Principles of
Fermentation
Technology

Principles of Fermentation Technology

THIRD EDITION

Peter F. Stanbury

Allan Whitaker

Stephen J. Hall

AMSTERDAM • BOSTON • HEIDELBERG • LONDON
NEW YORK • OXFORD • PARIS • SAN DIEGO
SAN FRANCISCO • SINGAPORE • SYDNEY • TOKYO
Butterworth-Heinemann is an imprint of Elsevier

Butterworth-Heinemann is an imprint of Elsevier
The Boulevard, Langford Lane, Kidlington, Oxford OX5 1GB, United Kingdom
50 Hampshire Street, 5th Floor, Cambridge, MA 02139, United States

Library of Congress Cataloging-in-Publication Data
A catalog record for this book is available from the Library of Congress

British Library Cataloguing-in-Publication Data
A catalogue record for this book is available from the British Library

ISBN: 978-0-08-099953-1

For information on all Butterworth-Heinemann publications
visit our website at https://www.elsevier.com/

Working together
to grow libraries in
developing countries

www.elsevier.com • www.bookaid.org

Publisher: Joe Hayton
Acquisition Editor: Fiona Geraghty
Editorial Project Manager: Maria Convey
Production Project Manager: Nicky Carter
Designer: Maria Inês Cruz

Typeset by Thomson Digital

This book is dedicated to all the staff, past and present,
of the Department of Biological and Environmental Sciences,
University of Hertfordshire.

Contents

Acknowledgments

The cover includes an image of the New Brunswick™ BioFlo® 610 fermenter Copyright © 2015 Courtesy of Eppendorf AG, Germany.

We wish to thank the authors, publishers, and manufacturing companies listed below for allowing us to reproduce either original or copyright material:

Authors
S. Abe (Fig. 3.15), A.W. Nienow (Figs. 7.10, 7.11, 9.15, 9.22, from *Trends in Biotechnology*, **8** (1990)); J.W. Richards (Figs. 5.3–5.6, 5.8, 7.18 and Table 5.2) from *Introduction to Industrial Sterilization*, Academic Press, London (1968), F.G. Shinskey (Fig. 8.11); R.M. Talcott (Figs. 10.11–10.13).

Publishers and Manufacturing Companies
Academic Press, London and New York: Figs. 1.2, 4.5, 7.1, 7.9, 7.14, 7.45, 7.52, 7.57, 9.25, 10.6, 10.28, and Table 8.3
Alfa Laval Ltd., Camberley: Figs. 5.11, 5.12, 5.14, 10.16, 10.17, and 10.20

American Chemical Society: Figs. 7.43, 7.50

American Society for Microbiology: Figs. 3.52, 9.19

American Society for Testing and Materials: Fig. 6.24. Copyright ASTM, reprinted with permission

Applikon Biotechnology, Tewkesbury, UK: Fig. 7.16 and Table 7.5

Bioengineered Bugs: Fig. 6.19

Bioprocess International: Fig. 6.11

Bio/Technology: Table 3.8

Blackwell Scientific Publications Ltd: Figs. 1.1 and 2.10

British Mycological Society: Fig. 7.49

British Valve and Actuator Manufacturers Association: Figs. 7.28–7.35, 7.37 and 7.38

Butterworth-Heinemann: Figs. 6.23, 7.22, 7.25, Table 3.9

Canadian Chemical News, Ottawa: Fig. 10.36

Celltainer Biotech BV, The Netherlands: Fig. 6.8

Chapman and Hall: Fig 7.47

Chemineer UK, Derby, UK: Figs. 9.6 and 9.23

Chilton Book Company Ltd., Radnor, Pennsylvania, USA: Figs. 8.2, 8.3, 8.4, 8.5, 8.8, and 8.9

Colder Products Company: Figs. 6.25b and 6.26b

EMD Millipore Corporation: Fig. 6.26a

Eppendorf AG, Germany: Figs 7.5 and 7.6

European Molecular Biology Laboratory: Fig. 12.5

Marcel Dekker Inc.: Figs. 6.16–6.18

Elsevier: Figs. 2.2, 2.14, 3.41, 3.47, 3.48, 3.53, 3.54, 3.56, 5.5, 5.6, 5.15, 5.20, 6.15, 7.10, 7.11, 8.14, 8.22, 8.24, 8.26, 9.2, 9.22, 10.4, 10.9, 10.30. Tables 2.5, 5.6, 6.4, 9.6, 9.10

Ellis Horwood: Figs. 9.18 and 10.5. Table 9.3

Fedegari Group: Table 5.3

GE Healthcare Life Sciences: Fig. 6.25a

Inceltech LH, Reading: Fig. 7.17

International Thomson Publishing Services: Figs. 5.16, 6.13, 7.24

Institute of Chemical Engineering: Fig. 11.7

Institute of Water Pollution Control: Fig. 11.6

IRL Press: Figs. 4.3, 6.5, 8.28

Japan Society for Bioscience, Biotechnology and Agrochemistry: Fig. 3.25

Kluwer Academic Publications: Fig. 7.53, reprinted with permission from Vardar-Sukan, F. and Sukan, S.S. (1992) *Recent Advances in Biotechnology*

MacMillan: Table 1.1

Marshall Biotechnology Ltd.: Fig. 7.23

McGraw Hill, New York: Fig. 7.27, 7.36, 8.23, 8.25, 10.10

Microbiology Research Foundation of Japan, Tokyo: Fig. 3.23

Microbiology Society: Figs. 3.27, 3.50, 3.51, and Tables 3.2 and 9.2

Nature Publishing Group: Fig. 3.3 and Table 3.10

New Brunswick Ltd., Hatfield, UK: Figs. 7.15, 7.26, and 7.56

New York Academy of Sciences: Figs. 2.14, 3.5, 3.6, 3.30

Oxford University Press: Figs. 3.34, 3.35, 12.7, 12.8, and Table 3.7

Pall Corporation, Portsmouth, UK: Figs. 5.19, 5.24, 5.25

Parker domnick hunter, Birtley, UK: Figs. 5.20, 5.21, 5.26, and 5.27

PubChem: Figs. 3.39, 3.40, and 3.55

Royal Society of Chemistry: Fig. 6.21

Sartorius Stedim UK Ltd., Epsom, UK: Figs. 6.7, 6.25c, 7.4, 7.7, 9.26

Science and Technology Letters, Northwood, UK: Fig. 9.24

Society for Industrial Microbiology, USA: Fig. 9.20

Southern Cotton Oil Company, Memphis, USA. Table 4.8

Spirax Sarco Ltd., Cheltenham, UK: Figs. 7.39–7.42

Springer. Figs. 3.31, 3.36, and 8.7. Tables 4.19 and 6.1

John Wiley and Sons: Figs. 5.17, 6.6, 6.10, 7.44, 7.51, 7.54, 8.11, 10.11–10.13, 10.21, 10.23, 10.24, 12.1, and Tables 2.6, 6.3, 12.3, 12.5

We also wish to thank Nick Hutchinson (Parker domnick hunter), Rob Smyth (Sartorius Stedim UK Ltd.), Geoff Simmons (Eppendorf UK Ltd.), Tom Watson (Pall Corporation), and particularly Maria Convey, our long-suffering Editorial Project Manager, and Nicky Carter, Production Project Manager.

Last but not least, we wish to express our thanks to Lesley Stanbury and Lorna Whitaker for their support, encouragement, and patience during the preparation of both this, and previous editions of "*Principles of Fermentation Technology*."

May 2016

An introduction to fermentation processes

The term "fermentation" is derived from the Latin verb *fervere,* to boil, thus describing the appearance of the action of yeast on the extracts of fruit or malted grain. The boiling appearance is due to the production of carbon dioxide bubbles caused by the anaerobic catabolism of the sugar present in the extract. However, fermentation has come to have with different meanings to biochemists and to industrial microbiologists. Its biochemical meaning relates to the generation of energy by the catabolism of organic compounds, whereas its meaning in industrial microbiology tends to be much broader.

The catabolism of sugar is an oxidative process, which results in the production of reduced pyridine nucleotides, which must be reoxidized for the process to continue. Under aerobic conditions, reoxidation of reduced pyridine nucleotide occurs by electron transfer, via the cytochrome system, with oxygen acting as the terminal electron acceptor. However, under anaerobic condition, reduced pyridine nucleotide oxidation is coupled with the reduction of an organic compound, which is often a subsequent product of the catabolic pathway. In the case of the action of yeast on fruit or grain extracts, NADH is regenerated by the reduction of pyruvic acid to ethanol. Different microbial taxa are capable of reducing pyruvate to a wide range of end products, as illustrated in Fig. 1.1. Thus, the term fermentation has been used in a strict biochemical sense to mean an energy-generation process in which organic compounds act as both electron donors and terminal electron acceptors.

The production of ethanol by the action of yeast on malt or fruit extracts has been carried out on a large scale for many years and was the first "industrial" process for the production of a microbial metabolite. Thus, industrial microbiologists have extended the term fermentation to describe any process for the production of product by the mass culture of a microorganism. Brewing and the production of organic solvents may be described as fermentation in both senses of the word but the description of an aerobic process as a fermentation is obviously using the term in the broader, microbiological, context and it is in this sense that the term is used in this book.

THE RANGE OF FERMENTATION PROCESSES

There are five major groups of commercially important fermentations:

1. Those that produce microbial cells (or biomass) as the product.
2. Those that produce microbial enzymes.

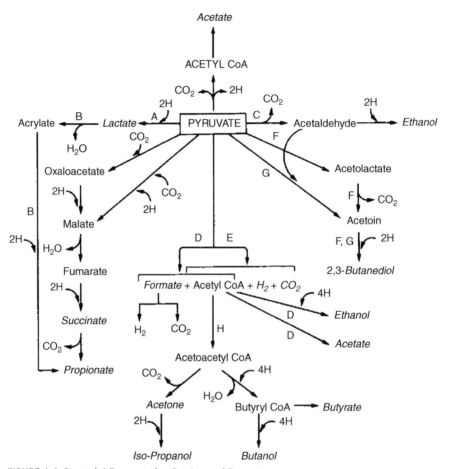

FIGURE 1.1 Bacterial Fermentation Products of Pyruvate

Pyruvate formed by the catabolism of glucose is further metabolized by pathways which are characteristic of particular organisms and which serve as a biochemical aid to identification. End products of fermentations are italicized (Dawes & Large, 1982).
A, Lactic acid bacteria (*Streptococcus, Lactobacillus*); B, *Clostridium propionicum*; C, Yeast, *Acetobacter, Zymomonas, Sarcina ventriculi, Erwinia amylovora*; D, Enterobacteriaceae (coli-aerogenes); E, Clostridia; F, *Klebsiella*; G, Yeast; H, Clostridia (butyric, butylic organisms); I, Propionic acid bacteria.

3. Those that produce microbial metabolites.
4. Those that produce recombinant products.
5. Those that modify a compound that is added to the fermentation—the transformation process.

The historical development of these processes will be considered in a later section of this chapter, but it is first necessary to include a brief description of the five groups.

MICROBIAL BIOMASS

The commercial production of microbial biomass may be divided into two major processes: the production of yeast to be used in the baking industry and the production of microbial cells to be used as human food or animal feed (single-cell protein). Bakers' yeast has been produced on a large scale since early 1900s and yeast was produced as human food in Germany during the First World War. However, it was not until the 1960s that the production of microbial biomass as a source of food protein was explored to any great depth. As a result of this work, reviewed briefly in Chapter 2, a few large-scale continuous processes for animal feed production were established in the 1970s. These processes were based on hydrocarbon feedstocks, which could not compete against other high protein animal feeds, resulting in their closure in the late 1980s (Sharp, 1989). However, the demise of the animal feed biomass fermentation was balanced by ICI plc and Rank Hovis McDougal establishing a process for the production of fungal biomass for human food. This process was based on a more stable economic platform and has been a significant economic success (Wiebe, 2004).

MICROBIAL ENZYMES

Enzymes have been produced commercially from plant, animal, and microbial sources. However, microbial enzymes have the enormous advantage of being able to be produced in large quantities by established fermentation techniques. Also, it is infinitely easier to improve the productivity of a microbial system compared with a plant or an animal one. Furthermore, the advent of recombinant DNA technology has enabled enzymes of animal origin to be synthesized by microorganisms (see Chapter 12). The uses to which microbial enzymes have been put are summarized in Table 1.1, from which it may be seen that the majority of applications are in the food and related industries. Enzyme production is closely controlled in microorganisms and in order to improve productivity these controls may have to be exploited or modified. Such control systems as induction may be exploited by including inducers in the medium (see Chapter 4), whereas repression control may be removed by mutation and recombination techniques. Also, the number of gene copies coding for the enzyme may be increased by recombinant DNA techniques. Aspects of strain improvement are discussed in Chapter 3.

MICROBIAL METABOLITES

The growth of a microbial culture can be divided into a number of stages, as discussed in Chapter 2. After the inoculation of a culture into a nutrient medium there is a period during which growth does not appear to occur; this period is referred as the lag phase and may be considered as a time of adaptation. Following a period during which the growth rate of the cells gradually increases, the cells grow at a constant maximum rate and this period is known as the log, or exponential, phase. Eventually, growth ceases and the cells enter the so-called stationary phase. After a further

Table 1.1 Commercial Applications of Enzymes

Industry	Application	Enzyme	Source
Baking and milling	Reduction of dough viscosity, acceleration of fermentation, increase in loaf volume, improvement of crumb softness, and maintenance of freshness	Amylase	Fungal
	Improvement of dough texture, reduction of mixing time, increase in loaf volume	Protease	Fungal/bacterial
Brewing	Mashing	Amylase	Fungal/bacterial
	Chill proofing	Protease	Fungal/bacterial
	Improvement of fine filtration	β-Glucanase	Fungal/bacterial
Cereals	Precooked baby foods, breakfast foods	Amylase	Fungal
Chocolate and cocoa	Manufacture of syrups	Amylase	Fungal/bacterial
Coffee	Coffee bean fermentation	Pectinase	Fungal
	Preparation of coffee concentrates	Pectinase, hemicellulase	Fungal
Confectionery	Manufacture of soft center candies	Invertase, pectinase	Fungal/bacterial
Cotton	Low temperature processing	Pectate lyase	Fungal
Corn syrup	Manufacture of high-maltose syrups	Amylase	Fungal
	Production of low D.E. syrups	Amylase	Bacterial
	Production of glucose from corn syrup	Amyloglycosidase	Fungal
	Manufacture of fructose syrups	Glucose isomerase	Bacterial
Dairy	Manufacture of protein hydrolysates	Protease	Fungal/bacterial
	Stabilization of evaporated milk	Protease	Fungal
	Production of whole milk concentrates, ice cream, and frozen desserts	Lactase	Yeast
	Curdling milk	Protease	Fungal/bacterial
Eggs, dried	Glucose removal	Glucose oxidase	Fungal
Fruit juices	Clarification	Pectinases	Fungal
	Oxygen removal	Glucose oxidase	Fungal
Laundry	Detergents	Protease, lipase	Bacterial
Leather	Dehairing, baiting	Protease	Fungal/bacterial
Meat	Tenderization	Protease	Fungal
Paper	Removal of wood waxes	Lipase	Fungal
Pharmaceutical	Digestive aids	Amylase, protease	Fungal

Table 1.1 Commercial Applications of Enzymes (*cont.*)

Industry	Application	Enzyme	Source
	Antiblood clotting	Streptokinase	Bacterial
	Various clinical tests	Numerous	Fungal/bacterial
	Biotransformations	Numerous	Fungal/bacterial
Photography	Recovery of silver from spent film	Protease	Bacterial
Protein hydrolysates	Manufacture	Proteases	Fungal/bacterial
Soft drinks	Stabilization	Glucose oxidase, catalase	Fungal
Textiles	Desizing of fabrics	Amylase	Bacterial
Vegetables	Preparation of purees and soups	Pectinase, amylase, cellulase	Fungal

Modified from Boing (1982).

period of time, the viable cell number declines as the culture enters the death phase. As well as this kinetic description of growth, the behavior of a culture may also be described according to the products that it produces during the various stages of the growth curve. During the log phase of growth, the products produced are either anabolites (products of biosynthesis) essential to the growth of the organism and include amino acids, nucleotides, proteins, nucleic acids, lipids, carbohydrates, etc. or are catabolites (products of catabolism) such as ethanol and lactic acid, as illustrated in Fig. 1.1. These products are referred as the primary products of metabolism and the phase in which they are produced (equivalent to the log, or exponential phase) as the trophophase (Bu'Lock et al., 1965).

Many products of primary metabolism are of considerable economic importance and are being produced by fermentation, as illustrated in Table 1.2. The synthesis of anabolic primary metabolites by wild-type microorganisms is such that their production is sufficient to meet the requirements of the organism. Thus, it is the task of the industrial microbiologist to modify the wild-type organism and to provide cultural conditions to improve the productivity of these compounds. This has been achieved very successfully, over many years, by the selection of induced mutants, the use of recombinant DNA technology, and the control of the process environment of the producing organism. This is exemplified by the production of amino acids where productivity has been increased by several orders of magnitude. However, despite these spectacular achievements, microbial processes have only been able to compete with the chemical industry for the production of relatively complex and high value compounds. In recent years, this situation has begun to change. The advances in metabolic engineering arising from genomics, proteomics, and metabolomics have provided new powerful techniques to further understand the physiology of "over-production" and to reengineer microorganisms to "over-produce" end products and intermediates of primary metabolism. Combined with the rising cost of petroleum and the desirability of environmentally friendly processes these advances are now facilitating the

Table 1.2 Some Primary Products of Microbial Metabolism and Their Commercial Significance

Primary Metabolite	Commercial Significance
Ethanol	"Active ingredient" in alcoholic beverages
	Used as a motor-car fuel when blended with petroleum
Organic acids	Various uses in the food industry
Glutamic acid	Flavor enhancer
Lysine	Feed supplement
Nucleotides	Flavor enhancers
Phenylalanine	Precursor of aspartame, sweetener
Polysaccharides	Applications in the food industry
	Enhanced oil recovery
Vitamins	Feed supplements

development of economic microbial processes for the production of bulk chemicals and feedstocks for the chemical industry (Otero & Nielsen, 2010; Van Dien, 2013). These aspects are considered later in this chapter and in Chapter 3.

During the deceleration and stationary phases, some microbial cultures synthesize compounds which are not produced during the trophophase and which do not appear to have any obvious function in cell metabolism. These compounds are referred to as the secondary compounds of metabolism and the phase in which they are produced (equivalent to the stationary phase) as the idiophase (Bu'Lock et al., 1965). It is important to realize that secondary metabolism may occur in continuous cultures at low growth rates and is a property of slow-growing, as well as nongrowing cells. When it is appreciated that microorganisms grow at relatively low growth rates in their natural environments, it is tempting to suggest that it is the idiophase state that prevails in nature rather than the trophophase, which may be more of a property of microorganisms in culture. The interrelationships between primary and secondary metabolism are illustrated in Fig. 1.2, from which it may be seen that secondary metabolites tend to be elaborated from the intermediates and products of primary metabolism. Although the primary biosynthetic routes illustrated in Fig. 1.2 are common to the vast majority of microorganisms, each secondary product would be synthesized by only a relatively few different microbial species. Thus, Fig. 1.2 is a representation of the secondary metabolism exhibited by a very wide range of different microorganisms. Also, not all microorganisms undergo secondary metabolism—it is common amongst microorganisms that differentiate such as the filamentous bacteria and fungi and the sporing bacteria but it is not found, for example, in the Enterobacteriaceae. Thus, the taxonomic distribution of secondary metabolism is quite different from that of primary metabolism. It is important to appreciate that the classification of microbial products into primary and secondary metabolites is a convenient, but in some cases, artificial system. To quote Bushell (1988), the classification "should not be allowed to act as a conceptual straitjacket, forcing the reader to consider all products

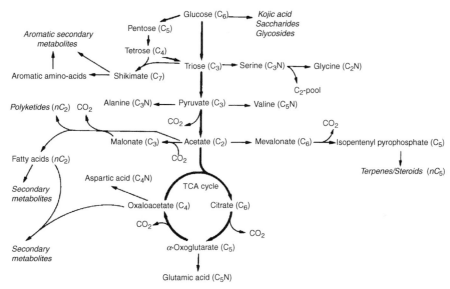

FIGURE 1.2 The Interrelationships Between Primary and Secondary Metabolism

Primary catabolic routes are shown in heavy lines and secondary products are italicized (Turner, 1971).

as either primary or secondary metabolites." It is sometimes difficult to categorize a product as primary or secondary and the kinetics of synthesis of certain compounds may change depending on the cultural conditions.

The physiological role of secondary metabolism in the producer organism in its natural environment has been the subject of considerable debate and their functions include effecting differentiation, inhibiting competitors, and modulating host physiology. However, the importance of these metabolites to the fermentation industry is the effects they have on organisms other than those that produce them. Many secondary metabolites have antimicrobial activity, others are specific enzyme inhibitors, some are growth promoters and many have pharmacological properties (Table 1.3). Thus, the products of secondary metabolism have formed the basis of a major section

Table 1.3 Some Secondary Products of Microbial Metabolism and Their Commercial Significance

Secondary Metabolite	Commercial Significance
Penicillin, cephalosporin, streptomycin	Antibiotics
Bleomycin, mitomycin	Anticancer agents
Lovastatin	Cholesterol-lowering agent
Cyclosporine A	Immunosuppressant
Avermectins	Antiparasitic agents

of the fermentation industry. As in the case for primary metabolites, wild-type microorganisms tend to produce only low concentrations of secondary metabolites, their synthesis being controlled by induction, quorum sensing, growth rate, feedback systems, and catabolite repression, modulated by a range of effector molecules (van Wezel & McDowall, 2011). The techniques which have been developed to improve secondary metabolite production are considered in Chapters 3 and 4.

RECOMBINANT PRODUCTS

The advent of recombinant DNA technology has extended the range of potential fermentation products. Genes from higher organisms may be introduced into microbial cells such that the recipients are capable of synthesizing "foreign" proteins. These proteins are described as "heterologous" meaning "derived from a different organism." A wide range of microbial cells has been used as hosts for such systems including *Escherichia coli, Saccharomyces cerevisiae,* and filamentous fungi. Animal cells cultured in fermentation systems are also widely used for the production of heterologous proteins. Although the animal cell processes were based on microbial fermentation technology, a number of novel problems had to be solved—animal cells were considered extremely fragile compared with microbial cells, the achievable cell density is very much less than in a microbial process and the media are very complex. These aspects are considered in detail in Chapters 4 and 7. Products produced by such genetically engineered organisms include interferon, insulin, human serum albumin, factors VIII and IX, epidermal growth factor, calf chymosin, and bovine somatostatin. Important factors in the design of these processes include the secretion of the product, minimization of the degradation of the product, and control of the onset of synthesis during the fermentation, as well as maximizing the expression of the foreign gene. These aspects are considered in more detail later in this chapter and in Chapters 4 and 12.

TRANSFORMATION PROCESSES

Microbial cells may be used to convert a compound into a structurally related, financially more valuable, compound. Because microorganisms can behave as chiral catalysts with high positional specificity and stereospecificity, microbial processes are more specific than purely chemical ones and enable the addition, removal, or modification of functional groups at specific sites on a complex molecule without the use of chemical protection. The reactions, which may be catalyzed include dehydrogenation, oxidation, hydroxylation, dehydration and condensation, decarboxylation, animation, deamination, and isomerization. Microbial processes have the additional advantage over chemical reagents of operating at relatively low temperatures and pressures without the requirement for potentially polluting heavy-metal catalysts. Although the production of vinegar is the oldest established microbial transformation process (conversion of ethanol to acetic acid), the majority of these processes involve the production of high-value compounds including steroids, antibiotics, and prostaglandins.

However, the conversion of acetonitrile to acrylamide by *Rhodococcus rhodochrous* is an example of the technology being used in the manufacturing of a bulk chemical—20,000 metric tons being produced annually (Demain & Adrio, 2008).

A novel application of microbial transformation is the use of microorganisms to mimic mammalian metabolism. Humans and animals will metabolize drugs such that they may be removed from the body. The resulting metabolites may be biologically active themselves—either eliciting a desirable effect or causing damage to the organism. Thus, in the development of a drug it is necessary to determine the activity of not only the administered drug but also its metabolites. These studies may require significant amount of the metabolites and while it may be possible to isolate them from tissues, blood, urine, or faeces of the experimental animal, their concentration is often very low resulting in such approaches being time-consuming, expensive, and far from pleasant. Sime (2006) discussed the exploitation of the metabolic ability of microorganisms to perform these biotransformations. Thus, drug metabolites have been produced in small-scale fermentation, facilitating the investigation of their biological activity and/or toxicity.

The anomaly of the transformation fermentation process is that a large biomass has to be produced to catalyze a single reaction. Thus, many processes have been streamlined by immobilizing either the whole cells, or the isolated enzymes, which catalyze the reactions, on an inert support. The immobilized cells or enzymes may then be considered as catalysts, which may be reused many times.

THE CHRONOLOGICAL DEVELOPMENT OF THE FERMENTATION INDUSTRY

The chronological development of the fermentation industry may be represented as five overlapping stages as illustrated in Table 1.4. The development of the industry prior to 1900 is represented by stage 1, where the products were confined to potable alcohol and vinegar. Although beer was first brewed by the ancient Egyptians, the first true large-scale breweries date from the early 1700s when wooden vats of 1500 barrels capacity were introduced (Corran, 1975). Even some process control was attempted in these early breweries, as indicated by the recorded use of thermometers in 1757 and the development of primitive heat exchangers in 1801. By the mid-1800s, the role of yeasts in alcoholic fermentation had been demonstrated independently by Cagniard-Latour, Schwann, and Kutzing but it was Pasteur who eventually convinced the scientific world of the obligatory role of these microorganisms in the process. During the late 1800s, Hansen started his pioneering work at the Carlsberg brewery and developed methods for isolating and propagating single yeast cells to produce pure cultures and established sophisticated techniques for the production of starter cultures. However, use of pure cultures did not spread to the British ale breweries and it is true to say that many of the small, traditional, ale-producing breweries still use mixed yeast cultures at the present time but, nevertheless, succeed in producing high quality products.

Table 1.4 The Stages in the Chronological Development of the Fermentation Industry

Stage	Main Products	Vessels	Process Control	Culture Method	Quality Control	Pilot Plant Facilities	Strain Selection
1 Pre-1900	Alcohol	Wooden, up to 1500 barrels capacity Copper used in later breweries	Use of thermometer, hydrometer and heat exchangers	Batch	Virtually nil	Nil	Pure yeast cultures used at the Carlsberg brewery (1886)
	Vinegar	Barrels, shallow trays, trickle filters		Batch	Virtually nil	Nil	Fermentations inoculated with 'good' vinegar
2 1900–1940	Bakers' yeast glycerol, citric acid, lactic acid and acetone/butanol	Steel vessels of up to 200 m³ for acetone/butanol Air spargers used for bakers' yeast Mechanical stirring used in small vessels	pH electrodes with off-line control Temperature control	Batch and fed-batch systems	Virtually nil	Virtually nil	Pure cultures used
3 1940–date	Penicillin, streptomycin, other antibiotics, gibberellin, amino acids, nucleotides, transformations, enzymes	Mechanically aerated vessels, operated aseptically—true fermenters	Sterilizable pH and oxygen electrodes. Use of control loops which were later computerized	Batch and fed-batch common Continuous culture introduced for brewing and some primary metabolites	Very important	Becomes common	Mutation and selection programmes essential

4 1964–date	Single-cell protein using hydrocarbon and other feedstocks	Pressure cycle and pressure jet vessels developed to overcome gas and heat exchange problems	Use of computer linked control loops	Continuous culture with medium recycle	Very important	Very important	Genetic engineering of producer strains attempted
5 1982–date	Production of heterologous proteins by microbial and animal cells Monoclonal antibodies produced by animal cells	Fermenters developed in stages 3 and 4. Animal cell reactors developed	Control and sensors developed in stages 3 and 4	Batch, fed-batch or continuous Continuous perfusion developed for animal cell processes	Very important	Very important	Introduction of foreign genes into microbial and animal cell hosts. In vitro recombinant DNA techniques used in the improvement of stage 3 products
6 2000–date	Use of "synthetic biology" to improve established fermentations and develop new bulk chemical processes	Fermenters developed in stages 3 and 4	Control and sensors developed in stages 3 and 4	Batch, fed-batch or continuous	Very important	Very important	Synthetic biology used to develop existing and novel fermentations

Vinegar was originally produced by leaving wine in shallow bowls or partially filled barrels where it was slowly oxidized to vinegar by the development of a natural flora. The appreciation of the importance of air in the process eventually led to the development of the "generator" which consisted of a vessel packed with an inert material (such as coke, charcoal, and various types of wood shavings) over which the wine or beer was allowed to trickle. The vinegar generator may be considered as the first "aerobic" fermenter to be developed. By the late 1800s to early 1900s, the initial medium was being pasteurized and inoculated with 10% good vinegar to make it acidic, and therefore resistant to contamination, as well as providing a good inoculum (Bioletti, 1921). Thus, by the beginning of the twentieth century the concepts of process control were well established in both the brewing and vinegar industries.

Between the years 1900 and 1940, the main new products were yeast biomass, glycerol, citric acid, lactic acid, acetone, and butanol. Probably the most important advances during this period were the developments in the bakers' yeast and solvent fermentations. The production of bakers' yeast is an aerobic process and it was soon recognized that the rapid growth of yeast cells in a rich medium (or wort) led to oxygen depletion that in turn, resulted in ethanol production at the expense of biomass formation. The problem was minimized by restricting the initial wort concentration, such that the growth of the cells was limited by the availability of the carbon source rather than oxygen. Subsequent growth of the culture was then controlled by adding further wort in small increments. This technique is now called fed-batch culture and is widely used in the fermentation industry to avoid conditions of oxygen limitation. The aeration of these early yeast cultures was also improved by the introduction of air through sparging tubes, which could be steam cleaned (De Becze & Liebmann, 1944).

The development of the acetone–butanol fermentation during the First World War by the pioneering efforts of Weizmann at Manchester University led to the establishment of the first truly aseptic fermentation. All the processes discussed so far could be conducted with relatively little contamination provided that a good inoculum was used and reasonable standards of hygiene employed. However, the anaerobic butanol process was susceptible to contamination by aerobic bacteria in the early stages of the fermentation, and by acid-producing anaerobic ones, once anaerobic conditions had been established in the later stages of the process. The fermenters employed were vertical cylinders with hemispherical tops and bottoms constructed from mild steel. They could be steam sterilized under pressure and were constructed to minimize the possibility of contamination. Two-thousand-hectoliter fermenters were commissioned which presented the problems of inoculum development and the maintenance of aseptic conditions during the inoculation procedure. The techniques developed for the production of these organic solvents were a major advance in fermentation technology and paved the way for the successful introduction of aseptic aerobic processes in the 1940s. In the late 1940s, fermentation still provided 65% of butanol and 10% of acetone produced in the United States of America (Jackson, 1958). However, the solvent fermentations became uneconomic with the development of competing processes based on the petrochemical feedstocks and they ceased to exist. It is interesting to note that approximately one hundred years after the development of the solvent fermentations that the competitiveness of fermentation and petrochemical processes for the production of some bulk chemicals may be reversed. As discussed earlier in this chapter, the rising cost of crude oil, the attractiveness of environmentally friendly processes, and the advances in metabolic engineering may lead to the resurrection of modern versions of these old processes.

The third stage of the development of the fermentation industry arose in the 1940s as a result of the wartime need to produce penicillin in submerged culture under aseptic conditions. The production of penicillin is an aerobic process that is very vulnerable to contamination. Thus, although the knowledge gained from the solvent fermentations was exceptionally valuable, the problems of sparging a culture with large volumes of sterile air and mixing a highly viscous broth had to be overcome. Also, unlike the solvent fermentations, penicillin was synthesized in very small quantities by the initial isolates and this resulted in the establishment of strain-improvement programs, which became a dominant feature of the industry in subsequent years. Process development was also aided by the introduction of pilot-plant facilities, which enabled the testing of new techniques on a semi-production scale. The development of a large-scale extraction process for the recovery of penicillin was another major advance at this time. The technology established for the penicillin fermentation provided the basis for the development of a wide range of new processes. This was probably the stage when the most significant changes in fermentation technology (as compared with genetic technology) took place resulting in the establishment of many new processes over the period, including other antibiotics, vitamins, gibberellin, amino acids, enzymes, and steroid transformations. From the 1960s onward, microbial products were screened for activities other than simply antimicrobial properties and screens became more and more sophisticated. These screens have given rise to those operating today utilizing miniaturized culture systems, robotic automation, and elegant assays.

While the products described earlier belonging to stage 3 are all biosynthetic compounds, one catabolic product gained significance in the mid-1970s. Brazil and the United States of America initiated programs for the manufacturing of ethanol as a motor fuel in 1975 and 1978 respectively. This was an attempt by both the countries to lessen their dependency on imported petroleum by using fermentation processes to convert carbohydrates to ethanol (bioethanol); the Brazilian process being based on sugarcane and the American process based on maize starch. Forty years after Brazil's initiative, energy independence is still the "holy grail" of most governments—the United States of America produced 56 billion liters of bioethanol in 2015 (Renewable Fuels Association, 2016). However, the political and social issues associated with the diversion of land from food production to fuel are obvious and thus the development of processes using cellulose and lignin feedstocks rather than sugar and starch is the next stage in this saga.

In the early 1960s, the decisions of several multinational companies to investigate the production of microbial biomass as a source of feed protein led to a number of developments, which may be regarded as the fourth stage in the progress of the industry. The largest mechanically stirred fermentation vessels developed during stage 3 were in the range 80,000–150,000 dm^3. However, the relatively low selling price of microbial biomass necessitated its production in much larger quantities than other fermentation products in order for the process to be profitable. Also, hydrocarbons were considered as the potential carbon sources that would result in increased oxygen demands and high heat outputs by these fermentations (see Chapters 4 and 9). These requirements led to the development of the pressure jet and pressure cycle fermenters that eliminated the need for mechanical stirring (see Chapter 7). Another feature of these potential processes was that they would have to be operated continuously if they were to be economic. At this time batch and fed-batch fermentations were common in the industry but the technique of growing an organism continuously by adding fresh medium to the vessel and removing culture fluid had been applied only to a very limited extent on a large scale. The brewers were also investigating the potential of continuous culture at this time, but its application in that industry was short-lived. Several companies persevered in the biomass field and a few processes came to fruition, of which the most long-lived was the ICI Pruteen animal feed process, which utilized a continuous 3,000,000-dm^3 pressure cycle fermenter for the culture of *Methylophilus methylotrophus* with methanol as carbon source (Smith, 1981; Sharp, 1989). The operation of an extremely large continuous fermenter for time periods in excess of 100 days presented a considerable aseptic operation problem, far greater than that faced by the antibiotic industry in the 1940s and 1950s. The aseptic operation of fermenters of this type was achieved as a result of the high standards of fermenter construction, the continuous sterilization of feed streams, and the utilization of computer systems to control the sterilization and operation cycles, thus minimizing the possibility of human error. However, although the Pruteen process was a technological triumph, it became an economic failure because the product was out-priced by soybean and fishmeal. Eventually, in 1989, the plant was demolished, marking the end of a short, but very exciting, era in the fermentation industry.

While biomass is a very low-value, high-volume product, the fifth stage in the progress of the industry resulted in the establishment of very high-value, low-volume products; a stage often referred as "new biotechnology." As mentioned earlier in this chapter, in vitro recombinant DNA technology enabled the expression of human and mammalian genes in cultured animal cells and microorganisms, thereby enabling the development of relatively large-scale fermentation processes for the production of human proteins, which could then be used therapeutically. These products have been classified by the regulatory authorities as biologicals, not as drugs, and thus come under the same regulatory controls as do vaccines, and along with vaccines are commonly referred as biopharmaceuticals. The exploitation of genetic engineering coincided approximately with another major development in biotechnology that influenced the progress of the fermentation industry—the production of monoclonal antibodies (mabs). The availability of monoclonals opened the door to sophisticated analytical techniques and raised hopes for their use as therapeutic agents. Although the practical use of mabs was initially limited to analytical applications (Webb, 1993), they are now well-established as therapeutic biopharmaceuticals and are produced in fermentation systems employing both mammalian cells and microorganisms as production agents (Walsh, 2012). Table 1.5 lists a selection of the recombinant proteins licensed for therapeutic use and the production systems.

In 1982, recombinant human insulin became the first heterologous protein to be approved for medical use. Eight more products were approved in the 1980s comprising two approvals for human growth hormone, two for interferons, one monoclonal antibody, one recombinant vaccine for hepatitis B, one for tissue plasminogen activator, and one for erythropoietin (Walsh, 2012). During this period, the pharmaceutical industry was also very active in developing "conventional" microbial processes which resulted in a number of new microbial products reaching the marketplace in the late 1980s and early 1990s. Buckland (1992) listed four secondary metabolites which were launched in the 1980s: cyclosporine, an immunoregulant used to control the

Table 1.5 The Major Groups of Recombinant Proteins Developed as Therapeutic Agents

Therapeutic Group	Recombinant Protein	Production System	Clinical Use, Treatment of
Blood factors	Factor VIII	Mammalian cells	Anemia
Thrombolytics,	Tissue plasminogen activator	Mammalian cells, *E.coli*	Clot lysis
Anticoagulants	Hirudin	*S. cerevisiae*	Anticoagulant
Hormones	Insulin	*E.coli, S. cerevisiae*	Diabetes
	Human growth hormone	*E.coli, S. cerevisiae*	Hypopituitary dwarfism
	Follicle stimulating hormone	Mammalian cells	Infertility
	Glucagon	*E.coli, S. cerevisiae*	Type 2 diabetes
Growth factors	Erythropoietin	Mammalian cells	Anemia
	Granulocyte-macrophage colony, stimulating factor	*E.coli*	Bone marrow transplantation
	Granulocyte colony stimulating Factor	*E.coli*, mammalian cells	Cancer
Cytokines	Interferon-alpha	*E.coli*	Cancers, hepatitis B leukemia
	Interferon-beta	*E.coli,*	Cancers, amytotrophic lateral sclerosis, genital warts

Modified from Dykes (1993) and Walsh (2012).

rejection of transplanted organs; imipenem, a modified carbapenem, which had the widest antimicrobial spectrum of any antibiotic; lovastatin, a drug used for reducing cholesterol levels, and ivermectin, an antiparasitic drug which has been used to prevent "African River Blindness" as well as in veterinary practice. The sale of these four products added together exceeded that of the totality of recombinant proteins at that time. However, the developments over the last twenty years have resulted in biopharmaceuticals reaching annual sales of 100 billion dollars—representing one third of the global pharmaceutical market (Nielsen, 2013). In 2010, thirty biopharmaceuticals recorded sales of more than 1 billion dollars each (Walsh, 2012).

By the end of 2012, approximately 220 biopharmaceuticals had been approved (Berlec & Strukelj, 2013; Reader, 2013) with 31% being produced in *E. coli*, 15% in yeast, 43% in mammalian cells (mainly Chinese Hamster Ovarian cell lines, CHO), the remaining 11% being produced by hybridoma cells, insect cells, and transgenic animals and plants. One of the 2012 recombinant products, Elelyso, (a human taliglucerase alfa, used for the treatment of the lysosomal storage disorder, Gaucher's disease) was approved to be produced in plant cell culture. This was the first approval of this production system and may pave the way for future processes. Whereas the approved products in the 1980s were predominantly hormones and cytokines, the approvals over the 2005–12 period were dominated by antibody-based products and engineered proteins, that is, proteins which have been modified postsynthesis. These modifications include:

- Antibody-drug conjugates in which, for example, anticancer drugs are linked to monoclonal antibodies which bind to the tumor, thus directing the drug to its target.
- Modifications which extend the half-life of the therapeutic protein in the body.
- Modifications which alter the pharmacokinetic properties of the therapeutic protein.

The commercial exploitation of recombinant proteins has necessitated the construction of production facilities designed to contain the engineered producing organism or cell culture. Thus, these processes are drawing on the experience of vaccine fermentations where pathogenic organisms have been grown on relatively large scales. Also, as indicated earlier, recombinant proteins are classified as biologicals, not as drugs, and thus come under the same regulatory authorities as do vaccines. The major difference between the approval of drugs and biologicals is that the process for the production of a biological must be precisely specified and carried out in a facility that has been inspected and licensed by the regulatory authority, which is not the case for the production of drugs (antibiotics, for example) (Bader, 1992). Thus, any changes that a manufacturer wishes to incorporate into a licensed process must receive regulatory approval. For drugs, only major changes require approval prior to implementation. The result of these containment and regulatory requirements is that the cost of developing a recombinant protein process is extremely high. Farid (2007) collated industry data which suggests that the investment costs for a 20,000 dm^3 plant was \$60 M and that for a 200,00 dm^3 plant was \$600 M, with validation costs accounting for approximately 10–20% of this expenditure. Earlier in the development of the heterologous protein sector Buckland (1992) claimed that it cost as much to build a 3000 dm^3 scale facility for Biologics as for a 200,000 dm^3 scale facility for an antibiotic.

While the development of recombinant DNA technology facilitated the production of heterologous proteins, the advances in genomics, proteomics, and metabolic flux analysis are the basis of the sixth stage in the progress of the industry. The sequencing of the complete genomes of a wide range of organisms and the development of computerized systems to store and access the data has enabled the comparison of genomes and the visualization of gene expression in terms of both mRNA and protein profiles, the transcriptome and proteome respectively. Metabolic flux analysis examines the flux (or flow) of intermediates through a pathway (or pathways) and enables the construction of mathematical models mimicking the metabolic networks of the cell. The combination of these approaches has enabled workers to take a more holistic view of the workings of an organism, such that the outlook of the molecular biologist and biochemist have coincided with that of the physiologist in attempting to understand the functioning of the whole organism rather than simply its component parts. The term given to this rediscovery of physiology is "systems biology" and its application to biotechnology, "synthetic biology." The adoption of a systems biology approach by fermentation scientists has enabled them to build upon established fermentation processes and take them to a further level. For example, Ikeda, Ohnishi, Hayashi, and Mitsuhashi (2006) compared the genome sequence of a lysine-producing industrial strain of *Corynebacterium glutamicum* with that of the wild type. The industrial strain had been manipulated by many rounds of mutation and directed selection (see Chapter 3) such that it contained not only the lesions giving overproduction but also undesirable mutations which had been inadvertently coselected during strain development. This comparison enabled the construction of a strain containing only the desirable lesions. Becker, Zelder, Hafner, Schroder, and Wittmann (2011) further developed this approach by constructing a lysine-producing strain by modifying the wild-type to optimize precursor supply, feedback control, metabolic flux, and NADPH supply.

Thus the goal of synthetic biology is to maximize the yield of the desired product while minimizing that of unwanted or unnecessary metabolites. Obviously, this has always been the goal of the fermentation scientist but the tools of synthetic biology may make this goal a reality. An exciting application of the approach is in the production of bulk chemicals and feedstocks for the chemical industry, in competition with the petrochemicals. Such products would be low value, high volume (as were the ill-fated biomass processes) necessitating very high yields. The USA company Genomatica has claimed a viable process for the production of 1,4-butanediol, an important chemical intermediate, from a manipulated strain of *E. coli*, a strain developed using the synthetic biology approach (Yim, Hasselbeck, Niu, & Pujol-baxley, 2011). The success of such processes will depend upon their economic competitiveness and it will be interesting to see whether they thrive or go the way of the biomass processes.

THE COMPONENT PARTS OF A FERMENTATION PROCESS

Regardless of the type of fermentation (with the possible exception of some transformation processes) an established process may be divided into six basic component parts:

1. The formulation of media to be used in culturing the process organism during the development of the inoculum and in the production fermenter.
2. The sterilization of the medium, fermenters, and ancillary equipment.
3. The production of an active, pure culture in sufficient quantity to inoculate the production vessel.
4. The growth of the organism in the production fermenter under optimum conditions for product formation.
5. The extraction of the product and its purification.
6. The disposal of effluents produced by the process.

The interrelationships between the six component parts are illustrated in Fig. 1.3.

However, one must also visualize the research and development program which is designed to gradually improve the overall efficiency of the fermentation. Before a fermentation process is established a producer organism has to be isolated, modified such that it produces the desired product in commercial quantities, its cultural requirements determined and the plant designed accordingly. Also, the extraction process has to be established. The development program would involve the continual improvement of the process organism, the culture medium, and the extraction process.

The subsequent chapters in this book consider the basic principles underlying the component parts of a fermentation. Chapter 2 considers growth, comprehension of which is crucial to understanding many aspects of the process, other than simply the

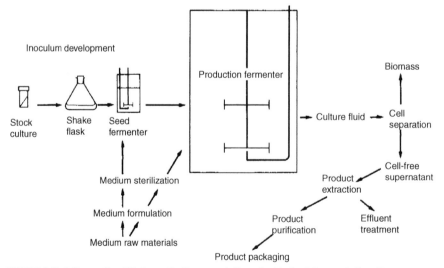

FIGURE 1.3 A Generalized Schematic Representation of a Typical Fermentation Process

growth of the organism in the production fermenter. The isolation and improvement of commercial strains is considered in Chapter 3 and the design of media in Chapter 4. The sterilization of the medium, fermenters, and air is considered in Chapter 5 and the techniques for the development of inocula are discussed in Chapter 6. Chapters 7, 8 and 9 consider the fermenter as an environment for the culture of microorganisms; Chapter 7 considers the design and construction of fermenters including contained systems and animal cell fermenters, Chapter 8 discusses the instrumentation involved in monitoring and maintaining a controlled environment in a fermenter, while the provision of oxygen to a culture is investigated in Chapter 9. The recovery of fermentation products is dealt with in Chapter 10 and the environmental impact of fermentation processes and the regulatory framework in which they must operate is addressed in Chapter 11. Finally, the production of heterologous proteins is discussed in Chapter 12. Throughout the book examples are drawn from a very wide range of fermentations to illustrate the applications of the techniques being discussed but it has not been attempted to give detailed considerations of specific processes as this is well covered elsewhere, for example, in the *Comprehensive Biotechnology* series edited by Moo-Young (2011). We hope that the approach adopted in this book will give the reader an understanding of the basic principles underlying the commercial techniques used for the large-scale production of microbial products.

REFERENCES

Bader, F. G. (1992). Evolution in fermentation facility design from antibiotics to recombinant proteins. In M. R. Ladisch, & A. Bose (Eds.), *Harnessing biotechnology for the 21st century* (pp. 228–231). Washington, DC: American Chemical Society.

Becker, J., Zelder, O., Hafner, S., Schroder, H., & Wittmann, C. (2011). From zero to hero – design-based systems metabolic engineering of *Corynebacterium glutamicum* for L-lysine production. *Metabolic Engineering*, *13*, 159–168.

Berlec, A., & Strukelj, B. (2013). Current state and recent advances in biopharmaceutical production in *Escherichia coli*, yeasts and mammalian cells. *Journal of Industrial Microbiology and Biotechnology*, *40*, 257–274.

Bioletti, F. T. (1921). The manufacture of vinegar. In C. E. Marshall (Ed.), *Microbiology* (pp. 636–648). London: Churchill.

Boing, J. T. P. (1982). Enzyme production. In G. Reed (Ed.), *Prescott and Dunn's industrial microbiology* (4th ed., pp. 634–708). New York: MacMillan.

Buckland, B. C. (1992). Reduction to practice. In M. R. Ladisch, & A. Bose (Eds.), *Harnessing biotechnology for the 21st century* (pp. 215–218). Washington, DC: American Chemical Society.

Bu'Lock, J. D., Hamilton, D., Hulme, M. A., Powell, A. J., Shepherd, D., Smalley, H. M., & Smith, G. N. (1965). Metabolic development and secondary biosynthesis in *Penicillium urticae*. *Canadian Journal of Microbiology*, *11*, 765–778.

Bushell, M. E. (1988). Application of the principles of industrial microbiology to biotechnology. In A. Wiseman (Ed.), *Principles of biotechnology* (pp. 5–43). New York: Chapman and Hall.

Corran, H. S. (1975). *A history of brewing*. David and Charles, Newton Abbott.

Dawes, I., & Large, P. J. (1982). Class 1 reactions: Supply of carbon skeletons. In J. Mandelstam, K. McQuillen, & I. Dawes (Eds.), *Biochemistry of bacterial growth* (pp. 125–158). Oxford: Blackwell.

De Becze, G. I., & Liebmann, A. J. (1944). Aeration in the production of compressed yeast. *Industrial Engineering Chemistry, 36*, 882–890.

Demain, A. L., & Adio, J. L. (2008). Contributions of microorganisms to industrial biology. *Molecular Biotechnology, 38*, 41–55.

Dykes, C. W. (1993). Molecular biology in the pharmaceutical industry. In J. M. Walker, & E. B. Gingold (Eds.), *Molecular biology and biotechnology* (pp. 155–176). Cambridge: Royal Society of Chemistry.

Farid, S. (2007). Process economics of industrial monoclonal antibody manufacture. *Journal of Chromatography B, 848*(1), 8–18.

Ikeda, M., Ohnishi, J., Hayashi, M., & Mitsuhashi, S. (2006). A genome-based approach to create a minimally mutated *Corynebacterium glutamcium* strain for efficient L-lysine production. *Journal of Microbiology and Biotechnology, 33*, 610–615.

Jackson, R. W. (1958). Potential utilization of agricultural commodities by fermentation. *Economic Botany, 12*(1), 42–53.

Moo-Young, M. (2011). *Comprehensive biotechnology* (Vols. 1–4). (2nd ed.). Amsterdam: Elsevier.

Nielsen, J. (2013). Production of biopharmaceutical proteins by yeast: advances through metabolic engineering. *Bioengineering, 4*(4), 207–211.

Otero, J. M., & Nielsen, J. (2010). Industrial systems biology. *Biotechnology and Bioengineering, 105*(3), 439–460.

Reader, R. A. (2013). FDA biopharmaceutical product approvals and trends in 2012. *BioProcess International, 11*(3), 18–27.

Renewable Fuels Association (2016). Fueling a High Octane Future: 2016 Ethanol Industry Outlook. Available from http://www.ethanolrfa.org/resources/publications/outlook/

Sharp, D. H. (1989). *Bioprotein manufacture: A critical assessment.* Chichester: Ellis Horwood.

Sime, J. (2006). Microbial systems to mimic mammalian metabolism. Speciality Chemicals Magazine (pp. 34–35).

Smith, S. R. L. (1981). Some aspects of ICI's single cell protein process. In H. Dalton (Ed.), *Microbial growth on C1 compounds* (pp. 342–348). London: Heyden.

Turner, W. B. (1971). *Fungal metabolites.* London: Academic Press.

Van Dien, S. (2013). From the first drop to the first truckload: commercialization of microbial processes for renewable chemicals. *Current Opinion in Biotechnology, 24*, 1061–1068.

van Wezel, G. P., & Mcdowall, K. M. (2011). The regulation of the secondary metabolism of *Streptomyces*: new links and experimental advances. *Natural Product Reports, 28*, 1311–1333.

Walsh, G. (2012). New biopharmaceuticals. *Biopharm International.* June, 1–5.

Webb, M. (1993). Monoclonal antibodies. In J. M. Walker, & E. B. Gingold (Eds.), *Molecular biology and biotechnology* (pp. 357–386). Cambridge: Royal Society of Chemistry.

Wiebe, M. G. (2004). Myco-protein—an overview of a successful fungal product. *Mycologist, 18*, 17–20.

Yim, H., Hasselbeck, R., Niu, W., Pujol-baxley, C., et al. (2011). Metabolic engineering of *Escherichia coli* for direct production of 1,4-butanediol. *Nature Chemical Biology, 7*, 445–452.

Microbial growth kinetics

As outlined in Chapter 1, fermentations may be carried out as batch, continuous, and fed-batch processes. The mode of operation is, to a large extent, dictated by the type of product being produced. This chapter will consider the kinetics and applications of batch, continuous, and fed-batch processes.

BATCH CULTURE

Batch culture is a closed culture system that contains an initial, limited amount of nutrient. The inoculated culture will pass through a number of phases, as illustrated in Fig. 2.1. After inoculation there is a period during which it appears that no growth takes place; this period is referred to as the lag phase and may be considered as a time of adaptation. In a commercial process, the length of the lag phase should be reduced as much as possible and this may be achieved by using a suitable inoculum, and cultural conditions as described in depth in Chapter 6.

EXPONENTIAL PHASE

Following a period during which the growth rate of the cells gradually increases, the cells grow at a constant, maximum rate and this period is known as the log, or exponential, phase and the increase in biomass concentration will be proportional to the initial biomass concentration.

$$\frac{dx}{dt} \propto x$$

where x is the concentration of microbial biomass (g dm^{-3}), t is time (h), d is a small change.

This proportional relationship can be transformed into an equation by introducing a constant, the specific growth rate (μ), that is, the biomass produced per unit of biomass and takes the unit per hours. Thus:

$$\frac{dx}{dt} = \mu x \tag{2.1}$$

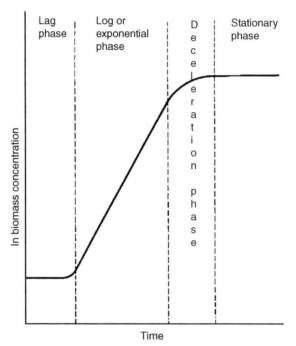

FIGURE 2.1 **Growth of a Typical Microbial Culture in Batch Conditions**

On integration Eq. (2.1) gives:

$$x_t = x_0 e^{\mu t} \tag{2.2}$$

where x_0 is the original biomass concentration, x_t is the biomass concentration after the time interval, t hours, e is the base of the natural logarithm.

On taking natural logarithms, Eq. (2.2) becomes:

$$\ln x_t = \ln x_0 + \mu t \tag{2.3}$$

Thus, a plot of the natural logarithm of biomass concentration against time should yield a straight line, the slope of which would equal to μ. During the exponential phase nutrients are in excess and the organism is growing at its maximum specific growth rate, μ_{max}. It is important to appreciate that the μ_{max} value is the maximum growth rate under the prevailing conditions of the experiment, thus the value of μ_{max} will be affected by, for example, the medium composition, pH, and temperature. Typical values of μ_{max} for a range of microorganisms are given in Table 2.1.

It is easy to visualize the exponential growth of single celled organisms that replicate by binary fission. Indeed, animal and plant cells in suspension culture will behave very similarly to unicellular microorganisms (Griffiths, 1986; Petersen & Alfermann, 1993). However, it is more difficult to appreciate that mycelial organisms,

Table 2.1 Some Representative Values of μ_{max} (Obtained Under the Conditions Specified in the Original Reference) for a Range of Organisms

Organism	μ_{max} (h^{-1})	References
Vibrio natriegens	4.24	Eagon (1961)
Methylomonas methanolytica	0.53	Dostalek et al. (1972)
Aspergillus nidulans	0.36	Trinci (1969)
Penicillium chrysogenum	0.12	Trinci (1969)
Fusarium graminearum Schwabe	0.28	Trinci (1992)
Plant cells in suspension culture	0.01–0.046	Petersen and Alfermann (1993)
Animal cells	0.01–0.05	Lavery (1990)

which grow only at the apices of the hyphae, also grow exponentially. The filamentous fungi and the filamentous bacteria (particularly the genus *Streptomyces*) are significant fermentation organisms and thus an understanding of their growth is important. Plomley (1959) was the first to suggest that filamentous fungi have a "growth unit" that is replicated at a constant rate and is composed of the hyphal apex (tip) and a short length of supporting hypha. Trinci (1974) demonstrated that the total hyphal length of a mycelium and the number of tips increased exponentially at approximately the same rate indicating that a branch is initiated when a certain hyphal length is reached. Robinson and Smith (1979) demonstrated that it is the volume of a fungal hypha rather than simply the length, that is, the branch initiation factor and Riesenberger and Bergter (1979) confirmed the same observation for *Streptomyces hygroscopicus*. Thus, branching in both fungi and streptomycetes is initiated when the biomass of the hyphal growth unit exceeds a critical level. This is equivalent to the division of a single celled organism when the cell reaches a critical mass. Hence, the rate of increase in hyphal mass, total length, and number of tips is dictated by the specific growth rate and:

$$\frac{dx}{dt} = \mu x,$$

$$\frac{dH}{dt} = \mu H,$$

$$\frac{dA}{dt} = \mu A$$

where H is total hyphal length and A is the number of growing tips. Although the growth of both filamentous fungi and streptomycetes are described by identical kinetics, the mechanisms associated with apical growth differ. The movement of materials to the fungal growing tip is dependent on a microtubule-based transport system (Egan, McClintock, & Reck-Peterson, 2012), whereas that in *Streptomyces* is facilitated by the coiled coil protein DivIVA that recruits other proteins to the growing site forming multiprotein assemblies termed polarisomes (Flardh, Richards, Hempel, Howard, & Butner, 2012).

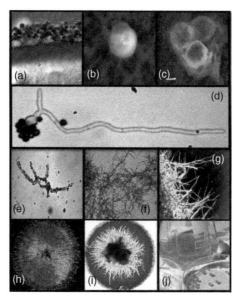

FIGURE 2.2 Morphological Forms of *Aspergillus* sp

(a) profile view of conidiophores (diameter 200 µm) on solid agar medium, (b) single spore, (c) spore package (spore diameter 5 µm), (d) germinated tube (length approx. 250 µm), (e) coagulated type of mycel, in which single ungerminated spores adhere to germinated hyphal tubes (length approx. 100 µm), (f) dispersed mycel, (g) exposed hyphae of a pellet (pellet hair) (length approx. 100 µm), (h) pellet slice (diameter approx. 1000 µm), (i) hairy biopellet (pellet diameter approx. 1000 µm), and (j) submerged biopellets. (Krull et al., 2013)

In submerged liquid culture (shake flask or fermenter), a mycelial organism may grow as dispersed hyphal fragments or as pellets (as shown in Fig. 2.2) and whether the culture is filamentous or pelleted can have a significant influence on the products produced by a mycelial organism (Krull et al., 2013). As discussed in more detail in Chapter 6, the key factors influencing hyphal morphology in submerged culture are the concentration of spores in the inoculum, medium design, and shear conditions. The influence of morphology on culture rheology and oxygen supply is discussed in Chapter 9. The growth of pellets will be exponential until the density of the pellet results in diffusion limitation. Under such limitation, the central biomass of the pellet will not receive a supply of nutrients, nor will potentially toxic products diffuse out. Thus, the growth of the pellet proceeds from the outer shell of biomass that is the actively growing zone and was described by Pirt (1975) as:

$$M^{1/3} = kt + M_0^{1/3}$$

where M_0 and M are the mycelium mass at time 0 and t, respectively. Thus, a plot of the cube root of mycelial mass against time will give a straight line, the slope of which equals k.

It is possible for new pellets to be generated by the fragmentation of old pellets and, thus, the behavior of a pelleted culture may be intermediate between exponential and cube root growth.

DECELERATION AND STATIONARY PHASES

Whether the organism is unicellular or mycelial, the foregoing equations predict that growth will continue indefinitely. However, growth results in the consumption of nutrients and the excretion of microbial products; events which influence the growth of the organism. Thus, after a certain time the growth rate of the culture decreases until growth ceases. The cessation of growth may be due to the depletion of some essential nutrient in the medium (substrate limitation), the accumulation of some autotoxic product of the organism in the medium (toxin limitation) or a combination of the two.

The nature of the limitation of growth may be explored by growing the organism in the presence of a range of substrate concentrations and plotting the biomass concentration at stationary phase against the initial substrate concentration, as shown in Fig. 2.3. From Fig. 2.3 it may be seen that over the zone A to B an increase in initial substrate concentration gives a proportional increase in the biomass produced at stationary phase, indicating that the substrate is limiting. The situation may be described by the equation:

$$x = Y(S_R - s) \tag{2.4}$$

where x is the concentration of biomass produced, Y is the yield factor (g biomass produced g^{-1} substrate consumed), S_R is the initial substrate concentration, and s is the residual substrate concentration.

Over the zone A to B in Fig. 2.3, s equals zero at the point of cessation of growth. Thus, Eq. (2.4) may be used to predict the biomass that may be produced from a certain amount of substrate. Over the zone C to D an increase in the initial substrate concentration does not give a proportional increase in biomass. This may be due to either the exhaustion of another substrate or the accumulation of toxic products. Over

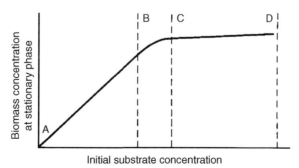

FIGURE 2.3 The Effect of Initial Substrate Concentration on the Biomass Concentration at the Onset of Stationary Phase, in Batch Culture

the zone B to C the utilization of the substrate is deleteriously affected by either the accumulating toxins or the availability of another substrate.

The yield factor (Y) is a measure of the efficiency of conversion of any one substrate into biomass and it can be used to predict the substrate concentration required to produce a certain biomass concentration. However, it is important to appreciate that Y is not a constant—it will vary according to growth rate, pH, temperature, the limiting substrate, and the concentration of the substrates in excess.

The decrease in growth rate and the cessation of growth, due to the depletion of substrate, may be described by the relationship between μ and the residual growth-limiting substrate, represented in Eq. (2.5) and in Fig. 2.4 (Monod, 1942):

$$\mu = \mu_{max}s/(K_s + s) \tag{2.5}$$

Where, s is the substrate concentration in the presence of the organism, K_s is the substrate utilization constant, numerically equal to substrate concentration, when μ is half μ_{max} and is a measure of the affinity of the organism for its substrate.

The zone A to B in Fig. 2.4 is equivalent to the exponential phase in batch culture where substrate concentration is in excess and growth is at μ_{max}. The zone C to A in Fig. 2.4 is equivalent to the deceleration phase of batch culture where the growth of the organism has resulted in the depletion of substrate to a growth-limiting concentration which will not support μ_{max}. If the organism has a very high affinity for the limiting substrate (a low K_s value), the growth rate will not be affected until the substrate concentration has declined to a very low level. Thus, the deceleration phase for such a culture would be short. However, if the organism has a low affinity for the substrate (a high K_s value) the growth rate will be deleteriously affected at a relatively high substrate concentration. Thus, the deceleration phase for such a culture would be relatively long. Typical values of K_s for a range of organisms and

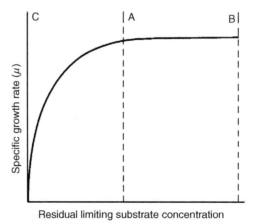

FIGURE 2.4 The Effect of Residual Limiting Substrate Concentration on the Specific Growth Rate of a Hypothetical Bacterium

Table 2.2 Some Representative Values of K_s for a Range of Microorganisms and Substrates

Organism	Substrate	K_s (mg dm^{-3})	References
Escherichia coli	Glucose	6.8×10^{-2}	Shehata and Marr (1971)
Saccharomyces cerevisiae	Glucose	25.0	Pirt and Kurowski (1970)
Pseudomonas sp.	Methanol	0.7	Harrison (1973)

substrates are shown in Table 2.2, from which it may be seen that such values are usually very small and the affinity for substrate is high. It will be appreciated that the biomass concentration at the end of the exponential phase is at its highest and, thus, the decline in substrate concentration will be very rapid so that the time period during which the substrate concentration is close to K_s is very short. While the concept of K_s facilitates the quantitative description of the relationship between specific growth rate and substrate concentration it should not be regarded as a true constant. There are many cases in the literature of microorganisms expressing different enzyme systems, achieving the same metabolic end point, depending on the concentration of substrate. Harder and Dijkhuizen's review (1983) and that of Ferenci (1999) cite many such examples for carbon and nitrogen metabolism in which high affinity (low K_s) systems are expressed under limitation and low affinity systems (high K_s) expressed under nutrient excess conditions, thus enabling organisms to "scavenge" for substrate under conditions of nutrient stress.

The stationary phase in batch culture is that point where the growth rate has declined to zero. However, it is important to appreciate that the cessation of growth is not the microbiological equivalent of a car running out of fuel. Although the two situations may be the result of fuel limitation, microorganisms have evolved strategies that avoid the consequences of coming to a halt in the fast lane. The kinetic descriptions discussed so far ignore the physiological adaptations that microorganisms undergo during a period of declining growth rate—adaptations that equip them to survive periods of nutrient starvation. Stationary phase cells are not simply exponential phase cells that have stopped growing—they are physiologically different.

Sigma factors are bacterial protein transcription factors that facilitate promoter recognition by RNA polymerase, thus enabling gene transcription and, ultimately, gene expression. Each RNA polymerase molecule consists of one sigma factor and a core enzyme (consisting of several units)—the nature of the sigma factor dictates the promoters that may be recognized. All bacteria have one sigma factor that recognizes the promoters of "housekeeping" genes enabling growth. However, they also have a range of sigma factors that recognize the promoters of other genes that may be switched on under specific circumstances. Thus, the deployment of particular sigma factors under specific prevailing circumstances enables the organism to adapt to its environment and change its gene expression profile and hence its phenotype. *E. coli* has seven sigma factors (see Table 2.3) one of which, σ^{38} or σ^S, recognizes genes transcribed uniquely during the stationary phase (Landini, Egli, Wolf, & Lacour, 2014). Bacteria have been shown to modify their physiology in response to both growth rate and biomass concentration. The response to biomass concentration is referred to as "quorum sensing"—a phenomenon in which the expression of certain genes only occurs when the culture reaches a threshold biomass. In this system, each cell produces a signal molecule, the concentration of which in the environment is then dependent on the number of bacteria producing it. Thus, as biomass concentration increases so does that of the signal molecule, until it reaches the threshold level and specific genes are induced. The nature of the signal molecules and some of the processes controlled by quorum sensing are shown in Table 2.4. An example of quorum sensing in the induction of secondary metabolism is discussed in detail in Chapter 6. However, in an elegant continuous culture experiment (see in later sections), Ihssen and Egli (2004) demonstrated that the level of σ^S in *E. coli* is controlled by growth

Table 2.3 The Sigma Factors of *Escherichia coli*

Sigma Factor	Function
σ^{70} or σ^D (RpoD)	Housekeeping sigma factor—recognizes genes required for growth
σ^{19} or σ^I (FecI)	The ferric citrate sigma factor, recognizes the *fec* gene for iron transport
σ^{24} or σ^E (RpoE)	Regulates and responds to extracytoplasmic functions
σ^{28} or σ^F (RpoF)	Control of flagella and pilli synthesis
σ^{32} or σ^H (RpoH)	Controls the production of heat shock proteins
σ^{38} or σ^s (RpoS)	Controls the general stress response of cells entering the stationary phase
σ^{54} or σ^N (RpoN)	Controls the response to nitrogen limitation

Table 2.4 Quorum Sensing Systems

Signal Molecule	Controlled Property	Taxonomic Group
Gamma-butyrolactones	Initiation of secondary metabolism and morphological differentiation	*Streptomyces* spp.
Acyl homoserine lactones	Bacterial bioluminescence	Gram negative bacteria
	Virulence	
	Antibiotic synthesis	
Oligopeptides	Biofilm formation	Gram positive bacteria
	Competence	
	Sporulation	
	Virulence	

rate and not by biomass concentration with σ^S levels being enhanced at low growth rates—that is, under conditions of nutrient depletion or toxin accumulation akin to the deceleration and stationary phases. The expression of the genes recognized by σ^S results in the expression of a raft of phenotypes, protecting the cells from a range of stresses that may be experienced in the stationary phase. The range of σ^S influenced characteristics include:

- cell size—stationary phase cells are smaller than those from the exponential phase, thus increasing the surface area to volume ratio and facilitating the enhanced uptake of limiting nutrients;
- production of detoxifying enzymes such as catalase and superoxide dismutase;
- repair and protection systems including DNA repair and protein protection by chaperonins;
- resistance to osmotic stress;
- resistance to high temperatures;
- resistance to adverse pH.

The σ^S governed responses involve approximately 500 genes, accounting for 10% of the genome and the overall process has been termed the "general stress response" (Hengge-Aronis, 1996). However, only about 140 genes are expressed simply as a result of enhanced σ^S levels—the control of the remainder is mediated by both σ^S and specific environmental stresses. Such an orchestrated wide-reaching process would have a significant energy demand—a requirement that is at odds with the energy status of stationary phase organisms. Landini et al. (2014) discusses the "general stress response" as an immediate reaction to nutrient deprivation by cells which still have the metabolic activity to take the necessary action to protect themselves from impending stress—that is, cells which have not yet entered the stationary phase but are experiencing growth rates less than the maximum. The ubiquitous nature of the response means that the organism is then protected against a range of adverse conditions that may develop. The control of σ^S synthesis and activity is a complex interaction of initiation of transcription, modulation of the mRNA transcripts and their translation and the regulation of the degradation of σ^S and its affinity for promoters. Landini et al. (2014) summarize these control systems in their excellent review.

While *E. coli* responds to nutrient limitation by modulating its physiology, other bacteria respond more dramatically by undergoing complex differentiation processes that enable the production of cell types capable of surviving adverse conditions. *Bacillus subtilis* produces a range of cell types including endospores (dormant cells), cannibal cells that prey on vegetative cells (of the same species), and thus overcome nutrient limitation, matrix producing cells that form biofilms and motile cells bearing flagella. The streptomycetes (filamentous bacteria) produce aerial hyphae bearing exospores. As in *E.coli*, sigma factors also play key controlling roles in the transition from exponential growth to stationary phase in these differentiating organisms. In *Bacillus subtilis,* there are at least 17 alternative sigma factors with sigma-H being paramount in a transcription cascade controlling the development of the endospore. Sigma-H has been shown to control the expression of 87 genes in *B. subtilis*

(Britton et al., 2002). While the "stationary phase response" in *E.coli* has been attributed to the organism's titration of its decreasing growth rate (due to nutrient limitation), in *B. subtilis* the transition to sporulation and other morphological types is a response to the complex interaction of the detection of both biomass level (quorum sensing) and nutrient limitation (Lazazzera, 2000; Britton et al., 2002). The degree of nutrient limitation modulates the quorum sensing response, again enabling the organism to undergo a series of energy-dependent transformations to adapt to imminent starvation conditions before the source of that energy is completely depleted.

The production of aerial hyphae and sporulation by the streptomycetes under nutrient limitation is a highly complex process that is responding to environmental conditions and accompanied by other stress responses such as protection against free radicals. *Streptomyces coelicolor* has 63 different sigma factors (Hopwood, 2007), 49 of which belong to the ECF family (extracytoplasmic function) and detect environmental change, including nutrient limitation and oxidative stress. It is interesting to note that morphological differentiation in *Streptomyces griseus* is governed by quorum sensing whereas that in *S. coelicolor* is not. Thus, closely related organisms have evolved different mechanisms to accomplish the same end point. The filamentous fungi also produce a range of taxonomically dependent spore types, again responding to environmental signals. However, it is important to appreciate that many fungi and streptomycetes will not undergo complete differentiation in submerged liquid culture, as this is not their natural habitat. Sporulation of mycelial organisms in submerged culture is considered in Chapter 6 (inoculum development) from which it can be appreciated that it has been observed far more frequently for fungal systems than for streptomycete ones. The differentiation of *Streptomyces* spp. has been studied extensively in the last 10 years and has been shown to involve a programmed cell death (PCD) process. As the name suggests, PCD is a carefully controlled process resulting in cell death that is actually beneficial to the development and survival of the colony as a whole. Although PCD is more associated with eukaryotic organisms (see later) it has been observed in a number of prokaryotes, particularly when grown on solidified medium enabling the development of defined colonies (Tanouchi, Pai, Buchler, & You, 2012). The development of *Streptomyces antibioticus* on solidified medium involves a number of distinct stages (Manteca, Fernandez, & Sanchez, 2005; Yague, Lopez-Garcia, Rioseras, Sanchez, & Manteca, 2012, 2013):

- Compartmentalized young mycelium (termed MI) develops from germinated spores.
- Selected compartments of the MI mycelium die, controlled by a highly ordered process; the remaining viable segments then develop into a multinucleated second (MII) mycelium, presumably utilizing substrate from the "sacrificed" cells.
- The MII mycelium develops in the agar medium until it undergoes a second PCD event and the surviving MII express the synthesis of an outer hydrophobic layer and grow into the air. Again, aerial mycelium develops at the expense of the dead cells.
- The aerial mycelium produces aerial spores.

Manteca, Alvarez, Salazar, Yague, and Sanchez, 2008; Manteca, Sanchez, Jung, Schwammle, and Jensen, 2010 demonstrated that *S. coelicolor* differentiated in submerged liquid culture from MI mycelium to MII, as a result of PCD. However, the MII mycelium did not develop a hydrophobic layer and sporulation did not occur. Interestingly, secondary metabolism was only expressed in MII mycelium. Thus, contrary to previous expectations, differentiation did occur in submerged culture and this was associated with secondary metabolism.

As indicated earlier, animal cell cultures also follow the same basic pattern of growth as microbial cultures and enter a stationary phase. Although cells may be physically damaged by extreme culture conditions, resulting in necrosis, the more common explanation of cessation of growth is apoptosis or PCD. This is a regulated cellular process in response to nutrient deprivation or metabolite accumulation. The phenomenon can be limited by the addition of fresh medium (see later discussion of continuous and fed-batch culture) or by supplementing the medium with antiapoptotic additives such as insulin growth factor (Butler, 2005, 2012). Chon and Zarbis-Papastoitsis (2011) reported that the development of media free of animal-derived components has resulted in significantly increased viable cell densities—with reports of more than 1.5×10^7 cells cm^{-3} being achieved. Cell lines may also be genetically modified by inserting antiapoptotic genes, thus reducing the expression of the programmed cell death cascade (Butler, 2005, 2012). These aspects are discussed in Chapters 4 and 12.

The concept of stationary phase cells being physiologically different from those in exponential phase is reinforced by the phenomenon of secondary metabolite production. The nature of primary and secondary metabolites was introduced in Chapter 1 and the point was made that microorganisms capable of differentiation are often, also, prolific producers of secondary metabolites, compounds not produced during the exponential phase. Bull (1974) pointed out that the stationary phase is a misnomer in terms of the physiological activity of the organism, and suggested that this phase be termed as the maximum population phase. The metabolic activity of the stationary phase has also been recognized in the physiological descriptions of microbial growth presented by Borrow et al. (1961) and Bu'Lock et al. (1965). Borrow et al. (1961) investigated the biosynthesis of gibberellic acid by *Gibberella fujikuroi* and divided the growth of the organism into several phases:

1. The balanced phase; equivalent to the early to middle exponential phase.
2. The storage phase; equivalent to the late exponential phase where the increase in mass is due to the accumulation of lipid and carbohydrate.
3. The maintenance phase; equivalent to the stationary phase.

Gibberellic acid (a secondary metabolite) was synthesized only toward the end of the storage phase and during the maintenance phase. As discussed in Chapter 1, Bu'Lock et al. (1965) coined the terms trophophase, to refer to the exponential phase, and idiophase to refer to the stationary phase where secondary metabolites are produced. The idiophase was depicted as the period subsequent to the exponential phase in which secondary metabolites were synthesized. However, it is now obvious that the culture conditions may be manipulated to induce secondary metabolism during logarithmic growth, for example, by the use of carbon sources, which support a reduced maximum growth rate (see Chapter 4).

Pirt (1975) has discussed the kinetics of product formation by microbial cultures in terms of growth-linked products and nongrowth linked products. Growth-linked may be considered equivalent to primary metabolites, which are synthesized by growing cells and nongrowth linked may be considered equivalent to secondary metabolites. The formation of a growth-linked product may be described by the equation:

$$\frac{dp}{dt} = q_p x \tag{2.6}$$

where p is the concentration of product, and q_p is the specific rate of product formation (mg product g^{-1} biomass h^{-1}).

Also, product formation is related to biomass production by the equation:

$$\frac{dp}{dx} = Y_{p/x} \tag{2.7}$$

where $Y_{p/x}$ is the yield of product in terms of biomass (g product g^{-1} biomass).

Multiply Eq. (2.7) by dx/dt, then:

$$\frac{dx}{dt} \cdot \frac{dp}{dx} = Y_{p/x} \cdot \frac{dx}{dt}$$

and

$$\frac{dp}{dt} = Y_{p/x} \cdot \frac{dx}{dt}.$$

But $\dfrac{dx}{dt} = \mu x$ and therefore:

$$\frac{dp}{dt} = Y_{p/x} \cdot \mu x$$

and

$$\frac{dp}{dt} = q_p x$$

and therefore:

$$\begin{aligned} q_p \cdot x &= Y_{p/x} \cdot \mu x, \\ q_p &= Y_{p/x} \cdot \mu \end{aligned} \tag{2.8}$$

From Eq. (2.8), it may be seen that when product formation is growth associated, the specific rate of product formation increases with specific growth rate. Thus, productivity in batch culture will be greatest at μ_{max} and improved product output will be achieved by increasing both μ and biomass concentration. Nongrowth linked product formation is related to biomass concentration and, thus, increased productivity in batch culture should be associated with an increase in biomass. However, it should be remembered that nongrowth related secondary metabolites are produced only under certain physiological conditions—primarily under limitation of a particular substrate so that the biomass must be in the correct "physiological state" before production can be achieved. The elucidation of the environmental conditions, which create the correct "physiological state" is extremely difficult in batch culture and this aspect is developed in a later section.

Thus, batch fermentation may be used to produce biomass, primary metabolites, and secondary metabolites. For biomass production, cultural conditions supporting the fastest growth rate and maximum cell population would be used; for primary metabolite production conditions to extend the exponential phase accompanied by product excretion and for secondary metabolite production, conditions giving a short exponential phase and an extended production phase, or conditions giving a decreased growth rate in the log phase resulting in earlier secondary metabolite formation.

CONTINUOUS CULTURE

Exponential growth in batch culture may be prolonged by the addition of fresh medium to the vessel. Provided that the medium has been designed such that growth is substrate limited (ie, by some component of the medium), and not toxin limited, exponential growth will proceed until the additional substrate is exhausted. This exercise may be repeated until the vessel is full. However, if an overflow device was fitted to the fermenter such that the added medium displaced an equal volume of culture from the vessel then continuous production of cells could be achieved (Fig. 2.5). If medium is fed continuously to such a culture at a suitable rate, a steady state is

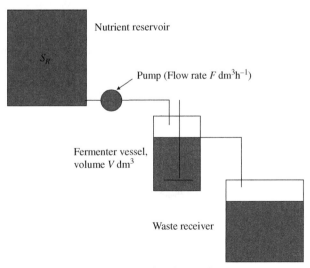

Nutrient reservoir

S_R

Pump (Flow rate F dm^{3}h^{-1})

Fermenter vessel, volume V dm^3

Waste receiver

FIGURE 2.5 A Schematic Representation of a Continuous Culture

achieved eventually, that is, formation of new biomass by the culture is balanced by the loss of cells from the vessel. The flow of medium into the vessel is related to the volume of the vessel by the term dilution rate, D, defined as:

$$D = \frac{F}{V} \tag{2.9}$$

where F is the flow rate ($dm^3\ h^{-1}$) and V is the volume (dm^3).

Thus, D is expressed in the unit h^{-1}.

The net change in cell concentration over a time period may be expressed as:

$$\frac{dx}{dt} = growth - output$$

or

$$\frac{dx}{dt} = \mu x - Dx. \tag{2.10}$$

Under steady-state conditions the cell concentration remains constant, thus $dx/dt = 0$ and:

$$\mu x = Dx \tag{2.11}$$

and

$$\mu = D. \tag{2.12}$$

Thus, under steady-state conditions the specific growth rate is controlled by the dilution rate, which is an experimental variable. It will be recalled that under batch culture conditions, an organism will grow at its maximum specific growth rate and, therefore, it is obvious that a continuous culture may be operated only at dilution rates below the maximum specific growth rate. Thus, within certain limits, the dilution rate may be used to control the growth rate of the culture.

The growth of the cells in a continuous culture of this type is controlled by the availability of the growth limiting chemical component of the medium and, thus, the system is described as a chemostat. The mechanism underlying the controlling effect of the dilution rate is essentially the relationship expressed in Eq. (2.5), demonstrated by Monod in 1942:

$$\mu = \mu_{max} s/(K_s + s)$$

At steady state, $\mu = D$, and, therefore,

$$D = \mu_{max} \bar{s}/(K_s + \bar{s})$$

where $\bar{s}$ is the steady-state concentration of substrate in the chemostat, and

$$\bar{s} = \frac{K_s D}{(\mu_{max} - D)} \tag{2.13}$$

Eq. (2.13) predicts that the substrate concentration is determined by the dilution rate. In effect, this occurs by growth of the cells depleting the substrate to a concentration that supports the growth rate equal to the dilution rate. If substrate is depleted below the level that supports the growth rate dictated by the dilution rate, the following sequence of events takes place:

1. The growth rate of the cells will be less than the dilution rate and they will be washed out of the vessel at a rate greater than they are being produced, resulting in a decrease in biomass concentration.
2. The substrate concentration in the vessel will rise because fewer cells are left in the vessel to consume it.
3. The increased substrate concentration in the vessel will result in the cells growing at a rate greater than the dilution rate and biomass concentration will increase.
4. The steady state will be reestablished.

Thus, a chemostat is a nutrient-limited self-balancing culture system that may be maintained in a steady state over a wide range of submaximum specific growth rates.

The concentration of cells in the chemostat at steady state is described by the equation:

$$\bar{x} = Y(S_R - \bar{s}) \tag{2.14}$$

Where, $\bar{x}$ is the steady-state cell concentration in the chemostat.

By combining Eqs. (2.13) and (2.14), then:

$$\bar{x} = Y\left[S_R - \left\{\frac{K_s D}{(\mu_{max} - D)}\right\}\right] \tag{2.15}$$

Thus, the biomass concentration at steady state is determined by the operational variables, S_R and D. If S_R is increased, $\bar{x}$ will increase but $\bar{s}$, the residual substrate concentration in the chemostat at the new steady state, will remain the same. If D is increased, μ will increase ($\mu = D$) and the residual substrate at the new steady state would have increased to support the elevated growth rate; thus, less substrate will be available to be converted into biomass, resulting in a lower biomass steady state value.

An alternative type of continuous culture to the chemostat is the turbidostat, where the concentration of cells in the culture is kept constant by controlling the flow of medium such that the turbidity of the culture is kept within certain, narrow limits. This may be achieved by monitoring the biomass with a photoelectric cell and feeding the signal to a pump-supplying medium to the culture such that the pump is switched on if the biomass exceeds the set point and is switched off if the biomass falls below the set point. Systems other than turbidity may be used to monitor the biomass concentration, such as CO_2 concentration or pH in which case it would be more correct to term the culture a biostat. The chemostat is the more commonly used system because it has the advantage over the biostat of not requiring complex control systems to maintain a steady state. However, the biostat may be advantageous in continuous enrichment culture in avoiding the total washout of the culture in its early stages and this aspect is discussed in Chapter 3.

The kinetic characteristics of an organism (and, therefore, its behavior in a chemostat) are described by the numerical values of the "constants" Y, μ_{max}, and K_s. It is important to recall the earlier discussion that Y and K_s are not true constants and their values may vary depending on cultural conditions. The value of Y affects the steady-state biomass concentration; the value of μ_{max} affects the maximum dilution rate that may be employed and the value of K_s affects the residual substrate concentration (and, hence, the biomass concentration) and also the maximum dilution rate that may be used. Fig. 2.6 illustrates the continuous culture behavior of a hypothetical bacterium with a low K_s value for the limiting substrate, compared with the initial limiting substrate concentration. With increasing dilution rate, the residual substrate concentration increases only slightly until D approaches μ_{max} when s increases significantly. The dilution rate at which x equals zero (ie, the cells have been washed out of the system) is termed as the critical dilution rate (D_{crit}) and is given by the equation:

$$D_{crit} = \frac{\mu_{max}S_R}{(K_s + S_R)} \tag{2.16}$$

Thus, D_{crit} is affected by the constants, μ_{max} and K_s, and the variable, S_R; the larger S_R the closer is D_{crit} to μ_{max}. However, μ_{max} cannot be achieved in a simple steady state chemostat because substrate limited conditions must always prevail.

Fig. 2.7 illustrates the continuous culture behavior of a hypothetical bacterium with a high K_s for the limiting substrate compared with the initial limiting substrate concentration. With increasing dilution rate, the residual substrate concentration increases significantly to support the increased growth rate. Thus, there is a gradual

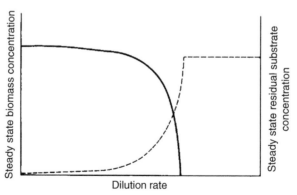

FIGURE 2.6 The Effect of Dilution Rate on the Steady-State Biomass and Residual Substrate Concentrations in a Chemostat Culture of a Microorganism with a Low K_s Value for the Limiting Substrate, Compared with the Initial Substrate Concentration

_____, Steady-state biomass concentration; — — —, Steady-state residual substrate concentration.

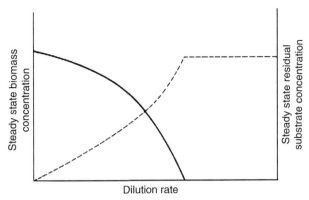

FIGURE 2.7 The Effect of Dilution Rate on the Steady-State Biomass and Residual Substrate Concentrations in a Chemostat of a Microorganism with a High K_s Value for the Limiting Substrate, Compared with the Initial Substrate Concentration

_____, Steady-state biomass concentration; — — —, Steady-state residual substrate concentration.

increase in s and a decrease in x as D approaches D_{crit}. Fig. 2.8 illustrates the effect of increasing the initial limiting substrate concentration on $\bar{x}$ and $\bar{s}$. As S_R is increased, so $\bar{x}$ increases, but the residual substrate concentration is unaffected. Also, D_{crit} increases slightly with an increase in S_R.

The results of chemostat experiments may differ from those predicted by the foregoing theory. The reasons for these deviations may be anomalies associated with the equipment or the theory not predicting the behavior of the organism under certain

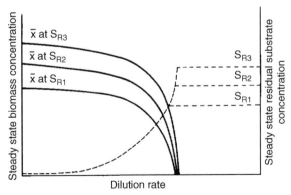

FIGURE 2.8 The Effect of the Increased Initial Substrate Concentration on the Steady-State Biomass and Residual Substrate Concentrations in a Chemostat

_____, Steady-state biomass concentration; — — —, Steady-state residual substrate concentration. S_{R1}, S_{R2}, and S_{R3} represent increasing concentrations of the limiting substrate in the feed medium.

circumstances. Practical anomalies include imperfect mixing and wall growth. Imperfect mixing would cause an increase in the degree of heterogeneity in the fermenter with some organisms being subject to nutrient excess while others are under severe limitation. This phenomenon is particularly relevant to very low dilution rate systems when the flow of medium is likely to be very intermittent. This problem may be overcome by the use of feedback systems, as discussed later in this chapter. Wall growth is another commonly encountered practical difficulty in which the organism adheres to the inner surfaces of the reactor resulting, again, in an increase in heterogeneity. The immobilized cells are not subject to removal from the vessel but will consume substrate resulting in the suspended biomass concentration being lower than predicted. Wall growth may be limited by coating the inner surfaces of the vessel with Teflon.

A frequent observation in carbon and energy limited chemostats is that the biomass concentration at low dilution rates is lower than predicted. This phenomenon has been explained by the concept of "maintenance"—defined by Pirt (1965) as energy used for functions other than the production of new cell material. Thus, non-growth related energy consumption is used to maintain viability—that is, viability is maintained by the consumption of a "threshold" level of energy source before further production of biomass is possible—a process also termed "endogenous metabolism" (Herbert, 1958). Under extreme carbon and energy limitation at very low growth rates (dilution rates), the proportion of carbon used for maintenance is greater than at higher dilution rates resulting in a lower biomass yield. The processes that would account for energy consumption in this context would be:

- osmoregulation,
- maintenance of the optimum internal pH,
- cell motility,
- defense mechanisms, for example, against oxygen stress,
- proofreading,
- synthesis and turnover of macromolecules such as RNA and proteins.

This concept of maintenance has been described by several kinetic "constants" accounting for the diversion of substrate from growth-related activities. Herbert (1958) proposed "endogenous metabolism," (a with units of h^{-1}), also termed the specific maintenance rate by Marr, Nilson, and Clark (1963):

$$dx/dt = (\mu - a)x$$

In this relationship the term ax represents the loss of biomass through maintenance which Pirt (1965) described as an artificial concept. Pirt built an analysis of maintenance on previous work by Schulze and Lipe (1964) who defined maintenance in terms of substrate consumed to maintain the biomass. Thus:

Overall rate of substrate use = Rate of substrate use for maintenance + rate of substrate use for growth

$$\frac{ds}{dt} = \left(\frac{ds}{dt}\right)_M + \left(\frac{ds}{dt}\right)_G$$

The overall observed yield factor (ie, including substrate consumed for maintenance) is given by the term Y_{app}, thus $-(ds/dt).Y_{app} = dx/dt$; remembering that $dx/dt = \mu x$, then:

$$\frac{ds}{dt} = \frac{-\mu x}{Y_{app}}$$

The conversion of substrate into biomass excluding maintenance is then given by the "true" growth yield Y_G, and:

$$\left(\frac{ds}{dt}\right)_G = -\frac{\mu x}{Y_G} \tag{2.17}$$

The substrate consumed for maintenance can then be represented using the term m (maintenance coefficient) that represents the substrate consumed per unit of biomass per hour, that is,

$$\left(\frac{ds}{dt}\right)_M = -mx$$

Eq. 2.17 then reduces to:

$$\frac{1}{Y_{app}} = \frac{m}{\mu} + \frac{1}{Y_G}$$

The limitations of these models have been discussed by van Bodegom (2007). He emphasized that the accumulation of storage products, extracellular losses and changes in metabolic pathways (resulting in changed efficiency) will all affect energy utilization, and hence biomass yield, but do not fit the concept of maintenance as "endogenous metabolism". Furthermore, cell death, while not being a component of "maintenance," would also contribute to a decrease in yield if it were not constant with growth rate. The consideration of maintenance as a constant, independent of growth rate, is also at odds with current knowledge of slow growing organisms. Stouthamer, Bulthius, and van Versveld (1990) showed that growth was energetically more expensive at low growth rates and thus the maintenance and the true growth yield were not biological constants. Indeed, a consideration of the progression of *E. coli* into the stationary phase (described earlier in this chapter) reveals additional resource expenditure involved in the initiation of the "general stress response"—approximately 10% of the genome being under the control of σ^s which is produced in response to low growth rates.

van Bodegom (2007) developed a conceptual model of maintenance m_{tot} that incorporated all the components that may contribute to a decline in conversion of carbon source to biomass:

$$m_{tot} = d\,\frac{\mu_r}{\mu_{max}}\left(\frac{\dfrac{1}{Y_G} + m_p}{1 - d/\mu_r}\right)$$

where d is relative death rate; μ_r is relative growth rate or the growth rate attributed to active cells, Y_G is the true growth yield, and m_p is the physiological maintenance requirement.

This equation shows that m_{tot} is not a constant, but is a nonlinear function of all the component variables. While it is difficult to allocate values to some of these variables, the approach highlights the complexity underlying a "steady-state" and that a variation in yield cannot be accounted for by a simple interpretation of the maintenance concept—it is influenced by the changes in the physiological state of an organism at different growth rates.

The basic chemostat may be modified in a number of ways, but the most common modifications are the addition of extra stages (vessels) and the feedback of biomass into the vessel.

MULTISTAGE SYSTEMS

A multistage system is illustrated in Fig. 2.9. The advantage of a multistage chemostat is that different conditions prevail in the separate stages. This may be advantageous in the utilization of multiple carbon sources and for the separation of biomass production from metabolite formation in different stages, for example, in production of secondary metabolites and biofuels. Harte and Webb (1967) demonstrated that when *Klebsiella aerogenes* was grown on a mixture of glucose and maltose only the glucose was utilized in the first stage and maltose in the second. Secondary metabolism may occur in the second stage of a dual system in which the second stage acts as a holding tank where the growth rate is much smaller than that in the first stage. Li et al. (2014) reported the use of an 11 stage continuous process for bioethanol production, each vessel having a capacity of 480,000 dm^3. The first fermenter was aerated and thus enabled rapid biomass development while the remaining fermenters

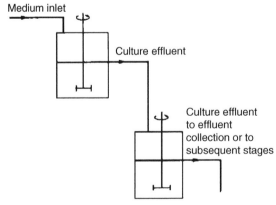

FIGURE 2.9 A Multistage Chemostat

were operated anaerobically, facilitating ethanol production. An earlier example of the industrial application of the technique is in continuous brewing which is described in a later section.

FEEDBACK SYSTEMS

A chemostat incorporating biomass feedback has been modified such that the biomass in the vessel reaches a concentration above that possible in a simple chemostat, that is, greater than $Y(S_R-s)$. Biomass concentration may be achieved by:

1. Internal feedback. Limiting the exit of biomass from the chemostat such that the biomass in the effluent stream is less concentrated than in the vessel.
2. External feedback. Subjecting the effluent stream to a biomass separation process, such as sedimentation or centrifugation, and returning a portion of the concentrated biomass to the growth vessel.

Pirt (1975) gave a full kinetic description of these feedback systems and this account summarizes his analysis.

Internal feedback

A diagrammatic representation of an internal feedback system is shown in Fig. 2.10a. Effluent is removed from the vessel in two streams, one filtered, resulting in a dilute effluent stream (and, thus, a concentration of biomass in the reactor) and one

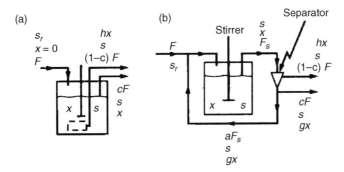

FIGURE 2.10 Diagrammatic Representations of Chemostats with Feedback (Pirt, 1975)

(a) Internal feedback. F, flow rate of incoming medium (dm^3 h^{-1}); c, fraction of the outflow which is not filtered; x, biomass concentration in the vessel and in the unfiltered stream; hx, biomass concentration in the filtered stream (b) External feedback. F, flow rate from the medium reservoir (dm^3 h^{-1}); F_s, flow rate of the effluent upstream of the separator; x, biomass concentration in the vessel and upstream of the separator; hx, biomass concentration in the dilute stream from the separator; g, factor by which the separator concentrates the biomass; a, proportion of the flow which is fed back to the fermenter; s, substrate concentration in the vessel and effluent lines; S_R, substrate concentration in the medium reservoir.

unfiltered. The proportion of the outflow leaving via the filter and the effectiveness of the filter then determines the degree of feedback. The flow rate of incoming medium is designated $F(\text{dm}^3 \text{ h}^{-1})$ and the fraction of the outflow which is not filtered is designated c; thus the outflow rate of the unfiltered stream is cF and that of the filtered stream is $(1-c)F$. The concentration of the biomass in the fermenter and in the unfiltered stream is x and the concentration of biomass in the filtered stream is hx. The biomass balance of the system is:

Change in biomass = Growth−Output in unfiltered stream−Output in filtered stream, which may be expressed as:

$$\frac{dx}{dt} = \mu x - cDx - (1-c)Dhx \tag{2.18}$$

or:

$$\frac{dx}{dt} = \mu x - D\{c(1-h)+h\}x.$$

At steady state $dx/dt = 0$, thus:

$$\mu \bar{x} = D\{c(1-h)+h\}\bar{x}.$$

and

$$\mu = D(c(1-h)+h)$$

If the term "$c(1-h) + h$" is represented by "A," then:

$$\mu = AD \tag{2.19}$$

When the filtered effluent stream is cell free then $h = 0$ and $A = c$. However, if the filter removes very little biomass then h will approach 1, and when $h = 1$ there is no feedback and $A = h$. Thus, the range of values of A is c to h and when feedback occurs A is less than 1, which means that μ is less than D.

The concentration of the growth limiting substrate in the vessel at steady state is then given by:

$$\bar{s} = \frac{K_s AD}{(\mu_{\text{max}} - AD)} \tag{2.20}$$

and the biomass concentration at steady state is given by:

$$\bar{x} = \frac{Y}{A(S_R - \bar{s})} \tag{2.21}$$

External feedback

A diagrammatic representation of an external feedback system is shown in Fig. 2.10b. The effluent from the fermenter is fed through a separator, such as a continuous centrifuge or filter, which produces two effluent streams—a concentrated biomass stream and a dilute one. A fraction of the concentrated stream is then returned to the vessel. The flow rate from the medium reservoir is $F(dm^3 h^{-1})$; the flow rate of the effluent upstream of the separator is $F_s(dm^3 h^{-1})$ and the concentration of biomass in the stream (and in the fermenter) is x; a is the proportion of the flow which is fed back to the fermenter and g is the factor by which the separator concentrates the biomass. Biomass balance in the system will be:

$$\text{Change} = \text{growth} - \text{output} + \text{feedback}$$

or

$$\frac{dx}{dt} = \mu x - \frac{F_s x}{V} + \frac{a F_s g x}{V} \tag{2.22}$$

The culture outflow (before separation), F_s, from the chemostat is:

$$F_s = F + a F_s$$

or

$$F_s = \frac{F}{(1-a)}$$

substituting $F/(1-a)$ for F_s in Eq. (2.22) and remembering that $D = F/V$:

$$\frac{dx}{dt} = \mu x - \frac{Dx}{(1-a)} + \frac{agDx}{(1-a)} \tag{2.23}$$

If all the cells are returned to the fermenter then biomass will continue to accumulate in the vessel. However, if the feedback is partial then a steady state may be achieved, $dx/dt = 0$ and:

$$\mu = BD \tag{2.24}$$

where $B = (1-ag)/(1-a)$.

The steady-state substrate and biomass concentrations in a fermenter with feedback are then given by the following equations:

$$\bar{s} = \frac{BDK_s}{(\mu_{max} - D)} \tag{2.25}$$

$$\bar{x} = \frac{Y}{B(S_R - s)} \tag{2.26}$$

From the equations describing μ, $\bar{s}$, and $\bar{x}$ (2.19–2.21, 2.24–2.26) in a fermenter with either external or internal feedback it can be appreciated that:

1. Dilution rate is greater than growth rate.
2. Biomass concentration in the vessel is increased.
3. The increased biomass concentration results in a decrease in the residual substrate compared with a simple chemostat.

4. The maximum output of biomass and products is increased.
5. Because D is less than μ, the critical dilution rate (the dilution rate at which washout occurs) is increased.

Biomass feedback is applied widely in effluent treatment systems where the advantages of feedback contribute significantly to the process efficiency. The outlet substrate concentration is considerably less and the feedback of biomass may improve stability in effluent treatment systems where mixed substrates of varying concentration are used. The system will also result in increased productivity of microbial products as illustrated by Major and Bull (1989) who reported very high lactic acid productivities in laboratory biomass recycle fermentations. Anaerobic processes are particularly suited to feedback continuous culture because the elevated biomass is not susceptible to oxygen limitation.

Feedback systems seem well suited for animal cell culture where low growth rates and low cell densities limit productivity and while it appears particularly attractive for immobilized animal cells, the advances in cell separation technology has also enabled the development of suspended culture processes. The potential for continuous animal-cell processes is considered in the next section of this chapter.

COMPARISON OF BATCH AND CONTINUOUS CULTURE IN INDUSTRIAL PROCESSES

Biomass productivity

The productivity of a culture system may be described as the output of biomass per unit time of the fermentation. Thus, the productivity of a batch culture may be represented as:

$$R_{batch} = \frac{(x_{max} - x_0)}{(t_i + t_{ii})} \tag{2.27}$$

where R is the output of the culture (g biomass dm^{-3} h^{-1}), x_{max} is the maximum cell concentration achieved at stationary phase, x_0 is the initial cell concentration at inoculation, t_i is the time during which the organism grows at μ_{max}, and t_{ii} is the time during which the organism is not growing at μ_{max} and includes the lag phase, the deceleration phase and the periods of batching, sterilizing, and harvesting.

The productivity of a continuous culture may be represented as:

$$R_{cont} = D\bar{x}\left(1 - \frac{t_{iii}}{T}\right) \tag{2.28}$$

where R_{cont} is the output of the culture (g biomass dm^{-3} h^{-1}), t_{iii} is the time period prior to the establishment of a steady state and includes vessel preparation, sterilization, and operation in batch culture prior to continuous operation, and T is the time period during which steady state conditions prevail.

The term $D\bar{x}$ increases with increasing dilution rate until it reaches a maximum value, after which any further increase in D results in a decrease in $D\bar{x}$, as illustrated

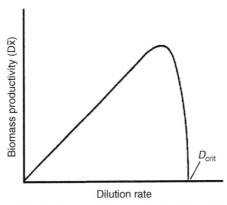

FIGURE 2.11 The Effect of Dilution Rate on Biomass Productivity in Steady-State Continuous Culture

in Fig. 2.11. Thus, maximum productivity of biomass may be achieved by the use of the dilution rate giving the highest value of $D\bar{x}$.

The output of a batch fermentation described by Eq. (2.27) is an average over the period of the fermentation and, because the rate of biomass production is dependent on initial biomass ($dx/dt = \mu x$), the vast proportion of the biomass is produced toward the end of the fermentation. Thus, productivity in batch culture is at its maximum only toward the end of the process. For a continuous culture operating at the optimum dilution rate, under steady-state conditions, the productivity will be constant and always maximum. Thus, the productivity of the continuous system must be greater than the batch. A continuous system may be operated for a very long time period (several weeks or months) so that the negative contribution of the unproductive time, t_{iii}, to productivity would be minimal. However, a batch culture may be operated for only a limited time period so that the negative contribution of the time, t_{ii}, would be very significant, especially when it is remembered that the batch culture would have to be reestablished many times during the time-course of a continuous run. Thus, the superior productivity of biomass by a continuous culture, compared with a batch culture, is due to the maintenance of maximum output conditions throughout the fermentation and the insignificance of the nonproductive period associated with a long-running continuous process.

The steady state achievable in a continuous process also adds to the advantage of improved biomass productivity. Cell concentration, substrate concentration, product concentration, and toxin concentration should remain constant throughout the fermentation. Thus, once the culture is established the demands of the fermentation, in terms of process control, should be constant. In a batch fermentation, the demands of the culture vary during the fermentation—at the beginning, the oxygen demand is low but toward the end the demand is high, due to the high biomass and the increased viscosity of the broth. Also, the amount of cooling required will increase during the process, as will the degree of pH control. In a continuous process oxygen demand, cooling requirements and pH control should remain constant. Thus, the use of continuous culture should allow for the easier introduction of process automation.

A batch process requires periods of intensive labor during medium preparation, sterilization, batching, and harvesting but relatively little during the fermentation itself. However, a continuous process results in a more constant labor demand in that medium is supplied continuously sterilized (see Chapter 5), the product is continuously extracted and the relative time spent on equipment preparation and sterilization is very small.

The argument against continuous biomass processes is that the duration of a continuous fermentation is very much longer than a batch one so that there is a greater probability of a contaminating organism entering the continuous process and a greater probability of equipment failure. However, problems of contamination and equipment reliability are related to equipment design, construction, and operation and, provided sufficiently rigorous standards are applied, these problems can be overcome. In fact, the fermentation industry has recognized the superiority of continuous culture for the production of biomass and several large-scale processes have been established. This aspect is considered in more detail in a later section of this chapter.

Metabolite productivity

Theoretically, a fermentation to produce a metabolite should also be more productive in continuous culture than in batch because a continuous culture may be operated at the dilution rate that maintains product output at its maximum, whereas in batch culture product formation may be a transient phenomenon during the fermentation. The kinetics of product formation in continuous culture have been reviewed by Pirt (1975) and Trilli (1990). Product formation in a chemostat may be described as:

Change in product concentration = production − output:

or:

$$\frac{dp}{dt} = q_p x - D_p$$

(2.29)

where p is the concentration of product and q_p is the specific rate of product formation (mg product g^{-1} biomass h^{-1}).

At steady state, $dp/dt = 0$, and thus:

$$\bar{p} = \frac{q_p \cdot x}{D}$$

(2.30)

where $\bar{p}$ is the steady-state product concentration. If q_p is strictly related to μ, then as D increases so will q_p; thus, the steady-state product concentration ($\bar{p}$) and product output (D_p) will behave in the same way as x and D_x, as shown in Fig. 2.12a. Thus, the production of a growth-related primary metabolite will be described by Eq. (2.30). If q_p is independent of μ then it will be unaffected by D and thus the

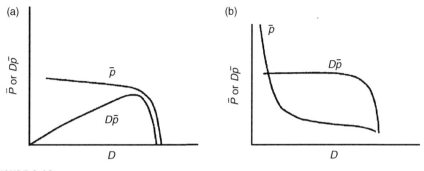

FIGURE 2.12

The effect of D on steady-state product concentration ($\bar{p}$) and product output ($D\bar{p}$) when (a) q_p is growth related. (b) q_p is independent of growth rate.

concentration will decline with increasing D but output will remain constant, as shown in Fig. 2.12b. If product formation occurs only within a certain range of growth rates (dilution rates) then a more complex relationship is produced.

Thus, from this consideration a chemostat process for the production of a product can be designed to optimize either output (g dm^{-3} h^{-1}) or product concentration. However, as Heijnen, Terwisscha van Scheltinga, and Straathof (1992) explained, when q_p is growth-related the advantage of high productivity obtained at high dilution rates must be balanced against the disadvantage of low product concentration resulting in increased downstream processing costs. The other arguments presented for the superiority of continuous culture for biomass production also hold true for product synthesis—ease of automation and the advantages of steady state conditions. The question that then arises is "Why has the fermentation industry not adopted continuous culture for the manufacture of microbial products?" It can be appreciated that the arguments cited previously against continuous culture (contamination and equipment reliability) are not valid, as these difficulties have been overcome in the large-scale continuous biomass processes. The answer to the question lies in the highly selective nature of continuous culture. We have already seen that μ is determined by D in a steady state chemostat and that μ and D are related to substrate concentration according to the equation:

$$D = \mu = \frac{\mu_{max}\bar{s}}{(K_s + \bar{s})}$$

The effect of substrate concentration on specific growth rate for two organisms, A and B, is shown in Fig. 2.13. A is capable of growing at a higher specific growth rate at any substrate concentration. The self-balancing properties of the chemostat mean that the organism reduces the substrate concentration to the value where $\mu = D$. Thus, at dilution rate X, organism A would reduce the substrate concentration

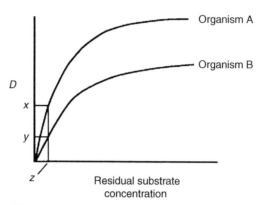

FIGURE 2.13 Competition Between two Organisms in a Chemostat

to Z. However, at this substrate concentration, organism B could grow only at a μ of Y. Therefore, if organisms A and B were introduced into a chemostat operating at dilution rate X, A would reduce the substrate concentration to Z at which B could not maintain a μ of X and would be washed out at a rate of $(X-Y)$ and a monoculture of A would be established eventually. The same situation would occur if A and B were mutant strains arising from the same organism. Commercial organisms have been selected and mutated to produce metabolites at very high concentrations (see Chapter 3) and, as a result, tend to grow inefficiently with low μ_{max} values and, possibly, high K_s values. Back mutants of production strains produce much lower concentrations of product and, thus, grow more efficiently. If such back mutants arise in a chemostat industrial process, then the production strain will be displaced from the fermentation as described in the foregoing scenario.

Calcott (1981) described this phenomenon as "contamination from within" and this type of "contamination" cannot be solved by the design of more "secure" fermenters. Thus, it is the problem of strain degeneration that has limited the application of large-scale continuous culture to biomass and, to a lesser extent, potable, and industrial alcohol. The production of alcohol by continuous culture is feasible because it is a byproduct of energy generation and, thus, is not a drain on the resources of the organism. Indeed, Monte Alegre, Rigo, and Joekes (2003) reported that 30% of Brazilian bioethanol is produced by continuous culture. However, it is possible that the technique could be exploited for other products provided strain degeneration is controlled; this may be possible in certain genetically engineered strains. Some processes have been developed using chemostat culture but the "industry standard" for large-scale culture of microbial metabolites is the fed-batch process, described later in this chapter.

The adoption of continuous culture for animal cell products is even more complex than for microbial systems. Continuous animal cell culture processes were considered in the 1980s and 90s and Griffiths (1992) compared the following process options for producing an animal cell product:

1. Batch culture.
2. Semicontinuous culture where a portion of the culture is harvested at regular intervals and replaced by an equal volume of medium.
3. Fed-batch culture where medium is fed to the culture resulting in an increase in volume (see in later sections).
4. Continuous perfusion where an immobilized cell population is perfused with fresh medium and is equivalent to an internal feedback continuous system.
5. Continuous culture.

The characteristics of all five modes of operation are shown in Table 2.5, from which it may be seen that the perfusion continuous system appears extremely attractive. However, it had been frequently reported at that time that the practicalities of running a large-scale continuous perfusion system presented considerable difficulties. The process had to be reliable and able to operate aseptically for the long periods necessary to exploit the advantage of a continuous process. Also, the licensing of a continuous process could present some difficulties where a consignment of product must be traceable to a batch of raw materials. In a long-term continuous process several different batches of media would have had to be used which presented the problem of associating product with raw material. Furthermore, it was difficult to monitor the genetic stability of cells immobilized in a large reactor system. In a thought-provoking paper Kadouri and Spier (1997) referred to these issues as "myths" and concluded that there were many examples of continuous processes that had been licensed at that time which were stable and engineered such that they were free of contamination. However, over the coming years the animal cell culture industry followed the route taken by the microbial fermentation industry and adopted fed-batch culture as its preferred option as discussed later in this chapter. Nevertheless, feedback continuous cell culture

Table 2.5 Comparison of the Performance of Different Operational Modes of an Animal Cell Fermentation (After Griffiths, 1992)

Operational Mode	Cell No. ($\times 10^{-6} cm^{-3}$)	Product Yield (mg day^{-1})	Product Yield (mg month^{-1})	Length of Run (days)
Batch	3	100	200	7
Semi-continuous	3	200	600	21
Fed-batch	6	200	500	14
Perfusion	30 +	3000	12000	>30
Continuous culture	2	300	1200	>100

(or perfusion culture) still featured as a production method to combat problems of low cell density, low product concentration, and product instability (Bonham-Carter and Shevitz, 2011). A key feature of perfusion culture is the design of the cell retention device. Early systems used spin filters—centrifugal filtration systems—located within the vessel to achieve internal feedback and such an approach could enable production targets to be met without resorting to large-scale vessels. However, a spin filter takes up significant space in the fermenter, is difficult to scale up and its reliability is questionable. Also, improvements in medium design and mammalian cell expression systems for heterologous proteins resulted in the increases in both cell count and product yield. These developments undermined the advantages of perfusion culture and made the fed-batch solution more attractive and scaling-up by increasing fermenter size, economically feasible. Thus, the adoption of perfusion culture has been predominantly influenced by the developments in the biology of the producing systems and the engineering of the equipment—the former enabling high productivity to be achieved without the frailty of the latter. However, two engineering developments in recent years have now enabled the exploitation of the biological advances in reliable perfusion culture—improved cell retention systems and the adoption of "single use" vessels.

As the name implies, single use vessels are disposable reactors used for a single fermentation run. These sterile bag or bottle vessels are available in volumes ranging from $1\,cm^3$ to $2000\,dm^3$ are constructed from regulatory approved plastics and can incorporate mixing and aeration equipment as well as disposable probes (Butler, 2012). These systems avoid the need for validated cleaning and sterilization regimes, reduce both risk and high capital investment in the stainless steel fermentation plant, retain flexibility in process development, and can be attached to perfusion equipment. The design of disposable bioreactors is considered in more detail in Chapter 7. Pollock, Ho, and Farid (2013) reviewed the development of industrial perfusion systems and Table 2.6 from Pollock's paper summarizes some key continuous perfusion cell culture processes, from which it can be seen that the retention system changed from the spin filter to gravity settling and finally alternating tangential flow filtration (ATF). ATF is a hollow fiber tangential flow filtration system (see Chapter 10) that incorporates a back-flushing cleaning process to prevent the filter from fouling. Pollock et al. (2013) devised a simulation tool to facilitate the choice between perfusion continuous culture (utilizing either spin filter or ATF technology) and fed-batch culture. Their model predicted that the spin filter perfusion approach would struggle to compete on economic, environmental, operational, and reliability criteria. The economic advantages of the ATF system, however, outweighed some reliability issues even based on only a threefold higher cell density than fed-batch culture. However, if environmental or operational factors were dominant then the fed-batch approach was preferable.

Continuous brewing and biomass production, which are the major industrial applications of continuous microbial culture, will now be considered in more detail.

Table 2.6 Perfusion Cell Culture Processes, Reported by Pollock et al. (2013)

Product Name	Protein or Monoclonal Antibody	Clinical Application	Company	Date of First Approval	Perfusion System	Reactor Size (dm³)
ReoPro®	mAb-abciximab	Angioplasty	Janssen Biotech*	1994	Spin filter	500
Cerezyme®	Beta-glucocerebrosidase	Gaucher disease	Genzyme	1994	Gravity settler	2,000
Gonal-f®	r.FSH	Anovulation	Merck-Serono	1997	ND	
Remicade®	mAb-infliximab	Rheumatoid arthritis and other autoimmune diseases	Janssen Biotech*	1998	Spin filter	500
Simulect®	mAb- basiliximab	Transplant rejection	Novartis	1998	Rotational sieve filtration	250
Rebif®	Interferon beta-1a	Multiple sclerosis	Merck-Serono	1998	Fixed bed	75
Kogenate-FS®	r.Factor VIII	Haemophilia A	Bayer	2000	Gravity settler	1,500
Xigris®	r.activated protein C	Sepsis	Eli Lilly	2001	Gravity settler	1,500
Fabrazyme®	Algasidase beta	Fabry disease	Genzyme	2003	Gravity settler	2,000
Myozyme®	Alglucosidase alfa	Pompe disease	Genzyme	2006	Gravity settler	4,000
Simponi®	mAb-golimumab	Rheumatoid arthritis and other autoimmune diseases	Janssen Biotech*	2009	Alternating tangential flow	500 and 1000
Stelara®	mAb-ustekinumab	Psoriasis	Janssen Biotech*	2009	Alternating tangential flow	500

*Microcarrier based process

Continuous brewing

The brewing industry in the United Kingdom has had a relatively brief "courtship" with continuous culture. Two types of continuous brewing have been used:

1. The cascade or multistage system.
2. The tower system.

Hough et al. (1976) described the cascade system utilized at Watneys' Mortlake brewery in London. The process utilized three vessels, the first two for fermentation and the third for separation of the yeast biomass. The specific gravity of the wort was reduced from 1040 to 1019 in the first vessel and from 1019 to 1011 in the second vessel. The residence time for the system was 15–20 h, using wort in the specific gravity range of 1035–1040, and it could be run continuously for 3 months. However, it is believed that the system was abandoned due to problems of excessive biomass production. This process appears to have been used widely in New Zealand with greater success (Kirsop, 1982), but apparently newer installations are of the batch type.

A typical tower fermenter for brewing is illustrated in Fig. 2.14. The system is partially closed in that relatively little yeast leaves the fermenter due to the highly flocculent nature of the strains employed. Thus, the system is a type of internal feedback. Wort is introduced into the base of the tower and passes through a porous plug of yeast. As the wort rises through the vessel it is progressively fermented and leaves the fermenter via a yeast-separation zone, which is twice the diameter of the rest of the tower. Hough et al. (1976) described the protocol employed in the establishment of a yeast plug in the tower and the subsequent operating conditions. Prior to the fermentation, the tower is thoroughly cleaned and steam sterilized, aseptic operation being more important for the continuous process than the batch one. The vessel is filled partially with sterile wort and inoculated with a laboratory culture. The initial stages are designed to encourage high biomass production by the periodical addition of wort over about a 9-day period. A porous plug of yeast develops at the base of the tower. The flow rate of the wort is then gradually increased over a further 9–12 days, by which time an approximate steady state may be achieved. Following the establishment of a high biomass in the fermenter the system is operated such that the wort is converted to beer with the formation of approximately the same amount of yeast as would be produced in the batch process. The beer produced during the 3-week start-up period is usually below specification and would have to be blended with high-quality beer. Thus, more than 3 months' continuous operation is necessary to compensate for the initial losses of the process.

The major advantage of the continuous tower process was that the wort residence time could be reduced from about 1 week to 4–8 h as compared with the batch system. However, the development of the cylindro-conical vessel (initially described by Nathan, 1930, but not introduced until the 1970s, see also Chapter 7) led to the shortening of the batch fermentation time to approximately 48 h. Although this is still considerably longer than the residence time in a tower fermenter, it should be remembered that beer conditioning and packaging takes considerably longer than the fermentation stage, so that the difference in the overall processing time between the

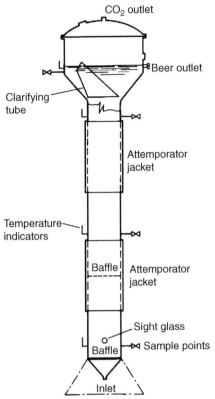

FIGURE 2.14 A Schematic Representation of a Tower Fermenter for the Brewing of Beer (Royston, 1966)

tower and the cylindro-conical batch process is not sufficient to justify the disadvantages of the tower. The major disadvantages of the tower system are the long start-up time, the technical complexity of the plant, the requirement for more highly skilled personnel than for a batch plant, the inflexibility of the system in that a long time delay would ensue between changing from one beer fermentation to another, and finally, the difficulty in matching the flavor of the continuously produced product with that of the traditional batch product. Thus, the continuous-tower system has fallen from use, with virtual universal adoption of the cylindro-conical batch process. However, there is evidence of the possible use of continuous processes for the production of alcohol free-beer employing immobilized yeast cells (Branyik et al., 2012).

Continuous culture and biomass production

Microbial biomass that is produced for human or animal consumption is referred to as single cell protein (SCP). Although yeast was produced as food on a large scale in Germany during the First World War (Laskin, 1977), the concept of utilizing microbial biomass as food was not thoroughly investigated until the 1960s. Since the 1960s, a large number of industrial companies have explored the potential of producing SCP from a wide range of carbon sources. Almost without exception, these investigations have been based on the use of continuous culture as the growth technique.

As discussed previously, continuous culture is the ideal method for the production of microbial biomass. The superior productivity of the technique, compared with that of batch culture, may be exploited fully and the problem of strain degeneration is not as significant as in the production of microbial metabolites. The selective pressure in the chemostat would tend to work to the advantage of the industrialist producing SCP, in that the most efficient strain of the organism would be selected, although this is not necessarily the case for mycelial processes. The development of SCP processes generated considerable research into large-scale chemostat design and the behavior of the production organism in these very large vessels. Many "novel" fermenters have been designed for SCP processes and these are considered in more detail in Chapter 7.

A very wide range of carbon sources have been investigated for the production of SCP. Whey has been used as a carbon source for biomass production since the 1940s and such fermentations have been shown to be economic in that they provide a high-grade feed product, while removing an, otherwise, troublesome waste product of the cheese industry (Meyrath & Bayer, 1979). Cellulose has been investigated extensively as a potential carbon source for SCP production and this work has been reviewed by Callihan and Clemmer (1979) and Woodward (1987). The major difficulty associated with the use of cellulose as a substrate is its recalcitrant nature.

An enormous amount of research has been conducted into the use of hydrocarbons as sources of carbon for biomass processes; the hydrocarbons investigated being methane, methanol, and n-alkanes. A large number of commercial firms were involved in this research field but very few created viable, commercial processes based on SCP production from hydrocarbons because of the economic difficulties involved (Sharp, 1989). At the start of this research, hydrocarbons were relatively cheap but, following the 1973 Middle East War, oil prices escalated and transformed the economic basis of biomass production from petroleum sources. ICI were successful in developing a commercial process for the production of bacterial biomass (Pruteen) from methanol at an annual rate of 54,000–70,000 tons. The process utilized a novel air-lift, pressure cycle fermenter, of 3000 m^3 capacity, and was the first commercial process to produce SCP from methanol (King, 1982). The fermentation was run successfully for periods in excess of 100 days without contamination (Howells, 1982). Regrettably, the economics of the process were such that when the price of soya and fishmeal declined, Pruteen could not compete as an animal feed. Selling in bulk ceased in 1985 (Sharp, 1989) and the 3000 m^3 vessel was eventually demolished.

The expertize developed by ICI during the Pruteen project and RHM's research into the use of a fungus, *Fusarium graminearum,* for the production of human food formed the basis of a joint venture between the two companies. Because the product was a human food rather than an animal feed, the process was economically viable. The ICI pressure cycle pilot-plant was used to produce the fungal biomass (mycoprotein, marketed as Quorn) in continuous culture. The advantage of fungal biomass is that it may be processed to give a textured protein that is acceptable for human consumption. The low shear properties of the air-lift vessel conserve the desirable morphology of the fungus. The process is operated at a dilution rate of between 0.17 and 0.20 h^{-1} (μ_{max} is 0.28 h^{-1}). The phenomenon of mutation and intense selection in the chemostat has proved to be problematical in mycoprotein fermentation, because highly branched mutants have arisen in the vessel resulting in the loss of the desirable morphology. However, the process was routinely operated in chemostat culture for 1000 h on the full scale and the isolation of morphologically stable mutants using continuous culture systems eventually enabled the system to be run for much longer periods (Trinci, 1992).

COMPARISON OF BATCH AND CONTINUOUS CULTURE AS INVESTIGATIVE TOOLS

Although the use of continuous culture on an industrial scale is limited, it is an invaluable investigative technique. The principle characteristic of batch culture is change. Even during the log phase, cultural conditions are not constant and it is only the constant maximum specific growth rate that gives the semblance of stability—biomass concentration, substrate concentration, and microbial products all change exponentially. During the deceleration phase, the onset of nutrient limitation causes the growth rate to decline from its maximum to zero in a very short time, so it is virtually impossible to study the physiological effects of nutrient limitation in batch culture. As Trilli (1990) pointed out, adaptation of an organism to change is not instantaneous, so that the activity of a batch culture is not in equilibrium with the composition of its environment. Physiological events in a batch culture may have been initiated by a change in the environment that took place some significant time before the change was observed. Thus, it is very difficult to relate "cause and effect." The major feature of continuous culture, on the other hand, is "the steady state"—biomass, substrate, and product concentration should remain constant over very long periods of time. Specific growth rate is controlled by dilution rate and growth is nutrient limited. However, it is important not to exaggerate the significance of the steady state because a constant biomass level does not necessarily indicate that the culture is physiologically stable (Malek, Votruba, & Ricica, 1988). Indeed, Ferenci (2006, 2008) emphasized that the concentration of limiting nutrient in an apparent "steady state" system can continue to decline for hundreds of hours. Also, a stable biomass concentration does not necessarily imply a genetically homogeneous population—nor a population genetically identical to the inoculum. The earlier discussions of competition in the chemostat, and the difficulty in its industrial-scale use, stressed the highly selective nature of the continuous culture environment that can result in the selection of mutants better adapted to the operating conditions. Thus, it is important to monitor the genetic variation of the culture and avoid extended culture periods when the study of mutant evolution is not the purpose of the investigation.

Provided the potential "pitfalls" of chemostat culture are appreciated and experiments designed accordingly, it is possible to separate the effects of growth rate and other environmental conditions, for example, temperature, pH, and dissolved oxygen concentration. Furthermore, because any of a wide range of substrates may be used to limit growth in the chemostat the effects of μ, limiting substrate concentration, and biomass concentration may be distinguished from each other. Although the "golden age" of continuous culture was in the 1960s and 1970s, a core of microbial physiologists continued to appreciate the unique contribution that the chemostat could make to the elucidation of the physiology of industrially relevant systems. An excellent example is the effect of growth rate and limiting substrate on metabolite formation. The interpretation of secondary metabolites as compounds produced in the idiophase of batch culture may lead one to suppose that the specific production rates (q_p) of such compounds are inversely linked to specific growth rate (μ). The testing of this supposition may be achieved in chemostat culture and it has been shown to be correct for cephamycin and thienamycin synthesis by *Streptomyces cattleya* (Lilley, Clark, & Lawrence, 1981) and gibberellin by *Gibberella fujikuroi* (Bu'lock, Detroy, Hostalek, & Munim-al-Shakarchi, 1974). However, different relationships have been demonstrated in other systems. Pirt and Righelato (1967) and Ryu and Hospodka (1980) showed that the q_p of penicillin is positively correlated with μ up to a specific growth rate of $0.013\ h^{-1}$, after which it is independent of μ. Pirt (1990) suggested that the apparent negative correlation may be related to penicillin degradation. These observations indicate that the growth rate in a commercial penicillin process should not decline below $0.013\ h^{-1}$. Positive correlations between μ and q_p have been obtained for Chlortetracycline production by *Streptomyces aureofaciens* (Sikyta, Slezak, & Herold, 1961), Oxytetracycline by *Streptomyces rimosus* (Rhodes, 1984) and erythromycin A by *Streptomyces erythraeus* (Trilli, Crossley, & Kontakou, 1987).

In recent years, continuous culture experimentation has again become fashionable and its use has increased significantly. This is due to the need for reproducible, homogeneous and fully controlled culture systems to exploit the advances in bioscience technology as we enter "the post-genomic era" (Hoskisson & Hobbs, 2005 and Bull, 2010).

The key advances that have been developed are:

- The sequencing of whole genomes, enabling comparison of microbial genomes and the preparation of DNA microarrays.
- The use of DNA microarrays to examine gene expression in growing organisms.
- The development of two dimensional gel electrophoresis for the separation of proteins and their identification using matrix assisted laser desorption ionization (MALDI) mass spectrometry.
- The development of metabolic flux analysis which uses stoichiometric models of metabolism, based on enzyme kinetics data, and C^{13} labeling to construct metabolic networks between biochemical pathways.

- The development of bioinformatics computer systems to store, access, and compare data generated from the above.

Thus, microbiologists now have the tools to investigate the mRNA, protein and metabolite profiles of whole organisms and, more importantly, how these profiles change under different physiological conditions—that is, the relationship between gene expression, phenotype, and the environment. The chemostat "steady state" is the ideal environment in which to examine these global changes—an environment in which the growth rate, limiting nutrient, nutrients in excess, and physical conditions can all be maintained at fixed, predetermined levels. Equally, the chemostat is ideal to investigate adaptation at a molecular level to changes in the environment of the organism—such as shifts in growth rate, substrate concentration, and physical conditions—when the transition from one steady state to another can be followed at the molecular level. The study of these processes in batch culture would be difficult in the extreme due to the variation in the cultural conditions and growth either at the maximum rate (in the exponential phase) or at rates continually diminishing to zero (in the deceleration phase). Both Hoskisson and Hobbs (2005) and Bull (2010) have reviewed the renaissance of the chemostat in the postgenomic era and cite literature that has significantly advanced our understanding of microbial physiology. Two studies will be discussed here as exemplars of these approaches.

Tai et al. (2005) argued that if transcriptional change could be reliably related to the fermenter environment then the detection of such change may be used to titrate the physiological state of the fermentation. Furthermore, the identification of key transcripts (termed "signature transcripts") related to specific, industrially significant, parameters would enable both process monitoring and optimization to be achieved by recording the change in gene expression of the process organism itself. However, changes in transcription may be due to more than one environmental parameter that, in batch culture, would be occurring simultaneously. The choice of continuous culture to explore this scenario enabled these workers to expose *Saccharomyces cerevisiae*, an important industrial model organism, to both aerobic and anaerobic conditions under various limitations at a fixed growth rate. Of the 877 genes that were differentially expressed in response to anaerobiosis, only 155 responded to oxygen deprivation regardless of nutrient limitation, that is, consistently linked to anaerobic conditions. Thus, continuous culture experimentation enabled the identification of transcripts reliably linked to a single environmental parameter.

Semisynthetic cephalosporin production by *Penicillium chrysogenum* requires supplementation of the growth medium with adipic acid, a side chain precursor. However, adipic acid is also used by the organism as a carbon and energy source resulting in loss of conversion to cephalosporin. Veiga et al. (2012) cultured the producer strain in a glucose-limited chemostat at a low (secondary metabolism enabling) dilution rate and compared the transcriptome in the presence and absence of adipic acid. This enabled the identification of upregulated genes coding for acyl-CoA oxidases and dehydrogenases, enzymes catalyzing the first step in the beta-oxidation pathway. Deletion of these genes from the production strain resulted in the redirection of adipic acid to cephalosporin synthesis giving an increased yield.

The fiercely selective nature of the chemostat, which is its major disadvantage for industrial production, makes it an excellent tool for the isolation and improvement of microorganisms. The use of continuous culture in this context is considered in Chapter 3, from which it may be seen that continuous enrichment culture offers considerable advantages over batch enrichment techniques and that continuous culture may be used very successfully to select strains producing higher yields of certain microbial enzymes.

FED-BATCH CULTURE

Yoshida, Yamane, and Nakamoto (1973) introduced the term fed-batch culture to describe batch cultures which are fed continuously, or sequentially, with medium, without the removal of culture fluid. A fed-batch culture is established initially in batch mode and is then fed according to one of the following feed strategies:

1. The same medium used to establish the batch culture is added, resulting in an increase in volume.
2. A solution of the limiting substrate at the same concentration as that in the initial medium is added, resulting in an increase in volume.
3. A concentrated solution of the limiting substrate is added at a rate less than in (1) and (2), resulting in an increase in volume.
4. A very concentrated solution of the limiting substrate is added at a rate less than in (1), (2), and (3), resulting in an insignificant increase in volume.

Fed-batch systems employing strategies (1) and (2) are described as variable volume, whereas a system employing strategy (4) is described as fixed volume. The use of strategy (3) gives a culture intermediate between the two extremes of variable and fixed volume.

The kinetics of the two basic types of fed-batch culture, variable volume, and fixed volume, will now be described.

VARIABLE VOLUME FED-BATCH CULTURE

The kinetics of variable volume fed-batch culture have been developed by Dunn and Mor (1975) and Pirt (1974, 1975, 1979). The following account is based on that of Pirt (1975). Consider a batch culture in which growth is limited by the concentration of one substrate; the biomass at any point in time will be described by the equation:

$$x_t = x_0 + Y(S_R - s) \tag{2.31}$$

where x_t is the biomass concentration after time, t hours, and x_0 is the inoculum concentration.

The final biomass concentration produced when $s = 0$ may be described as x_{max} and, provided that x_0 is small compared with x_{max}:

$$x_{max} \simeq Y \cdot S_R \qquad (2.32)$$

If, at the time when $x = x_{max}$, a medium feed is started such that the dilution rate is less than μ_{max}, virtually all the substrate will be consumed as fast as it enters the culture, thus:

$$FS_R \simeq \mu \left(\frac{X}{Y} \right) \qquad (2.33)$$

where F is the flow rate of the medium feed, and X is the total biomass in the culture, described by $X = xV$, where V is the volume of the culture medium in the vessel at time t.

From Eq. (2.33) it may be concluded that input of substrate is equalled by consumption of substrate by the cells. Thus, $(ds/dt) \simeq 0$. Although the total biomass in the culture (X) increases with time, cell concentration (x) remains virtually constant, that is, $(dx/dt) = 0$ and therefore $\mu = D$. This situation is termed a quasi steady state. As time progresses, the dilution rate will decrease as the volume increases and D will be given the expression:

$$D = \frac{F}{(V_0 + Ft)} \qquad (2.34)$$

where V_0 is the original volume. Thus, according to Monod kinetics, residual substrate should decrease as D decreases resulting in an increase in the cell concentration. However, over most of the range of μ that will operate in fed-batch culture, S_R will be much larger than K_s so that, for all practical purposes, the change in residual substrate concentration would be extremely small and may be considered as zero. Thus, provided that D is less than μ_{max} and K_s is much smaller than S_R, a quasi steady state may be achieved. The quasi steady state is illustrated in Fig. 2.15a. The major difference between the steady state of a chemostat and the quasi steady state of a fed-batch culture is that μ is constant in the chemostat but decreases in the fed-batch.

Pirt (1979) has expressed the change in product concentration in variable volume fed-batch culture in the same way as for continuous culture (see Eq. 2.29):

$$dp/dt = q_p x - Dp.$$

Thus, product concentration changes according to the balance between production rate and dilution by the feed. However, in the genuine steady state of a chemostat, dilution rate and growth rate are constant whereas in a fed-batch quasi steady state they change over the time of the fermentation. Product concentration in the chemostat will reach a steady state, but in a fed-batch system the profile of the product concentration over the time of the fermentation will be dependent on the relationship

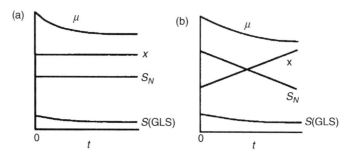

FIGURE 2.15 Time Profiles of Fed-Batch Cultures

μ, specific growth rate; x, biomass concentration; S(GLS), growth limiting substrate; S_N, any other substrate than S(GLS). (a) Variable volume fed-batch culture. (b) Fixed volume fed-batch culture (Pirt, 1979).

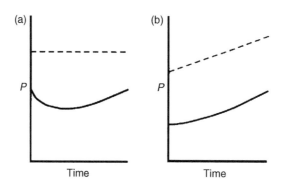

FIGURE 2.16

Product Concentration (p) in Fed-Batch Culture When q_p is Growth Related (-------) or Nongrowth Related, that is, q_p Constant (———). (a) Variable volume fed-batch culture. (b) Fixed volume fed-batch culture.

Modified from Pirt, 1979

between q_p and μ (hence D). If q_p is strictly growth related then it will change as μ changes with D and, thus, the product concentration will remain constant. However, if q_p is constant and independent of μ, then product concentration will decrease at the start of the cycle when D_p is greater than $q_p x$ but will rise with time as D decreases and $q_p x$ becomes greater than D_p. These relationships are shown in Fig. 2.16a. If q_p is related to μ in a complex manner, then the product concentration will vary according to that relationship. Thus, the feed strategy of a fed-batch system would be optimized according to the relationship between q_p and μ.

FIXED VOLUME FED-BATCH CULTURE

Pirt (1979) described the kinetics of fixed volume fed-batch culture as follows. Consider a batch culture in which the growth of the process organism has depleted the limiting substrate to a limiting level. If the limiting substrate is then added in a concentrated feed such that the broth volume remains almost constant, then:

$$\frac{dx}{dt} = GY \qquad (2.35)$$

where G is the substrate feed rate (g dm^{-3} h^{-1}) and Y is the yield factor.

But $dx/dt = \mu x$, thus substituting for dx/dt in Eq. (2.35) gives:

$\mu x = GY$, and thus:

$$\mu = \frac{GY}{x} \qquad (2.36)$$

Provided that GY/x does not exceed μ_{max} then the limiting substrate will be consumed as soon as it enters the fermenter and $ds/dt \simeq 0$. However, dx/dt cannot be equated to zero, as in the case of variable volume fed-batch, because the biomass concentration, as well as the total amount of biomass in the fermenter, will increase with time. Biomass concentration is given by the equation:

$$x_t = x_a + GYt \qquad (2.37)$$

where x_t is the biomass after operating in fed batch for t hours and x_a is the biomass concentration at the onset of fed-batch culture.

As biomass increases then the specific growth rate will decline according to Eq. (2.36). The behavior of a fixed volume fed-batch culture is illustrated in Fig. 2.15b from which it may be seen that μ declines (according to Eq. 2.36), the limiting substrate concentration remains virtually constant, biomass increases and the concentrations of the nonlimiting nutrients decline.

Pirt (1979) described the product balance in a fixed volume fed-batch system as:

$$\frac{dp}{dt} = q_p x$$

but substituting for x from Eq. (2.37) gives:

$$\frac{dp}{dt} = q_p(x_a + GYt)$$

If q_p is strictly growth-rate related then product concentration will rise linearly as for biomass. However, if q_p is constant then the rate of increase in product concentration will rise as growth rate declines, that is, as time progresses and x increases. These relationships are shown in Fig. 2.16b. If q_p is related to μ in a complex manner then the product concentration will vary according to that relationship. As in the case of variable volume fed-batch the feed profile would be optimized according to the relationship between q_p and μ.

FED-BATCH CULTURE AT A CONSTANT SPECIFIC GROWTH RATE

It is possible to operate both a variable volume and a fixed volume fed-batch culture at a constant specific growth rate by increasing the feed rate exponentially over the period of the fermentation. Lee (1996) gave the following equation, which can be used to program the control system:

$$G_t = F_t \cdot S_t = \left(\frac{\mu}{Y_{x/s}} + m \right) x_t \cdot V_t$$

where G_t is the mass flow rate of the feed at time t (g h^{-1}) F_t is the volumetric flow rate at time t (dm^3 h^{-1}), S_t is the substrate concentration in the feed medium at time t (g dm^{-3}), μ is the specific growth rate (h^{-1}), $Y_{x/s}$ is the yield factor (g substrate g^{-1} biomass), m is the maintenance coefficient (g substrate g^{-1} biomass h^{-1}), x_t is the biomass concentration at time t (g dm^{-3}), and V_t is the volume of culture in the fermenter at time t (dm^3).

During growth at the constant specific growth rate (μ), the total amount of biomass in the fermenter ($x_t \cdot V_t$) increases exponentially with time according to the equation:

$$x_t \cdot V_t = x_0 \cdot V_0 \cdot e^{\mu t}$$

where x_0 is the initial concentration of the biomass, t is the time, and V_0 is the initial volume of the culture. Thus:

$$G_t = F_t \cdot S_t = \left(\frac{\mu}{Y_{x/s}} + m \right) x_0 \cdot V_0 \cdot e^{\mu t}$$

The application of this equation assumes that the yield factor and maintenance energy are constant, an assumption which does not always hold true, as discussed earlier. If the maintenance energy is very low then this has been disregarded in the equation by some workers (Salehmin, Annuari, & Chisti, 2014).

CYCLIC FED-BATCH CULTURE

The life of a variable volume fed-batch fermentation may be extended beyond the time it takes to fill the fermenter by withdrawing a portion of the culture and using the residual culture as the starting point for a further fed-batch process. The decrease in volume results in a significant increase in the dilution rate (assuming that the flow rate remains constant) and thus, eventually, in an increase in the specific growth rate. The increase in μ is then followed by its gradual decrease as the quasi steady state is reestablished. Such a cycle may be repeated several times resulting in a series of fed-batch fermentations. Thus, the organism would experience a periodic shift-up in the growth rate followed by a gradual shift-down. This periodicity in growth rate may be achieved in fixed volume fed-batch systems by diluting the culture when the biomass reaches a concentration that cannot be maintained under aerobic conditions.

Dilution results in a decline in x and, thus, according to Eq. (2.36) an increase in μ. Subsequently, as feeding continues, the growth rate will decline gradually as biomass increases and approaches the maximum sustainable in the vessel once more, at which point the culture may be diluted again. Dilution would be achieved by withdrawing culture and refilling to the original level with sterile water or medium not containing the feed substrate.

APPLICATION OF FED-BATCH CULTURE

The use of fed-batch culture by the fermentation industry takes advantage of the fact that the concentration of the limiting substrate may be maintained at a very low level, thus avoiding both the repressive and toxic effects of high substrate concentration. Furthermore, the fed-batch system also gives control over the organisms' growth rate, which is related to a wide range of industrially important physiological properties including product formation, intracellular storage levels, RNA and protein levels, plasmid stability, overflow metabolite production (such as acetate and lactate), the substrate consumption rate, the specific oxygen uptake rate, and the specific product production rate (Schuler and Marison, 2012). Both variable and fixed volume systems result in low limiting substrate concentrations and growth rate control, and while the quasi steady state of the variable volume system has the advantage of maintaining the concentrations of the biomass and the nonlimiting nutrients constant, most commercial processes involve the addition of concentrated feeds resulting in the biomass concentration increasing with time. However, Pirt (1979) emphasized that the concentrations of substrates other than those, which limit growth can have a significant effect on biomass composition and product formation and this aspect needs to be recognized in the design of processes employing concentrated feeds.

The obvious advantage of cyclic fed-batch culture is that the productive phase of a process may be extended under controlled conditions. However, a further advantage lies in the controlled periodic shifts in growth rate that may provide an opportunity to optimize product synthesis. Dunn and Mor (1975) pointed out that changes in the rates of chemical processes can give rise to increases in intermediate concentrations and similar effects may be possible in microbial systems. This observation is particularly relevant to secondary metabolite production that is maximal in batch culture during the deceleration phase. Bushell (1989) suggested that optimum conditions for secondary metabolite synthesis may occur during the transition phase after the withdrawal of a volume of broth from the vessel and before the reestablishment of the steady-state following the resumption of the nutrient feed. During this period the dilution rate will be greater than the growth rate but, according to Bushell, the rate of uptake of the growth-limiting substrate should respond immediately to the increased substrate concentration. Thus, an imbalance results between the substrate uptake rate and the specific growth rate—this imbalance then contributing to a diversion of intermediates into secondary metabolism.

The advantages of cyclic fed-batch culture must be weighed against difficulties inherent in the system. It is pertinent to remember Pirt's cautionary comments (1979) regarding high concentration feeds, reiterated by Queener and Swartz (1979), that care has to be taken in the design of cyclic fed-batch processes to ensure that toxins do not accumulate to inhibitory levels and that nutrients other than those incorporated into the feed medium become limiting. Also, a prolonged series of fed-batch cycles may result in the accumulation of nonproducing or low-producing variants, as described for continuous culture.

The early fed-batch systems that were developed did not incorporate any form of feedback control of medium addition rate and relied on the inherent kinetics of the process to maintain stability. These systems were either based on a fixed feed rate (resulting in a decline in μ with time) or on a predetermined feed rate profile, based on knowledge of the process, such that the growth rate could, ostensibly, be maintained constant or varied during the fermentation (see the earlier section on "fed-batch culture at a constant specific growth rate"). However, the use of concentrated feeds resulting in very high biomass concentrations and the development of more sophisticated feeding programs has necessitated the introduction of feedback control techniques so that any perturbations in the process can be rectified. In such feedback controlled fermentations, a process parameter directly related to the organism's physiological state is monitored continuously by an online sensor, or intermittently by an offline assay process, necessitating the assay of a culture sample. The value generated by the sensor or assay may then be used *per se* in a control loop (see Chapter 8) to control the medium feed rate, or processed to calculate a derived value which may, in turn, be used in a control loop. In their 2012 review, Schuler and Marison explained that, ideally, the specific growth rate itself should be used as the derived value to achieve control. Parameters that have been utilized as control signals in their own right, or as a means of calculating μ include online measurements such as dissolved oxygen concentration, pH, and effluent gas composition; and offline measurements such as biomass, product concentration, and limiting substrate concentration. The beauty of online detection is that the value is being determined continuously, enabling direct control whereas the assay of samples is a discontinuous process resulting in intermittent control and the likelihood that the value of the control parameter has changed before the control intervention is activated. When it is considered that the period between sampling may be hourly (or longer) then it is appreciated that the effectiveness of the control may be seriously compromised by offline analysis (Soons et al., 2006). The "holy grail" of fermentation control is the ability to reliably and directly measure, online, the concentrations of biomass, substrate, and product, thereby making the real-time assessment of process performance a reality. While online detection of dissolved oxygen, pH and oxygen, and carbon dioxide in the effluent gas stream are routinely used for the indirect estimation of biomass. The use of dielectric, Fourier transform infrared (FTIR) and near-infrared (NIR) spectroscopy offer considerable potential for the online monitoring of not only biomass, but substrates and product as well. These approaches are considered in more detail in Chapter 8, and are reviewed by Schuler and Marison (2012) and Marison et al. (2013).

EXAMPLES OF THE USE OF FED-BATCH CULTURE

Fed-batch culture has been used in two major ways by the fermentation industry:

- To control the oxygen uptake rate of the process organism such that its oxygen demand does not exceed the oxygen supply capacity of the fermenter.
- To control the specific growth rate at an optimum value for product formation.

Both approaches can also feature in a single process, for example, if growth rate is controlled by the addition of a concentrated feed then biomass increases over time. As a result, oxygen demand increases and the fermentation is in danger of becoming oxygen-limited. Low dissolved oxygen concentration can then be used as an alarm signal to decrease the growth rate such that the oxygen demand is restrained to within the supply capacity of the fermenter (Nunez, Garelli, & De Battista, 2013).

The earliest example of the control of oxygen demand is given in the fed-batch process for the production of bakers' yeast in 1915. It was recognized that an excess of malt in the medium would lead to too high a growth rate resulting in an oxygen demand in excess of that which could be met by the equipment. This resulted in the development of anaerobic conditions and the formation of ethanol at the expense of biomass production (Reed & Nagodawithana, 1991). Thus, the organism was grown in an initially weak medium to which additional medium was added at a rate less than the maximum rate at which the organism could use it. The resulting process fulfilled the criteria stipulated in Pirt's (1975) kinetic description for the establishment of a quasi steady state, that is, a substrate limited culture and the use of a feed rate equivalent to a dilution rate less than μ_{max}. It is now recognized that bakers' yeast is very sensitive to free glucose and respiratory activity may be repressed at a concentration of about 5 μg dm^{-3}—a phenomenon known as the Crabtree effect (Crabtree, 1929). Thus, a high glucose concentration represses respiratory activity as well as giving rise to a high growth rate, the oxygen demand of which cannot be met. In modern fed-batch processes for yeast production feedback-control strategies have been introduced according to the principles discussed earlier. Fiechter (1982) described the feed of molasses being controlled by the automatic measurement of traces of ethanol in the exhaust gas of the fermenter. Although such systems may result in low growth rates, the biomass yield is near the theoretical maximum obtainable. Cannizzaro, Valentinotti, and von Stockar (2004) extended this approach by developing a control algorithm to control the flow rate (and hence growth rate and oxygen demand) such that the ethanol concentration was maintained at a minimum level. Several workers have utilized the derived variable, respiratory quotient (RQ) which is the ratio of carbon dioxide produced to oxygen consumed. The stoichiometry of aerobic carbohydrate utilization is:

$$(CH_2O) + O_2 = CO_2 + H_2O$$

and, thus, the RQ should be 1.0. An RQ above 1.0 indicates that system is oxygen-limited and some of the carbohydrate is being consumed anaerobically. The RQ may be calculated by comparing the composition of the inlet and outlet gasses.

Hussain and Ramachandran (2002) and Kiran and Jana (2009) used neural network control systems to link the growth rate to the difference between the specific oxygen uptake rate (m moles oxygen gram^{-1} biomass h^{-1}) and the specific carbon dioxide (m moles carbon dioxide gram^{-1} biomass h^{-1}) production rate, an approach claimed to give stronger control than utilizing the RQ itself.

It is interesting to note that the production of recombinant (heterologous) proteins from yeast may be achieved using fed-batch culture techniques very similar to that developed for the bakers' yeast fermentation. Gu, Park, and Kim (1991) reported the production of hepatitis B surface antigen (HBsAg) in a 0.9 dm^3 fed-batch reactor where the feed rate was increased exponentially to maintain a relatively high growth rate. HBsAg production was growth associated in this strain and good productivity was achieved by maintaining a high growth rate while keeping the glucose concentration below that which would repress respiratory activity. Ibba, Bonarius, Kuhla, Smith, and Kuenzi (1993) reported a cyclic fed-batch process for recombinant hirudin from *S. cerevisiae*, under the control of a constitutive promoter. The cyclic fed-batch process gave three times the hirudin activity of a continuous fermentation, the superior productivity being due to increased transcription although a genetic explanation of the results could not be offered. More recently, a number of investigators have devised fed-batch processes maximizing the biomass concentration in *S. cerevisiae* heterologous protein processes—concentrations ranging from 100–130 g dm^{-3} have been reported. Biener, Steinkamper, and Horn (2012) maintained such a fed-batch process at a constant growth rate by using calorimetric data as the control algorithm. Anane, van Ensburg, and Gorgens (2013) explored the production of recombinant alpha-glucuronidase by *S. cerevisiae* using the fed-batch approach to maintain a constant growth rate below the point at which oxygen limitation occurs. At growth rates below 0.12 h^{-1} a yield coefficient was achieved of 0.4 g biomass g^{-1} glucose consumed, close to the theoretical maximum of 0.51 g biomass g^{-1}.

Pichia pastoris is a methanol-utilizing yeast, which has been used as a host for heterologous protein production with fed-batch culture again being the preferred fermentation protocol. The expression of the heterologous protein is under the control of the promoter of the alcohol oxidase gene (AOX1), the expression of which is induced by methanol thereby giving a convenient system to switch on product formation. The process involves three stages (Potvin, Ahmad, & Zhang, 2012):

1. Batch culture with glycerol as the carbon source.
2. Fed-batch culture with glycerol as the carbon source.
3. Fed-batch culture with methanol as the carbon source.

Biomass is accumulated during the first two phases and the heterologous protein is synthesized in the third phase—thus growth and product formation are separated. Methanol is a highly reduced substrate and thus its utilization results in a high oxygen demand. Several strategies have been adopted to control the fed-batch fermentation including maintaining a constant growth rate by exponential feeding, controlling the oxygen uptake rate by limiting the growth rate and maintaining a constant methanol concentration by controlling the feed rate.

The penicillin fermentation provides an excellent example of the use of feed systems in the production of a secondary metabolite. The fermentation may be divided into two phases—the "rapid-growth" phase, during which the culture grows at μ_{max}, and the "slow-growth" or "production" phase. Glucose feeds may be used to control the metabolism of the organism during both phases. During the rapid-growth phase, an excess of glucose causes an accumulation of acid and a biomass oxygen demand greater than the aeration capacity of the fermenter, whereas glucose starvation may result in the organic nitrogen in the medium being used as a carbon source, resulting in a high pH and inadequate biomass formation (Queener & Swartz, 1979). In early versions of the penicillin fermentation, the accumulation of hexose was prevented by the use of a slowly hydrolyzed carbohydrate such as lactose in simple batch culture (Matelova, 1976). Considerable increase in productivity were then achieved by the use of computer controlled feeding of glucose during the rapid growth phase, such that the dissolved oxygen or pH was maintained within certain limits. Both control parameters essentially measure the same activity in that both oxygen concentration and pH will fall when glucose is in excess, due to an increased respiration rate and the accumulation of organic acids when the respiration rate exceeds the aeration capacity of the fermenter. Both systems appeared to work well in controlling feed rates during the rapid-growth phase (Queener & Swartz, 1979).

During the production phase of the penicillin fermentation, the feed rates utilized should limit the growth rate and oxygen consumption such that a high rate of penicillin synthesis is achieved, and sufficient dissolved oxygen is available in the medium. The control factor in this phase is normally dissolved oxygen because pH is less responsive to the effect of dissolved oxygen on penicillin synthesis than on growth. As the fed-batch process proceeds then the total biomass, viscosity, and oxygen demand increase until, eventually, the fermentation is oxygen limited. However, limitation may be delayed by reducing the feed rate as the fermentation progresses and this may be achieved by the use of computer-controlled systems. A number of workers have developed mathematical models of the penicillin process enabling more precise control of the fed-batch process (Rodrigues & Filho, 1996; Patnaik, 1999, 2000; Pico-Marco, Pico, & De Battista, 2005; Almquist, Cvijovic, Hatzimanikatis, Nielsen, & Jirstrand, 2014).

Many enzymes are subject to catabolite repression, where enzyme synthesis is prevented by the presence of rapidly utilized carbon sources (Aunstrup, Andresen, Falch, & Nielsen, 1979). It is obvious that this phenomenon must be avoided in enzyme fermentations and fed-batch culture is the major technique used to achieve this. Concentrated medium is fed to the culture such that the carbon source does not reach the threshold for catabolite repression. For example, Waki, Luga, and Ichikawa (1982) controlled the production of cellulase by *Trichoderma reesei* in fed-batch culture utilizing CO_2 production as the control factor and Suzuki, Mushiga, Yamane, and Shimizu (1988) achieved high lipase production from *Pseudomonas fluorescens* also using CO_2 production to control the addition of an oil feed. Salehmin et al. (2014) achieved success in optimizing the production of lipase by the yeast *Candida rugosa*. They employed fed-batch culture with an exponentially increasing feed rate of palm oil to achieve a fixed growth rate. Maximum enzyme yield was seen at a low specific growth rate of 0.05 h^{-1}, the authors speculating that at higher values of μ, carbon would be diverted to support the higher growth rate at the expense of enzyme production.

Significant success has been achieved in growing *E. coli* at high cell densities in fed-batch culture for the production of heterologous proteins. A wide range of approaches have been used including constant rate feeding, programmed feeding, exponential feeding, and feedback control commonly utilizing pH and dissolved oxygen as the control parameter. The key issues are the avoidance of overflow metabolism, and hence acetate excretion at the expense of the carbon source. Relatively low specific growth rates (between 0.2 and 0.35 h^{-1}) have been recommended to avoid acetate accumulation. Another feature of these processes is the separation of the growth phase from the protein production phase. Thus, the growth phase can be optimized to facilitate a high biomass concentration before the heterologous protein is induced. High heterologous protein levels can induce the stationary phase stress responses described earlier in this chapter and thus it is desirable to produce a large biomass before the fermentation is vulnerable to this type of perturbation (Lee, 1996; Choi, Keum, & Lee, 2006; Rosano & Ceccarelli, 2014).

Mammalian cell cultures have been extensively used for heterologous protein production and, although perfusion culture is increasing in popularity, fed-batch is still the most common process (Lu et al., 2013). As just described for *E. coli,* the whole range of fed-batch approaches have been pressed into service in mammalian cell culture—from simple processes in which feeding may range from manual intermittent (so-called bolus feeding) feeding to dynamic systems involving feedback control utilizing a range of feeding algorithms. However, Lu et al. (2013) commented that while dynamic systems have resulted in successful processes, the majority of systems rely on a predetermined feeding strategy rather than one that responds to the culture's changing requirements. It appears that the much lower rates of metabolism associated with mammalian cells has encouraged a perception that adequate control can be achieved using basic fed-batch approaches. Lu et al. (2013) used two CHO (Chinese hamster ovary) cell lines producing different IgG antibodies. Feedback control of the cultures was based on either capacitance measurements (of cell density) or automated glucose assay (of the limiting substrate). While one cell line more than doubled productivity under dynamic control, the other showed more modest improvement and the process had to be integrated with a medium development program.

REFERENCES

Almquist, J., Cvijovic, M., Hatzimanikatis, V., Nielsen, J., & Jirstrand, M. (2014). Kinetic models in industrial biotechnology—improving cell factory performance. *Metabolic Engineering, 24*, 39–60.

Anane, E., van Ensburg, E., & Gorgens, J. F. (2013). Optimisation and scale-up of alpha-glucosidase production by recombinant *Saccharomyces cerevisiae* in aerobic fed-batch culture with constant growth rate. *Biochemical Engineering Journal, 81*, 1–7.

Aunstrup, K., Andresen, O., Falch, E. A., & Nielsen, T. K. (1979). Production of microbial enzymes. In H. J. Peppler, & D. Perlman (Eds.), *Microbial technology* (pp. 282–309). (1). New York: Academic Press.

Biener, R., Steinkamper, A., & Horn, T. (2012). Calorimetric control of the specific growth rate during fed-batch cultures of *Saccharomyces cerevisiae*. *Journal of Biotechnology*, *160*, 195–201.

Bonham-Carter, J., & Shevitz, J. (2011). A brief history of perfusion manufacturing. *Bioprocess International*, *9*(9), 24–30.

Borrow, A., Jefferys, E. G., Kessel, R. H. J., Lloyd, E. C., Lloyd, P. B., & Nixon, I. S. (1961). Metabolism of *Gibberella fujikuroi* in stirred culture. *Canadian Journal of Microbiology*, *7*, 227–276.

Branyik, T., Silva, D. P., Baszczynski, M., Lehnert, R., & Almeida e Silva (2012). A review of methods of low alcohol and alcohol-free beer production. *Journal of Food Engineering*, *108*(4), 493–506.

Britton, R. A., Eichenberger, P., Gonzalez-Pastor, J. E., Fawcett, P., Monson, R., Losick, R., & Grossman, A. D. (2002). Genome-wide analysis of the tsationary-phase sigma factor (sigma-H) regulon of *Bacillus subtilis*. *Journal of Bacteriology*, *184*(17), 4881–4890.

Bu'lock, J. D., Hamilton, D., Hulme, M. A., Powell, A. J., Shepherd, D., Smalley, H. M., & Smith, G. N. (1965). Metabolic development and secondary biosynthesis in *Penicillium urticae*. *Canadian Journal of Microbiolgy*, *11*, 765–778.

Bu'lock, J. D., Detroy, R. W., Hostalek, Z., & Munim-al-Shakarchi, A. (1974). Regulation of biosynthesis in *Gibberella fujikuroi*. *Transactions of the British Mycological Society*, *62*, 377–389.

Bull, A. T. (1974). Microbial growth. In A. T. Bull, J. R. Lagnado, J. O. Thomas, & K. F. Tipton (Eds.), *Companion to biochemistry, selected topics for further study* (pp. 415–442). London: Longman.

Bull, A. T. (2010). The renaissance of continuous culture in the post-genomics era. *Journal of Industrial Microbiology and Biotechnology*, *37*, 993–1021.

Bushell, M. E. (1989). The process physiology of secondary metabolite production. In S. Baumberg, I. S. Hunter, & P. M. Rhodes (Eds.), *Microbial products: new approaches. Symposium of society for general microbiology Vol. 44* (pp. 95–120). Cambridge: Cambridge University Press.

Butler, M (2005). Animal cell cultures: recent advances and perspectives in the production of biopharmaceuticals. *Applied Microbiology and Biotechnology*, *68*, 283–291.

Butler, M (2012). Recent advances in technology supporting biopharmaceutical production from mammalian cells. *Applied Microbiology and Biotechnology*, *96*, 885–894.

Calcott, P. H. (1981). The construction and operation of continuous cultures. In P. H. Calcott (Ed.), *Continuous culture of cells* (pp. 13–26). (1). Boca Raton: CRC Press.

Callihan, C. D. & Clemmer, J. E. (1979). Biomass from cellulosic materials. In A. H. Rose (Ed.) *Economic microbiology Vol. 4*, Microbial biomass (pp. 208–270). London: Academic Press.

Cannizzaro, C., Valentinotti, S., & von Stockar, U. (2004). Control of yeast fed-batch process through regulation of extracellular ethanol concentration. *Bioprocess and Biosystems Engineering*, *26*, 377–383.

Choi, J. H. C., Keum, K. C., & Lee, S. Y. (2006). Production of recombinant protein by high cell density culture of *Escherichia coli*. *Chemical Engineering Science*, *61*, 876–885.

Chon, J. H., & Zarbis-Papastoitsis, G. (2011). Advances in the production and downstream processing of antibodies. *New Biotechnology*, *28*(5), 458–463.

Crabtree, H. G. (1929). Observations on the carbohydrate metabolism of tumours. *The Biochemical Journal*, *23*, 536–545.

Dostalek, M., Haggstrom, C., Molin, N. (1972). Optimisation of biomass production from methanol. Fermentation technology today. In G. Terui (Ed.) *Proceedings of fourth international fermentation symposium* (pp. 497–511).

Dunn, I. J., & Mor, J. -R. (1975). Variable-volume continuous cultivation. *Biotechnology and Bioengineering*, *17*, 1805–1822.

Eagon, R. G. (1961). *Pseudomonas natriegens,* a marine bacterium with a generation time of less than ten minutes. *Journal of Bacteriology*, *83*, 736–737.

Egan, M. J., McClintock, M. A., & Reck-Peterson, S. L. (2012). Microtubule-based transport in filamentous fungi. *Current Opinion in Microbiology*, *15*, 637–645.

Ferenci, T. (1999). Growth of bacterial cultures 50 years on: towards an uncertainty principle instead of constants in bacterial growth kinetics. *Research in Microbiology*, *150*, 431–438.

Ferenci, T. (2006). A cultural divide on the use of chemostats. *Microbiology*, *152*, 1247–1248.

Ferenci, T. (2008). Bacterial physiology, regulation and mutational adaptation in a chemostat environment. *Advances in Microbial Physiology.*, *53*, 22–169.

Fiechter, A. (1982). Regulatory aspects in yeast metabolism. In M. Moo-Young, C. W. Robinson, & C. Vezina (Eds.), *Advances in Biotechnology, 1. Scientific and Engineering Principles* (pp. 261–267). Toronto: Pergamon Press.

Flardh, K., Richards, D. M., Hempel, A. M., Howard, M., & Butner, M. (2012). Regulation of apical growth and hyphal branching in *Sreptomyces*. *Current Opinion in Microbiology*, *15*, 737–743.

Griffiths, J. B. (1986). Scaling-up of animal cell cultures. In R. I. Freshney (Ed.), *Animal cell culture, a practical approach* (pp. 33–70). Oxford: IRL Press.

Griffiths, J. B. (1992). Animal cell processes—batch or continuous? *Journal of Biotechnology*, *22*, 21–30.

Gu, M. B., Park, M. H., & Kim, D. -I. (1991). Growth rate control in fed-batch cultures of recombinant *Saccharomyces cerevisiae* producing hepatitis B surface antigen (HBsAg). *Applied Microbiology and Biotechnology*, *35*, 46–50.

Harder, W., & Dijkhuizen, L. (1983). Physiological responses to nutrient limitation. *Annual Review of Microbiology*, *37*, 1–23.

Harrison, D. E. F. (1973). Studies on the affinity of methanol and methane utilizing bacteria for their carbon substrates. *The Journal of Applied Bacteriology*, *36*, 301–308.

Harte, M. J., & Webb, F. C. (1967). Utilization of mixed sugars in continuous fermentations. *Biotechnology and Bioengineering*, *9*, 205–221.

Heijnen, J. J., Terwisscha van Scheltinga, A. H., & Straathof, A. J. (1992). Fundamental bottlenecks in the application of continuous bioprocesses. *Journal of Biotechnology*, *22*, 3–20.

Hengge-Aronis, R. (1996). Back to log phase: sigmaS as a global regulator in the osmotic control of gene expression. *Research in Microbiology*, *21*, 887–893.

Herbert, D. (1958). Some principles of continuous culture. In G. Tunevall (Ed.), *Recent progress in microbiology, seventh international congress for microbiology* (pp. 381–396). Sweden: Stockholm.

Hopwood, D. A. (2007). *Streptomyces in nature*. Oxford: Oxford University Press.

Hoskisson, P. A., & Hobbs, G. (2005). Continuous culture—making a comeback? *Microbiology*, *151*(10), 3153–3159.

Hough, J. S., Keevil, C. W., Maric, V., Philliskirk, G., & Young, T. W. (1976). Continuous culture in brewing. In A. C. R. Dean, D. C. Ellwood, C. G. T. Evans, & Meiling, J. (Eds.), *Continuous Culture, Vol. 6, Applications and New Fields* (pp. 226–238). Chichester: Ellis Horwood.

Howells, E. R. (1982). Single-cell protein and related technology. *Chemical Industry*, *7*, 508–511.

Hussain, M. A. H., & Ramachandran, K. B. (2002). Comparative evaluation of various control systems for fed-batch fermentation. *Bioprocess and Biosystems Engineering*, *24*, 309–318.

Ibba, M., Bonarius, D., Kuhla, J., Smith, A., & Kuenzi, M. (1993). Mode of cultivation is critical for the optimal expression of recombinant hirudin by *Saccharomyces cerevisiae*. *Biotechnology Letters*, *15*(7), 667–672.

Ihssen, J., & Egli, T. (2004). Specific growth rate and not cell density controls the general stress response in *Escherichia coli*. *Microbiology*, *150*, 1637–1648.

Kadouri, A., & Spier, R. E. (1997). Some myths and messages concerning the batch and continuous culture of animal cells. *Cytotechnology*, *24*, 89–98.

King, P. P. (1982). Biotechnology: an industrial view. *Journal of Chemical Technology and Biotechnology*, *32*, 2–8.

Kiran, A. U. M., & Jana, A. K. (2009). Control of continuous fed-batch fermentation process using neural network based model prediction controller. *Bioprocess and Biosystems Engineering*, *32*, 801–808.

Kirsop, B. H. (1982). Developments in beer fermentation. In A. Wiseman (Ed.), *Topics in enzyme and fermentation biotechnology* (pp. 79–131). (6). Chichester: Ellis Horwood.

Krull, R., Wucherpfennig, T., Esfandabadi, M. E., Walisco, R., Melzer, G., Hempel, D. C., Kampen, I., Kwade, A., & Wittmann, C. (2013). Characterization and control of fungal morphology for improved production performance in biotechnology. *Journal of Biotechnology*, *163*, 112–123.

Landini, P., Egli, T., Wolf, J., & Lacour, S. (2014). sigmaS, a major player in the response to environmental stress in *Escherichia coli*: role, regulation and mechanisms of promoter recognition. *Environmental Microbiology Reports*, *6*(1), 1–13.

Laskin, A. I. (1977). Single cell protein. In D. Perlman (Ed.), *Annual reports on fermentation processes* (pp. 151–180). (1). New York: Academic Press.

Lavery, M. (1990). Animal cell fermentation. In B. McNeil, & L. M. Harvey (Eds.), *Fermentation: A practical approach* (pp. 205–220). Oxford: IRL Press.

Lazazzera, B. A. (2000). Quorum sensing and starvation: signals for entry into stationary phase. *Current Opinion in Microbiology*, *3*, 177–182.

Lee, S. Y. (1996). High cell density culture of *Escherichia coli*. *Trends in Biotechnology*, *14*, 98–105.

Li, J., Chen, X., Qi, B., Luo, J., Zhang, Y., Su, Y., & Wan, Y. (2014). Efficient production of acetone-butanol-ethanol (ABE) from cassava by a fermentation-pervaporation coupled process. *Bioresource Technology*, *169*, 251–257.

Lilley, G., Clark, A. E., & Lawrence, G. C. (1981). Control of the production of cephamycin C and thienamycin by *Streptomyces cattleya* NRRL 8057. *Journal of Chemical Technology and Biotechnology*, *31*, 127–134.

Lu, F., Toh, P. C., Burnett, I., Li, F., Hudson, T., Amanullah, A., & Li, J. (2013). Automated dynamic fed-batch process and media optimization for high productivity cell culture process development. *Biotechnology and Bioengineering*, *110*, 191–205.

Major, N. C., & Bull, A. T. (1989). The physiology of lactate production by *Lactobacillus delbrueckii* in a chemostat with cell recycle. *Biotechnology and Bioengineering*, *34*, 592–599.

Malek, I., Votruba, J., & Ricica, J. (1988). The continuality principle and the role of continuous cultivation of microorganisms in basic research. In P. Kyslik, E. A. Dawes, V. Krumphanzl, & M. Novak (Eds.), *Continuous culture* (pp. 95–104). London: Academic Press.

Manteca, A., Fernandez, M., & Sanchez, J. (2005). A death round affecting a young compartmentalized mycelium precedes aerial mycelium dismantling in confluent surface cultures of *Streptomyces antibioticus*. *Microbiology*, *151*, 3689–3697.

Manteca, A., Alvarez, R., Salazar, N., Yague, P., & Sanchez, J. (2008). Mycelium differentiation and antibiotic production in liquid cultures of *Streptomyces coelicolor*. *Applied and Environmental Microbiology*, *74*, 3877–3886.

Manteca, A., Sanchez, J., Jung, H. R., Schwammle, V., & Jensen, O. N. (2010). Quantitative proteomic analysis of *Streptomyces coelicolor* development demonstrates the switch from primary to secondary metabolism associated with hyphae differentiation. *Molecular Cell Proteomics*, *9*, 1423–1436.

Marison, I., Hennessy, S., Foley, R., Schuler, M., Sivaprakasam, S., & Freeland, B. (2013). The choice of suitable online analytical techniques and data processing for monitoring of bioprocesses. *Advances in Biochemical Engineering/Biotechnology*, *132*, 249–280.

Marr, A. G., Nilson, E. H., & Clark, D. J. (1963). The maintenance requirements of *Escherichia coli*. *New York Academy of Science*, *102*, 536–548.

Matelova, V. (1976). Utilization of carbon sources during penicillin biosynthesis. *Folia Microbiologica*, *21*, 208–209.

Meyrath, J. and Bayer, K. (1979). Biomass from whey. In Economic Microbiology, vol. 4. Microbial Biomass, pp. 208–270. (Editor Rose, A. H). Academic Press, New York.

Monod, J. (1942). *Recherches sur les Croissances des Cultures Bacteriennes* (2nd ed.). Paris: Hermann and Cie.

Monte Alegre, R., Rigo, M., & Joekes, I. (2003). Ethanol fermentation of a diluted molasses medium by *Saccharomyces cerevisiae* immobilized on chrysotile. *Brazilian Archives of Biology and Technology*, *46*(4), 751–757.

Nathan, L. (1930). Improvements in the fermentation and maturation of beers. *Journal of the Institute of Brewing*, *36*, 538–550.

Nunez, S., Garelli, F., & De Battista, H. (2013). Decentralized control with minimum dissolved oxygen guarantees in fed-batch cultivations. *Industrial Engineering Chemistry Research*, *52*, 18014–18021.

Patnaik, P. R. (1999). Penicillin fermentation revisited: a topological analysis of kinetic multiplicity. *Process Biochemistry*, *34*, 737–743.

Patnaik, P. R. (2000). Penicillin fermentation: mechanisms and models for industrial-scale bioreactors. *Critical Reviews in Biotechnology*, *20*(1), 1–15.

Petersen, M., & Alfermann, A. W. (1993). Plant cell cultures. In H. -J. Rhem, & G. Reed (Eds.), *Biotechnology, biological fundamentals* (pp. 577–614). (1). Weinheim: VCH.

Pico-Marco, E., Pico, J., & De Battista, H. (2005). Sliding mode scheme for adaptive control in biotechnological fed-batch processes. *International Journal of Control*, *78*(2), 128–141.

Pirt, S. J. (1965). The maintenance energy of bacterial in growing cultures. *Proceedings of the Royal Society of London. Series B, Biological Sciences*, *163*(991), 224–231.

Pirt, S. J. (1974). The theory of fed batch culture with reference to the penicillin fermentation. *Journal of Applied Chemistry and Biotechnology*, *24*, 415–424.

Pirt, S. J. (1975). *Principles of microbe and cell cultivation*. Oxford: Blackwell.

Pirt, S. J. (1979). Fed-batch culture of microbes. *Annals of the New York Academy of Sciences*, *326*, 119–125.

Pirt, S. J. (1990). The dynamics of microbial processes: A personal view. In R. K. Poole, M. J. Bazin, & C. W. Keevil (Eds.), *Microbial growth dynamics* (pp. 1–16). Oxford: IRL Press.

Pirt, S. J., & Kurowski, W. M. (1970). An extension of the theory of the chemostat with feedback of organisms. Its experimental realization with a yeast culture. *Journal of General Microbiology*, *63*, 357–366.

Pirt, S. J., & Righelato, R. C. (1967). Effect of growth rate on the synthesis of penicillin by *Penicillium chrysogenum* in batch and continuous culture. *Applied Microbiology, 15,* 1284–1290.

Plomley, N. J. B. (1959). Formation of the colony in the fungus *Chaetomium. Australian Journal of Biological Sciences, 12,* 53–64.

Pollock, J., Ho, Sa. V., & Farid, S. S. (2013). Fed-batch and perfusion culture processes: economic, environmental, and operational feasibility under uncertainty. *Biotechnology and Bioengineering, 110*(1), 206–219.

Potvin, G., Ahmad, A., & Zhang, Z. (2012). Bioprocess engineering aspects of heterologous protein production in *Pichia pastoris*: a review. *Biochemical Engineering Journal, 64,* 91–105.

Queener, S., & Swartz, R. (1979). Penicillins; Biosynthetic and semisynthetic. In A. H. Rose (Ed.), *Economic Microbiology, Vol. 3. Secondary Products of Metabolism* pp. 35–123. Academic Press, London.

Reed, G., & Nagodawithana, T. W. (1991). *Yeast technology* (2nd ed.) (pp. 413–437). New York: Van Nostrand Reinhold.

Rhodes, P. M. (1984). The production of oxytetracycline in chemostat culture. *Biotechnology and Bioengineering, 26,* 382–385.

Riesenberger, D., & Bergter, F. (1979). Dependence of macromolecular composition and morphology of Streptomyces hygroscopicus on specific growth rate. *Zeitschrift fur allgemeine Mikrobiologie, 19,* 415–430.

Robinson, P. M., & Smith, J. M. (1979). Development of cells and huyphae of *Geotrichum candidum* in chemostat and batch culture. *Transactions of the British Mycological Society, 72,* 39–47.

Rodrigues, J. D. A., & Filho, R. M. (1996). Optimal feed rates strategies with operating constraints for the penicillin production process. *Chemical Engineering Science, 51*(11), 2859–2864.

Rosano, G. L., & Ceccarelli, E. A. (2014). Recombinant protein expression in *Escherichia coli*: advances and challenges. *Frontiers in Microbiology*(5), 1–17.

Royston, M. G. (1966). Tower fermentation of beer. *Process Biochemistry, 1*(4), 215–221.

Ryu, D. D., & Hospodka, J. (1980). Quantitative physiology of *Penicillium chrysogenum* in penicillin fermentation. *Biotechnology Bioengineering, 22,* 289–298.

Salehmin, M. N. I., Annuari, M. S. M., & Chisti, Y. (2014). High cell density fed-batch fermentation for the production of a microbial lipase. *Biochemical Engineering Journal, 85,* 8–14.

Schuler, M. M., & Marison, I. W. (2012). Real-time monitoring and control of microbial bioprocesses with focus on the specific growth rate: current state and future prospects. *Applied Microbiology and Biotechnology, 94*(6), 1469–1482.

Schulze, K. L., & Lipe, R. S. (1964). Relationship between substrate concentration, growth rate, and respiration rate of *Escherichia coli* in continuous culture. *Archiv fur Mikrobiologie, 48,* 1–20.

Sharp, D. H. (1989). *Bioprotein manufacture. A critical assessment.* Chichester: Ellis Horwood.

Shehata, T. E., & Marr, A. G. (1971). Effect of nutrient concentration on the growth of *Escherichia coli. Journal of Bacteriology, 107,* 210–216.

Sikyta, B., Slezak, J., & Herold, H. (1961). Growth of *Streptomyces aureofaciens* in continuous culture. *Applied Microbiology, 9,* 233–238.

Soons, Z. I. T. A., Voogt, J. A., van Straten, G., & van Boxtel, A. J. B. (2006). Constant specific growth rate in fed-batch cultivation of *Bordetella pertussis* using adaptive control. *Journal of Biotechnology, 125,* 252–268.

Stouthamer, A. H., Bulthius, B. A., & van Versveld, H. W. (1990). Energetics of low growth rates and its relevance for the maintenance concept. In R. K. Poole, M. J. Bazin, & C. W. Keevil (Eds.), *Microbial growth dynamics* (pp. 85–102). Oxford: IRL Press.

Suzuki, T., Mushiga, Y., Yamane, T., & Shimizu, S. (1988). Mass production of lipase by fed-batch culture of *Pseudomonas fluorescens*. *Applied Microbiology Biotechnology*, *27*, 417–422.

Tai, S. W., Boer, V. M., Daran-Lapujade, P., Walsh, M. C., et al. (2005). Two-dimensional transcriptome analysis in chemostat cultures. *Journal of Biological Chemistry*, *280*(1), 437–447.

Tanouchi, Y., Pai, A., Buchler, N. E., & You, L. (2012). Programmed stress-induced altruistic death in engineered bacteria. *Molecular Systems Biology*, *8*, 626–638.

Trilli, A. (1990). Kinetics of secondary metabolite production. In R. K. Poole, M. J. Bazin, & C. W. Keevil (Eds.), *Microbial growth dynamics* (pp. 103–126). Oxford: IRL Press.

Trilli, A., Crossley, M. V., & Kontakou, M. (1987). Relation between growth and erythromycin production in *Streptomyces erythraeus*. *Biotechnology Letters*, *9*, 765–770.

Trinci, A. P. J. (1969). A kinetic study of growth of *Aspergillus nidulans* and other fungi. *Journal of General Microbiology*, *57*, 11–24.

Trinci, A. P. J. (1974). A study of the kinetics of hyphal extension and branch initiation of fungal mycelia. *Journal of General Microbiology*, *81*, 225–236.

Trinci, A. P. J. (1992). Mycoprotein: a twenty year overnight success story. *Mycological Research*, *96*(1), 1–13.

van Bodegom, P. K. (2007). Microbial maintenance: a critical review on its quantification. *Microbial Ecology*, *53*, 513–523.

Veiga, T., Gomberi, A. K., Landes, N., Verhoven, M. D., et al. (2012). Metabolic engineering of β-oxidation in *Penicillium chrysogenum* for improved semi-synthetic cephalosporin biosynthesis. *Metabolic Engineering*, *14*(4), 437–448.

Waki, T., Luga, K., & Ichikawa, K. (1982). Production of cellulase in fed-batch culture. In M. Moo-Young, C. W. Robinson, & C. Vezina (Eds.), *Advances in biotechnology, scientific and engineering principles* (pp. 359–364). Toronto: Pergamon Press.

Woodward, J. (1987). Utilization of cellulose as a fermentation substrate: Problems and potential. In J. D. Stowell, A. J. Beardsmore, C. W. Keevil, & J. R. Woodward (Eds.), *Carbon substrates in biotechnology* (pp. 45–66). Oxford: IRL Press.

Yague, P., Lopez-Garcia, M. T., Rioseras, B., Sanchez, J., & Manteca, A. (2012). New insights on the development of *Streptomyces* and their relationships with secondary metabolite production. *Current Trends in Microbiology*, *8*, 65–73.

Yague, P., Lopez-Garcia, M. T., Rioseras, B., Sanchez, J., & Manteca, A. (2013). Presporulation stages of *Streptomyces* differentiation: state-of-the-art and future perspectives. *FEMS Microbiology Letters*, *342*(2), 79–88.

Yoshida, F., Yamane, T., & Nakamoto, K. (1973). Fed-batch hydrocarbon fermentations with colloidal emulsion feed. *Biotechnology and Bioengineering*, *15*, 257–270.

The isolation and improvement of industrially important microorganisms

ISOLATION OF INDUSTRIALLY IMPORTANT MICROORGANISMS

The remarkable biochemical diversity of microorganisms has been exploited by the fermentation industry by isolating strains from the natural environment able to produce products of commercial value. The so-called "golden era" of antibiotic discovery in the 1950s and 1960s, culturing soil microorganisms and screening their products, yielded the key groups of antibiotics currently in use (Davies, 2011). However, the lack of success in isolating producers of novel compounds in the late 1980s and 1990s led the pharmaceutical industry to the conclusion that the pool of microbial potential had been "overfished" and, as a result, the major policy decision was taken to cease this activity. The industry then looked to its chemists to produce the next generation of small molecule pharmaceuticals and put its faith in the new field of combinatorial chemistry—which enables the synthesis of many thousands of compounds in a single process. The organic chemist's prayer is attributed to John Bu'Lock, a chemist turned microbiologist, who made a major contribution to natural product science:

> *"Lord, I fall upon my knees*
> *And pray that all my syntheses*
> *Will prove superior*
> *To those produced by bacteria."*

Unfortunately, the application of combinatorial chemistry was not the answer to this prayer and the industry gained relatively little from enormous effort and expenditure. The details of this scenario will be considered in a later section as well as the contribution that microbial genomics made to the next stage of natural product exploration. However, the "golden age" gave rise to much of the industry and it should be remembered that antibiotics are not the only natural products produced by microorganism isolated from soil. Thus, the basic principles of strain isolation will be considered to give an appreciation of the origin of many current industrial producers and the basis of future developments.

The first stage in the screening for microorganisms of potential industrial application is their isolation. Isolation involves obtaining either pure or mixed cultures followed by their assessment to determine which carry out the desired reaction or produce the desired product. In some cases it is possible to design the isolation procedure in such a way that the growth of potential producers is encouraged or that they may be recognized at the isolation stage, whereas in other cases organisms must be isolated and producers recognized at a subsequent stage. However, it should be remembered that the isolate must eventually carry out the process economically and therefore the selection of the culture to be used is a compromise between the productivity of the organism and the economic constraints of the process. Bull, Ellwood, and Ratledge (1979) cited a number of criteria as being important in the choice of organism:

1. The nutritional characteristics of the organism. Depending on the value of the product, a process may have to be carried out using a cheap medium or a predetermined one, for example, the use of methanol as an energy source. These requirements may be met by the suitable design of the isolation medium.
2. The optimum temperature of the organism. The use of an organism having an optimum temperature above 40°C considerably reduces the cooling costs of a large-scale fermentation and, therefore, the use of such a temperature in the isolation procedure may be beneficial.
3. The reaction of the organism with the equipment to be employed and the suitability of the organism to the type of process to be used.
4. The stability of the organism and its amenability to genetic manipulation.
5. The productivity of the organism, measured in its ability to convert substrate into product and to give a high yield of product per unit time.
6. The ease of product recovery from the culture.

Points 3, 4, and 6 would have to be assessed in detailed tests subsequent to isolation and the organism most well suited to an economic process chosen on the basis of these results. However, before the process may be put into commercial operation the toxicity of the product and the organism has to be assessed.

The aforementioned account implies that cultures must be isolated, in some way, from natural environments. However, the industrial microbiologist may also "isolate" microorganisms from culture collections. A selection of the major collections is given in Table 3.1. Such collections may provide organisms of known characteristics but may not contain those possessing the most desirable features, whereas the environment contains a myriad of organisms, very few of which may be satisfactory. It is certainly cheaper to buy a culture than to isolate from nature, but it is also true that a superior organism may be found after an exhaustive search of a range of natural environments. However, it is always worthwhile to purchase cultures demonstrating the desired characteristics, however weakly, as they may be used as model systems to develop culture and assay techniques that may then be applied to the assessment of natural isolates.

The ideal isolation procedure commences with an environmental source (frequently soil) which is highly probable to be rich in the desired types, is so designed as to favor the growth of those organisms possessing the industrially important

Table 3.1 Major Culture Collections

Culture Collection	Website
National Collection of Type Cultures (NCTC)	http://www.phe-culturecollections.org.uk
National Collection of Industrial Food and Marine Bacteria (NCIMB Ltd)	http://www.ncimb.com
National Collection of Yeast Cultures (NCYC)	http://www.ncyc.co.uk
UK National Collection of Fungus Cultures	http://www.cabi.org
The British Antarctic Survey Culture Collection	http://www.cabi.org
National Collection of Plant Pathogenic Bacteria	http://ncppb.fera.defra.gov.uk
American Type Culture Collection (ATCC)	http://www.lgcstandards-atcc.org/?geo_country=gb
Deutsche Sammlung von Mikroorganismen und Zelkulturen (DSMZ)	https://www.dsmz.de
Centraalbureau voor Schimmelcultures (CBS)	http://www.cbs.knaw.nl
Czech Collection of Microorganisms (CCM)	http://www.sci.muni.cz/ccm/index.html
Collection Nationale de Cultures de Microorganismes (CNCM)	http://www.pasteur.fr/recherche/unites/Cncm/index-en.html
Japan Collection of Microorganisms (JCM)	http://jcm.brc.riken.jp/en/

characteristic (ie, the industrially useful characteristic is used as a selective factor) and incorporates a simple test to distinguish the most desirable types. Selective pressure may be used in the isolation of organisms that will grow on particular substrates, in the presence of certain compounds or under cultural conditions adverse to other types. However, if it is not possible to apply selective pressure for the desired character, it may be possible to design a procedure to select for a microbial taxon which is known to show the characteristic at a relatively high frequency, for example, the production of antibiotics by streptomycetes. Alternatively, the isolation procedure may be designed to exclude certain microbial "weeds" and to encourage the growth of more novel types. Indeed, as pointed out by Bull, Goodfellow, and Slater (1992), for screening programs to continue to generate new products, it is becoming increasingly more important to concentrate on lesser known microbial taxa or to utilize very specific screening tests to identify the desired activity. During the 1980s significant advances were made in the establishment of taxonomic databases describing the properties of microbial groups and these databases have been used to predict the cultural conditions that would select for the growth of particular taxa. Thus, the advances in the taxonomic description of taxa have allowed the rational design of procedures for the isolation of strains that may have a high probability of being productive or are representatives of unusual groups. Furthermore, whole-genome sequencing of microorganisms (developed in the 2000s) can give an insight into their cultural requirements (de Macedo Lemos, Alves, & Campanharo, 2003). The sequence can be used to construct a metabolic map and thus identify nutrients that the organism cannot synthesize (Song et al., 2008), thereby enabling the design of a defined medium. To assist this approach, Richards et al. (2014) constructed a database (MediaDB) of defined media that have been used for the cultivation of organisms with sequenced genomes. The advances in pharmacology and molecular biology have also enabled the design of screening tests to identify productive strains among the isolated organisms. However, as will be discussed later, this approach did not live up to expectations in the identification of novel antibiotics.

ISOLATION METHODS UTILIZING SELECTION OF THE DESIRED CHARACTERISTIC

In this section we consider methods that take direct advantage of the industrially relevant property exhibited by an organism to isolate that organism from an environmental source. Isolation methods depending on the use of desirable characteristics as selective factors are essentially types of enrichment culture. Enrichment culture is a technique resulting in an increase in the number of a given organism relative to the numbers of other types in the original inoculum. The process involves taking an environmental source (usually soil) containing a mixed population and providing conditions either suitable for the growth of the desired type, or unsuitable for the growth of the other organisms, for example, by the provision of particular substrates or the inclusion of certain inhibitors. Prior to the culture stage it is often advantageous to subject the soil to conditions that favor the survival of the organisms in question. For example, air-drying the soil will favor the survival of actinomycetes.

Enrichment liquid culture

Enrichment liquid culture is frequently carried out in shake flasks. However, the growth of the desired type from a mixed inoculum will result in the modification of the medium and therefore changes the selective force which may allow the growth of other organisms, still viable from the initial inoculum, resulting in a succession. The selective force may be reestablished by inoculating the enriched culture into identical fresh medium. Such subculturing may be repeated several times before the dominant organism is isolated by spreading a small inoculum of the enriched culture onto solidified medium. The time of subculture in an enrichment process is critical and should correspond to the point at which the desired organism is dominant.

The prevalence of an organism in a batch enrichment culture will depend on its maximum specific growth rate compared with the maximum specific growth rates of the other organisms capable of growth in the inoculum. Thus, provided that the enrichment broth is subcultured at the correct times, the dominant organism will be the fastest growing of those capable of growth. However, it is not necessarily true that the organism with the highest specific growth rate is the most useful, for it may be desirable to isolate the organism with the highest affinity for the limiting substrate.

The problems of time of transfer and selection on the basis of maximum specific growth rate may be overcome by the use of a continuous process where fresh medium is added to the culture at a constant rate. Under such conditions, the selective force is maintained at a constant level and the dominant organism will be selected on the basis of its affinity for the limiting substrate rather than its maximum growth rate.

The basic principles of continuous culture are considered in Chapter 2 from which it may be seen that the growth rate in continuous culture is controlled by the dilution rate and is related to the limiting substrate concentration by the equation:

$$\mu = \mu_{max} s/(K_s + s). \tag{3.1}$$

Eq. (3.1) is represented graphically in Fig. 3.1. A model of the competition between two organisms capable of growth in a continuous enrichment culture is represented in Fig. 3.2. Consider the behavior of the two organisms, A and B, in Fig. 3.2. In continuous culture the specific growth rate is determined by the substrate concentration and is equal to the dilution rate, so that at dilution rates below point Y in Fig. 3.2 strain B would be able to maintain a higher growth rate than strain A, whereas at dilution rates above Y strain A would be able to maintain a higher growth rate. Thus, if A and B were present in a continuous enrichment culture, limited by the substrate depicted in Fig. 3.2, strain A would be selected at dilution rates above

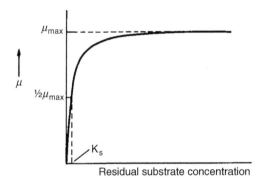

FIGURE 3.1 The Effect of Substrate Concentration on the Specific Growth Rate of a Microorganism

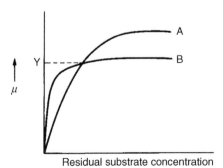

FIGURE 3.2 The Effect of Substrate Concentration on the Specific Growth Rates of Two Microorganisms A and B

Y and strain B would be selected at dilution rates below Y. Thus, the organisms that are isolated by continuous enrichment culture will depend on the dilution rate employed and may result in the isolation of organisms not so readily recovered by batch techniques.

Continuous enrichment techniques are especially valuable in isolating organisms to be used in a continuous-flow commercial process. Organisms isolated by batch enrichment and purification on solid media frequently perform poorly in continuous culture (Harrison, Wilkinson, Wren, & Harwood, 1976), whereas continuous enrichment provides an organism, or mixture of organisms, adapted to continuous culture. The enrichment procedure should be designed such that the predicted isolate meets as many of the criteria of the proposed process as possible and both Johnson (1972) and Harrison (1978) have discussed such procedures for the isolation of organisms to be used for biomass production. Johnson emphasized the importance of using the carbon source to be employed in the subsequent commercial process as the sole source of organic carbon in the enrichment medium, and that the medium should be carbon limited. The inclusion of other organic carbon sources, such as vitamins or yeast extract, may result in the isolation of strains adapted to using these, rather than the principal carbon source, as energy sources. The isolation of an organism capable of growth on a simple medium should also form the basis of a cheaper commercial process and should be more resistant to contamination—a major consideration in the design of a commercial continuous process. The use of as high as possible an isolation temperature should also result in the isolation of a strain presenting minimal cooling problems in the subsequent process.

The main difficulty in using a continuous-enrichment process is the washout of the inoculum before an adapted culture is established. Johnson (1972) suggested that the isolation process should be started in batch culture using a 20% inoculum and as soon as growth is observed, the culture should be transferred to fresh medium and the subsequent purification and stabilization of the enrichment performed in continuous culture. The continuous system should be periodically inoculated with soil or sewage that may not only be a source of potential isolates but should also ensure that the dominant flora is extremely resistant to contamination.

Harrison (1978) proposed two solutions to the problem of early washout in continuous isolation processes: The first uses a turbidostat and the second uses a two-stage chemostat (see Chapter 2). A turbidostat is a continuous-flow system provided with a photoelectric cell to determine the turbidity of the culture and maintain the turbidity between set points by initiating or terminating the addition of medium. Thus, washout is avoided, as the medium supply will be switched off if the biomass falls below the lower fixed point. The use of a turbidostat will result in selection on the basis of maximum specific growth rate as it operates at high levels of limiting substrate. Thus, although the use of the turbidostat removes the danger of washout it is not as flexible a system as the chemostat that may be used at a range of dilution rates. The two-stage chemostat described by Harrison (1978) is very similar to Johnson (1972) procedure. The first stage of the system was used as a continuous inoculum for the second stage and consisted of a large bottle containing a basic medium inoculated with a soil infusion. Continuous inoculation was employed until an increasing absorbance was observed in the second stage. Bull et al. (1992) advocated the use of feedback continuous systems for the isolation of strains with particularly high affinity for substrate and this approach would also guard against premature washout.

The use of continuous enrichment culture has frequently resulted in the selection of stable, mixed cultures presumably based on some form of symbiotic relationship. It is extremely unlikely that such mixed, stable systems could be isolated by batch techniques so that the adoption of continuous enrichment may result in the development of novel, mixed culture fermentations. Harrison, Topiwala, and Hamer (1972) isolated a mixed culture using methane as the carbon source in a continuous enrichment and demonstrated that the mixture contained one methylotroph and a number of nonmethylotrophic symbionts. The performance of the methylotroph in pure culture was invariably poorer than the mixture in terms of growth rate, yield, and culture stability.

Continuous enrichment has also been used for the isolation of organisms to be used in systems other than biomass production; Rowley and Bull (1977) used the technique to isolate an *Arthrobacter* sp. producing a yeast lysing enzyme complex. The technique has been used widely for the isolation of strains capable of degrading environmental pollutants. For example, Futamata, Nagano, Watanabe, and Hiraishi (2005) isolated a number of *Variovax* strains able to degrade trichloroethylene.

Enrichment cultures using solidified media

Solidified media have been used for the isolation of certain enzyme producers and these techniques usually involve the use of a selective medium incorporating the substrate of the enzyme that encourages the growth of the producing types. Aunstrup, Outtrup, Andresen, and Dambmann (1972) isolated species of *Bacillus* producing alkaline proteases. Soils of various pHs were used as the initial inoculum and, to a certain extent, the number of producers isolated correlated with the alkalinity of the soil sample. The soil samples were pasteurized to eliminate nonsporulating organisms and then spread onto the surface of agar media at pH 9–10, containing a dispersion of an insoluble protein. Colonies that produced a clear zone due to the digestion of the insoluble protein were taken to be alkaline protease producers. The size of the clearing zone could not be used quantitatively to select high producers, as there was not an absolute correlation between the size of the clearing zone and the production of alkaline protease in submerged culture. However, this example demonstrates the importance of choice of starting material, the use of a selective force in the isolation and the incorporation of a preliminary diagnostic test, albeit of limited use.

ISOLATION METHODS NOT UTILIZING SELECTION OF THE DESIRED CHARACTERISTIC—FROM THE "WAKSMAN PLATFORM" TO THE 1990s

It is exceptionally difficult to construct an enrichment isolation procedure that can give the producer of a secondary metabolite a selective advantage. Therefore, a pool of organisms is isolated and subsequently tested for the desired characteristic in a separate screening process. Waksman was the first to use extensive screening in his discovery of streptomycin in which the antibacterial activity of streptomycete isolates was assayed by detecting the inhibition of the growth of pathogenic bacteria as clear zones on an overlay plate (Schatz, Bugie, & Waksman, 1944). This approach, known as the Waksman platform, was adopted by pharmaceutical companies throughout the world and gave rise to the "golden age" of antibiotic discovery referred to earlier. However, the effectiveness of the platform declined as it resulted in the reisolation of strains that had already been screened many times before. However, this "reinvention of the wheel" syndrome has been minimized in two major ways:

1. Developing procedures to favor the isolation of unusual taxa that are less likely to have been screened previously.
2. Identifying selectable features correlated with the unselectable industrial trait thus enabling an enrichment process to be developed.

A very wide range of techniques have been developed to increase the probability of isolating novel organisms, including the pretreatment of soil (drying, heat-treatment, supplementation with substrates, irradiation, and ultrasonic waves, Tiwari & Gupta, 2013). An interesting approach is the use of a collection of bacteriophage as a pretreatment technique to eliminate common organisms or the detection of environmental bacteriophage as indicators of unusual actinomycetes (Kurtboke, 2011). Much use has been made of numerical taxonomic databases to design media selective for particular taxa. For example, Williams' group at Liverpool University (U.K.) used such databases to design isolation media to either encourage the growth of uncommon streptomycetes or to discourage the growth of *Streptomyces albidoflavus* (Vickers, Williams, & Ross, 1984; Williams & Vickers, 1988). Several groups of workers took advantage of the antibiotic sensitivity information stored in taxonomic databases to design media selective for particular taxa, as shown in Table 3.2. The incorporation of particular antibiotics into isolation media may result in the selection of the resistant taxa. Such techniques have been reviewed by Goodfellow and O'Donnell (1989); Bull et al. (1992); Goodfellow (2010); and Tiwari and Gupta (2013). The isolates from an isolation procedure would be screened for activity and then the growth requirements of the positive cultures determined. This knowledge can then be used to optimize the isolation medium and the cycle begins again.

While taxonomic databases are a convenient source of information, it is important to appreciate the observation made by Huck, Porter, and Bushell (1991) that these systems were designed to provide information for taxonomic differentiation within a group. Thus, some of the diagnostic data may not be applicable to isolation systems. More significantly, the reactions of organisms outside the taxon in question would not be listed. Of course, an environmental sample contains a vast variety of organisms and the design of isolation media based on knowledge of only one taxon may inadvertently result in the preferential isolation of undesirable types. Huck et al. (1991) used the statistical stepwise discrimination analysis (SDA) technique to design media for the positive selection of antibiotic producing soil isolates. This was

Table 3.2 Antibacterial Compounds Used in Selective Media for the Isolation of Actinomycetes (Goodfellow & O'Donnell, 1989)

Selective Agent	Target Organism	References
Bruneomycin	*Micromonospora*	Preobrazhenskaia et al. (1975)
Dihydroxymethylfuratriazone	*Microtetraspora*	Tomita et al. (1980)
Gentamycine	*Micromonospora*	Bibikova, Ivanitskaya, and Singal (1981)
Kanamycin	*Actinomadura*	Chormonova (1978)
Nitrofurazone	*Streptomyces*	Yoshiokova (1952)
Novobiocin	*Micromonospora*	Sveshnikova, Chormonova, Lavrova, Trekhova, and Preobrazhenskaya (1976)
Tellurite	*Actinoplanes*	Willoughby (1971)
Tunicamycin	*Micromonospora*	Wakisaka et al. (1982)

Table 3.3 Selective Substrates for the Isolation of Actinomycetes and Antibiotic-Producing Actinomycetes (Huck et al., 1991)

Substrates Selective for Actinomycetes	Substrates Selective for Antibiotic-Producing Actinomycetes
Proline	Proline
Glucose (1.0%)	Glucose (1.0%)
Glycerol	Glycerol
Starch	Starch
Humic acid (0.1%)	Humic acid (0.1%)
Propionate (0.1%)	Zinc
Methanol	Alanine
Nitrate	Potassium
Calcium	Vitamins
	Cobalt (0.05%)
	Phenol (0.01%)
	Asparagine

achieved by characterizing a collection of eubacterial and actinomycete soil isolates according to 43 physiological and nutritional tests. Certain features were identified which, when used as selective factors, enhanced the probability of either isolating actinomycetes or antibiotic-producing actinomycetes. These features are shown in Table 3.3. Using this approach several media were developed which enhanced the isolation of antibiotic producers.

The most desirable isolation medium would be one that selects for the desired types and also allows maximum genetic expression. Cultures grown on such media could then be used directly in a screen. However, it is more common that, once

Table 3.4 Guidelines for "Overproduction Media" (Nisbet, 1982)

1.	Prepare a range of media in which different types of nutrients become growth-limiting, for example, C, N, P, O.
2.	For each type of nutrient depletion use different forms of the growth-sufficient nutrient.
3.	Use a polymeric or complexed form of the growth-limiting nutrient.
4.	Avoid the use of readily assimilated forms of carbon (glucose) or nitrogen (NH_4^+) that may cause catabolite repression.
5.	Ensure that known cofactors are present (Co^{3+}, Mg^{2+}, Mn^{2+}, Fe^{2+}).
6.	Buffer to minimize pH changes.

isolated, the organisms are grown on a range of media designed to enhance productivity. Nisbet (1982) put forward some guidelines for the design of such media and these are summarized in Table 3.4. Although the issue of maximizing gene expression has been a guiding principle in drug discovery its significance was fully appreciated when the model streptomycete, *Streptomyces coelicolor*, was completely sequenced. It was discovered that its genome coded for twenty secondary metabolites, yet the organism is only credited with producing three antimicrobials (Bentley, Chater, & Cerdeno-Tarranga, 2002). Similarly the genome sequence of *Streptomyces avermitilis* revealed thirty secondary metabolite gene clusters (Ikeda, Ishikawa, & Hanamoto, 2003).

The difficulties of isolating novel organisms and avoiding the "rediscovery" of microbial products were compensated by the development of sophisticated screening methods to make the most of the potential hidden in both microbial broths and synthesized chemical compounds. The next section addresses the design of such screens and also puts into context the policy decisions taken by the pharmaceutical industry, faced with diminishing returns from exceptionally expensive discovery programs. We will then consider the recent advances in the isolation of novel organisms and the way forward in the development of the next generation "discovery platform."

SCREENING METHODS AND HIGH THROUGHPUT SCREENING

The early screening strategies based on the Waksman platform were empirical, labor intensive and showed diminishing success rates, as the number of commercially important compounds isolated increased. Thus, new screening methods were developed that were more precisely targeted to identify the desired activity.

Antibiotics were initially detected by growing the potential producer on an agar plate in the presence of an organism (or organisms) against which antimicrobial action was required. Production of the antibiotic was detected by inhibition of the test organism(s). Alternatively, the microbial isolate could be grown in liquid culture and the cell-free broth tested for activity. This approach was extended by using a range of organisms to detect antibiotics with a defined antibacterial spectrum. For example, Zahner (1978) discussed the use of test organisms to detect the production of antibiotics with confined action spectra. The use of *Bacillus subtilis* and *Streptomyces viridochromogenes* or *Clostridium pasteurianum* allowed the identification of anti-biotics with a low activity against *B. subtilis* and a high activity against the other test organisms. Such antibiotics may be new because they would not have been isolated by the more common tests using *B. subtilis* alone. The kirromycin group of antibiot-ics was discovered using methods of this type.

In the 15 years prior to 1971, no novel naturally occurring β-lactams were discov-ered (Nisbet & Porter, 1989). However, the advances made in the understanding of cell-wall biosynthesis and the mode of action of antibiotics allowed the development of mode of action screens in the 1970s that resulted in a very significant increase in the discoveries of new β-lactam antibiotics. Nagarajan et al. (1971) discovery of the cephamycins was based on the detection of compounds that induced morphological changes in susceptible bacteria. The appreciation that the mode of action of penicillins was the inhibition of the transpeptidase enzyme crosslinking mucopeptide molecules led to the development of enzyme inhibitor assays, which were particularly attractive because they could be automated. Fleming, Nisbet, and Brewer (1982) described the development of an automated screen for the detection of carboxypeptidase inhibitors that led to the detection of several novel cephamycin and carbapenem compounds.

The increasing frequency of penicillin and cephalosporin resistance among clini-cal bacteria led to the development of mechanism-based screens for the isolation of more effective antibacterials. The logic of the Beechams Pharmaceuticals group of Brown et al. (1976) was to search for a compound that would inhibit β-lactamase and could be incorporated with ampicillin as a combination therapeutic agent. Sam-ples were tested for their ability to increase the inhibitory effect of ampicillin on a β-lactamase-producing *Klebsiella aerogenes* and this strategy resulted in the discov-ery of clavulanic acid, produced by *Streptomyces clavuligerus*. Clavulanic acid was combined with amoxicillin (a semisynthetic penicillin) and marketed as "augmentin" which Lewis described in 2013 as "one of the most successful antibiotics on the market." This approach has also resulted in recent success with the approval in 2015 of the AstraZeneca/Actavis drug, avibactam, which protects ceftazidime (a third gen-eration cephalosporin) from degradation by β-lactamases (Ehmann et al., 2012).

The concept of using the inhibition of enzymes as a screening mechanism was pioneered by Umezawa (1972) in his search for microbial products inhibiting key enzymes of human metabolism. His approach was based on the logic that if a com-pound inhibits a key human enzyme in vitro, it may have a pharmacological action in vivo. Such screens have been applied to a wide range of pharmacological targets and have resulted in the isolation of several important drugs. Endo, Kurooda, and Tsujita (1976), of the Japanese pharmaceutical company Sankyo, screened fungal cultures for inhibitors of cholesterol synthesis—specifically, hydroxy-methyl-glutaryl-CoA reductase, the rate-limiting enzyme in the biosynthesis of sterols. They isolated "compactin" but, although it was highly effective in reducing cholesterol levels in patients with hypercholesterolemia, it was discontinued due to safety fears. Merck, in association with Sankyo, screened for similar compounds and Alberts et al. (1980) isolated mevinolin (later called lovastatin) from *Aspergillus terreus*. This was the first statin to reach the clinic and resulted in the development of a range of semisynthetic

Table 3.5 Enzyme Inhibitors Produced by Members of the Actinobacteria (Manivasagan et al., 2014)

Inhibited Enzyme Group	Use/Potential Use of Inhibitors
α-Amylase	Controlling blood glucose and serum insulin levels
β-Glucosidase	Antiviral; control of obesity
Serine proteases	Antifungal
Aspartic proteases	Control of gastroesophageal reflux disease (GERD)
Calpain—a cysteine proteases	Potential in control of neurodegenerative disorders
Pyroglutamyl peptidase	Elucidation of the mechanism of diseases caused by pyroglutamyl peptidase
Angiotensin converting enzyme (ACE)	Antihypertensive
Chitinase	Antifungal
Lipase	Antiobesity
Monoamine oxidase	Antidepressant
Tyrosinase	Skin whitener cosmetic

and synthetic statins, the total worldwide sales of which have been predicted to reach a trillion US dollars by 2020 (Ioannidis, 2014). Bull et al. (1992) listed the following clinical situations in which microbial products have been shown to inhibit key enzymes: hypercholesterolaemia, hypertension, gastric inflammation, muscular dystrophy, benign prostate hyperplasia, and systemic lupus erythematosus. Manivasagan, Venkatesan, Sivakumar, and Kim (2014) reviewed the extensive range of enzyme inhibitors produced from the Actinobacteria, summarized in Table 3.5. An interesting specific example is provided by the work of Hashimoto, Mural, Ezaki, Morikawa, and Hatanaka (1990). The activity of carbapenem antibiotics is lost in therapy due to renal dehydropeptidase activity. These workers isolated microbial products capable of inhibiting the enzyme that could then be administered along with the antibiotic to maintain its clinical activity.

The detection of pharmacological agents by receptor ligand binding assays has been developed rapidly by pharmaceutical companies (Fang, 2012). These are extensions of the enzyme inhibitor approach but agents that block receptor sites are likely to be more effective at very low concentrations. The gastrointestinal hormone cholecystokinin (CCK) controls a range of digestive activities such as pancreatic secretion and gall bladder contraction. Receptor screening identified a fungal metabolite, aperlicin (from *Aspergillus alliaceus*) that had a very high affinity for CCK receptors. Although the fungal product did not prove to be a suitable drug it was used as a model for the design of analogs which were receptor binders and pharmacologically acceptable.

The progress in molecular biology, genetics, and immunology also contributed extensively to the development of innovative screens in the 1990s, by enabling the construction of specific detector strains, increasing the availability of enzymes and receptors, and constructing extremely sensitive assays. Furthermore, these advances were paralleled by the development of robotic automation systems that facilitated an enormous increase in the rate of screening. Thus, combination of sensitive assays with robotic automation gave rise to what has become know as high throughput screening (HTS). The process is usually carried out using microtiter plates that have been scaled-up from the original 96 well devices to 384, 1536, or 3456 wells. Obviously, the considerable investment in such HTS systems can only be justified if there are sufficient potential agents to test. Agents for testing would include microbial cell-free broths and synthesized chemical compounds. As mentioned previously, the diminishing returns from microbial cultures and a belief that the microbial world had been "over–fished" for pharmaceuticals led the industry to concentrate on compounds synthesized using combinatorial chemistry. Such compounds were available in sufficient numbers to feed the high throughput screens and were deemed a better investment than the isolation of microorganisms and their associated culture and purification of their products.

The determination of the complete gene sequence of *Haemophilus influenzae* in 1995, followed by the sequencing of other key pathogens, gave a means of identifying potential antibiotic targets at a genome level. Payne, Gwynn, Holmes, and Pompliano (2008) provided an invaluable insight into the rationale of GlaxoSmithKline's (GSK) use of this technology in selecting screening targets for antibiotic discovery. GSK scientists hypothesized that a gene that was highly conserved in a range of both Gram negative and positive bacteria was highly likely to be essential to microbial growth. Thus, they compared the genomes of *H. influenza*, *Moraxella catarrhalis* (both Gram-negative), *Streptococcus pneumoniae*, *Staphylococcus aureus*, and *Enterococcus faecalis* (all Gram-positive) and identified genes that were not only highly conserved but present as only single copies (to reduce the probability of resistance) and had no human homologues (to focus on bacterial-specific genes). To test the hypothesis that these genes were essential for growth, the genes were replaced (using allelic replacement mutagenesis) with an antibiotic resistance marker. Failure of the mutant to grow strongly suggested (but did not guarantee) the essential nature of the gene. However, growth in the absence of the gene indicated it to be nonessential and thus did not merit further consideration. A gene had to be shown to be essential in at least two of the Gram-positive species to be considered a target as well as its level of expression (when artificially modulated) being commensurate with strain viability. More than 350 potential target genes were identified from the genome sequence analyses of which 127 were judged as essential by the allelic replacement test. The next step in the process was to clone each target gene into a bacterial host to express their protein that could then be purified and incorporated into a high throughput screen. Most of the proteins were enzymes, the assays of which were available in the literature and only 10% of the targets failed due to an inability to develop a screening assay. Between 1995 and 2001, 67 high-throughput screens were run against the company's compound library of up to half a million compounds, predominantly synthesized chemicals. Only 16 of the screens gave positive results (hits) and of these only 5

Table 3.6 GSK High Throughput Screens Generating Antibacterial Lead Compounds (Payne et al., 2008)

Target Protein	Function in the Bacterium
β-Keto-acyl carrier protein synthase III (FabH)	Fatty acid synthesis
Enoyl-acyl carrier protein reductase (Fab1)	Fatty acid synthesis
Peptide deformylase	Protein modification
Methionyl-tRNA synthetase	tRNA synthetase
Phenylalanine-tRNA synthetase	tRNA synthetase

gave rise to "lead" compounds (Table 3.6). Leads were defined as hits that also had antibacterial activity along with evidence that the mode of action was commensurate with the inhibition of the target protein. In all cases, the lead was produced by chemical modification of the hit compound. Of the successful screens shown in Table 3.6, only the two related to fatty acid synthesis are actually novel targets, the others are proven targets for antibiotics produced by actinomycetes and fungi. Each of the 16 high-throughput screens cost approximately US $1 million. The workers concluded: "The level of success was unsustainably low in relation to the large effort invested."

GSK were not alone in their lack of success in screening for antibacterials—Chan, Holmes, and Payne (2004) reported that more than 125 antibacterial screens, using 60 different targets, were operated by 34 different companies between 1996 and 2004, with no promising candidates coming anywhere near fruition. It is important to appreciate that the HTS approach has been extremely successful in the discovery of drugs affecting human targets. Thus, why was it so difficult to discover novel antibiotics? A number of reviewers have addressed this anomaly (Baltz, 2008; Lewis, 2013, 2015; Demain, 2014) and two key issues have been identified:

1. The high-throughput antibiotic screens relied predominantly on enzyme assays rather than viable bacteria and one of the major issues in the effectiveness of an antibiotic is its ability to be taken up by the bacterium. This is particularly relevant for Gram negative organisms, which have an outer membrane of negatively charged lipopolysaccharide, thereby blocking access to hydrophobic compounds, and an inner membrane which blocks the uptake of hydrophilic ones. Furthermore, if a compound does pass both barriers, they may be "deported" by multidrug pumps. Thus, the most common reason for the later failure of a "hit" molecule in an in vitro screen is its inability to penetrate the bacterial cell.

2. In their analysis of GSK's genomics-driven HTS, Payne et al. (2008) cited the lack of chemical diversity of its compound library as a major contributory factor to the lack of success. Berdy (2012) estimated that 1.6% of the approximately 70,000 known microbial products have given rise to drugs whereas only 0.005% of the 8–10 million synthetic compounds have done so. The explanation for this difference is that the microbiological molecules are the result of evolution

and natural selection such that nature has already done the prescreening and provided us with a "prescreened" library. Synthetic libraries not only contain a less diverse range of compounds compared with natural products but they have also been compiled to facilitate their eventual use as orally administered drugs. In an elegant analysis, Lipinski, Lombardo, Dominy, and Feeney (2001) formulated a set of "rules" (the rule of 5) to define the physicochemical characteristics that define the solubility and permeability of potential drugs that can be administered by mouth: Poor absorption is more likely when a compound has more than 5 hydrogen donors, more than 10 hydrogen bond acceptors, a molecular weight of greater than 500 Da and a CLogP (a measure of hydrophilicity) greater than 5. However, obeying the rule of 5 does not define a compound that will penetrate a prokaryote and thus an antibiotic HTS based on a chemical library obeying the rule is operating at a disadvantage.

In addition to these issues of discovery, the search for new antibiotics has obviously been influenced by the economics of the process. The clinical dose of an oral antibiotic is in the range 250–1000 mg administered up to 3 times daily for the period of the infection—this is approximately two to three orders of magnitude greater than the dose for a drug targeted at a human receptor. Thus, not only must the antibiotic be manufactured in larger quantities it is also more likely than other drugs to fail toxicity testing in the development. Drugs that control or prevent physiological (chronic) disease are also taken over a very long period compared with the short duration of antibiotic therapy. Furthermore, a novel antibiotic would be kept "in reserve" and only used against multiply resistant strains. As a result, the predicted financial returns of an antibiotic are far less favorable than other drugs. For example, Lewis (2013) quoted the annual sales of the cholesterol-lowering statin, atorvastatin, as $12 billion whereas that of the best-selling antibiotic, levofloxacin, is $2.5 billion. The time period from the initiation of a discovery program for a drug through to the marketplace is estimated as 14–21 years (Demain, 2014) and between 1999 and 2003, the total cost rose from $500–600 million to $900 million. In Europe and the United States of America, a patent gives the inventor protection for 20 years. However, the patent application must be made long before the drug is marketable, thus the effective protected time is often only 7–12 years. Thus, the economics of operating an antibiotic discovery program is at odds with the importance of the compounds to humankind. The development of antibiotic resistance necessitates the introduction of new antibiotics and there is a real fear of the world returning to the "preantibiotic era" in which an apparently minor infection could develop into a life-threatening situation. This raises questions of the role of government in commercial research funding. However, antibiotic resistance and the quest for new compounds is an international problem requiring governmental cooperation on a world stage. The Ebola virus outbreak in 2014/15 precipitated what has been called a "minor miracle" of public and private sector collaboration (Keller, 2015). The threat of a global *Ebola* pandemic running out of control in Africa eventually led to a World Health Organization initiative involving major governments, pharmaceutical companies and regulatory agencies to take an effective vaccine to human trials in less than a year (Henao-Restrepo, Longini, Egger, & Dean, 2015). It may take a similar initiative to gain the upper hand over the evolution of bacterial antibiotic resistance.

RETURN OF NATURAL PRODUCTS

It is all very well to quote the proportion of microbial products that have given rise to drugs compared with synthetic compounds but it must be remembered that the industry was faced with the very real decline in successful hits from microbial sources. As discussed previously, rediscovery was a major problem as well as the complexity of culture broths and the frequency of false-positives. Thus, a return to natural product screening must address these issues to make it an economic proposition. It has been well known for many years that the number of microbial taxa cultured from environmental samples represents the "tip of the iceberg" of microbial communities (Bull et al., 1992). However, the perceived size of the "tip" has been decreasing as a fuller understanding of natural microbial communities has emerged. This understanding is a result of what Bull, Ward, and Goodfellow (2000) described as a "paradigm shift" in search and discovery strategies—attributed to the development of metagenomics, the study of genetic material recovered directly from environmental samples. The process involves the extraction of DNA from an environmental sample (eg, soil) followed by PCR amplification of 16s ribosomal RNA genes using 16s rRNA gene primers. The amplified fragments are cloned into *E. coli*, thereby generating a library of 16s rRNA genes representative of the microbial community in the sampled environment. Such approaches have shown that the diversity of the environmental 16s rRNA genes far exceeds that shown by such genes from organisms cultured from the same environment, with Overmann (2013) suggesting that cultured bacterial species may represent only 0.1–0.001% of the actual total number. This compares with previous estimates of 1–20%. Thus, the problem of natural product discovery is not that the pool has been overfished but that we have been fishing in a very small pond. The extent of this untapped resource is exemplified by the existence of "candidate divisions"—lineages of prokaryotes which have not been cultured but their existence has been demonstrated by 16s rRNA metagenomics. The Bacteria and Archaea domains are composed of at least 60 major taxons (phyla or divisions), approximately half of which are candidate divisions, that is, no representatives have been cultured (Gasc, Ribiere, Parisot, Beugnot, & Defois, 2015). It will be recalled that whole genome sequencing of two streptomycetes (*S. coelicolor* and *S. avermitilis*) revealed that these organisms were capable of synthesizing far more secondary metabolites than had previously been realized. Thus, not only has screening hardly scratched the surface of the potential range of microorganisms but the individual repertoires of those isolated has also been under-exploited.

A number of approaches have been proposed to take advantage of this new appreciation of microbial diversity: the exploration of different environments and the development of new culture techniques to isolate novel organisms, thus broadening the base of the discovery process; increasing the gene expression of isolated organisms such that their full potential can be realized; screening for the genes coding for products rather than screening for the products themselves; developing novel screens to detect activity; and minimizing the anomalies previously associated with natural product screening. Obviously, these approaches are not mutually exclusive and their combination may give rise to synergy in discovery.

BROADENING THE BASE OF THE DISCOVERY PROCESS AND MAXIMIZING GENE EXPRESSION

The strategy of searching unusual environments for novel microorganisms is hardly a new approach but the development of the methodologies and an extension of the screened environments still have significant potential in product discovery. Tiwari and Gupta (2013) cited the discovery of 11 new families and 120 new genera of actinomycetes between 2001 and 2013. Notably productive searches have isolated strains from desert soils, marine environments, and from plant-microbe associations, as reviewed by Tiwari and Gupta (2013). A fuller appreciation of microbial ecology has begun to reveal the extent of the interactions between organisms in their natural environment that may be exploited in their isolation. Kaeberlein, Lewis, and Epstein (2002) developed a cultural system that simulated the natural marine environment. Microorganisms present in the upper layer of intertidal marine sediment were separated from the sediment, serially diluted and mixed with warm agar made with seawater. The agar, containing the organisms was placed in the center of a ring sitting on a 0.03 μm pore-size polycarbonate membrane and another membrane placed on the top—thus entrapping the embedded organisms in a diffusion membrane sandwich. The diffusion chamber was then incubated in a seawater aquarium. Thus, nutrients from the seawater were able to diffuse into the chamber creating an oligotrophic environment. As a result, microcolonies developed on the surface of the agar. This system was miniaturized such that 384×1.25 μL diffusion chambers could be accommodated on a 72 by 19 mm plate, termed an isolation chip or "ichip," as shown in Fig. 3.3 (Nichols et al., 2010). The ichip was dipped in warm agar containing a diluted suspension of organisms such that there was a high probability that each chamber contained one cell. As before, the plate containing the chambers (the ichip) was incubated in its simulated natural environment, in this case either seawater or a soil suspension. After incubation for two weeks, the mini agar plugs in each chamber could be examined for microcolony growth. A control using 500 μL volumes of seeded agar (the same volume contained in an ichip) in a 24 well cell culture plate was incubated conventionally. Organisms were identified using 16s rRNA gene sequencing. There was virtually no overlap of species isolated from the ichip with those from the culture dish, thus indicating the very different environment provided in the diffusion system. Ling, Schneider, Peoples, and Spoering (2015) used the ichip methodology to isolate strains that were then cultured in conventional systems and screened for antibacterial activity. A new species of β-proteobacteria, *Eleftheria terrae*, was shown to produce a novel antibiotic (named teixobactin) active against Gram-positive bacteria by binding to lipid precursors of the cell wall components,

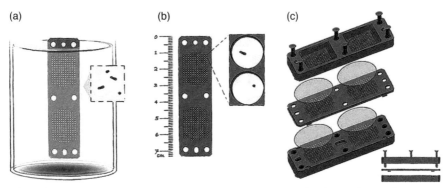

FIGURE 3.3 Isolation Chip, or Ichip, for High-Throughput Microbial Cultivation in Situ

The iChip (a) consists of a central plate (b) which houses growing microorganisms, semi-permeable membranes on each side of the plate, which separate the plate from the environment, and two supporting side panels (c). The central plate and side panels have multiple matching through-holes. When the central plate is dipped into suspension of cells in molten agar, the through-holes capture small volumes of this suspension, which solidify in the form of small agar plugs. Alternatively, molten agar can be dispensed into the chambers. The membranes are attached and the iChip is then placed in soil from which the sample originated.

Ling et al. (2015)

peptidoglycan and teichoic acid, the same site of action as vancomycin. The compound is a nonribosomal depsipeptide (contains an ester bond as well as peptide bonds, Fig. 3.4) composed of nonprotein amino acids and was effective in animal infection models and nontoxic in preliminary trials. Crucially, it took 30 years before vancomycin resistance was detected, which is thought to have originated from the Gram-positive producer organism (*Amycolatopsis orientalis*), indicating the robust nature of the target site. Ling et al. (2015) failed to develop teixobactin resistance in Gram-positive organisms and resistance in Gram-negatives is due to the impermeability of the outer membrane, a mechanism that cannot be transferred to Gram-positive bacteria. *E. terrae* is a Gram-negative bacterium and thus its resistance to its own antibiotic is also due to its outer membrane, thus eliminating a possible mechanism of teixobactin resistance equivalent to that of vancomycin. Thus, the isolation of novel organisms using an isolation technique that mimicked their natural environment resulted in the detection of a novel molecule with a mode of action that precludes the rapid development of resistance.

The recognition of "silent" antibiotic genes referred to earlier has stimulated research into the development of cultural techniques that will switch them on and thus realize the potential of isolates (Zhu, Sandiford, & van Wezel, 2014). The differentiation of streptomycetes involves a complex process of programmed cell death in which the autolysis of vegetative mycelium provides substrates for the development of aerial, spore-bearing hyphae, as discussed in Chapter 2. This differentiation

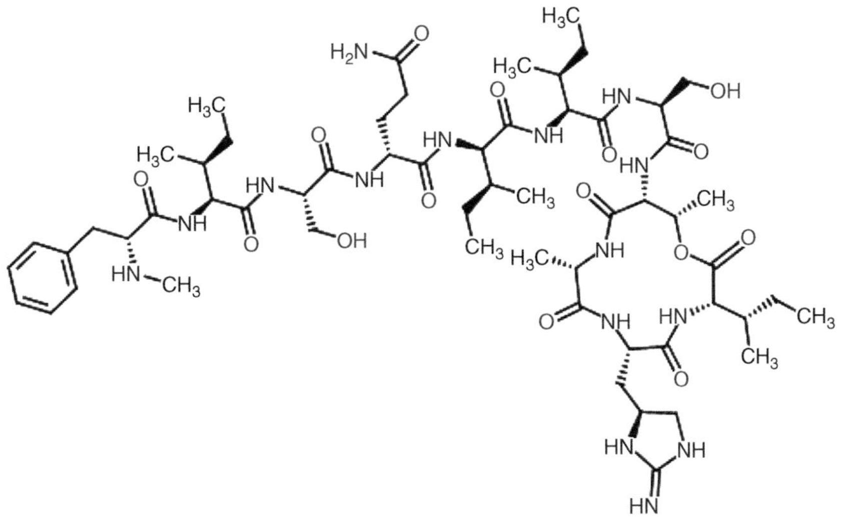

FIGURE 3.4 The Structure of Teixobactin

corresponds with the onset of secondary metabolism and antibiotic production, thus inhibiting the growth of bacteria competing for the released nutrients. Thus, the inclusion of molecules that result from the degradation of polymers may mimic this situation and stimulate antibiotic synthesis. For example, N-acetylglucoseamine (a peptidoglycan component) has been shown to be a key regulator of the streptomycete transcription factor DasR, a pleiotropic regulator orchestrating the control of primary and secondary metabolism. The presence of N-acetylglucoseamine in poor growth conditions has been shown to stimulate secondary metabolism. Interestingly, the rare earth metals scandium and lanthanum are also known to enhance antibiotic synthesis and their inclusion in screening media may also be beneficial.

The recognition of microbial interaction in natural ecosystems has led to the investigation of signaling between different organisms as well as between members of the same species with Romero, Traxler, Lopez, and Kolter (2011) debating the concept of antibiotics as signal molecules in their natural environment. Thus, cocultivation of different organisms has resulted in the expression of previously unexpressed systems. The first report by Watanabe, Okubo, Suzuki, and Izaki (1992) related to the discovery of the polyketide enacyloxin that was produced by a *Gluconobacter* sp. in combination with either *Neurospora crassa* or *Aspergillus oryzae*. The field has expanded considerably with the huge improvement in analytical techniques (Wu, Kim, van Wezel, & Choi, 2015a). For example, Wu et al. (2015b) grew *S. coelicolor* and *Aspergillus niger* together in submerged liquid culture and identified their products using NMR analysis. *A. niger* formed 100–500 μm pellets to which the streptomycete adhered and eventually covered. As a result of the interaction, *A. niger* produced a cyclic dipeptide, cyclo(-Phe-Phe) and 2-hydroxyphenylacetic acid. Interestingly,

A. niger could produce these compounds when cultured in the cell-free broth of the *Streptomyces*, indicating that the mechanism involved signaling or the provision of precursors rather than physical contact. Furthermore, biotransformation studies with *o*-coumaric acid and caffeic acid yielded a number of novel products. This work demonstrates the potential of such studies and may be valuable to further exploit the biosynthetic potential of established secondary metabolite producers.

The identification of quorum sensing type signal molecules as initiators of secondary metabolism in the actinomycetes has opened another possible approach to awaken silent genes (discussed in more detail later in this chapter). The discovery of A-factor in the control of streptomycin synthesis in *Streptomyces griseus* soon resulted in a wider appreciation of the role of γ-butyrolactones in secondary metabolism. Thus the inclusion of such molecules in culture systems may result in stimulation of secondary metabolite gene expression.

The approaches to gene awakening considered so far involve manipulating the environment of the organism in an attempt to maximize its gene expression without necessarily knowing which genes are present. These approaches have been described as "looking harder" rather than "looking differently." The "look differently" philosophy has exploited the advances in genome sequencing and bioinformatics to search directly for the genes of secondary metabolism, a process termed genome mining. Two approaches have been used to exploit this approach: (1) the production of the product in the organism encoding the genes and (2) the heterologous production of the product, that is, the incorporation of the genes into a host producer (Bachmann, Van Lanen, & Baltz, 2014). The most well studied secondary metabolism genes and associated enzyme systems are those for the synthesis of the polyketides and nonribosomal peptides. Both groups are synthesized by multimodular enzyme complexes, the polyketide synthases (PKSs) and the nonribosomal peptide synthases (NRPSs), respectively. The genes for each product group are grouped together in a tight locus and the enzyme complexes are composed of enzyme modules, each module catalyzing a particular reaction. Some of the modules are universal while others are "optional extras" resulting in a huge range of potential products depending on the precise composition of the complex. Knowing the PKS or NRPS genes in an organism allows the prediction of the enzyme complex and thus the structure that the complex will synthesize. The first example of this approach is the discovery of the novel polyketide antifungal farnesylated benzodiazepinone by Ecopia Bioscience (McAlpine et al., 2005). These workers identified promising, but unexpressed, gene clusters in *Streptomyces aizunensis* using genome scanning, a shotgun DNA sequencing method in which DNA fragments were cloned into a cosmid library and screened with genome sequence tags of known biosynthetic genes. The identified genes could then be completely sequenced which enabled the prediction of the product structure. Expression of the polyketide was achieved by growing the organism in almost fifty different media. Thus, the use of genome mining identified an organism that had the capacity to produce a novel compound and justified the devotion of resources to unlock its synthesis. The advances in DNA sequencing superseded the early techniques of genome scanning and it is now possible to sequence the whole genome of an organism both rapidly and economically. Equally, the advances in bioinformatics have also revolutionized the processing of sequence information. The development of the "antibiotics and secondary metabolism analysis shell" (antiSMASH) software (Medema et al., 2011b; Weber et al., 2015) not only enabled the rapid identification of secondary metabolism genes but also the partial prediction of the chemical structures for which they coded (Helfrich, Eiter, & Piel, 2014).

As the understanding of the control of secondary metabolism has increased, the approaches to gene awakening have become more sophisticated. Most secondary metabolism genes are controlled by regulator genes, both negative and positive (Zotchev, Sekura, & Katz, 2012), and thus the removal of repressors or the over production of positive regulators may stimulate gene expression. Laureti et al. (2011) identified a 150 kb span in the genome sequence of *Streptomyces ambofaciens* comprising 25 genes commensurate with polyketide synthesis. Bioinformatic analysis revealed that the predicted product was unknown and also demonstrated the presence of three potential regulatory genes. Cloning of one of these genes and its over expression on a plasmid in *S. ambofaciens* resulted in the production of a group of very closely related 51-membered macrolides, which they named stambomycins. Biological activity screening revealed that the stambomycins had potential as anticancer lead compounds. However, not all organisms bearing the gifts of silent genes are amenable to this type of genetic manipulation and a solution to this problem is the expression of the genes in a heterologous host; Gomez-Escribano and Bibb (2011) modified *S. coelicolor* for precisely this purpose. The model streptomycete was modified by deleting its four endogenous antibiotic gene clusters and introducing point mutations into two pleiotropic genes. These manipulations ensured that a heterologous pathway did not have to compete with endogenous secondary metabolite pathways for precursors; endogenous products did not complicate the purification of the heterologous products; and the heterologous product was produced at a high level.

The genome mining discussed so far relates to the analysis of cultured organisms and the expression of silent genes. However, this strategy does not address the issue of the untapped potential of the microbial population that remains uncultured. The field of metagenomics was referred to earlier in this discussion in the context of screening DNA extracted from the environment for 16sRNA gene sequences in an attempt to assess the true extent and diversity of uncultured microorganisms. Metagenomics has also been applied to the discovery of natural products by searching DNA extracted from the environment (environmental DNA or eDNA) for secondary metabolism genes and cloning these genes into a heterologous host, thus bypassing the need for cultivation. Two approaches have been used, termed function-based and sequence-based (Milshteyn, Schneider, & Brady, 2014). The function-based approach involves the direct screening of metagenomic clones for desired, new biological activity while the sequence-based approach involves the sequencing and bioinformatic analysis of eDNA to identify potential secondary metabolism genes. The aim of the functional approach is to provide culture broths for activity screening. The first novel molecule discovered by this approach was that of Wang et al. (2000) who expressed a soil DNA library in *Streptomyces lividans* (1020 clones) and isolated terragine

A using an HPLC detection system. However, much simpler screens are required to process large libraries and workers have tended to use antibacterial activity or pigment production to focus on clones of potential interest. A number of refinements of the function-based approach are needed to fully realize its potential—the cloning of larger fragments such that the probability of incorporating a whole secondary metabolism gene cluster into a clone is increased; the development of a wider range of hosts to facilitate efficient expression; and the enrichment of the libraries to facilitate screening (Milshteyn et al., 2014).

The most promising sequence-based approach has involved a PCR-based "sequence tag" methodology focusing on the secondary metabolite classes incorporating the most diversity, primarily the polyketides and the nonribosomal peptides (Milshteyn et al., 2014). This philosophy relies upon a homology-based screen in which gene sequences in eDNA are recognized as similar to known sequences of secondary metabolism genes, for example, polyketide synthesis. The process uses PCR amplification based on degenerative primers designed to recognize the desired gene. A degenerative primer is a mixture of primers each of which has a slightly different sequence; thereby extending the range of sequences that may be amplified. The amplified DNA may not only be cloned into a host to produce the product but also sequenced thus enabling an in silico search to predict the structure using bioinformatics databases. This approach has been used to screen DNA both from relatively restricted microbial populations and eDNA from soil or marine samples. For example, Wilson, Mori, Ruckert, and Uria (2014) used this approach to demonstrate that virtually the entire spectrum of secondary metabolites produced by the sponge, *Theonella swinhoei*, was actually produced by a single bacterial endosymbiont *Entotheonella factor* TSY1.

The development of novel approaches to screening natural products has been discussed by Lewis (2013) and Harvey, Edrada-Ebel, and Quinn (2015). The first problem to be addressed in a natural product screening program is the removal of compounds that cause artifacts, for example, polyphenols that give positive results against a wide range of screens. Separating a crude extract into fractions and testing each separately may not only reduce interference in an assay but may also enable the earlier elimination of "rediscovery." The Wyeth pharmaceutical company developed a fractionation scheme in which the microbial cultures were first screened for secondary metabolites using liquid chromatography linked to mass spectrometry employing an infinity diode array detector and an infinity evaporative light scattering detector (LCMS-DAD-ELSD) (Wagenaar, 2008). Those cultures that produced secondary metabolites were then methanol extracted and fractionated using reverse-phase HPLC. Early fractions contained media ingredients and highly polar compounds while late fractions contained lipophilic compounds, both of which were discarded. Ten middle-order HPLC fractions were retained and subjected to screening and compared with the screening of the crude extract. Interestingly, 80% of the active cultures only showed the activity in the fractionated extracts.

As discussed earlier, the lack of success in screening synthetic chemical libraries for antibiotic activity may be due to the limitation of their diversity by the application of Lipinski's "rule of five." While these rules describe the characteristics required for an orally administered drug they are not commensurate with compounds that must be accessible to prokaryotic cells. Lewis (2013) argued that an empirical analysis of the physicochemical properties of known antibacterials would give the broad characteristics of a molecule that would penetrate a prokaryotic cell. Multidrug resistance efflux is an active transport system in which a wide range of structurally unrelated antibiotics may be pumped out of a bacterium. These pumps are particularly effective against hydrophobic cations but not against anions and hydrophilic compounds. Relatively hydrophilic compounds with a molecular mass of less than 600 Da penetrate prokaryotes well as do ones containing atoms rarely found in natural products, such as fluorine and boron. Interestingly, fluoroquinones, the only successful synthetic antibiotic group, are hydrophilic, small and have both an anionic group and a fluoride atom. A wider examination of antimicrobials may enable the formulation of set of characteristics to select relevant members of a synthetic chemical library.

Wang, Soisson, and Young (2006) and Young, Jayasuriya, and Ondeyka (2006) of Merck Research Laboratories assessed inhibitors of bacterial type II fatty-acid synthesis as potential antibiotics. These workers developed *S. aureus* strains that were either downregulated or upregulated in FabF protein production; FabF being a key enzyme in type II fatty acid synthesis. Both strains were challenged with natural products in a well-diffusion assay to identify those that inhibited only the downregulated strain. Such activity predicted the mode of action to be the inhibition of FabF. This approach resulted in the discovery of a new class of antibiotics, the platensimycins, following a screen of 250,000 fermentation broths. The broths tested were not new and had been tested previously in a variety of screens. However, the discriminatory nature of the whole cell assay could not only identify specific inhibitors but, crucially, also do so at low concentrations, due to the debilitated nature of the FabF downregulated strain.

Lewis (2013) also raises the interesting issue of the search for broad-spectrum antibiotics. Many screens were designed to identify compounds that would inhibit both Gram positive and Gram-negative organisms and thus those that only inhibited Gram positives were discarded. A broad-spectrum antibiotic is beneficial if the identity of the pathogen is unknown but also have significant side effects by inhibiting the natural gut flora of the patient. The discovery of more specific antibiotics, directed at particular pathogens, might be an easier task than finding agents that are active against a broad range of pathogens. However, such products would only be appropriate in the clinic if the identity of the pathogen were known. Thus, it would require a change of practice in the diagnostic laboratory—the replacement of cultural techniques with much more rapid molecular methods.

Richard Baltz has been active and influential in industrial microbiology for over 40 years and in his authoritative review (2016) he predicted that we are entering a "New Golden Age" for natural product discovery and development. He based this prediction on three major advances, coupled with the developments in bioinformatics and mass spectrometry: the discovery, due to inexpensive gene sequencing, of an "unprecedented" number of secondary metabolism gene clusters;

the revelation by metagenomics of the extent of microbial diversity yet to be exploited; and the increased understanding of the biosynthesis of polyketides and nonribosomal proteins.

IMPROVEMENT OF INDUSTRIAL MICROORGANISMS

Natural isolates usually produce commercially important products in very low concentrations and therefore every attempt is made to increase the productivity of the chosen organism. Increased yields may be achieved by optimizing the culture medium and growth conditions, but this approach will be limited by the organism's maximum ability to synthesize the product. The potential productivity of the organism is controlled by its genome and, therefore, the genome must be modified to increase the potential yield. The cultural requirements of the modified organism would then be examined to provide conditions that would fully exploit the increased potential of the culture, while further attempts are made to beneficially change the genome of the already improved strain. Thus, the process of strain improvement involves the continual genetic modification of the culture, followed by reappraisals of its cultural requirements.

Genetic modification may be achieved by selecting natural variants, by selecting induced mutants and by selecting recombinants. There is a small probability of a genetic change occurring each time a cell divides and when it is considered that a microbial culture will undergo a vast number of such divisions, it is not surprising that the culture will become more heterogeneous. The heterogeneity of some cultures can present serious problems of yield degeneration because the variants are usually inferior producers compared with the original culture. However, variants have been isolated which are superior producers and this has been observed frequently in the early stages in the development of a natural product from a newly isolated organism. An explanation of this phenomenon for mycelial organisms may be that most new isolates are probably heterokaryons (contain more than one type of nucleus) and the selection of the progeny of uninucleate spores results in the production of homokaryons (contain only one type of nucleus) that may be superior producers. However, the phenomenon is also observed with unicellular isolates that are certainly not heterokaryons. Therefore, it is worthwhile to periodically plate out the producing culture and screen a proportion of the progeny for productivity; this practice has the added advantage that the operator tends to become familiar with morphological characteristics associated with high productivity and, by selecting "typical" colonies, a strain subject to yield degeneration may still be used with consistent results.

Therefore, selection of natural variants may result in increased yields but it is not possible to rely on such improvements, and thus techniques have been employed to increase the chances of improving the culture. The first approach used in strain improvement was the induction and isolation of mutants, an approach that was enormously successful (Baltz, 2016) and was combined with the exploitation of recombination systems such as protoplast fusion. The development of recombinant DNA technology increased the armory of the industrial microbiologist such that more directed approaches to strain improvement could be applied and also opened the door to the production of heterologous products. Genome sequencing and the development of computer systems to store and access data gave rise to the "postgenomic" era. The influence of these powerful tools led to the development of "new" subject areas—"systems biology" and its application to biotechnology, "synthetic biology"; effectively new terms for the discipline of microbial physiology enhanced by the availability of very powerful tools. At this point it is important to appreciate that the chronological development of techniques and knowledge did not entirely replace previous approaches, but built upon them. The industrial strains developed by mutation and selection were the starting point for the application of both recombinant DNA technology and genomics to strain improvement and these strains are the basis of many of those used in current processes.

These tools for strain improvement have been used for the development of both primary and secondary metabolite fermentations. However, there are some fundamental differences in the ways in which the tools have been applied, which is due to both the nature of the products and the knowledge of the pathways concerned. Thus, the improvement of primary product biosynthesis will be considered separately from that of secondary metabolite biosynthesis.

IMPROVEMENT OF STRAINS PRODUCING PRIMARY BIOSYNTHETIC PRODUCTS

Primary biosynthetic pathways convert the intermediates of carbon catabolic pathways to end products required for the production of biomass. These pathways are regulated such that the supply of end products is commensurate with the needs of the organism and are generally reductive in nature resulting in the consumption of NADPH. The improvement in the yield of these products has been achieved predominantly by targeting three sites:

- The control of the terminal pathway
- The supply of precursors from central metabolism
- The supply of NADPH by the recycling of NADP.

The targeting of these three sites corresponds approximately to the chronological development of the strain improvement process. The first approach was the induction and selection of mutants, which focused primarily on the removal of the control of the terminal pathway. The supply of precursors was addressed by recombinant DNA technology and the upregulation of NADP recycling was achieved using a systems biology approach. This is a simplification of the developments over more than 50 years because there was considerable overlap between the application of the evolving technologies to the three target sites and other targets were also relevant to strain improvement. The evolution of the strain improvement process as applied to primary metabolism will now be reviewed by considering the selection of mutants, the application of recombination (natural systems, protoplast fusion, and recombinant DNA technology) and finally the developments in the postgenomic era.

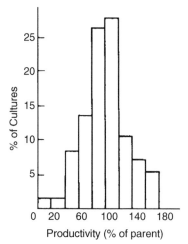

FIGURE 3.5 The Spread in Productivity of Chortetracycline of Natural Variants
of *Streptomyces viridifaciens* (Dulaney & Dulaney, 1967)

Selection of induced mutants synthesizing improved levels of primary metabolites

An example from secondary metabolism helps to place in context the challenge of the isolation of hyper-producing mutants. Dulaney and Dulaney (1967) compared the spread in productivity of chlortetracycline of natural variants of *Streptomyces viridifaciens* with the spread in productivity of the survivors of an ultraviolet treatment. The results of their comparison are shown in Figs. 3.5 and 3.6, from which it may be seen that although there are more inferior producers among the survivors of the ultraviolet treatment there are also strains producing more than twice the parental level, far greater than the best of the natural variants. The use of ultraviolet light is only one of a large number of physical or chemical agents that increase the mutation-rate—such agents are termed mutagens. The reader is referred to Baltz (1986) and Birge (1988) for accounts of the modes of action of mutagens. The vast majority of induced mutations are deleterious to the yield of the desired product but, as shown in Fig. 3.6, a minority is more productive than the parent. The problem of obtaining the high-yielding mutants has been addressed by designing cultural techniques that enable the separation of the few desirable types from the large number of mediocre producers.

The separation of desirable mutants from the other survivors of a mutation treatment is similar to the isolation of desirable organisms from nature. Where possible, the mutant isolation procedure should use the improved characteristic of the desired mutant as a selective factor. Presumably, superior productivity is a result of a diversion of precursors into the product and/or a modification of the control mechanisms limiting the level of production. Thus, knowledge of the biosynthetic route and the mechanisms of control of the biosynthesis of the product should enable the

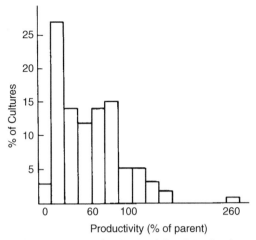

FIGURE 3.6 The Spread in Chlortetracycline Productivity of the Survivors of a UV-Treated Population of *Streptomyces viridifaciens* (Dulaney & Dulaney, 1967)

prediction of a "blueprint" of the desirable mutant. Such a "blueprint" might then enable the design of isolation techniques that would give the desired mutants a selective advantage over the other types present. Knowledge of biosynthetic routes and control mechanisms are more detailed for primary metabolites and, therefore, the use of selective pressure in mutant isolation is common in the fields of amino acid, nucleotide, and vitamin production. Before considering the methods used for the selection of mutants producing improved levels of primary metabolites, it is necessary to study the mechanisms of control of their biosynthesis such that the "blueprints," referred to earlier, may be drawn accurately. The levels of primary metabolites in bacteria are regulated by feedback control systems such that, when the product is present in sufficient quantity its biosynthesis is switched off. The major systems involved are feedback inhibition, feedback repression, and attenuation. Feedback inhibition is the situation where the end product of a biochemical pathway inhibits the activity of an enzyme catalyzing one of the reactions (normally the first reaction) of the pathway. Inhibition acts by the end product binding to the enzyme at an allosteric site that results in interference with the attachment of the enzyme to its substrate. Feedback repression acts at the gene level and is the situation where the end product of a biochemical pathway acts as a corepressor. The aporepressor is coded by the regulator gene but is only active when combined with the corepressor (the pathway end product). The active complex of co- and aporepressor will bind to the operator site and prevents the transcription of messenger RNA by RNA polymerase, thus repressing the synthesis of an enzyme (or enzymes) catalyzing a reaction (or reactions) of the pathway. Attenuation occurs in the control of the biosynthesis of some amino acids and also acts on transcription, but the controlling factor is not the end product (amino acid) but the charged tRNA molecule (the amino acid chemically bonded to its transfer RNA) that delivers the amino acid to the ribosome in the synthesis of a protein. The ratio of charged/uncharged tRNA acts as a means to measure the level of the amino acid. Although transcription will be initiated regardless of the charged/uncharged tRNA levels, it will be terminated if there are a significant number of tRNA molecules in the charged state (ie, there is sufficient amino acid present). Termination occurs between the operator and the first structural gene. Attenuation controls the synthesis of threonine, isoleucine, valine, leucine, phenylalanine, and histidine, while tryptophan is controlled by both attenuation and repression. Attenuation acts first, responding to an increasing level of tryptophan, followed by repression as the amino acid becomes in excess (Yanofsky, Kelley, & Horn, 1984).

Feedback inhibition and repression/attenuation frequently act in concert in the control of biosynthetic pathways, where inhibition may be visualized as a rapid control that switches off the biosynthesis of an end product and repression/attenuation as a mechanism to then switch off the synthesis of temporarily redundant enzymes. The control of pathways giving rise to only one product (ie, unbranched pathways) is normally achieved by the first enzyme in the sequence being susceptible to inhibition by the end product and the synthesis of all the enzymes being susceptible to repression/attenuation by the end product, as shown in Fig. 3.7.

The control of biosynthetic pathways giving rise to a number of end products (branched pathways) is more complex than the control of simple, unbranched sequences. The end products of the same, branched biosynthetic pathway are rarely required by the microorganism to the same extent, so that if an end product exerts control over a part of the pathway common to two, or more, end products then the organism may suffer deprivation of the products not participating in the control. Thus, mechanisms have evolved which enable the level of end products of branched pathways to be controlled without depriving the cell of essential intermediates. The following descriptions of these mechanisms are based on the effect of the control, which may be arrived at by inhibition, repression, or a combination of both systems.

Concerted or multivalent feedback control. This control system involves the control of the pathway by more than one end product—the first enzyme of the pathway is inhibited or repressed only when all end products are in excess, as shown in Fig. 3.8.

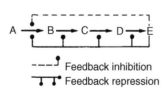

---⦙ Feedback inhibition
⊤ ⊤⦙ Feedback repression

FIGURE 3.7 The Control of a Biosynthetic Pathway Converting precursor A to End Product E Via the Intermediates B, C, and D

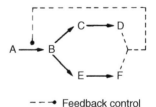

– – –→ Feedback control

FIGURE 3.8 The Control of a Biosynthetic Pathway by the Concerted Effects of Products D and F on the First Enzyme of the Pathway

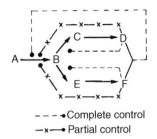

FIGURE 3.9 The Control of a Biosynthetic Pathway by the Cooperative Control by End Products D and F

Cooperative feedback control. The system is similar to concerted control except that weak control may be effected by each end product independently. Thus, the presence of all end products in excess results in a synergistic repression or inhibition. The system is illustrated in Fig. 3.9 and it may be seen that for efficient control to occur when one product is in excess there should be a further control operational immediately after the branch point to the excess product. Thus, the reduced flow of intermediates will be diverted to the product that is still required.

Cumulative feedback control. Each of the end products of the pathway inhibits the first enzyme by a certain percentage independently of the other end products. In Fig. 3.10 both D and F independently reduce the activity of the first enzyme by 50%, resulting in total inhibition when both products are in excess. As in the case of cooperative control, each end product must exert control immediately after the branch point so that the common intermediate, B, is diverted away from the pathway of the product in excess.

Sequential feedback control. Each end product of the pathway controls the enzyme immediately after the branch point to the product. The intermediates that then build up as a result, control earlier enzymes in the pathway.

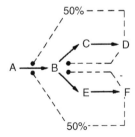

FIGURE 3.10 The Control of a Biosynthetic Pathway by the Cumulative Control of Products D and F

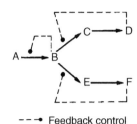

- - -• Feedback control

FIGURE 3.11 The Control of a Biosynthetic Pathway by Sequential Feedback Control

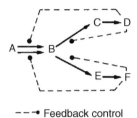

- - -• Feedback control

FIGURE 3.12 The Control of two Isoenzymes (Catalyzing the Conversion of A to B) by End Products D and F

Thus, in Fig. 3.11, D inhibits the conversion of B to C, and F inhibits the conversion of B to E. The inhibitory action of D, F, or both, would result in an accumulation of B, which, in turn, would inhibit the conversion of A to B. This control is common in the genus *Bacillus*.

Isoenzyme control. Isoenzymes are enzymes that catalyze the same reaction but differ in their control characteristics. Thus, if a critical control reaction of a pathway is catalyzed by more than one isoenzyme, then the different isoenzymes may be controlled by the different end products. Such a control system should be very efficient, provided that control exists immediately after the branch point so that the reduced flow of intermediates is diverted away from the product in excess. An example of the system is shown in Fig. 3.12 and is commonly found in members of the *Enterobacteriaceae*.

Thus, the levels of microbial metabolites may be controlled by a variety of mechanisms, such that end products are synthesized in amounts not greater than those required for growth. However, the ideal industrial microorganism should produce amounts far greater than those required for growth and, as suggested earlier, an understanding of the control of production of a metabolite may enable the construction of a "blueprint" of the most useful industrial mutant, that is, one where the production of the metabolite is not restricted by the organism's control system. Such postulated mutants may be modified in three ways:

1. The organism may be modified such that the end products that control the key enzymes of the pathway are lost from the cell due to some abnormality in the permeability of the cell membrane.
2. The organism may be modified such that it does not produce the end products that control the key enzymes of the pathway.
3. The organism may be modified such that it does not recognize the presence of inhibiting or repressing levels of the normal control metabolites.

Isolation of *Corynebacterium glutamicum* and the role of permeability in the glutamate fermentation

The birth of the amino acid fermentation industry occurred in 1957 with the isolation of the glutamic acid producing bacterium, *Corynebacterium glutamicum* (Kinoshita, Udaka, and Shimono, 1957a). *Dashi* is one of the fundamental ingredients of *washoku*

(Japanese cuisine) and is a stock made from *konbu*, an edible seaweed (*Laminaria saccharina*), and *bonito* flakes (dried fish). The distinct flavor of *dashi* is due to a combination of glutamate extracted from the *Laminaria* and inosine monophosphate, a nucleotide extracted from the *bonito*. Ikeda (1909) identified the sodium salt of glutamic acid as one of the flavor ingredients and it was produced commercially by the acid hydrolysis of wheat gluten, a process that established the Ajinomoto Company. The increasing demand for glutamate resulted in a race for a chemical or fermentation method to produce monosodium glutamate and this was won by Kyowa Hakko Kagyo Ltd and reported by Kinoshita et al. (1957a). Kinoshita's group isolated a biotin-requiring, glutamate-producing organism, subsequently named *C. glutamicum*. Thus, *C. glutamicum* is a natural biotin auxotroph and provided that the level of biotin in the production medium was below 5 μg dm^{-3} then the organism would excrete glutamate, but at concentrations of biotin optimum for growth it produced lactate (Kinoshita & Nakayama, 1978).

Biotin is a cofactor of acetyl-CoA carboxylase, a key enzyme in the synthesis of fatty acids. Thus, the effect of biotin limitation was attributed to a partial disruption of the cell membrane's permeability resulting in the passive excretion of glutamate from the cell—the "leak model." Glutamate controls its own biosynthesis by the feedback repression of phosphoenolpyruvate carboxylase and citrate synthase and both repression and inhibition of NADP-glutamate dehydrogenase with glutamate being maintained at a level of 25–36 μg mg^{-1} dry weight of cells in the presence of excess biotin. However, under biotin limitation loss of glutamate from the cell prevents its accumulation to inhibiting and repressing levels, resulting in glutamate accumulation in the medium up to a level of 50 g dm^{-3} (Demain & Birnbaum, 1968). Further evidence of the permeability hypothesis is provided by the observations of Somerson and Phillips (1961) that penicillin would induce glutamate excretion under excess biotin, as would the surfactant polyoxyethylene sorbitan monooleate (Tween 80) (Udagawa, Abe, & Kinoshita, 1962). The effect of penicillin on growing cells is to weaken the cell wall and thus impart a physical stress on the cell membrane and Tween 80 would also cause a physical abuse of the cell envelope, thus explaining the effect of both conditions on glutamate production in terms of membrane function. Glycerol auxotrophs of corynebacteria have also been shown to over produce glutamate under glycerol limitation. Despite this portfolio of evidence the "leak model" was challenged by the discovery that glutamate excretion occurs via a special efflux carrier system (Hoischen & Kraemer, 1989; Gutmann, Hoischen, & Kramer, 1992) and would not "leak" from the cells in a passive manner. Furthermore, it is difficult to explain the high glutamate production of 60–80 g dm^{-3} by this model simply in terms of the balance of intra- and extracellular glutamate (Shimizu & Hirasawa, 2007). However, the evidence all falls into place in the light of an elegant piece of work by Nakamura, Hirano, Ito, and Wachi (2007) who demonstrated that the glutamate efflux system is a mechanosensitive channel. Such channels are membrane proteins that can respond to mechanical stress and have been implicated in bacteria as safety valves to prevent lysis. This finding is entirely compatible with the effects of biotin limitation, penicillin, Tween, and glycerol on glutamate production all of which cause changes in membrane tension. As a result, the channel responds to the mechanical stress, opens due to a conformational change in the protein and releases glutamate from the cell.

α-Ketoglutarate is the branch point between the TCA cycle and glutamate biosynthesis (Fig. 3.13). The enzyme α-ketoglutarate dehydrogenase (αKGDH) converts α-ketoglutarate to succinyl coenzyme A in the TCA cycle and it was originally

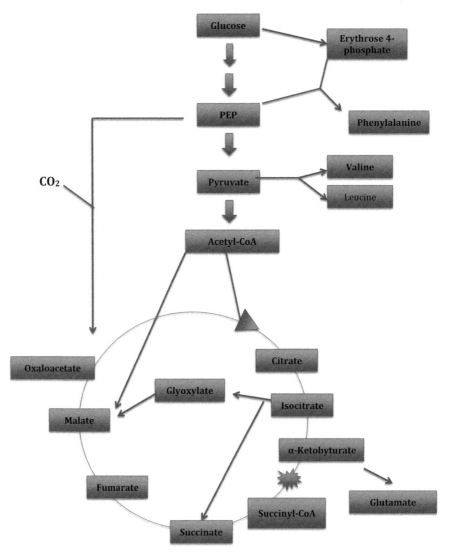

FIGURE 3.13 Glycolysis and the TCA and Glyoxylate Cycles and Glutamate Biosynthesis in *Corynebacterium glutamicum*

PEP, Phosphenol pyruvate; ✦ , inactive α-ketobutyrate dehydrogenase.

thought that *C. glutamicum* was deficient in this enzyme and thus α-ketoglutarate would be diverted to glutamate. However, Shiio and Ujigawa-Takeda (1980) demonstrated the presence of αKDH in the glutamate producer and this raised the question of the control of the enzyme and its role in glutamate synthesis. Niebisch, Kabus, Schultz, Weil, and Bott (2006) identified a novel protein, OdhI, as a regulator of αKDH. The influence of OdhI on αKDH depends on its phosphorylated state—the unphosphorylated form binds to a subunit of αKDH and inhibits its activity whereas the phosphorylated form will not bind. Phosphorylation is catalyzed by a protein kinase G and dephosphorylation is catalyzed by a phospho-serine/threonine protein phosphatase. Proteome analysis showed a significant increase in OdhI after penicillin exposure, that is, a treatment that induces glutamate synthesis. In their review of amino acid production, Ikeda and Takeno (2013) put forward an outline scenario for glutamate over production:

- Biotin limitation, penicillin or Tween 80 exposure induces the synthesis of the regulator OdhI in its unphosphorylated form.
- Unphosphorylated OdhI inhibits αKDH, diverting α-ketoglutarate to glutamic acid.
- Biotin limitation etc. also causes an increase in membrane tension resulting in the opening of the mechanosensitive channel, which allows the export of glutamate, thus avoiding it reaching inhibiting or repressing intracellular levels.

The mechanisms linking the environmental conditions with OdhI synthesis and the control of its phosphorylation/dephosphorylation have still to be established at the time of writing.

It is important to appreciate that *C. glutamicum* is a natural biotin auxotroph that was isolated from the natural environment and was not an artificially induced and selected mutant. However, *C. glutamicum* is an exceptionally important fermentation organism that is used not only for the commercial production of glutamic acid but also for the production of many other primary metabolites. The significance of *C. glutamicum* is exemplified by its use to produce in excess of 2 million metric tons of monosodium glutamate and almost 1 million metric tons of lysine per year (Sanchez & Demain, 2008).

Isolation of mutants that do not produce feedback inhibitors or repressors

Mutants that do not produce certain feedback inhibitors or repressors have proved useful for the production of intermediates of unbranched pathways; and intermediates and end products of branched pathways. Demain (1972) presented several "blue-prints" of hypothetical mutants producing intermediates and end products of biosynthetic pathways and these are illustrated in Fig. 3.14. The mutants illustrated in Fig. 3.14 do not produce some of the inhibitors or repressors of the pathways considered and, thus, the control of the pathway is lifted, but, because the control factors are also essential for growth, they must be incorporated into the medium at concentrations that will allow growth to proceed but will not evoke the normal control reactions.

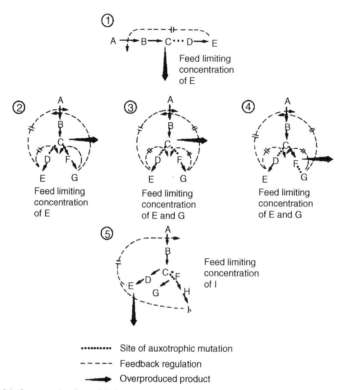

FIGURE 3.14 Overproduction of Primary Metabolites by Decreasing the Concentration of a Repressing or Inhibiting End Product (Demain, 1972)

In the case of Fig. 3.14(1) the unbranched pathway is normally controlled by feedback inhibition or repression of the first enzyme of the pathway by the end product, E. However, the organism represented in Fig. 3.14(1) is auxotrophic for E due to the inability to convert C to D so that control of the pathway is lifted and C will be accumulated provided that E is included in the medium at a level sufficient to maintain growth but insufficient to cause inhibition or repression.

Fig. 3.14(2) is a branched pathway controlled by the concerted inhibition of the first enzyme in the pathway by the combined effects of E and G. The mutant illustrated is auxotrophic for E due to an inability to convert C to D, resulting in the removal of the concerted control of the first enzyme. Provided that E is included in the medium at a level sufficient to allow growth but insufficient to cause inhibition then C will be accumulated due to the control of the end product G on the conversion of C to F. The example shown in Fig. 3.14(3) is similar to that in Fig. 3.14(2) except that it is a double auxotroph and requires the feeding of both E and G. Fig. 3.14(4) is, again, the same pathway and illustrates another double mutant with the deletion for the production of G occurring between F and G, resulting in the accumulation of F.

Fig. 3.14(5) illustrates the accumulation of an end product of a branched pathway that is normally controlled by the feedback inhibition of the first enzyme in the pathway by the concerted effects of E and I. The mutant illustrated is auxotrophic for I and G due to an inability to convert C to F and, thus, provided G and I are supplied in quantities which will satisfy growth requirements without causing inhibition, the end product, E, will be accumulated.

All the hypothetical examples discussed earlier are auxotrophic mutants and, under certain circumstances, may accumulate relatively high concentrations of intermediates or end products. Therefore, the isolation of auxotrophic mutants may result in the isolation of high-producing strains, provided that the mutation for auxotrophy occurs at the correct site, for example, between C and D in Figs. 3.14(1) and (2). The recovery of auxotrophs is a simpler process than is the recovery of high producers, as such, so that the best approach is to design a procedure to select relevant auxotrophs from the survivors of a mutation and subsequently screen the selected auxotrophs for productivity. Productive strains among the auxotrophs may be detected by overlayering colonies of the mutants with agar suspensions of bacteria auxotrophic for the required product. The high-producing mutants may be identified by the growth of the overlay around the producer. The most commonly used methods for the recovery of auxotrophic mutants are the use of some form of enrichment culture or the use of a technique to visually identify the mutants.

The enrichment processes employed are based on the provision of conditions that adversely affect the prototrophic cells but do not damage the auxotrophs. Such conditions may be achieved by exposing the population, in minimal medium, to an antimicrobial agent that only affects dividing cells and should result in the death of the growing prototrophs but the survival of the nongrowing auxotrophs. Several techniques have been developed using different antimicrobials suitable for use with a range of microorganisms.

Davis (1949) developed an enrichment technique utilizing penicillin as the inhibitory agent. The survivors of a mutation treatment were first cultured in liquid complete medium, harvested by centrifugation, washed, and resuspended in minimal medium plus penicillin. Only the growing prototrophic cells were susceptible to the penicillin and the nongrowing auxotrophs survived. The cells were harvested by centrifugation, washed (to remove the penicillin and products released from lysed cells) and resuspended in complete medium to allow the growth of the auxotrophs, which could then be purified on solidified medium. The nature of the auxotrophs isolated may be determined by the design of the so-called complete medium; if only one addition is made to the minimal medium then mutants auxotrophic for the additive should be isolated. Weiner, Voll, and Cook (1974) developed a very similar method utilizing nalidixic acid as the selective agent and showed it to be a useful alternative for bacteria resistant to penicillin.

Abe (1972) described the use of Davis' technique to isolate auxotrophic mutants of the glutamic acid producing organism *C. glutamicum*. The procedure is outlined in Fig. 3.15.

Advantage has also been taken of the fact that the ungerminated spores of some organisms are more resistant to certain compounds than are the germinated spores.

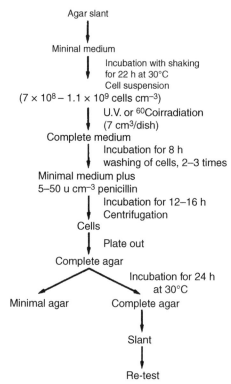

FIGURE 3.15 The Use of the Penicillin Selection Method for the Isolation of Auxotrophic Mutants of *C. glutamicum* (Abe, 1972)

Thus, by culturing mutated spores in minimal medium only the prototrophs will germinate and subsequent treatment of the spore suspension with a suitable compound would kill the germinated prototrophic spores but leave the ungerminated auxotrophic spores unharmed. The auxotrophic spores may then be isolated by washing, to remove the inhibitor, and cultured on supplemented medium. Ganju and Iyengar (1968) developed a technique of this type using sodium pentachlorophenate against the spores of *Penicillium chrysogenum, Streptomyces aureofaciens, S. olivaceus,* and *B. subtilis.*

The mechanical separation of auxotrophic and prototrophic spores of filamentous organisms has been achieved by the "filtration enrichment method" (Catcheside, 1954). Liquid minimal medium is inoculated with mutated spores and shaken for a few hours, during which time the prototrophs will germinate but the auxotrophs will not. The suspension may then be filtered through a suitable medium, such as sintered glass, which will tend to retain the germinated spores resulting in a concentration of auxotrophic spores in the filtrate.

The visual identification of auxotrophs is based on the alternating exposure of suspected colonies to supplemented and minimal media. Colonies that grow on supplemented media, but not on minimal, are auxotrophic. The alternating exposure of colonies to supplemented and minimal medium has been achieved by replica plating (Lederberg & Lederberg, 1952). The technique consists of allowing the survivors of a mutation treatment to develop colonies on petri dishes of supplemented medium and then transferring a portion of each colony to minimal medium. The transfer process may be "mechanized" by using some form of replicator. For bacteria the replicator is a sterile velvet pad attached to a circular support and replication is achieved by inverting the petri dish on to the pad, thus leaving an imprint of the colonies on the pad which may be used to inoculate new plates by pressing the plates on to the pad. It may be possible to replicate fungal and streptomycete cultures using a velvet pad, but, if unsatisfactory results are obtained, a steel pin replicator may be more appropriate.

Visual identification of auxotrophs has also been achieved by the so-called "sandwich technique." The survivors of a mutation treatment are seeded in a layer of minimal agar in a petri dish. The plate is incubated for 1 or 2 days and the colonies developed are marked on the base of the plate, after which a layer of supplemented agar is poured over the surface. The colonies that then appear after a further incubation period are auxotrophic, as they were unable to grow on the minimal medium.

Many auxotrophic mutants have been produced from *C. glutamicum* for the synthesis of both amino acids and nucleotide related compounds. As discussed previously, *C. glutamicum* is a biotin-requiring organism that will produce glutamic acid under biotin-limited conditions but it is important to remember that mutants of this organism, employed for the production of other amino acids, must be supplied with levels of biotin optimum for growth. Biotin-limited conditions will result in these mutants producing glutamate and not the desired amino acid.

Auxotrophic mutants of *C. glutamicum* have been used for the production of lysine. The control of the production of the aspartate family of amino acids in *C. glutamicum* is shown in Fig. 3.16. Aspartokinase, the first enzyme in the pathway (1 in Fig. 3.16), is controlled by the concerted feedback inhibition of lysine and threonine. Homoserine dehydrogenase (2 in Fig. 3.16) is subject to feedback inhibition by threonine and repression by methionine. The first enzyme in the route from aspartate semialdehyde to lysine is not subject to feedback control. Thus, the control system found in *C. glutamicum* is a relatively simple one. Nakayama, Kituda, and Kinoshita (1961) selected a homoserine auxotroph of *C. glutamicum,* by the penicillin selection and replica plating method, which produced lysine in a medium containing a low level of homoserine, or threonine plus methionine. The mutant lacked homoserine dehydrogenase that allowed aspartic semialdehyde to be converted solely to lysine and the resulting lack of threonine removed the concerted feedback inhibition of aspartokinase. Kinoshita and Nakayama (1978) quoted the homoserine auxotroph, *C. glutamicum 901,* as producing 44 g dm^{-3} lysine and Sanchez and Demain refer to such mutants producing up to 70 g dm^{-3}.

The control of the production of arginine in *C. glutamicum* is shown in Fig. 3.17. The major control of the pathway is the feedback inhibition of the second enzyme in the sequence, acetylglutamic acid phosphorylating enzyme, although the first enzyme may also be subject to regulation. Kinoshita, Nakayama, and Udaka (1957b)

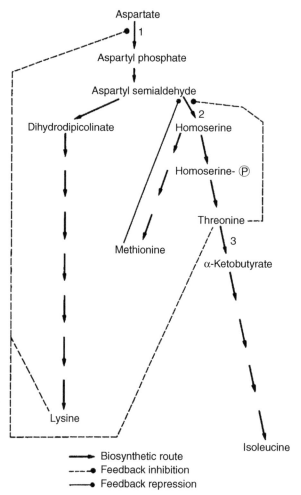

FIGURE 3.16 The Control of the Aspartate Family of Amino Acids in *C. glutamicum*

1, Aspartokinase; *2,* homoserine dehydrogenase; *3,* threonine deaminase.

isolated a citrulline requiring auxotroph of *C. glutamicum* that would accumulate ornithine at a molar yield of 36% from glucose, in the presence of limiting arginine and excess biotin. The mutant lacked the enzyme converting ornithine to citrulline that resulted in the cessation of arginine synthesis and, therefore, the removal of the control of the pathway.

Inosine monophosphate (IMP) was identified as a flavor-enhancing agent, along with glutamate, in *dashi* (Japanese stock), as discussed earlier. Whereas glutamate originated from the *Laminaria* used to prepare the stock, IMP came from the other ingredient in *dashi*, *bonito* flakes. The combination of glutamate and IMP results in a synergistic flavor enhancement 30 times that of glutamate alone (Sanchez and

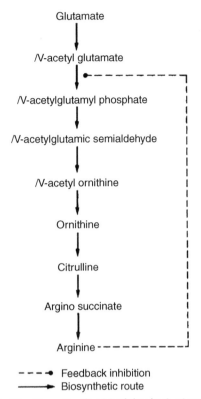

FIGURE 3.17 The Control of the Biosynthesis of Arginine in *C. glutamicum*

Demain, 2008). IMP is produced commercially by the chemical phosphorylation of inosine (Shibai, Enei, & Hirose, 1978) that is produced from auxotrophic strains of *B. subtilis.* The control of the production of purine nucleotides is shown in Fig. 3.18. The main sites of control shown in Fig. 3.18 are:

1. Phosphoribosyl pyrophosphate (PRPP) amidinotransferase (the first enzyme in the sequence) is feedback inhibited by AMP but only very slightly by GMP.
2. The synthesis of PRPP amidinotransferase is repressed by the cooperative action of AMP and GMP, as are the syntheses of the other enzymes in the pathway to IMP all of which are coded by the *pur* operon in *B. subtilis.*
3. IMP dehydrogenase is feedback inhibited and repressed by GMP.
4. Adenylosuccinate synthase is repressed by AMP but is not significantly inhibited.

Mutants that are auxotrophic for AMP or doubly auxotrophic for AMP and GMP have been isolated which will excrete inosine at levels of up to 15 g dm^{-3} (Sanchez and Demain, 2008). AMP auxotrophs, lacking adenylosuccinate synthase activity, require the feeding of small quantities of adenosine but will accumulate inosine due to the removal of the inhibition and cooperative repression of PRPP amidinotransferase,

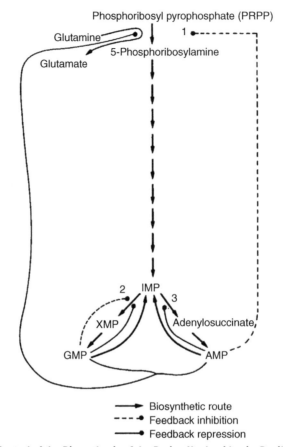

FIGURE 3.18 Control of the Biosynthesis of the Purine Nucleotides in *Bacillus subtilis*

1, Reaction catalyzed by PRPP amidinotransferase; *2*, reaction catalyzed by IMP dehydrogenase; *3*, reaction catalyzed by adenylosuccinate synthase; *AMP*, adenosine monophosphate; *IMP*, inosine monophosphate; *XMP*, xanthine monophosphate; *GMP*, guanosine monophosphate.

as shown in Fig. 3.19. AMP and GMP double auxotrophs will produce inosine due to the removal of the controls normally imposed by the two end products, as illustrated in Fig. 3.20. Such double auxotrophs require the feeding of both adenosine and guanosine, in small concentrations.

Isolation of mutants that do not recognize the presence of inhibitors and repressors

The use of auxotrophic mutants has resulted in the production of many microbial products in large concentrations, but, obviously, such mutants are not suitable for the synthesis of products that control their own synthesis independently. A hypothetical

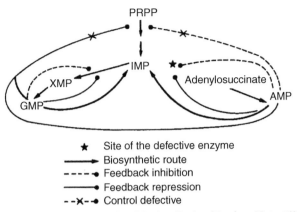

FIGURE 3.19 The Control of the Synthesis of Purine Nucleotides in a Mutant With a Defective Adenylosuccinate Synthase

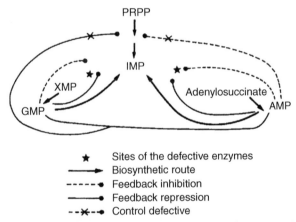

FIGURE 3.20 The Control of the Synthesis of Purine Nucleotides in a Mutant Defective in Adenylosuccinate Synthase and IMP Dehydrogenase Activities

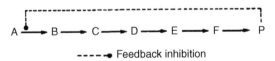

FIGURE 3.21 The Control of the Production of an End Product P

example is shown in Fig. 3.21 where the end product P controls its own biosynthesis by feedback inhibition of the first enzyme in the pathway. If it is required to produce the intermediate F in large concentrations then this may be achieved by the isolation of a mutant auxotrophic for P, blocked between F and P. However, if P is required to be synthesized in large concentrations it is quite useless to produce an auxotrophic mutant. The solution to this problem is to modify the organism such that the first enzyme in the pathway no longer recognizes the presence of inhibiting levels of P. The isolation of mutants altered in the recognition of control factors has been achieved principally by the use of two techniques:

1. The isolation of analog resistant mutants.
2. The isolation of revertants.

An analog is a compound that is very similar in structure to another compound. Analogs of amino acids and nucleotides are frequently growth inhibitory, and their inhibitory properties may be due to a number of possible mechanisms. For example, the analog may be used in the biosynthesis of macromolecules resulting in the production of defective cellular components. In some circumstances the analog is not incorporated in place of the natural product but interferes with its biosynthesis by mimicking its control properties. For example, consider the pathway illustrated in Fig. 3.21 where the end product, P, feedback inhibits the first enzyme in the pathway. If P* were an analog of P (which could not substitute for P in biosynthesis) and were to inhibit the first enzyme in a similar way to P, then the biosynthesis of P may be prevented by P* which could result in the inhibition of the growth of the organism.

Mutants may be isolated which are resistant to the inhibitory effects of the analog and, if the site of toxicity of the analog is the mimicking of the control properties of the natural product, such mutants may overproduce the compound to which the analog is analogous. To return to the example of the biosynthesis of P where P* is inhibitory due to its mimicking the control properties of P; a mutant may be isolated which may be capable of growing in the presence of P* due to the fact that the first enzyme in the pathway is no longer susceptible to inhibition by the analog. The modified enzyme of the resistant mutant may not only be resistant to inhibition by the analog but may also be resistant to the control effects of the natural end product, P, resulting in the uninhibited production of P. If the control system were the repression of enzyme synthesis, then the resistant mutant may be modified such that the enzyme synthesis machinery does not recognize the presence of the analog. However, the site of resistance of the mutant may not be due to a modification of the control system; for example, the mutant may be capable of degrading the analog, in which case the mutant would not be expected to overproduce the end product. Thus, analog resistant mutants may be expected to overproduce the end product to which the analog is analogous provided that:

1. The toxicity of the analog is due to its mimicking the control properties of the natural product.
2. The site of resistance of the resistant mutant is the site of control by the end product.

Resistant mutants may be isolated by exposing the survivors of a mutation treatment to a suitable concentration of the analog in growth medium and purifying any colonies that develop. Sermonti (1969) described a method to determine the suitable concentration. The organism was exposed to a range of concentrations of the toxic analog by inoculating each of a number of agar plates containing increasing levels of the analog with 10^6–10^9 cells. The plates were incubated for several days and examined to determine the lowest concentration of analog which allowed only a very few isolated colonies to grow, or completely inhibited growth. The survivors of a mutation treatment may then be challenged with the predetermined concentration of the analog on solid medium. Colonies that develop in the presence of the analog may be resistant mutants.

Szybalski (1952) constructed a method of exposing the survivors of a mutation to a range of analog concentrations on a single plate. Known as the gradient plate technique, it consists of pouring 20 cm^3 of molten agar medium, containing the analog, into a slightly slanted petri dish and allowing the agar to set at an angle. After the agar has set, a layer of medium not containing the analog is added and allowed to set with the plate level. The analog will diffuse into the upper layer giving a concentration gradient across the plate and the survivors of a mutation treatment may be spread over the surface of the plate and incubated. Resistant mutants should be detected as isolated colonies appearing beyond a zone of confluent growth, as indicated in Fig. 3.22. Whichever method is used for the isolation of analog-resistant mutants, great care should be taken to ensure that the isolates are genuinely resistant to the analog by streaking them, together with analog-sensitive controls, on both analog-supplemented and analog-free media. The resistant isolates should then be screened for the production of the desired compound by over layering them with a bacterial strain requiring the compound; producers may then be recognized by a halo of growth of the indicator strain.

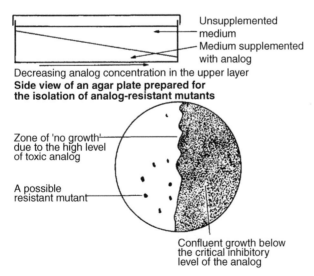

Unsupplemented medium

Medium supplemented with analog

Decreasing analog concentration in the upper layer

Side view of an agar plate prepared for the isolation of analog-resistant mutants

Zone of 'no growth' due to the high level of toxic analog

A possible resistant mutant

Confluent growth below the critical inhibitory level of the analog

Surface view of a gradient plate after inoculation and incubation.

FIGURE 3.22 The Gradient Plate Technique for the Isolation of Analog-Resistant Mutants

Sano and Shiio (1970) investigated the use of lysine analog-resistant mutants of *Brevibacterium flavum* for the production of lysine. The control of the biosynthesis of the aspartate family of amino acids in *B. flavum* is as illustrated for *C. glutamicum* in Fig. 3.16. The main control of lysine synthesis is the concerted feedback inhibition of aspartokinase by lysine and threonine. Sano and Shiio demonstrated that *S*-(2 aminoethyl) cysteine (AEC) completely inhibited the growth of *B. flavum* in the presence of threonine, but only partially in its absence. Also, the inhibition by AEC and threonine could be reversed by the addition of lysine. This evidence suggested that the inhibitory effect of AEC was due to its mimicking lysine in the concerted inhibition of aspartokinase. AEC-threonine-resistant mutants were isolated by plating the survivors of a mutation treatment on minimal agar containing 1 mg cm^{-3} of both AEC and threonine. A relatively large number of the resistant isolates accumulated lysine, the best producers synthesizing more than 30 g dm^{-3}. Investigation of the lysine producers indicated that their aspartokinases had been desensitized to the concerted inhibition by lysine and threonine.

The development of an arginine-producing strain of *B. flavum* by Kubota, Onda, Kamijo, Yoshinaga, and Oka-mura (1973) provides an excellent example of the selection of a series of mutants resistant to increasing levels of an analog. The control of the biosynthesis of arginine in *B. flavum* is similar to that shown for *C. glutamicum* in Fig. 3.17. Kubota et al. selected mutants resistant to the arginine analog, 2-thiazolealanine, and the genealogy of the mutants is shown in Fig. 3.23. Strain number 352 produced 25.3 g dm^{-3} arginine. Presumably, the mutants were altered in the susceptibility of the second enzyme in the pathway to inhibition by arginine.

The second technique used for the isolation of mutants altered in the recognition of control factors is the isolation of revertant mutants. Auxotrophic mutants may revert to the phenotype of the mutant "parent." Consider the hypothetical pathway illustrated in Fig. 3.21 where P controls its own production by feedback inhibiting the first enzyme (*a*) of the pathway. A mutant does not produce the enzyme, *a*, and is, therefore, auxotrophic for P. However, a revertant of the mutant produces large

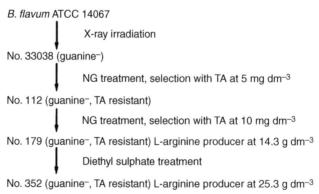

FIGURE 3.23 The Genealogy of L-Arginine-Producing Mutants of *B. flavum*

TA, Thiazolealanine; *NG*, *N*-methyl-*N*′-nitro-*N*-nitroso-guanidine (Kubota et al., 1973).

concentrations of P. The explanation of the behavior of the revertant is that, with two mutations having occurred at loci concerned with the production of enzyme *a,* the enzyme of the revertant is different from the enzyme of the original prototrophic strain and is not susceptible to the control by P. Revertants may occur spontaneously or mutagenic agents may be used to increase the frequency of occurrence, but the recognition of the revertants would be achieved by plating millions of cells on medium which would allow the growth of only the revertants, that is, in the earlier example, on medium lacking P.

Shiio and Sano, 1969 investigated the use of prototrophic revertants of *B. flavum* for the production of lysine. These workers isolated prototrophic revertants from a homoserine dehydrogenase-defective mutant. The revertants were obtained as small-colony forming strains and produced up to 23 g dm^{-3} lysine. The overproduction of lysine was shown to be due to the very low level of homoserine dehydrogenase in the revertants that presumably, resulted in the synthesis of threonine and methionine in quantities sufficient for some growth, but insufficient to cause inhibition or repression.

Mutant isolation programs for the improvement of strains producing primary metabolites did not rely on the use of only one selection technique. Most projects employed a number of methods including the selection of natural variants and the selection of induced mutants by a variety of means. The selection of bacteria over-producing threonine provides a good example of the use of a variety of selection techniques. Attempts to isolate auxotrophic mutants of *C. glutamicum* producing threonine were unsuccessful despite the fact that productive auxotrophic strains of *Escherichia coli* had been isolated. Huang (1961) demonstrated threonine production at a level of 2–4 g dm^{-3} by a diaminopimelate and methionine double auxotroph of *E. coli*. Kase, Tanaka, and Nakayama (1971) isolated a triple auxotrophic mutant of *E. coli* that required diaminopimelate, methionine, and isoleucine and produced between 15 and 20 g dm^{-3} threonine. The control of the production of the aspartate family of amino acids in *E. coli* is shown in Fig. 3.24 and that in *C. glutamicum* in Fig. 3.16. The mechanism of control in *E. coli* involves a system of isoenzymes, three isoenzymic forms of aspartokinase and two of homoserine dehydrogenase, under the influence of different end products. However, in *C. glutamicum* control is effected by the concerted inhibition of a single aspartokinase by threonine and lysine; by the inhibition of homoserine dehydrogenase by threonine and the repression of homoserine dehydrogenase by methionine. Thus, the control of homoserine dehydrogenase may not be removed by auxotrophy without the loss of threonine production. However, in *E. coli* methionine auxotrophy would remove control of the methionine-sensitive homoserine dehydrogenase and aspartokinase that would still allow threonine production, despite the control of the threonine-sensitive isoenzymes by threonine. *E. coli* mutants also lacking lysine and isoleucine would be relieved of the control of the lysine-sensitive aspartokinase and the degradation of threonine to isoleucine.

The production of threonine by *C. glutamicum* was achieved by the use of combined auxotrophic and analog resistant mutants. A good example of the approach is given by Kase and Nakayama (1972) who obtained stepwise improvements in productivity by the imposition of resistance to α-amino-β-hydroxyvaleric acid

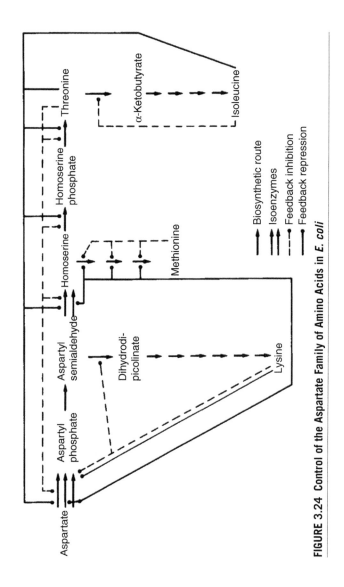

FIGURE 3.24 Control of the Aspartate Family of Amino Acids in *E. coli*

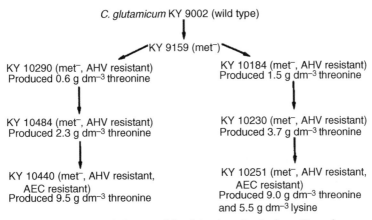

FIGURE 3.25 **The Genealogy of Mutants of C. *glutamicum* Producing L-Threonine or L-Threonine Plus L-Lysine**

AHV, α-Amino-β-hydroxy valeric acid; *AEC,* S-(β-aminoethy)-L-cysteine (Kase & Nakayama, 1972).

(a threonine analog) and S-(β-aminoethyl)-L-cysteine (a lysine analog) on a methionine auxotroph of *C. glutamicum.* The genealogy of the mutants is shown in Fig. 3.25. The analog-resistant strains were altered in the susceptibility of aspartokinase and homoserine dehydrogenase to control, and the lack of methionine removed the repression control of homoserine dehydrogenase. The use of transduction and recombinant DNA technology has resulted in the construction of far more effective threonine producers and these strains are considered in later sections of this chapter.

The use of recombination systems for the improvement of primary metabolite production
Transduction, transformation, and protoplast fusion
Hopwood (1979) defined recombination, in its broadest sense, as "any process which helps to generate new combinations of genes that were originally present in different individuals." The early use of recombination in strain improvement was extremely limited and took little advantage of natural systems (the parasexual processes of conjugation, transformation, and transduction in bacteria and sexual and parasexual processes in fungi). This was mainly due, according to Adrio and Demain (2010) to the very poor recombination frequencies in industrial microorganisms. However, there are a number of early examples of the use of transduction and transformation to improve amino acid producers. Komatsubara, Kisumi, and Chibata (1979, 1983) investigated the over production of threonine by *Serratia marcescens,* which is very similar to *E. coli* K-12 in its control of the aspartate family of amino acids (Fig. 3.24). Thus, as discussed in the previous section, the control of this pathway involves three aspartokinase and two homoserine dehydrogenase isoenzymes susceptible to control by the different end products as shown in Fig. 3.24.

These workers isolated three different analog resistant mutants; (1) showing feedback inhibition-resistance of threonine-sensitive aspartokinase and homoserine dehydrogenase; (2) showing constitutive synthesis of all threonine synthesizing enzymes and lack of threonine degradation; (3) showing both feedback inhibition and repression resistance of lysine-sensitive aspartokinase. Using two transductional crosses, these characteristics were combined in the one strain that then produced 25 g dm^{-3} threonine. A further analog-resistant mutant was isolated that was constitutive for methionine-sensitive aspartokinase and homoserine dehydrogenase and this mutation was also incorporated into the production strain by another transduction cross. Thus, all the aspartokinase and homoserine dehydrogenase enzymes in the production strain were resistant to both feedback inhibition and repression control and as a result produced 40 g dm^{-3} threonine. A similar example of the early use of recombination is provided by Yoneda (1980) who used transformation to assemble five different mutations for superior α-amylase production into one strain of *B. subtilis*. The final strain showed an enhancement of 250-fold compared with the original.

It was the development of protoplast fusion techniques in the 1970s and 1980s that transformed the landscape of strain improvement and enabled the properties of different strains or families of mutants to be combined together (Adrio and Demain, 2010). Up until this time strain improvement had concentrated on yield improvement—but often at the expense of other characteristics, particularly the loss of vigor commonly found in strains that have undergone several generations of mutation and selection. Protoplasts are cells devoid of their cell walls and may be prepared by subjecting cells to the action of wall degrading enzymes in isotonic solutions. Thus, lysozyme is used for the production of bacterial protoplasts and chitinase for fungal. Protoplasts may regenerate their cell walls and are then capable of growth as normal cells. Cell fusion, followed by genetic recombination, may occur between protoplasts of strains that would otherwise not fuse and the resulting fused protoplast may regenerate a cell wall and grow as a normal cell (Fig. 3.26). Thus, protoplasts may be used to overcome some recombination barriers. Protoplast fusion has been demonstrated in a large number of industrially important organisms including *Streptomyces* spp. (Hopwood, Wright, Bibb, & Cohen, 1977), *Bacillus* spp. (Fodor & Alfoldi, 1976), corynebacteria (Karasawa, Tosaka, Ikeda, & Yoshii, 1986), filamentous fungi (Ferenczy, Kevei, & Zsolt, 1974), and yeasts (Sipiczki & Ferenczy, 1977). Karasawa et al. (1986) used the technique to improve the fermentation rates of *B. flavum* lysine producers developed using repeated mutation and directed selection. Such strains were good lysine producers but showed low glucose consumption and growth rates, undesirable features that had been inadvertently introduced during the mutation and selection program. A protoplast fusion was performed between the lysine producer and a fast growing strain; a fusant was isolated displaying the desirable characteristics of high lysine production and high glucose consumption rate resulting in a much faster fermentation. The same authors used protoplast fusion to produce a superior threonine producing *Brevibacterium lactofermentum* strain. Lysine auxotrophy was introduced into a threonine and lysine overproducer by fusing it with a lysine auxotroph—the recombinant produced higher levels of threonine due to its lysine auxotrophy.

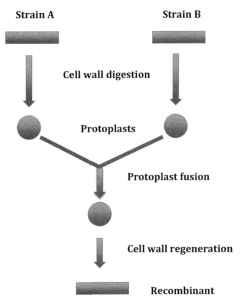

FIGURE 3.26 Protoplast Fusion Between Two Strains, A and B

Recombinant DNA technology

The work of Komatsubara et al. (1979, 1983), using transduction to combine desirable mutations in one strain of *S. marcescens*, discussed earlier is an example of the in vivo transfer of DNA between bacteria. The development of in vitro recombinant DNA technology hugely enhanced the ability to transfer genetic material derived from one species or strain to another. Transduction techniques make use of phage particles which will pick up genetic information from the chromosome of one bacterial species, infect another bacterial species and in so doing introduce the genetic information from the first host. The information from the first host may then be expressed in the second host. Whereas, transduction depends on vectors collecting information from one cell and incorporating it into another, the in vitro techniques involve the insertion of the information into the vector by in vitro manipulation followed by the insertion of the carrier and its associated "extra" DNA into the recipient cell. Because the DNA is incorporated into the vector by in vitro methods the source of the DNA is not limited to that of the host organism of the vector. Thus, DNA from human or animal cells may be introduced into the recipient cell. The basic requirements for the in vitro transfer and expression of foreign DNA in a host microorganism as follows:

1. A "vector" DNA molecule (plasmid or phage) capable of entering the host cell and replicating within it. Ideally the vector should be small, easily prepared and must contain at least one site where integration of foreign DNA will not destroy an essential function.
2. A method of splicing foreign genetic information into the vector.

3. A method of introducing the vector/foreign DNA recombinants into the host cell and selecting for their presence. Commonly used simple characteristics include drug resistance, immunity, plaque formation, or an inserted gene recognizable by its ability to complement a known auxotroph.

4. A method of assaying for the "foreign" gene product of choice from the population of recombinants created.

The initial work focused on *E. coli* but subsequently techniques have been developed for the insertion of foreign DNA into a range of bacteria, yeasts, filamentous fungi, and animal cells. The range of vectors has been discussed by Whitehouse (2015) and Harbron (2015) for bacteria, Curran and Bugeja (2015) for yeasts, Kieser, Bibb, Butner, Chater, and Hopwood (2000) for streptomycetes, Zhang, Daubaras, and Suen (2004) for filamentous fungi, Twyman and Whitelaw (2010) for mammalian cells, and Hitchman et al. (2010) for insect cells. The insertion of information into the vector molecule is achieved by the action of restriction endonucleases and DNA ligase. Site-specific endonucleases produce specific DNA fragments that may be joined to another similarly treated DNA molecule using DNA ligase. The modified vector is then normally introduced into the recipient cell by transformation. Because the transformation process is an inefficient one, selectable genes must be incorporated into the vector DNA so that the transformed cells may be cultured preferentially from the mixture of transformed and parental cells. This is normally accomplished by the use of drug-resistant markers so that those cells containing the vector will be capable of growth in the presence of a certain antimicrobial agent. The process is shown diagrammatically in Fig. 3.27.

Recombinant DNA technology has been used widely for the improvement of native microbial products. Frequently, this has involved "self-cloning" work where a chromosomal gene is inserted into a plasmid and the plasmid incorporated into the original strain and maintained at a high copy number. Thus, this is not an example of recombination because the engineered strain is altered only in the number of copies of the gene and does not contain genes that were present originally in a different organism. However, the techniques employed in the construction of these strains are the same as those used in the construction of chimeric strains, so it is logical to consider this aspect here.

The first application of gene amplification to industrial strains was for the improvement of enzyme production. Indeed, some regulatory mutants isolated by conventional means owed their productivity to their containing multiple copies of the relevant gene as well as the regulatory lesion. For example, the *E. coli* β-galactosidase constitutive mutants isolated by Horiuchi, Horiuchi, and Novick (1963) in chemostat culture also contained up to four copies of the *lacZ* gene. According to Demain (1990), during the 1960s and early 1970s, the number of gene copies was increased by using plasmids or transducing phage in the same species. The production of β-galactosidase, penicillinase, chloramphenicol transacetylase, and aspartate transcarbamylase were all increased by transferring plasmids containing the structural gene into recipient cultures, especially when the plasmid replicated faster than the host chromosome. The advent of recombinant DNA technology increased the applicability of this approach

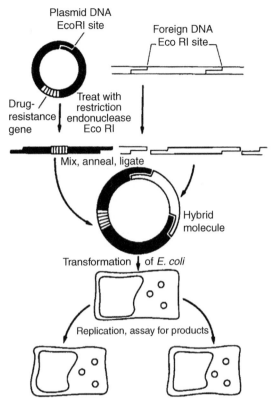

FIGURE 3.27 A Summary of the Steps in in Vitro Genetic Recombination

Both plasmid vector and foreign DNA are cut by the restriction endonuclease, EcoRI, producing linear double-stranded DNA fragments with single-stranded cohesive projections. EcoRI recognizes the oligonucleotide sequence $^{GAATTC}_{CTTAAG}$ and will cut any double-stranded DNA molecule to yield fragments with the same cohesive ends $^{GAATT}_C$, $^C_{TTAAG}$. On mixing vector and foreign DNA, hybrids form into circular molecules which can be covalently joined using DNA ligase. Transformation of *E. coli* results in the low-frequency uptake of hybrid molecules whose presence can be detected by the ability of the plasmid to confer drug resistance on the host (Atherton, Byrom, & Dart, 1979).

by allowing the construction of vectors containing the desired gene and enabling the transfer of DNA to other species. Adrio and Demain (2010, 2014) summarized the following uses of recombinant DNA technology in enzyme manufacture:

- Production of enzymes in industrial microorganisms that were originally produced in organisms that were difficult to grow, difficult to genetically manipulate or were pathogenic or toxigenic.

- Increasing the yield by the use of multiple gene copies, strong promoters, and efficient signal sequences.
- The use of protein engineering to improve the stability, activity or specificity of an enzyme.

The use of multicopy plasmids in the 1980s resulted in the development of a number of strains producing elevated enzyme levels. The development of the first commercial recombinant lipase for use in washing powders was achieved by Novo Nordisk (Sharma, Christi, & Banerjee, 2001). A lipase was isolated from *Humicola lanuginosa* (formerly *Thermomyces lanuginosus*) that was unsuitable as an industrial organism. A cDNA clone coding for the lipase was prepared and transformed into the industrial fungus *A. oryzae*. The recombinant *Aspergillus* produced the lipase at high levels. As well as achieving high levels of production and production in a heterologous host, microbial enzymes have been modified to improve their catalytic characteristics. For example, van den Burg, Vriend, Veltman, Venema, and Eijsink (1998) modified a protease from *Bacillus* (now *Geothermus*) *stearothermophilus* using site-directed mutagenesis such that it was active at 100°C in the presence of denaturing agents.

The first successful application of genetic engineering techniques to the production of amino acids was obtained in threonine production with *E. coli*. Debabov (1982) investigated the production of threonine by a threonine analog resistant mutant of *E. coli* K12. The entire threonine operon was introduced into a plasmid that was then incorporated into the organism by transformation. The plasmid copy number in the cell was approximately twenty and the activity of the threonine operon enzymes (measured as homoserine dehydrogenase activity) was increased 40–50 times. The manipulated organism produced 30 g dm^{-3} threonine, compared with 2–3 g dm^{-3} by the nonmanipulated strain. Miwa et al. (1983) utilized similar techniques in constructing an *E. coli* strain capable of synthesizing 65 g dm^{-3} threonine. It is important to appreciate that the genes which were amplified in these production strains were already resistant to feedback repression so that the multicopies present in the modified organism were expressed and not subject to control. Thus, the recombinant DNA techniques built on the achievements made with directed mutant isolation.

The application of genetic engineering to the industrially important corynebacteria was hampered for some years by the lack of suitable vectors. However, vectors have been constructed from corynebacterial plasmids and transformation and selective systems developed. The first patents for suitable vectors were registered by the two Japanese companies Ajinomoto (1983) and Kyowa Hakko Kogyo (1983) and now a range of vectors is available with kanamycin, chloramphenicol, and hygromycin as common selectable resistance markers. This subject has been reviewed in detail by Patek and Nesvera (2013). *C. glutamicum—E. coli* shuttle vectors have been developed and these are maintained in *C. glutamicum* at 10–50 copies per chromosome. Thus, these plasmids are especially useful in enabling increased product formation due to the higher gene dosage. Much of the work published in the early 1990s used this approach (see examples later in the text). Integrating vectors are essential for the disruption and replacement of *C. glutamicum* genes and these have been based on

E. coli plasmid vectors that do not replicate in *C. glutamicum*. As well as plasmid copy number, the efficiency of expression of a cloned gene will determine the level of product. Thus, *C. glutamicum* expression vectors are available in which promoter and ribosome binding sites are positioned just before the restriction sites (for gene insertion) thus placing the inserted gene under the control of the plasmid promoter. Most of the *E. coli—C. glutamicum* shuttle vectors incorporate this feature, for example, using the inducible lactose promoter such that expression can be induced by the addition of the gratuitous inducer, isopropyl-β-D-thiogalactopyranoside (IPTG). Although useful as a laboratory technique, IPTG induction is usually considered too expensive to use on a production scale and thus a temperature-sensitive repressor is frequently incorporated. In this system, the repressor binds to the promoter/operator (thus preventing expression) but at 40°C the binding becomes unstable and expression occurs. Expression is therefore induced by a heat shock. Transformation was originally first achieved in *C. glutamicum* using protoplasts but subsequently electroporation techniques were developed to introduce the required DNA (Dunican & Shivnan, 1989). These systems have enabled not only the use of recombinant DNA technology for strain improvement but have also facilitated the detailed investigation of the molecular biology of these important amino acid and nucleotide producers.

It is not surprising that the improvement of threonine production was the first reported use of recombinant DNA technology with amino acid producing corynebacteria (Shiio & Nakamori, 1989). It may be recalled from the discussion of the development of the early threonine producers that *C. glutamicum* was not particularly amenable for threonine over production using auxotrophs and analog resistant mutants. Over production was achieved by incorporating a DNA fragment coding for homoserine dehydrogenase from a *B. lactofermentum* threonine producer into a plasmid and introducing the modified plasmid back into the *Brevibacterium*. A similar approach was used for homoserine kinase and a strain was developed with remarkably increased homoserine kinase and dehydrogenase activities that produced 33 g dm^{-3} threonine (Morinaga et al., 1987).

The application of recombinant DNA technology to the development of tryptophan producing strains has been particularly valuable. Relatively low productivity was achieved using the classical mutation/selection approach despite attempts over a 40-year period (Ikeda, 2006). The control of the aromatic family of amino acids in *C. glutamicum* is shown in Fig. 3.28. The first step of the pathway is catalyzed by two isoenzymic forms of 3-deoxy-D-arabino-heptulosonate 7-phosphate synthase (DS); one susceptible to feedback inhibition by tyrosine (product of the *aro* I gene) and the other by concerted feedback inhibition by both tyrosine and phenylalanine (product of the *aro* II gene). The tyrosine and phenylalanine sensitive DS forms a protein complex with chorismic mutase, the first enzyme after the branch point to phenylalanine and tyrosine. Each of the three end products controls its own biosynthesis by inhibiting the first enzyme of its pathway. Additionally, the synthesis of all the enzymes for tryptophan synthesis is repressed by attenuation control.

The manipulation of *C. glutamicum* became possible with the development of cloning vectors in the 1980s. Earlier attempts had involved the stepwise isolation of

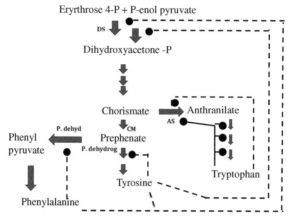

FIGURE 3.28 Control of the Aromatic Family of Amino Acids in _C. glutamicum_

--•, Feedback inhibition control; _AS,_ anthranilate synthase; _CM,_ chorismate mutase; _DS,_ DAHP synthase; _P. dehyd,_ prephenate dehydratase; —•, feedback repression; _P. dehydrog,_ prephenate dehydrogenase.

auxotrophic mutants for phenylalanine and tyrosine and/or resistance to structural analogs of tryptophan, phenylalanine, and tyrosine (Hagino & Nakayama, 1975). Using a mutation and selection approach, Shiio, Sugimoto, and Kawamura (1984) also added resistance to the sulfonamide antibiotic, sulfaguanidine, to the organism's portfolio of mutations and increased productivity of tryptophan to 19 g dm^{-3}. The logic of sulfonamide resistance is that the antibiotic inhibits the production of folic acid that is synthesized from para-aminobenzoic acid that, in turn, is produced from chorismate. Mutants resistant to sulfonamides were known to over-produce chorismate and thus such mutants may also have had the potential for tryptophan production. Shiio's strain produced 19.5 g dm^{-3} tryptophan and also over-produced chorismate, as predicted by the rationale of the selection process.

Katsumata and Ikeda (1993) used Shiio's strain as a basis for a recombinant DNA approach and cloned the first enzyme of the pathway (DS) and the tryptophan operon into a plasmid that was maintained at a high copy number in the producing organism. This strain yielded 43 g dm^{-3} tryptophan, a 54% increase over the parent. However, two problems were encountered with the engineered strain—sugar utilization and viability declined late in the fermentation and the plasmid was lost from the cells in the absence of selective pressure. Although antibiotic selective pressure could have been used to maintain plasmid stability, this would have been a problematic approach for a large-scale fermentation. The loss of viability was attributed to the accumulation of indole in the culture broth. The final step in tryptophan synthesis is the combining together of indole and serine. Thus, the accumulation of indole appeared to be due to the lack of sufficient serine to match the increased flux of indole through the tryptophan pathway from chorismate. The addition of serine to the fermentation led to a decline in indole, enhanced sugar utilization and increased viability. This observation gave rise to the development of a new plasmid that addressed both the problem of plasmid stability as well as serine availability. The *serA* gene, coding for phosphoglycerate dehydrogenase, the first enzyme in the serine pathway, was added to the plasmid. The host strain was then mutated to loss of *serA*, thus making the organism auxotrophic for serine and dependent on the presence of the plasmid for serine synthesis. This strain produced 50 g dm^{-3} tryptophan and the new multicopy plasmid was stably maintained. Further modification of this strain concentrated on engineering central metabolism as it was predicted that limitation could have been due to the supply of the pathway's precursors, phophoenolpyruvate (from glycolysis) and erythrose-4-phosphate (from the pentose phosphate pathway). Incorporation of the pentose-phosphate pathway gene for phophoketolase into the *C. glutamicum* plasmid resulted in a further increase in tryptophan yield to 58 g dm^{-3}. Thus, the elegant application of recombinant DNA technology developed a traditionally mutated strain into one in which the modifications to the final tryptophan pathway were integrated with changes in precursor supply that could sustain the increased potential of the industrial organism.

The application of recombinant DNA technology to the development of processes for the production of phenylalanine is an excellent illustration of the interrelationship between mutant development and genetic engineering. Phenylalanine is a precursor of the sweetener, aspartame, and is thus an exceptionally important fermentation product. Backman et al. (1990) described the rationale used in the construction of an *E. coli* strain capable of synthesizing commercial levels of phenylalanine. *E. coli* was chosen as the producer because of its rapid growth, the availability of recombinant DNA techniques and the extensive genetic database.

The control of the biosynthesis of the aromatic family of amino acids in *E. coli* is shown in Fig. 3.29. The first step in the pathway is catalyzed by three isoenzymes of dihydroxyacetone phosphate (DAHP) synthase, each being susceptible to one of the three end products of the aromatic pathway, phenylalanine, tyrosine, or tryptophan. Control is achieved by both repression of enzyme synthesis and inhibition of enzyme activity. Within the common pathway to chorismic acid, the production of shikimate kinase is also susceptible to repression. The conversion of chorismic acid to prephenic acid is catalyzed by two isoenzymes of chorismate mutase, each being susceptible to feedback inhibition and repression by one of either tyrosine or phenylalanine. Each isoenzyme also carries an additional activity associated with either the phenylalanine or tyrosine branch. The tyrosine sensitive isoenzyme carries prephenate dehydrogenase activity (the next enzyme in the route to tyrosine) while the phenylalanine sensitive enzyme carries prephenate dehydratase (the next enzyme in the route to phenylalanine).

The regulation of gene expression was modified by:

1. Both tyrosine sensitive and phenylalanine sensitive DAHP synthase and shikimic kinase are regulated by the repressor protein coded by the *tyr*R gene. The *tyr*R gene had been cloned and was subjected to in vitro mutagenesis and the wild type gene replaced. This one mutation then resulted in the derepression of both tyrosine and phenylalanine DAHP synthase as well as shikimic kinase.
2. The chorismic mutase/prephenate dehydratase protein is under both repression and attenuation control by phenylalanine. These controls were eliminated

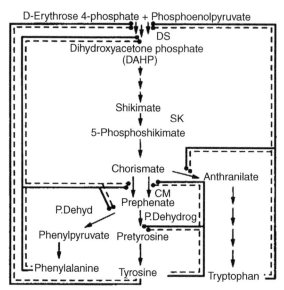

FIGURE 3.29 Control of the Aromatic Amino Acid Family in *E. coli*

--◆, Feedback inhibition control; *AS*, anthranilate synthase; *CM*, chorismate mutase; *DS*, DAHP synthase; *P. dehyd*, prephenate dehydratase; →◆, feedback repression; *P. dehydrog*, prephenate dehydrogenase; *SK*, shikimate kinase.

by replacing the normal promoter with one not containing the regulatory sequences, thus giving a 10 times increase in gene expression.

The regulation of enzyme inhibition was modified by:

1. Chorismic mutase/prephenate dehydratase is subject to feedback inhibition by phenylalanine. It was shown that a certain tryptophan residue was particularly important in the manifestation of feedback inhibition. In vitro mutagenesis was used to delete the tryptophan codon and the modified gene introduced along with the substituted promoter referred to earlier. The enzyme produced by the modified gene was no longer susceptible to phenylalanine.

2. The rationale was to limit the availability of tyrosine such that the tyrosine-sensitive DAHP synthase isoenzyme would not be inhibited. Although the other two isoenzymes would still be susceptible to phenylalanine and tryptophan inhibition, sufficient DAHP would be synthesized by the third isoenzyme (in the absence of tyrosine) to facilitate overproduction of phenylalanine. Traditionally, this objective would have been achieved by using a tyrosine auxotroph fed with limiting tyrosine. However, these workers developed an excision vector system. An excision vector is a genetic element that can carry a cloned gene and can both integrate into, and be excised from, the bacterial chromosome. The vector is based on the bacteriophage lambda and the technology is explained in Fig. 3.30. The excision of the vector can be induced by a temperature

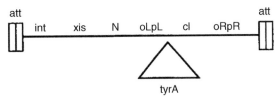

FIGURE 3.30 Excision Vector Technology

An excision vector is represented as a line between two boxes. Above the line and boxes are indicated those genes from bacteriophage lambda that are carried on the vector. oRpR is required for the expression of a repressor determined by *cl*. This repressor binds to oRpR and oLpL to prevent the expression of other lambda genes. The other genes collectively act to form a recombination activity that allows the vector to integrate into or excise from the bacterial chromosome at the sites (indicated by the boxes) named *att*. Other genes, such as *tyr*A (indicated below the excision vector), can be cloned into the excision vector prior to its introduction into a target cell and can thereby be present or absent in the cell in coordination with the vector. Upon entering a cell, the genes for recombination (*N, xis,* and *int*) are expressed because there is not yet any repressor. As repressor accumulates, it shuts off the expression of those genes. In a fraction of the recipient cells, the recombination enzymes cause the vector to integrate into the cell chromosome before those enzymes decay away. That cell and all of its progeny inherit the vector and any gene it might carry (such as *tyr*A). If the repressor is inactivated, such as by high temperature, new recombination enzymes are formed that excise the vector from the chromosome. In such a cell and all of its progeny, the vector and the gene(s) it might carry are lost (Backman et al., 1990).

shock. Thus, when the vector is excised the progeny of the cell will lose the inserted DNA. The *tyr*A gene was deleted from the production strain, inserted into the excision vector and transformed back into the organism where the vector became integrated into the chromosome. The fermentation could then be conducted using a cheap medium (the organism was not auxotrophic at inoculation) and allowing growth to an acceptable density. A heat shock may then be used to initiate vector excision, tyrosine auxotrophy and, hence, phenylalanine synthesis.

These efforts should have generated a high-producing strain. However, the tyrosine-sensitive DAHP synthase was susceptible to inhibition by high concentrations of phenylalanine. It will be recalled that the flow of DAHP was intended to come from the deregulated tyrosine sensitive isoenzyme. Thus, the final step in the development of the strain was to render this isoenzyme resistant to phenylalanine inhibition. This was achieved by the selection of mutants resistant to phenylalanine analogs. Thus, the strain was improved using a combination of gene cloning, in vitro mutagenesis and analog resistance, indicating the importance of the contribution of a range of techniques to strain development. The final strain was capable of producing 50 g dm^{-3} phenylalanine at a yield of 0.23 g g^{-1} glucose and 2 g g^{-1} biomass. The organism produced very low amounts of the other products and intermediates of the pathway which the workers claimed was due to the very precise manipulation of the strain which avoided the concomitant adverse characteristics associated with many highly mutated organisms.

Ikeda and Katsumata (1992) redesigned a tryptophan producing *C. glutamicum* strain such that it overproduced either phenylalanine or tyrosine. The regulation of the aromatic pathway in *C. glutamicum* is shown in Fig. 3.28. Phenylalanine producers that were resistant to phenylalanine analogs were used as sources of genes coding for enzymes resistant to control. Thus, a plasmid was constructed containing genes coding for the deregulated forms of DAHP synthase, chorismate mutase, and prephenate dehydratase. The vector was introduced into a tryptophan overproducer having the following features:

1. Chorismate mutase deficient and, therefore, auxotrophic for both tyrosine and phenylalanine.
2. Wild type DAHP synthase.
3. Anthranilate synthase partially desensitized to inhibition by tryptophan.

The transformed strain was capable of producing 28 g dm^{-3} phenylalanine and the levels of all three enzymes coded for by the vector were amplified approximately seven fold. If the original tryptophan producing strain was transformed with a plasmid containing only DAHP synthase and chorismate mutase, then tyrosine was overproduced (26 g dm^{-3}). Thus, genetic engineering techniques allowed a tryptophan producer to be redesigned into either a phenylalanine or tyrosine producer.

A different application of recombinant DNA technology is seen in the modification of the ICI plc. Pruteen organism, *Methylomonas methylotrophus*. The efficiency of the organisms' ammonia utilization was improved by the incorporation of a plasmid containing the glutamate dehydrogenase gene from *E. coli* (Windon et al., 1980). The wild type *M. methylotrophus* contained only the glutamine synthetase/glutamate synthase system that although having a lower K_m value than glutamate dehydrogenase, consumes a mole of ATP for every mole of NH_3 incorporated. Glutamate dehydrogenase, on the other hand, has a lower affinity for ammonia but does not consume ATP. In the commercial process ammonia was in excess because methanol was the limiting substrate so the expenditure of ATP in the utilization of ammonia was wasteful of energy. The manipulated organism was capable of more efficient NH_3 metabolism, which resulted in a 5% yield improvement in carbon conversion. However, the strain was not used in the industrial process due to problems of scale-up.

Postgenomic era—the influence of genomics, transcriptomics, and fluxomics on the improvement of primary metabolite producers
Application of genomics
Genomics uses the tools of gene sequencing and bioinformatics to study the biology of an organism at the chromosomal level. The knowledge obtained from a whole

genome sequence enables the development of a raft of new information on the functioning of an organism both at, and below, the level of the genome itself. Comparison of the sequence with gene databases enables the prediction of the role of each gene and the protein each produces. Thus, the development of information systems, and the means to interrogate them, has been as crucial to the success of genome investigation as has been the sequencing science. Three major DNA databases have been established: GenBank (USA), EMBL-BANK (Europe), and DDBJ (Japan) that receive information from laboratories, and share it with each other, on a daily basis. The information stored in these data depositories is available to the public and thus enables laboratories all over the world to benefit from, and contribute to, the development of the subject. The searching of both DNA and protein databases for matching sequences is enabled by a number of algorithms such as BLAST (Basic Local Allignment Search Tool). Thus, the term in silico has been added to in vivo and in vitro, describing a new era of biological exploration.

The complete genome sequence of *C. glutamicum* ATCC 13032 was first elucidated in 2001 by the Japanese company Kyowa Hakko Kogyo Co., Ltd. (Nakagawa et al., 2001) and deposited in the public database (GenBank NC_003450). Kyowa's competitor, Ajinomoto, sequenced the genome of a closely related species, *Corynebacterium efficiens*, in 2002 (Fudou et al., 2002) and deposited it in 2003 (GenBank database, NC_004369). Quite independently, Kalinowski et al., 2003 published the sequence of *C. glutamicum* ATCC 13032 in 2003 and in 2007 the sequence of *C. glutamicum* strain R was published by Yukawa, Omumasaba, Nonaka, Kos, and Okai (2007). Ohnishi et al. (2002) was the first to apply the knowledge of *C. glutamicum*'s genome sequence in an attempt to produce a "minimum mutation strain." Amino acid producing strains that have been developed by mutation and selection have proved to be highly successful commercial organisms. However, the selection of desirable traits using, for example, analog resistance, does not prevent the coselection of other mutations that negatively affect strain performance. Thus, strains that have undergone multiple mutation/selection procedures may have accumulated a range of undesirable mutations resulting in their being less vigorous, slower growing, and less resistant to stressful conditions. Also, the presence of background mutations may confuse the interpretation of the mechanism of over production that may, in fact, not be due to a "selected" mutation, thus making further logical, directed strain improvement problematic. As discussed earlier, protoplast fusion was used in an attempt to remove deleterious markers by generating recombinants between high producing strains (that lacked vigor) and wild types that grew well but did not over produce. Ohnishi et al.'s more direct strategy was to compare key gene sequences of a high lysine producing strain of *C. glutamicum* (B6) with that of the fully sequenced wild type to identify any mutated genes that could be responsible for over production. The influence of the mutated genes on lysine production could then be assessed by their sequential introduction into the wild-type by allelic replacement (Fig. 3.31). Ohnishi et al. focused their initial attention on 16 genes of the terminal lysine pathway (Fig. 3.32) and, knowing the sequence of the wild-type, were able to prepare PCR primers based on the nucleotide sequences flanking each intact gene. The PCR products derived from

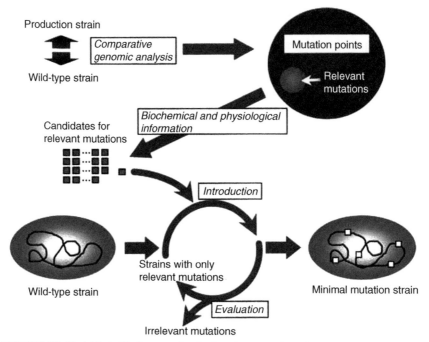

FIGURE 3.31 Ohnishi et al.'s Strategy for the Development of a Minimal Mutation L-Lysine Producing Strain of *Corynebacterium glutamicum* by the Sequential Addition to the Wild-Type of Mutations Identified From the Production Strain (Ohnishi et al., 2002)

the high producer were then sequenced and compared with the wild type genes. Each of the following five genes were shown to contain a point mutation:

hom—coding for homoserine dehydrogenase
lysC—coding for aspartokinase
dapE—coding for succinyl-L-diaminopimelate desuccinylase
dapF—coding for diaminopimelate epimerase
pyc—pyruvate carboxylase

The two *dap* mutations (*E* and *F*) were considered negligible because they resulted in neither amino acid substitution nor change to a rare codon. It can be seen from Fig. 3.16 that the control of the aspartate family of amino acids in *C. glutamicum* is achieved by the concerted inhibition of aspartokinase by lysine and threonine and the inhibition of homoserine dehydrogenase by threonine. The mutant alleles of *lysC* and *hom* (designated *lysC311* and *hom59* respectively) were introduced individually into the wild type strain by allelic replacement. The presence of *lysC311* gave the phenotype of resistance to the lysine analog *S*-(2-aminoethyl)-L-cysteine (AEC) and *hom59* resulted in a partial requirement for homoserine, observations commensurate with the history of the original producer strain (B6). Analog resistance of

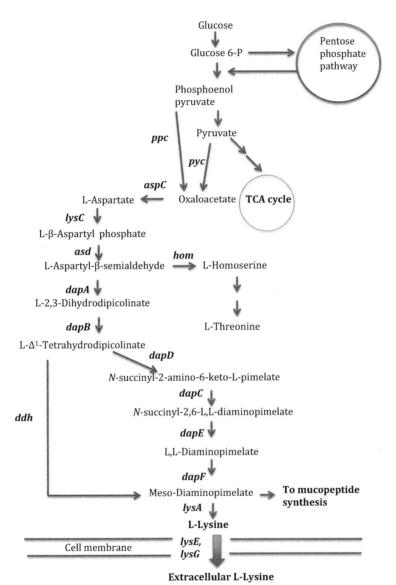

FIGURE 3.32 L-Lysine Biosynthetic Pathway in *C. glutamicum*

Enzymes encoded by genes: *asd,* aspartate semialdehyde dehydrogenase; *aspC,* aspartate aminotransferase; *dapA,* dihydrodipicolinate synthase; *dapB,* dihydrodipicolinate reductase; *dapC,* succinyl-L-diaminopimelate aminotransferase; *dapD,* tetrahydodipicolinate succinylase; *dapE,* succinyl-L-diaminopimelate desuccinylase; *dapF,* diaminopimelate epimerase; *ddh,* diaminopimelate dehydrogenase; hom, homoserine dehydrogenase; *lysA,* diaminopimelate decarboxylase; *lysC,* aspartokinase; *lysE,* lysine exporter; *lysG,* lysine exporter regulator; *ppc,* phosphenolpyruvate carboxylase; *pyc,* pyruvate carboxylase.

Modified from Ohnishi et al. (2002). Further details are given in Fig. 3.37 and Fig. 3.38.

aspartokinase may be expected to release feedback inhibition; and partial auxotrophy for homoserine would result in depleted threonine, thereby lifting inhibition of homoserine dehydrogenase. The *lysC311* single mutant produced 50 g dm^{-3} lysine; the *hom59* mutant produced 10 g dm^{-3} lysine whereas a wild type background transformed with both mutations resulted in a synergistic production of 75 g dm^{-3} lysine. Crucially, this reconstructed strain resembled the wild-type in its high growth rate and rate of glucose consumption, indicating that the background of deleterious mutations introduced by the many rounds of mutation and selection had been circumvented.

The final mutation revealed in this work was that of *pyc* coding for pyruvate carboxylase (*pyc458*), an anaplerotic enzyme fixing carbon dioxide in the synthesis of oxaloacetate, the immediate precursor of the aspartate family. Previous work on the lysine fermentation had concentrated on the terminal pathway and had not addressed the supply of precursors. The further incorporation of *pyc458* along with *lysC311* and *hom59* into the wild-type resulted in a strain (designated AHP-3) producing 80 g dm^{-3} lysine and, importantly, the highest production rate of 3.0 g dm^{-3} h^{-1} reported at that time; the high production rate being due to the high growth rate of the strain. Pyruvate carboxylase had not been a target in the strain improvement process used in the development of strain B6 and no selection mechanism existed for its isolation. Thus, the mutation had been coselected along with selectable markers during the process, illustrating that the undefined background of the industrial strain included both desirable and undesirable lesions. It may be recalled from our earlier discussion that the three key focal points for yield improvement are—control of the terminal pathway, provision of precursors, and the provision of NADPH. Thus, Ohnishi, Katahira, Mitsuhashi, Kakita, and Ikeda (2005) turned their attention to the supply of NADPH by investigating the genes associated with the pentose phosphate pathway, the major source of NADPH. Again, the gene sequence of the wild-type was used to prepare PCR primers based on the nucleotide sequences flanking each intact gene of the pentose phosphate pathway. Following comparison of the sequences of the PCR products with the wild-type, a point mutation was identified in the *gnd* gene, coding for 6-phosphogluconate dehydrogenase. Using the same allelic replacement methodology described earlier the mutated allele was added to the manipulated wild-type containing *pyc458*, *lysC311*, and *hom59*. The yield of this strain improved by 15% and again retained the vitality of the wild-type such that the fermentation was completed in 30 h, compared with 50 for the industrial B6 producer. Thus, this mutation had also been coselected and its addition to the other three mutations in a background free of undesirable lesions led to the development of a high-producing vigorous strain.

Application of transcriptomics

Despite the success of the strategy of Ohnishi et al. (2005), the final strain still produced less lysine than did B6—albeit in a much shorter time. Thus, strain B6 contained additional, undetected, lesions that had been co-selected and were contributing to its lysine synthesizing capability. To investigate this phenomenon, the group utilized the knowledge of the complete gene sequence of the organism to construct a DNA microarray, thus opening the door to the world of transcriptomics (Hayashi et al., 2006).

The transcriptome of an organism is its total mRNA profile at a particular time and under particular conditions. A microarray is a collection of DNA oligonucleotides (probes), each representative of a single gene, immobilized on a solid surface such as glass, plastic, or silica. This set of immobilized probes may then be used to hybridize with another DNA strand to detect the presence of complimentary sequences. The array may represent the entire genome or it may be limited to a particular range of genes under investigation. The oligonucleotides can be synthesized and then "spotted" onto the surface using a robotic micropipettting device to create microscopic zones of DNA. Alternatively, the DNA may actually be synthesized in situ on the surface of the array—a highly sophisticated approach restricted to commercial manufacture. Total mRNA is extracted from the organism and complementary DNA (cDNA) prepared from it using reverse transcriptase. The cDNA is then labeled with a fluorescent dye (a fluorophore, such as cyanine), which enables visualization by a laser beam at a particular wavelength—commonly Cy2 (fluorescing at 510 nm, the green zone of the spectrum) or Cy5 (fluorescing at 670 nm, the red zone of the spectrum). The array is exposed to the labeled DNA, washed and then exposed to laser excitation at the relevant wavelength. Two transcriptomes can be compared by labeling each with a different fluorophore (green or red) and exposing the array to a mixture of the labeled transcriptomes. Thus, transcription of a gene is indicated by the response of the array spot to laser stimulation—green, red or (if the gene has been transcribed equally in both transcriptomes) yellow. The relative expression is indicated by the spectrum of color images, (from green to red) and is visualized and analyzed using computer software. The process is illustrated in Fig. 3.33.

The comparison of the transcriptome of the lysine industrial strain (B6) with the wild type enabled the identification of genes that had been upregulated in B6. Three aspects of the B6 transcriptome stood out—high expression of the pentose phosphate genes, low expression of the TCA genes and global induction of amino acid biosynthesis genes. The changes in the central metabolic pathways were commensurate with previous findings on the flux through these pathways—but the upregulation of the amino acid genes was unexpected. In particular, the expression of *lysC* (aspartokinase) and *asd* (aspartate semialdehyde dehydrogenase) were significantly elevated in B6, yet no mechanism was known for the control of the synthesis of these genes. Arrays comparing the transcriptome of the Ohnishi et al. (2005) AHP-3 strain with the wild-type did not show the elevated transcription of either central metabolism or amino acid genes. This result was in line with the modifications incorporated into AHP-3 as it only carried mutations related to enzyme behavior and not gene expression. Thus, it appeared that the increased transcription seen in B6 might have been involved in lysine over production—that is, desirable mutations that had been co-selected. The changes in the transcription of TCA and pentose phosphate genes may have enhanced NADPH supply and the global induction of amino acid genes may have enhanced the flux (flow) to lysine due to the particular effect on *lysC* and *asd*. In *E. coli*, the expression of amino acid synthesizing genes are globally upregulated when amino acid levels are low—a mechanism termed the stringent response, initiated by the nucleotide ppGpp. Although such a system has not been found in

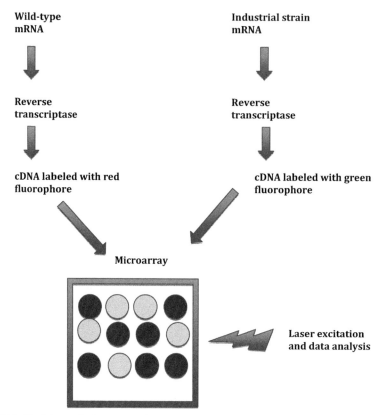

FIGURE 3.33 Schematic Representation of the Comparison of the Transcriptome of a Wild-Type with that of an Industrial Strain Using a DNA Microarray

C. glutamicum it appears that a form of global regulation has a significant role to play in lysine over production.

Application of metabolic flux analysis and a systems biology approach

Metabolism in a microorganism cannot be considered simply as composed of metabolic pathways operating independently because in reality they are combined into a complex network that has evolved to respond to environmental change and enable efficient biomass production. Thus, technology that reveals the flow of metabolites through an organism's metabolic network can significantly enhance a rational approach to strain improvement. Metabolic flux analysis (MFA) is a method that has the potential to achieve this objective as it quantitatively describes the network of an organism. Although MFA indicates what is happening in a fermentation, it does not intrinsically explain the control of the system. However, the generated data can give a valuable insight into the operation of a fermentation and facilitate the design of a blueprint of an over–producing strain. The first approach to elucidate the flux of metabolites through the network was a stoichiometric one that was comprehensively described in Holms (1997, 2001) excellent reviews on which the following discussion is based. The first requirement is that the metabolic network of the organism is completely described. This can be assembled form the literature but has been made very much easier in recent years by the availability of complete genome sequences. The sequence provides the information to deduce the full enzyme complement of the organism and thus the reactions that it may potentially catalyze. The next stage is the complete description of everything that the organism consumes as its carbon source(s) (the input) and everything that it produces (the output), from which the metabolic flux can be constructed. This is achieved by growth experiments in which a variation of the yield factor is calculated, that is, the glucose required to produce a certain biomass. The biomass is then analyzed to determine its monomer composition, for example, as described by Mousdale (1997). These monomers are the core building blocks that have been produced from the central metabolism pathways and exist predominantly in a cell as polymers—nucleic acids, proteins, carbohydrates, and lipids and represent the output of metabolism.

Although medium ingredients are normally measured as mass, in MFA it is more convenient and simple to express amount by moles. Thus, for example, in glycolysis 1 mole of glucose becomes 1 mole of glucose 6-phosphate that is converted to a mole of each of glyceraldehyde 3-phosphate and dihydroxyacetone phosphate. If this were expressed in mass then 100 g of glucose would be converted to 144 g of glucose 6-phosphate and a total of 198 g of triose phosphates. The output of the process is composed of the moles of monomers synthesized and incorporated into biomass. Both input and output are then expressed as specific units, related to the moles consumed, or produced, by a kilogram dry weight of biomass. Thus, a carbon source is measured in terms of the moles consumed to generate a kilogram of dry biomass and the output is measured as the moles of a product produced per kilogram of dry biomass (moles kg^{-1} dry wt.). Likewise all intermediates between carbon source and monomer product are described in terms of moles kg^{-1} dry weight of biomass. The progress of the carbon source through the central pathways of the organism to the eventual monomer products (and thus the production of biomass) is called "throughput." The flux through the pathway is the rate at which the carbon source is consumed and intermediates and products are produced, and it is calculated by multiplying the throughput by the specific growth rate of the culture (μ), measured in h^{-1}. Thus, the units of flux are moles kg^{-1} dry wt. h^{-1}.

The central metabolic pathways (CMPs) of an organism are the pathways that are both catabolic and anabolic (hence amphibolic) and produce both energy and precursors of anabolism. Thus, in an aerobic organism the CMPs are glycolysis, the pentose phosphate pathway, phosphoenolpyruvate carboxylase, pyruvate dehydrogenase, and the TCA cycle. There are approximately 30 intermediates of the CMPs, between 7 and 9 of which act as precursors for biosynthesis. Having determined the monomer composition of the organism the next stage in completing a MFA is to deduce the amount of each precursor that is converted into monomers (and eventually into biomass). An example of this approach is shown in Table 3.7 for *E. coli*

Table 3.7 The Monomer Content of *E. coli* ML308 Biomass and the Precursors Consumed in its Synthesis

Monomer Content of Biomass (Moles kg⁻¹ dry wt.)		Precursors Used for Biosynthesis (Moles kg⁻¹ dry wt.)							
		G6P	TP	PG	PEP	PYR	OAA	OGA	AcCoA
Ala	0.454					0.454			
Arg	0.252							0.252	
Asp	0.201						0.201		
Asn	0.101						0.101		
Cys	0.101			0.302					
Glu	0.353							0.353	
Gln	0.201							0.201	
Gly	0.430			0.430					
His	0.050	0.050							
Ile	0.252					0.252	0.252		
Leu	0.403					0.806			0.404
Lys	0.403					0.403	0.403		
Met	0.201					0.201ᵃ	0.201		
Phe	0.151	0.151			0.302				
Pro	0.252							0.252	
Ser	0.302			0.302					
Thr	0.252						0.252		
Trp	0.050	0.100		0.050	0.050				
Tyr	0.101	0.101			0.202				
Val	0.302					0.604			
AMP	0.115	0.115		0.115					
dAMP	0.024	0.024		0.024					
GMP	0.115	0.115		0.115					
dGMP	0.024	0.024		0.024					
CMP	0.115	0.115					0.115		
dCMP	0.024	0.024					0.024		
UMP	0.115	0.015					0.115		
dTMP	0.024	0.024					0.024		
C₁₆ fatty acid	0.280								2.240
Glycerophosphate	0.140		0.140						
Carbohydrate as glucose	1.026	1.026							
Total		1.98	0.14	1.36	0.55	2.32	1.69	1.06	2.64

Abbreviations: G6P, glucose 6-phosphate; TP, triose-phosphate; PG, phosphoglycerate; PEP, phosphoenolpyruvate; PYR, pyruvate; OAA, oxaloacetate; OGA, oxoglutarate; AcCoA, acetyl-coenzyme A.
ᵃNegative
From Holms (1997), with permission of Oxford University Press

ML308 (Holms, 1997). The monomer content of the biomass (moles kg^{-1} dry wt.) is shown in the first column followed by the intermediates of central metabolism used as precursors for these monomers. Thus, remembering the biosynthetic route to each monomer, it is possible to calculate simply the amount of precursor consumed in its synthesis. Thus, to produce 0.454 moles of alanine, 0.454 moles of pyruvate (its sole precursor) will be required. Similarly, 0.201 moles of methionine will require the consumption of 0.201 moles of both its precursors, pyruvate, and oxaloacetate. The amount of each precursor that is diverted into biomass can then be obtained by adding all their contributions to the synthesis of each monomer. For example, phosphoenolpyruvate (PEP) is a precursor of the three aromatic amino acids (tyrosine, tryptophan, and phenylalanine) resulting in 0.55 moles being incorporated into a kg of biomass. The throughput diagram for *E. coli* growing on glucose is given in Fig. 3.34. This shows the uptake of glucose via the posphotransferase system and its route through the central metabolism pathways. The arrows leading out from the diagram as "Biomass" are precursors contributing to biosynthesis, thus biomass production—for example, PEP giving rise to the aromatic amino acids. Holms (1997) quoted that *E. coli* ML308 required 11.24 moles of glucose to produce a kg of biomass (dry weight) in batch culture at a specific growth rate of 0.94 h^{-1}. Thus, this value is included in Fig. 3.34, as the glucose input. The various "outputs" of the precursors are now incorporated—for example, 0.55 for phosphoenolpyruvate and 1.69 for oxaloacetate. It will also be noticed that a significant amount of acetate is excreted from the cells. Holms demonstrated that acetate is only excreted in chemostat culture when the growth rate exceeds 0.72 h^{-1} indicating the flux of carbon through the pathways is in excess of requirements at higher growth rates. Under the growth-limiting conditions of lower growth rates, only 8.64 moles of glucose are consumed in the production of a kg of biomass, compared with 11.24 at higher growth rates. Thus, growth at the lower growth rate is more efficient than that in batch culture—a situation also seen in fed-batch culture (see Chapter 2).

The next stage in the construction of the throughput diagram is to connect the input (glucose) and outputs in Fig. 3.34. *E. coli* takes up glucose via the glucose posphotransferase system (PTS), which transfers the phosphate moiety from phosphoenolpyruvate to glucose, generating glucose-6-phosphate and pyruvate. Thus, 11.24 moles of glucose give 11.24 moles of glucose-6-phosphate. The outputs from each precursor (as calculated in Table 3.7) are then included in Fig. 3.34, for example, 1.98 moles of glucose-6-phosphate, 0.14 moles of triose phosphate, and 1.34 moles of phosphoglycerate. The loss of glucose-6-phosphate to biosynthesis leaves 9.26 moles entering glycolysis and the production of a total of 18.52 moles of triose phosphate, 0.14 of which goes to biosynthesis leaving 18.38 to be converted to 3-phosphoglycerate. 1.34 moles of 3-phosphoglycerate are accounted for as precursor leaving 17.04 moles to be converted to phosphoenolpyruvate (PEP). The node at PEP branches four ways. Biosynthesis accounts for 0.55 moles. The output to biosynthesis from the TCA cycle is via oxoglutarate and oxaloacetate but this output must be replaced by anaplerosis—that is, by the conversion of PEP to oxaloacetate by the carbon dioxide fixing reaction catalyzed by phosphoenolpyruvate carboxylase.

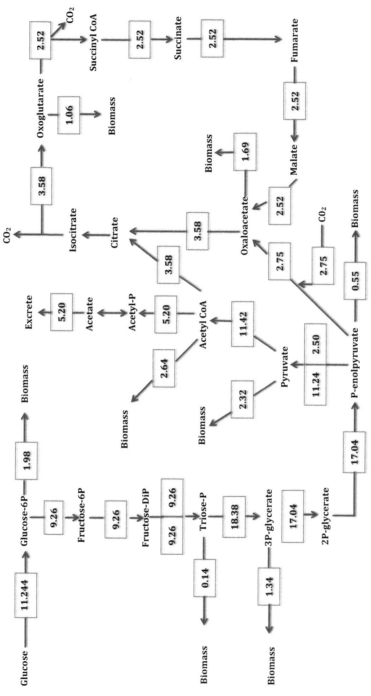

FIGURE 3.34 Throughput Diagram for *E. coli* ML308 Grown on Glucose as Sole Carbon Source in Batch Culture (units, moles kg⁻¹ dry Weight Biomass)

Modified from Holms (1997), with permission of Oxford University Press.

The total output from the TCA cycle is equivalent to 2.75 moles of PEP, which is thus introduced into the TCA cycle by its conversion to oxaloacetate. It will be recalled that 11.24 moles of PEP would have been converted to pyruvate by the PTS system in the uptake of glucose, thus leaving only a contribution of 2.5 moles of pyruvate to the pyruvate pool by pyruvate kinase. Of this pyruvate pool, 2.32 moles are diverted to biosynthesis and 11.42 converted to acetyl-CoA. A three-way node then distributes acetyl-CoA to biosynthesis (2.64 moles), acetate excretion (5.20 moles), and the TCA cycle (3.58 moles). This throughput analysis can then be converted to a flux analysis by multiplying all the throughputs by the specific growth rate.

The throughput analysis shown in Fig. 3.34 represents the broad picture of the organism's metabolism with the output from each precursor being bulked together. Also, the throughput of glucose to the pentose phosphate pathway is bulked together in the precursor role of glucose-6-phosphate. However, Fig. 3.34 can be expanded to show every biosynthetic enzyme or to amplify a component part. For example, the throughput from oxoglutarate to its individual end products is shown in Fig. 3.35. Carbon dioxide is also shown in the overall throughput diagram and it must be taken into account in an audit of the carbon balance. The gas is released by the action of both isocitrate dehydrogenase (isocitrate to oxoglutarate) and oxoglutarate dehydrogenase (oxoglutarate to succinyl CoA), the former releasing 3.58 moles kg^{-1} biomass and the latter 2.52 moles kg^{-1} biomass. It will also be recalled that the anaplerotic reaction catalyzed by phosphoenolpyruvate carboxylase produces 2.75 moles kg^{-1} oxaloacetate and thus also consumes 2.75 moles kg^{-1} CO_2. Thus, the net production of CO_2 is 3.35 moles kg^{-1} (3.58 + 2.52 − 2.75). As well as generating precursors for biosynthesis, the CMPs are obviously also generating sufficient reducing power and ATP to convert the precursors into monomers and finally into polymers

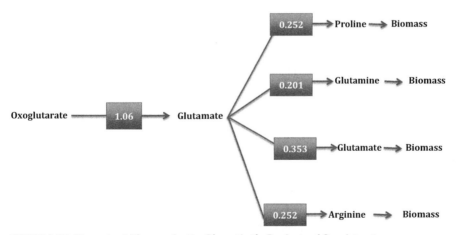

FIGURE 3.35 Throughput Diagram for the Biosynthetic Products of Oxoglutarate in *E. coli* ML308

Units are moles kg^{-1} dry weight of biomass. Modified from Holms (1997), with permission of Oxford University Press.

and biomass. However, stoichiometric flux analysis does not give a direct appreciation of the provision of ATP and reducing power and relies upon balances for ATP, NADH, and NADPH and may not represent the true in vivo reality (Becker & Wittmann, 2013). However, the use of a stoichiometric approach to elucidate flux in the *C. glutamicum* lysine fermentation demonstrated high pentose phosphate pathway activity compared with *E. coli*, primarily due to the lack of NADH/NADP transhydrogenase in the Corynebacterium (Vallino & Stephanopoulos, 1991; Dominguez, Nezondet, Lindley, & Cocaign, 1993). Thus, a significant flux of glucose is required through the pentose phosphate pathway to produce sufficient NADPH for lysine synthesis.

Some of the limitations of the solely stoichiometric approach to MFA have been overcome by using substrates labeled with the stable isotope, ^{13}C, reviewed by Wittmann (2007) and Becker and Wittmann (2013, 2014). The procedure is normally performed on exponentially growing cells and although chemostat cultures would seem the most desirable, due to their steady-state properties, most work has employed batch cultures. Also, it is important that the study is done using a defined medium, that is, lacking any complex carbon source that would confuse the interpretation of the consumption of the labeled substrate. The labeled substrate is taken up by the organism and assimilated into all its metabolites—the labeling pattern of the metabolites being dictated by the route taken by the label through the cellular pathways (Becker and Wittmann, 2013). Amino acids are derived from a wide range of metabolic precursors and thus represent the outcome of flux from a carbon substrate through all the central pathways of an organism. When that carbon substrate is labeled then the labeling patterns of the amino acids will be representative of the route that has given rise to them. As a result, the amino acids from total biomass hydrolysates have been widely used as the means to access flux data (Dauner & Sauer, 2000) with gas chromatography coupled with mass spectrometry (GC-MS) being the most widely used methodology for analysis (Wittmann, 2007). Gas chromatography separates the components of the hydrolysate and mass spectrometry determines the ^{13}C labeling. The ^{13}C labeled compounds may differ in the number of incorporated heavy atoms as well as their position, the final labeling of the end products being determined by the route of their precursors through the metabolic pathways. The labeling will obviously also be affected by the labeling pattern of the substrate. For example, labeling with [1-^{13}C] glucose can distinguish between the flux of glucose-6-phosphate through glycolysis or the pentose phosphate pathway. The 1-C of glucose is decarboxylated in the pentose phosphate pathway and thus pyruvate emanating from this route is unlabeled, whereas half the pyruvate moieties generated via glycolysis will be labeled, as shown in Fig. 3.36. The use of a mixture of universally labeled and unlabeled glucose has been shown to be useful in resolving fluxes downstream of phosphoenol pyruvate (Wittmann, 2007). The data generated from these studies are challenging to interpret and, thus, it was this aspect that limited the wider application of the heavy isotope labeling approach. However, sophisticated computer programs have now been developed to assist in this analysis. For example, OpenFLUX, developed by Quek, Wittmann, Nielsen, and Kromer (2009), is open access software that

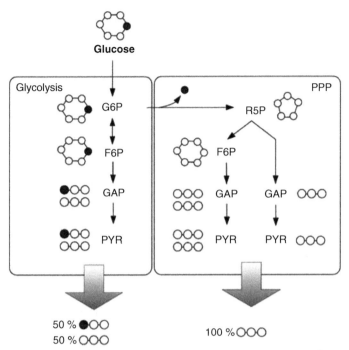

FIGURE 3.36 The Flux Partitioning of Glucose-6-Phosphate Between Glycolysis and the Pentose Phosphate Pathway (PPP) as Revealed by Feeding [1-^{13}C] Glucose

The heavy atom is decarboxylated in the PPP resulting in the production of unlabelled pyruvate. ● ^{13}C label. Becker and Wittmann (2013).

utilizes a spreadsheet-based interface and is supported by a comprehensive manual. The program enables the operator to incorporate the complete flux model of the organism containing all the relevant reactions and quantitative descriptions of the transfer of ^{13}C atoms through each reaction. The software runs a random initial flux through the model (in silico metabolism) and computes a simulated set of GC-MS labeling data. These simulated data are then compared with the experimental results and an optimization algorithm is applied to vary the predicted flux until the in silico data match the experimental.

Thus, the knowledge generated from genome sequencing gave rise to the development of genomics and transcriptomics, contributed to the field of proteomics and laid a firm foundation on which to base more comprehensive metabolic flux analyses. This explosion of knowledge, combined with the ready availability of enormous computer power, has resulted in increasingly sophisticated mathematical modeling of biological systems and given rise to what has been called "systems biology." Systems biology is, in effect, the science of the physiologist and considers the organism as a whole rather than adopting a reductionist approach that concentrates on individual aspects of an organism's performance. An excellent example of the application of this approach to strain improvement is given by the work of Becker, Zelder, Hafner, Schroder, and Wittmann (2011) who succeeded in developing a strain of *C. glutamicum* producing the highest published yield of lysine. Rather than start with a high producing industrial strain which contained both desirable and undesirable genetic lesions, their rationale was to begin with the wild-type organism and, from a knowledge of its metabolic flux, construct a blueprint of a fully genetically defined lysine overproducer. The ^{13}C flux analysis of the wild type showed an equal distribution of carbon from glucose into glycolysis and the pentose phosphate pathway resulting in high flux through the TCA cycle and efficient energy generation. The following key modifications were predicted to form the basis of an "ideal" producer:

- Increased flux of aspartate through the lysine biosynthetic pathway and greater bias toward the diaminopimelate dehydrogenase branch.
- Increased flux through carbon dioxide fixing, anaplerotic reactions that supplement the level of oxaloacetate—the key lysine precursor.
- Reduced flux through the oxaloacetate decarboxylating reactions associated with gluconeogenesis.
- Reduced TCA cycle flux.
- Increased pentose phosphate flux, thus enhancing NADH supply.
- Decreased flux through other anabolic routes.

These modifications were achieved by the sequential development of eleven strains. Modified genes were incorporated using the *sacB* recombination system (Jager, Schafer, Puhler, Labes, & Wohlleben, 1992). The control of the biosynthesis of lysine in *C. glutamicum* is shown in Fig. 3.16 and the overall route to lysine from glucose, giving associated enzymes and their coding genes, is shown in Fig. 3.32. The branch from glycolysis to the pentose phosphate pathway is illustrated in Fig. 3.37 and the inter-relationships between oxaloacetate, glycolysis, gluconeogenesis, and the TCA cycle are shown in Fig. 3.38. The development of the strains was as follows, with each subsequent strain retaining the modifications of the previous:

Strain LYS-1. The first point of attack was the introduction of a modification to the *lysC* gene coding for aspartokinase, the first enzyme in the route to lysine and the aspartate family of amino acids. This modification (nucleotide substitution C932T) had been shown previously to lift the threonine and lysine concerted feedback inhibition of aspartokinase (Cremer, Eggeling, & Sahm, 1991). An alternative approach would have been to disrupt the synthesis of threonine but this would have resulted in the production of an auxotroph for threonine, methionine and isoleucine that would have placed a major limitation on the design of the fermentation medium. This strain produced a slight increase in lysine production.

Strain LYS-2. As shown in Fig. 3.32, there are two alternative routes for the conversion of terahydrodipicolinate to meso-diaminopimelate in *C. glutamicum*, with the diaminopimelate dehydrogenase (coded for by the *ddh* gene) route being more effective (Schrumpf et al., 1991; Melzer, Esfandabadi, Franco-Lara, & Wittmann, 2009). To facilitate the flux of tetrahydrodipicolinate via the energetically more favorable

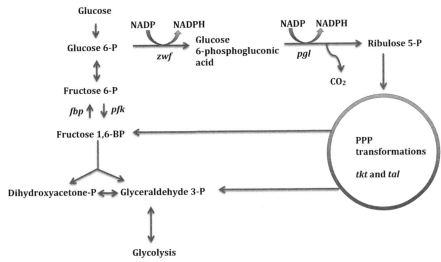

FIGURE 3.37 Relationship Between the Early Stages of Glycolysis and the Pentose Phosphate Pathway in *C. glutamicum*

Genes coding for the enzymes at the indicated steps: *fbp*, fructose bisphosphatase; *pfk*, phosphofructokinase; *pgl*, 6-phosphogluconate dehydrogenase; *tal*, transaldolase; *tkt*, transketolase; *zwf*, glucose 6-phosphate dehydrogenase

route, a second copy of the *ddh* gene was introduced. As a result, diaminopimelate dehydrogenase activity doubled and lysine yield improved by 25%.

Strain LYS-3. Measurement of flux in LYS-2 indicated a complex rearrangement in the flows through the carboxylating and decarboxylating reactions between oxaloacetate, pyruvate, and PEP shown in Fig. 3.38. In particular, the flux through PEP carboxykinase resulted in a diversion of oxaloacetate away from lysine. Thus, the gene encoding PEP carboxykinase, *pck*, was deleted from LYS-2. This resulted in a 30% increase in lysine yield.

Strains LYS-4, 5, 6, and 7. This stage of the improvement process concentrated on the lysine pathway (Fig. 3.32) and included four sequential modifications. It will be recalled that LYS-1 was modified by lifting the threonine and isoleucine inhibition of aspartokinase (*lysc* gene). The expression of this modified gene was increased by substituting its promoter with that of the *sod* (superoxide dismutase) gene. SOD is a vital antioxidant defensive enzyme and thus its expression is controlled by a very strong promoter, the characteristics of which make it eminently suitable as a replacement promoter to enable over expression of desirable genes. The intermediates were diverted to the *ddh* branch by the increased expression of both dihydrodipicolinate synthase (*dapB*) and diaminopimelate decarboxylase (*lysA*). The former was achieved by promoter replacement with *sod*, the latter by incorporating an extra gene copy. Meso diaminopimelic acid is not only the immediate precursor of lysine but also a key component of mucopeptide, the cell wall polymer responsible for mechanical

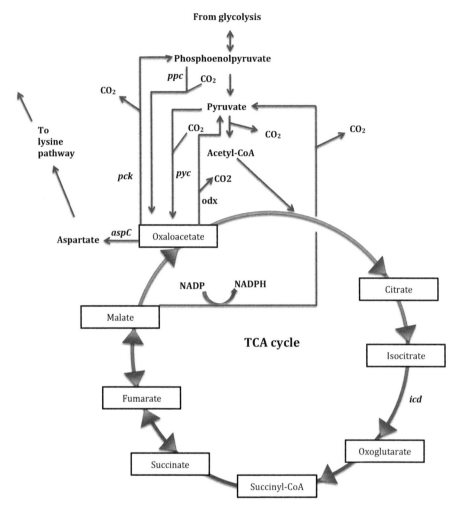

FIGURE 3.38 Inter-Relationships Between Oxaloacetate, Glycolysis, Gluconeogenesis, the TCA Cycle and the Biosynthetic Pathway to Lysine in *C. glutamicum*

Enzymes encoded by genes: *aspc,* aspartate amino transferase; *icd,* isocitrate dehydrogenase; *odx,* oxaloacetate decarboxylase; *pck,* phosphoenolpyruvate carboxykinase; *ppc,* phosphoenolpyruvate carboxylase; *pyc,* pyruvate carboxylase.

strength (Fig. 3.32). Thus, to increase flux to lysine, and divert it away from muco-peptide synthesis, an extra copy of the *lys A* gene was incorporated. Finally, the last stage in this sequence was to address the loss of aspartyl semialdehyde to the threonine pathway due to the action of homoserine dehydrogenase (coded by gene *hom*). The modification was based on Ohnishi et al. (2002) finding that a specific nucleotide exchange within *hom* gave an enzyme with a reduced affinity for homoserine, thus favoring the distribution of this intermediate to dihydrodipicolinate. Thus, *hom* was replaced with this modified gene using allelic replacement technology and the addition of this feature to those of LYS-4, 5, and 6 resulted in an increase in lysine yield of 18%. Flux analysis of strain LYS-7 indicated that the increased lysine yield was attributable to a decrease in growth due to the redirection of aspartate semialdehyde away from threonine and the limitation placed on mucopeptide synthesis by the lack of meso-diaminopimelate.

Strain LYS-8 and 9. The next steps focused on the manipulation of the pyruvate node to achieve an elevated flux to oxaloacetate. The major route to oxaloacetate is via the anaplerotic (CO_2 fixing) reaction catalyzed by pyruvate carboxylase, coded for by the gene *pyc*. The manipulation was twofold; LYS-8, allelic replacement with a *pyc* derivative incorporating a nucleotide substitution revealed by Ohnishi et al. (2002) and LYS-9, the replacement of the wild-type promoter with that of the *sod* gene. The former improved the kinetics of the enzyme and the latter enhanced the expression of *pyc*. However, these manipulations did not have the desired effect on lysine yield, with LYS-8 yield actually decreasing. Flux analysis of both strains showed that the flow to oxaloacetate was increased as predicted but the oxaloacetate was not directed to aspartate and then to lysine. Oxaloacetate was directed round the TCA cycle to malate that was then decarboxylated to pyruvate by the malic enzyme, thus effectively diverting oxaloacetate back to glycolysis via the TCA cycle (Fig. 3.38). Removing the malate enzyme was not an option as interplay between the TCA cycle and glycolysis is required and, as the strain lacks PEP carboxykinase, the malic enzyme is the only remaining route. Thus, the solution was to manipulate the flux through the TCA cycle in strain LYS-10.

Strain LYS-10. The flux through the TCA cycle was reduced by changing the start codon of the *icd* gene, coding for isocitrate dehydrogenase, from ATG to the rare codon GTG; a technique previously used to downregulate protein synthesis in *C. glutamicum* (Becker, Buschke, Bucker, & Wittmann, 2010). Flux analysis of this strain revealed a reduction in flux to the TCA cycle from 48%–40% and a 40% increase in the flux through the anaplerotic (CO_2 fixing) pyruvate carboxylase to oxaloacetate. However, flux through the malic enzyme actually increased and this was attributed to its use as a NADPH generating system to provide sufficient reducing power for the increased lysine production. The pentose phosphate pathway (PPP) normally supplies NADPH for biosynthesis (Fig. 3.37) but the glucose flux through the pathway was consistent at about 60% in strains LYS-2–LYS-9. This observation suggested that the PPP was limiting the NADPH supply resulting in the activation of the malic enzyme, the loss of CO_2 in the conversion of oxaloacetate to pyruvate and the increased flux of oxaloacetate through the TCA cycle (Fig. 3.38). Thus, the final stage in the development of the strain was the manipulation of the flux through the PPP.

Strains LYS-11 and 12. The relationship between the early stages of glycolysis and the PPP is shown in Fig. 3.37 from which it can be seen that the distribution of glucose 6-phosphate between glucose 6-phosphogluconate and fructose 6-phosphate determines the relative fluxes through glycolysis and the PPP. Gluconeogenesis is effectively the reverse of glycolysis, converting pyruvate to glucose when the organism is growing on noncarbohydrate carbon source. The key enzyme in this process is phosphofructokinase, catalyzing the conversion of fructose 1,6-bisphosphate to fructose 6-phosphate. Becker, Heinzle, Klopprogge, Zelder, and Wittmann (2005) had previously demonstrated that flux through the PPP could be stimulated by increasing the level of fructose 1,6-bisphosphatase, achieved by replacing the promoter of its gene (*fbp*) with that of elongation factor tu (*eftu*). Thus, the level of fructose 6-P is effectively increased and because the reaction between fructose 6-P and glucose 6-P is reversible, the balance of the reaction is then in the reverse direction and glucose 6-P is directed into the PPP. This approach was applied to develop strain LYS-11, which increased lysine production by a further 18%. Finally, strain LYS-12 was developed by replacing the promoter of the *tkt* operon (coding for all the genes of the PPP pathway) with the *sod* promoter.

Flux through the PPP in strain LYS-12 increased to 85% accompanied by a reduction in flux through the TCA cycle and the malic enzyme, increased anaplerotic production of oxaloacetate via pyruvate carboxylase and increased flux to lysine. Under fed-batch conditions, the strain was able to produce 120 g dm^{-3} lysine at a rate of 4.0 g dm^{-3} h^{-1}, a conversion of glucose to lysine of 55% in a fermentation lasting only 30 h. The concentration of lysine and the glucose conversion figures were at least equivalent to the highest reported industrial strains but, crucially, the genetic modifications that resulted in these remarkable statistics were in a wild-type background, devoid of inadvertently introduced deleterious mutations. Thus, the organism still grew vigorously, consumed glucose efficiently and, most importantly, its genetics and metabolism were fully explained and understood thereby making its scale-up to an industrial process very much more manageable.

Becker et al.'s work on the lysine pathway in *C. glutamicum* is an excellent case study of the application of metabolic engineering to the improvement of fermentation strains. Further examples are given by Kind, Becker, and Wittmann (2013)—lysine; Zhang, Zhang, Kang, Du, and Chen (2015)—phenylalanine; Ikeda, Mitsuhashi, Tanaka, and Hayahsi (2009)—arginine; and Hasegawa et al. (2012)—valine.

IMPROVEMENT OF STRAINS PRODUCING SECONDARY BIOSYNTHETIC PRODUCTS

As discussed in Chapters 1 and 2, secondary metabolites have been traditionally described as produced by slow-growing or nongrowing microorganisms, limited in their taxonomic distribution and playing no obvious function in cell growth. However, it is now appreciated that secondary metabolites have roles in effecting differentiation, inhibiting competitors, and modulating host physiology. These compounds are synthesized from the intermediates and end products of primary metabolism and exhibit a diverse range of structure and biological activity—it is the wide range of biological activity that makes secondary metabolites attractive targets for commercial production. For example, polyketides and macrolides are synthesized from acetate moieties, polyisoprenoids from isoprene units, β-lactams from amino acids and a wide range of compounds by nonribosomal peptide synthesis (Fig. 3.39). The control of primary

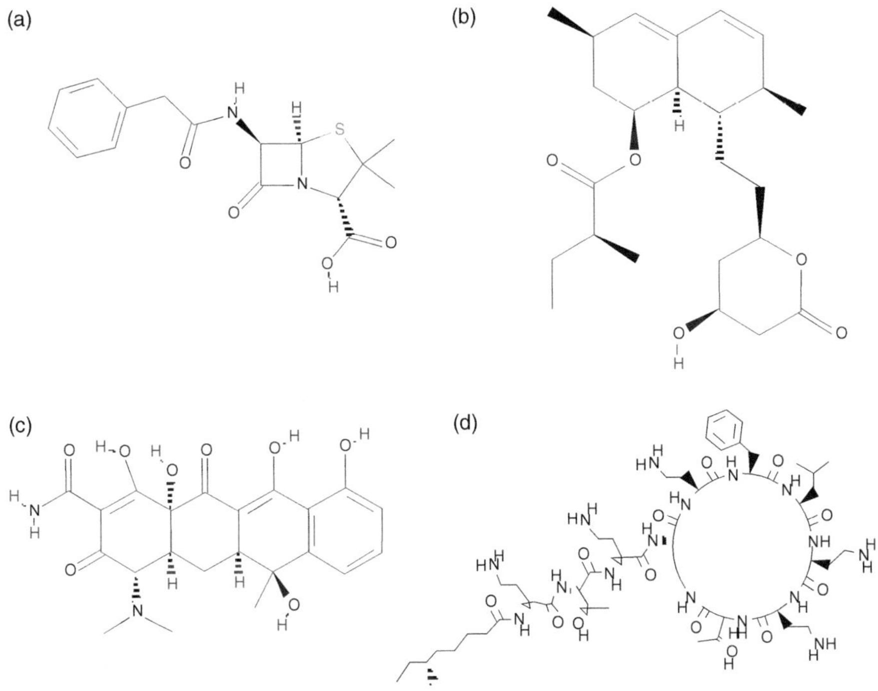

(a)

(b)

(c)

(d)

Polymyxin B

FIGURE 3.39 The Structures of Selected Secondary Metabolites

(a) Penicillin, G (b), lovastatin, (c) tetracycline, (d) polymyxin.

Data deposited in or computed by PubChem. URL: https://pubchem.ncbi.nlm.nih.gov

biosynthetic products is dominated by feedback systems that have evolved to ensure that the level of synthesis is commensurate with the requirements of growth. However, the control of secondary metabolism is dominated by systems that switch on biosynthesis such that, in the natural environment, the products are produced only when appropriate. In recent years it has become very clear that sophisticated genetic systems have evolved that initiate these control systems and emphasize the fact that these compounds are not "waste products" or "evolutionary relics," but metabolites central to the life-style of the producer. As discussed earlier in this chapter, each streptomycete that has been sequenced has been shown to be capable of the production of a wide range of secondary metabolites, many of which are antibiotics. This diverse productivity may be related to the wide range of organisms in soil with which a producing organism has to compete or to the variety of environmental conditions that may prevail in soil, thus enabling the producer organism to manufacture different defensive compounds should the availability of raw materials change (Challis & Hopwood, 2003; van Wezel & McDowall, 2011). The ability to respond to the diverse stimuli of a changing soil environment may also begin to explain the diversity of systems that control the onset of secondary metabolism.

The most prolific secondary metabolite producers are the Actinobacteria (the actinomycetes), particularly the genus *Streptomyces* and much of the material discussed in this section has come from work on this genus. The genes of secondary metabolism tend to be clustered together and many contain only a single regulatory gene that processes the signals that stimulate the transcription of their associated genes. These regulators have been termed "cluster-situated (transcriptional) regulators" or CSRs (Huang et al., 2003; van Wezel and McDowall, 2011) and it is their protein products that act as transcription factors and switch on the secondary metabolism cluster. For example, the actinorhodin gene cluster is controlled by the protein transcription factor ActII-ORF4 (coded by the gene *actII-ORF4*) and the streptomycin gene cluster by StrR (coded by gene *strR*), both these CSRs being members of the *Streptomyces* Antibiotic Regulatory Protein (SARP) family. The CSRs ultimately respond to an environmental stimulus relayed to it by a signal molecule via a signaling pathway and it is common that several CSRs may be under the control of a pleiotropic regulator (one that causes several responses) that is switched on upstream of the CSRs. The factors that stimulate the onset of secondary metabolism and influence the CSRs are associated with the shift into substrate limitation, and include the concentration of biomass, starvation, and growth rate (Barka et al., 2016).

The response to biomass concentration is referred to as quorum sensing—a phenomenon in which expression of certain genes only occurs when the culture reaches a threshold biomass (see Chapters 2 and 6). The control of antibiotic synthesis by quorum sensing would result in a "corporate reaction" from the mycelium resulting in the simultaneous synthesis of an antibiotic resulting in an ecologically significant concentration of the antimicrobial. Khokhlov et al. (1967) discovered the first quorum sensing molecule, a γ-butyrolactone named A-factor (Fig. 3.40), which was synthesized by *S. griseus* at a very low concentration (10^{-9} M) and was required for both the production of streptomycin and sporulation. The exceptionally high specific activity of A-factor suggested it played a "hormonal" role in controlling the differentiation of *S. griseus* and eventually Beppu and his coworkers at the University of Tokyo resolved its mode of action (Horinouchi & Beppu, 2007). A-factor is synthesized by exponential-phase mycelium and its production is amplified as the culture enters the stationary phase. Miyake, Kuzuyama, Horinouchi, and Beppu (1989) isolated a protein named A-factor receptor protein (ArpA, coded by the gene *arpA*) that would preferentially bind A-factor. The key experimental evidence in elucidating the mode of action of A-factor was that mutants unable to synthesize the receptor protein produced elevated levels of streptomycin. The conclusion from this observation was that the receptor protein was acting as a repressor of gene expression and its binding with A-factor released its repressive effects, enabling the synthesis of streptomycin. It was subsequently shown that ArpA represses the expression of the regulator gene, *adpA*, and the binding of A-factor to ArpA results in derepression, that is, the transcription of the *adpA* gene and the expression of the AdpA protein. AdpA is a transcription factor that enables the transcription of *strR*, the CSR that, in turn, enables the transcription of the streptomycin structural genes. However, AdpA not only controls the expression of *strR* but also another two families of antibiotics, aerial mycelium

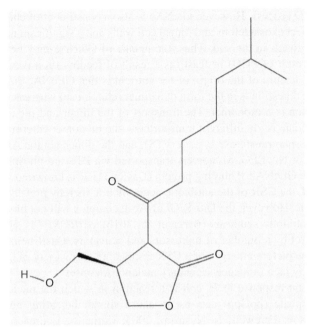

FIGURE 3.40 The Structure of A Factor, a γ-Buturolactone Synthesized by *Streptomyces griseus*

Data deposited in or computed by PubChem. URL: *https://pubchem.ncbi.nlm.nih.gov.*

formation and sporulation. Thus, AdpA is a pleiotropic regulator, controlling a wide range of developmental systems in *S. griseus*, each via a local transcription factor. However, other factors have been identified that interact with the promoter of *strR* (van Wezel and McDowall, 2011)—in particular, AtrA that closely resembles the cAMP receptor of *E. coli*. Thus, CSRs may be the focal point of a number of environmental stimuli involved in the onset of the synthesis of a secondary metabolite or group of metabolites.

The behavior of streptomycetes entering the stationary phase was discussed in Chapter 2 (Microbial Growth Kinetics). Yague, Lopez-Garcia, Rioseras, Sanchez, and Manteca (2012, 2013) identified two programmed cell death (PCD) events in *S. antibioticus* when grown on solidified medium—the first when mycelial segments of basal mycelium (MI) died and the remaining viable segments gave rise to a secondary mycelium (MII), presumably growing on substrate from the "sacrificed" cells. The developing MII mycelium eventually underwent a similar PCD—this time the cannibalism giving rise to aerial mycelium and spore production. Antibiotic production is frequently associated with these stages, presumably to prevent the invasion of marauding competitors benefitting from the sacrificed mycelium. The stimulus for antibiotic synthesis (secondary metabolism) under these conditions appears to be a component of the cell wall, *N*-acetyl glucosamine (GlcNAc) monomers released during the PCD process. However, GlcNAc is also a constituent of chitin, the major polymer of insect exoskeletons and fungal cell walls and a significant nutrient source for bacterial growth in the soil. Thus, the uptake of GlcNAc may be indicative of famine (as a result of PCD) or feast (as a result of feeding on a dead insect). The distinguishing feature of the origin of the nutrient is that GLcNAc emanating from external chitin digestion is in the form of a dimer (chitobiose) whereas that produced by autodigestion is a monomer. The transport of the dimer and monomer through the cell membrane is via different transporters—the monomer entering via the PEP dependent phosphotransferase system (PTS), and the dimer via the ABC transporters, DasABC or NgcEFG. Monomers transported via PTS are phosphorylated and can bind to the GLcNAc-P binding protein (DasR). DasR is known to be a repressor of *actII-ORF4*, the CSR of the antibiotic actinorhodin, thereby preventing synthesis of the antibiotic. However, the Das-R/GLcNAc-P complex will not bind to the CSR, resulting in antibiotic synthesis (Barka et al., 2016). *actIIORF4* is subject to transcription control by a number of repressors and activators, DasR being only one of them. Thus, the picture emerges of a CSR responding to a variety of environmental stimuli that may then influence secondary metabolism onset.

The stringent response in *E. coli* is a reaction in which the nucleotide guanosine tetraphosphate (ppGpp) acts as an alarm signal indicating nutrient starvation (Magnusson, Farewell, & Nystrom, 2005; Carneiro, Lourenco, Ferreira, & Rocha, 2011). The nucleotide is synthesized by two different pathways, named after the enzymes catalyzing the conversion of GTP and ATP to ppGpp—the RelA (ppGpp synthetase 1) pathway and the SpoT (ppGpp synthetase II) pathway. RelA is located at the ribosome and is activated (thus synthesizing ppGpp) when amino acid starvation is detected by the binding of uncharged t-RNA. SpoT carries both ppGpp synthesizing and hydrolyzing activity and is responsible for its synthesis in response to any nutrient starvation other than amino acid depletion. ppGpp acts by binding to RNA polymerase and both inhibits the synthesis of stable RNA moieties (rRNA and tRNA) and stimulates the expression of a large number of stress-related genes. The process is summarized in Fig. 3.41. *S. coelicolor* has homologues of both RelA and SpoT, named RelA and RshA, respectively, which respond broadly in similar ways to their *E. coli* counterparts. A number of reports have linked increased ppGpp levels with both the onset of secondary metabolism and morphological differentiation in streptomycetes. Clavulanic acid is an unusual antibiotic in that it is produced during the exponential phase and its production is increased if *relA* (the gene coding for RelA) is inactivated, that is, ppGpp lowers clavulanic production (Gomez-Escribano, Martin, Hesketh, Bibb, & Liras, 2008). Thus, the control of clavulanic acid biosynthesis by ppGpp is commensurate with its phase of production. Although the details of the mechanism of ppGpp involvement in secondary metabolism are not clear, the involvement of the nucleotide is very well documented and, as will be discussed later, is a target for strain improvement.

Further control mechanisms associated with secondary metabolism involve two-component systems (TCSs). A TCS is a signaling system incorporating a sensor and a response regulator. The sensor detects environmental information at the outer

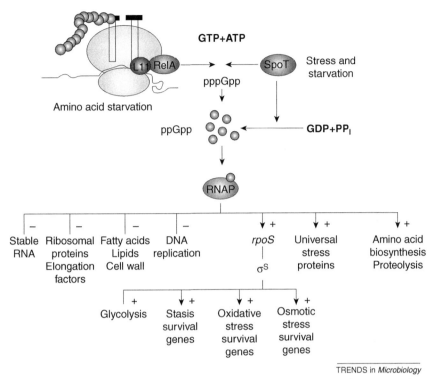

TRENDS in *Microbiology*

FIGURE 3.41 The Stringent Response in *E. coli* in Which ppGpp is Synthesized by RelA in Response to Amino Acid Starvation and by SpoT in Response to Stress or Other Nutrient Depletion

RNAP, RNA polymerase.

Magnusson et al., 2005.

surface of the cell membrane and relays information to the response regulator in the cytoplasm that can then initiate a change in gene expression. The detection system is a membrane-bound histidine kinase that phosphorylates itself (using ATP as a phosphate donor) when activated by its environmental stimulus. The phosphoryl group is then transferred to an aspartate residue in the response regulator protein. The activated response regulator may then interact with a specific gene promoter and initiate, or prevent, transcription. The phosphorylated response regulator may be dephosphorylated, thereby returning it to its basal state. The half-life of the phosphorylated response regulator may vary from seconds to hours and thus determines the sensitivity of the control system. The genome of *S. coelicolor* contains 67 two-component systems (Martin & Liras, 2010) but the molecular mechanism of action of most is not known. However, it is well known that high phosphate levels can repress secondary metabolism and Sola-Landa, Moura, and Martin (2003) showed that this control in *S. lividans* is mediated by the PhoR-PhoP two-component system.

Probably the most widely reported phenomenon associated with secondary metabolism is its repression by rapidly utilized carbon sources, particularly glucose. The phenomenon of carbon catabolite repression is a very common characteristic in bacteria and is manifested when glucose is used preferentially over other, less favorable, carbon sources. However, the mechanism of catabolite repression is not well understood in the streptomycetes. Uptake of glucose by the streptomycetes does not involve the phosphoenolpyruvate system, despite the fact that this system operates for the uptake of N-acetylglucoseamine. van Wezel et al. (2005) demonstrated that glucose uptake by S. coelicolor is facilitated by the permease, GlcP. The key protein associated with glucose mediated catabolite repression appears to be glucokinase (catalyzes the phosphorylation of glucose to glucose-6-phosphate) and its mechanism seems to be linked to its interaction with GlcP.

Our consideration of primary metabolites discussed the control of the terminal pathway, the supply of precursors and the supply of NADPH as targets for strain improvement. The targeting of these sites corresponded approximately with the development of techniques for the directed selection of mutants, the application of recombinant DNA technology and finally the development of systems biology. The improvement of secondary metabolite producers has followed a similar trend in that early advances were due to the isolation of mutants and the application of the more sophisticated approaches enabled a more focused strategy on key targets to evolve. However, the nature of secondary metabolites and their different control resulted in the adoption of different approaches, particularly in the isolation of mutants.

Isolation of mutants producing improved yields of secondary metabolites
The empirical approach
The discussion of the directed selection of mutants considered primary products whose biosynthesis and control had been sufficiently understood to prepare "blueprints" of the desirable mutants that then enabled the construction of suitable selection procedures. In contrast, important secondary metabolites were being produced long before their biosynthetic pathways, and certainly the control of those pathways, had been elucidated. Thus, strain improvement programs had to be developed without this fundamental knowledge that meant that they depended on the random selection of the survivors of mutagen exposure. Elander and Vournakis (1986) described these techniques as "hit or miss methods that require brute force, persistence and skill in the art of microbiology." However, despite the limited knowledge underlying these approaches they were extremely effective in increasing the yields of antibiotics. Chain's original Oxford *Penicillium notatum* isolate produced 5 mg dm^{-3} of penicillin whereas Demain cited commercial strains in 2006 producing over 70 g dm^{-3}. More rational approaches have been developed which reduce the empirical nature of strain-improvement programs. These developments include streamlining the empirical techniques and the use of more directed selection methods. Before discussing the attempts at directed selection the empirical approach will be considered along with the attempts made to improve this approach, including miniaturized programs.

The empirical approach to strain improvement involves subjecting a population of the micro-organism to a mutation treatment, frequently nitrosoguanidine (NTG), and then screening a proportion of the survivors of the treatment for improved productivity. The assessment of the chosen survivors was usually carried out in shake flasks, resulting in the procedure being costly, both in terms of time and personnel. According to Fantini (1966) the two questions that arise in the design of such programs are:

1. How many colonies from the survivors of a mutation treatment should be isolated for testing?
2. Which colonies should be isolated?

In attempting to answer the first question Fantini dismissed statistical approaches as impractical and claimed that the number of colonies isolated for testing is determined by the practical limiting factors of personnel, incubator and shaker space, and time. However, Davies (1964) demonstrated that a statistical approach could give valuable guidelines for the efficient utilization of physical resources in strain-development programs. Davies based a computer simulation of a mutation and screening program on the availability of 200 shaker spaces and practical results of error variance and the distribution of yield likely to occur among the mutants. He assumed that the majority of the progeny of a mutation would give a small, rather than a large increase in productivity and it would, therefore, be more feasible to screen a small number in the hope of obtaining a small increase rather than screen a large number in the hope of obtaining a large increase. The major difficulty inherent in this approach is the error involved in determining small increases, which implies the replication of screening tests and, therefore, the use of more facilities.

Davies used the computer simulation to investigate the merits of replication in screening programs and finally proposed the use of a two-stage scheme where mutants were tested singly in the first stage and then the better producers were tested in quadruplicate in the second stage. Davies concluded that such two-stage schemes were adequate over a wide range of conditions, although a one-stage screen could be used if the testing error were small and the frequency of occurrence of favorable types high, and a three-stage screen if the testing error were high and the frequency of occurrence of favorable types low. A screening program based on Davies' proposals is shown in Fig. 3.42.

The answer to Fantini's second question (which colonies should be isolated?) is extremely difficult in the field of secondary metabolites and in Davies' scheme the colonies were chosen at random. The selection of colonies on the basis of changed morphology has been considered by a number of workers and it appears that this is an undesirable technique. Elander (1966) demonstrated that it was preferable to isolate normal morphological types as, although a morphological variant may be a superior producer, it might require considerable fermentation development to materialize the increased production. Also, Alikhanian (1962) claimed that the vast majority of morphological mutants of the antibiotic producing actinomycetes tested were inferior producers. The most common type of shake flask program quoted is similar to that

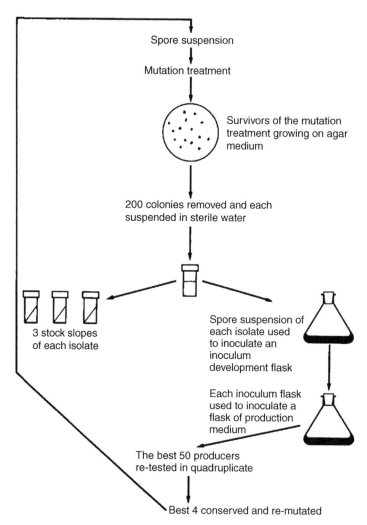

Spore suspension

Mutation treatment

Survivors of the mutation treatment growing on agar medium

200 colonies removed and each suspended in sterile water

3 stock slopes of each isolate

Spore suspension of each isolate used to inoculate an inoculum development flask

Each inoculum flask used to inoculate a flask of production medium

The best 50 producers re-tested in quadruplicate

Best 4 conserved and re-mutated

FIGURE 3.42 A Strain-Improvement Program for a Secondary Metabolite Producing Culture (Davies, 1964)

of Davies where the choice of colonies is random. However, there are now many reports of screens that miniaturize the procedure for the improvement of secondary metabolite producing strains. Such miniaturized systems are designed to enable the productivity of all (or a significant proportion) of the survivors to be assessed, which should eliminate (or reduce) the problem of choice of colonies for assessment and increase the throughput of cultures.

The basis of the miniaturized techniques is to grow the survivors of the mutation treatment either in a very low volume of liquid medium or on solidified (agar) medium. If the product is an antibiotic, the agar-grown colonies may be overlayed with an indicator organism sensitive to the antibiotic produced, allowing the assay to be done in situ. The level of antibiotic is assessed by the degree of inhibition of the overlayed indicator. The system is simple to apply to strains producing low levels of antibiotic but must be modified to allow the screening of high producers where very large zones of inhibition would be obtained. Also a system should be used to free the superior producers from contaminating indicator organisms. Dulaney and Dulaney (1967) used overlay techniques in the isolation of mutants producing chlortetracycline. Mutated spores were cultured on an agar medium in petri dishes for 6 days and then covered with pieces of sterile cellophane. An overlay of agar containing the indicator organism was then added and the plates incubated overnight. The mutant colonies were kept free from contamination by the cellophane and the size of the inhibition zone could be controlled by the depth of the base layer, the age of the colonies when overlayed, the depth of the overlay and the temperature of incubation. The system was designed such that a single colony of a nonmutated strain would not produce an inhibition zone but that two adjacent colonies would. In practice, the depth of the overlay controlled the size of the inhibition zone. Dulaney and Dulaney (1967) obtained a far greater enrichment in the number of desired phenotypes by the overlay technique than by random selection and testing in liquid medium.

Ichikawa, Date, Ishikura, and Ozaki (1971) screened for the increased production of the antibiotic kasugamycin by a *Streptomyces* sp., using a miniaturized technique termed the agar piece method. In order to prevent interference between colonies, mutated spores were grown on plugs of agar that were then placed on assay plates containing agar seeded with the indicator organism, levels of the antibiotic being determined by the size of inhibition zones. By combining this technique with a medium improvement program, the authors obtained a tenfold increase in the productivity. Ichikawa's method was modified by Ditchburn, Giddings, and MacDonald (1974) for the isolation of *Aspergillus nidulans* mutants synthesizing improved levels of penicillin. These workers obtained promising results using the technique and claimed that its potential for the recovery of higher-yielding mutants, for a given expenditure of labor time, was greater than the shake-flask method. The level of production of penicillin by *A. nidulans* is very small compared with *P. chrysogenum* and the technique would have to be modified considerably for commercial use.

Ball and McGonagle (1978) developed a miniaturized technique suitable for the assay of penicillin production by "high-yielding" strains of *P. chrysogenum* (producing up to 6000 units cm^{-3}). These workers highlighted the design of the solidified medium as a critical factor in the optimization of the method. They claimed that the growth of a colony on agar-solidified medium would be unlikely to modify the medium to the same extent as the growth of the organism on the same medium in submerged liquid culture. Thus, the nutrient-limiting conditions that favor the onset of antibiotic production might not be achieved by the growth of the culture on solid medium which would not allow the full production potential of the culture to be detected. However, nutrient-limiting conditions were achieved in the solidified medium by omitting the main carbon source and reducing the corn steep-liquor content. Mutated spores were plated over the surface of petri dishes of the nutrient limited medium such that ten to twenty colonies developed per plate. The time of incubation of the plates was not quoted but "when the colonies were of a size suitable for accurate measurement," a pasteurized spore suspension of *B. subtilis* containing 0.16 units cm^{-3} of penicillinase (to limit the size of the inhibition zones) was dispersed over the surface of the dish. The plates were then incubated for 18–24 h and the inhibition zones were examined. Suitable colonies were freed of contaminating *B. subtilis* by culturing on nutrient agar containing sodium penicillin and streptomycin sulfate. The use of this technique enabled three operators to scan 15,000 survivors from ultraviolet irradiated spore populations in 3 months.

The major disadvantages of the miniaturized solidified medium technique approach is that productivity expressed on the solid medium may not be expressed in the subsequent liquid culture and conversely, colonies not showing activity on solid media may be highly productive in liquid medium. Despite these limitations the aforementioned workers have demonstrated that the approach has considerable merit and Ball (1978) claimed that the increase in throughput might be as much as 20 times that of a conventional shake-flask program. Work by Bushell's group (Pickup, Nolan, & Bushell, 1993) provides an interesting insight into the problem of certain streptomycete isolates producing a secondary metabolite on agar but not in liquid culture. It is important to appreciate that the strains were natural isolates and not the survivors of mutation treatments but the conclusions do have relevance to mutant development. Working on the premise that nonproduction in liquid medium may be due to a more fragmented morphology, these strains were subjected to a filtration enrichment procedure. Liquid cultures were filtered through a linen filter and the filtrate and retentate mycelium recultured in fresh medium. The procedure was then repeated sequentially. The enriched retentate mycelium of several isolates which were previously unable to produce antibiotic in liquid culture gave rise to stable filamentous types synthesizing in liquid medium. Although this approach could not be used routinely in a high-throughput mutant screen it may be applied to a few high-producing strains which have not fulfilled their promise (detected in an agar screen) in liquid media.

An excellent example of a miniaturized screening program is given by the work of Dunn-Coleman et al. (1991) of Genencor International. These workers developed a low-volume liquid medium system to isolate mutants of *A. niger* var. *awamori* capable of improved secretion of bovine chymosin. The *Aspergillus* strain had been constructed using recombinant DNA technology. Mutated spore suspensions were diluted and inoculated into 96-well microtiter plates using robots. The dilutions were such that each well should have contained one viable spore. The plates were incubated in a static incubator and then harvested using a robotic pipetting station and assayed for product. Typically, 50–60,000 mutated viable spores were assessed in each screen. The most promising 10–50 strains from the miniaturized screen were then tested in shake flask culture. The best producer from the shake-flask screen was then used as the starter for a further round of mutagenesis. The results of seven rounds of

Table 3.8 Chymosin Production by NTG Mutated Strains of *Aspergillus niger* var. *awamori* (Dunn-Coleman et al., 1991)

Rounds of NTG Mutagenesis and Recurrent Selection	Chymosin Concentration ($\mu g\ cm^{-3}$)	
	Microtiter Plate	Shake Flask
Parent	42	286
First	50	273
Second	64	280
Third	136	377
Fourth	170	475
Fifth	200	510
Sixth	256	646

mutation and selection are shown in Table 3.8. Investigation of the improved strains showed that they were capable of secreting a range of enzymes and, thus, the strains were presumed to be modified in their secretion mechanisms. It is interesting to note that an improved secreter of a heterologous protein has been isolated using a traditional mutation and selection approach that has been miniaturized and automated.

In several cases the examination of randomly selected high-yielding mutants has indicated that their superiority may have been due to modifications of their control systems. Goulden and Chataway (1969) demonstrated that a mutant of *P. chrysogenum,* producing high levels of penicillin, was less sensitive to the control of acetohydroxyacid synthase by valine. Valine is one of the precursors of penicillin and its synthesis is controlled by its feedback inhibition of acetohydroxyacid synthase. Thus, removal of the control of valine synthesis may result in the production of higher valine levels and, hence, greater production of penicillin. Pruess and Johnson (1967) demonstrated that higher-yielding strains of *P. chrysogenum* also contained higher levels of acyltransferase, the enzyme that catalyzes the addition of phenylacetic acid to 6-aminopenicillanic acid. Dulaney (1954) reported that the best producer of streptomycin among *S. griseus* mutants was auxotrophic for vitamin B_{12} and Demain (1973) claimed that this auxotrophic mutation was still a characteristic of production strains being used in 1969. Thus, it appears that some strains isolated by random selection, and overproducing secondary metabolites, are altered in ways similar to those strains isolated by directed selection techniques and overproducing primary metabolites.

Use of directed selection in the isolation of improved secondary metabolite producers

There are many examples in the literature where a more directed selection approach has been adopted for the improvement of secondary metabolite producers. The techniques used include the isolation of auxotrophs, revertants, and analog-resistant mutants. Although mutation of secondary metabolite producers to auxotrophy for a primary biosynthetic product frequently results in their producing lower yields, cases of improved productivity have also been demonstrated. For example, Alikhanian, Mindlin, Goldat, and Vladimizov (1959) investigated the tetracycline producing abilities of 53 auxotrophic mutants, all of which produced significantly less tetracycline than the parent strain in normal production medium. Supplementation of the medium with the growth requirement resulted in one mutant expressing productivity superior to that of the parent.

It is sometimes difficult to explain the precise reason for the effect of mutation to auxotrophy on the production of secondary metabolites, but in the majority of cases it has been demonstrated to be an effect on the secondary metabolic system rather than, simply, an effect on the growth of the organism. The simplest explanation for the deleterious effect on secondary metabolite yield is that the auxotroph is blocked in the biosynthesis of a precursor of the end product, for example, Polsinelli, Albertini, Cassani, and Cifferi (1965) demonstrated that auxotrophs of *Streptomyces antibioticus* which required any of the precursors of actinomycin (isoleucine, valine, or threonine) were poor producers of the antibiotic.

Many secondary metabolites may be considered as end products of branched pathways that also give rise to primary metabolites. Thus, a mutation to auxotrophy for the primary end product may also influence the production of the secondary product. In *P. chrysogenum,* lysine and penicillin share the same common biosynthetic route to α-aminoadipic acid, as shown in Fig. 3.43. The role of lysine in the penicillin fermentation has also led to the investigation of lysine auxotrophs as potential superior penicillin producers. Demain (1957) demonstrated that lysine was inhibitory to the biosynthesis of penicillin. The explanation of this phenomenon is considered to be the inhibition of homocitrate synthase by lysine resulting in the depletion of α-aminoadipic acid required for penicillin synthesis (Demain & Masurekar, 1974). It may be postulated that lysine auxotrophs blocked immediately after α-aminoadipic acid would produce higher levels of penicillin due to the diversion of the intermediate toward penicillin synthesis and the removal of any control of homocitrate synthase by endogenous lysine. O'Sullivan and Pirt (1973) investigated the production of penicillin by lysine auxotrophs of *P. chrysogenum* in continuous culture but were unable to demonstrate improved productivity in a range of lysine-supplemented media. Luengo, Revilla, Villanueva, and Martin (1979) examined the lysine regulation of penicillin biosynthesis in low-producing and industrial strains of *P. chrysogenum.* The industrial strain was capable of producing up to 12,000 units cm^{-3} penicillin. These workers demonstrated that although the onset of penicillin synthesis in the industrial strain was less sensitive to lysine than in the low-producing strain, the empirical industrial selection procedures had not completely removed the mechanisms of lysine regulation of penicillin biosynthesis. Luengo et al. (1979) expressed the possibility that overproduces of penicillin may be obtained by the selection of mutants to lysine regulation. However, it should be noted that the industrial strain employed was a representative of only one series of penicillin producers and that other series may have already been modified with respect to the regulatory effects of lysine. Nevertheless, this study does indicate the possibility that selection of strains resistant to lysine control may be overproduces of penicillin.

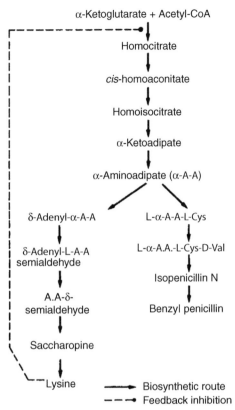

α-Ketoglutarate + Acetyl-CoA

Homocitrate

cis-homoaconitate

Homoisocitrate

α-Ketoadipate

α-Aminoadipate (α-A-A)

δ-Adenyl-α-A-A   L-α-A-A-L-Cys

δ-Adenyl-L-A-A   L-α-A.A.-L-Cys-D-Val
semialdehyde

A.A-δ-   Isopenicillin N
semialdehyde
                        Benzyl penicillin

Saccharopine

Lysine   ——→ Biosynthetic route
              ---→ Feedback inhibition

FIGURE 3.43 Biosynthesis of Benzyl Penicillin and Lysine in *Penicillium chrysogenum*

It is far more difficult to explain the effect of auxotrophy for factors not associated with the biosynthesis of the secondary metabolite. Polsinelli et al. (1965) demonstrated that seven out of 27 auxotrophs of *S. griseus* produced more actinomycin than did the prototrophic parent. Dulaney and Dulaney (1967) demonstrated increased chlortetracycline yields in eight out of eleven auxotrophs of *S. viridifaciens* when grown in supplemented media. In neither of these cases were the auxotrophic requirements directly involved in the biosynthesis of the secondary metabolites. Demain (1973) put forward two possible explanations to attempt to account for the behavior of such auxotrophs. The first explanation is that the auxotrophic factors were involved in "cross-pathway" regulation with the secondary metabolite or its precursors. Demain quoted several examples of cross-pathway regulation in primary metabolism where the activity of one pathway is affected by the product of an apparently unrelated sequence. The alternative explanation is that the effect on secondary metabolism is not due to the auxotrophy but to a second mutation accompanying the auxotrophy, that is, a double mutation. Demain cited two attempts to determine whether the effects of auxotrophy on secondary metabolism were due to double mutations or to the auxotrophy. MacDonald, Hutchinson, and Gillett (1963) reverted a low-producing thiosulfate-requiring mutant of *P. chrysogenum* to thiosulfate independence and examined penicillin productivity by the revertants. Approximately half of the revertants reacquired their "grandparents" production level, whereas the other half retained their poor productivity. Polsinelli et al. (1965) reverted five isoleucine–valine auxotrophs of *S. antibioticus* (which also produced low levels of actinomycin compared with the parent strain) to prototrophy and discovered that some were returned to normal production and others to higher production levels than the "grandparent." Thus, in the case of Polsinelli et al. (1965) mutants, it is unlikely that the effect on secondary metabolism was due to a double mutation, but it is possible that this was the case for some of MacDonald et al.'s strains. However, it should be remembered that both these groups of auxotrophs were poor secondary metabolite producers blocked in routes directly involved with the secondary biosynthetic pathway. It may be more relevant to examine the nature of auxotrophic strains blocked in apparently unrelated pathways and produce improved levels of the secondary metabolite. It is not possible to say whether any of the auxotrophic mutants previously discussed produced superior levels of the secondary metabolite as a result of double mutations, but it may be possible to exploit this possibility in the future.

The treatment of bacterial cells with NTG results in clusters of mutations around the replicating fork of the chromosome (Guerola, Ingraham, & Cerda-Olmedo, 1971). Thus, if one of the mutations were selectable (eg, auxotrophy) it may be possible to isolate a strain containing the selectable mutation along with other nonselectable mutations that map close by. The efficient application of this technique would depend on the accurate mapping of the gene involved in producing the secondary metabolite so that neighboring mutations may be selected. This technique may be valuable for the selection of mutants of the bacilli and streptomycetes producing high levels of antibiotics where mutations affecting synthesis may be mapped for each biosynthetic step. Comutation by NTG may then be followed by selection for changes to genes adjacent to those loci known to be involved in production of the particular secondary metabolite.

The isolation of analog-resistant mutants has already been discussed in the field of primary metabolism, where the rationale was that a mutant resistant to the inhibitory effects of an analog, which mimics the control characteristics of the natural metabolite, might overproduce the natural metabolite. This approach has been adopted, or may be adopted, in a number of guises in the field of secondary metabolism. The most obvious application is the enhancement of the production of a precursor of the secondary metabolite. This approach was adopted by Elander, Mabe, Hamill, and Gorman (1971) in the isolation of mutants of *Pseudomonas aureofaciens* overproducing the antibiotic pyrrolnitrin. Tryptophan is a precursor of pyrrolnitrin and although it is stimulatory to production it is uneconomic to use as an additive in an industrial process. Thus, Elander isolated mutants resistant to tryptophan analogs using the gradient plate technique described previously. A strain was eventually isolated that produced 2–3 times more antibiotic than the parent and was resistant to feedback inhibition by tryptophan. Addition of tryptophan to the improved strain would not result in higher pyrrolnitrin synthesis, indicating that tryptophan supply was no longer the limiting factor and the organism was producing sufficient endogenous tryptophan for antibiotic synthesis.

Godfrey (1973) isolated mutants of *Streptomyces lipmanii* resistant to the valine analog, trifluoroleucine. These mutants produced higher levels of cephamycin than the parent strain, and appeared to be deregulated for the isoleucine, leucine, valine biosynthetic pathway, indicating that valine may have been a rate-limiting step in cephamycin synthesis.

Methionine has been demonstrated to stimulate the biosynthesis of cephalosporin by *Acremonium chrysogenum* and superior producers have been isolated in the form of methionine analog-resistant mutants (Chang and Elander, 1982). Lysine analog resistant mutants have yielded a greater frequency of superior β-lactam antibiotic producers compared with random selection (Elander, 1989).

There are many cases in the literature of a secondary metabolite preventing its own synthesis. Martin (1978) cited the following examples: chloramphenicol, auro-dox, cycloheximide, staphylomycin, ristomycin, puromycin, fungicidin, candihexin, mycophenolic acid, and penicillin. The precise mechanisms of these controls are not clear but they appear to be specific against the synthesis of the secondary metabolite and not against the general metabolism of the cells. The mechanism of the control of its own synthesis by chloramphenicol appears to be the repression of arylamine synthetase (the first enzyme in the pathway from chorismic acid to chloramphenicol) by chloramphenicol. Jones and Vining (1976) demonstrated that arylamine synthetase was fully repressed by 100 mg dm^{-3} chloramphenicol, a level of antibiotic which neither affected cell growth nor the activities of the other enzymes of the chloramphenicol pathway.

Martin (1978) quoted several examples of correlations between the level of secondary metabolite accumulation and the level that causes "feedback" control, which may imply that the factor limiting the yields of some secondary metabolites is the feedback inhibition by the end product. Much more recently, Liu, Chater, Chandra, Niu, and Tan (2013) cited evidence of the involvement feedback control of secondary metabolism in their review of molecular regulation of antibiotic biosynthesis. The selection of mutants resistant to feedback inhibition by a secondary metabolite is a far more difficult task than the isolation of strains resistant to primary metabolic control. It is extremely unlikely that a toxic analog of the secondary metabolite could be found where the toxicity lay in the mimicking of the feedback control of the secondary metabolite, a compound which would not be necessary for growth. However, the detection of mutants resistant to feedback inhibition by antibiotics may be achieved by the use of solidified media screening techniques, similar to the miniaturized screening techniques previously described. The technique would involve culturing the survivors of a mutation treatment on solidified medium containing hitherto repressing levels of the antibiotic and detecting improved producers by overlaying the colonies with an indicator organism. The difficulty inherent in this technique is that the incorporated antibiotic, itself, will inhibit the development of the indicator organism. This problem may be overcome by adjusting the depth of the overlay or the concentration of the indicator such that an inhibition zone could be produced only by a level of antibiotic greater than that incorporated in the original medium. However, it is unlikely that this would be a satisfactory solution for the selection of a high-producing commercial strain. Another approach would be to utilize an analog of the antibiotic which mimicked the feedback control by the natural product but which did not have antimicrobial properties. Inhibition of antibiotic synthesis by analogs has been demonstrated in the cases of aurodox (Liu, McDaniel, & Schaffner, 1972) and penicillin (Gordee & Day, 1972). An alternative may be to use a mutant indicator bacterium for the overlay that is only sensitive to levels of the antibiotic in excess of that originally incorporated into the medium. In cases where it has been demonstrated that feedback control by the end product plays an important role in limiting productivity it would probably be worthwhile to design such procedures.

Many secondary metabolites have been shown to be toxic to the producing cell when it is present in the trophophase (growth phase) (Demain, 1974). Thus, it appears that a "switch" in the metabolism of the organism in the idiophase enables it to produce an otherwise "autotoxic" product. Furthermore, it has been demonstrated that, in some cases, the higher the resistance to the secondary metabolite in the growth phase, the higher is productivity in the production phase. Dolezilova, Spizek, Vondracek, Paleckova, and Vanek (1965) demonstrated that the level of production of nystatin by various strains of *S. noursei* was related to the resistance of the strain to the antibiotic in the growth phase; a nonproducing mutant was inhibited by 20 units cm^{-3}, the parent strain produced 6000 units cm^{-3} and was inhibited by 2000 units cm^{-3} in the growth phase and a mutant producing 15,000 units cm^{-3} was found to be resistant to 20,000 units cm^{-3}. The possible relationship between antibiotic resistance and productivity may be used to advantage in the selection of high-producing mutants by culturing the survivors of a mutation treatment in the presence of a high level of the end product. Those strains capable of growth in the presence of a high level of the antibiotic may also be capable of high productivity in the idiophase. This approach has been used successfully for antifungal azasterols (Bu'Lock, 1980), streptomycin (Woodruff, 1966), and ristomycin (Trenina & Trutneva, 1966) but without success for novobiocin (Hoeksema & Smith, 1961).

A similar rationale was used by McGuire, Glotfelty, and White (1980) for the improvement of daunorubicin production, an anthracycline antitumor agent synthesized by the red-pigmented streptomycete *Streptomyces peuceticus*. The red pigment is presumably anthracycline, which shares its polyketide origin with daunorubicin. The directed selection was based on the ability of the antibiotic cerulenin to suppress polyketide synthesis. Cerulenin is an antibiotic that targets fatty acid synthase, thus ultimately disrupting membrane function. Fatty acid synthase and polyketide synthase are very similar complexes and thus resistance to cerulenin may result in an associated modification in polyketide synthesis. This was indicated in *S. peuceticus* by the lack of the red pigment on cerulenin agar. Thus, mutants that were still capable of oligoketide synthesis in the presence of the inhibitor would remain red. Approximately a third of the resistant mutants were superior daunorubicin producers. Miyake et al. (2006) used a similar approach in isolating cerulenin-resistant mutants of *Monascus pilosus* producing an elevated level of the polyketide lovastatin. Zhang, Hunter, and Tham (2012) reported that the mechanism of cerulenin resistance was the production of increased levels of the enzyme that "titrates out" the inhibitor.

A potentially toxic compound may be rendered harmless by an organism converting it to a secondary metabolite or a secondary metabolite complexing with it. Ions of heavy metals such as Hg^{2+}, Cu^{2+} and related organometallic ions are known to complex with β-lactam antibiotics (Elander and Vournakis, 1986) and such agents have been used to select resistant mutants. The logic of this selection process is that overproduction of a β-lactam may be the mechanism for increased resistance to the heavy metal. Elander and Vournakis (1986) reported that the frequency of superior cephalosporin C producers was greater amongst mutants resistant to mercuric chloride than among random samples of the survivors of ultraviolet treatment.

The conversion of the penicillin precursor, phenylacetic acid, to penicillin is thought to be an example of detoxification by conversion to a secondary metabolite. It appears that strains capable of withstanding higher concentrations of the precursor may be able to synthesize higher levels of the end product. Polya and Nyiri (1966) applied this hypothesis in selecting phenylacetic acid-resistant mutants of *P. chrysogenum* and demonstrated that 7% of the isolates showed enhanced penicillin production. Barrios-Gonzalez, Montenegro, and Martin (1993) investigated the same phenomenon and developed methods for the enrichment of both spores and early idiophase mycelium resistant to phenyl acetic acid. Of the resistant spore population, 16.7% were superior penicillin producers as compared with 50% of the resistant idiophase population. This suggests that the selective force may be more "directed" by using a population already committed to penicillin synthesis. Although the best mutants contained elevated levels of acyltransferase (the enzyme that directly detoxifies PAA), they also showed higher levels of isopenicillin N synthetase (cyclase), which the authors claimed to be the limiting step. Thus, it was claimed that the screening method was not specific to select for acyltransferase elevation but could select for strains having a faster carbon flow through the pathway enabling them to use PAA faster.

Ball (1978) stressed that the major difficulty in the selection of toxic precursor resistant mutants is that the site of resistance of a mutant may not result in the organism overproducing the end product—for example, the resistance may be due to an alteration in the permeability of the mutant or due to the mutants' ability to degrade the precursor to a harmless metabolite unrelated to the desired end product. Barrios-Gonzalez's approach of using mycelial fragments already committed to penicillin synthesis decreases the likelihood of such "false selection" occurring.

The repression of secondary metabolism has been attributed to the high concentrations of key nutrients present in the exponential phase—particularly the carbon, phosphate, and nitrogen sources. Thus, analogs of repressing media components have been used to select resistant mutants. For example, arsenate and vanadate have been used as selective agents to isolate phosphate resistant strains (DeWitt, Jackson, & Paulus, 1989) and deoxyglucose to isolate glucose catabolite repression resistant strains (Allen, McNally, Lowendorf, Slayman, & Free, 1989). Deoxyglucose is a toxic analog of glucose and mutants resistant to the compound having lower glucose uptake rates resulting in the lifting of catabolite repression. Repression by ammonium ion can be reduced by the isolation of mutants resistant to bialaphos—an inhibitor of glutamine synthetase, the key enzyme in ammonium ion assimilation (Zhang et al., 2012).

The involvement of ppGpp in the onset of secondary metabolism is very well documented, although the details of the mechanism are far from clear. This scenario was debated earlier in our consideration of the control of secondary metabolism. This perceived role of ppGpp has been exploited in strain improvement by the development of an approach termed "ribosome engineering" (Ochi et al., 2004). The involvement of the ribosome in ppGpp synthesis caused some workers to explore the relationship between mutations giving resistance to antibiotics acting at the ribosome and the production of secondary metabolites. An excellent example of the approach is given by Wang, Hoasaka, and Ochi's (2008) investigation of actinorhodin production by *S. coelicolor*. These workers sequentially introduced mutations ultimately giving resistance to eight different ribosome-acting antibiotics. The process was very simple and relied on the detection of naturally occurring mutations by plating more than 10^9 spores on the surface of agar plates containing antibiotics. Thus, no mutagens or recombinant DNA technology techniques were used. The best producer, resistant to seven antibiotics, produced 1.63 g dm^{-3} actinorhodin—a 180-fold increase over the wild type. This elevated productivity was associated with an increase in ppGpp synthesis and *relA* transcription and the maintenance of protein synthesis into the stationary phase compared with the wild type. Importantly, the improved actinorhodin synthesis was lost in a *relA*⁻ multiply resistant strain indicating the necessity of ppGpp synthesis for hyper-production. The molecular mechanisms involved in this process are unclear and the authors pointed out that the ppGpp and ribosome characteristics did not necessarily occur in each successive mutant strain. However, remarkable results were achieved in a short time using very simple techniques that can be easily applied to different secondary metabolite producers.

As discussed in the section on primary metabolites, a mutant may revert to the phenotype of its "parent," but the genotype of the revertant may not, necessarily, be the same as the original "parent." Some revertant auxotrophs have been demonstrated to accumulate primary metabolites (Shiio and Sano, 1969) and attempts have been made to apply the technique to the isolation of mutants overproducing secondary metabolites. Two approaches have been used in the field of secondary metabolites with respect to the isolation of revertants:

- The reversion of mutants auxotrophic for primary metabolites that may influence the production of a secondary metabolite.
- The reversion of mutants that have lost the ability to produce the secondary metabolite.

As previously discussed, it is difficult to account for some of the effects of auxotrophy on secondary metabolism and it appears that some may be due to as yet unresolved cross-pathway phenomena and others due to the expression of other mutations associated with the auxotrophy. Similarly, revertant mutants may affect secondary metabolism in a number of ways—direct effects on the pathway, cross-pathway effects, and effects due to mutations other than the detected reversion.

Dulaney and Dulaney (1967) investigated the tetracycline productivities of a population of prototroph revertants of *S. viridifaciens* derived from each of five auxotrophs. These workers predicted that some revertants might be productive due to direct influence of the mutations on tetracycline biosynthesis but that others may be so because they contained other lesions. Superior producers were obtained in all the prototroph-revertant populations apart from those derived from a homocysteine auxotroph. However, the frequency of the occurrence of the superior producers was similar to that obtained by the random selection of the survivors of a mutation treatment in all but one of the populations. The exceptional population was the revertants of a methionine auxotroph, 98% of which produced between 1.2 and 3.2 times as much tetracycline as the original prototrophic culture. A possible explanation of the very favorable titres of the population may be the role of methionine as the methyl donor in tetracycline biosynthesis and that methionine availability limited the production of the secondary metabolite in the original prototroph. However, addition of exogenous methionine to the prototroph did not result in superior productivity.

Polsinelli et al. (1965) also demonstrated that reversion of five mutants of an actinomycin producing strain of *S. antibioticus* blocked in the isoleucine–valine pathway resulted in the isolation of some superior mutants. Godfrey (1973) reported that the reversion of a cysteine auxotrophic mutant of *S. lipmanii* resulted in improved production of cephamycin. These studies provide promising evidence that the selection of revertants of auxotrophs of primary metabolites involved in secondary metabolism may yield a high number of productive mutants.

The reversion of nonproducing strains may result in the detection of a high-producing mutant as that mutant would have undergone at least two mutations associated with the production of the secondary metabolite. Dulaney and Dulaney (1967) plated the progeny of a mutation of a nonproducing strain of *S. viridifaciens* onto solidified production medium and screened for superior tetracycline producers by an overlay technique. A mutant was isolated which produced 9 times the tetracycline yield of the original "parent." The major difficulty inherent in this technique is that nonproducing mutants of high-yielding strains may be incapable of being reverted due to extreme deficiencies in their metabolism. Indeed, Rowlands (1992) suggested that strains which are non (or low) producers and illustrate other effects such as poor growth and sporulation are best discarded, but that revertant mutants not showing pleiotropic effects are perhaps more likely to possess genuine increases in biosynthetic activity than those produced by any other technique, having been mutated twice in genes directly affecting product formation.

Use of recombination systems for the improvement of secondary metabolite production

The vast majority of commercial secondary metabolite producers are either fungi or actinomycetes and recombination has been used for strain improvement in both groups. The first approaches used the parasexual cycle in industrial fungi and conjugation in the streptomycetes (often referred to as spore mating) to achieve recombination. The development of protoplast fusion significantly increased the use of recombination for both groups of organisms and the subsequent application of recombinant DNA technology resulted in another significant step in strain development.

The application of the parasexual cycle

Many industrially important fungi do not possess a sexual stage and therefore it would appear difficult to achieve recombination in these organisms. However, Pontecorvo, Roper, Hemmons, MacDo-nald, and Bufton (1953) demonstrated that nuclear fusion and gene segregation could take place outside, or in the absence of, the sexual organs. The process was termed the parasexual cycle and has been demonstrated in the imperfect fungi, *A. niger* and *P. chrysogenum,* as well as the sexual fungus *A. nidulans.* In order for parasexual recombination to take place in an imperfect fungus, nuclear fusion must occur between unlike nuclei in the vegetative hyphae of the organism. Thus, recombination may be achieved only in an organism in which at least two different types of nuclei coexist, that is, a heterokaryon. The heterozygous diploid nucleus resulting from the fusion of the two different haploid nuclei may give rise to a diploid clone and, in rare cases, a diploid nucleus in the clone may undergo an abnormal mitosis resulting in mitotic segregation and the development of recombinant clones which may be either diploid or haploid.

Recombinant clones may be detected by their display of recessive characteristics not expressed in the heterokaryon. Analysis of the recombinants normally demonstrates them to be segregant for only one, or a few linked, markers and culture of the segregants results in the development of clones displaying more recessive characters than the initial segregant. The process of recombination during the growth of the heterozygous diploid may occur in two ways: mitotic crossing over, which results in diploid recombinants, and haploidization, which results in haploid recombinants.

Mitotic crossing over is the result of an abnormal mitosis. The normal mitosis of a heterozygous diploid cell (where $2n = 2$) is shown in Fig. 3.44. During mitosis, each pair of homologous chromosomes replicate to produce two pairs of chromatids and a chromatid of one pair migrates to a pole of the cell with a chromatid of the other pair. Division of the cell at the equator results in the production of two cells, both of which are heterozygous for all the genes on the chromosome. Mitotic crossing over involves the exchange of distal segments between chromatids of homologous chromosomes as shown in Fig. 3.45. This process may result in the production of daughter nuclei homozygous for a portion of one pair of chromosomes and in the expression of any recessive alleles contained in that portion. Thus, the clone arising from the partial homozygote will be recombinant and further mitotic crossing over in the recombinant will result in the expression of more recessive alleles.

Haploidization is a process that results in the unequal distribution of chromatids between the progeny of mitosis. Thus, of the four chromatids of a homologous chromosome pair, three may migrate to one pole and one to another resulting in the

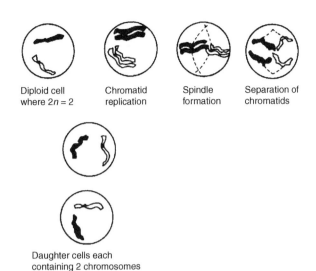

FIGURE 3.44 Diagrammatic Representation of the Mitotic Division of a Eukaryotic Cell Containing Two Chromosomes

The nuclear membrane has not been portrayed in the figure.

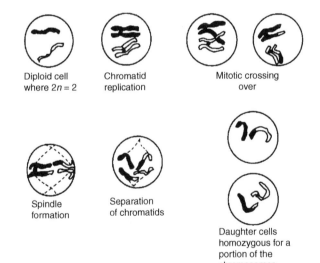

FIGURE 3.45 Diagrammatic Representation of Mitosis Including Mitotic Crossing Over

formation of two nuclei, one containing $2n + 1$ chromosomes and the other containing $2n - 1$ chromosomes. The $2n - 1$ nuclei tend toward the haploid state by the progressive random loss of further chromosomes. Thus, the resulting haploid nucleus will contain a random assortment of the homologues of the chromosomes of the organism. A representation of the process is shown in Fig. 3.46.

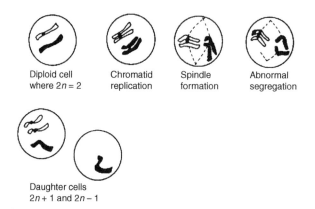

FIGURE 3.46 **Diagrammatic Representation of Mitosis Involving Haploidization**

Therefore, the major components of the parasexual cycle are the establishment of a heterokaryon, vegetative nuclear fusion, and mitotic crossing over or haploidization resulting in the formation of a recombinant. In practice, the occurrence and detection of these stages may be enhanced by the use of auxotrophic markers. The parents of the cross are made auxotrophic for different requirements and cultured together on minimal medium. The auxotrophs will grow very slightly due to the carryover of their growth requirements from the previous media, but if a heterokaryon is produced, by anastomoses (mycelial connections) forming between the two parents, then it will grow rapidly. The frequency of vegetative nuclear fusion in the heterokaryon may be enhanced by the use of agents such as camphor vapor or ultraviolet light; mitotic segregation may be enhanced by the use of agents such as X-rays, nitrogen mustard, p-fluorophenylalanine, and ultraviolet light (Sermonti 1969).

The application of the parasexual cycle to industrially important fungi has been hindered by a number of problems (Elander, 1980). A major difficulty is the influence of the auxotrophic markers (used for the selection of the heterokaryon) on the synthesis of the desired product. As discussed earlier, auxotrophic mutations have been observed to have quite unpredictable results on the production of some secondary metabolites. Even conidial color markers have been shown to have deleterious effects on product synthesis (Elander, 1980). Even when suitable markers are available, the induction of heterokaryons in some industrial fungi has been demonstrated to be a difficult process and specialized techniques had to be employed to increase the probabilities of heterokaryon formation (MacDonald & Holt, 1976). However, the development of protoplast fusion methods in the late 1970s enabled efficient heterokaryon formation to be achieved and removed the major barrier to the application of the parasexual cycle to strain improvement. These developments are considered in more detail in the next section on protoplast fusion.

Despite the early difficulties of inducing the parasexual cycle in some industrial fungi, it was used in two ways to study these organisms. The cycle was used to investigate the genetics of the producing strains as well as to develop recombinant superior producers. Information obtained on the basic genetics of the industrial fungi using the parasexual cycle included the number of chromosomes (or linkage groups), the allocation of genes, important in product synthesis, to particular chromosomes, and the mapping of important genes on a chromosome. Sermonti (1969) and MacDonald and Holt (1976) have described the techniques that were used to achieve these objectives.

Initial studies on haploid strains of *P. chrysogenum* derived from parasexual crosses demonstrated that most of the progeny exhibited the genotype of one of the parents, that is, no recombination had occurred (MacDonald, 1968). MacDonald suggested that one of the reasons for this lack of success might have been the difference in gross chromosomal morphology between the parents caused by certain mutagens used in the development of the strains. Thus, one precaution to be adopted in the development of strains that may be used in a parasexual cross is the avoidance of mutagens that may cause gross changes in chromosomal morphology. Subsequently, Ball (1978) suggested that the careful choice of markers and the use of strains giving similar titers and being "not too divergent" could result in achieving more recombinants.

Although diploids produced by the parasexual cycle are frequently unstable, stable diploids have been used for the industrial production of penicillin. Elander (1967) isolated a diploid from a sister cross of *P. chrysogenum,* which was shown to be a better penicillin producer than the parents and was morphologically more stable (a sister cross is one where the two strains differ only in the markers used to induce the heterokaryon). One explanation of the superior performance of the diploid may have been its resistance to strain degeneration caused by deleterious recessive mutations. Such mutations would only have been expressed in the diploid if both alleles had been modified. Calam, Daglish, and McCann (1976) demonstrated that a diploid strain of *P. chrysogenum* was more stable than haploid mutants and mutation and selection of the diploid gave rise to a diploid strain producing higher levels of penicillin than the parents. Ball (1978) claimed that the usefulness of diploids may only be short term, presumably implying problems of degeneration, and may best be used as stepping-stones to a recombinant haploid.

The advantages to be gained from the industrial use of parasexual recombinants are not confined to the amalgamation of different yield improving mutations. Equally advantageous would be the introduction of characteristics that would make the process more economic, for example, low viscosity, sporulation, and the elimination of unwanted products. The development of these issues will be considered in the next section on protoplast fusion.

Protoplast fusion

Fusion of fungal protoplasts appears to be an excellent technique to obtain heterokaryons between strains where conventional techniques have failed, or, indeed, as the method of choice. Thus, this approach has allowed the use of the parasexual cycle for breeding purposes in situations where it had not been previously possible. This situation is illustrated by the work of Peberdy, Eyssen, and Anne (1977) who succeeded in obtaining heterokaryons between *P. chrysogenum* and *P. cyaneofulvum* and demonstrated the formation of diploids which gave rise to recombinants after treatment with *p*-fluorophenylalanine or benomyl. Although it has been claimed that *P. chrysogenum* and *P. cyaneofulvum* are not different species of *Penicillium* (Samson, Hadlock, & Stolk, 1977), Peberdy et al. (1977) still demonstrated that protoplast fusion could be successful where conventional techniques had failed.

A demonstration of the use of protoplast fusion for an industrial fungus is provided by the work of Hamlyn and Ball (1979) on the cephalosporin producer, *C. acremonium*. These workers compared the effectiveness of conventional techniques of obtaining nuclear fusion between strains of C. *acremonium* with the protoplast fusion technique. The results from conventional techniques suggested that nuclear fusion was difficult to achieve. Electron microscopic examination of fused protoplasts indicated that up to 1% underwent immediate nuclear fusion. Recombinants were obtained in both sister and divergent crosses. A cross between an asporulating, slow-growing strain with a sporulating fast-growing strain, which only produced one-third of the cephalosporin level of the first strain eventually resulted in the isolation of a recombinant which combined the desirable properties of both strains, that is, a strain which demonstrated good sporulation, a high growth rate, and produced 40% more antibiotic than the higher-yielding parent. Chang and Elander (1982) utilized protoplast fusion to combine the desirable qualities of two strains of *P. chrysogenum*. Protoplasts from two strains, differing in colony morphology and the abilities to produce penicillin V (the desired product) and OH-V penicillin (an undesirable product), were fused, followed by plating on a nonselective medium. Out of 100 stable colonies that were scored, two possessed the desirable morphology, high penicillin V and low OH-V penicillin productivities.

Lein (1986) reported the penicillin strain improvement program adopted by Panlabs, Inc. This program included random and directed selection as well as protoplast fusion. Table 3.9 illustrates the properties of the two strains used in a protoplast fusion and one of the recombinants selected. To avoid any adverse effects no selective genetic markers were used and the regenerated colonies were screened on the basis of colony morphology and spore color. A total of 238 colonies judged to be recombinants were screened for penicillin V production and the culture with the best combination of properties is shown in the table. Thus, the desirable characteristics of each strain were combined in the recombinant.

Protoplasts are also useful in the filamentous fungi for manipulations other than cell fusion. Rowlands (1992) suggested that they might be used in mutagenesis of nonsporulating fungi. Spores are the cells of choice for the mutagenesis of filamentous fungi but this is obviously impossible for nonsporulating strains. Mycelial fragments may be used but these will be multinucleate and very high mutagen doses are required. Although some protoplasts will be nonnucleate or multinucleate at least some will be uninucleate, which will express any modified genes after mutation. Also, protoplasts will take up DNA in in vitro genetic manipulation experiments and this aspect will be discussed in a later section of this chapter.

Recombination can take place between actinomycetes by conjugation (Hopwood, 1976) and phage transduction (Studdard, 1979). However, both these

Table 3.9 The Use of Protoplast Fusion for the Improvement of a Penicillin V Producer (Lein, 1986)

Characteristics	Parent A	Parent B	Best Recombinant
Spores per slant ($\times 10^8$)	2.2	2.5	7.5
Germination frequency (%)	99	40	49
Color of sporulating colonies	Green	Pale green	Deep green
Seed growth	Good	Poor	Good
Penicillin V yield (mg cm^{-3})	11.7	18.5	18.0
Phenylacetic oxidation	Yes	No	No

mechanisms involve the transfer of only small regions of the bacterial chromosome. Furthermore, the low frequencies at which recombination occurs necessitate the use of selectable genetic markers such as auxotrophy or antibiotic resistance. The introduction of such markers is time-consuming but also they can detrimentally affect the synthetic capacity of the organism. Protoplast fusion has particular advantages over conjugation in that the technique involves the participation of the entire genome in recombination. Also, Hopwood (1979) developed techniques that resulted in the recovery of a very high proportion of recombinants from the fusion products of *S. coelicolor* protoplasts. By subjecting protoplasts to an exposure of ultraviolet light, sufficient to kill 99% of them prior to fusion, Hopwood claimed a 10-fold increase in recombinant detection for strains normally giving a low yield of recombinants (1%) and a doubling of the recombination frequency for a cross normally yielding 20% recombinants. Such yields of recombinants means that they would be detectable by simply screening a random proportion of the progeny of a protoplast fusion and the use of selectable markers to "force" out the recombinants would not be necessary. Examples of the application of the technique to actinomycete strain improvement include cephamycin C yield enhancement in *Nocardia lactamdurans* (Wesseling, 1982) and the improvement of lignin degradation in *Streptomyces viridosporus* (Petty & Crawford, 1984).

The impact of protoplast fusion was taken to a new level with the development of a technique given the term "genome shuffling." Two publications in 2002 used this approach to improve the acid tolerance of an industrial *Lactobacillus* strain (Patnaik et al., 2002) and the production of tylosin by *Streptomyces fradiae* (Zhang et al., 2002). The philosophy of this method was to combine mutation and selection with recombination using protoplast fusion. Zhang et al. (2002) subjected spores of *S. fradiae* (strain SF1) to NTG mutagenesis and cultured 22,000 survivors in microtiter plates, selecting 11 mutants for further study. Protoplasts were prepared from the mutants, mixed in equal quantities and fused. Fused protoplasts were allowed to regenerate cell walls and 1,000 progeny screened for tylosin production. Seven progeny from the fusion process, producing more than the original parent, were selected, protoplasted, fused, regenerated, and screened. A further seven strains with improved production were selected as before and the best two examined in detail.

The productivity of the two selected strains were compared with the productivity of SF1 and strain SF21 that had been produced from SF1 by 20 rounds of conventional mutagenesis and selection, over a period of 20 years. The results were quite dramatic—the two rounds of genome shuffling, taking 1 year, produced strains giving yields equivalent to those of mutants produced from 20 rounds of mutation and selection over a period of 20 years. The random mutagenesis involved over a million assays whereas the genome shuffling approach involved only 24,000. Yields improved by eightfold in both systems—but genome shuffling reduced the time involved from 20 years to 1. Thus, the shuffling of genes between multiple parental genomes, an unusually promiscuous cross, achieved the equivalent of a conventional program in a 20th of the time and, crucially, employed simple techniques requiring no knowledge of the pathways involved, nor of their control.

The adoption of genome shuffling as a means of achieving whole-genome engineering has been reported for a number of systems, for example, natamycin production by *Streptomyces gilvosporeus* (Luo et al., 2012); nosiheptide production by *Streptomyces actuosus* (Wang et al., 2014) and spinosad production by *Saccharopolyspora spinosa* (Wang, Xue, He, Peng, & Yao, 2015).

Recombinant DNA technology

The basic principles of the use of recombinant DNA technology has been considered in the previous section of this chapter dealing with its application to the improvement of primary metabolite production. The application of in vitro recombinant DNA technology to the improvement of secondary metabolite production may not be as advanced as it is for primary metabolites, but it has made a very significant contribution. Techniques have been developed for the genetic manipulation of the filamentous fungi (Zhang et al., 2004) and the streptomycetes (Kieser et al., 2000 and Bekker, Dodd, Brady, & Rumbold, 2014). As discussed earlier, the genes for the biosynthesis of a secondary metabolite are clustered both in the streptomycetes and the fungi. Furthermore, these clusters also contain the genes for regulation, resistance, export, and extracellular processing. Work of this type has not only increased the basic understanding of the molecular genetics of secondary metabolism, but it has also facilitated strain improvement to be conducted in a rational manner, with four broad approaches being used: amplifying the copy number of genes coding for key enzymes or regulatory elements; increasing precursor availability; removing competing reactions that divert intermediates away from the desired end product, and modifying the pathways to produce novel products.

An excellent early example of amplifying gene copy number is provided by the work of the Lilly Research Laboratories group (Skatrud, 1992) in attempting to increase cephalosporin C synthesis by the fungus, *Cephalosporium acremonium*. Four critical stages were involved:

- Identifying the biochemical rate-limiting step in the cephalosporin C industrial fermentation.
- Cloning the gene coding for the enzyme catalyzing the rate-limiting step.

- Constructing a vector containing the cloned gene and introducing it into the production strain.
- Screening the transformants for increased productivity.

The biosynthetic route to cephalosporin C is shown in Fig. 3.47. Analysis of fermentation broths showed an accumulation of penicillin N that indicated a bottleneck at the next reaction, that is, the ring expansion step where the 5-membered penicillin N ring is expanded to the 6-membered ring of deacetoxycephalosporin C. This step is catalyzed by deacetoxycephalosporin C synthase (DAOCS, commonly called expandase) coded for by the gene *cef*EF (already previously cloned by Samson et al., 1987). *cef*EF also codes for the next enzyme in the route, deacetylcephalosporin C synthase (DACS, commonly called hydroxylase). Thus, if the expandase step were rate limiting, introduction of extra copies of cefEF should relieve the limitation and eliminate the accumulation of penicillin N.

The production strain of *C. acremonium* was transformed with a plasmid containing an exact copy of *cef*EF and the transformants screened for increased productivity. Approximately one in four transformants were superior producers and one produced almost 50% more antibiotic in a laboratory-scale fermentation. Analysis of the transformant showed that a single copy of the transforming DNA had integrated into chromosome III, whereas native *cef*EF resides in chromosome II. In pilot scale fed-batch fermentations the transformant showed a 15% increase in yield, still a very significant increase for an industrial strain. As predicted, the transformant did not accumulate penicillin N and the bottleneck appeared to have been relieved. The superior strain was among the first eight transformants examined, whereas 10,000 survivors of a mutagen exposure rendered no improved types. This work illustrated the enormous potential of recombinant DNA technology for secondary metabolite yield improvement.

The amplification of a regulatory gene is exemplified by Perez-Llarena, Liras, Rodriguez-Garcia, and Martin (1997) work on the gene *ccaR* of *S. clavuligerus*. The protein encoded by the gene resembles those produced by "cluster-situated transcriptional regulators" (CSRs). Disruption of the gene resulted in the loss of both cephamycin and clavulanic acid; restoration of the gene restored productivity and amplification gave a two- to threefold increase in production of both compounds.

Malmberg, Hu, and Sherman's (1993) early application of the technology in the study of cephamycin synthesis in wild-type *S. clavuligerus* illustrates the second approach—increasing precursor availability. The three precursors of cephamycin synthesis are the three amino acids L-cysteine, D-valine, and α-aminoadipic acid (α-AAA). However, the synthesis of α-AAA is different in the fungi and streptomycetes—in the former, it is a precursor of lysine whereas in the latter it is a catabolic product. The observations that supplementation with α-AAA and resistance to lysine analogs stimulated cephamycin synthesis led these workers to increase α-AAA availability by amplifying the copy number of the *lat* gene, coding for lysine ε-aminotransferase (LAT), the first enzyme in the conversion of lysine to α-AAA. This was achieved using an engineered plasmid containing the *lat* gene that integrated into

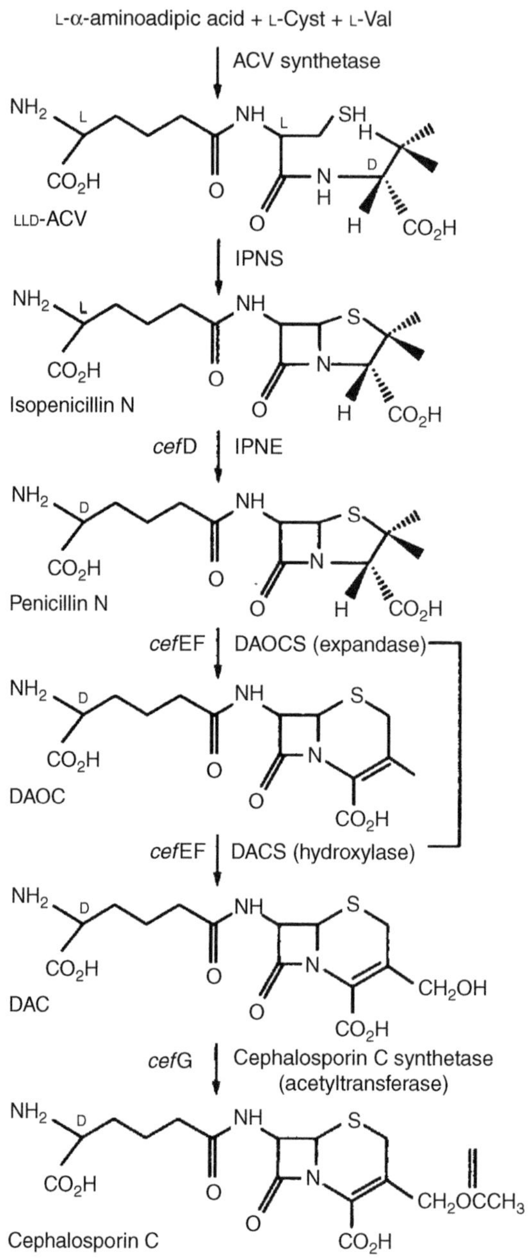

FIGURE 3.47 The Biosynthetic Route to Cephalosporin C in *Cephalosporium acremonium* (Skatrud, 1992)

Abbreviations: *ACV*, delta-(L-2-aminoadipyl)-L-cysteinyl-D-valine; *cef*, Genes coding for cephalosporin synthesis; *DAC*, Deacetylcephalosporin C; *DACS*, DAC synthase (commonly known as hydroxylase); *DAOC*, Deacetoxycephalosporin C; *DAOCS*, DAOC synthase (commonly known as expandase); *IPNS*, Isopenicillin-N-synthase.

the cephamycin C gene cluster. The transformed strain produced 5 times the wild-type concentration of both cephamycin C and its precursor, *O*-carbamoyl deacetyl-cephalosporin C, but crucially in identical proportions, indicating an early influence on the pathway in the transformant. The increased productivity of the engineered strain correlated with its elevated activity of LAT.

Li and Townsend's paper in 2006 also describes increasing the synthesis of a precursor of a secondary metabolite—this time, the production of clavulanic acid by *S. clavuligerus*. The biosynthetic route to clavulanic acid is shown in Fig. 3.48, from which it can be seen that its two precursors are D-glyceraldehyde 3-phosphate and L-arginine. Supplementation of the culture with arginine simply increased the concentration of the arginine pool without stimulating product synthesis while glycerol supplementation enhanced clavulanic production, suggesting glyceraldehyde 3-phosphate as the limiting factor. The key intermediate in the pathway is D-glyceraldehyde-3-phosphate (D-G3P), located at the three-way branch point to clavulanic acid, glycolysis, and gluconeogenesis, and thus glyceraldehyde-3-phosphate dehydrogenase (GAPDH) is a key enzyme influencing flux at this node. Two different GAPDH enzymes are produced by *S. clavuligerus*, coded by the genes *gap1*

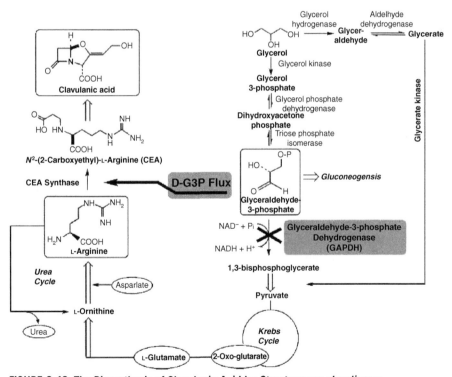

FIGURE 3.48 The Biosynthesis of Clavulanic Acid by *Streptomyces clavuligerus*

Where D-G3P is D-glyceraldehyde 3-phosphate (Li & Townsend, 2006).

and *gap2*. Each of these genes was inactivated in separate strains by targeted gene disruption. Clavulanic acid production was doubled as a result of disruption of *gap1* but disruption of *gap2* had little effect. Supplementation of the *gap1*-disrupted strain gave the reverse results seen in the wild type, that is, arginine supplementation stimulated product formation while glycerol did not, indicating a sufficiency of D-G3P and arginine limitation. Biomass production was unaffected by the manipulation and thus glycolysis was still operational in the enhanced strain. This could have been due to the activity of Gap2 or to the alternative glycerol pathway, shown in Fig. 3.48, involving glycerol hydrogenase, aldehyde dehydrogenase, and glycerate kinase. The disrupted gene was incorporated into the genome by double-crossover through homologous recombination and, thus, the construct was stable and not subject to reversion due to plasmid loss. Further work on this strain (Banos, Perez-Redondo, Koekman, & Liras, 2009) increased the gene dosage of the *glp* operon, comprising the genes coding for glycerol transporter, glycerol kinase, and glycerol-3-phosphate dehydrogenase. This strain elevated clavulanic acid yield by a further 4.5-fold and with glycerol supplementation, 7.5-fold.

The ability of individual streptomycetes to produce a range of different secondary metabolites has been addressed several times in this chapter. This characteristic is manifested in two ways—the production of closely related metabolites of the same pathway and the ability to produce a range of secondary metabolites originating from different pathways. Paradkar et al. (2001) investigated both aspects in their study of the production of clavulanic acid by *S. clavuligerus*. As well as clavulanic acid, this organism produces a number of closely related compounds that have an inverted stereochemistry in their ring structure, the antipodal clavams (Fig. 3.49) and the un-related *β*-lactam, cephamycin C, produced by a separate biosynthetic pathway. These workers disrupted both cephamycin C and antipodal clavam synthesis in an industrial strain of *S. clavuligerus* by replacing the *lat* gene and the *cvm1* gene, respectively, with cloned copies of these genes interrupted by the insertion of an apramycin resistance gene. The double mutant produced about a 10% increase in clavulanic acid production—a significant result when it is considered that the strain was already producing several orders of magnitude more product than the wild-type. However, the improved production was not the only asset of the modified strain—removal of the antipodal clavans (that are relatively toxic) is advantageous from a regulatory point of view and also facilitated the downstream extraction process (Paradkar, 2013).

The isolation of organisms from the natural environment synthesizing commercially useful metabolites is an expensive and laborious process. Therefore, other approaches have been used to produce novel compounds that may be of some industrial significance. Probably the most successful alternative approach has been the semisynthetic one where microbial products have been chemically modified, for example, the semisynthetic penicillins. The advent of readily available recombination techniques resulted in attempts to produce novel compounds from recombinants, particularly streptomycetes. The rationale behind these experiments was that by mixing the genotypes of two organisms, synthesizing different metabolites, then new combinations of biosynthetic genes, and hence pathways, may be produced. Little

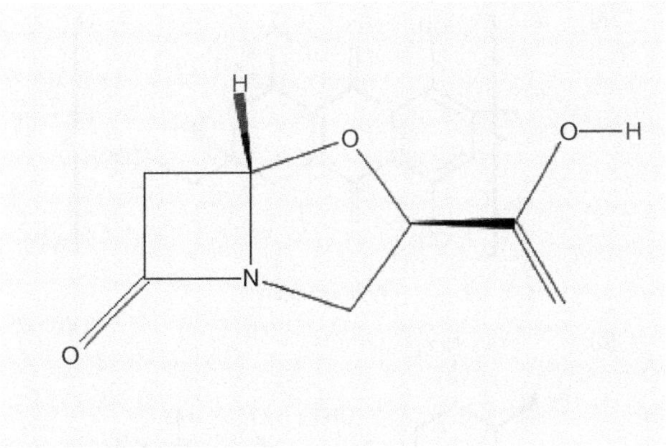

(a) 5S or antipodal clavam (Clavam-2-carboxylate)

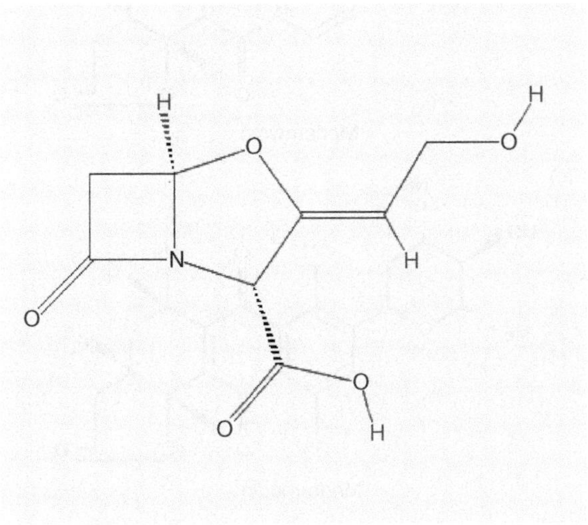

(b) Clavulanic acid

FIGURE 3.49

Antipodal clavam (a) and clavulanic acid (b), both synthesized by *Streptomyces clavuligerus*.

Data deposited in or computed by PubChem. URL: *https://pubchem.ncbi.nlm.nih.gov.*

progress was achieved using protoplast fusion but the exploitation of recombinant DNA technology yielded some significant successes.

S. coelicolor produces the polyketide actinorhodin (Fig. 3.50) while *Streptomyces* sp. AM 7161 produces medermycin. Hopwood et al. (1985) transformed some of the cloned genes coding for actinorhodin into *Streptomyces* sp. AM7161. The recombinant

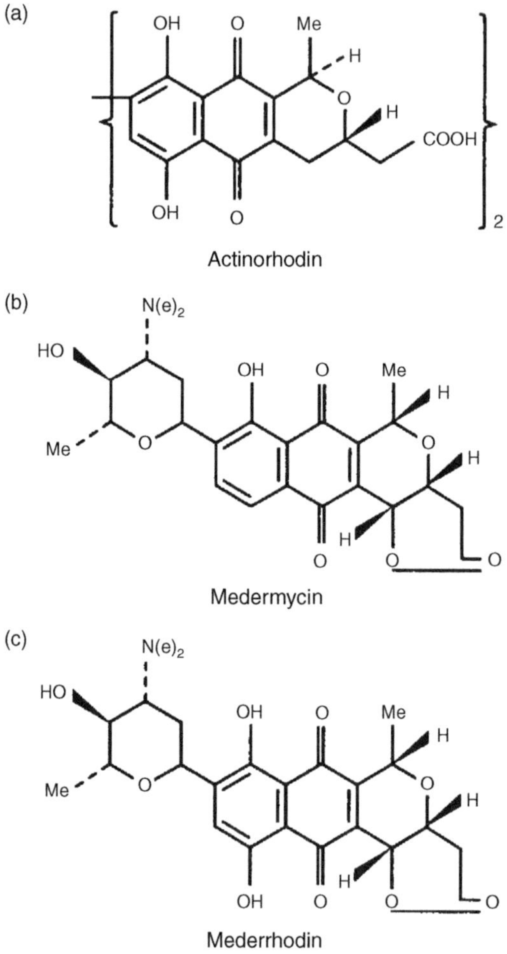

FIGURE 3.50

(a) Structure of actinorhodin, (b) structure of medermycin, (c) structure of mederrhodin (Hunter & Baumberg, 1989).

produced another antibiotic, mederrhodin A (Fig. 3.50). The modified strain contained the *acftV* gene from *S. coelicolor* coding for the p-hydroxylation of actinorhodin; the enzyme was also capable of hydroxylating medermycin. Hopwood et al. (1985) also introduced the entire actinorhodin gene cluster into *Streptomyces violaceoruber* that produces granaticin or dihydrogranaticin (Fig. 3.51). The recombinant synthesized the novel antibiotic, dihydrogranatirhodin that has the same structure as dihydrogranaticin apart from the stereochemistry at one of its chiral centers (Fig. 3.51).

The enzymes responsible for polyketide biosynthesis have been extensively studied and significant advances have been made in the understanding of both their biochemistry and genetics. Polyketide synthases (PKSs) are multifunctional

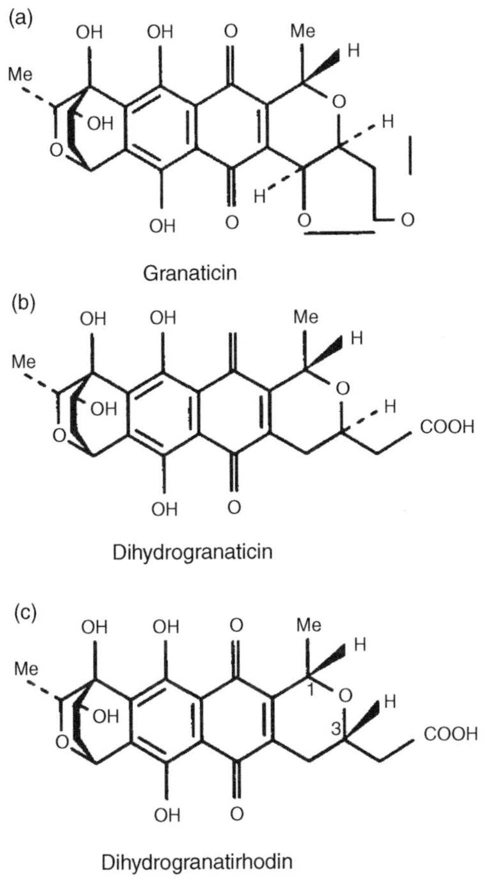

FIGURE 3.51

(a) Structure of granaticin, (b) structure of dihydrogranaticin, (c) structure of dihydrogranatirhodin (Hunter & Baumberg, 1989).

enzymes that catalyze repeated decarboxylative condensations between coenzyme A thioesters and are very similar to the fatty acid synthases. An enormous range of microbial polyketides is known and the variation is due to the chain length, the nature of the precursors and the subsequent modification reactions that occur after cyclization. McDaniel, Ebert-khosla, Hopwood, and Khosla (1993) constructed plasmids coding for recombinant PKSs and achieved expression in *S. coelicolor*. Five novel polyketides were synthesized and the system showed great promise for the production of new, potentially valuable, compounds as well as providing carbon skeletons amenable to derivatization by organic chemists.

A commercially relevant example is provided by the Lilly Research Laboratories group (Beckman, Cantwell, Whiteman, Queener, & Abraham, 1993) who produced a strain of *Penicillium chrysogenum* capable of accumulating deacetoxycephalosporin C (DAOC), a compound that can be biotransformed into 7-aminodesacetoxycephalosporanic acid (7ADAOC) which is a precursor of at least three chemically synthesized clinically important cephalosporins. The conventional route to 7ADAOC is by the chemical ring expansion of benzylpenicillin that is far more complex than the biotransformation of DAOC. DAOC is produced as an intermediate in the synthesis of cephalosporin C by *Cephalosporium acremonium,* as shown in Fig. 3.47. However, it would be very difficult to manufacture the compound from *C. acremonium* by blocking the conversion of DAOC to deacetylcephalosporin C because a single bifunctional enzyme catalyzes both the conversion of penicillin N to DAOC (expandase activity) and the hydroxylation of DAOC to deacetylcephalosporin C (hydroxylase activity). The bifunctional protein is encoded by the gene *cef* EF. In *S. clavuligerus* the expandase and hydroxylase enzymes are separate proteins encoded by the genes *cef* E and *cef* F, respectively. Thus, *S. clavuligerus* could be modified for the commercial production of DAOC. However, the Lilly group did not have a strain capable of producing high cephalosporin levels available to them, but they did have a *P. chrysogenum* that overproduced isopenicillin N (IPN) as an intermediate in phenoxymethylpenicillin synthesis as shown in Fig. 3.52. Thus, to enable the *Penicillium* to produce DAOC the insertion of two genes was necessary, that is, those encoding isopenicillin epimerase and expandase. These workers had already cloned *cef*E (coding for expandase) from *S. clavuligerus* and *cef*D (coding for epimerase) from *S. lipmanii*. Thus, using sophisticated vector technology, a *P. chrysogenum* was transformed with a plasmid containing both *cef*D and *cef*E. One of the isolated transformants was capable of synthesizing up to 2.5 g dm^{-3} DAOC and both the *cef*D and *cef*E genes were integrated into the organism's genome. Thus, the recombinant produced by the Lilly group contained a branched biosynthetic pathway giving rise to the two end products, DAOC and phenoxymethylpenicillin.

A novel application of recombinant DNA technology to strain improvement is the development of reporter-guided mutant selection. Askenazi et al. (2003) were the first to develop the system that combined random mutagenesis with a directed

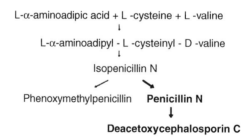

FIGURE 3.52 β-Lactam Biosynthetic Pathway in Wild-Type *P. chrysogenum* (Pathway in Non-bold Font) and in *P. chrysogenum* Transformed with pBOB13 and pZAZ6 (Branched Pathway Which Includes Steps in Bold Font)

Enzyme steps (non-bold font, top to bottom): ACV synthetase, IPN synthase, acyl coenzyme A: IPN acyltransferase. Enzyme steps (bold font, top to bottom): IPN epimerase, penicillin N expandase (= DAOC synthase). (Beckman et al., 1993.)

selection process based on a reporter gene (Fig. 3.53). As discussed previously, the control of secondary metabolite pathways involves a regulatory cascade in which a transcription factor, coded by a regulatory gene, switches on a lower-level regulatory gene that in turn, enables transcription of the structural genes of a pathway. A plasmid is constructed in which a reporter gene (usually antibiotic resistance) is linked to the promoter of the pathway regulatory gene. Thus, the reporter gene is under the

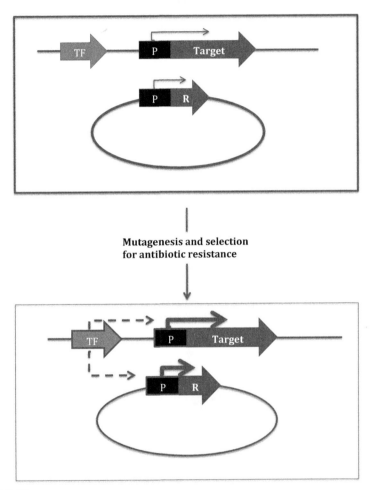

FIGURE 3.53

Reporter-guided mutant selection representing a bacterial cell showing a portion of the chromosome (with the target secondary metabolite gene and its promoter (P) controlled by a transcription factor, TF) and an engineered plasmid containing an antibiotic resistance gene (R) controlled by the secondary metabolism promoter. Mutagenesis, followed by selection for antibiotic resistance selects expression for the up-regulated TF via the expression of the reporter.

Modified from Xiang et al. (2009).

control of the same cascade as are the secondary metabolism genes. The producer organism is transformed with the plasmid and subjected to mutagenesis. Mutants that have been altered in the control cascade such that the regulator gene is over expressed will also express the reporter gene, resulting in the expression of a selectable characteristic (antibiotic resistance). A disadvantage of the system is the number of false positive antibiotic-resistant mutants that are generated that are not due to expression of the reporter gene. Xiang et al. (2009) addressed this problem by linking the desired promoter to two, different, reporter genes. Thus, a false positive could only be generated by two, unrelated, mutations (Fig. 3.54). The basis of the reporter system is that the desired mutation is functioning in *trans*. Thus, a *cis* mutation in the reporter

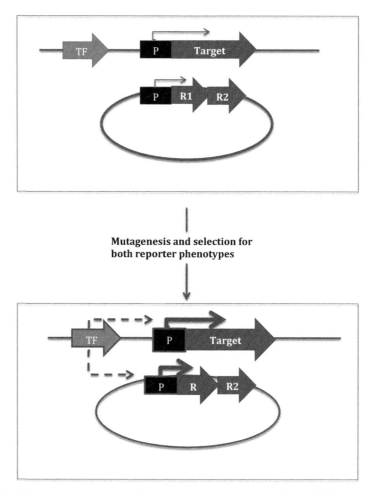

FIGURE 3.54 Dual Reporter-Guided Mutant Selection, as for Fig. 3.53 but With an Additional Reporter Gene (R2)—See Text for Explanation

Modified from Xiang et al. (2009).

construct promoter that gives elevated reporter gene expression will also be a false positive, so it is not possible to fully eliminate the possibility.

Xiang et al. (2009) applied the approach to the improvement of clavulanic acid production by *S. clavuligerus*. The reporter construct consisted of the promoter of the clavulanic acid regulator gene, *ccaR*, linked to the kanamycin resistance gene, *neo*, and *xylE*, coding for catechol 2,3-dioxygenase. Colonies expressing *xylE* convert catechol to its yellow derivative, 2-hydroxymuconic semialdehyde. The advantage of a color-related phenotype is that the degree of expression of the reporter can be estimated by the intensity of the color reaction. Using this methodology, Xiang et al. (2009) showed that 90% of mutants selected using the double-reporter system produced elevated clavulanic acid levels. Interestingly, the mutants were genetically diverse, thus giving an additional resource in attempting to elucidate the factors controlling productivity.

Postgenomic era—the influence of genomics, transcriptomics, and proteomics on the improvement of secondary metabolite producers

The impact of genome sequencing on the evolution of industrial strains of *C. glutamicum* was considered earlier in this chapter and the reader is referred to that discussion for an introduction to the basic principles. A complete genome sequence not only enables the identification of key mutations in industrial strains but also opens the door to the fields of transcriptomics, proteomics, and metabolomics. The progress made in these fields has been made possible by the advances in both molecular biology and bioinformatics, the latter facilitating the processing of vast amounts of data and the development of metabolic models. Key contributions to the understanding of the physiology of secondary metabolism were the complete genome sequencing of the "model organisms" of secondary metabolism—*S. coelicolor* (Bentley et al., 2002) and *P. chrysogenum* (van den Berg et al., 2008). The *S. coelicolor* genome contained the largest number of predicted genes (7825) of any bacterium sequenced at that time—commensurate, perhaps, with the morphological, physiological, and chemical complexity of the organism. A surprisingly large number of proteins (965) was attributed a regulatory role, with an abundance of sigma factors (65) and two-component regulators (53 sensor-regulator pairs). Also, the large number of genes coding for extracellular enzymes is typical of soil microorganisms. Most surprisingly, and discussed in an earlier section of this chapter on strain isolation, was the identification of 18 clusters of genes predicted to code for enzymes of secondary metabolism—a discovery that resuscitated the search for novel metabolites. Another feature, also shown in other streptomycete genomes, is the occurrence of several genes coding for proteins with the same activity in central carbon metabolism (Hiltner, Hunter, & Hoskison, 2015). For example, in our earlier discussion of clavulanic acid production by *S. clavuligerus* two isoenzymic forms of glyceraldehyde-3-phosphate dehydrogenase were active. This duplication of enzyme activity confuses the results of flux analysis as the flux through a reaction is measured as the combined activity of all participating enzymes and may not reflect the sophisticated control that may be happening. A dedicated database,

StrepDB, has been developed for the processing of information associated with the *S. coelicolor* sequence and its coverage has now been expanded to include a wide range of streptomycete genomes and products (Lucas et al., 2013). Mining both fungal and bacterial genomes for secondary metabolism genes has also been facilitated by the "antibiotics and secondary metabolism analysis shell" (antiSMASH), a web server and stand-alone tool for the automatic genomic identification and analysis of biosynthetic gene clusters, now available in its third iteration (Weber et al., 2015).

The sequenced *P. chrysogenum* strain was Wisconsin 54-1255, an early descendent of the original isolate NRRL 1951 (from which all industrial strains were derived), producing approximately 2000 μg cm^{-3} penicillin (Barreiro, Martin, & Garcia-Estrada, 2012). The genome contained 12,943 genes (similar in size to other filamentous fungi) and, in common with sequenced streptomycetes, contained secondary metabolism genes other than those coding for its known secondary metabolite products. Interestingly, two penicillin biosynthetic genes, *pcbAB* and *pcbC* lacked introns, suggesting that the penicillin gene cluster originated by horizontal gene transfer from a prokaryote.

The application of genomics in the field of secondary metabolism includes the direct comparison of the genomes of high and low producers and the use of both transcriptomics and proteomics to investigate the physiology and molecular biology of secondary metabolite production. An excellent example of the "omics" approach is the investigation of the penicillin fermentation. The Wisconsin 54-1255 *P. chrysogenum* sequence elucidated by van den Berg et al. enabled the development of a microarray to investigate gene expression. It was already appreciated that the penicillin gene cluster was amplified in high producers (Fierro et al., 1995) and van den Berg et al.'s sequence showed their strain to have only one—thus, the amplification occurred later in the strain improvement programs. The comparison of the transcriptomes of Wisconsin 54-1255 with an industrial strain enabled a greater understanding of the basis of hyper-productivity. The strains were investigated under penicillin producing conditions and the differential expression of their genes compared. The results are summarized in Table 3.10. Particularly noteworthy are the increased expression in the high producer of genes associated with precursor synthesis—valine, cysteine, and α-aminoadipic acid—and both up- and downregulated genes related to the peroxisome. The peroxisome is an important organelle in penicillin synthesis as it is the site for the final two steps in the pathway.

Further insight into the development of penicillin industrial strains has been given by Barreiro et al. (2012) review of the comparison of the proteomes of the wild-type strain, the moderate producer Wisconsin 54-1255 and a high producing strain, AS-P-78. The following key observations shed light on the changes inherent in high penicillin producers:

1. Increasing penicillin yield was commensurate with a decrease in the catabolic pathways and an increase in the biosynthetic pathways of the three precursor amino acids—valine, cysteine, and α-aminoadipic acid.

Table 3.10 Upregulated and Downregulated Genes of *P. chrysogenum* Under Penicillin Producing Conditions (van den Berg et al., 2008)

Gene Group	Number of Genes Upregulated	Number of Genes Downregulated
Growth and metabolism	106	75
Pathway from precursors to penicillin G	6	1
Penicillin G secretion	13	4
Peroxisome related	2	4
β-Lactam metabolism	4	2
Penylacetic acid metabolism	6	0
Cell differentiating and morphology	14	11
Other	156	174
Total	307	271

2. Despite the fact that AS-P-78 contains up to six copies of the penicillin gene cluster, the levels of associated proteins were not proportionately increased. Thus, the superior productivity of AS-P-78 cannot be explained entirely by the amplification of the gene cluster.
3. The expression of proteins involved in the degradation or modification of penicillin were downregulated in the high producer.
4. The original wild-type produced a number of proteins involved in pigment synthesis while this characteristic was absent in the moderate and high producers. This was an expected observation as the loss of pigmentation was an early feature in the strain development programs. Also, proteins associated with the invasion of fruit (the organism's natural habitat) were expressed at lower levels in the penicillin producers.
5. Phenylacetic acid (PAA) is fed to the penicillin fermentation, forming the side chain of penicillin G. However, PAA can be degraded via the homogentisate pathway, thus making it unavailable for antibiotic synthesis. The degradatory pathway enzymes were expressed at much lower levels in the penicillin-producing strains and phenylacetyl-CoA ligase and isopenicillin N acyltransferase, which catalyze the conversion of isopenicillin N to penicillin G, were overexpressed. These enzymes are present in the peroxisome as well as a wide repertoire of ligases, specific for a range of side-chains.
6. The synthesis of one mole of penicillin synthesis requires 8–10 moles of NADPH (Kleijn et al., 2007). The major source of NADPH is the pentose phosphate pathway, and the production of key enzymes such as ribose-5-phosphate isomerase and transketolase were upregulated in the high producer.

No penicillin pathway-specific regulatory genes have been elucidated in the penicillin gene cluster and control appears to be exerted at a global level by regulators such as VeA and LaeA. The "velvet" morphological mutation was first recorded in *A. nidulans* by Kafer (1965). The name described its velvety appearance due to its prolific production of conidia (asexual spores). The velvet complex of proteins consists of VeA, VelB, and LaeA and couple secondary metabolism with differentiation including sexual and asexual reproduction (Palmer et al., 2013). LaeA was the most recent protein to be added to this group by Butchko, Adams, and Keller (1999) who isolated *A. nidulans* mutants that had lost the ability to produce the secondary metabolite, sterigmatocystin, but retained the ability to sporulate. A gene named *laeA* was shown to complement one of these mutants and its protein product, LaeA, was shown to be a regulator of secondary metabolism. Deletion of *laeA* blocked the expression of secondary metabolism gene clusters (such as sterigmatocystin, penicillin, and lovastatin) whereas its over expression increased transcription of these gene clusters and their biosynthetic products. The combination of VeA-VelB was required for the production of sexual spores whereas VeA-LaeA was required for the production of sterigmatocystin. The mode of action has been attributed to LaeA's interaction with histone protein that may result in making concealed portions of the genome accessible to transcription (Bok & Keller, 2004; Sanchez, Somoza, Keller, & Wang, 2012).

Thus, elucidation of the genome sequence and the accompanying transcriptomic and proteomic investigations have revealed many key lesions in penicillin high-yielding strains that contribute to their productivity. However, it is interesting to note form Table 3.10 that the considerable majority of genes that were differentially expressed in the low and medium producers were classified under "growth and metabolism" or "other." Some of these may be making a real contribution to penicillin synthesis but many will have been introduced during the mutation/selection process and actually be deleterious in terms of the organism's vitality.

Lum, Huang, Hutchinson, and Kao (2004) took advantage of the *S. coelicolor* DNA array in an early example of the comparison of the transcriptome of a high producer with that of a wild-type. By adding the erythromycin gene cluster to the *S. coelicolor* array they were able to produce a functioning array to investigate erythromycin production by *Saccharoployspora erythrea*. They demonstrated that production of erythromycin was accompanied by the sustained, coordinated expression of the erythromycin genes suggesting control by a regulator and the absence of cluster-associated regulator genes implied orchestration at a global level. Subsequently, using DNA binding studies of the promoter regions of the five erythromycin genes, Chng, Lumm, Vroom, and Kao (2008) identified the regulator protein as BldD (product of gene *bldD*). BldD is a regulator of aerial hyphal development in *Streptomyces* and acts as a repressor. Transcriptome studies showed *bldD* expression was elevated in the production strain and its deletion caused a sevenfold reduction in erythromycin production and blocked aerial development on solidified medium.

More recently, an example of the application of transcriptomics to the elucidation of potential genetic targets for strain improvement is provided by the work of Medema et al. (2011a) who elucidated the genome sequence of the clavulanic acid (CA) producer, *S. clavuligerus*. Knowing the genome sequence enabled Medema et al. to construct a DNA array that was then used to compare the transcriptome of the wild-type with that of an industrial strain, DS48802, developed using conventional mutation and selection. DS48802 produced CA at approximately 100 times the level of the wild type. As is typical of secondary metabolite producers, *S. clavuligerus* produces a range of secondary metabolite products. Thus, as well as CA, it synthesizes at least four different clavam metabolites, the methionine antimetabolite, alanylclavam, and cephamycin C. The gene clusters coding for CA and cephamycin are adjacent on the chromosome and, together, are referred to as a "super-cluster." The transcriptome of DS48802 showed the complete super-cluster, including the regulator genes *claR and ccaR,* to be upregulated by a factor of two- to eightfold, as was the two-component system that induced the expression of the alanylclavam cluster. This response suggests that a common "higher-level" regulator was operational, with AdpA the prime candidate as its transcription was elevated 2.5 times that of the wild-type. The pathway to CA is given in Fig. 3.48, and was discussed in our earlier consideration of its rational modification by Li and Townsend (2006). Glycerol uptake and metabolism was upregulated in the industrial strain, while aconitase and citrate synthase were downregulated, resulting in a reduced flux of the CA precursor, glyceraldehyde-3-phosphate, to the TCA cycle and its diversion to the synthesis of CA. The outcome of this scenario is very similar to that of Li and Townsend's modification of glyceraldehyde-3-phosphate dehydrogenase discussed earlier, but still enabling reduced TCA flux to facilitate arginine synthesis (the other precursor of CA). Other upregulated primary metabolism genes included glutamine and glutamate synthetases (possibly involved in the synthesis of arginine via the urea cycle) and transporters for both ammonia and phosphate. Thus, these insights into the physiology of the industrial producer reveal potential leads to develop a strain by molecular engineering, devoid of damaging lesions introduced by the random mutation and selection approach.

The remarkable progress made in sequencing technology, and the associated massive reduction in cost, has enabled the secondary metabolism community to adopt the approach as a "routine" methodology and resulted in a "mushrooming" of sequenced strains. At the time of writing 229 streptomycete and 659 fungal sequences had been completed and were publicly available. Thus, it is perfectly feasible to genome sequence the strains in the family tree of a high producing strain in an attempt to correlate increased productivity with particular mutations. However, in reality, the processing of such data would be challenging as each round of mutation is likely to have introduced multiple mutations, many of which will have no impact on secondary metabolism or may damage the vigor of the strain (Baltz, 2016). Nevertheless, there are examples in the literature of this approach and they are likely to be representative of a multitude of confidential "inhouse" studies done in the industry. Peano et al. (2014) used this methodology to rationalize the productivity of a strain of *Amycolatopsis mediterranei* over producing rifamycin B. Rifamycin B is a polyketide antibiotic synthesized from an aromatic starter unit, 3-amino-5-hydroxybenzoic acid (AHBA), extended by two molecules of malonyl CoA and eight molecules of (s)-methylmalonyl-CoA (Figs. 3.55 and 3.56). Three strains derived from a common ancestor were sequenced. The highest producer, HP-130, had been produced from

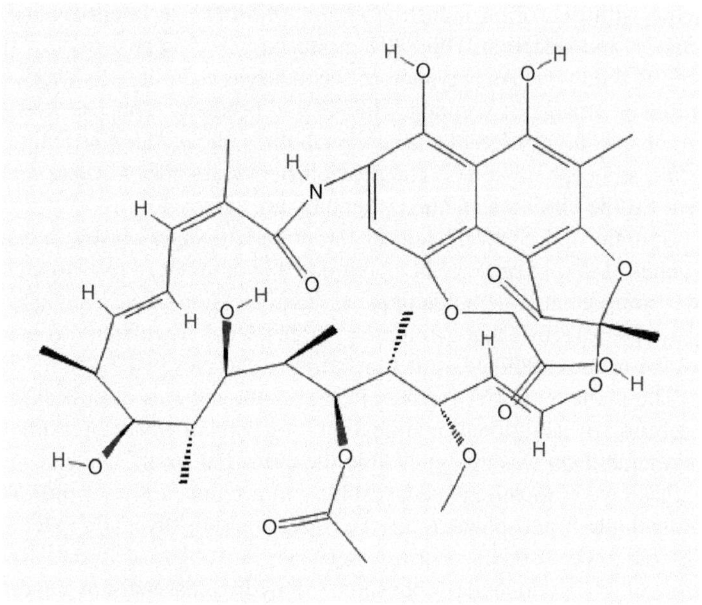

FIGURE 3.55 The Structure of Rifamycin

Data deposited in or computed by PubChem. URL: *https://pubchem.ncbi.nlm.nih.gov.*

its ancestral strain by 12 successive rounds of mutation with UV light and methylnitronitrosoguanidine (MNNG). Its sequence revealed that the strain, compared with its ancestor, contained 250 mutations in the coding regions, indicating that each round of mutagenesis had contributed more than 20 mutations to the genome. Of the 250 lesions, 109 were unique to HP-130—54 missense, 4 frameshift, 3 nonsense, 2 in-frame insertions, and 45 silent mutations that did not lead to amino acid substitutions. Work was then focused on two mutations, in *mut B2* and *argS2*. Gene *mutB2* is one of two *mutB* paralogs catalyzing the isomerization of succinyl-CoA to methylmalonyl-CoA, a key precursor of rifamycin B (see Fig. 3.56). The pool concentration of methylmalonyl-CoA in HP-130 was twice that in a lower producing strain (S699) whereas S699 had higher succinyl-CoA levels than HP-130. Having identified this potential target, the hypothesis was tested by disrupting the *mutB2* gene in S699, thereby increasing rifamycin B to that of HP-130. Investigation of the *argS2* gene revealed that it coded for one of two arginyl-tRNA synthetases and this was commensurate with the reduced arginyl-tRNA synthetase activity in HP130 compared with S699, as well as an increased ppGpp level. It will be recalled that ppGpp is a key signal molecule in the control of the onset of secondary metabolism and its synthesis can be induced by amino acid shortage (the stringent response) detected by the binding of uncharged tRNA molecules to the ribosome. Thus, it was postulated that the phenotype of the mutation was manifested by the reduced arginyl-tRNA synthetase activity resulting in uncharged arginine-tRNA binding at the ribosome,

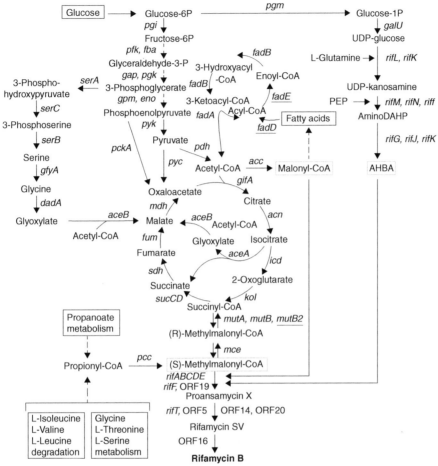

FIGURE 3.56 The Biosynthetic Route to Rifamycin in *Amycolatopsis mediterranei* (Peano et al., 2014)

initiation of the stringent response and the synthesis of ppGpp. Disruption of *argS2* in S699 resulted in elevated levels of both ppGpp and rifamycin but not to the extent seen in HP-130.

Thus, the comparison of the genomes of high and low producers of secondary metabolites can reveal key mutations to facilitate strain construction as was exemplified in Ohnishi's work on the development of a lysine producing strain of *C. glutamicum* (Ohnishi et al., 2002). The next stage in this strategy would be to express mutations into a wild-type background, as was done by Ohnishi, to eliminate the deleterious lesions introduced by mutagenesis. However, the challenge is the identification of mutations that may be relevant to secondary metabolism over production in a system that has evolved to react to a multitude of environmental stimuli.

While the advances in genomics have contributed significantly to the improvement of secondary metabolite producers, probably the greatest impact is the revelation of "silent" secondary metabolism gene clusters. The exploitation of this discovery was discussed in the first section of this chapter on the isolation of natural product producers and a key issue raised was the development of heterologous hosts in which to express these silent clusters. As more and more fungi and actinomycetes are sequenced, more silent clusters will be revealed in organisms already "in captivity." Thus, the availability of reliable heterologous expression hosts and the technology to introduce large DNA sequences is critical to the industry being able to capitalize on this new resource. Baltz (2016) has addressed both the technology and the range of available hosts in an authoritative review in which he distinguished the need for two different host types: discovery hosts, genetically amenable and capable of expressing a wide range of gene clusters without being affected by the products produced, nor interfering with expression by producing their own products; manufacturing hosts, capable of producing high concentrations of product and amenable to the stresses incurred in large-scale manufacture. Currently, work has focused on the development of discovery hosts with the streptomycetes *S. avermitilis*, *S. coelicolor*, *S. albus* J1074, and *S. ambofaciens* being the most promising.

SUMMARY

The tasks of both discovering new microbial compounds and improving the synthesis of known ones have become more and more challenging. Early work on isolation and improvement relied on a "blunderbuss" approach yet, due to the ingenuity and resourcefulness of the individuals involved, resulted in the establishment of a wide range of highly successful processes. The rational improvement programs developed in the amino-acid industry pointed the way to the adoption of such approaches for both secondary metabolite discovery and improvement and the development of miniaturized screening systems allowed the industry to take full advantage of robotic systems to revolutionize screening throughput. However, it is the gigantic developments in molecular biology that have allowed the industry to enter the next stage in its evolution. Recombinant DNA technology and genome sequencing has enabled not only the synthesis of industrial strains but also the rebirth of natural product discovery.

REFERENCES

Abe, S. (1972). Mutants and their isolation. In K. Yamada, S. Kinoshita, T. Tsunoda, & K. Aida (Eds.), *The Microbial Production of Amino Acids* (pp. 39–66). New York: Halstead Press.

Adrio, J. L., & Demain, A. L. (2010). Recombinant organisms for production of industrial products. *Bioengineered Bugs*, *1*(2), 116–131.

Adrio, J. L., & Demain, A. L. (2014). Microbial enzymes: tools for biotechnological processes. *Biomolecules*, *4*(1), 117–139.

Ajinomoto, 1983. European patent application 71023.

Alberts, A. W., Chen, J., Kuron, G., Hunt, V., Huff, J., Hoffman, C., Rothtock, J., Lopez, M., Joshua, H., Harris, E., Patchett, A., Monoghan, R., Currie, S., Stapley, E., Albers-Schonberg, G., Hesens, O., Hirsh-field, J., Hoogsteen, K., Liesch, J., & Springer, J. (1980). Mevinolin, a highly potent inhibitor of HMG-CoA reductase and cholesterol lowering agent. *Proceedings of National Academy of Sciences of the United States of America*, 7, 3957–3961.

Alikhanian, S. I. (1962). Induced mutagenesis in the selection of micro-organisms. *Advances in Applied Microbiology*, 4, 1–5.

Alikhanian, S. I., Mindlin, S. Z., Goldat, S. V., & Vladimizov, A. V. (1959). Genetics of organisms producing tetracyclines. *Annals of New York Academy of Sciences*, 81, 914.

Allen, K. E., McNally, M. T., Lowendorf, H. S., Slayman, C. W., & Free, S. J. (1989). Deoxyglucose resistant mutants of *Neurospora crassa*: isolation, mapping and biochemical characterisation. *Journal of Bacteriology*, 171, 53–58.

Askenazi, M., Driggers, E. M., Holtzman, D. A., Norman, T. C., Iverson, S., Zimmer, D. P., et al. (2003). Integrating transcription and metabolite profiles to direct the engineering of lovastatin-producing fungal strains. *Nature Biotechnology*, 21, 150–156.

Atherton, K. T., Byrom, D., & Dart, E. C. (1979). Genetic manipulation for industrial processes. *Society for General Microbiology Symposium.*, 29, 379–406.

Aunstrup, K., Outtrup, H., Andresen, O., Dambmann, C., 1972. Proteases from alkalophilic *Bacillus* sp. *Fermentation Technology Today: Proceedings of the Fourth International Fermentation Symposium* (pp. 299–305).

Bachmann, B. O., Van Lanen, S. G., & Baltz, R. H. (2014). Microbial genome mining for accelerated natural product discovery: is a renaissance in the making? *Journal of Industrial Microbiology and Biotechnology*, 41, 175–184.

Backman, K., O'Connor, M. J., Maruya, A., Rudd, E., McKay, D., Balakrishnan, R., Radjai, M., DiPasquan-Tonio, V., Shoda, D., Hatch, R., & Venkatasubramanian, K. (1990). Genetic engineering of metabolic pathways applied to the production of phenylalanine. *Annals of the New York Academy of Sciences*, 589, 16–24.

Ball, C. (1978). Genetics in the development of the penicillin process. In R. Hutter, T. Leisinger, J. Nuesch, & W. Wehrlin (Eds.), *Antibiotics and Other Secondary Metabolites, Biosynthesis and Production* (pp. 163–176). London: Academic Press.

Ball, C., & McGonagle, M. P. (1978). Development and evaluation of a potency index screen for detecting mutants of *P. chrysogenum* having increased penicillin yields. *Journal of Applied Bacteriology*, 45, 67–74.

Baltz, R. H. (1986). Mutagenesis in *Streptomyces* spp. In A. L. Demain, & N. A. Solomon (Eds.), *Industrial Microbiology, Biotechnology* (pp. 184–190). Washington: American Microbiological Society.

Baltz, R. H. (2008). Renaissance in antibacterial discovery from actinomycetes. *Current Opinion in Pharmacology*, 8, 557–563.

Baltz, R. H. (2016). Genetic manipulation of secondary metabolite biosynthesis for improved production in *Streptomyces* and other actinomycetes. *Journal of Industrial Microbiology and Biotechnology*, 43(2–3), 343–370.

Banos, S., Perez-Redondo, R., Koekman, B., & Liras, P. (2009). Glycerol utilization gene cluster in *Streptomyces clavuligerus*. *Applied and Environmental Microbiology*, 75(9), 2991–2995.

Barka, E. A., Vatsa, P., Sanchez, L., Gaveau-Vaillant, N., Jacquard, C., Klenk, H. -P., Clement, C., Ouhdouch, Y., & van Wezel, G. (2016). Taxonomy, physiology, and natural products of *Actinobacteria*. *Microbiology and Molecular Biology Reviews*, 80(1), 1–43.

Barreiro, C., Martin, J. F., Garcia-Estrada, C., 2012. Proteomics shows new faces for the old penicillin producer *Penicillium chrysogenum*. *Journal of Biomedicine and Biotechnology,* 2012, 15 pages.

Barrios-Gonzalez, J., Montenegro, E., & Martin, J. F. (1993). Penicillin production by mutants resistant to phenylacetic acid. *Journal of Fermentation and Bioengineering, 76*(6), 455–458.

Becker, J., & Wittmann, C. (2013). Pathways at work: metabolic flux analysis of the industrial cell factory *Corynebacterium glutamicum*. In H. Yakawa, & M. Inui (Eds.), *Corynebacterium Glutamicum Biology and Biotechnology* (pp. 217–238). Heidelberg: Springer.

Becker, J., & Wittmann, C. (2014). GC-MS-Based ^{13}C metabolic flux analysis. In J. O. Kromer, L. K. Nielsen, & L. M. Blank (Eds.), *Metabolic Flux Analysis Methods and Protocols* (pp. 165–174). New York: Humana Press.

Becker, J., Heinzle, E., Klopprogge, C., Zelder, O., & Wittmann, C. (2005). Amplified expression of fructose 1,6-bisphosphate in *Corynebacterium glutamicum* increases in vivo flux through the pentose phosphate pathway and lysine production on different carbon sources. *Applied and Environmental Microbiology, 71*, 8587–8596.

Becker, J., Buschke, N., Bucker, R., & Wittmann, C. (2010). Systems level engineering of *Corynebacterium glutamicum*—reprogramming translational efficiency for superior production. *Engineering and Life Sciences, 10*, 430–438.

Becker, J., Zelder, O., Hafner, S., Schroder, H., & Wittmann, C. (2011). From zero to hero–design-based systems metabolic engineering of *Corynebacterium glutamicum* for L-lysine production. *Metabolic Engineering, 13*, 159–168.

Beckman, R., Cantwell, C., Whiteman, P., Queener, S. W., & Abraham, E. P. (1993). Production of deaceto-cephalosporin C by transformants of *Penicillium chrysogenum*: antibiotic biosynthetic pathway engineering. In R. H. Baltz, G. D. Hegeman, & P. L. Skatrud (Eds.), *Industrial Microorganisms: Basic and Applied Molecular Genetics* (pp. 177–182). Washington: American Society for Microbiology.

Bekker, V., Dodd, A., Brady, D., & Rumbold, K. (2014). Tools for metabolic engineering in *Streptomyces*. *Bioengineered, 5*(5), 293–299.

Bentley, S. d., Chater, K. F., Cerdeno-Tarranga, A. -M., et al. (2002). Complete genome sequence of the model actinomycete *Streptomyces coelicolor* A3(2). *Nature, 417*, 141–147.

Berdy, J. (2012). Thoughts and facts about antibiotics: where we are now and where we are going. *The Journal of Antibiotics, 65*, 385–395.

Bibikova, M. V., Ivanitskaya, L. P., & Singal, S. M. (1981). Direct screening on selective media with gentamycin of organisms which produce aminoglycoside antibiotics. *Antibiotiki, 26*, 488–492.

Birge, E. A. (1988). *Bacterial and Bacteriophage Genetics*. New York: Springer Verlag, 58–88.

Bok, J. W., & Keller, N. (2004). LaeA, a regulator of secondary metabolism in *Aspergillus* spp. *Eukaryotic Cell, 3*(2), 527–535.

Brown, A. G., Butterworth, D., Cole, M., Hanscombe, G., Hood, J. D., & Reading, C. (1976). Naturally occurring β-lactamase inhibitors with antibacterial activity. *The Journal of Antibiotics, 29*, 668–669.

Bu'Lock, J. D. (1980). Resistance of a fungus to its own antifungal metabolites and the effectiveness of resistance selection in screening for higher yielding mutants. *Biotechnology Letters, 3*, 285–290.

Bull, A. T., Ellwood, D. C., & Ratledge, C. (1979). The changing scene in microbial technology. *Society of General Microbiology Symposium, 29*, 1–28.

Bull, A. T., Goodfellow, M., & Slater, J. H. (1992). Biodiversity as a source of innovation in biotechnology. *Annual Review of Microbiology, 46*, 219–252.

Bull, A. T., Ward, A. C., & Goodfellow, M. (2000). Search and discovery strategies for bio-technology: the paradigm shift. *Microbiology and Molecular Biologgy Reviews*, *64*(3), 573–606.

Butchko, R. A., Adams, T. H., & Keller, N. P. (1999). *Aspergillus nidulans* mutants defective in *stc* gene cluster regulation. *Genetics*, *153*, 715–720.

Calam, C. T., Daglish, L. B., & McCann, E. P. (1976). Penicillin: Tactics in strain improve-ment. In K. D. MacDonald (Ed.), *Second International Symposium on the Genetics of Industrial Microorganisms* (pp. 273–287). London: Academic Press.

Carneiro, S., Lourenco, A., Ferreira, E. C., & Rocha, I. (2011). Stringent response of *Esch-erichia coli*: revisiting the bibliome usimg literature mining. *Microbial Informatics and Experimentation*, *1*(14), 1–24.

Catcheside, D. G. (1954). Isolation of nutritional mutants of *Neurospora crassa* by filtration enrichment. *Journal of General Microbiology*, *11*, 34–36.

Challis, G. L., & Hopwood, D. A. (2003). Synergy and contingency as driving forces for the evolution of multiple secondary metabolite production by *Streptomyces* species. *Proceed-ings of the National Academy of Sciences of the United States of America*, *100*(Suppl. 2), 14555–14561.

Chan, P. F., Holmes, D. J., & Payne (2004). Finding the gems using genomic discovery: antibacterial drug discovery strategies—the success and the challenges. *Drug Discovery Today: Therapeutic Strategies*, *1*(4), 519–527.

Chang, L. T., & Elander, R. P. (1982). Rational selection for improved cephalosporin C production in strains of *Acremonium chrysogenum*. *Developments in Industrial Microbiol-ogy*, *20*, 367–380.

Chng, C., Lumm, A. M., Vroom, A., & Kao, C. M. (2008). A key developmental regulator controls the synthesis of the antibiotic erythromycin in *Sacchsorpolyspora erythraea*. *Proceedings of the National Academy of Sciences United States of America*, *105*, 11346–11351.

Chormonava, N. T. (1978). Isolation of *Actinomadura* from soil samples on selective media with kanamycin and rifampicin. *Antibiotiki*, *23*, 22–26.

Cremer, J., Eggeling, L., & Sahm, H. (1991). Control of the lysine biosynthetic sequence in *Corynebacterium glutamicum* as analyzed by overexpression of the individual overex-pressing genes. *Applied and Environmental Microbiology*, *57*(6), 1746–1752.

Curran, B. P. G., & Bugeja, V. C. (2015). The biotechnology and molecular biology of yeast. In R. Rappley, & D. Whitehouse (Eds.), *Molecular Biology, Biotechnology* (6th ed., pp. 40–75). Cambridge: Royal Society of Chemistry.

Dauner, M., & Sauer, U. (2000). GC-MS analysis of amino acids rapidly provides rich infor-mation for isotopomer balancing. *Biotechnology Progress*, *16*(4), 642–649.

Davies, O. L. (1964). Screening for improved mutants in antibiotic research. *Biometrics*, *20*, 576–591.

Davies, J. (2011). How to discover new antibiotics: harvesting the pravome. *Current Opinion in Chemical Biology*, *15*, 5–10.

Davis, B. D. (1949). Nutritionally deficient bacterial mutants isolated by penicillin. *Proceed-ings of the National Academy of Sciences, United States of America*, *35*, 1–10.

de Macedo Lemos, E. G., Alves, L. M. C., & Campanharo, J. C. (2003). Genomics-based design of defined media for the plant pathogen *Xylella fastidiosa*. *FEMS Microbiology Letters*, *219*, 39–45.

Debabov, V. G. (1982). Gene engineering and microbiological industry. In V. Krumphanzl, B. Sikyta, & Z. Vanek (Eds.), *Overproduction of Microbial Products* (pp. 345–352). London: Academic Press.

Demain, A. L. (1957). Inhibition of penicillin formation by lysine. *Archives of Biochemistry and Biophysics*, *67*, 244–245.

Demain, A. L. (1972). Cellular and environmental factors affecting the synthesis and excretion of metabolites. *Journal of Applied Chemistry and Biotechnology*, *22*, 346–362.

Demain, A. L. (1973). Mutation and production of secondary metabolites. *Advances in Applied Microbiology*, *16*, 177–202.

Demain, A. L. (1974). How do antibiotic-producing organisms avoid suicide? *Annals of the New York Academy of Sciences*, *235*, 601–612.

Demain, A. L. (1990). Regulation and exploitation of enzyme biosynthesis. In W. M. Fogarty, & C. T. Kelly (Eds.), *Microbial Enzymes and Biotechnology* (2nd ed., pp. 331–368). Barking, UK: Elsevier.

Demain, A. l. (2014). Importance of microbial natural products and the need to revitalize their discovery. *Journal of Industrial Microbiology*, *41*, 185–201.

Demain, A. L., & Birnbaum, J. (1968). Alteration of permeability for the release of metabolites from the microbial cell. *Current Topics in Microbiology and Immunology*, *46*, 1–25.

Demain, A. L., & Masurekar, P. S. (1974). Lysine inhibition of *in vivo* homocitrate synthesis in *Penicillium chrysogenum*. *Journal of General Microbiology*, *82*, 143–151.

DeWitt, J. P., Jackson, J. V., & Paulus, T. J. (1989). Actinomycetes. In J. O. Neway (Ed.), *Fermentation Process Development of Industrial Organisms* (pp. 1–72). New York: Marcel Dekker.

Ditchburn, P., Giddings, B., & MacDonald, K. D. (1974). Rapid screening for the isolation of mutants in *A. nidulans* with increased penicillin yields. *The Journals of Applied Bacteriology*, *37*, 515–523.

Dolezilova, L., Spizek, J., Vondracek, M., Paleckova, F., & Vanek, Z. (1965). Cycloheximide producing and fungicidin producing mutants of *Streptomyces noursei*. *Journal of General Microbiology*, *39*, 305–310.

Dominguez, H., Nezondet, C., Lindley, N. D., & Cocaign, M. (1993). Modified carbon flux during oxygen limited growth of *Corynebacterium glutamicum* and the consequences for amino acid overproduction. *Biotechnology Letters*, *15*(5), 449–454.

Dulaney, E. L. (1954). Induced mutation and strain selection in some industrially important micro-organisms. *Annals of the New York Academy of Sciences*, *60*, 155–167.

Dulaney, E. L., & Dulaney, D. D. (1967). Mutant populations of *Streptomyces viridifaciens*. *Transactions of the New York Academy of Sciences*, *29*, 782–799.

Dunican, L. K., & Shivnan, E. (1989). High frequency transformation of whole cells of amino acid producing coryne-form bacteria using high voltage electropolation. *Biotechnology*, *7*, 1067–1070.

Dunn-Coleman, N. S., Bloebaum, P., Berka, R. M., Bodie, E., Robinson, N., Armstrong, G., Ward, M., Przetak, M., Carter, G. L., LaCost, R., Wislon, L. J., Kodama, K. H., Baliu, E. F., Bower, B., Lamsa, M., & Heinsohn, H. (1991). Commercial levels of chymosin production by *Aspergillus*. *Biotechnology*, *9*(10), 976–981.

Ehmann, D. E., Jahic, H., Ross, P. L., Gu, R. -F., Hu, J., Kern, G., Walkup, G. K., & Fisher, S. L. (2012). Avibactam is a covalent, reversible, non-β-lactam β-lactamase inhibitor. *Proceedings of the National Academy of Sciences United States of America*, *109*(29), 11663–11668.

Elander, R. P. (1966). Two decades of strain development in antibiotic-producing micro-organisms. *Developments in Industrial Microbiology*, *7*, 61–73.

Elander, R. P. (1967). Enhanced penicillin synthesis in mutant and recombinant strains of *P. chrysogenum*. In H. Stubble (Ed.), *Induced Mutations and their Utilisation* (pp. 403–423). Berlin: Academic Verlag.

Elander, R. P. (1980). New genetic approaches to industrially important fungi. *Biotechnology and Bioengineering, 22*(Suppl. 1), 49–61.

Elander, R. P. (1989). Bioprocess technology in industrial fungi. In J. O. Neway (Ed.), *Fermentation Process Development of Industrial Organisms* (pp. 169–220). New York: Marcel Dekker.

Elander, R. P., & Vournakis, J. N. (1986). Genetic aspects of overproduction of antibiotics and other secondary metabolites. In Z. Vanek, & Z. Hostalek (Eds.), *Overproduction of Microbial Metabolites. Strain Improvement and Process Control Strategies* (pp. 63–80). Boston: Butter-worths.

Elander, R. P., Mabe, J. A., Hamill, R. L., & Gorman, M. (1971). Biosynthesis of pyrrolnitrin by analogue resistant mutants of *Pseudomonas fluorescens. Folia Microbiologica, 16,* 157–165.

Endo, A., Kurooda, M., & Tsujita, Y. (1976). ML-236A, ML-236B and ML236C, new inhibitors of cholesterogenesis produced by *Penicillium citrinum. Journal of Antibiotics (Japan), 29,* 1346–1348.

Fang, Y. (2012). Ligand-receptor interaction platfora and their application for drug discovery. *Expert Opinion on Drug Discovery, 7*(10), 969–988.

Fantini, A. A. (1966). Experimental approaches to strain improvement. *Developments in Industrial Microbiology, 7,* 79–87.

Ferenczy, L., Kevei, F., & Zsolt, J. (1974). Fusion of fungal protoplasts. *Nature (London), 248,* 793–794.

Fierro, F., Barredo, J. L., Diez, B., Gutierrez, S., Fernandez, F. J., & Martin, J. F. (1995). The penicillin gene cluster is amplified in tandem repeats linked by conserved hexanucleotide sequences. *Proceedings of the National Academy of Sciences United States of America, 92,* 6200–6204.

Fleming, I. D., Nisbet, L. J., & Brewer, S. J. (1982). Target directed antimicrobial screens. In J. D. Bu'Lock, L. J. Nisbet, & D. J. Winstanley (Eds.), *Bioactive Microbial Products: Search and Discovery* (pp. 107–130). London: Academic Press.

Fodor, K., & Alfoldi, L. (1976). Fusion of protoplasts of *Bacillus megatherium. Proceedings of the National Academy of Sciences, United States of America, 73,* 2147–2150.

Fudou, R., Jojima, Y., Seto, A., Yamada, K., Kimura, E., Nakamutsa, T., Hiraishi, A., & Yamanaka, S. (2002). *Corynebacterium efficiens* sp. nov., a glutamic-acid producing species from sopil and vegetables. *International Journal of Systematic and Evolutionary Microbiology, 52,* 1127–1131.

Futamata, H., Nagano, Y., Watanabe, K., & Hiraishi, A. (2005). Unique kinetic properties of phenol-degrading *Variovax* strains responsible for efficient trichloroethylene degradation in a chemostat enrichment culture. *Applied and Environmental Microbiology, 71*(2), 904–911.

Ganju, P. L., & Iyengar, M. R. S. (1968). An enrichment technique for isolation of auxotrophic micro-organisms. *Hindustan Antibiotics Bulletin, 11,* 12–21.

Gasc, C., Ribiere, C., Parisot, N., Beugnot, R., Defois, C., et al. (2015). Capturing prokaryotic dark matter. *Research in Microbiology, 166,* 814–830.

Godfrey, O. W. (1973). Isolation of regulatory mutants of the aspartic and pyruvic families and their effects on antibiotic production in *Streptomyces lipmanii. Antimicrobial Agents and Chemotherapy, 4,* 73–79.

Gomez-Escribano, J. P., & Bibb, M. J. (2011). Engineering *Streptomyces coelicolor* for heterologous expression of secondary metabolite gene clusters. *Microbial Biotechnology, 4*(2), 207–215.

Gomez-Escribano, J. P., Martin, J. F., Hesketh, A., Bibb, M. J., & Liras, P. (2008). *Streptomyces clavuligerus relA*-null mutants overproduce clavulanic acid and cephamycin C: negative regulation of secondary metabolism by (p)ppGpp. *Microbiology, 154*, 744–755.

Goodfellow, M. (2010). Selective Isolation of Actinobacteria. In R. H. Davies, A. L. Demain (Eds.), A. T. Bull, J. E. Davies (Section Eds.), *Manual of Industrial Microbiology, Biotechnology,* Section 1 *Isolation, Screening of Secondary Metabolites, Enzymes* (pp. 13–27). Washington: American Society for Microbiology Press.

Goodfellow, M., & O'Donnell, A. G. (1989). Search and discovery of industrially significant actinomycetes. In S. Baumberg, I. S. Hunter, & P. M. Rhodes (Eds.), *Microbial Products New Approaches (Society for General Microbiology Symposia)* (pp. 343–383). (Vol. 44). Cambridge: Cambridge University Press.

Gordee, E. R., & Day, L. E. (1972). Effect of exogenous penicillin on penicillin synthesis. *Antimicrobial Agents and Chemotherapy, 1*, 315–322.

Goulden, S. A., & Chataway, F. W. (1969). End product control of acetohydroxy acid synthase by valine in *P. chrysogenum* Q176 and a high yielding mutant. *Journal of General Microbiology, 59*, 111–118.

Guerola, N., Ingraham, J. L., & Cerda-Olmedo, E. (1971). Introduction of closely linked multiple mutations by nitrosoguanidine. *Nature New Biology, London, 230*, 122–125.

Gutmann, M., Hoischen, C., & Kramer, R. (1992). Carrier-mediated glutamate secretion by *Corynebacterium glutamicum* under biotin limitation. *Biochimica Biophysica Acta, 1112*, 115–123.

Hagino, H., & Nakayama, K. (1975). L-Tryptophan production by analog-resistant mutants derived from a phenylalanine and tyrosine double auxotroph of *Corynebacterium glutamicum*. *Agricultural and Biological Chemistry, 39*, 343–349.

Hamlyn, P. F., & Ball, C. (1979). Recombination studies with *Cephalosporium acremonium*. In O. K. Sebek, & A. I. Laskin (Eds.), *Genetics of Industrial Micro-organisms* (pp. 185–191). Washington: American Society for Microbiology.

Harbron, S. (2015). Protein expression. In R. Rapley, & D. Whitehouse (Eds.), *Molecular Biology and Biotechnology* (pp. 76–108). Cambridge: Royal Society of Chemistry.

Harrison, D. E. F. (1978). Mixed cultures in industrial fermentation processes. *Advances in Applied Microbiology, 24*, 129–162.

Harrison, D. E. F., Topiwala, H. H., Hamer, G. (1972). Yield and productivity in SCP production from methane and methanol. In *Fermentation Technology Today, Proceedings of the fourth international fermentation symposium,* (pp. 491–495).

Harrison, D. E. F., Wilkinson, T. G., Wren, S. J., & Harwood, J. H. (1976). Mixed bacterial cultures as a basis for continuous production of SCP from C_1 compounds. In A. C. R. Dean, D. C. Ellwood, C. G. T. Evans, & J. Melling (Eds.), *Continuous Culture 6, Applications and New Fields* (pp. 122–134). Chichester: Ellis Horwood.

Harvey, A. L., Edrada-Ebel, R., & Quinn, R. J. (2015). The re-emrgence of natural products for drug discovery in the genomics era. *Nature Reviews Drug Discovery, 14*, 111–129.

Hasegawa, S., Uematsu, K., Natsuma, Y., Suda, M., Hiraga, K., Tojima, T., Inui, M., & Yukawa, H. (2012). Improvement of the redox balance increases L-valine production by *Corynebacterium glutamicum* under oxygen deprivation conditions. *Applied and Environmental Microbiology, 78*(3), 865–875.

Hashimoto, S., Mural, H., Ezaki, M., Morikawa, N., Hatanaka, H., et al. (1990). Studies on new dehydrogenase inhibitors. 1. Taxonomy, fermentation, isolation and physico-chemical properties. *Journal of Antibiotics, 43*, 29–35.

Hayashi, M., Ohnishi, J., Mitsuhashi, S., Yonetani, Y., Hashimoto, S., & Ikeda, M. (2006). Transcriptome analysis reveals global expression changes in an industrial L-lysine producer of *Corynebacterium glutamicum*. *Bioscience Biotechnology and Biochemistry*, *70*(2), 546–550.

Helfrich, E. J. N., Eiter, S., & Piel, J. (2014). Recent advances in genome-based polyketide discovery. *Current Opinion in Biotechnology*, *29*, 107–115.

Henao-Restrepo, A. M., Longini, I. M., Egger, M., Dean, N. E., et al. (2015). Efficacy and effectiveness of an rVSV-vectored vaccine expressing Ebola surface glycoprotein: interim results from the Guinea ring vaccination cluster-randomised test. *The Lancet*, *386*(9996), 857–866.

Hiltner, J. K., Hunter, I. S., & Hoskison, P. A. (2015). Tailoring specialized metabolite production in *Streptomyces*. *Advances in Applied Microbiology*, *91*, 237–255.

Hitchman, R. B., Posse, R. D., & King, L. A. (2010). Protein expression in insect cells. In M. Moo-Young, (Editor in Chief), M. Butler, (Vol. Ed.), *Comprehensive Biotechnology, 2nd ed., Vol. 1, Scientific Fundamentals of Biotechnology* (pp 323–340). Amsterdam: Elsevier.

Hoeksema, H., & Smith, C. G. (1961). Novobiocin. *Progress in Industrial Microbiology*, *3*, 93–139.

Hoischen, C., & Kraemer, R. (1989). Evidence for an efflux mechanism involved in the secretion of glutamate by *Corynebacterium glutamicum*. *Archives of Microbiology*, *151*, 342–347.

Holms, W. H. (2001). Flux analysis—a basic tool of microbial physiology. *Advances in Microbial Physiology*, *45*, 271–340.

Holms, W. H. (1997). Metabolic flux analysis. In P. M. Rhodes, & P. F. Stanbury (Eds.), *Applied Microbial Physiology—A practical Approach* (pp. 213–248). Oxford: IRL Press at Oxford University Press.

Hopwood, D. A. (1976). Genetics of antibiotic production in Streptomycetes. In D. Schlessinger (Ed.), *Microbiology—1976* (pp. 558–562). Washington: American Society for Microbiology.

Hopwood, D. A. (1979). The many faces of recombination. In O. K. Sebek, & A. I. Laskin (Eds.), *Genetics of Industrial Micro-organisms* (pp. 1–9). Washington: American Society for Microbiology.

Hopwood, D. A., Wright, H. M., Bibb, M. J., & Cohen, S. N. (1977). Genetic recombination through protoplast fusion in Streptomycetes. *Nature (London)*, *268*, 171–174.

Hopwood, D. A., Malpartida, F. F., Kieser, H. M., Ikeda, H., Duncan, J., Fujii, I., Rudd, B. A. M., Floss, H. G., & Omura, S. (1985). Production of hybrid antibiotics by genetic engineering. *Nature (London)*, *314*, 642–646.

Horinouchi, S., & Beppu, T. K. (2007). Hormonal control by A-factor of morphological development and secondary metabolism in *Streptomyces*. *Proceedings of the Japan Academy, Series B*, *83*, 277–295.

Horiuchi, T., Horiuchi, S., & Novick, A. (1963). The genetic basis of hyper-synthesis of β Galactosidase. *Genetics*, *48*, 157–169.

Huang, H. T. (1961). Production of l-threonine by auxotrophic mutants of *E. coli*. *Applied Microbiology*, *9*, 419–424.

Huang, J., Shi, J., Molle, V., Sohlberg, B., Weaver, D., Bibb, M. J., Karoonuthaisiri, N., Lih, C. -J., Kao, C. M., Buttner, M. J., & Cohen, S. N. (2003). Cross-regulation among dispratae antibiotic biosynthetic pathways of *Streptomyces coelicolor*. *Molecular Microbiology*, *58*(5), 1276–1287.

Huck, T. A., Porter, N., & Bushell, M. E. (1991). Positive selection of antibiotic producing soil isolates. *Journal of General Microbiology, 137*(10), 2321–2329.

Hunter, I. S., & Baumberg, S. (1989). Molecular genetics of antibiotic formation. In S. Baumberg, I. S. Hunter, & P. M. Rhodes (Eds.), *Microbial Products New Approaches (Society for General Microbiology Symposia)* (pp. 121–162). (Vol. 44). Cambridge: Cambridge University Press.

Ichikawa, T., Date, M., Ishikura, T., & Ozaki, A. (1971). Improvement of kasugamycin producing strains by the agar piece method and prototroph method. *Folia Microbiologica, 16*, 218–224.

Ikeda, K. A. (1909). A production method of seasoning consists of salt of L-glutamic acid. Japanese Patent 14804.

Ikeda, M. (2006). Towards bacterial strains overproducing L-tryptophan and other aromatics by metabolic engineering. *Applied Microbiology and Biotechnology, 69*(6), 615–626.

Ikeda, M., & Katsumata, R. (1992). Metabolic engineering to produce tyrosine or phenylalanine in a tryptophan-producing *Corynebacterium glutamicum* strain. *Applied and Environmental Microbiology, 58*(3), 781–785.

Ikeda, M., & Takeno, S. (2013). Amino acid production by *Corynebacterium glutamicum*. In H. Yukawa, & M. Inui (Eds.), *Corynebacterium glutamicum Biology and Biotechnology* (pp. 107–148). Heidelberg: Springer.

Ikeda, H., Ishikawa, J., Hanamoto, A., et al. (2003). Complete genome sequence and comparative analysis of the industrial microorganism *Strepromyces avermitilis. Nature Biotechnology, 21*, 526–531.

Ikeda, M., Mitsuhashi, S., Tanaka, K., & Hayahsi, M. (2009). Reengineering of a *Corynebacterium glutamicum* L-arginine and L-citrulline producer. *Applied and Environmental Microbiology, 75*(6), 1635–1641.

Ioannidis, J. P. (2014). More than a billion people taking statins?: Potential implications of the new cardiovascular guidelines. *Journal of the American Medical Association, 311*, 463–464.

Jager, W., Schafer, A., Puhler, A., Labes, B., & Wohlleben, W. (1992). Expression of the *Bacillus subtilis sacB* gene leads to sucrose sensitivity in the Gram-positive bacterium *Corynebacterium glutamicum* but not in *Streptomyces lividans. Journal of Bacteriology, 174*(16), 5462–5465.

Johnson, M. J. (1972). Techniques for selection and evaluation of culture for biomass production. *Fermentation Technology Today, Proceedings of the Fourth International Fermentation Symposium* (pp. 473–478).

Jones, A., & Vining, L. C. (1976). Biosynthesis of chloramphenicol in Streptomycetes: identification of p-amino-l-phenylalanine as a product from the action of arylamine synthetase on chorismic acid. *Canadian Journal of Microbiology, 22*, 237–244.

Kaeberlein, T., Lewis, K., & Epstein, S. S. (2002). Isolating "uncultivable" microorganisms in pure culture in a simulated natural environment. *Science, 296*, 1127–1128.

Kafer, E. (1965). Origins of translocations in *Aspergillus nidulans. Genetics, 52*, 217–232.

Kalinowski, J., Bathe, B., Bartels, D., Bischoff, N., Bott, M., Burkovski, A., et al. (2003). The complete *Corynebacterium glutamicum* ATCC 13032 gene sequence and its impact on the production of L-aspartate-derived amino acids and vitamins. *Journal of Bacteriology, 104*, 5–25.

Karasawa, M., Tosaka, O., Ikeda, S., & Yoshii, H. (1986). Application of protoplast fusion to the development of l-threonine and l-lysine producers. *Agricultural and Biological Chemistry, 50*(2), 339–346.

Kase, H., & Nakayama, K. (1972). Production of l-threonine by analogue resistant mutants. *Agricultural and Biological Chemistry, 36*, 1611–1621.

Kase, H., Tanaka, H., & Nakayama, K. (1971). Studies on L-threonine fermentation: production of L-threonine by auxotrophic mutants. *Agricultural and Biological Chemistry, 35*, 2089–2096.

Katsumata, R., & Ikeda, M. (1993). Hyperproduction of tryptophan in *Corynebacterium glutamicum* by pathway engineering. *Nature Biotechnology, 11*, 921–925.

Keller, J. (2015). Why the new Ebola vaccine is a minor miracle. *Pacific Standard* August 10, 2015.

Khokhlov, A. S., Tovarova, I. I., Borisova, L. N., Pliner, S. A., Schevenko, Kornitskaya, E. Y., Ivkina, N. S., & Rapoport, I. A. (1967). A-factor responsible for the biosynthesis of streptomycin by a mutantstrain of *Actinomyces streptomycini*. *Doklady Akademii Nauk SSSR, 177*, 283–299.

Kieser, T., Bibb, M. J., Butner, M. J., Chater, K. F., & Hopwood, D. A. (2000). *Practical Streptomycete Genetics*. Norwich: John Innes Foundation.

Kind, S., Becker, J., & Wittmann, C. (2013). Increased lysine production by flux coupling of the tricarboxylic acid cycle and the lysine biosynthetic pathway – metabolic engineering of the availability of succinyl-CoA in *Corynebacterium glutamicum*. *Metabolic Engineering, 15*, 184–195.

Kinoshita, S., & Nakayama, K. (1978). Amino acids. In A. H. Rose (Ed.), *Primary Products of Metabolism. Economic Microbiology* (pp. 210–262). (Vol. 2). London: Academic Press.

Kinoshita, S., Udaka, S., & Shimono, M. (1957a). Studies on the amino acid fermentations, part 1. Production of l-gtutamic acid by various micro-organisms. *The Journal of General and Applied Microbiology, 3*, 193–205.

Kinoshita, S., Nakayama, K., & Udaka, S. (1957b). Fermentative production of l-Orn. *The Journal of General and Applied Microbiology, 3*, 276–277.

Kleijn, R. J., Liu, F., van Winden, W. M., van Gulik, W. M., Ras, C., & Heijnen, J. J. (2007). Cytostolic NADPH metabolism in penicillin-G producing and non-producing chemostat cultures of *Penicillium chrysogenum*. *Metabolic Engineering, 9*(1), 112–123.

Komatsubara, S., Kisumi, M., & Chibata, I. (1979). Transductional construction of a threonine-producing strain of *Serratia marcescens*. *Applied and Environmental Microbiology, 38*(6), 1045–1051.

Komatsubara, S., Kisumi, M., & Chibata, I. (1983). Transductional construction of a threonine-hyperproducing strain of *Serratia marcescens*: lack of feedback controls of three aspartokinases and two homoerine dehydrogenases. *Applied and Environmental Microbiology, 45*(5), 1445–1452.

Kubota, K., Onda, T., Kamijo, H., Yoshinaga, F., & Oka-mura, S. (1973). Microbial production of l-arginine by mutants of glutamic acid-producing bacteria. *Jorunal of General Microbiology, 19*, 339–352.

Kurtboke, D. I. (2011). Exploitation of phage battery in the search for bioactive actinomycetes. *Applied Microbiology and Biotechnology, 89*, 931–937.

Kyowa Hakko Kogyo (1983). European patent application 73062.

Laureti, L., Song, L., Huang, S., Corre, C., Lebold, P., Challis, G. L., & Aigle, B. (2011). Indentification of a 51-membered macrolide complex by activation of a silent polyketide synthase in *Streptomyces ambofaciens*. *Proceedings of the National Academy of Sciences of the United States of America, 108*(15), 6258–6263.

Lederberg, J., & Lederberg, E. M. (1952). Replica plating and indirect selection of bacterial mutants. *Journal of Bacteriology, 63*, 399–406.

Lein, J. (1986). The Panlabs strain improvement program. In Z. Vanek, & Z. Hostalek (Eds.), *Overproduction of Microbial Metabolites. Strain Improvement and Process Control Strategies* (pp. 105–140). Boston: Butterworths.

Lewis, K. (2013). Platforms for antibiotic discovery. *Nature Reviews Drug Discovery, 12,* 371–387.

Lewis, K. (2015). Challenges of antibiotic discovery. *Microbe, 10*(9), 363–369.

Li, R., & Townsend, C. A. (2006). Rational strain improvement for enhanced clavulanic acid production by genetic engineering of the glycolytic pathway in *Streptomyces clavuligerus*. *Metabolic Engineering, 8,* 240–252.

Ling, L. L., Schneider, T., Peoples, A. J., Spoering, A. L., et al. (2015). A new antibiotic kills pathogens without detectable resistance. *Nature, 517,* 455–459.

Lipinski, C. A., Lombardo, F., Dominy, B. W., & Feeney, P. J. (2001). Experimental and computational approaches to estimate solubility and permeability in drug discovery and development settings. *Advances in Drug Delivery Reviews, 64,* 4–17.

Liu, C. M., McDaniel, L. E., & Schaffner, C. P. (1972). Candicidin biogenesis. *Journal of Antibiotics, 25,* 116–121.

Liu, G., Chater, K. F., Chandra, G., Niu, G., & Tan, H. (2013). Molecular regulation of antibiotic biosynthesis in *Streptomyces*. *Microbiology and Molecular Biology Reviews, 77*(1), 112–143.

Lucas, X., Senger, C., Erxleben, A., Gruning, B. A., Dorimg, K., Mosch, J., et al. (2013). StreptomeDB: a resource for natural compounds isolated from *Streptomyces* species. *Nucleic Acid Research, 41,* D1130–D1136.

Luengo, J. M., Revilla, G., Villanueva, J. R., & Martin, J. F. (1979). Lysine regulation of penicillin biosynthesis in low-producing and industrial strains of *Penicillium chrysogenum*. *Journal of General Microbiology, 115,* 207–211.

Lum, A. M., Huang, J., Hutchinson, C. R., & Kao, C. M. (2004). Reverse engineering of industrial pharmaceutical-producing actinomycete strains using DNA microarrays. *Metabolic Engineering., 6,* 186–196.

Luo, J. -M., Li, J. -S., Liu, D., Liu, F., Wang, Y. -T., Song, X. -R., & Wang, M. (2012). Genome shuffling of *Streptomyces gilvosporeus* for improving natamycin production. *Journal of Agricultural and Food Chemistry, 60,* 6026–6036.

MacDonald, K. D. (1968). The persistence of parental genome segregation in *P. chrysogenum* after nitrogen mustard treatment. *Mutation Research, 5,* 302–305.

MacDonald, K. D., & Holt, G. (1976). Genetics of biosynthesis and overproduction of penicillin. *Science Progress, 63,* 547–573.

MacDonald, K. D., Hutchinson, J. M., & Gillett, W. A. (1963). Formation and segregation of heterozygous diploids between a wild strain and derivative of high yield in *Penicillium chrysogenum*. *Journal of General Microbiology, 33,* 385–394.

Magnusson, L. U., Farewell, A., & Nystrom, T. (2005). ppGpp: a global regulator in *Escherichia coli*. *Trends in Microbiology, 13*(5), 236–242.

Malmberg, L. -H., Hu, W. -S., & Sherman, D. H. (1993). Precursor flux control through targeted chromosomal insertion of the lysine ε-aminotreansferase (*lat*) gene in cephamycin C biosynthesis. *Journal of Bacteriology, 175,* 6916–6924.

Manivasagan, P., Venkatesan, J., Sivakumar, K., & Kim, S. -K. (2014). Actinobacterial enzyme inhibitors—A review. *Critical Reviews in Microbiology, 41*(2), 261–272.

Martin, J. F. (1978). Manipulation of gene expression in the development of antibiotic production. In R. Hutter, T. Leisinger, J. Nuesch, & W. Wehrlin (Eds.), *Antibiotics and Other Secondary Metabolites. Biosynthesis and Production*. London: Academic Press.

Martin, J. -F., & Liras, P. (2010). Engineering of regulatory cascades and networks controlling antibiotic biosynthesis in *Streptomyces. Current Opinion in Microbiology, 13*, 263–273.

McAlpine, J. B., Bachmann, B. O., Piraee, M., Tremblay, S., Alarco, A. -M., Zazopoulos, E., & Farnet, C. M. (2005). Microbial genomics as a guide to drug discovery and structural elucidation; ECO-02301, a novel antifungal agent, as an example. *Journal of Natural products, 68*, 493–496.

McDaniel, R., Ebert-khosla, S., Hopwood, D. A., & Khosla, C. (1993). Engineered biosynthesis of novel polyketides. *Science, 262*, 1546–1550.

McGuire, J. C., Glotfelty, G., & White, R. J. (1980). *FEMS Microbiol. Lett., 9*, 141–143.

Medema, M. H., Alam, M. T., Heijne, W. H. M., van den Berg, M. A., Muller, U., Trefzer, A., Bovenberg, R. A. L., Breitling, R., & Takano, E. (2011a). Genome-wide gene expression changes in an industrial clavulanic acid overproduction strain of *Streptomyces clavuligerus. Microbial Biotechnology, 4*(2), 300–305.

Medema, M. H., Blin, K., Cimermanclc, P., deJager, V., Zakrzewski, P., Fischbach, M. A., Weber, T., Takano, E., & Breitling, R. (2011b). AntiSMASH: rapid identification, annotation and analysis of secondary metabolite biosynthesis gene clusters in bacterial and fungal genome sequences. *Nucleic Acids Research, 39*, W339–W346.

Melzer, G., Esfandabadi, M. E., Franco-Lara, E., & Wittmann, C. (2009). Flux design: *in silico* design of cell factories based on correlation of pathway fluxes to desired properties. *BMC Systems Biology, 3*, 120.

Milshteyn, A., Schneider, J. S., & Brady, S. F. (2014). Mining the metabolome: identifying novel natural products from microbial communities. *Chemical Biology, 21*(9), 1211–1223.

Miwa, K., Tsl chida, Y., Kurahashi, O., Nakamori, S., Sano, K., & Momose, H. (1983). Construction of L-threonine overproducing strains of *Escherichia coli* K12 using recombinant DNA techniques. *Agricultural and Biological Chemistry, 47*, 2329–2334.

Miyake, K., Kuzuyama, T., Horinouchi, S., & Beppu, T. (1989). Detection and properties of A-factor binding protein from *Streptomyces griseus. Journal of Bacteriology, 171*, 4298–4302.

Miyake, T., Uchitomi, K., Zhang, M. -Y., Kono, I., Nozaki, N., Sammoto, H., & Inagaki, K. (2006). Effects of principal nutrients on lovastatin production by *Monascus pilosus. Bioscience Biotechnology and Biochemistry, 70*(5), 1154–1159.

Morinaga, Y., Takagi, H., Ishida, M., Miwa, K., Sato, T., Nakamoui, S., & Sano, K. (1987). Threonine production by co-existence of cloned genes coding homoserine dehydrogenase and homoserine kinase in *Breuibacterium lactofermentum. Agricultural and Biological Chemistry, 51*, 93–100.

Mousdale, D. M. (1997). The analytical chemistry of microbial cultures. In P. M. Rhodes, & P. F. Stanbury (Eds.), *Applied Microbial Physiology—A Practical Approach* (pp. 165–192). Oxford: IRL Press at Oxford University Press.

Nagarajan, R., Boeck, L. D., Gorman, M., Hamill, R. L., Higgins, C. E., Hoehn, M. M., Stark, W. H., & Whitney, J. G. (1971). β-Lactam antibiotics from *Streptomyces. Journal of the American Chemical Society, 93*, 2308–2310.

Nakagawa, S., Mizoguchi, H., Ando, S., Hayashi, M., Ochiai, A., Yokoi, H., Tateishi, N., Senoh, A., Ikeda, M., & Ozaki, A. (2001). Novel polynucleotides. *European Patent* 1,108,790.

Nakamura, J., Hirano, S., Ito, H., & Wachi, M. (2007). Mutations of the *Corynebacterium glutamicum* NCgl1221 gene, encoding a mechanosensitive channel homolog, induce L-glutamic acid production. *Applied and Environmental Microbiology, 73*, 4491–4498.

Nakayama, K., Kituda, S., & Kinoshita, K. (1961). Induction of nutritional mutants of glutamic acid bacteria and their amino acid accumulation. *The Journal of General and Applied Microbiology, 7*(1), 41–51.

Nichols, D., Cahoon, N., Trakhtenberg, E. M., Pham, L., Mehta, A., Belanger, A., Kanigan, T., Lewis, K., & Epstein, S. S. (2010). Use of Ichip for high-throughput in situ cultivation of "uncultivable" microbial species. *Applied and Environmental Microbiology*, *76*(8), 2445–2450.

Niebisch, A., Kabus, A., Schultz, C., Weil, B., & Bott, M. (2006). Corynebacterial protein kinase G controls 2-oxoglutarate dehydrogenase acivity via the phosphorylation status of the OdhI protein. *Journal of Biological Chemistry*, *281*, 12300–12307.

Nisbet, L. J. (1982). Current strategies in the search for bioactive products. *Journal of Chemical Technology and Biotechnology*, *32*, 251–270.

Nisbet, L. J., & Porter, N. (1989). The impact of pharmacology and molecular biology on the exploitation of microbial products. In S. Baumberg, I. S. Hunter, & P. M. Rhodes (Eds.), *Microbial Products New Approaches (Society for General Microbiology Symposia)* (pp. 309–342). (Vol. 44). Cambridge: Cambridge University Press.

O'Sullivan, C. Y., & Pirt, S. J. (1973). Penicillin production by lysine auxotrophs of *P. chrysogenum*. *Journal of General Microbiology*, *76*, 65–75.

Ochi, K., Okamoto, S., Tozawa, Y., Inaoka, T., Hosaka, T., Xu, J., & Kurosawa, K. (2004). Ribosome engineering and secondary metabolite production. *Advances in Applied Microbiology*, *56*, 155–184.

Ohnishi, J., Mitsuhashi, S., Hayashi, M., Ando, S., Yokoi, H., Ochiai, K., & Ikeda, M. (2002). A novel methodology employing *Corynebacterium glutamicum* genome information to generate a new L-lysine-producing mutant. *Applied Microbiology and Biotechnology*, *58*, 217–223.

Ohnishi, J., Katahira, R., Mitsuhashi, S., Kakita, S., & Ikeda, M. (2005). A novel *gnd* mutation leading to increased L-lysine production in *Corynebacterium glutamicum*. *FEMS Microbiology Letters*, *242*, 265–274.

Overmann, J. (2013). Principles of enrichment, isolation, cultivation, and preservation of prokaryotes. In E. F. DeLong, S. Lory, E. Stackebrandt, & F. Thompson (Eds.), *The Prokaryotes: Prokaryotic Biology and Symbiotic Associations* (4th ed.). Heidelburg: Springer.

Palmer, J. M., Theisen, J. M., Duran, R. M., Grayburn, W. S., Calvo, A. M., & Keller, N. P. (2013). Secondary metabolism and development is mediated by LlmF control of VEA subcellular localization in *Aspergillus nidulans*. *PLOS Genetics*, *9*(1), e1003193.

Paradkar, A. (2013). Clavulanic acid production by *Streptomyces clavuligerus*: biogenesis, regulation and strain improvement. *The Journal of Antibiotics*, *66*, 411–420.

Paradkar, A. S., Mosher, R. H., Anders, C., Griffin, A., Griffin, J., Hughes, C., Greaves, P., Barton, B., & Jensen, S. E. (2001). Applications of gene replacement technology to *Streptomyces clavuligerus* strain development for clavulanic acid production. *Applied and Environmental Microbiology*, *67*(5), 2292–2297.

Patek, M., & Nesvera, J. (2013). Promoters and plasmid vectors of *Corynebacterium glutamicum*. In H. Yukawa, & M. Inui (Eds.), *Corynebacterium Glutamicum: Biology and Biotechnology* (pp. 51–88). Heidelberg: Springer.

Patnaik, R., Louie, S., Gavrilovic, V., Perry, K., Stemmer, W. P. C., Ryan, C. M., & del Cardayre, S. (2002). Genome shuffling of *Lactobacillus* for improved acid tolerance. *Nature Biotechnology*, *20*, 707–712.

Payne, D. J., Gwynn, M. N., Holmes, D. J., & Pompliano, D. L. (2008). Drugs for bad bugs: confronting the challenges of antibacterial discovery. *Nature Reviews Drug Discovery*, *6*, 29–40.

Peano, C., Damiano, F., Forcato, M., Pietrelli, A., Palumbo, C., Corti, G., et al. (2014). Comparative genomics revealed key molecular targets to rapidly convert a reference rifamycin-producing bacterial strain into an overproducer by genetic engineering. *Metabolic Engineering*, *26*, 1–16.

Peberdy, J. F., Eyssen, H., & Anne, J. (1977). Interspecific hybridisation between *Penicillium chrysogenum* and *Penicillium cyaneofulvum* following protoplast fusion. *Molecular and General Genetics, 157*, 281–284.

Perez-Llarena, F. J., Liras, P., Rodriguez-Garcia, A., & Martin, J. F. (1997). A regulatory gene (*ccaR*) required for cephamycin and clavulanic acid production in *Streptomyces clavuligerus*: amplification results in overproduction of both β-lactam compounds. *Journal of Bacteriology, 179*(6), 2053–2059.

Petty, T. M., & Crawford, D. L. (1984). Enhancement of lignin degradation in *Streptomyces* spp. by protoplast fusion. *Applied Environmental Microbiology, 47*, 439–440.

Pickup, K., Nolan, R. D., & Bushell, M. E. (1993). A method for increasing the success rate of duplicating antibiotic activity in agar and liquid culture of *Streptomyces* isolates in new antibiotic screens. *Journal of Fermentation and Bioengineering, 76*(2), 89–93.

Polsinelli, M., Albertini, A., Cassani, G., & Cifferi, O. (1965). Relation of biochemical mutations to actinomycin synthesis in *Streptomyces antibioticus*. *Journal of General Microbiology, 39*, 239–246.

Polya, K., & Nyiri, L. (1966). *Abstracts of Ninth International Congress of Microbiology,* p. 172.

Pontecorvo, G., Roper, J. A., Hemmons, L. M., MacDonald, K. D., & Bufton, A. W. J. (1953). The genetics of *A. nidulans*. *Advances in Genetics, 5*, 141–238.

Preobrazhenskaia, T. P., Lavrova, N. V. M., Ukholina, R. S., & Nechaeva, N. P. (1975). Isolation of new species of *Actinomadura* on selectice media with streptomycin and bruneomycin. *Antibiotiki, 30*, 404–408.

Pruess, D. L., & Johnson, M. J. (1967). Penicillin acyltransferases in *P. chrysogenum*. *Journal of Bacteriology, 94*, 1502–1509.

Quek, L. -E., Wittmann, C., Nielsen, L. K., & Kromer, J. O. (2009). OpenFLUX: efficient modeling software for 13C-based metabolic flux analysis. *Microbial Cell Factories, 8*, 25.

Richards, M. A., Cassen, V., Heavner, B. D., Ajami, N. E., Herrmann, A., Simeonidis, E., & Price, N. D. (2014). Media DB: a database of microbial growth conditions in defined media. *PLOS One, 9*(8), e10358.

Romero, D., Traxler, M. F., Lopez, D., & Kolter, R. (2011). Antibiotics as signal molecules. *Chemical Reviews, 111*(9), 5492–5505.

Rowlands, R. T. (1992). Strain improvement and strain stability. In D. B. Finkelstein, & C. Ball (Eds.), *Biotechnology of Filamentous Fungi. Technology and Products* (pp. 41–64). Stoneham: Butterworth-Heinemann.

Rowley, B. I., & Bull, A. T. (1977). Isolation of a yeast lysing *Arthrobacler* species and the production of lytic enzyme complex in batch and continuous culture. *Biotechnology and Bioengineering, 19*, 879–900.

Samson, R. A., Hadlock, R., & Stolk, A. C. (1977). A taxonomic study of the *Penicillium chrysogenum* series. *Antonie van Leuwenhoek, 43*, 169–175.

Samson, S. M., Dotzlav, J. E., Becker, G. W., van Frank, R. M., Veal, L. E., Yeh, W. -K., Miller, J. R., Queener, S. W., & Ingolia, T. D. (1987). *Biotechnology, 5*, 1207–1214.

Sanchez, S., & Demain, A. L. (2008). Metabolic regulation and overproduction of primary metabolites. *Microbial Biotechnology, 1*(4), 283–319.

Sanchez, J. F., Somoza, A. D., Keller, N. P., & Wang, C. C. C. (2012). Advances in *Aspergillus* secondary metabolite research in the post-genomic era. *Natural Product Reports, 29*(3), 351–371.

Sano, K., & Shiio, I. (1970). Microbial production of l-lysine III—Production by mutants resistant to *S*-(l-aminethyl)-l-cysteine. *The Journal of General and Applied Microbiology, 16*, 373–391.

Schatz, A., Bugie, E., & Waksman, S. A. (1944). Streptomycin, a substance exhibiting antibiotic activity against Gram-positive and Gram-negative bacteria. *Proceedings of the Society for Experimental Biology and Medicine*, *55*, 66–69.

Schrumpf, B., Schwarzer, A., Kalinowski, J., Puhler, A., Eggeling, L., & Sahm, H. (1991). A functionally split pathway for lysine synthesis in *Corynebacterium glutamicum*. *Journal of Bacteriology*, *173*(14), 4510–4516.

Sermonti, G. (1969). *Genetics of Antibiotic Producing Organisms*. London: Wiley-Interscience.

Sharma, R., Christi, Y., & Banerjee, U. C. (2001). Production, purification, characterization and applications of lipases. *Biotechnology Advances*, *19*, 627–662.

Shibai, H., Enei, H., & Hirose, Y. (1978). Purine nucleoside fermentations. *Process Biochemistry*, *13*(11), 6–8.

Shiio, I., & Nakamori, S. (1989). Coryneform bacteria. In J. O. Neway (Ed.), *Fermentation Process Development of Industrial Organisms* (pp. 133–168). New York: Marcel Dekker.

Shiio, I., & Sano, K. (1969). Microbial production of l-lysine II—Production by mutants sensitive to threonine or methionine. *The Journal of General and Applied Microbiology*, *15*, 267–287.

Shiio, I., & Ujigawa-Takeda, K. (1980). Presence and regulation of α-ketoglutarate dehydrogenase complex in a gliutamate producing bacterium *Brevibacterium flavum*. *Agricultural and Biological Chemistry*, *44*, 1897–1904.

Shiio, I., Sugimoto, S.-I., & Kawamura, K. (1984). Production of L-tryptophan by sulphonamide-resistant mutants. *Agricultural and Biological Chemistry*, *48*(8), 2073–2080.

Shimizu, H., & Hirasawa, T. (2007). Production of glutamate and glutamate-related amino acids: molecular mechanism analysis and metabolic engineering. In V. Wendisch (Ed.), *Amino Acid Biosynthesis~ Pathways, Regulation and Metabolic Engineering* (pp. 1–38). Berlin: Springer.

Sipiczki, M., & Ferenczy, L. (1977). Protoplast fusion of *Schizosaccharomyces pombe* auxotrophic mutants of identical mating type. *Molecular and General Genetics*, *157*, 77–83.

Skatrud, P. L. (1992). Genetic engineering of β-lactam antibiotic biosynthetic pathways in filamentous fungi. *Trends in Biotechnology*, *10*, 324–329.

Sola-Landa, A., Moura, R. S., & Martin, J. F. (2003). The two-component PhoR-PhoP system controls both primary metabolism and secondary metabolite biosynthesis in *Streptomyces lividans*. *Proceedings of the National Academy of Sciences of the United States of America*, *100*, 6133–6318.

Somerson, N. and Phillips (1961). *Precede d'obtention d'acide glutamique*. Belgian Patent No. 593,807.

Song, H., Kim, T. Y., Choi, B.-K., Choi, S. J., Nielson, L. K., Chang, H. N., & Lee, S. Y. (2008). Development of chemically defined medium for *Mannheimia succiniciproducens* based on its genome sequence. *Applied Microbiology and Biotechnology*, *79*, 263–272.

Studdard, C. (1979). Transduction of auxotrophic markers in a chloramphenicol-producing strain of *Streptomyces*. *Journal of General Microbiology*, *110*, 479–483.

Sveshnikova, M. A., Chormonova, N. T., Lavrova, N. V., Trekhova, L. P., & Preobrazhenskaya, T. P. (1976). Isolation of soil actinomycetes on selective media with novobiocin. *Antibiotiki*, *21*, 784–787.

Szybalski, W. (1952). Microbial selection. Part 1. Gradient plate technique for study of bacterial resistance. *Science*, *116*, 46–48.

Tiwari, K., & Gupta, R. K. (2013). Diversity and isolation of rare actinomycetes: an overview. *Critical Reviews in Microbiology*, *39*(3), 256–294.

Tomita, K., Hoshino, Y., Sashira, T., Hasegawa, K., Akiyama, M., Tsukiura, H., & Kawaguichi, H. (1980). Taxonomy of the antibiotic Bu 2313-producing organism *Microtetraspora caesia* sp. nov. *The Journal of Antibiotics*, *33*, 1491–1501.

Trenina, G. A., & Trutneva, E. M. (1966). Use of ristomycin in selection of active variants of *Proactinomyces fructiferi* var. *ristomycini*. *Antibiotiki*, *11*, 770–774.

Twyman, R. M., & Whitelaw, B. (2010). Gene expression in recombinant animal cells and transgenic animals. In M. C. Flickinger (Ed.), *Encyclopedia of Industrial Biotechnology* (pp. 213–295). New York: Wiley.

Udagawa, K., Abe, S., & Kinoshita, K. (1962). *Journal of Fermentation Technology (Osaka) (Japanese)*, *40*, 614.

Umezawa, H. (1972). In *Enzyme Inhibitors of Microbial Origin*. Tokyo: University of Tokyo Press.

Vallino, J. J., & Stephanopoulos, G. (1991). In S. K. Sidkar, M. Bier, & P. Todd (Eds.), *Frontiers in Bioprocessing* (pp. 205–219). Boca Raton: CRC Press.

van den Berg, M. A., Albang, R., Albermann, K., Badger, J. H., Daran, J. -M., et al. (2008). Genome sequencing and analysis of the filamentous fungus *Penicillium chrysogenum*. *Nature Biotechnology*, *26*(10), 1161–1168.

van den Burg, B., Vriend, G., Veltman, O. R., Venema, G., & Eijsink, V. G. H. (1998). Engineering an enzyme to resist boiling. *Proceedings of the National Academy of Sciences of the United States of America*, *95*, 2056–2060.

van Wezel, G. P., & McDowall, K. J. (2011). The regulation of the secondary metabolism of *Streptomyces*: new links and experimental advances. *Natural Product Reports*, *28*, 1311–1333.

van Wezel, G. P., Mahr, K., Konig, M., Traag, B. A., Pimentel-Schmitt, E. F., Willimek, A., & Titgemeyer, F. (2005). GlcP constitutes the major glucose uptake system of *Streptomyces coelicolor* A3(2). *Molecular Microbiology*, *55*(2), 624–636.

Vickers, J. C., Williams, S. T., & Ross, G. W. (1984). A taxonomic approach to selective isolation of streptomycetes from soil. In L. Oritz-Oritz, L. F. Bojalil, & V. Yakoleff (Eds.), *Biological, Biochemical and Biomedical Aspects of Actinomycetes* (pp. 553–561). Orlando: Academic Press.

Wagenaar, M. M. (2008). Pre-fractionated microbial samples—the second generation natural products library at Wyeth. *Molecules*, *13*, 1406–1426.

Wakisaka, Y., Kawamura, Y., Yasuda, Y., Koizuma, K., & Nishimoto, Y. (1982). A selection procedure for *Micromonospora*. *The Journal of Antibiotics*, *35*, 36–82.

Wang, J., Soisson, S. M., Young, K., et al. (2006). Platensimycin is a selective FabF inhibitor with potent antibiotic properties. *Nature*, *441*, 358–361.

Wang, G., Hoasaka, T., & Ochi, K. (2008). Dramatic activation of antibiotic production in *Streptomyces coelicolor* by cumulative drug resistance mutations. *Applied and Environmental Microbiology*, *74*(9), 2834–2840.

Wang, Q., Zhang, D., Li, Y., Zhang, F., Wang, C., & Liang, X. (2014). Genome shuffling and ribosome engineering of *Streptomyces actuosus* for high-yield nosiheptide production. *Applied Biochemistry and Biotechnology*, *173*, 1553–1563.

Wang, H., Xue, W., He, Y. -M., Peng, R. -H., & Yao, Q. -H. (2015). Improvement of the ability to produce spinosad in *Saccharopolyspora spinosa* through the acquisition of drug resistance and genome shuffling. *Annals of Microbiology*, *65*(2), 771–777.

Wang, G., Graziani, E., Waters, B., Pan, W., Li, X., McDermott, J. et al. (2000). Novel natural products from soil libraries in a streptomycete host. *Organic Letters*, *2*(16), 2401–2404.

Watanabe, T., Okubo, N., Suzuki, T. K., & Izaki, K. (1992). New polyenic antibiotics active against Gram-positive and Gram-negative bacteria vi. Non-lactonic polyene antibiotic, enacyloxin IIa, inhibits binding of aminoacyl-tRNA to a site of ribosomes. *The Journal of Antibiotics*, *45*(4), 572–574.

Weber, T., Biln, K., Duddela, S., Krug, D., Kim, H. U., Bruccoleru, R., et al. (2015). antiSMASH 3.0-a comprehensive resource for the genome mining of biosynthetic gene clusters. *Nucleic Acids Research, 43*(Web Server issue), W237–W243.

Weiner, R. M., Voll, M. J., & Cook, T. M. (1974). Nalidixic acid for enrichment of auxotrophs in cultures of *Salmonella typhimurium*. *Applied Microbiology, 28*(4), 579–581.

Wesseling, A. C. (1982). Protoplast fusion among the actinomycetes and its indusrial applications. *Developments in Industrial Microbiology, 23*, 31–40.

Whitehouse, D. (2015). Genes and genomics. In R. Rapley, & D. Whitehouse (Eds.), *Molecular Biology and Biotechnology* (pp. 40–75). Cambridge: Royal Society of Chemistry.

Williams, S. T., & Vickers, J. C. (1988). Detection of actinomycetes in natural habitats—problems and perspectives. In Y. Okami, T. Beppu, & H. Ogawara (Eds.), *Biology of Actinomycetes 88* (pp. 265–270). Tokyo: Japan Science Society.

Willoughby, L. G. (1971). Observations on some aquatic actinomycetes of streams and rivers. *Freshwater Biology, 7*, 23–27.

Wilson, M. C., Mori, T., Ruckert, C., Uria, A. R., et al. (2014). An environmental bacterial taxon with a large and distinct metabolic repertoire. *Nature, 506*(7486), 58–62.

Windon, J. D., Worsey, M. J., Pioli, E. M., Pioli, D., Barth, P. T., Atherton, K. T., Dart, E. C., Byrom, D., Powell, K., & Senior, P. J. (1980). *Nature (London), 287*, 396.

Wittmann, C. (2007). Fluxome analysis using GC-MS. *Microbial Cell Factories, 6*(6), 1–17.

Woodruff, H. B. (1966). The physiology of antibiotic production: role of producing organisms. *Society for General Microbiology Symposia, 16*, 22–46.

Wu, C., Kim, H. K., van Wezel, G. P., & Choi, Y. H. (2015a). Metabolomics in the natural product field—a gateway to novel antibiotics. *Drug Discovery Today Technologies, 13*, 12–16.

Wu, C., Zacchetti, B., Ram, A. F. J., van Wezel, G. P., Claessen, D., & Choi, Y. H. (2015b). Expanding the chemical space for natural products by *Aspergillus-Streptomyces* co-cultivation and biotransformation. *Scientific Reports, 5*, Article number: 10868.

Xiang, S. -H., Li, J., Yin, H., Zheng, J. -T., Yang, X., Wang, H. -B., Luo, J. -L., Bai, H., & Yang, K. -Q. (2009). Application of a double-reporter-guided mutant selection method to improve clavulanic acid production in *Streptomyces clavuligerus*. *Metabolic Engineering, 11*, 310–318.

Yague, P., Lopez-Garcia, M. T., Rioseras, B., Sanchez, J., & Manteca, A. (2012). New insights on the development of *Streptomyces* and their relationships with secondary metabolite production. *Current Trends in Microbiology, 8*, 65–73.

Yague, P., Lopez-Garcia, M. T., Rioseras, B., Sanchez, J., & Manteca, A. (2013). Presporulation stages of *Streptomyces* differentiation: state-of-the-art and future perspectives. *FEMS Microbiology Letters, 342*(2), 79–88.

Yanofsky, C., Kelley, R. L., & Horn, V. (1984). Repression is relieved before attenuation in the *trp* operon of *Escherichia coli* as tryptophan starvation becomes increasingly severe. *Journal of Bacteriology, 158*(3), 1018–1024.

Yoneda, Y. (1980). Increased production of extracellular enzymes by the synergistic effect of genes introduced into *Bacillus subtilis* by stepwise transformation. *Applied and Environmental Microbiology, 39*(1), 274–276.

Yoshiokova, H. (1952). A new rapid isolation procedure of soil streptomycetes. *The Journal of Antibiotics, 5*, 559–561.

Young, K., Jayasuriya, H., Ondeyka, J. G., et al. (2006). Discovery of FabH/FabF inhibitors from natural products. *Antimicrobial Agents and Chemotherapy, 50*(2), 519–526.

Yukawa, H., Omumasaba, C. A., Nonaka, H., Kos, P., Okai, N., et al. (2007). Comparative analysis of the *Corynebacterium glutamicum* group and complete gene sequence of strain R. *Microbiology, 153*, 1042–1058.

Zahner, H. (1978). The search for new secondary metabolites. In R. Hutter, T. Leisenger, Nuesch, & W. Wehrli (Eds.), *In: FEMS Symposium on Antibiotics and Other Secondary Metabolites. Biosynthesis and Production* (pp. 1–18). (Vol. 5). London: Academic Press.

Zhang, Y. -X., Perry, K., Vinci, V. A., Powell, K., Stemmer, W. P. C., & del Cardayre, S. B. (2002). Genome shuffling leads to rapid phenotypic improvement in bacteria. *Nature, 415*, 644–646.

Zhang, N., Daubaras, D. L., & Suen, W. -C. (2004). Heterologous protein expression. In Z. An (Ed.), *Handbook of Industrial Mycology* (pp. 667–688). New York: Marcel Dekker.

Zhang, W., Hunter, I. S., & Tham, R. (2012). Microbial and plant cell synthesis of secondary metabolites and strain improvement. In E. M. H. El-Mansi, C. F. A. Bryce, B. Dahhou, S. Sanchez, A. L. Demain, & A. R. Allman (Eds.), *Fermentation Microbiology and Biotechnology* (3rd ed., pp. 101–136). Boca-Raton: CRC Press.

Zhang, C., Zhang, J., Kang, Z., Du, G., & Chen, J. (2015). Rational engineering of multiple module pathways for the production of L-phenylalanine in *Corynebacterium glutamicum*. *Journal of Industrial Microbiology and Biotechnology, 42*, 787–797.

Zhu, H., Sandiford, S. K., & van Wezel, G. P. (2014). Triggers and cues that activate antibiotic production by actinomycetes. *Journal of Industrial Microbiology and Biotechnology, 41*, 371–386.

Zotchev, S. B., Sekura, O. N., & Katz, L. (2012). Genome-based bioprospecting of microbes for new therapeutics. *Current opinion in Biotechnology, 23*, 941–947.

Media for industrial fermentations

INTRODUCTION

Detailed investigation is needed to establish the most suitable medium for an individual fermentation process, but certain basic requirements must be met by any such medium. All microorganisms require water, sources of energy, carbon, nitrogen, mineral elements, and possibly vitamins plus oxygen if aerobic. On a small scale it is relatively simple to devise a medium containing pure compounds, but the resulting medium, although supporting satisfactory growth, may be unsuitable for use in a large scale process.

On a large scale one must normally use sources of nutrients to create a medium which will meet as many as possible of the following criteria:

1. It will produce the maximum yield of product or biomass per gram of substrate used.
2. It will produce the maximum concentration of product or biomass.
3. It will permit the maximum rate of product formation.
4. There will be the minimum yield of undesired products.
5. It will be of a consistent quality and be readily available throughout the year.
6. It will cause minimal problems during media making and sterilization.
7. It will cause minimal problems in other aspects of the production process particularly aeration and agitation, extraction, purification, and waste treatment.

The use of cane molasses, beet molasses, cereal grains, starch, glucose, sucrose, and lactose as carbon sources, and ammonium salts, urea, nitrates, corn steep liquor, soya bean meal, slaughter-house waste, and fermentation residues as nitrogen sources, have tended to meet most of the above criteria for production media because they are cheap substrates. However, other more expensive pure substrates may be chosen if the overall cost of the complete process can be reduced because it is possible to use simpler procedures. Other criteria are used to select suitable sporulation and inoculation media and these are considered in Chapter 6.

It must be remembered that the medium selected will affect the design of fermenter to be used. For example, the decision to use methanol and ammonia in the single cell protein process developed by ICI plc necessitated the design of a novel fermenter design (MacLennan, Gow, & Stringer, 1973; Sharp, 1989). The microbial oxidation of hydrocarbons is a highly aerobic and exothermic process. Thus, the production fermenter had to have a very high oxygen transfer capacity coupled with excellent cooling facilities. ICI plc solved these problems by developing an air lift fermenter (Chapter 7). Equally, if a fermenter is already available this will obviously influence the composition of the medium. Rhodes, Crosse, Ferguson, and Fletcher (1955) observed that the optimum concentrations of available nitrogen for griseoful-vin production showed some variation with the type of fermenter used. Some aspects of this topic are considered in Chapter 7. Solid-state fermentation (Chapter 7) is an alternative to deep aqueous (submerged) culture methods, which can have significant advantages such as higher yields of enzymes and secondary metabolites. Indeed some enzymes and secondary metabolites can only be produced economically by solid-state fermentation though the reasons for this are not fully understood (Barrios-Gonzalez, 2012).

The problem of developing a process from the laboratory to the pilot scale, and subsequently to the industrial scale, must also be considered. A laboratory medium may not be ideal in a large fermenter with a low gas-transfer pattern. A medium with a high viscosity will also need a higher power input for effective agitation. This will become more significant as the scale of the fermentation increases. Besides meeting requirements for growth and product formation, the medium may also influence pH variation, foam formation, the oxidation–reduction potential, and the morphological form of the organism. It may also be necessary to provide precursors or metabolic inhibitors. The medium will also affect product recovery and effluent treatment (Chapters 10 and 11).

Historically, undefined complex natural materials have been used in fermentation processes because they are much cheaper than pure substrates. However, there is often considerable batch variation because of variable concentrations of the component parts and impurities in natural materials which cause unpredictable biomass and/or product yields. For example, yeast extract is commonly used in many fermentations as a complex but inexpensive source of carbon and nitrogen which is rich in various amino acids, peptides, vitamins, growth factors, trace elements, and carbohydrates. However, batch-to-batch variations in composition of yeast extract can significantly affect the productivity of a fermentation (Zhang, Reddy, Buckland, & Greasham, 2003). As a consequence of these variations in composition small yield improvements are difficult to detect. Undefined media often make product recovery and effluent treatment more problematical because not all the components of a complex nutrient source will be consumed by the organism. The residual components may interfere with recovery (Chapter 10) and contribute to the biochemical oxygen demand of the effluent (Chapter 11).

Thus, although manufacturers have been reluctant to use defined media components because they are more expensive, pure substrates give more predictable yields from batch to batch and recovery, purification, and effluent treatment are much simpler and therefore cheaper. Process improvements are also easier to detect when pure substrates are used. Zhang, Sinha, and Meagher (2006) report the use of glycerophosphate in a defined media for the commercially important yeast *Pichia pastoris*. Glycerophosphate as a phosphorous source overcomes problems caused when a basal salts media containing bi- or tri-cationic phosphate (which precipitate at pH's above 5.5) is used.

Collins (1990) has given an excellent example of a process producing recombinant protein from *S. cerevisiae* instead of just biomass. The range of growth conditions which can be used is restricted because of factors affecting the stability of the recombinant protein. The control of pH and foam during growth in a fermenter were identified as two important parameters. Molasses would normally be used as the cheapest carbohydrate to grow yeast biomass in a large scale process. However, this is not acceptable for the recombinant protein production because of the difficulties, and incurred costs caused in subsequent purification which result from using crude undefined media components. Collins and coworkers therefore used a defined medium with glucose, sucrose, or another suitable carbon source of reasonable purity plus minimal salts, trace elements, pure vitamins, and ammonia as the main nitrogen source and for pH control. Other impurities in molasses might have helped to stabilize foams and led to the need to use antifoams. While chemically defined media have traditionally been preferred in laboratory fermentations (eg, the development of a defined media for the production of a glycopeptide antibiotic reported by Gunnarsson, Bruheim, & Nielsen, 2003) the use of chemically defined media has been gaining popularity in industrial fermentations for some time and has become particularly important in the production of biologicals for the reasons outlined above. The development and use of defined media in the industrial scale fermentations have been reviewed by Zhang and Greasham (1999). The effects of complex, defined, and industrial media on heterologous protein production by Saccharomyces cerevisiae are reviewed by Hahn-Hagerdal et al. (2005).

Aspects of microbial media have also been reviewed by Suomalainen and Oura (1971), Martin and Demain (1978), Iwai and Omura (1982), DeTilley, Mou, and Cooney (1983), Kuenzi and Auden (1983), Miller and Churchill (1986), Smith (1986) and Priest and Sharp (1989).

Media for culture of animal cells will be discussed later in this chapter.

TYPICAL MEDIA

Table 4.1 gives the recipes for some typical media for submerged culture fermentations. These examples are used to illustrate the range of media in use, but are not necessarily the best media in current use.

MEDIUM FORMULATION

Medium formulation is an essential stage in the design of successful laboratory experiments, pilot-scale development, and manufacturing processes. The constituents of a medium must satisfy the elemental requirements for cell biomass and metabolite production and there must be an adequate supply of energy for biosynthesis and cell

Table 4.1 Some Examples of Fermentation Media

Itaconic Acid (Nubel & Ratajak, 1962)		Clavulanic Acid (Box, 1980)	
Cane molasses (as sugar)	150 g dm^{-3}	Glycerol	1%
		Soybean flour	1.5%
$ZnSO_4$	1.0 g dm^{-3}	KH_2PO_4	0.1%
$ZnSO_4 \cdot 7H_2O$	3.0 g dm^{-3}	10% Pluronic L81 antifoam in soya bean oil	0.2%(v/v)
$CuSO_4 \cdot 5H_2O$	0.01 g gm^{-3}	Oxytetracycline (Anonymous, 1980)	
Amylase (Underkofler, 1966)		Starch	12% + 4% (Additional feeding)
		Technical amylase	0.1%
Ground soybean meal	1.85%	Yeast (dry weight)	1.5%
Autolyzed Brewers yeast fractions	1.50%	$CaCO_3$	2%
		Ammonium sulfate	1.5%
Distillers dried solubles	0.76%	Lactic acid	0.13%
NZ-amine (enzymatic casein hydrolysate)	0.65%	Lard oil	2%
		Total inorganic salts	0.01%
Lactose	4.75%		
$MgSO_4 \cdot 7H_2O$	0.04%	Gibberellic acid (Calam & Nixon, 1960)	
Hodag KG-1 antifoam	0.05%		
Avermectin (Stapley & Woodruff, 1982)		Glucose monohydrate	20 g dm^{-3}
		$MgSO_4$	1 g dm^{-3}
Cerelose	45 g	$NH_4H_2HPO_4$	2 g dm^{-3}
Peptonized milk	24 g	KH_2PO_4	5 g dm^{-4}
Autolyzed yeast	2.5 g	$FeSO_4 \cdot 7H_2O$	0.01 g dm^{-3}
Polyglycol P-2000	2.5 cm^3	$MnSO_4 \cdot 4H_2O$	0.01 g dm^{-3}
Distilled water	1 dm^3	$ZnSO_4 \cdot 7H_2O$	0.01 g dm^{-3}
pH	7.0	$CuSO_4 \cdot 5H_2O$	0.01 g dm^{-3}
		Corn steep liquor (as dry solids)	7.5 g dm^{-3}
		Glutamic acid (Gore, Reisman, & Gardner)	
		Dextrose	270 g dm^{-3}
Endotoxin from *Bacillus thuringiensis* (Holmberg, Sievanen, & Carlberg, 1980)		$NH_4H_2PO_4$	2 g dm^{-3}
Molasses	0–4%	$(NH_4)_2HPO_4$	2 g dm^{-3}
Soy flour	2–6%	K_2SO_4	2 g dm^{-3}
KH_2PO_4	0.5%	$MgSO_4 \cdot 7H_2O$	0.5 g dm^{-3}
KH_2PO_4	0.5%	$MnSO_4 \cdot H_2O$	0.04 g dm^{-3}

Table 4.1 Some Examples of Fermentation Media (*cont.*)

Itaconic Acid (Nubel & Ratajak, 1962)		Clavulanic Acid (Box, 1980)	
		$FeSO_4 \cdot 7H_2O$	0.02 g dm^{-3}
$MgSO_4 \cdot 7H_2O$	0.005%	Polyglycol 2000	0.3 g dm^{-3}
$MnSO_4 \cdot 4H_2O$	0.003%	Biotin	12 µg dm^{-3}
		Penicillin	11 µg dm^{-3}
$FeSO_4 \cdot 7H_2O$	0.001%	Penicillin (Perlman, 1970)	
$CaCl_2$	0.005%		
$Na(NH_4)_2PO_4 \cdot 4H_2O$	0.15%	Glucose or molasses (by continuous feed)	10% of total
Lysine (Nakayama, 1972a)			
Cane blackstrap molasses	20%	Corn-steep liquor	4–5% of total
Soybean meal hydrosylate (as weight of meal before hydrolysis with 6N H_2SO_4 and neutralized with ammonia water)	1.8%	Phenylacetic acid (by continuous feed)	0.5–0.8% of total
$CaCO_3$ or $MgSO_4$ added to buffer medium		Lard oil (or vegetable oil) antifoam by continuous addition pH 6.5–7.5 by acid or alkali addition	0.5% of total
Antifoam agent			

Note. The choice of constituents in the ten media is not a haphazard one. The rationale for medium design will be detailed in the remainder of the chapter.

maintenance. The first step to consider is an equation based on the stoichiometry for growth and product formation. Thus for an aerobic fermentation:

$$\text{carbon and energy source} + \text{nitrogen source} + O_2 + \text{other requirements} \rightarrow$$
$$\text{biomass} + \text{products} + CO_2 + H_2O + \text{heat}$$

This equation should be expressed in quantitative terms, which is important in the economical design of media if component wastage is to be minimal. Thus, it should be possible to calculate the minimal quantities of nutrients which will be needed to produce a specific amount of biomass. Knowing that a certain amount of biomass is necessary to produce a defined amount of product, it should be possible to calculate substrate concentrations necessary to produce required product yields. There may be medium components which are needed for product formation which are not required for biomass production. Unfortunately, it is not always easy to quantify all the factors precisely.

A knowledge of the elemental composition of a process microorganism is required for the solution of the elemental balance equation. This information may not be available so that data which is given in Table 4.2 will serve as a guide to the absolute minimum quantities of N, S, P, Mg, and K to include in an initial medium recipe. Trace elements (such as Fe, Zn, Cu, Mn, Co, Mo, B, W) may also be needed

Table 4.2 Element Composition of Bacteria, Yeasts, and Fungi (% by Dry Weight)

Element	Bacteria (Luria, 1960; Herbert, 1976; Aiba, Humphrey, & Millis, 1973)	Yeasts (Aiba et al., 1973; Herbert, 1976)	Fungi (Lilly, 1965; Aiba et al., 1973)
Carbon	50–53	45–50	40–63
Hydrogen	7	7	
Nitrogen	12–15	7.5–11	7–10
Phosphorus	2.0–3.0	0.8–2.6	0.4–4.5
Sulfur	0.2–1.0	0.01–0.24	0.1–0.5
Potassium	1.0–4.5	1.0–4.0	0.2–2.5
Sodium	0.5–1.0	0.01–0.1	0.02–0.5
Calcium	0.01–1.1	0.1–0.3	0.1–1.4
Magnesium	0.1–0.5	0.1–0.5	0.1–0.5
Chloride	0.5	—	—
Iron	0.02–0.2	0.01–0.5	0.1–0.2

in much smaller quantities. An analysis of relative concentrations of individual elements in bacterial cells and commonly used cultivation media quoted by Cooney (1981) showed that some nutrients are frequently added in substantial excess of that required, for example, P, K; however, others are often near limiting values, for example, Zn, Cu. The concentration of P is deliberately raised in many media to increase the buffering capacity. These points emphasize the need for considerable attention to be given to medium design.

Some microorganisms cannot synthesize specific nutrients, for example, amino acids, vitamins, or nucleotides. Once a specific growth factor has been identified it can be incorporated into a medium in adequate amounts as a pure compound or as a component of a complex mixture.

The carbon substrate has a dual role in biosynthesis and energy generation. The carbon requirement for biomass production under aerobic conditions may be estimated from the cellular yield coefficient (Y) which is defined as:

$$\frac{\text{Quantity of cell dry matter produced}}{\text{Quantity of carbon substrate utilized}}$$

Some values are given in Table 4.3. Thus for bacteria with a Y for glucose of 0.5, which is 0.5 g cells per gram glucose, the concentration of glucose needed to obtain 30 g dm^{-3} cells will be 30/0.5 = 60 g dm^{-3} glucose. One liter of this medium would also need to contain approximately 3.0 g N, 1.0 g P, 1.0 g K, 0.3 g S, and 0.1 g Mg. More details of Y values for different microorganisms and substrates are given by Atkinson and Mavituna (1991b).

An adequate supply of the carbon source is also essential for a product-forming fermentation process. In a critical study, analyses are made to determine how the

Table 4.3 Cellular Yield Coefficients (Y) of Bacteria on Different Carbon Substrates (Data from Abbott & Clamen, 1973)

Substrate	Cellular Yield Coefficient (g biomass dry wt. g^{-1} substrate)
Methane	0.62
n-Alkanes	1.03
Methanol	0.40
Ethanol	0.68
Acetate	0.34
Malate	0.36
Glucose (molasses)	0.51

observed conversion of the carbon source to the product compares with the theoretical maximum yield. This may be difficult because of limited knowledge of the biosynthetic pathways. Cooney (1979) has calculated theoretical yields for penicillin G biosynthesis on the basis of material and energy balances using a biosynthetic pathway based on reaction stoichiometry. The stoichiometry equation for the overall synthesis is:

$$a_2 C_6H_{12}O_6 + b_2 NH_3 + c_2 O_2 + d_2 H_2SO_4 + e_2 PAA$$
$$\rightarrow n_2 \text{ Pen G} + p_2 CO_2 + q_2 H_2O$$

where a_2, b_2, c_2, d_2, e_2, n_2, p_2, and q_2 are the stoichiometric coefficients and PAA is phenylacetic acid. Solution of this equation yields:

$$10/6 \, C_6H_{12}O_6 + 2NH_3 + 1/2 \, O_2 + H_2SO_4$$
$$+ \, C_8H_8O_2 \rightarrow C_{16}H_{18}O_4N_2S + 2CO_2 + 9H_2O$$

In this instance, it was calculated that the theoretical yield was 1.1 g penicillin G g^{-1} glucose (1837 units mg^{-1}).

Using a simple model for a batch-culture penicillin fermentation it was estimated that 28, 61, and 11% of the glucose consumed was used for cell mass, maintenance, and penicillin respectively. When experimental results of a fed-batch penicillin fermentation were analyzed, 26% of the glucose has been used for growth, 70% for maintenance, and 6% for penicillin. The maximum experimental conversion yield for penicillin was calculated to be 0.053 g per gram glucose (88.5 units mg^{-1}). Thus, the theoretical conversion value is many times higher than the experimental value. Hersbach, Van Der Beek, and Van Duck (1984) concluded that there were six possible biosynthetic pathways for penicillin production and two possible mechanisms for ATP production from NADH and FADH$_2$. They calculated that conversion yields by different pathways varied from 638 to 1544 units of penicillin per mg glucose. At that time the best quoted yields were 200 units penicillin per mg glucose. This gives a production of 13–29% of the maximum theoretical yield.

The other major nutrient which will be required is oxygen which is provided by aerating the culture, and this aspect is considered in detail in Chapter 9. The design of a medium will influence the oxygen demand of a culture in that the more reduced carbon sources will result in a higher oxygen demand. The amount of oxygen required may be determined stoichiometrically, and this aspect is also considered in Chapter 9. Optimization of the media is dealt with later in this chapter.

WATER

Water is the major component of almost all fermentation media, and is needed in many of the ancillary services such as heating, cooling, cleaning, and rinsing. Clean water of consistent composition is therefore required in large quantities from reliable permanent sources. When assessing the suitability of a water supply it is important to consider pH, dissolved salts, and effluent contamination.

The mineral content of the water is very important in brewing, and most critical in the mashing process, and historically influenced the siting of breweries and the types of beer produced. Hard waters containing high $CaSO_4$ concentrations are better for the English Burton bitter beers and Pilsen type lagers, while waters with a high carbonate content are better for the darker beers such as stouts. Nowadays, the water may be treated by deionization or other techniques and salts added, or the pH adjusted, to favor different beers so that breweries are not so dependent on the local water source. Detailed information is given by Hough, Briggs, and Stevens (1971) and Sentfen (1989).

The reuse or efficient use of water is normally of high priority. When ICI plc and John Brown Engineering developed a continuous-culture single cell protein (SCP) process at a production scale of 60,000 tons year^{-1}, it was realized that very high costs would be incurred if fresh purified water was used on a once-through basis, since operating at a cell concentration of 30 g biomass (dw) dm^{-3} would require 2700×10^6 dm^3 of water per annum (Ashley & Rodgers, 1986; Sharp, 1989). Laboratory tests to simulate the process showed that the *Methylophilus methylotrophus* could be grown successfully with 86% continuous recycling of supernatant with additions to make up depleted nutrients. This approach was therefore adopted in the full scale process to reduce capital and operating costs and it was estimated that water used on a once through basis without any recycling would have increased water costs by 50% and effluent treatment costs 10-fold.

Water reusage has also been discussed by Topiwala and Khosrovi (1978), Hamer (1979), and Levi, Shennan, and Ebbon (1979).

ENERGY SOURCES

Energy for growth comes from either the oxidation of medium components or from light. Most industrial microorganisms are chemoorganotrophs, therefore the commonest source of energy will be the carbon source such as carbohydrates, lipids, and proteins. Some microorganisms can also use hydrocarbons or methanol as carbon and energy sources (Bauchop and Elsden, 1960).

CARBON SOURCES
FACTORS INFLUENCING THE CHOICE OF CARBON SOURCE

It is now recognized that the rate at which the carbon source is metabolized can often influence the formation of biomass or production of primary or secondary metabolites. Fast growth due to high concentrations of rapidly metabolized sugars is often associated with low productivity of secondary metabolites. This has been demonstrated for a number of processes (Table 4.4). At one time the problem was overcome by using the less readily metabolized sugars such as lactose (Johnson, 1952), but many processes now use semicontinuous or continuous feed of glucose or sucrose, discussed in Chapter 2, and later in this chapter (Table 4.15). Alternatively, carbon catabolite regulation might be overcome by genetic modification of the producer organism (Chapter 3).

The main product of a fermentation process will often determine the choice of carbon source, particularly if the product results from the direct dissimilation of it. In fermentations such as ethanol or single-cell protein production where raw materials are 60–77% of the production cost, the selling price of the product will be determined largely by the cost of the carbon source (Whitaker, 1973; Moo-Young, 1977). It is

Table 4.4 Carbon Catabolite Regulation of Metabolite Biosynthesis

Metabolite	Microorganism	Interfering Carbon Source	References
Griseofulvin	*Penicillium griseofulvin*	Glucose	Rhodes (1963); Rhodes et al. (1955)
Penicillin	*P. chrysogenum*	Glucose	Pirt and Righelato (1967)
Cephalosporin	*Cephalosporium acremonium*	Glucose	Matsumura, Imanaka, Yoshida, and Taguchi (1978)
Aurantin	*Bacillus aurantinus*	Glycerol	Nishikiori et al. (1978)
α-Amylase	*B. licheniformis*	Glucose	Priest and Sharp (1989)
Bacitracin	*B. licheniformis*	Glucose	Weinberg (1967)
Puromycin	*Streptomyces alboniger*	Glucose	Sankaran and Pogell (1975)
Actinomycin	*S. antibioticus*	Glucose	Marshall, Redfield, Katz, and Weissback (1968)
Cephamycin C	*S. clavuligerus*	Glycerol	Aharonowitz and Demain (1978)
Neomycin	*S. fradiae*	Glucose	Majumdar and Majumdar (1965)
Cycloserine	*S. graphalus*	Glycerol	Svensson, Roy, and Gatenbeck (1983)
Streptomycin	*S. griseus*	Glucose	Inamine et al. (1969)
Kanamycin	*S. kanamyceticus*	Glucose	Basak and Majumdar (1973)
Novobiocin	*S. niveus*	Citrate	Kominek (1972)
Siomycin	*S. sioyaensis*	Glucose	Kimura (1967)

often part of a company development program to test a range of alternative carbon sources to determine the yield of product and its influence on the process and the cost of producing biomass and/or metabolite. This enables a company to use alternative substrates, depending on price and availability in different locations, and remain competitive. Upto ten different carbon sources have been or are being used by Pfizer Ltd for an antibiotic production process depending on the geographical location of the production site and prevailing economics (Stowell, 1987).

The purity of the carbon source may also affect the choice of substrate. For example, metallic ions must be removed from carbohydrate sources used in some citric acid processes (Karrow & Waksman, 1947; Woodward, Snell, & Nicholls, 1949; Smith, Nowakowska-Waszczuk, & Anderson, 1974).

The method of media preparation, particularly sterilization, may affect the suitability of some carbohydrates for individual fermentation processes. Starch suffers from the handicap that when heated in the sterilization process it gelatinizes, giving rise to very viscous liquids, so that only concentrations of upto 2% can be used without modification (Solomons, 1969).

Local laws may also dictate the substrates which may be used to make a number of beverages. In the Isle of Man, the Manx Brewers Act (1874) forbids the use of ingredients other than malt, sugar, and hops in the brewing of beer. There are similar laws applying to beer production in Germany. Scotch malt whisky may be made only from barley malt, water, and yeast. Within France, many wines may be called by a certain name only if the producing vineyard is within a limited geographical locality.

EXAMPLES OF COMMONLY USED CARBON SOURCES

Carbohydrates

It is common practice to use carbohydrates as the carbon source in microbial fermentation processes. The most widely available carbohydrate is starch obtained from maize grain. It is also obtained from other cereals, potatoes, and cassava. Analysis data for these substrates can be obtained from Atkinson and Mavituna (1991a). Maize and other cereals may also be used directly in a partially ground state, for example, maize chips. Starch may also be readily hydrolyzed by dilute acids and enzymes to give a variety of glucose preparations (solids and syrups). Hydrolyzed cassava starch is used as a major carbon source for glutamic acid production in Japan (Minoda, 1986). The use of starch particles and glucose as an inexpensive medium for ethanol production has been reported by Bawa et al. (2010). Syrups produced by acid hydrolysis may also contain toxic products, which may make them unsuitable for particular processes.

Barley grains may be partially germinated and heat treated to give the material known as malt, which contains a variety of sugars besides starch (Table 4.5). Malt is the main substrate for brewing beer and lager in many countries. Malt extracts may also be prepared from malted grain.

Sucrose is obtained from sugar cane and sugar beet. It is commonly used in fermentation media in a very impure form as beet or cane molasses (Table 4.6), which are the residues left after crystallization of sugar solutions in sugar refining. Molasses

Table 4.5 Carbohydrate Composition of Barley Malt (Harris, 1962) (Expressed as % Dry Weight of Total)

Starch	58–60
Sucrose	3–5
Reducing sugars	3–4
Other sugars	2
Hemicellulose	6–8
Cellulose	5

Table 4.6 Analysis of Beet and Cane Molasses (Rhodes & Fletcher, 1966) (Expressed as % of Total w/v)

	Beet	Cane
Sucrose	48.5	33.4
Raffinose	1.0	0
Invert sugar	1.0	21.2

Remainder is nonsugar.

is used in the production of high-volume/low-value products such as ethanol, SCP, organic and amino acids, and some microbial gums. In 1980, 300,000 tons of cane molasses were used for amino acid production in Japan (Minoda, 1986). Molasses or sucrose also may be used for the production of higher value/low-bulk products such as antibiotics, speciality enzymes, vaccines, and fine chemicals (Calik, Pehlivan, Ozcelik, Calik, & Ozdamar, 2004; Cheng, Demirci, & Catchmark, 2011; Papagianni & Papamichael, 2014). The cost of molasses will be very competitive when compared with pure carbohydrates. However, molasses contains many impurities and molasses-based fermentations will often need a more expensive and complicated extraction/purification stage to remove the impurities and effluent treatment will be more expensive because of the unutilized waste materials which are still present in the fermentation broth. Some new processes may require critical evaluation before the final decision is made to use molasses as the main carbon substrate.

The use of lactose and crude lactose (milk whey powder) in media formulations is now extremely limited since the introduction of continuous-feeding processes utilizing glucose, discussed in a later section of this chapter. However, Lukondeh, Ashbolt, and Rogers (2005) have report the use of a lactose based media for biomass production. Xylose may also provide a useful carbon source in fermentation media. Silva, Mussato, Roberto, and Teixeira (2012) report on the conversion of xylose to ethanol in a media supplemented with urea, magnesium sulfate, and yeast extract. It was found that K_La's of between 2.3 and 4.9 h^{-1} resulted in highest ethanol production and virtually complete xylose metabolism.

Corn steep liquor (Table 4.7) is a by-product after starch extraction from maize. Although primarily used as a nitrogen source, it does contain lactic acid, small

Table 4.7 Partial Analysis of Corn-Steep Liquor

Total Solids	51 %w/v
Acidity as lactic acid	15%w/v
Free reducing sugars	5.6%w/v
Free reducing sugars after hydrolysis	6.8%w/v
Total nitrogen	4%w/v
Amino acids as % of nitrogen	
Alanine	25
Arginine	8
Glutamic acid	8
Leucine	6
Proline	5
Isoleucine	3.5
Threonine	3.5
Valine	3.5
Phenylalanine	2.0
Methionine	1.0
Cystine	1.0
Ash	1.25%w/v
Potassium	20%
Phosphorus	1–5%
Sodium	0.3–1%
Magnesium	0.003–0.3%
Iron	0.01–0.3%
Copper	0.01–0.03%
Calcium	
Zinc	0.003–0.08%
Lead	
Silver	0.001–0.003%
Chromium	
B Vitamins	
Aneurine	41–49 μg g^{-1}
Biotin	0.34–0.38 μg g^{-1}
Calcium pantothenate	14.5–21.5 μg g^{-1}
Folic acid	0.26–0.6 μg g^{-1}
Nicotinamide	30–40 μg g^{-1}
Riboflavine	3.9–4.7 μg g^{-1}

Also niacin and pyridoxine
(Belik, Herold, & Doskocil, 1957; Misecka & Zelinka, 1959; Rhodes and Fletcher, 1966)

amounts of reducing sugars, and complex polysaccharides. Kadam and Newman (1997) describe the use of corn steep liquor in a low cost media in ethanol production. Certain other materials of plant origin, usually included as nitrogen sources, such as soyabean meal and Pharmamedia, contain small but significant amounts of carbohydrate.

Oils and fats

Oils were first used as carriers for antifoams in antibiotic processes (Solomons, 1969). Vegetable oils (olive, maize, cotton seed, linseed, soya bean, etc.) may also be used as carbon substrates, particularly for their content of the fatty acids: oleic, linoleic, and linolenic acid, because costs are competitive with those of carbohydrates. Bader, Boekeloo, Graham, and Cagle (1984) discussed factors favoring the use of oils instead of carbohydrates. A typical oil contains approximately 2.4 times the energy of glucose on a per weight basis. Oils also have a volume advantage as it would take $1.24 \, dm^3$ of soya bean oil to add 10 kcal of energy to a fermenter, whereas it would take $5 \, dm^3$ of glucose or sucrose assuming that they are being added as 50% w/w solutions. Ideally, in any fermentation process, the maximum working capacity of a vessel should be used. Oil based fed-batch fermentations permit this procedure to operate more successfully than those using carbohydrate feeds where a larger spare capacity must be catered for to allow for responses to a sudden reduction in the residual nutrient level (Stowell, 1987). Oils also have antifoam properties which may make downstream processing simpler in some cases, but normally they are not used solely for this purpose. However, residual (unutilized) oil can increase broth viscosity and hence decrease oxygen transfer efficiency and can create problems in some downstream processing operations.

Stowell, 1987 reported the results of a Pfizer antibiotic process operated with a range of oils and fats on a laboratory scale. On a purely technical basis glycerol trioleate was the most suitable substrate. In the United Kingdom however, when both technical and economic factors are considered, soyabean oil or rapeseed oil are the preferred substrates Papapanagiotou, Quinn, Molitor, Nienow, and Hewitt (2005) investigated the use of a microemulsion of rapeseed oil for oxytetracycline production by *Streptomyces rimosus*. More biomass was produced, oil utilization was increased threefold and oxytetracycline production was increased. Glycerol trioleate is known to be used in some fermentations where substrate purity is an important consideration. Methyl oleate has been used as the sole carbon substrate in cephalosporin production (Pan, Speth, McKillip, & Nash, 1982). Junker, Mann, Gailliot, Byrne, and Wilson (1998) report the successful use of soybean oil (with ammonium sulfate addition) in secondary metabolite production by *Streptomyces hygroscopicus*.

Hydrocarbons and their derivatives

There has been considerable interest in hydrocarbons. Development work has been done using *n*-alkanes for the production of organic acids, amino acids, vitamins and cofactors, nucleic acids, antibiotics, enzymes, and proteins (Fukui & Tanaka, 1980).

Methane, methanol, and *n*-alkanes have all been used as substrates for biomass production (Hamer, 1979; Levi et al., 1979; Drozd, 1987; Sharp, 1989).

Drozd (1987) discussed the advantages and disadvantages of hydrocarbons and their derivatives as fermentation substrates, particularly with reference to cost, process aspects, and purity. In processes where the feedstock costs are an appreciable fraction of the total manufacturing cost, cheap carbon sources are important. In the 1960s and early 1970s there was an incentive to consider using oil or natural gas derivatives as carbon substrates as costs were low and sugar prices were high. On a weight basis *n*-alkanes have approximately twice the carbon and three times the energy content of the same weight of sugar. Although petroleum-type products are initially impure, they can be refined to obtain very pure products in bulk quantities which would reduce the effluent treatment and downstream processing. At this time the view was also held that hydrocarbons would not be subject to the same fluctuations in cost as agriculturally derived feedstocks because it would be a stable priced commodity and might be used to provide a substrate for the conversion to microbial protein (SCP) for economic animal and/or human consumption. Sharp (1989), gives a very good account of market considerations of changes in price and how this would affect the price of SCP. The SCP would have to be cheaper, or as cheap as, soya meal to be marketed as an animal feed supplement. It is evident that both ICI plc and Shell plc made very careful assessments of likely future prices of soya meal during process evaluation.

SCP processes were developed by BP plc (Toprina from yeast grown on *n*-alkanes), ICI plc (Pruteen from bacteria grown on methanol), Hoechst/UBHE (Probion from bacteria on methanol), and Shell plc (bacteria on methane). Only BP plc and ICI plc eventually developed SCP at a production scale as an animal feed supplement (Sharp, 1989). BP's product was produced by an Italian subsidiary company, but rapidly withdrawn from manufacture because of Italian government opposition and the price of feed stock quadrupling in 1973. At this time the crude oil exporting nations (OPEC) had collectively raised the price of crude oil sold in the world market. In spite of the significant increase in the cost of crude oil and its derivatives, as well as recognizing the importance of competition from soya bean meal, the ICI plc directorate gave approval to build a full scale plant in 1976. Pruteen was marketed in the 1980s but eventually withdrawn because it could not compete with soya bean meal prices as an animal feed supplement.

Drozd (1987) has made a detailed study of hydrocarbon feedstocks and concluded that the cost of hydrocarbons does not make them economically attractive bulk feedstocks for the production of established products or potential new products where feedstock costs are an appreciable fraction of manufacturing costs of low-value bulk products. In SCP production, raw materials account for three quarters of the operating or variable costs and about half of the total costs of manufacture (Sharp, 1989). It was considered that hydrocarbons and their derivatives might have a potential role as feedstocks in the microbial production of higher value products such as intermediates, pharmaceuticals, fine chemicals, and agricultural chemicals (Drozd, 1987).

Other carbon sources

Biodiesel produced from animal fats and vegetable oils generates approximately 10% by weight glycerol as a by-product, which could be used as a carbon and energy source in microbial fermentations to produce valuable products such as 1,3-propanediol, dihydroxyactetone, and ethanol (da Silva, Mack, & Contiero, 2009). Shin, Lee, Jung, and Kim (2010) report the use of glycerol to gain a 12-fold increase (via upregulation of transcription and transporter genes) in cephalosporin C production by *Acremonium chrysogenum* M35.

Many agricultural waste products, by-products, and residues may also be utilized as fermentation substrates (Chapter 11). For example, Somrutai, Takagi, and Yoshida (1996) report the potential utilization of palm oil mill effluent (POME), a rapidly increasing waste from palm oil production in tropical countries, in acetone-butanol fermentation. The use of wheat bran and sunflower oil cake for α-amylase production has been reported by Rajagopalan and Krishnan (2009).

Phillips et al. (2015) have investigated alcohol (ethanol, butanol, and hexanol) production by *Clostridium carboxidivorans* from syngas (a mixture of hydrogen, carbon monoxide, and carbon dioxide in the ratio 20:70:10). The remaining media was a mixture of nitrogenous compounds, phosphate, and trace metals designed to enhance higher alcohol production. Carbon monoxide has been reported as a sole energy source in a phosphate buffered basal medium for the growth of *Eubacterium limosum* KIST612 (Chang, Kim, Kim, & Lovitt, 2007). De Coninck et al. (2000) report the use of skimmed milk and yeast extract as the carbon and nutrient source in enzyme production via the protozoa *Tetrahymena thermophila*.

NITROGEN SOURCES
EXAMPLES OF COMMONLY USED NITROGEN SOURCES

Most industrially used microorganisms can utilize inorganic or organic sources of nitrogen. Inorganic nitrogen may be supplied as ammonia gas, ammonium salts, or nitrates (Hutner, 1972). Ammonia has been used for pH control and as the major nitrogen source in a defined medium for the commercial production of human serum albumin by *Saccharomyces cerivisiae* (Collins, 1990). Ammonium salts such as ammonium sulfate will usually produce acid conditions as the ammonium ion is utilized and the free acid will be liberated. On the other hand nitrates will normally cause an alkaline drift as they are metabolized. Ammonium nitrate will first cause an acid drift as the ammonium ion is utilized, and nitrate assimilation is repressed. When the ammonium ion has been exhausted, there is an alkaline drift as the nitrate is used as an alternative nitrogen source (Morton & MacMillan, 1954). One exception to this pattern is the metabolism of *Gibberella fujikuroi* (Borrow et al., 1961, 1964). In the presence of nitrate the assimilation of ammonia is inhibited at pH 2.8–3.0. Nitrate assimilation continues until the pH has increased enough to allow the ammonia assimilation mechanism to restart.

Organic nitrogen may be supplied as amino acid, protein or urea, or in a complex media as yeast extract. In many instances growth will be faster with a supply of organic nitrogen, and a few microorganisms have an absolute requirement for amino acids. It might be thought that the main industrial need for pure amino acids would be in the deliberate addition to amino acid requiring mutants used in amino acid production. However, amino acids are more commonly added as complex organic nitrogen sources, which are nonhomogeneous, cheaper, and readily available. In lysine production, methionine and threonine are obtained from soybean hydrolysate since it would be too expensive to use the pure amino acids (Nakayama, 1972a).

Other proteinaceous nitrogen compounds serving as sources of amino acids include corn-steep liquor (see also carbon sources), soya meal, peanut meal, cottonseed meal (Pharmamedia, Table 4.8; and Proflo), Distillers' solubles meal, and yeast extract. Analysis of many of these products which include amino acids, vitamins, and minerals are given by Miller and Churchill (1986) and Atkinson and Mavituna (1991a). In storage these products may be affected by moisture, temperature changes, and ageing.

Table 4.8 The Composition of Pharmamedia (Traders Protein, Southern Cotton Oil Company, Division of Archer Dariels Midland Co.)

Component	Quantity
Total solids	99%
Carbohydrate	24.1%
Reducing sugars	1.2%
Nonreducing sugars	1.2%
Protein	57%
Amino nitrogen	4.7%
Components of amino nitrogen	
Lysine	4.5%
Leucine	6.1%
Isoleucine	3.3%
Threonine	3.3%
Valine	4.6%
Phenylalanine	5.9%
Tryptophan	1.0%
Methionine	1.5%
Cystine	1.5%
Aspartic acid	9.7%
Serine	4.6%
Proline	3.9%
Glycine	3.8%
Alanine	3.9%
Tyrosine	3.4%

Table 4.8 The Composition of Pharmamedia (Traders Protein, Southern Cotton Oil Company, Division of Archer Dariels Midland Co.) (*cont.*)

Component	Quantity
Histidine	3.0%
Arginine	12.3%
Mineral components	
Calcium	2,530 ppm
Chloride	685 ppm
Phosphorus	13,100 ppm
Iron	94 ppm
Sulphate	18,000 ppm
Magnesium	7,360 ppm
Potassium	17,200 ppm
Fat	4.5%
Vitamins	
Ascorbic acid	32.0 mg kg^{-1}
Thiamine	4.0 mg kg^{-1}
Riboflavin	4.8 mg kg^{-1}
Niacin	83.3 mg kg^{-1}
Pantothenic acid	12.4 mg kg^{-1}
Choline	3,270 mg kg^{-1}
Pyidoxine	16.4 mg kg^{-1}
Biotin	1.5 mg kg^{-1}
Folic acid	1.6 mg kg^{-1}
Inositol	10,800 mg kg^{-1}

Chemically defined amino acid media devoid of protein are necessary in the production of certain vaccines when they are intended for human use.

FACTORS INFLUENCING THE CHOICE OF NITROGEN SOURCE

Control mechanisms exist by which nitrate reductase, an enzyme involved in the conversion of nitrate to ammonium ion, is repressed in the presence of ammonia (Brown, MacDonald, & Meers, 1974). For this reason ammonia or the ammonium ion is the preferred nitrogen source. In fungi that have been investigated, the ammonium ion represses uptake of amino acids by general and specific amino acid permeases (Whitaker, 1976). In *Aspergillus nidulans,* ammonia also regulates the production of alkaline and neutral proteases (Cohen, 1973). Therefore, in mixtures of nitrogen sources, individual nitrogen components may influence metabolic regulation so that there is preferential assimilation of one component until its concentration has diminished.

It has been shown that antibiotic production by many microorganisms is influenced by the type and concentration of the nitrogen source in the culture medium

(Aharonowitz, 1980). Antibiotic production may be inhibited by a rapidly utilized nitrogen source (NH_4^+, NO_3^-, and certain amino acids). The antibiotic production only begins to increase in the culture broth after most of the nitrogen source has been consumed.

In shake flask media experiments, salts of weak acids (eg, ammonium succinate) may be used to serve as a nitrogen source and eradicate the source of a strong acid pH change due to chloride or sulfate ions which would be present if ammonium chloride or sulfate were used as the nitrogen source. This procedure makes it possible to use lower concentrations of phosphate to buffer the medium. High phosphate concentrations inhibit production of many secondary metabolites (see section: Minerals).

The use of complex nitrogen sources for antibiotic production has been a common practice. They are thought to help create physiological conditions in the trophophase, which favor antibiotic production in the idiophase (Martin & McDaniel, 1977). For example, in the production of polyene antibiotics, soybean meal has been considered a good nitrogen source because of the balance of nutrients, the low phosphorus content, and slow hydrolysis. It has been suggested that this gradual breakdown prevents the accumulation of ammonium ions and repressive amino acids. These are probably some of the reasons for the selection of ideal nitrogen sources for some secondary metabolites (Table 4.9).

In gibberellin production the nitrogen source has been shown to have an influence on directing the production of different gibberellins and the relative proportions of each type (Jefferys, 1970).

Other predetermined aspects of the process can also influence the choice of nitrogen source. Rhodes (1963) has shown that the optimum concentration of available nitrogen for griseofulvin production showed some variation depending on the form of inoculum and the type of fermenter being used. Obviously these factors must be borne in mind in the interpretation of results in media-development programs.

Table 4.9 Best Nitrogen Sources for Some Secondary Metabolites

Product	Main Nitrogen Source(s)	References
Penicillin	Corn-steep liquor	Moyer and Coghill (1946)
Bacitracin	Peanut granules	Inskeep, Benett, Dudley, and Shepard (1951)
Riboflavin	Pancreatic digest of gelatine	Malzahn, Phillips, and Hanson (1959)
Novobiocin	Distillers' solubles	Hoeksema and Smith (1961)
Rifomycin	Pharmamedia	Sensi and Thiemann (1967)
	Soybean meal, $(NH_4)_2SO_4$	
Gibberellins	Ammonium salt and natural plant nitrogen source	Jefferys (1970)
Butirosin	Dried beef blood or haemoglobin with $(NH_4)_2SO_4$	Claridge, Bush, Defuria, and Price (1974)
Polyenes	Soybean meal	Martin and McDaniel (1977)

Some of the complex nitrogenous material may not be utilized by a microorganism and create problems in downstream processing and effluent treatment. This can be an important factor in the final choice of substrate.

MINERALS

All microorganisms require certain mineral elements for growth and metabolism (Hughes & Poole, 1989, 1991). In many media, magnesium, phosphorus, potassium, sulfur, calcium, and chlorine are essential components, and because of the concentrations required, they must be added as distinct components. Others such as cobalt, copper, iron, manganese, molybdenum, and zinc are also essential but are usually present as impurities in other major ingredients. There is obviously a need for batch analysis of media components to ensure that this assumption can be justified, otherwise there may be deficiencies or excesses in different batches of media. See Tables 4.7 and 4.8 for the analysis of corn steep liquor and Pharmamedia, and Miller and Churchill (1986) for the analysis of other media ingredients of plant and animal origin. When synthetic media are used, the minor elements will have to be added deliberately. The form in which the minerals are usually supplied, and the concentration ranges, are given in Table 4.10. As a consequence of product composition analysis, as outlined earlier in this chapter, it is possible to estimate the amount of a specific mineral for medium design, for example, sulfur in penicillins and cephalosporins, chlorine in chlortetracycline.

The concentration of phosphate in a medium, particularly laboratory media in shake flasks, is often much higher than that of other mineral components. Part of this phosphate is being used as a buffer to minimize pH changes when external control of the pH is not being used.

Table 4.10 The Range of Typical Concentrations of Mineral Components $(g\ dm^{-3})$

Component	Range
[a]KH_2PO_4	1.0–4.0 (part may be as buffer)
$MgSO_4 \cdot 7H_2O$	0.25–3.0
KCl	0.5–12.0
$CaCO_3$	5.0–17.0
$FeSO_4 \cdot 4H_2O$	0.01–0.1
$ZnSO_4 \cdot 8H_2O$	0.1–1.0
$MnSO_4 \cdot H_2O$	0.01–0.1
$CuSO_4 \cdot 5H_2O$	0.003–0.01
$Na_2MoO_4 \cdot 2H_2O$	0.01–0.1

[a]*Complex media derived from plant and animal materials normally contain a considerable concentration of inorganic phosphate.*

In specific processes, the concentration of certain minerals may be very critical. Some secondary metabolic processes have a lower tolerance range to inorganic phosphate than vegetative growth. This phosphate should be sufficiently low as to be assimilated by the end of trophophase. In 1950, Garner et al. (1950) suggested that an important function of calcium salts in fermention media was to precipitate excess inorganic phosphates, and suggested that the calcium indirectly improved the yield of streptomycin. The inorganic phosphate concentration also influences the production of bacitracins, citric acid (surface culture), ergot, monomycin, novobiocin, oxytetracycline, polyenes, ristomycin, ri-famycin Y, streptomycin, vancomycin, and viomycin (Sensi & Thiemann, 1967; Demain, 1968; Liu, McDaniel, & Schaffner, 1970; Mertz & Doolin, 1973; Weinberg, 1974). However, pyrrolnitrin (Arima, Imanara, Kusaka, Fukada, & Tamura, 1965), bicyclomycin (Miyoshi et al., 1972), thiopeptin (Miyairi et al., 1970), and methylenomycin (Hobbs et al., 1992) are produced in a medium containing a high concentration of phosphate. Two monomycin antibiotics are selectively produced by *Streptomyces jamaicensis* when the phosphate is 0.1 mM or 0.4 mM (Hall & Hassall, 1970). Phosphate regulation has also been discussed by Weinberg (1974), Aharonowitz and Demain (1977), Martin and Demain (1980), Iwai and Omura (1982) and Demain and Piret (1991).

In a review of antibiotic biosynthesis, Liras, Asturias, and Martin (1990) recognized target enzymes which were (1) repressed by phosphate, (2) inhibited by phosphate, or (3) repression of an enzyme occurs but phosphate repression is not clearly proved. A phosphate control sequence has also been isolated and characterized from the phosphate regulated promoter that controls the biosynthesis of candicidin.

Weinberg (1970) has reviewed the nine trace elements of biological interest (Atomic numbers 23–30, 42). Of these nine, the concentrations of manganese, iron, and zinc are the most critical in secondary metabolism. In every secondary metabolic system in which sufficient data has been reported, the yield of the product varies linearly with the logarithmic concentration of the "key" metal. The linear relationship does not apply at concentrations of the metal which are either insufficient, or toxic, to cell growth. Some of the primary and secondary microbial products whose yields are affected by concentrations of trace metals greater than those required for maximum growth are given in Table 4.11.

Chlorine does not appear to play a nutritional role in the metabolism of fungi (Foster, 1949). It is, however, required by some of the halophilic bacteria (Larsen, 1962). Obviously, in those fermentations where a chlorine-containing metabolite is to be produced, the synthesis will have to be directed to ensure that the nonchloro-derivative is not formed. The most important compounds are chlortetracycline and griseofulvin. In griseofulvin production, adequate available chloride is provided by the inclusion of at least 0.1% KCl (Rhodes et al., 1955), as well as the chloride provided by the complex organic materials included as nitrogen sources. Other chlorine containing metabolites are caldriomycin, nornidulin, and mollisin.

Table 4.11 Trace Elements Influencing Primary and Secondary Metabolism

Product	Trace Element(s)	References
Bacitracin	Mn	Weinberg and Tonnis (1966)
Protease	Mn	Mizusawa, Ichishawa, and Yoshida (1966)
Gentamicin	Co	Tilley, Testa, and Dorman(1975)
Riboflavin	Fe, Co	Hickey (1945)
	Fe	Tanner, Vojnovich, and Van Lanen (1945)
Mitomycin	Fe	Weinberg (1970)
Monensin	Fe	Weinberg (1970)
Actinomycin	Fe, Zn	Katz, Pienta, and Sivak (1958)
Candicidin	Fe, Zn	Weinberg (1970)
Chloramphenicol	Fe, Zn	Gallicchio and Gottlieb (1958)
Neomycin	Fe, Zn	Majumdar and Majumdar (1965)
Patulin	Fe, Zn	Brack (1947)
Streptomycin	Fe, Zn	Weinberg (1970)
Citric acid	Fe, Zn, Cu	Shu and Johnson (1948)
Penicillin	Fe, Zn, Cu	Foster, Woodruff, and McDaniel (1943)
		Koffler, Knight, and Frazier (1947)
Griseofulvin	Zn	Grove (1967)

CHELATORS

Many media cannot be prepared or autoclaved without the formation of a visible precipitate of insoluble metal phosphates. Gaunt, Trinci, and Lynch (1984) demonstrated that when the medium of Mandels and Weber (1969) was autoclaved, a white precipitate of metal ions formed, containing all the iron and most of the calcium, manganese, and zinc present in the medium.

The problem of insoluble metal phosphate(s) may be eliminated by incorporating low concentrations of chelating agents such as ethylene diamine tetraacetic acid (EDTA), citric acid, polyphosphates, etc., into the medium. These chelating agents preferentially form complexes with the metal ions in a medium. The metal ions then may be gradually utilized by the microorganism (Hughes & Poole, 1991). Gaunt et al. (1984) were able to show that the precipitate was eliminated from Mandels and Weber's medium by the addition of EDTA at 25 mg dm^{-3}. It is important to check that a chelating agent does not cause inhibition of growth of the microorganism which is being cultured.

In many media, particularly those commonly used in large scale processes, there may not be a need to add a chelating agent as complex ingredients such as yeast extracts or proteose peptones will complex with metal ions and ensure gradual release of them during growth (Ramamoorthy & Kushner, 1975).

GROWTH FACTORS

Some microorganisms cannot synthesize a full complement of cell components and therefore require preformed compounds called growth factors. The growth factors most commonly required are vitamins, but there may also be a need for specific amino acids, fatty acids, or sterols. Many of the natural carbon and nitrogen sources used in media formulations contain all or some of the required growth factors (Atkinson & Mavituna, 1991a). When there is a vitamin deficiency it can often be eliminated by careful blending of materials (Rhodes & Fletcher, 1966). It is important to remember that if only one vitamin is required it may be occasionally more economical to add the pure vitamin, instead of using a larger bulk of a cheaper multiple vitamin source. Calcium pantothenate has been used in one medium formulation for vinegar production (Beaman, 1967). In processes used for the production of glutamic acid, limited concentrations of biotin must be present in the medium (Chapter 3). Some production strains may also require thiamine (Kinoshita & Tanaka, 1972). Xi et al. (2012) investigated the effects of biotin as a growth factor in succinic acid production by *Actinobacillus succinogenes* showing it's importance to succinate productivity at low biotin concentrations. A second growth factor (5-aminolevulinate) was also examined and similar results to biotin supplementation were obtained.

NUTRIENT RECYCLE

The need for water recycling in ICI plc's continuous-culture SCP process has already been discussed in an earlier section of this chapter. It was shown that *M. methylotrophus* could be grown in a medium containing 86% recycled supernatant plus additional fresh nutrients to make up losses. This approach made it possible to reduce the costs of media components, media preparation, and storage facilities (Ashley & Rodgers, 1986; Sharp, 1989).

BUFFERS

The control of pH may be extremely important if optimal productivity is to be achieved. A compound may be added to the medium to serve specifically as a buffer, or may also be used as a nutrient source. Many media are buffered at about pH 7.0 by the incorporation of calcium carbonate (as chalk). If the pH decreases the carbonate is decomposed. Obviously, phosphates which are part of many media also play an important role in buffering. However, high phosphate concentrations are critical in the production of many secondary metabolites (see section: Minerals earlier in this chapter).

The balanced use of the carbon and nitrogen sources will also form a basis for pH control as buffering capacity can be provided by the proteins, peptides, and amino acids, such as in corn-steep liquor. The pH may also be controlled externally by the addition of ammonia or sodium hydroxide and sulfuric acid (Chapter 8).

THE ADDITION OF PRECURSORS AND METABOLIC REGULATORS TO MEDIA

Some components of a fermentation medium help to regulate the production of the product rather than support the growth of the microorganism. Such additives include precursors, inhibitors, and inducers, all of which may be used to manipulate the progress of the fermentation.

PRECURSORS

Some chemicals, when added to certain fermentations, are directly incorporated into the desired product. Probably the earliest example is that of improving penicillin yields (Moyer & Coghill, 1946, 1947). A range of different side chains can be incorporated into the penicillin molecule. The significance of the different side chains was first appreciated when it was noted that the addition of corn-steep liquor increased the yield of penicillin from 20 to 100 units cm^{-3}. Corn-steep liquor was found to contain phenylethylamine, which was preferentially incorporated into the penicillin molecule to yield benzyl penicillin (Penicillin G). Having established that the activity of penicillin lay in the side chain, and that the limiting factor was the synthesis of the side chain, it became standard practice to add side-chain precursors to the medium, in particular phenylacetic acid. Smith and Bide (1948) showed that addition of phenylacetic acid and its derivatives to the medium were capable of both increasing penicillin production threefold and to directing biosynthesis toward increasing the proportion of benzyl penicillin from 0% to 93% at the expense of other penicillins. Phenylacetic acid is still the most widely used precursor in penicillin production. Some important examples of precursors are given in Table 4.12.

INHIBITORS

When certain inhibitors are added to fermentations, more of a specific product may be produced, or a metabolic intermediate which is normally metabolized is accumulated. One of the earliest examples is the microbial production of glycerol (Eoff, Linder, & Beyer, 1919). Glycerol production depends on modifying the ethanol fermentation by removing acetaldehyde. The addition of sodium bisulfite to the broth leads to the formation of the acetaldehyde bisulfite addition compound (sodium hydroxy ethyl sulfite). Since acetaldehyde is no longer available for reoxidation of $NADH_2$, its place as hydrogen acceptor is taken by dihydroacetone phosphate, produced during glycolysis. The product of this reaction is glycerol-3-phosphate, which is converted to glycerol.

Table 4.12 Precursors Used in Fermentation Processes

Precursor	Product	Microorganism	References
Phenylacetic-acid related compounds	Penicillin G	*Penicillium chrysogenum*	Moyer and Coghill (1947)
Phenoxy acetic acid	Penicillin V	*Penicillium chrysogenum*	Soper, Whitehead, Behrens, Corse, and Jones (1948)
Chloride	Chlortetracycline	*Streptomyces aureofaciens*	Van Dyck and de Somer (1952)
Chloride	Griseofulvin	*Penicillium griseofulvin*	Rhodes et al. (1955)
[a]Propionate	Riboflavin	*Lactobacillus bulgaricus*	Smiley and Stone (1955)
Cyanides	Vitamin B12	*Proprianobacterium, Streptomyces* spp.	Mervyn and Smith (1964)
β-Tononones	Carotenoids	*Phycomyces blakesleeanus*	Reyes, Chichester, and Nakayama (1964)
α-Amino butyric acid	L-Isoleucine	*Bacillus subtilis*	Nakayama (1972b)
D-Threonine	L-Isoleucine	*Serratia marcescens*	
Anthranilic acid	L-Tryptophan	*Hansenula anomala*	
Nucleosides and bases	Nikkomycins	*Streptomyces tendae*	Vecht-Lifshitz and Braun (1989)
Dihydronovobionic acid	Dihydronovo-biocin	*Streptomyces* sp.	Walton, McDaniel, and Woodruff (1962)
p-Hydroxycinnamate	Organomycin A and B	*Streptomyces organonensis*	Eiki, Kishi, Gomi, and Ogawa (1992)
DL-α-Amino butyric acid	Cyclosporin A	*Tolypocladium inflatum*	Kobel and Traber (1982)
L-Threonine	Cyclosporin C		
Tyrosine or *p*-hydroxy-phenylglycine	Dimethylvanco-mycin	*Nocardia orienlalis*	Boeck, Mertz, Wolter, and Higgens (1984)

[a]*Yields are not so high as by other techniques.*

The application of general, and specific, inhibitors are illustrated in Table 4.13. In most cases, the inhibitor is effective in increasing the yield of the desired product and reducing the yield of undesirable related products. A number of studies have been made with potential chlorination inhibitors, for example, bromide, to minimize chlortetracycline production during tetracycline fermentation (Gourevitch, Misiek, & Lein, 1956; Le petit, 1957; Goodman, Matrishin, Young, & McCormick, 1959; Lein, Sawmiller, & Cheney, 1959; Szumski, 1959).

Inhibitors have also been used to affect cell-wall structure and increase permeability for the release of metabolites. The best example is the use of penicllin and surfactants in glutamic acid production (Phillips & Somerson, 1960).

Table 4.13 Specific and General Inhibitors Used in Fermentations

Product	Inhibitor	Main Effect	Microorganism	References
Glycerol	Sodium bisulfite	Acetaldehyde production repressed	*Saccharomyces cenutsiae*	Eoff et al. (1919)
Tetracycline	Bromide	Chlortetracycline formation repressed	*Streptomyces aureofaciens*	Le petit (1957)
Glutamic acid	Penicillin	Cell wall permeability	*Micrococcus glutamicus*	Phillips and Somerson (1960)
Citric acid	Alkali metal/ phosphate, pH below 2.0	Oxalic acid repressed	*Aspergillus niger*	Batti (1967)
Valine	Various inhibitors	Various effects with different inhibitors	*Breuibacterium roseum*	Uemura, Sugisaki, and Takamura (1972)
Rifamycin B	Diethyl barbiturate	Other rifamycins inhibited	*Nocardia mediterranei*	Lancini and White (1973)
7-Chloro-6 de- methyltetracycline	Ethionine	Affects one- carbon transfer reactions	*Streptomyces aureofaciens*	Neidleman, Bienstock, and Bennett (1963)

INDUCERS

The majority of enzymes which are of industrial interest are inducible. Induced enzymes are synthesized only in response to the presence in the environment of an inducer. Inducers are often substrates such as starch or dextrins for amylases, maltose for pullulanase, and pectin for pectinases. Some inducers are very potent, such as isovaleronitrile inducing nitralase (Kobayashi, Nagasawa, & Yamada, 1992). Substrate analogs that are not attacked by the enzyme may also serve as enzyme inducers. Most inducers which are included in microbial enzyme media (Table 4.14) are substrates or substrate analogs, but intermediates and products may sometimes be used as inducers. For example, maltodextrins will induce amylase and fatty acids induce lipase. However, the cost may prohibit their use as inducers in a commercial process. Reviews have been published by Aunstrup, Andresen, Falch, and Nielsen (1979) and Demain (1990).

One unusual application of an inducer is the use of yeast mannan in streptomycin production (Inamine, Lago, & Demain, 1969). During the fermentation varying amounts of streptomycin and mannosidostreptomycin are produced. Since mannosidostreptomycin has only 20% of the biological activity of streptomycin, the former is an undesirable product. The production organism *Streptomyces griseus* can be induced by yeast mannan to produce β-mannosidase which will convert mannosidostreptomycin to streptomycin.

Table 4.14 Some Examples of Industrially Important Enzyme Inducers

Enzyme	Inducer	Microorganism	References
α-Amylase	Starch	*Aspergillus* spp.	Windish and Mhatre (1965)
	Maltose	*Bacillus subtilis*	
Pullulanase	Maltose	*Aerobacler aerogenes*	Wallenfels, Bender, and Rached (1966)
α-Mannosidase	Yeast mannans	*Streptomyces griseus*	Inamine et al. (1969)
Penicillin acylase	Phenylacetic acid	*Escherichia coli*	Carrington (1971)
Proteases	Various proteins	*Bacillus* spp.	Keay (1971)
		Streptococcus spp.	Aunstrup (1974)
		Streptomyces spp.	
		Asperigillus spp.	
		Mucor spp.	
Cellulase	Cellulose	*Trichoderma viride*	Reese (1972)
Pectinases	Pectin (beet pulp, apple pomace, citrus peel)	*Aspergillus* spp.	Fogarty and Ward (1974)
Nitralase	Isovaleronitrile	*Rhodococcus rhodochrous*	Kobayashi et al. (1992)

It is now possible to produce a number of heterologous proteins in yeasts, fungi, and bacteria. These include proteins of viral, human, animal, plant, and microbial origin (Peberdy, 1988; Wayne Davies, 1991). However, heterologous proteins may show some degree of toxicity to the host and have a major influence on the stability of heterologous protein expression. As well as restricting cell growth as biomass, the toxicity will provide selective conditions for segregant cells which no longer synthesize the protein at such a high level (Goodey, Doel, Piggott, Watson, & Carter, 1987). Therefore, optimum growth conditions may be achieved by not synthesizing a heterologous protein continuously and only inducing it after the host culture has grown up in a vessel to produce sufficient biomass (Piper & Kirk, 1991). In cells of *S. cerevisiae* where the *Gal*1 promoter is part of the gene expression system, product formation may be induced by galactose addition to the growth medium which contains glycerol or low nonrepressing levels of glucose as a carbon source.

One commercial system that has been developed is based on the *alc* A promoter in *Aspergillus nidulans* to express human interferon α2 (Wayne Davies, 1991). This can be induced by volatile chemicals, such as ethylmethyl ketone, which are added when biomass has increase to an adequate level and the growth medium contains a nonrepressing carbon source or low nonrepressing levels of glucose.

Methylotrophic yeasts such as *Hansenula polymorpha* and *Pichia pastoris* may be used as alternative systems because of the presence of an alcohol oxidase promoter (Veale & Sudbery, 1991). During growth on methanol, which also acts as an inducer, the promoter is induced to produce about 30% of the cell protein. In the presence of glucose or ethanol it is undetectable. Expression systems have been developed with *P. pastoris* for tumor necrosis factor, hepatitis B surface antigen, and α-galactosidase. Hepatitis B surface antigen and other heterologous proteins can also be expressed by *H. pofymorpha*.

OXYGEN REQUIREMENTS

It is sometimes forgotten that oxygen, although not added to an initial medium as such, is nevertheless a very important component of the medium in many processes, and its availability can be extremely important in controlling the growth rate and metabolite production. This will be discussed in detail in Chapter 9.

The medium may influence the oxygen availability in a number of ways including the following:

1. *Fast metabolism.* The culture may become oxygen limited because sufficient oxygen cannot be made available in the fermenter if certain substrates, such as rapidly metabolized sugars which lead to a high oxygen demand, are available in high concentrations.
2. *Rheology.* The individual components of the medium can influence the viscosity of the final medium and its subsequent behavior with respect to aeration and agitation.
3. *Antifoams.* Many of the antifoams in use will act as surface active agents and reduce the oxygen transfer rate. This topic will be considered in a later section of this chapter.

FAST METABOLISM

Nutritional factors can alter the oxygen demand of the culture. *Penicillium chrysogenum* will utilize glucose more rapidly than lactose or sucrose, and it therefore has a higher specific oxygen uptake rate when glucose is the main carbon source (Johnson, 1946). Therefore, when there is the possibility of oxygen limitation due to fast metabolism, it may be overcome by reducing the initial concentration of key substrates in the medium and adding additional quantities of these substrates as a continuous or semicontinuous feed during the fermentation (Tables 4.1 and 4.15; Chapters 2 and 9). It can also be overcome by changing the composition of the medium, incorporating higher carbohydrates (lactose, starch, etc.) and proteins which are not very rapidly metabolized and do not support such a large specific oxygen uptake rate.

RHEOLOGY

Deindoerfer and West (1960) reported that there can be considerable variation in the viscosity of compounds that may be included in fermentation media. Polymers in solution, particularly starch and other polysaccharides, may contribute to the rheological behavior of the fermentation broth (Tuffile & Pinho, 1970). As the

Table 4.15 Some Processes Using Batch Feed or Continuous Feed or in Which They Have been Tried

Product	Additions	References
Yeast	Molasses, nitrogen sources, P and Mg	Harrison (1971)
		Reed and Peppler (1973)
Glycerol	Sugar, Na_2CO_3	Eoff et al. (1919)
Acetone-butyl alcohol	Additions and withdrawals of wort	Soc Richard and Allente et Cle (1921)
Riboflavin	Carbohydrate	Moss and Klein (1946)
Penicillin	Glucose and NH_3	Hosler and Johnson (1953)
Novobiocin	Various carbon and nitrogen sources	Smith (1956)
Griseofulvin	Carbohydrate	Hockenhull (1959)
Rifamycin	Glucose, fatty acids	Pan, Bonanno, and Wagman (1959)
Gibberellins	Glucose	Borrow et al. (1960)
Vitamin B_{12}	Glucose	Becher et al. (1961)
Tetracyclines	Glucose	Avanzini (1963)
Citric acid	Carbohydrates, NH_3	Shepherd (1963)
Single-cell protein	Methanol	Harrison, Topiala, and Hamer (1972)
Candicidin	Glucose	Martin and McDaniel (1975)
Streptomycin	Glucose, ammonium sulfate	Singh, Bruzelius, and Heding (1976)
Cephalosporin	Fresh medium addition	Trilli, Michelini, Mantovani, and Pirt (1977)

polysaccharide is degraded, the effects on rheological properties will change. Allowances may also have to be made for polysaccharides being produced by the microorganism (Banks, Mantle, & Syczyrbak, 1974; Leduy, Marsan, & Coupal, 1974). This aspect is considered in more detail in Chapter 9.

ANTIFOAMS

In most microbiological processes, foaming is a problem. It may be due to a component in the medium or some factor produced by the microorganism. The most common cause of foaming is due to proteins in the medium, such as corn-steep liquor, Pharmamedia, peanut meal, soybean meal, yeast extract, or meat extract (Schugerl, 1985). These proteins may denature at the air–broth interface and form a skin which does not rupture readily. The foaming can cause removal of cells from the medium which will lead to autolysis and the further release of microbial cell proteins will probably increase the stability of the foam. If uncontrolled, then numerous changes may occur and physical and biological problems may be created. These include reduction in the working volume of the fermenter due to oxygen-exhausted gas bubbles circulating in the system (Lee & Tyman, 1988), changes in bubble size, lower mass, and heat transfer rates, invalid process data due to interference at sensing electrodes and incorrect monitoring and control (Vardar-Sukan, 1992). The biological problems include deposition of cells in upper parts of the fermenter, problems of sterile operation with the air filter exits of the fermenter becoming wet, and there is danger of microbial infection, and the possibility of siphoning leading to loss of product.

Hall, Dickinson, Pritchard, and Evans (1973) have recognized five patterns of foaming in fermentations:

1. Foaming remains at a constant level throughout the fermentation. Initially it is due to the medium and later due to microbial activity.
2. A steady fall in foaming during the early part of the fermentation, after which it remains constant. Initially it is due to the medium but there are no later effects caused by the microorganism.
3. The foaming falls slightly in the early stages of the fermentation then rises. There are very slight effects caused by the medium but the major effects are due to microbial activity.
4. The fermentation has a low initial foaming capacity which rises. These effects are due solely to microbial activity.
5. A more complex foaming pattern during the fermentation which may be a combination of two or more of the previously described patterns.

If excessive foaming is encountered there are three ways of approaching the problem:

1. To try and avoid foam formation by using a defined medium and a modification of some of the physical parameters (pH, temperature, aeration, and agitation). This assumes that the foam is due to a component in the medium and not a metabolite.
2. The foam is unavoidable and antifoam should be used. This is the more standard approach.
3. To use a mechanical foam breaker (Chapter 7).

Antifoams are surface active agents, reducing the surface tension in the foams and destabilizing protein films by: (1) hydrophobic bridges between two surfaces, (2) displacement of the absorbed protein, and (3) rapid spreading on the surface of the film (Van't Riet & Van Sonsbeck, 1992). Other possible mechanisms have been discussed by Ghildyal, Lonsane, and Karanth (1988), Lee and Tyman (1988), and Vardar-Sukan (1992).

An ideal antifoam should have the following properties:

1. Should disperse readily and have fast action on an existing foam.
2. Should be active at low concentrations.
3. Should be long acting in preventing new foam formation.
4. Should not be metabolized by the microorganism.
5. Should be nontoxic to the microorganism.
6. Should be nontoxic to humans and animals.
7. Should not cause any problems in the extraction and purification of the product.
8. Should not cause any handling hazards.

9. Should be cheap.
10. Should have no effect on oxygen transfer.
11. Should be heat sterilizable.

The following compounds which meet most of these requirements have been found to be most suitable in different fermentation processes (Solomons, 1969; Ghildyal et al., 1988):

1. Alcohols; stearyl and octyl decanol.
2. Esters.
3. Fatty acids and derivatives, particularly glycerides, which include cottonseed oil, linseed oil, soy-bean oil, olive oil, castor oil, sunflower oil, rapeseed oil, and cod liver oil.
4. Silicones.
5. Sulfonates.
6. Miscellaneous; alkaterge C, oxazaline, polypropylene glycol.

These antifoams are generally added when foaming occurs during the fermentation. Because many antifoams are of low solubility they need a carrier such as lard oil, liquid paraffin, or castor oil, which may be metabolized and affect the fermentation process (Solomons, 1967).

Unfortunately, the concentrations of many antifoams which are necessary to control fermentations will reduce the oxygen-transfer rate by as much as 50%; therefore antifoam additions must be kept to an absolute minimum. There are also other antifoams which will increase the oxygen-transfer rate (Ghildyal et al., 1988). If the oxygen-transfer rate is severely affected by antifoam addition then mechanical foam breakers may have to be considered as a possible alternative. Vardar-Sukan (1992) concluded that foam control in industry is still an empirical art. The best method for a particular process in one factory is not necessarily the best for the same process on another site. The design and operating parameters of a fermenter may affect the properties and quantity of foam formed.

MEDIUM OPTIMIZATION

At this stage it is important to consider the optimization of a medium such that it meets as many as possible of the seven criteria given in the introduction of this chapter. The meaning of optimization in this context does need careful consideration (Winkler, 1991). When considering the biomass growth phase in isolation, it must be recognized that efficiently grown biomass produced by an "optimized" high productivity growth phase is not necessarily best suited for its ultimate purpose, such as synthesizing the desired product. Different combinations and sequences of process conditions need to be investigated to determine the growth conditions, which produce the biomass with the physiological state best constituted for the product formation. There may be a sequence of phases each with a specific set of optimal conditions.

Medium optimization by the classical method of changing one independent variable (nutrient, antifoam, pH, temperature, etc.) while fixing all the others at a certain level can be extremely time consuming and expensive for a large number of variables. To make a full factorial search which would examine each possible combination of independent variable at appropriate levels could require a large number of experiments, x', where x is the number of levels and n is the number of variables. This may be quite appropriate for three nutrients at two concentrations (2^3 trials) but not for six nutrients at three concentrations. In this instance 3^6 (729) trials would be needed. Industrially the aim is to perform the minimum number of experiments to determine optimal conditions. Other alternative strategies must therefore be considered which allow more than one variable to be changed at a time. These methods have been discussed by Stowe and Mayer (1966), McDaniel, Bailey, Ethiraj, and Andrews (1976), Hendrix (1980), Nelson (1982), Dey (1985), Greasham and Inamine (1986), Bull, Huck, and Bushell (1990), and Hicks (1993). Kennedy and Krouse (1999) provide a very comprehensive review of media optimization strategies from component swapping through factorial design and on to more recent developments such as artificial neural networks and fuzzy logic.

When more than five independent variables are to be investigated, the Plackett–Burman design may be used to find the most important variables in a system, which are then optimized in further studies (Plackett & Burman, 1946). These authors give a series of designs for up to one hundred experiments using an experimental rationale known as balanced incomplete blocks. This technique allows for the evaluation of $X - 1$ variables by X experiments. X must be a multiple of 4, for example, 8, 12, 16, 20, 24, etc. Normally one determines how many experimental variables need to be included in an investigation and then selects the Plackett–Burman design, which meets that requirement most closely in multiples of four. Any factors not assigned to a variable can be designated as a dummy variable. Alternatively, factors known to not have any effect may be included and designated as dummy variables. As will be shown shortly in a worked example (Table 4.16), the incorporation of dummy variables into an experiment makes it possible to estimate the variance of an effect (experimental error).

Table 4.16 Plackett–Burman Design for Seven Variables (Nelson, 1982)

Trial	Variables							Yield
	A	**B**	**C**	**D**	**E**	**F**	**G**	
1	H	H	H	L	H	L	H	1.1
2	L	H	H	H	L	H	L	6.3
3	L	L	H	H	H	L	H	1.2
4	H	L	L	H	H	H	L	0.8
5	L	H	L	L	H	H	H	6.0
6	H	L	H	L	L	H	H	0.9
7	H	H	L	H	L	L	H	1.1
8	L	L	L	L	L	L	L	1.4

H, denotes a high level value; L, denotes a low level value.

Table 4.16 shows a Plackett–Burman design for seven variables (A–G) at high and low levels in which two factors, E and G, are designated as "dummy" variables. These can then be used in the design to obtain an estimate of error. Normally three dummy variables will provide an adequate estimate of the error. However, more can be used if fewer real variables need to be studied in an investigation (Stowe & Mayer, 1966). Each horizontal row represents a trial and each vertical column represents the H (high) and L (low) values of one variable in all the trials. This design (Table 4.16) requires that the frequency of each level of a variable in a given column should be equal and that in each test (horizontal row), the number of high and low variables should be equal. Consider the variable A; for the trials in which A is high, B is high in two of the trials and low in the other two. Similarly, C will be high in two trials and low in two, as will be for all the remaining variables. For those trials in which A is low, B will be high two times and low two times. This will also apply to all the other variables. Thus, the effects of changing the other variables cancel out when determining the effect of A. The same logic then applies to each variable. However, no changes are made to the high and low values for the E and G columns. Greasham and Inamine (1986) state that although the difference between the levels of each variable must be large enough to ensure that the optimum response will be included, caution must be taken when setting the level differential for sensitive variables, since a differential that is too large could mask the other variables. The trials are carried out in a randomized sequence.

The effects of the dummy variables are calculated in the same way as the effects of the experimental variables. If there are no interactions and no errors in measuring the response, the effect shown by a dummy variable should be 0. If the effect is not equal to 0, it is assumed to be a measure of the lack of experimental precision plus any analytical error in measuring the response (Stowe & Mayer, 1966).

This procedure will identify the important variables and allow them to be ranked in order of importance to decide which to investigate in a more detailed study to determine the optimum values to use.

The stages in analyzing the data (Tables 4.16 and 4.17) using Nelson's (1982) example are as follows:

Table 4.17 Analysis of the Yields Shown in Table 4.16 (Nelson, 1982)

	Factor						
	A	**B**	**C**	**D**	**E**	**F**	**G**
Σ(H)	3.9	14.5	9.5	9.4	9.1	14.0	9.2
Σ (L)	14.9	4.3	9.3	9.4	9.7	4.8	9.6
Difference	−11.0	10.2	0.2	0.0 −	0.6	9.2	−0.4
Effect	−2.75	2.55	0.05	0.00 −	0.15	2.30	−0.10
Mean square	15.125	13.005	0.005	0.000	0.045	10.580	0.020

Mean square for "error" $= \dfrac{0.045 + 0.020}{2} = 0.0325$

1. Determine the difference between the average of the H (high) and L (low) responses for each independent and dummy variable.

   Therefore the difference =

   $$\Sigma A\,(H) - \Sigma A\,(L).$$

   The effect of an independent variable on the response is the difference between the average response for the four experiments at the high level and the average value for four experiments at the low level.

   Thus the effect of

   $$A = \frac{\Sigma A\,(H)}{4} - \frac{\Sigma A\,(L)}{4}$$
   $$= \frac{2(\Sigma A\,(H) - \Sigma A\,(L))}{8}$$

   This value should be near zero for the dummy variables.

2. Estimate the mean square of each variable (the variance of effect).

   For A the mean square will be =

   $$\frac{(\Sigma A(H) - \Sigma A(L))^2}{8}$$

3. The experimental error can be calculated by averaging the mean squares of the dummy effects of E and G.

   Thus, the mean square for error =

   $$\frac{0.045 + 0.020}{2} = 0.0325$$

   This experimental error is not significant.

4. The final stage is to identify the factors which are showing large effects. In the example, this was done using an F-test for

   $$\frac{\text{Factor mean square}}{\text{Error mean square}}$$

   This gives the following values:

   $$A = \frac{15.125}{0.0325} = 465.4,$$
   $$B = \frac{13.005}{0.0325} = 400.2,$$
   $$C = \frac{0.0500}{0.0325} = 1.538,$$
   $$D = \frac{0.0000}{0.0325} = 0.00,$$
   $$F = \frac{10.580}{0.0325} = 325.6.$$

When Probability Tables are examined it is found that factors *A*, *B*, and *F* show large effects which are very significant, whereas *C* shows a very low effect which is not significant and *D* shows no effect. *A*, *B*, and *F* have been identified as the most important factors. The next stage would then be the optimization of the concentration of each factor, which will be discussed later.

Nelson (1982) has also referred to the possibility of two factor interactions which might occur when designing Table 4.16. This technique has also been discussed by McDaniel et al. (1976), Greasham and Inamine (1986), Bull et al. (1990), and Hicks (1993). El-Naggar, El-Bindary, and Nour (2013) report the use of Plackett-Burman design on process variables such as agitation rate and media components to optimize the production of antimicrobial metabolites by *Streptomyces anulatus* NEAE-94. It was found that of the variables screened agitation rate, inoculum size, and inoculum-age had significant effects on the productivity. These factors were further optimized using the three level Box-Behnken statistical design (Box & Behnken, 1960). The optimal media contained 20 g dm^{-3} starch, 2 g dm^{-3} KNO$_3$, 0.5 g dm^{-3} K$_2$HPO$_4$, 0.1 g dm^{-3} NaCl, and MgSO$_4$.7H$_2$O and 3 g dm^{-3} FeSO$_4$. This application of the combination of Plackett-Burman and Box-Behnken design and optimization have been reported by many authors for a wide range of fermentation applications. For example, Zhang et al. (2010) report their use in the production of human like collagen (HLC III) by recombinant *Escherichia coli* and Chen, Chiang, and Chao (2010) report their use in luciferase production by *Bacillus subtilis*.

The next stage in medium optimization would be to determine the optimum level of each key independent variable which has been identified by the Plackett–Burman design. This may be done using response surface optimization techniques which were introduced by Box and Wilson (1951). Hendrix (1980) has given a very readable account of this technique and the way in which it may be applied. Response surfaces are similar to contour plots or topographical maps. While topographical maps show lines of constant elevation, contour plots show lines of constant value. Thus, the contours of a response surface optimization plot show lines of identical response. In this context, response means the result of an experiment carried out at particular values of the variables being investigated.

The axes of the contour plot are the experimental variables and the area within the axes is termed the response surface. To construct a contour plot, the results (responses) of a series of experiments employing different combinations of the variables are inserted on the surface of the plot at the points delineated by the experimental conditions. Points giving the same results (equal responses) are then joined together to make a contour line. In its simplest form two variables are examined and the plot is two dimensional. It is important to appreciate that both variables are changed in the experimental series, rather than one being maintained constant, to ensure that the data are distributed over the response surface. In Fig. 4.1, the profile generated by fixing X_1 and changing X_2 and then using the best X_2 value and changing X_1 constitutes a cross, which may not encroach upon the area in which the optimum resides.

The technique may be applied at different levels of sophistication. Hendrix applied the technique at its simplest level to predict the optimum combination of two

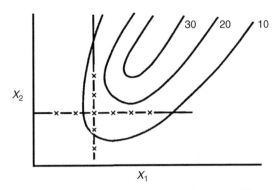

FIGURE 4.1 Optimal Point of a Response Surface by One Factor at a Time

variables. The values of the variables for the initial experiments are chosen randomly or with the guidance of previous experience of the process. There is little to be gained from using more than 15–20 experiments. The resulting contour map gives an indication of the area in which the optimum combination of variables resides. A new set of experiments may then be designed within the indicated zone. Hendrix proposed the following strategy to arrive at the optimum in an incremental fashion:

1. Define the space on the plot to be explored.
2. Run five random experiments in this space.
3. Define a new space centred upon the best of the five experiments and make the new space smaller than the previous one, perhaps by cutting each dimension by one half.
4. Run five more random experiments in this new space.
5. Continue doing this until no further improvement is observed, or until you cannot afford any more experiments.

The more sophisticated applications of the response surface technique use mathematical models to analyze the first round of experimental data and to predict the relationship between the response and the variables. These calculations then allow predictive contours to be drawn and facilitate a more rapid optimization with fewer experiments. If three or more variables are to be examined, then several contour maps will have to be constructed. Hicks (1993) gives an excellent account of the development of equations to model the different interactions which may take place between the variables. Several computer software packages are now available which allow the operator to determine the equations underlying the responses and, thus, to determine the likely area on the surface in which the optimum resides. Some examples of the types of response surface profiles that may be generated are illustrated in Fig. 4.2.

The following examples illustrate the application of the technique:

1. McDaniel et al. (1976), Fig. 4.3. The variables under investigation were cerelose and soybean level, with the analysis indicating the optimum to be 6.2% cerelose and 3.2% soybean.

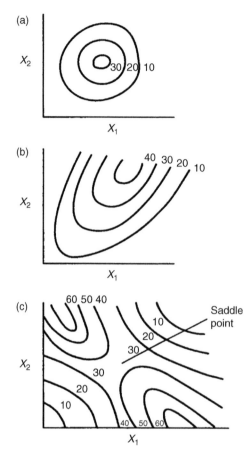

FIGURE 4.2 Typical Response Surfaces in Two Dimensions

(a) mound, (b) rising ridge, (c) saddle.

2. Saval, Pablos, and Sanchez (1993). The medium for streptomycin production was optimized for four components resulting in a 52% increase in streptomycin yield, a 10% increase in mycelial dry weight, and a 48% increase in specific growth rate (Table 4.18).

When further optimization experiments are necessary for medium development in large vessels, the number of experiments will normally be restricted because of the cost and the lack of spare large vessels (Spendley, Hext, & Himsworth, 1992). The simplex search method attempts to optimize n variables by initially performing $n + 1$ experimental trials. The results of this initial set of trials are then used to predict the conditions of the next experiment and the situation is repeated until the optimum combination is attained. Thus, after the first set of trials the optimization proceeds as individual experiments. The prediction is achieved using a graphical representation of the trials where the experimental variables are the axes. Using this procedure, the

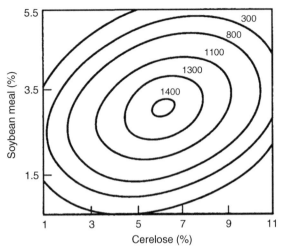

FIGURE 4.3 Contour Plot of Two Independent Variables, Cerelose, and Soybean Meal, for Optimization of the Candidin Fermentation

(Redrawn from McDaniel et al., 1976; Bull et al., 1990).

Table 4.18 Concentration of Nutrients in an Original and Optimized Medium for Streptomycin Production (Saval et al., 1993)

Nutrient (g dm⁻³)	Original medium	Optimized Medium
Glucose	10	23
Beer-yeast autolysate	25	27
NaCl	10	8
K$_2$HPO$_4$	1	1

experimental variables are plotted and not the results of the experiments. The initial experimental conditions are chosen such that the points on the graph are equidistant from one another and form the vertices of a polyhedron described as the simplex. Thus, with two variables the simplex will be an equilateral triangle. The results of the initial set of three experiments are then used to predict the next experiment enabling a new simplex to be constructed. The procedure will be explained using an example to optimize the concentrations of carbon and nitrogen sources in a medium for antibiotic production.

In our example, a graph is constructed in which the *x*-axis represents the concentration range of the carbon source (the first variable) and the *y*-axis represents the concentration range of the nitrogen source (the second variable). The first vertex A (experimental point) of the simplex represents the current concentrations of the two variables which are producing the best yield of the antibiotic. The experiment for the second vertex B is planned using a new carbon–nitrogen mixture and the position of

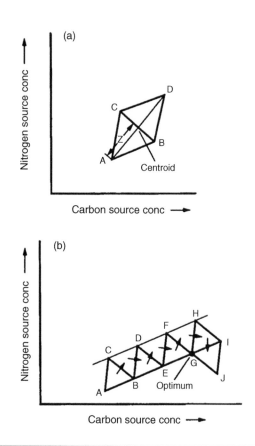

FIGURE 4.4

(a) Simplex optimization for a pair of independent variables (with reflection), (b) Simplex optimization of pair of independent variables which has reached the optimum.

the third vertex C can now be plotted on the graph using lengths AC and BC equal to AB (the simplex equilateral triangle, Fig. 4.4a). The concentrations of the carbon and nitrogen sources to use in the third experiment can now be determined graphically and the experiment can be undertaken to determine the yield of antibiotic. The results of the three experiments are assessed and the worst response to antibiotic production identified. In our example, experiment A was the worst and B the best. The simplex design is now used to design the next experiment. A new simplex (equilateral triangle) BCD is constructed opposite the worst response (ie, A) using the existing vertices B and C. A line is drawn from A through the centroid (mid point) of BC. D (the next experiment) will be on this line and the sides BD and CD will be the same length as BC. This process of constructing the new simplex is described as reflection. Once the position of D is known, the concentrations of the carbon and nitrogen sources can be determined graphically, the experiment performed and the production of antibiotic assayed. Thus, a series of simplexes can be constructed moving in a crabwise way. The procedure is continued until the optimum is located. At this point the simplex begins to circle on its self, indicating the optimum concentration (Fig 4.4b; Greasham & Inamine, 1986). However, if a new vertex exhibits the lowest response, the simplex would reflect back on to the previous one, halting movement toward the optimum. In this case, the new simplex is constructed opposite the second least desirable response using the method previously described.

If it is decided that the supposed optimum should be reached more rapidly then the distance z between the centroid and D may be increased (expanded) by a factor which is often two. If the optimum is thought to have been nearly reached then the distance z may be decreased by a factor of 0.5 (contraction). This modified simplex optimization was first proposed by Nelder and Mead (1965) and has been discussed by Greasham and Inamine (1986).

The simplex method may also be used in small scale media development experiments to help identify the possible optimum concentration ranges to test in more extensive multifactorial experiments.

ANIMAL CELL MEDIA

The use of animal cell cultures in biotechnology was initially confined to the production of vaccines, with the polio injectable vaccine being the first to be produced in an animal cell culture system in 1955. However, in 2014, 56% of biopharmaceuticals were manufactured using mammalian cell cultures (Walsh, 2014) and the current use of the technology for heterologous protein production by the pharmaceutical industry generates in excess of US$120 billion per annum (Chapters 1 and 12) (Bandaranayake & Almo, 2014). Thus, the development of safe, high yielding, and consistently performing cell culture media is crucial to the pharmaceutical industry.

THE DEVELOPMENT OF BASAL MEDIA

The pioneering work on animal cell culture began in the early 1900s (Landecker, 2007) and various biological fluids were combined with isotonic salt solutions to provide an environment for animal cells to grow in vitro. The most successful biological fluid used in this approach was serum (either fetal calf or calf). Blood consists of red and white blood cells suspended in a liquid (plasma). Serum is the supernatant remaining after the removal of cellular components of blood and allowing the resulting cell-free plasma to clot. It was not until the 1950s that the precise chemical requirements of cells in culture were examined in any depth, with the aim of designing chemically defined media (Butler, 2015). As a result of this work, a number of basal media were developed that form the foundation of most formulations to support the growth of animal cells in vitro and these are summarized in Table 4.18 (Jayme, Watanabe, & Shimada, 1997). The first medium of this type was Eagle's basal medium (BME) followed by a modified version, Eagle's minimal essential medium (EMEM) containing inorganic salts, amino acids, and vitamins as well as glucose as the energy source and glutamine as a nitrogen and additional energy source (Eagle, 1959). However, to support cell growth, rather than simply maintain cell viability, this medium still required supplementation with serum.

Eagle's pioneering work laid the foundation for two developmental approaches—one concentrating on the replacement of serum, with the objective of achieving chemically defined media, and the other focusing on the optimization of nutrient levels to support high cell densities. A completely synthetic medium (F12) was developed in 1965 that supported the growth of Chinese hamster ovary (CHO) cells. Compared with EMEM, this medium contained a wider range of amino acids and vitamins as well as other ingredients listed in Table 4.19. Dulbecco's group (Dulbecco & Freeman, 1959) modified Eagle's medium by increasing the concentration of the amino acids and adding further, nonessential, amino acids and trace elements. The resulting Dulbecco's modified Eagles medium (DMEM) supported higher cell densities and viability. Barnes and Sato (1980) combined the features of F12 and DMEM by combining them in a 1:1 mixture by volume to create DMEM/F12, a medium that has proved to be the most widely used basal synthetic medium, and when supplemented with serum, was suitable for the production of biopharmaceuticals (Jayme et al., 1997). RPMI (Roswell Park Memorial Institute) 1640 medium was developed by Moore, Gerner and Franklin (1967) and supported high cell density lymphocyte cultures by incorporating elevated nutrient levels as shown in Table 4.19. Murakami et al. (1984) adopted Sato's approach and achieved a higher performing medium (Rat Dermal Fibroblasts medium or RDF) by combining DMEM/F12 in a 1:1 volumetric ratio with RPMI 1640. RDF was further improved by substantially increasing the concentrations of the amino acids and glucose; the revised medium being named enriched RDF or eRDF (Murakami & Yamada, 1987). These media shown in Table 4.19 were generally used in conjunction with serum supplementation to achieve higher growth rates and cell concentrations. The serum could provide a range of unknown factors that would stimulate growth as well as augmenting the levels of nutrient already provided in the basal medium.

SERUM-FREE, ANIMAL-COMPONENT FREE, PROTEIN-FREE, AND CHEMICALLY DEFINED MEDIA

As well as developing eRDF, which has been very widely used as a basal medium in biotechnology processes, Murakami and others from Sato's group identified four components that were required for serum-free cultivation of hybridomas: insulin, transferrins, ethanolamine, and selenium—a cocktail referred to as ITES (Murakami et al., 1982). Insulin regulates the cellular uptake and metabolism of glucose, amino acids, and lipids but is also a very important antiapoptopic molecule such that its presence in cell culture media prevents the onset of apoptosis. Transferrins are a group of iron-binding glycoproteins involved with iron uptake by the cells. Ethanolamine is a precursor of phospholipid synthesis and selenium is an essential trace element that functions as a cofactor for glutathione peroxidases and thioredoxin reductase.

Table 4.19 Biochemical Composition of Basal Medium Formulations (mg dm^{-3}) (Jayme et al., 1997)

Component	EMEM (Eagle, 1959)	F-12	DMEM (Dulbecco & Freeman, 1959)	DMEM/F12 (Barnes and Sato, 1980)	RPMI 1640 (Moore et al., 1967)	RDF (Murakami et al., 1984)	eRDF (Murakami and Yamada, 1987)
Inorganic salts							
CaCl$_2$ (anhyd)	200.00	33.22	200.00	116.60			
CaCl$_2$.2H$_2$O						77.90	108.77
Ca(NO$_3$)$_3$.4H$_2$O					100.00	49.58	
CuSO$_4$.5H$_2$O		0.0024		0.0013		0.00062	0.00075
FeSO$_4$.7H$_2$O		0.83		0.417		0.208	0.222
Fe(NO$_3$)$_3$.9H$_2$O		0.10		0.05		0.025	
KCl	400.00	223.6	400.00	311.80	400.00	358.00	373.00
MgSO$_4$ (anhyd)	97.67	57.22	97.67	48.84	48.84	49.35	66.20
MgCl$_2$ (anhyd)				26.84			
MgCl$_2$.6H$_2$O						30.48	
NaCl	6800.00	7599.00	6400.00	6995.50	6000.00	6505.00	6435.00
NaHCO$_3$	2200.00	1176.00	3700.00	2438.00	2000.00	1050.00	1050.00
NaH$_2$PO$_4$.H$_2$O	140.00	142.00	125.00	62.5		31.2	
Na$_2$HPO$_4$ (anhyd)				71.02			
Na$_2$HPO$_4$.12H$_2$O					800.00	1100.00	1220.00
ZnSO$_4$.7H$_2$O		0.86		0.43		0.22	0.33
Subtotal	9837.67	9232.83	10922.67	10073.80	9348.84	9251.96	9253.42
Amino acids							
L-alanine		8.90		4.45		2.23	6.68
L-arginine HCl	126.00	211.00	84.00	147.50	200.00	194.00	582.00
L-aspartic acid		13.30		6.65	20.00	13.30	39.90
L-asparagine.H$_2$O		15.01		7.50	50.00	31.50	94.5

(continued)

Table 4.19 Biochemical Composition of Basal Medium Formulations (mg dm^{-3}) (Jayme et al., 1997) (cont.)

Component	EMEM (Eagle, 1959)	F-12	DMEM (Dulbecco & Freeman, 1959)	DMEM/F12 (Barnes and Sato, 1980)	RPMI 1640 (Moore et al., 1967)	RDF (Murakami et al., 1984)	eRDF (Murakami and Yamada, 1987)
L-cysteine HCL.H$_2$O		35.12		17.56		8.80	105.40
L-cystine	31.00					36.80	
L-cystine.2HCl		14.70	63.00	31.29	65.00		
L-glutamic acid				7.35	20.00	13.20	39.70
L-glutamine	2292.00	146.00	584.00	365.00	300.00	333.00	1000.00
Glycine		7.50	30.00	18.75	10.00	14.30	42.80
L-histidine HCl.H$_2$O	42.00	21.00	42.00	31.48	15.00	25.00	75.00
L-hydroxyproline					20.00	10.50	31.50
L-isoleucine	52.00	4.00	105.00	54.47	50.00	52.50	157.50
L-leucine	52.00	13.10	105.00	59.05	50.00	55.10	165.30
L-lysine HCl	73.00	36.50	146.00	91.25	40.00	65.80	197.30
L-methionine	15.00	4.50	30.00	17.24	15.00	16.40	49.20
L-phenylalanine	32.00	5.00	66.00	35.48	15.00	24.80	74.30
L-proline		34.50		17.25	20.00	18.40	55.30
L-serine		10.50	42.00	26.25	30.00	28.40	85.10
L-threonine	48.00	11.90	95.00	53.45	20.00	36.90	110.80
L-tryptophan	10.00	2.00	16.00	9.02	5.00	6.10	18.40
L-tyrosine						29.00	87.00
L-tyrosine.2Na.2H$_2$O	52.00	7.81	104.00	55.79	29.00		
L-valine	46.00	11.70	94.00	52.85	20.00	36.30	108.90
Subtotal	871.00	614.04	1606.00	1110.02	994.00	1051.53	3126.58

Vitamins

p-Aminobenzoic acid	1.00				1.00	0.51	0.51
Biotin	1.00	0.0073		0.0035	0.20	0.10	0.10
D-Ca pantothenate	1.00	0.50	4.00	2.24	0.25	0.67	0.67
Folic acid	1.00	1.30	4.00	2.65	1.00	1.81	1.81
Niacinamide	1.00	0.036	4.00	2.02	1.00	150	150
Pyridoxal HCl	1.00		4.00	2.00		1.00	1.00
Pyridoxine HCl		0.06		0.03	1.00	0.50	0.50
Riboflavin	0.10	0.037	0.40	0.22	0.20	0.21	0.21
Thiamine HCl	1.00	0.30	4.00	2.17	1.00	1.60	1.60
Vitamin B12		1.40		0.68	0.005	0.34	0.34
Subtotal	5.1	3.64	20.40	12.02	5.66	8.24	8.24

Miscellaneous

Choline chloride	1.00	14.00	4.00	8.98	3.0	6.14	12.29
D-Glucose	1000.00	1802.00	4500.00	3151.00	2000.00	1700.00	3423.00
Glutathione (reduced)					1.00	0.50	0.5
HEPES						1190.00	1190.00
Hypoxanthine (Na salt)		4.77	7.20	2.39		1.00	1.00
I-Inositol	2.00	18.00		12.60	35.00	23.40	46.80
Linoleic acid		0.084		0.042		0.021	0.021
Lipoic acid		0.21		0.105		0.052	0.052
Phenol red	10.00	1.20	15.00	8.10	5.00	6.56	5.00
Putrescine-2HCl		0.161		0.081		0.04	0.04
Pyruvate (sodium salt)		110.00		55.00		55.00	110.0
Thymidine		0.70		0.365		0.18	0.18
Subtotal	1013.00	1951.13	4526.20	3238.67	2044.00	2982.89	4788.88

Selected media represent the most frequently used, glutamine containing derivative of the referenced formulation.

The motivation for the development of serum-free media for biopharmaceutical production became acute with the increased perception of the risk of pathogen contamination emanating from animal products in cell culture media. The possible presence of prions, the agents of transmissible encephalopathies such as bovine spongiform encephalopathy (BSE), was a particular concern and led to most regulatory authorities demanding the use of serum-free medium for biopharmaceutical processes. However, this regulatory requirement was not the only reason for the development of such media as serum containing media also have the disadvantages associated with the use of a complex medium for the production of a protein, including:

1. Batch to batch variation. Any biological material is subject to variation in quality and composition and serum is no exception. Thus, such variation can give rise to unpredictable process performance that also makes the process more difficult to control.
2. Problems with purification. A culture medium supplemented with 10% serum can have a protein concentration approaching 10 g dm^{-3}, while the recombinant protein in a process under development may accumulate to only 0.1 g dm^{-3} (Butler, 2015). Thus, the isolation of the desired protein becomes a significant challenge.
3. Economics of the process. The industry standard serum was fetal calf serum but its cost was between \$500 and \$1000 dm^{-3}, which meant that the serum accounted for up to 95% of the medium costs (Butler, 2015). Also, the serum could only be sourced from countries that were designated BSE-free, in particular New Zealand. This could give rise to shortages that could threaten production schedules and further price inflation.
4. Ethical considerations. The harvesting of fetal calf serum causes considerable distress to the unborn calf. Two international workshops were organized in 2003 and 2009 that drew attention to the suffering caused to the live unborn calf during the harvesting of blood for serum manufacture. As a result of these workshops, recommendations were made that serum-free media should be developed for cell culture in basic and applied research (van der Valk et al., 2010).

The development of serum-free media is not a trivial task as serum provides a multitude of factors required for growth, contains protective compounds such as antitoxins, antioxidants, and protease inhibitors and the presence of albumen gives buffering capacity as well as physical protection against shear forces. Butler (2015) stressed the importance of statistical methods such as those described earlier in this chapter for the optimization of these media.

In order to conform to the requirements of the regulatory authorities, cell culture media must be free of not only serum but also any animal-derived components and, thus, can be described as animal-component free. Theoretically, this can be achieved by replacing the components of the serum that are required by the organism with pure chemical ingredients, resulting in a chemically defined medium, for example, the ITES cocktail referred earlier. Such a medium would be free of the disadvantages associated with serum-based and complex media in general. However, it has often been observed that chemically defined cell culture media support neither the cell concentration nor the productivities of that of their complex counterparts. A very common approach to replace the proteinaceous components of serum is to use a complex protein source of plant or microbial origin. However, the medium is then still effectively a complex one and may suffer from problems associated with variability and product purification. This issue has been partly addressed by the use of hydrolysates of plant and microbial proteins, which, while still complex, the original proteins have been hydrolyzed to peptides and present fewer down-stream concerns. Thus, all media for biopharmaceutical production must be serum-free and animal component-free, they may be also protein-free but they may not achieve the goal of being chemically defined. Gupta, Hageman, Wierenga, Boots, and Gruppen (2014) reported that the adoption of the "semidefined" approach is common in the biopharmaceutical industry with defined basal media being supplemented with plant protein hydrolysates. The following section addresses in more detail the key components that can be added to a basal medium to replace serum yet maintain high cell densities and productivities.

Protein hydrolysates

Protein hydrolysates (peptones) are widely used in microbial culture media but can also be used as a substitute for serum in mammalian cell cultures. Since that time, plant or microbial protein hydrolysates have been widely used as part of the strategy to replace serum in commercial media (Pasupuleti, Holmes, & Demain, 2010; Lobo-Alfonso, Price, & Jayme, 2010). Commonly used hydrolysates include yeast, soy, rapeseed, wheat, cotton, and pea. By hydrolyzing plant or yeast proteins, the resulting medium will be protein-free but rich in lower molecular weight peptides, thus reducing some of the downstream processing problems associated with whole protein substrates. However, these hydrolysates are highly complex mixtures, containing minerals, carbohydrates, and lipids as well as peptides and amino acids derived from proteins in the original microbial or plant source. Thus, these products have the advantage of supplying a number of possible requirements originally present in serum but have the inherent disadvantage of natural products, namely, variability between batches. The quality and characteristics of a protein hydrolysate are determined by the nature of the hydrolytic process. Yeast extract is prepared by the autolysis of yeast cells by osmotic shock followed by hydrolysis by enzymes from the yeast itself. A cocktail of enzymes is normally used to hydrolyze plant proteins designated for cell culture use. As a result, the degree of degradation, and thus the size distribution of the resulting peptides, can be controlled by the nature of the cocktail (Siemensma et al., 2010).

A number of publications have addressed the nature of the growth-stimulating components of hydrolysates. For example, Frages-Haddani et al. (2006) used membrane filtration fractionation to separate rapeseed hydrolysates and identified peptide fractions that had an antiapoptopic effect as well as increasing γ-interferon production by CHO cells. Chun, Kim, Lee, and Chung (2007) demonstrated that < 3000 molecular weight fractions of soy hydrolysates supported CHO cell growth and productivity to the same extent as the complete hydrolysate. Thus, removal of the larger components resulted in an effective medium that was protein-free and devoid of many of the down-stream processing problems associated with protein-rich media. Similarly, Michiels et al. (2011) obtained a fraction from soy peptone that caused the same stimulatory effect on CHO cells, as did the complete, far more complex, hydrolysate. Gupta et al. (2014) identified 410 compounds in a soy hydrolysate, 253 of which were peptides. However, the most important compounds that contributed to performance variation between batches were phenyl lactate and ferulate.

Thus, the key issues to be addressed in the improvement of hydrolysates are those of fractionation and variation. Fractionation may remove components deleterious to cell performance or to downstream processing or ensure the presence of advantageous components. Variation may be addressed by the strict control of the hydrolyzing process and there is promise in the use of recombinant enzymes produced by fermentation to achieve a more predictable product. An interesting development is the production of specific peptides by microbial fermentation. Both the Japanese fermentation companies, Ajnomoto and Kyowa Hakko produce such products under animal component-free conditions and such products may combine the stimulatory effects of protein hydrolysates with the lack of variability associated with chemically defined media (Pasupuleti et al., 2010). Finally, it is essential that the end-user of the medium runs performance tests to ensure that each batch of medium is acceptable before it is used in the commercial process.

Insulin

Insulin is a normal constituent of serum and has many effects on cell physiology including glucose metabolism, membrane transport, nucleic acid biosynthesis, and fatty acid biosynthesis as well as having antiapoptotic effects. Thus, insulin must be incorporated into serum-free media. The anomaly of having to replace an animal product (serum) with another (insulin) is resolved by the availability of recombinant insulin manufactured by microbial fermentation (Chapter 12). Butler (2015) discussed the phenomenon of the insulin concentration required by animal cell cultures in serum-free medium (approximately 5 µg cm^{-3}) being a thousand times that found in serum and concluded that it was due to insulin instability under nonphysiological conditions. This instability has been attributed to the high redox potential of cell culture media due to the presence of cysteine and can be addressed by replacing cysteine with cystine. The problem of concentration and stability of insulin has also been addressed by the use of insulin-like growth factors (IGF). IGF are naturally synthesized in the liver, are close structural relatives of insulin and are very effective cellular growth promoters. Morris and Schmid (2000) demonstrated that a recombinant IGF (Long R^{3}IGF-1) was the preferred growth factor for recombinant protein production in two cell lines of CHO cells. Long R^{3}IGF-1 is 200 times more potent and three times more stable than insulin (Butler, 2015).

Transferrins

It will be recalled that Murakami et al. (1982) demonstrated that transferrin was an essential component of serum in mammalian cell culture. Transferrins are a group of glycoproteins that facilitate iron transport under physiological conditions. They are strong ferric iron (iron III) chelators and as well as enabling iron transport. Thus, the presence of a transferrin in a serum-free medium binds iron, making it available to the cells but also removing it from the medium such that none is available to catalyze the production of free radicals and thus protects the cells against oxidative stress and protein damage. Recombinant transferrins available are compatible with animal component-free media.

Albumin

Albumin is the major protein present in serum, making up about 60% of the total protein and its role in cell culture has been reviewed by Francis (2010). The major roles of albumin are binding and transport of lipids, metal ions, amino acids, and other factors, pH control and its function as an antioxidant. Albumin also appears to play a role in cell culture as a protectant against the physical stresses of aeration and agitation although the explanation of the effect is far from clear (Francis, 2010). The replacement of the albumin in serum can be achieved using recombinant albumin but, as argued by Francis, it is hardly compatible with the objectives of a protein-free medium. Thus, the inclusion of recombinant albumin in production scale mammalian cell fermentations will depend on an analysis of the economic benefits of their inclusion and will, therefore, be process dependent.

Osmolality

The optimum range of osmotic pressure for growth is often quite narrow and varies with the type of cell and the species from which it was isolated. It may be necessary to adjust the concentration of NaCl when major additions are made to a medium.

pH

The normal buffer system in tissue culture media is the CO_2-bicarbonate system. This is a weak buffering system and can be improved by the use of a zwitterionic buffer such as Hepes, either in addition to or instead of the CO_2-bicarbonate buffer. Continuous pH control is achieved by the addition of sodium bicarbonate or sodium hydroxide (with fast mixing) when too acidic. The pH does not normally become too alkaline so acid additions are not required but provision may be made for CO_2 additions (Fleischaker, 1987).

Nonnutritional media supplements

Sodium carboxy methyl cellulose may be added to media at 0.1% to help to minimize mechanical damage caused by the shear force generated by the stirrer impeller. The problems of foam formation and subsequent cell damage and losses can affect animal cell growth. Pluronic F-68 (polyglycol) can provide a protective effect to animal cells in stirred and sparged vessels. In media which are devoid of Pluronic

F-68, cells may become more sensitive to direct bubble formation in the presence of an antifoam agent being used to supress foam formation (Zhang, Handacorrigan, & Spier, 1992).

REFERENCES

Abbott, B. J., & Clamen, A. (1973). The relationship of substrate, growth rate and mainte-nance coefficient to single cell protein production. *Biotechnology and Bioengineering*, *15*, 117–127.

Aharonowitz, Y. (1980). Nitrogen metabolite regulation of antibiotic biosynthesis. *Annual Review of Microbiology*, *34*, 209–233.

Aharonowitz, Y., & Demain, A. L. (1977). Influence of inorganic phosphate and organic buf-fers on cephalosporin production by *Streptomyces clauuligerus. Archives of Microbiol-ogy*, *115*, 169–173.

Aharonowitz, Y., & Demain, A. L. (1978). Carbon catabolite regulation of cephalosporin produc-tion in *Streptomyces clauuligerus. Antimicrobial Agents and Chemotherapy*, *14*, 159–164.

Aiba, S., Humphrey, A. E., & Millis, N. F. (1973). Scale-up. In *Biochemical engineering* (2nd ed., pp. 195–217). New York: Academic Press.

Anonymous (1980). Research Report. Research Institute of Antibiotics and Biotransformations. Roztoky, Czechoslovakia. Cited by Podojil, M., Blumauerova, M., Vanek, Z., & Culik, K. (1984). The tetracyclines; properties, biosynthesis and fermentation. In E. J. Vandamme, *Biotechnology of Industrial Antibiotics* (pp. 259–279). New York: Marcel Dekker.

Arima, K., Imanara, H., Kusaka, M., Fukada, A., & Tamura, G. (1965). Studies on pyrrolni-trin, a new antibiotic. I. Isolation of pyrrolnitrin. *The Journal of Antibiotics*, *18*, 201–204.

Ashley, M. H. J., & Rodgers, B. L. F. (1986). The efficient use of water in single cell protein production. In D. I. Alani, & M. Moo-Young (Eds.), *Perspectives in biotechnology and applied microbiology* (pp. 71–79). London: Elsevier.

Atkinson, B., & Mavituna, F (1991a). Process biotechnology. In *Biochemical engineering and biotechnology handbook* (2nd ed., pp. 43–81). London: Macmillan.

Atkinson, B., & Mavituna, F. (1991b). Stoichiometric aspects of microbial metabolism. In *Biochemical engineering and biotechnology handbook* (2nd ed., pp. 115–167). London: Macmillan.

Aunstrup, K. (1974). Industrial production of proteolytic enzymes. In B. Spencer (Ed.), *Industrial aspects of biochemistry, part A* (pp. 23–46). Amsterdam: North Holland.

Aunstrup, K., Andresen, O., Falch, E. A., & Nielsen, T. K. (1979). Production of microbial enzymes. In H. J. Peppier, & D. Perlman (Eds.), *Microbial Technology* (pp. 281–309). (Vol. 1). New York: Academic Press.

Avanzini, F. (1963). Preparation of tetracyline antibiotics. Chlortetracycline from *Streptomy-ces aureofaciens.* British Patent 939,476.

Bader, F. G., Boekeloo, M. K., Graham, H. E., & Cagle, J. W. (1984). Sterilization of oils: data to support the use of a continuous point-of-use sterilizer. *Biotechnology and Bioengineering*, *26*, 848–856.

Bandaranayake, A. D., & Almo, S. C. (2014). Recent advances in mammalian protein produc-tion. *FEBS Letters*, *588*, 253–260.

Banks, G. T., Mantle, P. G., & Syczyrbak, C. A. (1974). Large scale production of clavine alkaloids by *Claviceps fusiformis. Journal of General Microbiology*, *82*, 345–361.

Barrios-Gonzalez, J. (2012). Solid-state fermentation: physiology of solid medium, its molecular basis and applications. *Process Biochemistry*, *47*, 175–185.

Basak, K., & Majumdar, S. K. (1973). Utilization of carbon and nitrogen sources by *Streptomyces kanamyceticus* for kanamycin production. *Antimicrobial Agents and Chemotherapy*, *4*, 6–10.

Batti, M. R. (1967). Process for producing citric acid. U.S. Patent 3,335,067.

Bauchop, T., & Elsden, S. R. (1960). The growth of microorganisms in relation to their energy supply. *Journal of General Microbiology*, *23*, 457–466.

Bawa, N., Bear, D., Hill, G., Niu, C., & Roesler, W. (2010). Fermentation of glucose and starch particles using an inexpensive medium. *Journal of Chemical Technology and Biotechnology*, *85*, 441–446.

Beaman, R. G. (1967). Vinegar fermentation. In H. J. Peppier (Ed.), *Microbial technology* (pp. 344–359). New York: Reinhold.

Becher, E., Bernhauer, K., and Wilhann, G. (1961). Process for the conversion of benzimimidazole containing vitamin B12. factors, particularly Factor III to vitamin B12. U.S. Patent 2,976,220.

Belik, E., Herold, M., & Doskocil, J. (1957). The determination of B complex vitamins in corn-steep extracts by microbiological tests. *Chemical Papers*, *11*, 51–56.

Boeck, L. D., Mertz, F. D., Wolter, R. K., & Higgens, C. E. (1984). *N*-Diethyrvancomycin, a novel antibiotic produced by a strain of *Nocardia orientalis*. Taxonomy and fermentation. *The Journal of Antibiotics*, *37*, 446–453.

Borrow, A., Jefferys, E. G., and Nixon, I. S. (1960). Gibberellic acid. British Patent 838,033.

Borrow, A., Jefferys, E. G., Kessel, R. H. J., Lloyd, E. C., Lloyd, P. B., & Nixon, I. S. (1961). Metabolism of *Gibberella fujikuroi* in stirred culture. *Canadian Journal of Microbiology*, *7*, 227–276.

Borrow, A., Jefferys, E. G., Kessel, R. H. J., Lloyd, E. C., Lloyd, P. D., Rothwell, A., Rothwell, B., & Swait, J. C. (1964). The kinetics of metabolism of *Gibberella fujikuroi* in stirred culture. *Canadian Journal of Microbiology*, *10*, 407–444.

Box, S. J. (1980). Clavulanic acid and its salts. British Patent 1,563,103.

Box, G. E. P., & Behnken, D. W. (1960). Some new three-level designs for the study of quantitative variables. *Technometrics*, *2*, 455–475.

Box, G. E. P., & Wilson, K. B. (1951). On the experimental attainment of optimum conditions. *Journal of the Royal Statistical Society*, *13*, 1–45.

Brack, A. (1947). Antibacterial compounds. 1. The isolation of genistyl alcohol in addition to patulin from the filtrate of a penicillin culture. Some derivatives of genistyl alcohol. *Helvetica Chimica Acta*, *30*, 1–8.

Brown, C. M., MacDonald, D. S., & Meers, J. F. (1974). Physiological aspects of microbial inorganic nitrogen metabolism. *Advances in Microbial Physiology*, *11*, 1–62.

Bull, A. T., Huck, T. A., & Bushell, M. E. (1990). Optimization strategies in microbial process development and operation. In R. K. Poole, M. J. Bazin, & C. W. Keevil (Eds.), *Microbial growth dynamics* (pp. 145–168). Oxford: IRL Press.

Butler, M. (2015). Serum and protein free media. In M. Al-Rubeai (Ed.), *Animal cell culture* (pp. 223–236). Heidelberg: Springer.

Calam, C. T., & Nixon, I. S. (1960). Gibberellic acid. British Patent 839,652.

Calik, G., Pehlivan, N., Ozcelik, I. S., Calik, P., & Ozdamar, T. H. (2004). Fermentation and oxygen transfer characteristics in serine alkaline protease production by recombinant *Bacillus subtilis* in molasses based complex media. *Journal of Chemical Technology and Biotechnology*, *79*, 1243–1250.

Carrington, T. R. (1971). The development of commercial processes for the production of 6-aminopenicillanic acid (6-APA). *Proceedings of the Royal Society of London, 179*, 321–333.

Chang, I. S., Kim, D., Kim, B. H., & Lovitt, R. W. (2007). Use of an industrial grade medium and medium enhancing effects on high cell density CO fermentation by *Eubacterium limosum* KIST612. *Biotechnology Letters, 29*, 1183–1187.

Chen, P. T., Chiang, C. -J., & Chao, Y. -P. (2010). Medium optimization and production of secreted renilla luciferase in *Bacillus subtilis* by fed-batch fermentation. *Biochemical Engineering Journal, 49*, 395–400.

Cheng, K. -C., Demirci, A., & Catchmark, J. M. (2011). Evaluation of medium composition and fermentation parameters on pullulan production by *Aureobasidium pullulans*. *Food Science and Technology International, 17*(2), 99–109.

Chun, B. -H., Kim, J. -H., Lee, H. -J., & Chung, N. (2007). Usability of size-excluded fraction of soy protein hydrolysates for growth and viability of Chinese hamster ovary cells in protein-free suspension culture. *Bioresource Technology, 98*, 1000–1005.

Claridge, C. A., Bush, J. A., Defuria, M. D., & Price, K. E. (1974). Fermentation and mutation studies with a butirosin producing strain of *Bacillus circulons*. *Developments in Industrial Microbiology, 15*, 101–113.

Cohen, B. L. (1973). The neutral and alkaline proteases of *Aspergillus nidulans*. *Journal of General Microbiology, 77*, 21–28.

Collins, S. H. (1990). Production of secreted proteins in yeast. In T. J. R. Harris (Ed.), *Protein production by biotechnology* (pp. 61–78). London: Elsevier.

Cooney, C. L. (1979). Conversion yields in penicillin production: theory vs. practice. *Process Biochemistry, 14*(1), 31–33.

Cooney, C. L. (1981). Growth of micro-organisms. In H. J. Rehm, & G. Reed (Eds.), *Biotechnology* (pp. 73–112). (Vol. 1). Weinheim: Verlag Chemie.

da Silva, G. P., Mack, M., & Contiero, J. (2009). Glycerol: a promising and abundant carbon source for industrial microbiology. *Biotechnology Advances, 27*, 30–39.

De Coninck, J., Bouquelet, S., Dumortier, V., Duyme, F., & Verdier-denantes, I. (2000). Industrial media and fermentation processes for improved growth and protease production by *Tetrahymena thermophila* B111. *Journal of Industrial Microbiology and Biotechnology, 24*, 285–290.

DeTilley, G., Mou, D. G., & Cooney, C. L. (1983). Optimization and economics of antibiotics production. In J. E. Smith, D. R. Berry, & B. Kristiansen (Eds.), *The Filamentous Fungi* (pp. 190–209). (Vol. 4). London: Arnold.

Deindoerfer, F. H., & West, J. M. (1960). Rheological properties of fermentation broths. *Advances in Applied Microbiology, 2*, 265–273.

Demain, A. L. (1968). Regulatory mechanisms and the industrial production of microbial metabolites. *Lloydia, 31*, 395–418.

Demain, A. L. (1990). Regulation and exploitation of enzyme biosynthesis. In W. M. Fogarty, & K. T. Kelly (Eds.), *Microbial enzymes and biotechnology* (2nd ed., pp. 331–368). London: Elsevier.

Demain, A. L., & Piret, J. M. (1991). Cephamycin production by *Streptomyces clavuligerus*. In S. Baumberg, H. Krugel, & D. Noack (Eds.), *Genetics and product formation in streptomyces* (pp. 87–103). New York: Plenum.

Dey, A. (1985). *Orthogonal fractional factorial designs*. New York: Wiley, 1–25.

Drozd, J. W. (1987). Hydrocarbons as feedstocks for biotechnology. In J. D. Stowell, A. J. Beardsmore, C. W. Keevil, & J. R. Woodward (Eds.), *Carbon substrates in biotechnology* (pp. 119–138). Oxford: IRL Press.

Dulbecco, R., & Freeman, G. (1959). Plaque production by the polynoma virus. *Virology, 8*, 396–397.

Eagle, H. (1959). Nutrition needs of mammalian cells in tissue culture. *Science, 130*, 432–437.

Eiki, H., Kishi, I., Gomi, T., & Ogawa, M. (1992). 'Lights out' production of cephamycins in automated fermentation facilities. In M. R. Ladisch, & A. Bose (Eds.), *Harnessing biotechnology for the 21st century* (pp. 223–227). Washington: American Chemical Society.

El-Naggar, N. E. -A., El-Bindary, A. A., & Nour, N. S. (2013). Statistical optimization of process variables for antimetabolite production by *Streptomyces anulatus* NEAE-94 against some multidrug resistant strains. *International Journal of Pharmacology, 9*(6), 322–334.

Eoff, J. R., Linder, W. V., & Beyer, G. F. (1919). Production of glycerine from sugar by fermentation. *Industrial & Engineering Chemistry, 11*, 82–84.

Fleischaker, H. (1987). Microcarrier cell culture. In B. K. Lydersen (Ed.), *Large scale cell culture technology* (pp. 59–79). Munich: Hanser.

Fogarty, W. M., & Ward, O. P. (1974). Pectinases and pectic polysaccharides. *Progress in Industrial. Microbiology, 13*, 59–119.

Foster, J. W. (1949). *Chemical activities of the fungi.* New York: Academic Press.

Foster, J. W., Woodruff, H. B., & McDaniel, L. E. (1943). Microbiological aspects of penicillin. 3. Production of penicillin in subsurface cultures of *Penicillium notatum. Journal of Bacteriology, 46*, 421–432.

Frages-Haddani, B., Tessier, B., Chenu, S., Chevalot, I., Harscoat, C., Marc, I., Goergen, J. L., & Marc, A. (2006). Peptide fractions of reapeseed hydrolysates as an alternative to animal proteins in CHO cell culture media. *Process Biochemistry, 41*, 2297–2304.

Francis, G. L. (2010). Albumin and mammalian cell culture: implications for biotechnology applications. *Cytotechnology, 62*, 1–16.

Fukui, S., & Tanaka, A. (1980). Production of useful compounds from alkane media in Japan. *Advances in Biochemical Engineering, 17*, 1–35.

Gallicchio, V., & Gottlieb, D. (1958). The biosynthesis of chloramphenicol. III. Effects of micronutrients on synthesis. *Mycologia, 50*, 490–500.

Garner, H. R., Fahmy, M., Phillips, R. I., Koffler, H., Tetrault, P. A., & Bohonos, N. (1950). Chemical changes in the submerged growth of *Streptomyces griseus. Bacteriology Proceedings*, 139–140.

Gaunt, D. M., Trinci, A. P. J., & Lynch, J. M. (1984). Metal ion composition and physiology of *Trichoderma reesei* grown on a chemically defined medium prepared in two different ways. *Transactions of the British Mycological Society, 83*, 575–581.

Ghildyal, N. P., Lonsane, B. K., & Karanth, N. G. (1988). Foam control ID submerged fermentation: state of the art. *Advances in Applied Microbiology, 33*, 173–221.

Goodey, A. R., Doel, S., Piggott, J. R., Watson, M. E. E., & Carter, B. L. A. (1987). Expression and secretion of foreign polypeptides in yeast. In D. R. Berry, I. Russell, & G. G. Stewart (Eds.), *Yeast biotechnology* (pp. 401–429). London: Allen and Unwin.

Goodman, J. J., Matrishin, M., Young, R. W., & McCormick, J. R. D. (1959). Inhibition of the incorporation of chlorine into the tetracycline molecule. *Journal of Bacteriology, 78*, 492–499.

Gore, J. H., Reisman, H. B., & Gardner, C. H. (1968). L-Glutamic acid by continuous fermentation. U.S. Patent 3,402,102.

Gourevitch, A., Misiek, M., & Lein, J. (1956). Competitive inhibition by bromide of incorporation of chloride into the tetracycline molecule. *Antibiotics & Chemotherapy, 5*, 448–452.

Greasham, R., & Inamine, E. (1986). Nutritional improvement of processes. In A. L. Demain, & N. A. Solomon (Eds.), *Manual of industrial microbiology and biotechnology* (pp. 41–48). Washington: American Society for Microbiology.

Grove, J. F. (1967). Griseofulvin. In D. Gottlieb, & P. D. Shaw (Eds.), *Antibiotics, II* (pp. 123–213). Berlin: Springer-Verlag.

Gunnarsson, N., Bruheim, P., & Nielsen, J. (2003). Production of the glycopeptide antibiotic A40926 by *Nonomuraea* sp. ATTC 39727: influence of medium composition in batch fermentation. *Journal of Industrial Microbiology and Biotechnology*, *30*, 150–156.

Gupta, A. J., Hageman, J. A., Wierenga, P. A., Boots, J. -W., & Gruppen, H. (2014). Chemometric analysis of soy protein hydrolysates used in animal cell culture for IgG production—N untargeted metabolomics approach. *Process Biochemistry*, *49*, 309–317.

Hahn-Hagerdal, B., Karhumaa, K., Larsson, C. U., Gorwa-Grauslund, M., Gorgens, J., & Van zyl, W. H. (2005). Role of cultivation media in the development of yeast strains for large scale industrial use. *Microbial Cell Factories*, *4*(31).

Hall, M. J., & Hassall, C. H. (1970). Production of the monomycins, novel depsipeptide antibiotics. *Applied Microbiology*, *19*, 109–112.

Hall, M. J., Dickinson, S. D., Pritchard, R., & Evans, J. I. (1973). Foams and foam control in fermentation processes. *Progress in Industrial Microbiology*, *12*, 169–231.

Hamer, G. (1979). Biomass from natural gas. In A. H. Rose (Ed.), *Economic microbiology* (pp. 315–360). (Vol. 4). London: Academic Press.

Harris, G. (1962). The structural chemistry of barley and malt. In A. H. Cook (Ed.), *Barley and malt, biology, biochemistry and technology* (pp. 431–582). London: Academic Press.

Harrison, J. S. (1971). Yeast production. *Progress in Industrial Microbiology*, *10*, 129–177.

Harrison, D. E. F., Topiala, H., & Hamer, G. (1972). Yield and productivity in single cell protein production from methane and methanol. In G. Temi (Ed.), *Fermentation technology today* (pp. 491–495). Japan: Society of Fermentation Technology.

Hendrix, C. (1980). Through the response surface with test tube and pipe wrench. *Chemtech*, 488–497.

Herbert, D. (1976). Stoichiometric aspects of microbial growth. In C. R. Dean, D. C. Elhvood, C. G. T. Evans, & J. Melling (Eds.), *Continuous culture 6: applications and new fields* (pp. 1–30). Chichester: Ellis Horwood.

Hersbach, G. J. M., Van Der Beek, C. P., & Van Duck, P. W. M. (1984). The penicillins: properties, biosynthesis and fermentation. In E. J. Vandamme (Ed.), *Biotechnology of industrial antibiotics* (pp. 45–140). New York: Marcel Dekker.

Hickey, R. J. (1945). The inactivation of iron by 2,2"-bipyridine and its effect on riboflavine synthesis by *Clostridium acetobutylicum*. *Archives of Biochemistry*, *8*, 439–447.

Hicks, C. R. (1993). *Fundamental concepts in the design of experiments* (4th ed.). New York: Saunders.

Hobbs, G., et al. (1992). An integrated approach to study regulation of production of the antibiotic methylenomycin by *Streptomyces coelicolor* A3(2). *Journal of Bacteriology*, *174*, 1487–1494.

Hockenhull, J. H. (1959). Improvements in or relating to antibiotics. British Patent 868,958.

Hoeksema, H., & Smith, C. G. (1961). Novobiocin. *Progress in Industrial Microbiology*, *3*, 91–139.

Holmberg, A., Sievanen, R., & Carlberg, G. (1980). Fermentation of *Bacillus thuringiensis* for endotoxin production. *Biotechnology and Bioengineering*, *22*, 1707–1724.

Hosler, P., & Johnson, M. J. (1953). Penicillin from chemically defined media. *Industrial & Engineering Chemistry*, *45*, 871–874.

Hough, J. S., Briggs, D. E., & Stevens, R. (1971). *Malting and brewing science*. London: Chapman and Hall, Chapter 7.

Hughes, M. N., & Poole, R. K. (1989). *Metals and microorganisms*. London: Chapman and Hall.

Hughes, M. N., & Poole, R. K. (1991). Metal speciation and microbial growth—the hard (and soft) facts. *Journal of General Microbiology, 137*, 725–734.

Hutner, S. H. (1972). Inorganic nutrition. *Annual Review of Microbiology, 26*, 313–346.

Inamine, E., Lago, B. D., & Demain, A. L. (1969). Regulation of mannosidase, an enzyme of streptomycin biosynthesis. In D. Perlman (Ed.), *Fermentation advances* (pp. 199–221). New York: Academic Press.

Inskeep, G. C., Benett, R. E., Dudley, J. F., & Shepard, M. W. (1951). Bacitracin, product of biochemical engineering. *Industrial & Engineering Chemistry, 43*, 1488–1498.

Iwai, Y., & Omura, S. (1982). Culture conditions for screening of new antibiotics. *The Journal of Antibiotics, 35*, 123–141.

Jayme, D., Watanabe, T., & Shimada, T. (1997). Basal medium development for serum-free culture: a historical perspective. *Cytotechnology, 23*, 95–101.

Jefferys, E. G. (1970). The gibberellin fermentation. *Advances in Applied Microbiology, 13*, 283–316.

Johnson, M. J. (1946). Metabolism of penicillin producing moulds. *Annals of the New York Academy of Sciences, 48*, 57–66.

Johnson, M. J. (1952). Recent advances in penicillin fermentations. *Bulletin of the World Health Organization, 6*, 99–121.

Junker, B., Mann, Z., Gailliot, P., Byrne, K., & Wilson, J. (1998). Use of soybean oil and ammonium sulfate additions to optimize secondary metabolite production. *Biotechnology and Bioengineering, 60*(5), 580–588.

Kadam, K. L., & Newman, M. M. (1997). Development of a low cost fermentation medium for ethanol production from biomass. *Applied Microbiology and Biotechnology, 47*, 625–629.

Karrow, E. O., & Waksman, S. A. (1947). Production of citric acid in submerged culture. *Industrial & Engineering Chemistry, 39*, 821–825.

Katz, E., Pienta, P., & Sivak, A. (1958). The role of nutrition in the synthesis of actinomycin. *Applied Microbiology, 6*, 236–241.

Keay, L. (1971). Microbial proteases. *Process Biochemistry, 6*(8), 17–21.

Kennedy, M., & Krouse, D. (1999). Strategies for improving fermentation media performace: a review. *Journal of Industrial Microbiology and Biotechnology, 23*, 456–475.

Kimura, A. (1967). Biochemical studies on *Streptomyces sioyaensis*. 11. Mechanism of the inhibitory effect of glucose on siomycin formation. *Agricultural and Biological Chemistry, 31*, 845–852.

Kinoshita, S., & Tanaka, K. (1972). Glutamic acid. In K. Yamada, S. Kinoshita, T. Tsunoda, & K. Aida (Eds.), *The microbial production of amino acids* (pp. 263–324). New York: Halsted Press-Wiley.

Kobayashi, M., Nagasawa, T., & Yamada, H. (1992). Enzymatic synthesis of acrylamide; a success story not yet over. *Trends in Biotechnology, 10*, 402–408.

Kobel, H., & Traber, R. (1982). Directed synthesis of cyclosporins. *European Journal of Applied Microbiology and Biotechnology, 14*, 237–240.

Koffler, H., Knight, S. G., & Frazier, W. C. (1947). The effect of certain mineral elements on the production of penicillin in shake flasks. *The Journal of Bacteriology, 53*, 115–123.

Kominek, L. A. (1972). Biosynthesis of novobiocin by *Streptomyces niveus*. *Antimicrobial Agents and Chemotherapy, 1*, 123–134.

Kuenzi, M. T., & Auden, J. A. L. (1983). Design and control of fermentation processes. In L. J. Nisbet, & D. J. Winstanley (Eds.), *Bioactive Microbial Products* (pp. 91–116). (Vol. 2). London: Academic Press.

Lancini, G., & White, R. J. (1973). Rifamycin fermentation studies. *Process Biochemistry*, *8*(7), 14–16.

Landecker (2007). *Culturing life—how cells became technologies*. Cambridge, Massachusetts: Harvard University Press.

Larsen, H. (1962). Halophilism. In I. C. Gunsalus, & R. Y. Stanier (Eds.), *The Bacteria* (pp. 297–342). (Vol. 4). New York: Academic Press.

Le petit, S.p.A. (1957). Bromotetracycline. British Patent 772, 149.

Leduy, J., Marsan, A. A., & Coupal, B. (1974). A study of the theological properties of a non-newtonian fermentation broth. *Biotechnolgy and Bioengineering*, *16*, 61–76.

Lee, J. C., & Tyman, K. J. (1988). Antifoams and their effects on coalesence between protein stabilized bubbles. In R. Kine (Ed.), *Second international conference on bioreactor fluid dynamics* (pp. 353–377). London: Elsevier.

Lein, J., Sawmiller, L. F., & Cheney, L. C. (1959). Chlorina-tion inhibitors affecting the biosynthesis of tetracycline. *Applied Microbiology*, *7*, 149–157.

Levi, J. D., Shennan, J. L., & Ebbon, G. P. (1979). Biomass from liquid n-alkanes. In A. H. Rose (Ed.), *Economic Microbiology* (pp. 361–419). (Vol. 4). London: Academic Press.

Lilly, V. G. (1965). The chemical environment for growth. 1. Media, macro and micronutrients. In G. C. Ainsworth, & A. S. Sussman (Eds.), *The Fungi* (pp. 465–478). (Vol. 1). New York: Academic Press.

Liras, P., Asturias, J. A., & Martin, J. F. (1990). Phosphate control sequences involved in transcriptional regulation of antibiotic biosynthesis. *Trends in Biotechnology*, *8*, 184–189.

Liu, C. M., McDaniel, L. E., & Schaffner, C. P. (1970). Factors affecting the production of candicidin. *Antimicrobial Agents and Chemotherapy*, *7*, 196–202.

Lobo-Alfonso, J., Price, P., & Jayme, D. (2010). Hydrolysates as components of serum-free media for animal cell culture applications. In V. H. C. Pasupuleti (Ed.), *Protein hydrolysates in biotechnology* (pp. 55–78). Dordecht: Springer.

Lukondeh, T., Ashbolt, N. J., & Rogers, P. L. (2005). Fed-batch fermentation for production of Kluyveromyces marxianus Fll 510700 cultivated on a lactose based media. *Journal of Industrial Microbiology and Biotechnology*, *32*, 284–288.

Luria, S. E. (1960). The bacterial protoplasm: composition and organisation. In I. C. Gunsalus, & R. Y. Stanier (Eds.), *The Bacteria* (pp. 1–34). (Vol. 1). New York: Academic Press.

MacLennan, D. G., Gow, J. S., & Stringer, D. A. (1973). Methanol-bacterium process for SCP. *Process Biochemistry*, *8*(6), 22–24.

Majumdar, M. K., & Majumdar, S. K. (1965). Effects of minerals on neomycin production by *Streptomyces fradiae*. *Applied Microbiology*, *13*, 190–193.

Malzahn, R. C, Phillips, R. F., & Hanson, A. M. (1959). Riboflavin. U.S. Patent 2, 876, 169.

Mandels, M., & Weber, J. (1969). The production of cellulases. In R. F. Gould (Ed.), *Celluloses and their applications. Advances in chemistry series* (pp. 391–414). (Vol. 95). Washington: American Chemical Society.

Marshall, R., Redfield, B., Katz, E., & Weissback, H. (1968). Changes in phenoxazinone synthetase activity during the growth cycle of *Streptomyces antibioticus*. *Archives of Biochemistry and Biophysics*, *123*, 317–323.

Martin, J. F., & Demain, A. L. (1978). Fungal development and metabolite formation. In J. E. Smith, & D. R. Berry (Eds.), *The filamentous fungi* (pp. 426–450). (Vol. 3). London: Arnold.

Martin, J. F., & Demain, A. L. (1980). Control of antibiotic biosynthesis. *Microbiological Reviews*, *44*, 230–251.

Martin, J. F., & McDaniel, L. E. (1975). Kinetics of biosynthesis of polyene macrolide antibiotics by batch cultures: cell maturation time. *Biotechnology and Bioengineering.*, *17*, 925–938.

Martin, J. F., & McDaniel, L. E. (1977). Production of polyene macrolide antibiotics. *Advances in Applied Microbiology*, *21*, 1–52.

Matsumura, M., Imanaka, T., Yoshida, T., & Taguchi, H. (1978). Effect of glucose and methionine consumption rates on cephalosporin C production by *Cephalosporium acremonium*. *Journal of Fermentation Technology*, *56*, 345–353.

McDaniel, L. E., Bailey, E. G., Ethiraj, S., & Andrews, H. P. (1976). Application of response surface optimization techniques to polyene macrolide fermentation studies in shake flasks. *Developments in Industrial Microbiology*, *17*, 91–98.

Mertz, F. P., & Doolin, L. E. (1973). The effect of phosphate on the biosynthesis of vancomycin. *Canadian Journal of Microbiology*, *19*, 263–270.

Mervyn, I., & Smith, E. L. (1964). The biochemistry of vitamin B 12 fermentation. *Progress in Industrial Microbiology*, *5*, 105–201.

Michiels, J. -F., Barbau, J., De Boel, S., Dessy, S., Agathos, S. N., & Schneider, Y. -J. (2011). Characterisation of beneficial and detrimental effects of a soy peptone, as an additive for CHO cell cultivation. *Process Biochemistry*, *46*, 671–681.

Miller, T. L., & Churchill, B. W. (1986). Substrates for large scale fermentations. In A. L. Demain, & N. A. Solomons (Eds.), *Manual of industrial microbiology* (pp. 122–136). Washington, DC: American Society for Microbiology.

Minoda, Y. (1986). Raw materials for amino acid fermentation. *Developments in Industrial Microbiology*, *24*, 51–66.

Misecka, J., & Zelinka, J. (1959). The effect of some elements on the quality of corn steep with rererence to the production of penicillin. *Biologia (Bratislava)*, *14*, 591–596.

Miyairi, N., Miyoshi, T., Oki, H., Kohsaka, M., Ikushima, H., Kunugita, K., Sakai, H. H., & Imanaka, H. (1970). Studies on thiopeptin antibiotics. 1. Characteristics of thiopeptin B. *The Journal of Antibiotics*, *23*, 113–119.

Miyoshi, N., Miyairi, H., Aoki, H., Kohsaka, M., Sakai, H., & Imanaka, H. (1972). Bicyclomycin, a new antibiotic. I. Taxonomy, isolation and characterization. *The Journal of Antibiotics*, *25*, 269–275.

Mizusawa, K., Ichishawa, E., & Yoshida, F. (1966). Proteolytic enzymes of the thermophilic *Streptomyces*. II. Identification of the organism and some conditions of protease formation. *Agricultural and Biological Chemistry*, *30*, 35–41.

Moore, G. E., Gerner, R. E., & Franklin, H. A. (1967). Culture of human leukocytes. *The Journal of the American Medical Association*, *199*(8), 519–524.

Moo-Young, M. (1977). Economics of SCP production. *Process Biochemistry*, *12*(4), 6–10.

Morris, A. E., & Schmid, J. (2000). Effects of insulin and LongR(3) on serum-free Chniese hamster ovary cells expressing two recombinant proteins. *Biotechnology Progress*, *16*, 693–697.

Morton, A. G., & Macmillan, A. (1954). The assimilation of nitrogen from ammonium salts and nitrate by fungi. *Journal of Experimental Botany*, *5*, 232–252.

Moss, A. R., & Klein, R. (1946). Improvements in or relating to the manufacture of riboflavin. British Patent 615, 847.

Moyer, A. J., & Coghill, R. D. (1946). Penicillin. IX. The laboratory scale production of penicillin by submerged culture by *Penicillium notatum* Westling (NRRL832). *Journal of Bacteriology*, *51*, 79–93.

Moyer, A. J., & Coghill, R. D. (1947). Penicillin. IX. The effect of phenylacetic acid on penicillin production. *Journal of Bacteriology*, *53*, 329–341.

Murakami, H., & Yamada, K. (1987). Production of cancer-specific monoclonal antibodies with human-human hybridomas and their serum-free, high density, perfusion culture. In R. E. Spier, & J. B. Griffiths (Eds.), *Modern approaches to animal cell technology* (pp. 52–76). London: Butterworths.

Murakami, H., Masui, H., Sato, G. H., Sueoka, N., Chow, T. P., & Konosueoka, T. (1982). Growth of hybridoma cells in serum-free medium: ethanolamine is an essential component. *Proceedings of the National Academy of Sciences, USA, 79,* 1158–1162.

Murakami, H., Shimomura, T., Nakamura, T., Ohashi, H., Shinohara, K., & Omura, H. (1984). Development of a basal medium for serum-free cultivation of hybridoma cells in high density. *Journal of the Agricultural Chemical Society of Japan, 56,* 575–583.

Nakayama, K. (1972a). Lysine and diaminopimelic acid. In K. Yamada, S. Kinoshita, T. Tsunoda, & K. Aiba (Eds.), *The microbial production of amino acids* (pp. 369–397). New York: Halsted Press.

Nakayama, K. (1972b). Micro-organisms in amino acid fermentation. In G. Temi (Ed.), *Fermentation technology today* (pp. 433–438). Japan: Society of Fermentation Technology.

Neidleman, S. L., Bienstock, E., & Bennett, R. E. (1963). Biosynthesis of 7-chloro-6-demethyltetracycline in the presence of aminopterin and ethionine. *Biochimica et Biophysica Acta, 71,* 199–201.

Nelder, J. A., & Mead, R. (1965). A simplex method for function minimization. *The Computer Journal, 7,* 308–313.

Nelson, L. S. (1982). Technical aids. *Journal of Quality Technology, 14*(2), 99–100.

Nishikiori, T., Masuma, R., Oiwa, R., Katagiri, M., Awaya, J., Iwai, Y., & Omura, S. (1978). Aurantinin, a new antibiotic of bacterial origin. *The Journal of Antibiotics, 31,* 525–532.

Nubel, R. D., & Ratajak, E. J. (1962). Process for producing itaconic acid. U.S. Patent 3,044,941.

Pan, S. C., Bonanno, S., & Wagman, G. H. (1959). Efficient utilization of fatty oils as energy source in penicillin fermentation. *Applied Microbiology, 7,* 176–180.

Pan, C. H., Speth, S. V., McKillip, E., & Nash, C. H. (1982). Methyl oleate-based medium for cephalosporin C production. *Developments in Industrial Microbiology, 23,* 315–323.

Papagianni, M., & Papamichael, E. M. (2014). Production of pediocin SM-1 by *Pediococcus pentosaceus* Mees 1934 in fed-batch fermentation: effects of sucrose concentration in a complex medium and process modeling. *Process Biochemistry, 49,* 2044–2048.

Papapanagiotou, P. A., Quinn, H., Molitor, J. -P., Nienow, A. W., & Hewitt, C. J. (2005). The use of phase inversion temperature (PIT) microemulsion technology to enhance oil utilization during *Streptomyces rimosus* fed-batch fermentations to produce oxytetracycline. *Biotechnology Letters, 27,* 1579–1585.

Pasupuleti, V. H. C., Holmes, C., & Demain, A. L. (2010). Applications of protein hydrolysates in biotechnology. In V. H. C. Pasupuleti (Ed.), *Protein hydrolysates in biotechnology* (pp. 1–9). Dordecht: Springer.

Peberdy, J. F. (1988). Genetic manipulation. In D. R. Berry (Ed.), *Physiology of industrial fungi* (pp. 187–218). Oxford: Blackwell.

Perlman, D. (1970) The evolution of penicillin manufacturing processes. In A.L. Elder (Ed.), The History of Penicillin Production, *Chemical Engineering Progress Symposium Series, Vol. 60(100),* pp. 25-30, Oxford: Blackwell Publishing.

Phillips, T., & Somerson, L. (1960). Glutamic acid. U.S. Patent 3,080,297.

Phillips, J. R., Atiyeh, H. K., Tanner, R. S., Torres, J. R., Saxena, J., Wilkins, M. R., & Huhnke, R. L. (2015). Butanol and hexanol production in *Clostridium carboxidivorans* syngas fermentation: medium development and culture techniques. *Bioresource Technology, 190,* 114–121.

Piper, P. W., & Kirk, N. (1991). Inducing heterologous gene expression in yeast as fermentations approach maximal biomass. In A. Wiseman (Ed.), *Genetically engineered proteins and enzymes from yeast: process control* (pp. 147–184). Chichester: Ellis Horwood.

Pirt, S. J., & Righelato, R. C. (1967). Effect of growth rate on the synthesis of penicillin by *Penicillium chrysogenum* in batch and chemostat cultures. *Applied Microbiology*, *15*, 1284–1290.

Plackett, R. L., & Burman, J. P. (1946). The design of multifactorial experiments. *Biometrika*, *33*, 305–325.

Priest, F. G., & Sharp, R. J. (1989). Fermentation of bacilli. In J. O. Neway (Ed.), *Fermentation process development of industrial organisms* (pp. 73–112). London: Plenum.

Rajagopalan, G., & Krishnan, C. (2009). Optimization of agro-residual medium for α-amylase production from a hyper-producing *Bacillus subtilis* KCC103 in submerged fermentation. *Journal of Chemical Technology and Biotechnology*, *84*, 618–625.

Ramamoorthy, S., & Kushner, D. J. (1975). Binding of mecuric and other heavy metal ions by microbial growth media. *Microbial Ecology*, *2*, 162–176.

Reed, G., & Peppler, H. J. (1973). In *Yeast Technology*. Westport: AVI Publishing Company, Chapter 5.

Reese, E. T. (1972). Enzyme production from insoluble substrates. *Biotechnology and Bioengineering Symposium*, *3*, 43–62.

Reyes, P., Chichester, C. O., & Nakayama, T. O. M. (1964). Mechanism of *β*-ionone stimulation of carotenoid and ergosterol biosynthesis in *Phycomyces blakeseeanus*. *Biochimica et Biophysica Acta*, *90*, 578–592.

Rhodes, A. (1963). Griseofulvin: production and biosynthesis. *Progress in Industrial Microbiology*, *4*, 165–187.

Rhodes, A., & Fletcher, D. L. (1966). *Principles of industrial microbiology*. Oxford: Pergamon Press.

Rhodes, A., Crosse, R., Ferguson, T. P., & Fletcher, D. L. (1955). Improvements in or relating to the production of the antibiotic griseofulvin. British Patent 784, 618.

Sankaran, L., & Pogell, B. M. (1975). Biosynthesis of puromycin in *Streptomyces alboniger*: Regulation and properties of *O*-dimethyl puromycin *O*-methyl transferase. *Antimicrobial Agents and Chemotherapy*, *8*, 727–732.

Saval, S., Pablos, L., & Sanchez, S. (1993). Optimization of a culture medium for streptomycin production using response-surface methodology. *Bioresource Technology*, *43*, 19–25.

Schugerl, K. (1985). Foam formation, foam suppression and the effect of foam on growth. *Process Biochemistry*, *20*(4), 122–123.

Sensi, P., & Thiemann, J. E. (1967). Production of rifamycins. *Progress in Industrial Microbiology*, *6*, 21–60.

Sentfen, H. (1989). Brewing liquor: quality requirements and corrective measures. *Brauerei Rundschau*, *100*, 53–56.

Sharp, D. H. (1989). *Bioprotein manufacture: a critical assessment*. Chichester: Ellis Horwood.

Shepherd, J. (1963). Citric acid. U.S. Patent 3,083,144.

Shin, H. Y., Lee, J. Y., Jung, Y. R., & Kim, S. W. (2010). Stimulation of cephalosporin C production in *Acremonium chrysogenum* M35 by glycerol. *Bioresource Technology*, *101*, 4549–4553.

Shu, P., & Johnson, M. J. (1948). Interdependence of medium constituents in citric acid production by submerged fermentation. *Journal of Bacteriology*, *56*, 577–585.

Siemensma, A., Babcock, J., Wilcox, C., & Huttinga, H. (2010). Towards an understanding of how protein hydrolysates stimulate more efficient biosynthesis in cultured cells.

In V. J. Pasupuleti, & A. L. Demain (Eds.), *Protein Hydrolysates in Biotechnology* (pp. 33–54). Netherlands: Springer.

Silva, J. P. A., Mussato, S. I., Roberto, I. C., & Teixeira, J. A. (2012). Fermentation media and oxygen transfer conditions that maximize the xylose conversion to ethanol by *Pichia stipsis. Renewable Energy, 37*, 259–265.

Singh, A., Bruzelius, E., & Heding, H. (1976). Streptomycin, a fermentation study. *European Journal of Applied Microbiology, 3*, 97–101.

Smiley, K. L., & Stone, L. (1955). Production of riboflavine by *Ashbya gossypii.* U.S. Patent 2,702, 265.

Smith, C. G. (1956). Fermentation studies with *Streptomyces niueus. Applied Microbiology, 4*, 232–236.

Smith, J. E. (1986). Concepts of industrial antibiotic production. In D. I. Alani, & M. Moo-Young (Eds.), *Perspectives in biotechnology and applied microbiology* (pp. 106–142). London: Elsevier.

Smith, E. L., & Bide, A. E. (1948). Penicillin salts. *Biochemical Journal, 42*, xvii–xvii10.

Smith, J. E., Nowakowska-Waszczuk, K., & Anderson, J. G. (1974). Organic acid production by mycelial fungi. In B. Spencer (Ed.), *Industrial aspects of biochemistry, part A* (pp. 297–317). Amsterdam: North Holland.

Soc Richard, & Allente et CIe (1921). Acetone and butyl alcohol. British Patent 176, 284.

Solomons, G. L. (1967). Antifoams. *Process Biochemistry, 2*(10), 47–48.

Solomons, G. L. (1969). *Materials and methods in fermentation.* London: Academic Press.

Somrutai, W., Takagi, M., & Yoshida, T. (1996). Acetone-butanol fermentation by *Clostridium aurantibutyricum* ATTC 17777 from a model medium for palm oil mill effluent. *Journal of Fermentation and Bioengineering, 81*(6), 543–547.

Soper, Q. F., Whitehead, C., Behrens, O. K., Corse, J. J., & Jones, R. G. (1948). Biosynthesis of penicillins. VII. Oxy- and mercapto acetic acids. *Journal of Chemical Society, 70*, 2849–2855.

Spendley, W., Hext, G. R., & Himsworth, F. R. (1962). Sequential application of simplex designs in optimization and evolutionary operation. *Technometrics, 4*, 441–461.

Stapley, E. O., & Woodruff, H. B. (1982). Avermectin, antiparasitic lactones produced by *Streptomyces avermitilis* isolated from a soil in Japan. In H. Umezawa, A. L. Demain, T. Mata, & C. R. Hutchinson (Eds.), *Trends in antibiotic research* (pp. 154–170). Tokyo: Japanese Antibiotic Research Association.

Stowe, R. A., & Mayer, R. P. (1966). Efficient screening of process variables. *Industrial Engineering and Chemistry, 56*, 36–40.

Stowell, J. D. (1987). The application of oils and fats in antibiotic processes. In J. D. Stowell, A. J. Beardsmore, C. W. Keevil, & J. R. Woodward (Eds.), *Carbon Substrates in Biotechnology* (pp. 139–159). Oxford: IRL Press.

Suomalainen, H., & Oura, A. (1971). Yeast nutrition and solute uptake. In A. H. Rose, & J. S. Harrison (Eds.), *The Yeasts* (pp. 3–74). (Vol. 2). London: Academic Press.

Svensson, M. L., Roy, P., & Gatenbeck, S. (1983). Glycerol catabolite regulation of D-cycloserine production in *Streptomyces garyphalus. Archives of Microbiology, 135*, 191–193.

Szumski, S. A. (1959). Chlorotetracycline fermentation. U.S. Patent 2,871,167.

Tanner, F., Vojnovich, C., & Van Lanen, J. (1945). Riboflavin production by *Candida* species. *Science, 101*, 180–181.

Tilley, B. C, Testa, R.T., & Dorman, E. (1975). A role of cobalt ions in the biosythesis of gentamicin. *Abstracts of papers of 31st Meeting for Soc. Indust. Microbiol.*

Topiwala, H. H., & Khosrovi, B. (1978). Water recycle in biomass production processes. *Biotechnology and Bioengineering, 20*, 73–85.

Trilli, A., Michelini, V., Mantovani, V., & Pirt, S. J. (1977). Estimation of productivities in repeated fed batch cephalosporin fermentation. *Journal of Applied Chemistry and Biotechnology, 27*, 219–224.

Tuffile, C. M., & Pinho, F. (1970). Determination of oxygen-transfer coefficients in viscous streptomycete fermentations. *Biotechnology and Bioengineering, 12*, 849–871.

Uemura, T., Sugisaki, Z., & Takamura, Y. (1972). Valine. In K. Yamada, S. Kinoshita, T. Tsunoda, & K. Aida (Eds.), *The microbial production of amino acids* (pp. 339–368). New York: Halsted Press–Wiley.

Underkofler, L. A. (1966). Production of commercial enzymes. In G. Reed (Ed.), *Enzymes in food processing*. New York: Academic Press, Chapter 10.

van der Valk, J., Brunner, D., De Smet, K., Fex Svenningsen, A., Honegger, P., Knudsen, L. E., et al. (2010). Optimization of chemically defined cell culture media—replacing fetal bovine serum in mammalian in vitro methods. *Toxicologi in Vitro, 24*, 1053–1063.

Van Dyck, P., & De Somer, P. (1952). Production and extraction methods of aureomycin. *Antibiotics and Chemotherapy, 2*, 184–198.

Van't Riet, K., & Van Sonsbeck, H. M. (1992). Foaming, mass transfer and mixing interactions in large-scale fermenters. In M. R. Ladisch, & A. Bose (Eds.), *Harnessing biotechnology for the 21st century* (pp. 189–192). Washington: American Chemical Society.

Vardar-Sukan, F. (1992). Foaming and its control in bioprocesses. In F. Vardar-Sukan, & S. Suha-Sukan (Eds.), *Recent advances in biotechnology* (pp. 113–146). Dordrecht: Kluwer.

Veale, R. A., & Sudbery, P. E. (1991). Methylotrophic yeasts as gene expression systems. In J. F. Perberdy, C. E. Caten, J. E. Ogden, & J. W. Bennett (Eds.), *Applied molecular genetics of fungi* (pp. 118–128). Cambridge: Cambridge University.

Vecht-Lifshitz, S. E., & Braun, S. (1989). Fermentation broth of *Bacillus thuringiensis* as a source of precursors for production of nikkomycins. *Letters in Applied Microbiology, 9*, 79–81.

Wallenfels, K., Bender, H., & Rached, J. R. (1966). Pullulanase from *Aerobacter aerogenes*. *Biochemical and Biophysics Research Communication, 22*, 254–261.

Walsh, G. (2014). Biopharmaceutical benchmarks 2014. *Nature Biotechnology, 32*(10), 992–1000.

Walton, R. B., McDaniel, L. E., & Woodruff, H. B. (1962). Biosynthesis of novobiocin analogues. *Developments in Industrial Microbiology, 3*, 370–375.

Wayne Davies, R. (1991). Expression of heterologous genes in filamentous fungi. In J. F. Perberdy, C. E. Caten, J. E. Ogden, & J. W. Bennett (Eds.), *Applied molecular genetics of fungi* (pp. 103–117). Cambridge: Cambridge University.

Weinberg, E. D. (1967). Bacitracin, gramicidin and tyrocidin. In D. Gottlieb, & P. D. ShaWwl (Eds.), *Antibiotics* (pp. 240–253). (Vol. 2). Berlin: Springer-Verlag.

Weinberg, E. D. (1970). Biosynthesis of secondary metabolites: roles of trace metals. *Advances in Microbial Physics, 4*, 1–44.

Weinberg, E. D. (1974). Secondary metabolism: control by temperature and inorganic phosphate. *Developments in Industrial Microbiology, 15*, 70–81.

Weinberg, E. D., & Tonnis, S. M. (1966). Action of chloramphenicol and its isomers on secondary biosynthesis processes of *Bacillus*. *Applied Microbiology, 14*, 850–856.

Whitaker, A. (1973). Fermentation economics. *Process Biochemistry, 8*(9), 23–26.

Whitaker, A. (1976). Amino acid transport into fungi: an essay. *Transactions of the British Mycological Society, 67*, 365–376.

Windish, W. W., & Mhatre, N. S. (1965). Microbial amylases. *Advances in Applied Microbiology, 7*, 273–304.

Winkler, M. A. (1991). Environmental design and time-profiling in computer controlled fermentation. In A. Wiseman (Ed.), *Genetically engineered protein and enzymes from yeast: production and control.* Chichester: Ellis Horwood, Chapter 4.

Woodward, J. C., Snell, R. L., & Nicholls, R. S. (1949). Conditioning molasses and the like for the production of citric acid. U.S. Patent 2,492,673.

Xi, Y. -L., Chen, K. -Q., Xu, R., Zhang, J. -H., Bai, X. -F., Jiang, M., Wei, P., & Chen, J. -Y. (2012). Effect of biotin and a similar compound on succinic acid fermentation by *Actinobacillus succinogenes* in a chemically defined media. *Biochemical Engineering Journal, 69,* 87–92.

Zhang, J., & Greasham, R. (1999). Chemically defined media for commercial fermentations. *Applied Microbiology and Biotechnology, 51,* 407–421.

Zhang, S., Handacorrigan, A., & Spier, R. E. (1992). Foaming and medium surfactant effects on the cultivation of animal cells in sirred and sparged bioreactors. *Journal of Biotechnology, 25,* 289–306.

Zhang, J., Reddy, J., Buckland, B., & Greasham, R. (2003). Towards consistent and productive complex media for industrial fermentations: Studies on yeast extract for a recombinant yeast fermentation process. *Biotechnolgy and Bioengineering, 82*(6), 640–652.

Zhang, W., Sinha, J., & Meagher, M. M. (2006). Glycerophosphate as a phosphorous source in a defined medium for *Pichia pastoris* fermentation. *Applied Microbiology and Biotechnology, 72,* 139–144.

Zhang, C., Fan, D., Shang, L., Ma, X., Luo, Y., Xue, W., & Gao, P. (2010). Optimization of fermentation process for human like collagen production of recombinant *Escherichia coli* using response surface methodology. *Chinese Journal of Chemical Engineering, 18*(1), 137–142.

Sterilization

5

INTRODUCTION

A fermentation product is produced by the culture of a certain organism, or animal cell line, in a nutrient medium. If a foreign microorganism invades the fermentation then the following consequences may occur:

1. The medium would have to support the growth of both the production organism and the contaminant, resulting in a loss of productivity.
2. If the fermentation is a continuous one then the contaminant may "outgrow" the production organism and displace it from the fermentation.
3. The foreign organism may contaminate the final product, for example, single-cell protein where the cells, separated from the broth, constitute the product.
4. The contaminant may produce compounds that make subsequent extraction of the final product difficult.
5. The contaminant may degrade the desired product; this is common in bacterial contamination of antibiotic fermentations where the contaminant would have to be resistant to the normal inhibitory effects of the antibiotic and degradation of the antibiotic is a common resistance mechanism, for example, the degradation of β-lactam antibiotics by β-lactamase-producing bacteria.
6. Contamination of a bacterial fermentation with phage could result in the lysis of the culture.

Avoidance of contamination may be achieved by:

1. Effective design and construction of the fermentation plant.
2. Using a pure inoculum to start the fermentation, as discussed in Chapter 6.
3. Sterilizing the medium to be employed.
4. Sterilizing the fermenter vessel.
5. Sterilizing all materials to be added to the fermentation during the process, for example, air, nutrient feeds, antifoams, and pH titrants.
6. Maintaining aseptic conditions during the fermentation.
7. Putting in place detailed operating procedures for sterilization, aseptic maintenance, and staff training.

The extent to which these procedures are adopted is determined by the likely probability of contamination and the nature of its consequences. Some fermentations are described as "protected"—that is, only a very limited range of microorganisms may utilize the medium, or the growth of the process organism may result in the development of selective growth conditions, such as a reduction in pH. The brewing of beer falls into this category; hop resins tend to inhibit the growth of many microorganisms and the growth of brewing yeasts tends to decrease the pH of the medium. Thus, brewing worts are boiled, but not necessarily sterilized, and the fermenters are thoroughly cleaned with disinfectant solution but are not necessarily sterile. Also, the precautions used in the development of inoculum for brewing are far less stringent than, for example, in an antibiotic fermentation. However, the vast majority of fermentations are not "protected" and, if contaminated, would suffer some of the consequences previously listed. A contaminated "classical" microbial fermentation would only be deemed suitable for down-stream processing, if it conformed with the characteristics described by Stockbridge et al. (2001). These authors considered the nature of the contaminant, the level of contamination, its ability to survive down-stream processing and the likelihood of the contaminant being a potential pathogen or producing a toxin (Fig. 5.1). Junker, Lester, Brix, Wong, and Nuechterlein (2006) reported that fermentations contaminated by a fungus are normally discarded due to the high probability of fungi producing toxic products. At the other end of the spectrum from protected fermentations are those processes (either microbial or cell culture) producing injectable biologics or vaccines. If contamination is detected in these processes then the batch is rejected, the equipment shut down and an investigation into the cause of contamination initiated (Pollard, 2011). Animal cell cultures are particularly susceptible to contamination due to their low growth rate, and hence the duration of the inoculum development programme as well as the production fermentation. Such processes are also vulnerable to viral and mycoplasma infection. Mycoplasmas lack cell walls and can quickly dominate an animal cell culture, producing up to 10^6-10^7 colony forming units (CFU) cm^{-3} but contributing virtually no turbidity to the culture (Stacey & Stacey, 2000).

In a wide-reaching review of aseptic operation, Pollard (2011) summarized the microbial taxa causing contamination of fermentations and their likely origin. The most common contaminants are Gram-positive spore-forming rods (*Bacillus* spp.) usually linked with inadequate medium sterilization, due to the presence of large insoluble medium particles, or poor cleaning of the fermenter prior to sterilization resulting in accumulation of dried broth in crevices and joints. The presence of Gram-negative rods is indicative of cooling water leaks, water in the inlet air, or inadequate filter sterilization while contamination by Gram-positive cocci is usually linked with a failure in air filtration. The economically acceptable frequency of contamination has been cited as one fermentation in a hundred (Sharma & Gurtu, 1993).

The approaches adopted to avoid contamination will be discussed in more detail, apart from the development of aseptic inocula, which is considered in Chapter 6, and the design and aseptic operation and containment of fermentation vessels, which are discussed in Chapters 6, and 7.

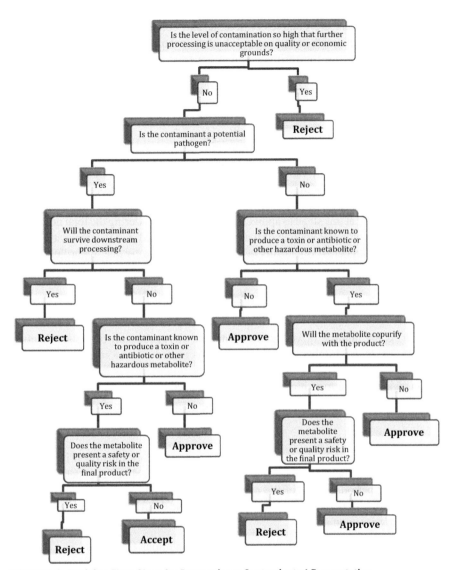

FIGURE 5.1 Decision Flow-Chart for Processing a Contaminated Fermentation

Modified from Stockbridge et al., 2001

MEDIUM STERILIZATION

As pointed out by Corbett (1985), media may be sterilized by filtration, radiation, ultrasonic treatment, chemical treatment, or heat. However, for practical reasons, steam is used almost universally for the sterilization of fermentation media. The major exception is the use of filtration for the sterilization of media for animal-cell culture—such media are completely soluble and contain heat labile components making filtration the method of choice. Filtration techniques will be considered later in this chapter. Before the techniques that are used for the steam sterilization of culture media are discussed, it is necessary to discuss the kinetics of sterilization. The destruction of microorganisms by steam (moist heat) may be described as a first-order chemical reaction and, thus, may be represented by the following equation:

$$-\frac{dN}{dt} = kN \tag{5.1}$$

where N is the number of viable organisms present, t is the time of the sterilization treatment (minutes), k is the reaction rate constant of the reaction, or the specific death rate (min^{-1}).

It is important at this stage to appreciate that we are considering the total number of organisms present in the volume of medium to be sterilized, *not* the concentration—the minimum number of organisms to contaminate a batch is one, regardless of the volume of the batch. On integration of Eq. (5.1), the following expression is obtained:

$$\frac{N_t}{N_0} = e^{-kt} \tag{5.2}$$

where N_0 is the number of viable organisms present at the start of the sterilization treatment, N_t is the number of viable organisms present after a treatment period, t minutes.

On taking natural logarithms, Eq. (5.2) is reduced to:

$$\ln\left(\frac{N_t}{N_0}\right) = -kt \tag{5.3}$$

The graphical representations of Eqs. (5.1) and (5.3) are illustrated in Fig. 5.2, from which it may be seen that viable organism number declines exponentially over the treatment period. A plot of the natural logarithm of N_t/N_0 against time yields a straight line, the slope of which equals-k. This kinetic description makes two predictions that appear anomalous:

1. An infinite time is required to achieve sterile conditions (ie, $N_t = 0$).
2. After a certain time, there will be less than one viable cell present.

Thus, in this context, a value of N_t of less than one is considered in terms of the probability of an organism surviving the treatment. For example, if it was predicted that a particular treatment period reduced the population to 0.1 of a viable organism, this implies that the probability of one organism surviving the treatment is one in ten. This may be better expressed in practical terms as a risk of one batch in ten becoming contaminated. This aspect of contamination will be considered later.

The relationship displayed in Fig. 5.2 would be observed only with the sterilization of a pure culture in one physiological form, under ideal sterilization conditions. The value of k is not only species dependent, but dependent on the physiological form of the cell; for example, the endospores of the genus *Bacillus* are far more

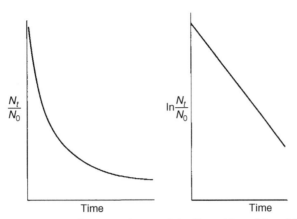

FIGURE 5.2 Plots of the Proportion of Survivors and the Natural Logarithm of the Proportion of Survivors in a Population of Microorganisms Subjected to a Lethal Temperature Over a Time Period

heat resistant than the vegetative cells. Richards (1968) produced a series of graphs illustrating the deviation from theory that may be experienced in practice. Fig. 5.3a,b and c illustrate the effect of the time of heat treatment on the survival of a population of bacterial endospores. The deviation from an immediate exponential decline in viable spore number is due to the heat activation of the spores, that is the induction of spore germination by the heat and moisture of the initial period of the sterilization process. In Fig. 5.3a, the activation of spores is significantly more than their destruction during the early stages of the process and, therefore, viable numbers increase

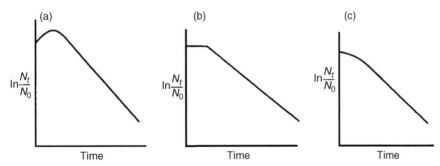

FIGURE 5.3 Plots of the Behavior of a Population of Bacterial Spores Subjected to a Lethal temperature Over a Time Period

(a) Initial population increase resulting from the heat activation of spores in the early stages of a sterilization process (Richards, 1968). (b) An initial stationary period observed during a sterilization treatment due to the death of spores being completely compensated by the heat activation of spores (Richards, 1968). (c) Initial population decline at a sub-maximum rate during a sterilization treatment due to the death of spores being compensated by the heat activation of spores (Richards, 1968).

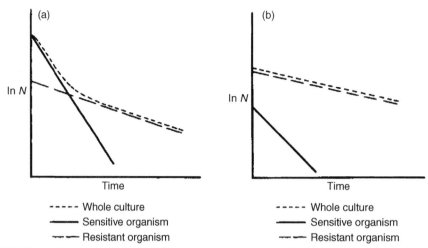

FIGURE 5.4 Plots of the Behavior of Mixed Microbial Cultures Subjected to a Lethal Temperature Over a Time Period

(a) The effect of a sterilization treatment on a mixed culture consisting of a high proportion of a very sensitive organism (Richards, 1968). (b) The effect of a sterilization treatment on a mixed culture consisting of a high proportion of a relatively resistant organism (Richards, 1968).

before the observation of exponential decline. In Fig. 5.3b activation is balanced by spore death and in Fig. 5.3c activation is less than spore death.

Fig. 5.4a and b illustrate typical results of the sterilization of mixed cultures containing two species with different heat sensitivities. In Fig. 5.4a the population consists mainly of the less-resistant type where the initial decline is due principally to the destruction of the less-resistant cell population and the later, less rapid decline, is due principally to the destruction of the more resistant cell population. Fig. 5.4b represents the reverse situation where the more resistant type predominates and its presence disguises the decrease in the number of the less resistant type.

As with any first-order reaction, the reaction rate increases with increase in temperature due to an increase in the reaction rate constant, which, in the case of the destruction of microorganisms, is the specific death rate (k). Thus, k is a true constant only under constant temperature conditions. The relationship between temperature and the reaction rate constant was demonstrated by Arrhenius and may be represented by the equation:

$$\frac{d\ln k}{dT} = \frac{E}{RT^2} \tag{5.4}$$

where E is the activation energy, R is the gas constant, T is the absolute temperature.

On integration Eq. (5.4) gives:

$$k = Ae^{-E/RT} \tag{5.5}$$

where A is the Arrhenius constant.

On taking natural logarithms, Eq. (5.5) becomes:

$$\ln k = \ln A - \frac{E}{RT} \tag{5.6}$$

From Eq. (5.6), it may be seen that a plot of $\ln k$ against the reciprocal of the absolute temperature will give a straight line. Such a plot is termed an Arrhenius plot and enables the calculation of the activation energy and the prediction of the reaction rate for any temperature. By combining Eqs. (5.3) and (5.5) together, the following expression may be derived for the heat sterilization of a pure culture at a constant temperature:

$$\ln N_0/N_t = A \cdot t \cdot e^{-(E/RT)}. \tag{5.7}$$

Deindoerfer and Humphrey (1959) used the term $\ln N_0/N_t$ as a design criterion for sterilization, which has been variously called the Del factor, Nabla factor, and sterilization criterion represented by the term ∇. Thus, the Del factor is a measure of the fractional reduction in viable organism count produced by a certain heat and time regime. Therefore:

$$\nabla = \ln(N_0/N_t)$$

but

$$\ln(N_0/N_t) = kt$$

and

$$kt = A \cdot t \cdot e^{-(E/RT)}$$

thus

$$\nabla = A \cdot t \cdot e^{-(E/RT)}. \tag{5.8}$$

On taking natural logarithms and rearranging, Eq. (5.8) becomes:

$$\ln t = \frac{E}{RT} + \ln\left(\frac{\nabla}{A}\right) \tag{5.9}$$

Thus, a plot of the natural logarithm of the time required to achieve a certain ∇ value against the reciprocal of the absolute temperature will yield a straight line, the slope of which is dependent on the activation energy, as shown in Fig. 5.5. From Fig. 5.5 it is clear that the same degree of sterilization (∇) may be obtained over a wide range of time and temperature regimes; that is, the same degree of sterilization may result from treatment at a high temperature for a short time as from a low temperature for a long time.

This kinetic description of bacterial death enables the design of procedures (giving certain ∇ factors) for the sterilization of fermentation broths. By choosing a value for N_t, procedures may be designed having a certain probability of achieving sterility, based upon the degree of risk that is considered acceptable. According to Deindoerfer

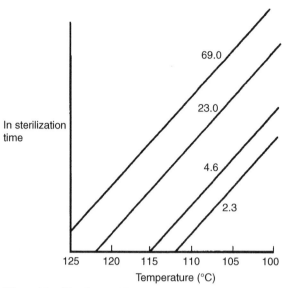

FIGURE 5.5 The Effect of Sterilization and Temperature on the Del Factor Achieved in the Process

The figures on the graph indicate the Del factors for each straight line.

Modified after Richards, 1966

and Humphrey (1959), Richards (1968), Banks (1979), and Corbett (1985) a risk factor of one batch in a thousand being contaminated is frequently used in the fermentation industry—that is, the final microbial count in the medium after sterilization should be 10^{-3} viable cells. However, to apply these kinetics, it is necessary to know the thermal death characteristics of all the taxa contaminating the fermenter and unsterile medium. This is an impossibility and, therefore, the assumption may be made that the only microbial contaminants present are spores of *Geobacillus stearothermophilus*—that is, one of the most heat-resistant microbial types known. *G. stearothermophilus* was previously included in the genus *Bacillus* as *Bacillus stearothermophilus* until its reclassification in 2001 (Nazina et al., 2001). Thus, by adopting *G. stearothermophilus* as the design organism a considerable safety factor should be built into the calculations. It should be remembered that *G. stearothermophilus* is not always adopted as the design organism. If the most heat-resistant organism contaminating the medium ingredients is known, then it may be advantageous to base the sterilization process on this organism. Deindoerfer and Humphrey (1959) determined the thermal death characteristics of *G. stearothermophilus* spores as:

$$\text{Activation energy} = 283.3\,\text{kJ}\,\text{mol}^{-1}$$
$$\text{Arrhenius constant} = 1 \times 10^{36.2}\,\text{second}^{-1}$$

However, it should be remembered that these kinetic values will vary according to the medium in which the spores are suspended, and this is particularly relevant

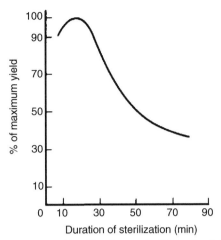

FIGURE 5.6 The Effect of the Time of Sterilization on the Yield of a Subsequent Fermentation (Richards, 1966)

when considering the sterilization of fats and oils (which are common fermentation substrates) where the relative humidity may be quite low. Bader, Boekeloo, Graham, and Cagle (1984) demonstrated that spores of *Bacillus macerans* suspended in oil were ten times more resistant to sterilization if they were dry than if they were wet.

A regime of time and temperature may now be determined to achieve the desired Del factor. However, a fermentation medium is not an inert mixture of components, and deleterious reactions may occur in the medium during the sterilization process, resulting in a loss of nutritive quality. Thus, the choice of regime is dictated by the requirement to achieve the desired reduction in microbial content with the least detrimental effect on the medium. Fig. 5.6 illustrates the deleterious effect of increasing medium sterilization time on the yield of product of subsequent fermentations. The initial rise in yield is due to some components of the medium being made more available to the process microorganism by the "cooking effect" of a brief sterilization period (Richards, 1966).

Two types of reaction contribute to the loss of nutrient quality during sterilization:

1. *Interactions between nutrient components of the medium.* A common occurrence during sterilization is the Maillard-type browning reaction that results in discoloration of the medium as well as loss of nutrient quality and the accumulation of growth-inhibitory compounds (Helou et al., 2014). These highly complex reactions normally occur between carbonyl groups, usually from reducing sugars, with the amino groups of amino acids and proteins. An example of the effect of sterilization time on the availability of glucose in a corn-steep liquor medium is shown in Table 5.1 (Corbett, 1985). Problems of this type can be resolved by sterilizing the sugar separately from the rest of the medium and recombining the two after cooling—as is done at a laboratory

Table 5.1 The Effect of Sterilization Time on Glucose Concentration and Product Accretion Rate in an Antibiotic Fermentation (Corbett, 1985)

Time at 121°C (min)	Amount of Added Glucose Remaining (%)	Relative Accretion Rate
60	35	90
40	46	92
30	64	100

scale (when sterilizing medium using an autoclave) and effectively in fed-batch fermentations. Marshall, Lilly, Gbewonyo, and Buckland (1995) measured the interactions between glucose and proteins caused by sterilization using changes in the absorption spectra and pH of a fermentation medium pre- and poststerilization. They demonstrated that a recombinant *E. coli* strain producing acidic fibroblast growth factor performed significantly better when glucose and protein were sterilized separately and the better performance correlated with decreased interaction between the two ingredients.

2. *Degradation of heat labile components.* Certain vitamins, amino acids, and proteins may be degraded during a steam sterilization regime. In extreme cases, such as the preparation of media for animal-cell culture, filtration may be used and this aspect will be discussed later in the chapter.

Both these problems may be resolved for most microbial fermentations by the judicious choice of steam sterilization regime. The thermal destruction of essential media components conforms approximately with first order reaction kinetics and, therefore, may be described by equations similar to those derived for the destruction of bacteria:

$$\frac{x_t}{x_0} = e^{-kd.t} \tag{5.10}$$

Where x_t is the concentration of nutrient after a heat treatment period, t; x_0 is the original concentration of nutrient at the onset of sterilization; kd is the reaction rate constant for the destruction of the medium component.

It is important to appreciate that we are considering the decline in the concentration of the nutrient component, whereas when we consider the decline in contaminants it is the absolute number of contaminating organisms that is quantified and not the concentration. The effect of temperature on the reaction rate constant may be expressed by the Arrhenius equation:

$$\ln kd = \ln A - E/RT.$$

Therefore, a plot of the natural logarithm of the reaction rate against $1/T$ will give a straight line, slope—(E/R). As the value of R, the gas constant, is fixed, the slope of the graph is determined by the value of the activation energy (E). The

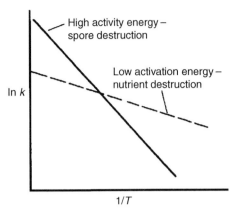

High activity energy –
spore destruction

Low activation energy –
nutrient destruction

ln *k*

1/*T*

FIGURE 5.7 The Effect of Activation Energy on Spore and Nutrient Destruction Quality

activation energy for the thermal destruction of *G. stearothermophilus* spores has been cited as 283.3 kJ mol^{-1}, whereas that for thermal destruction of nutrients is 40–125 kJ mol^{-1} (Richards, 1968). Li et al. (1998) quote the range of activation energy for microbial inactivation as 270–335 kJ mol^{-1} and that for chemical degradation as 40–105 kJ mol^{-1}. The activation energy for the Maillard reaction is also much lower than that for microbial destruction, for example, Nursten (2005) quotes values of between 40 and 190 kJ mol^{-1}. Fig. 5.7 is an Arrhenius plot for two reactions—one with a lower activation energy than the other. From this plot it may be seen that as temperature is increased, the reaction rate rises more rapidly for the reaction with the higher activation energy. Thus, considering the difference between activation energies for spore destruction and both nutrient degradation and the Maillard reaction, an increase in temperature would accelerate spore destruction more than medium denaturation and interaction.

In our earlier consideration of Del factors, it was evident that the same Del factor could be achieved over a range of temperature/time regimes. Thus, it would appear to be advantageous to employ a high temperature for a short time to achieve the desired probability of sterility, yet cause minimum nutrient degradation. Thus, the ideal technique would be to heat the fermentation medium to a high temperature, at which it is held for a short period, before being cooled rapidly to the fermentation temperature. However, it is obviously impossible to heat a batch of many thousands of liters of broth in a tank to a high temperature, hold for a short period and cool without the heating and cooling periods contributing considerably to the total sterilization time. The only practical method of materializing the objective of a short-time, high-temperature treatment is to sterilize the medium in a continuous stream. Moreover, the Maillard reactions between medium ingredients can be completely and conveniently, avoided by sterilizing reducing sugars and proteins (or protein derivatives) separately. This is achieved by incorporating several mixing tanks for the different components upstream of the sterilizer and processing the streams from them in sequence through the process, separated by a sterile water flush through the system (Soderberg, 2014). Continuous sterilization also gives saving in the design of the fermenter. If medium is sterilized batchwise in the fermenter, it must be agitated without aeration. As discussed in Chapter 9, aeration reduces the agitation power draw and thus the highest power draw occurs during batch sterilization and it is this requirement that dictates the capacity of the motor. Thus, the adoption of continuous sterilization enables a lower capacity agitator motor to be used in the production fermenter—saving both capital and running costs. In the past, the fermentation industry was reluctant to adopt continuous sterilization due to a number of disadvantages outweighing the advantage of nutrient quality. The relative merits of batch and continuous sterilization may be summarized as follows (Junker et al., 2006; Soderberg, 2014):

Advantages of continuous sterilization over batch sterilization:

1. Superior maintenance of medium quality.
2. Ability to sterilize medium components separately.
3. Superior energy efficiency, consuming 60–80% less steam and cooling water.
4. Ease of scale-up—discussed later.
5. Easier automatic control.
6. The reduction of surge capacity for steam.
7. The reduction of sterilization cycle time and hence the reduction in fermenter turnaround time, thus increasing productivity.
8. Under certain circumstances, the reduction of fermenter corrosion.
9. Enables the use of a lower capacity agitator in the fermenter giving economies in both capital and running costs.

Advantages of batch sterilization over continuous sterilization

1. Lower capital equipment costs.
2. Lower risk of contamination—continuous processes require the aseptic transfer of the sterile broth to the sterile vessel.
3. Easier manual control.
4. Easier to use with media containing a high proportion of solid matter.

The early continuous sterilizers were constructed as plate heat exchangers and these were unsuitable on two accounts:

1. Failure of the gaskets between the plates resulted in the mixing of sterile and unsterile streams.
2. Particulate components in the media would block the heat exchangers.

However, modern continuous sterilizers use double spiral heat exchangers in which the two streams are separated by a continuous steel division. Also, the spiral exchangers are far less susceptible to blockage. However, a major limitation to the adoption of continuous sterilization was the precision of control necessary for its success. This precision has been achieved with the development of sophisticated computerized monitoring and control systems resulting in continuous sterilization being very widely used and it is now the method of choice. However, batch sterilization is still used in many fermentation plants and, thus, it will be considered here before continuous sterilization is discussed in detail.

DESIGN OF BATCH STERILIZATION PROCESSES

Although a batch sterilization process is less successful in avoiding the destruction of nutrients than a continuous one, the objective in designing a batch process is still to achieve the required probability of obtaining sterility with the minimum loss of nutritive quality. The highest temperature, which appears to be feasible for batch sterilization is 121°C so the procedure should be designed such that exposure of the medium to this temperature is kept to a minimum. This is achieved by taking into account the contribution made to the sterilization by the heating and cooling periods of the batch treatment. Deindoerfer and Humphrey (1959) presented a method to assess the contribution made by the heating and cooling periods. The following information must be available for the design of a batch sterilization process:

1. A profile of the increase and decrease in the temperature of the fermentation medium during the heating and cooling periods of the sterilization cycle.
2. The number of microorganisms originally present in the medium.
3. The thermal death characteristics of the "design" organism. As explained earlier this may be G. *stearothermophilus* or an alternative organism relevant to the particular fermentation.

Knowing the original number of organisms present in the fermenter and the risk of sterilization failure considered acceptable, the required Del factor may be calculated. A frequently adopted risk of sterilization failure is 1 in 1000. This value should not be confused with the acceptable rate of contamination previously quoted as 1 in 100 as contamination results from failures of processes other than medium sterilization, including air sterilization, cooling equipment leaks, and aseptic operation failure. A sterilization failure rate of 1 in 1000 is commensurate with N_t being equal to 10^{-3} of a viable cell. It is worth reinforcing at this stage that we are considering the absolute total number of organisms present in the volume of medium being sterilized and *not* the concentration. If a specific case is considered where the unsterile broth was shown to contain 10^{11} viable organisms, then the Del factor may be calculated, thus:

$$\nabla = \ln(10^{11}/10^{-3})$$
$$\nabla = \ln 10^{14}$$
$$= 32.2.$$

Therefore, the overall Del factor required is 32.2. However, the destruction of cells occur during the heating and cooling of the broth as well as during the period at 121°C, thus, the overall Del factor may be represented as:

$$\nabla_{overall} = \nabla_{heating} + \nabla_{holding} + \nabla_{cooling}.$$

Knowing the temperature–time profile for the heating and cooling of the broth (prescribed by the characteristics of the available equipment) it is possible to determine the contribution made to the overall Del factor by these periods. Thus, knowing the Del factors contributed by heating and cooling, the holding time may be calculated to give the required overall Del factor.

CALCULATION OF THE DEL FACTOR DURING HEATING AND COOLING

The relationship between Del factor, the temperature and time is given by Eq. (5.8):

$$\nabla = A \cdot t \cdot e^{-(E/RT)}.$$

However, during the heating and cooling periods the temperature is not constant and, therefore, the calculation of ∇ would require the integration of Eq. (5.8) for the time–temperature regime observed. Deindoerfer and Humphrey (1959) produced integrated forms of the equation for a variety of temperature–time profiles, including linear, exponential, and hyperbolic. However, the regime observed in practice is frequently difficult to classify, making the application of these complex equations problematic. Richards (1968) demonstrated the use of a graphical method of integration and this is illustrated in Fig. 5.8. The time axis is divided into a number of equal increments t_1, t_2, t_3, etc., Richards suggesting 30 as a reasonable number.

For each increment, the temperature corresponding to the midpoint time is recorded. It may now be approximated that the total Del factor of the heating-up period is equivalent to the sum of the Del factors of the midpoint temperatures for each time increment. The value of the specific death rate of *G. stearothermophilus* spores at each midpoint temperature may be deduced from the Arrhenius equation using the

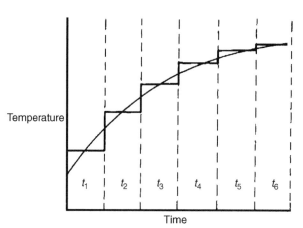

FIGURE 5.8 The Graphical Integration Method Applied to the Increase in Temperature Over a Time Period t_1, t_2 etc Represent Equal Time Intervals (Richards, 1968)

thermal death characteristic published by Deindoerfer and Humphrey (1959). The value of the Del factor corresponding to each time increment may then be calculated from the equations:

$$\nabla_1 = k_1 t,$$
$$\nabla_2 = k_2 t,$$
$$\nabla_3 = k_3 t,$$
etc.

The sum of the Del factors for all the increments will then equal the Del factor for the heating-up period. The Del factor for the cooling-down period may be calculated in a similar fashion.

CALCULATION OF THE HOLDING TIME AT CONSTANT TEMPERATURE

From the previous calculations the overall Del factor, as well as the Del factors of the heating and cooling parts of the cycle, have been determined. Therefore, the Del factor to be achieved during the holding time may be calculated by difference:

$$\nabla_{holding} = \nabla_{overall} - \nabla_{heating} - \nabla_{cooling}$$

Using our example where the overall Del factor is 32.2 and if it is taken that the heating Del factor was 9.8 and the cooling Del factor 10.1, the holding Del factor may be calculated:

$$\nabla_{holding} = 32.2 - 9.8 - 10.1,$$
$$\nabla_{holding} = 12.3.$$

But $\nabla = kt$, and from the data of Deindoerfer and Humphrey (1961) the specific death rate of G. stearothermophilus spores at 121°C is 2.54 min^{-1}.

$$\text{Therefore}, t = \nabla / k \text{ or } t = 12.3 / 2.54 = 4.84 \text{ min}.$$

If the contribution made by the heating and cooling parts of the cycle were ignored then the holding time would be given by the equation:

$$t = \nabla_{overall} / k = 32.2 / 2.54 = 12.68 \text{ min}.$$

Thus, by considering the contribution made to the sterilization process by the heating and cooling parts of the cycle a considerable reduction in exposure time is achieved.

RICHARDS' RAPID METHOD FOR THE DESIGN OF STERILIZATION CYCLES

Richards (1968) proposed a rapid method for the design of sterilization cycles avoiding the time-consuming graphical integrations. The method assumes that all spore destruction occurs at temperatures above 100°C and that those parts of the heating and cooling cycle above 100°C are linear. Both these assumptions appear reasonably valid and the technique loses very little in accuracy and gains considerably in simplicity. Furthermore, based on these assumptions, Richards has presented a table of Del factors for *G. stearothermophilus* spores which would be obtained in heating and cooling a broth up to (and down from) holding temperatures of 101–130°C, based on a temperature change of 1°C per min. This information is presented in Table 5.2, together with the specific death rates for *G. stearothermophilus* spores over the temperature range. If the rate of temperature change is 1°C per min, the Del factors for heating and cooling may be read directly from the table; if the temperature change deviates from 1°C per min, the Del factors may be altered by simple proportion. For example, if a fermentation broth were heated from 100 to 121°C in 30 min and cooled from 121 to 100°C in 17 min, the Del factors for the heating and cooling cycles may be determined as follows:

From Table 5.2, if the change in temperature had been 1°C per min, the Del factor for both the heating and cooling cycles would be 12.549. But the temperature change in the heating cycle was 21°C in 30 min; therefore,

$$\text{Del}_{\text{heating}} = (12.549 \times 30) / 21 = 17.93$$

and the temperature change in the cooling cycle was 21°C in 17 min, therefore,

$$\text{Del}_{\text{cooling}} = (12.549 \times 17) / 21 = 10.16.$$

Having calculated the Del factors for the heating and cooling periods the holding time at the constant temperature may be calculated as before.

SCALE UP AND OPTIMIZATION OF A BATCH STERILIZATION PROCESS

The use of the Del factor in the scale up of batch sterilization processes has been discussed by Banks (1979). It should be appreciated by this stage that the Del factor does not include a volume term, that is, absolute numbers of contaminants and survivors are considered, *not* their concentration. Thus, if the size of a fermenter is increased the initial number of spores in the medium will also be increased, but if the same probability of achieving sterility is required the final spore number should remain the same, resulting in an increase in the Del factor. For example, if a pilot sterilization were carried out in a 1000-dm^3 vessel with a medium containing 10^6 organisms cm^{-3} requiring a probability of contamination of 1 in 1000, the Del factor would be:

$$\nabla = \ln\{(10^6 \times 10^3 \times 10^3)/10^{-3}\}$$
$$= \ln(10^{12}/10^{-3})$$
$$= \ln 10^{15} = 34.5.$$

If the same probability of contamination were required in a 10,000-dm^3 vessel using the same medium the Del factor would be:

$$\nabla = \ln\{(10^6 \times 10^3 \times 10^4)/10^{-3}\}$$
$$= \ln(10^{13}/10^{-3})$$
$$= \ln 10^{16} = 36.8.$$

Table 5.2 Del Values for *B. Stearothermophilus* Spores for the Heating-Up Period Over a Temperature Range of 100–130°C, Assuming a Rate of Temperature Change of 1°C min^{-1} and Negligible Spore Destruction at Temperatures Below 100°C (Richards, 1968)

T(°C)	k (min⁻¹)	∇
100	0.019	–
101	0.025	0.044
102	0.032	0.076
103	0.040	0.116
104	0.051	0.168
105	0.065	0.233
106	0.083	0.316
107	0.105	0.420
108	0.133	0.553
109	0.168	0.720
110	0.212	0.932
111	0.267	1.199
112	0.336	1.535
113	0.423	1.957
114	0.531	2.488
115	0.666	3.154
116	0.835	3.989
117	1.045	5.034
118	1.307	6.341
119	1.633	7.973
120	2.037	10.010
121	2.538	12.549
122	3.160	15.708
123	3.929	19.638
124	4.881	24.518
125	6.056	30.574
126	7.506	38.080
127	9.293	47.373
128	11.494	58.867
129	14.200	73.067
130	17.524	90.591

Thus, the Del factor increases by 2.3 with a ten times increase in the fermenter volume. The holding time in the large vessel may be calculated by the graphical integration method or by the rapid method of Richards (1968), as discussed earlier, based on the temperature-time profile of the sterilization cycle in the large vessel. However, it must be appreciated that extending the holding time on the larger scale (to achieve the increased ∇ factor) will result in increased nutrient degradation. Also, the contribution of the heating-up and cooling-down periods to sterilization and nutrient destruction will be greater as scale increases because larger fermenters have less heat exchange surface per unit volume. Thus, the medium will spend a greater proportion of the time at lower temperatures that will contribute further to nutrient destruction.

Deindoerfer and Humphrey also produced a mathematical model of nutrient destruction based on the same logic as the Del factor and introduced the concept of a nutrient quality criterion (Q), given by the term:

$$Q = \ln\left(\frac{x_0}{x_t}\right) \qquad (5.11)$$

where x_0 is the concentration of essential heat labile nutrient in the original medium, x_t is the concentration of essential heat labile nutrient in the medium after a sterilization time, t.

Whereas the Del factor is based on absolute numbers of contaminants, the nutrient quality criterion is based on the concentration of the nutrient and, thus, the nutrient quality criterion is scale independent. Hence, ideally, Q should remain constant with an increase in scale. This presents us with the anomaly of having to increase the Del factor with scale (and thus the heat exposure) and yet maintain Q the same under the more severe sterilization conditions.

As considered earlier, the destruction of a nutrient may be considered a first-order reaction:

$$x_t/x_0 = e^{-kd.t}$$

where kd is the reaction rate constant for the destruction of the nutrient:
or

$$x_0/x_t = e^{kd \cdot t}$$

Thus, taking natural logarithms,

$$\ln(x_0/x_t) = kd.t$$

But

$$Q = \ln(x_0/x_t)$$

Therefore

$$Q = kd.t$$

The relationship between kd and absolute temperature is described by the Arrhenius equation:

$$kd = A \cdot e^{-(E/RT)}$$

therefore, substituting for *kd:*

$$Q = A \cdot t \cdot e^{-(E/RT)}.$$

This equation describes the relationship between the nutrient quality criterion (*Q*), and the time (*t*) and temperature (*T*) of the sterilization process and the thermodynamic constants for the reaction (*A* and *E*) and is directly equivalent to the Del factor Eq. (5.8):

$$\nabla = A \cdot t \cdot e^{-(E/RT)}.$$

Singh, Hensler, and Fuchs (1989) applied Deindoerfer and Humphrey's models to the optimization and scale-up of the batch sterilization of a large-scale (100,000 dm^3) antibiotic fermentation. The difficulty in applying the nutrient quality criterion is that the precise nutrient destruction reactions occurring in a complex medium are not known and thus neither are the kinetic constants of the reactions, *E* and *A*. These workers addressed this problem by adopting a "design" activation energy for nutrient destruction based on a "typical" value of 96.2 kJ mol^{-1}. The first step in their approach was to determine whether nutrient degradation was a key factor in the performance of their fermentation. A number of fermentations were conducted in four different fermenters sterilized at holding temperatures between 117 and 120°C. The heating and cooling profiles of the tanks were analyzed over the whole sterilization cycle and the following functions calculated:

$e^{-(E/RT)}$ for sterilization using Deindoerfer and Humphrey's value of *E* for *G. stearothermophilus* (283.3 kJ mol^{-1})
$e^{-(E/RT)}$ for nutrient destruction using a "typical" value of *E* for nutrient destruction of 96.2 kJ mol^{-1}.

These values were plotted against the sterilization cycle time, the areas under the two plots then giving the values of the terms ∇/A and Q/A. This empirical approach means that the Arrhenius constants for both reactions are incorporated into the derived terms using the experimental data. Statistical comparison of antibiotic yield with Q/A for each fermentation showed large Q/A values associated with poor antibiotic production. Indeed, a threshold value of Q/A was linked with low yield—fermentations above this value gave, on average, a 43% drop in product concentration. The worst performing fermenter (low antibiotic yield and a high Q/A value) also had the poorest heat transfer profile, thus linking product yield, nutrient quality, and fermenter characteristics. Mathematical models of the sterilization cycle of the fermenter were developed which enabled the effects of sterilization temperature and holding time to be predicted and optimized using a Simplex optimization procedure (discussed in more detail in Chapter 4). As a result, the temperature/holding time for the fermenter was changed from 119°C for 45 min to 122°C for 12 min, almost doubling the antibiotic yield. These workers also suggested that an experimental, rather than a predictive modeling, approach may be used to generate data for a Simplex analysis by calculating Q/A and ∇/A

values over a range of time/temperature regimes. This work demonstrates the number of important issues:

1. The ability to demonstrate that a decrease in yield on scale-up, or in different fermenters, can be attributed to the sterilization regime.
2. The magnitude of the effect of medium degradation on product yield.
3. The impact of the heat transfer characteristics of a fermenter on the sterilization cycle and hence the product yield.
4. The positive effect that a small increase in holding temperature can have on holding time and medium degradation.

AN ALTERNATIVE APPROACH TO STERILIZATION KINETICS: *D*, *Z*, AND *F* VALUES

In the earlier discussion of the destruction of microorganisms by steam (moist heat), the decline in the number of viable organisms over a treatment period was described as a first order reaction represented by Eq. (5.1), where k is the specific death rate:

$$-dN/dt = kN$$

and integration of this equation gave Eq. (5.2):

$$N_t/N_0 = e^{-kt}$$

which, on taking natural logarithms, reduces to Eq. (5.3):

$$\ln(N_t/N_0) = -kt$$

and a plot of this equation as natural logarithm of N_t/N_0 against time, giving a straight line of slope $-k$, is shown in Fig. 5.2. In Bigelow's (1921) original presentation of these kinetics, Eq. (5.3) was represented as a plot of $\log_{10} N/N_0$ against time, which could be conveniently and quickly achieved using semilogarithmic graph paper. The slope of this plot is $-k/2.303$, due to the change in the logarithm base (2.303 being the natural logarithm of 10). It has been common practice in the food industry to use the $\log_{10}$ graphical representation and this has resulted in the use of the constant "decimal reduction time" (D) which is the time required, at a specific temperature, to reduce the population to one tenth of its original value, that is, to kill 90% of the population. A simple example is given in Fig. 5.9 in which the value of D can be seen to be 5 min.

The relationship between D and k (the specific death rate) can be deduced as follows:

We have seen that:

$$\ln(N_t/N_0) = -kt$$

and thus:

$$\ln(N_0/N_t) = kt$$

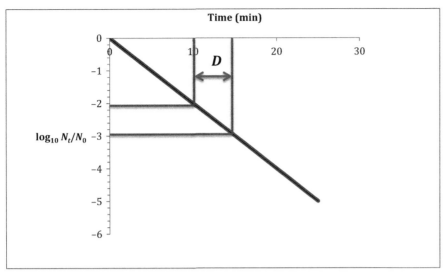

FIGURE 5.9 A Plot of the $\log_{10}$ of the Proportion of Survivors in a Population of Spores of a Heat-Resistant Bacterium

The decimal reduction time (D) is the time period required to achieve a 90% reduction in the population, that is, the time period for one "log reduction"; in this example, $D = 5$ min.

If t equals the decimal reduction time, then N_t will be $0.1N_0$, thus;

$$\ln(N_0/0.1N_0) = kD$$

$$N_0/0.1N_0 = 10$$

and the natural logarithm of 10 is 2.303, thus:

$$2.303 = kD$$

and:

$$D = 2.303/k$$

The relationship between D and temperature is a logarithmic one, as is to be expected from our earlier consideration of the effect of temperature on specific death rate, which is governed by the Arrhenius equation. However, in this analysis a simplified version of the Arrhenius equation is used and represented as a plot of $\log_{10} D$ against temperature (°C), rather than the natural logarithm of rate against the reciprocal of the absolute temperature. Such a plot is shown in Fig. 5.9, giving a straight-line relationship over the temperature range shown. This representation is then used to determine another term, the "temperature sensitivity" (or Z-value), the temperature change that results in a 10-fold change in the D-value. The Z-value of

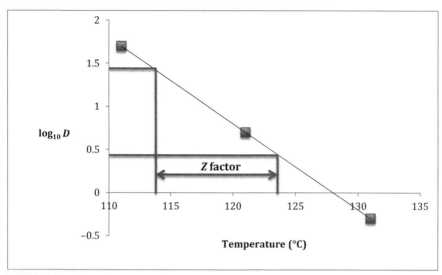

FIGURE 5.10 A Plot of the log$_{10}$ of the Decimal Reduction Time (*D*) of the Spores of a Heat-Resistant Bacterium

The *Z* factor is shown as the temperature change required (in °C) to cause a 10-fold change in the decimal reduction time, that is, a "log change". In this example the *Z* value = 10°C.

the hypothetical bacterium shown in Fig. 5.10 is 10°C, that is, an increase of 10°C results in a 10-fold decrease in the decimal reduction time. It is important to appreciate that the two models for predicting the effects of temperature on lethality are based on different approaches and have been shown to give different predictions in certain circumstances. Jones (1968) showed that over the temperature range 110–140°C, the two methods show reasonable agreement for *G. stearothermophilus*. Mann, Jiefer, and Leuenberger (2001) demonstrated that, the *Z* value gave a more accurate prediction of lethality in the high temperature range of 140–155°C. However, when biological variation was taken into account they showed that there was more variation in the individual data, whether based on *Z* or the Arrhenius relationship, than the difference between the two models.

The *D* and *Z* values can be used to design a sterilization regime in a similar manner to that described utilizing specific death rate and the Arrhenius constant. It will be recalled that a value of N_t of 10^{-3} is used as the acceptable failure rate of the sterilization of a fermentation medium and the kinetics of destruction are based on the assumption that the contaminants are spores of *G. stearothermophilus*, enabling the prediction of the required holding times for particular Del factors at particular temperatures. The *D* and *Z* values approach also uses the same failure rate of 10^{-3} and *G. stearothermophilus* as the design organism. The values of *D* and *Z* for spores of *G. stearothermophilus* in water are 3 min (at 121°C) and 10°C, respectively (Junker, Beshty, & Wilson, 1999). The number of log reductions to achieve N_t can be determined, knowing N_0, and thus the holding time can be calculated in terms of multiples of the decimal reduction time. As in the description of the use of Del factor, N_0 increases as the volume to be sterilized increases requiring an increase in the number of log reductions. The Z factor is then used to determine the effect of temperature on holding time by representing the lethality of the sterilization regime in terms of the equivalence of its exposure time to that of a regime held at 121.1°C. This comparative value is termed F_0 and is defined as the exposure time (minutes) at 121.1°C equivalent to the actual exposure time at the process temperature. The F_0 concept was originally developed in the food industry (based on the reference temperature of 250°F, hence the anomalous use of 121.1°C) and then adopted by the pharmaceutical industry for the sterilization of parenteral (injectable) products. Boeck, Alford, Pieper, and Huber (1989) applied the concept to the sterilization of fermentation media using 121°C as the reference temperature.

At its simplest level, F_0 can represented by the equation:

$$F_0 = t.10^{(T-121)/Z}$$

where t is time (min); T is temperature (°C); Z is the temperature sensitivity, which for *G. stearothermophilus* is considered equal to 10°C (Junker et al., 1999).

Thus, for a sterilization regime of 15 min at a constant temperature of 111°C, the F_0 would be:

$$F_0 = 15 \times 10^{(111-121)/10}$$

$$F_0 = 15 \times 10^{-10/10}$$

and
$$F_0 = 15 \times 10^{-1} = 1.5 \text{ min}$$

Thus, a holding time of 15 min at 111°C is equivalent in lethality to 1.5 min at 120°C. If the holding temperature were 131°C, then:

$$F_0 = 15 \times 10^{(131-121)/10}$$
$$F_0 = 15 \times 10^{10/10} = 15 \times 10$$
$$F_0 = 150 \text{ min.}$$

Thus, a holding time of 15 min at 131°C is equivalent in lethality to 150 min at 121°C. These two simple examples reinforce the importance of understanding the logarithmic effect of temperature on microbial destruction. However, as the heating and cooling cycles also contribute to the death of contaminating organisms, the calculation of F_0 should take into account for the whole sterilization cycle. The equation for F_0 is then modified to:

$$F_0 = t.\sum_{10} (T-121)/Z$$

The term Σ is used to indicate that F_0 is determined by adding together the cumulative effects of temperature over the whole cycle. A similar approach can be adopted

Table 5.3 Table of F_0 Values for a Range of Temperatures at a Holding Time of One Minute, Based on a Z Value of 10 and a Reference Temperature of 121.1°C

T(°C)	F_0 (min)
100	0.008
101	0.010
102	0.012
103	0.015
104	0.019
105	0.024
106	0.031
107	0.039
108	0.049
109	0.062
110	0. 077
111	0.097
112	0.123
113	0.154
114	0.194
115	0.245
116	0.308
117	0.388
118	0.489
119	0.615
120	0.774
121	0.975
122	1.227
123	1.545
124	1.945
125	2.448
126	3.082
127	3.881
128	4.885
129	6.150
130	7.743

Modified from Pistolesi and Mascherpa, (2014).

to that used by Richards (1968) to calculate the contribution of heating and cooling cycles to Del factors. Pistolesi and Mascherpa (2014) produced a table of F_0 values (based on a Z value of 10°C) for a range of temperatures held for one minute. An abbreviated version of their table is given in Table 5.3. Assuming that spore destruction is limited to temperatures above 100°C, incremental F_0 values for whole cycle (heating, holding, and cooling) may be calculated by dividing the profile into one minute increments, calculating the mean temperature for each increment, obtaining the F_0 for each increment from Table 5.3 and adding the values together. The resulting F_0 value takes into account any deviation of temperature from the set point during the holding time as well as the contribution of heating and cooling. Thus, the lethality of a process may be assessed after sterilization is complete and before inoculation. Although this manual postcalculation approach provides valuable information, it is tedious. However, it can be replaced by a computerized control system. Boeck et al. (1988) reported such a system in which the cycle was controlled to achieve the desired F_0 by online monitoring, F_0 calculation and compensatory temperature adjustment through the process.

Boeck et al. (1989) also developed a term, R_0 based on the Arrhenius equation to represent medium degradation:

$$R_0 = dt \sum_{t0}^{t} A.e^{-E/RT}$$

where t is sterilization time, A is the Arrhenius constant, E is the activation energy, R is the universal gas constant and T is the absolute temperature. A sterilization regime resulting in no medium degradation would give an R_0 value of zero and an increase in R_0 indicates a decrease in nutrient quality. Because R_0 is based on the Arrhenius equation, it is directly equivalent to Deindoerfer and Humphrey's Nutrient Quality Criterion (Q), discussed earlier:

$$Q = A \cdot t \cdot e^{-(E/RT)}.$$

The different representation of the two equations is due to R_0 explicitly including the whole sterilization cycle in the relationship, whereas Q is expressed in a more generic format. It will be recalled that the difficulty in applying of Deindoerfer and Humphrey's equation is that the precise nutrient destruction reactions occurring in a complex medium are not known and thus neither are the kinetic constants, E and A. Singh et al. (1989) approached this problem by adopting a "design" or "typical" value for E, the activation energy of nutrient degradation enabling them to apply Deindoerfer and Humphrey's Nutrient Quality Criterion over the whole sterilization cycle. Boeck et al. (1989) adopted a similar approach by making the following two assumptions to derive A and E and apply them to the calculation of R_0: (1) The reaction rate doubles for each temperatures increase of 10°C, and (2) R_0 increases by 1.0 unit $\min^{-1}$ at 121°C. This enabled the computation of the design values of E and $\ln A$ as 86.8 kJ mol^{-1} and 26.596 respectively. As in their work on F_0, Boeck et al. (1989) developed a computer control system to give an online calculation of the accumulating R_0 value during the sterilization cycle. Such a system, in combination with an online calculation of the accumulating F_0 value, would enable the sterilization process to be controlled to give the predetermined degrees of lethality and medium quality. It must be understood that the overriding criterion to be achieved in the process is the desired lethality—any change in the process cannot compromise the required F_0 value.

Boeck et al. (1989) produced similar results to those of Singh for Q. Increasing R_0 values resulted in a decrease in biomass and phosphate uptake in the fermentations. The same biomass yield was obtained in fermentations where the same R_0 factor had been achieved, but at different temperatures and an increase in temperature from 110 to 125°C gave an almost 10-fold increase in F_0 while R_0 was virtually unaffected. Junker et al. (1999) reported that raising the holding by 5 min had about a twofold greater effect on F_0 than R_0. Thus, these approaches can be used to identify medium degradation issues with variable fermentation performance both between different fermenters and different fermentation runs as well as the identification of medium degradation as the cause of reduced yield on scaling-up a fermentation.

Several workers have compared the use of the Arrhenius equation and the Z factor to predict specific death rates at different temperatures. The consensus of findings is that the two approaches give very similar results within the range 121°C–134°C (van Doornmalen & Kopinga, 2009) whereas at higher temperatures use of the Z-value gives a higher predicted specific death rate than the Arrhenius equation (Ling, Tang, Kong, & Mitcham, 2015).

VARIATION IN THE VALUES OF STERILIZATION KINETIC "CONSTANTS"

The values of specific death rate (k) and decimal reduction time (D) used in the previous section have been taken from the original literature specified. For example, the specific death rate of *G. stearothermophilus* quoted by Deindoerfer and Humphrey (1961) at 121°C was 2.54 min^{-1} and the decimal reduction time for the same organism quoted by Junker et al. (1999) was 3 min. However, a specific death rate of 2.54 min^{-1} corresponds to a decimal reduction time of 0.9 min. Thus, the value quoted by Deindoerfer and Humphrey indicates a greater susceptibility of *G. stearothermophilus* to wet heat than the data cited by Junker et al. (1999). However, it should not be surprising that a biological system shows such variation. Indeed, a range of death rates for *G. stearothermophilus* may be found in the literature, as shown in Table 5.4. The thermal resistance of spores is influenced by the environmental conditions pertaining during both the spore induction and spore destruction process. Guizelini, Vandenberghe, Sella, and Soccol (2012) investigated the heat resistance of *G. stearothermophilus* spores produced under a

Table 5.4 Published Values of Wet Heat Specific Death Rate, k (min^{-1}) and D (min) at 121°C for *G. Stearothermophilus* Spores

Strain	K (min^{-1})	D (min)	References
FS 1518	2.54	0.91	Deindoerfer and Humphrey, 1959
FS 1518	0.77	3.0	Bader, 1986, Junker, 2001
FS 617	2.9	0.8	Bader, 1986
Not specified	1.15	2.0	Pistolesi and Mascherpa, 2014

Table 5.5 Ratio of *D*-Values (121°C) of *G. Stearothermophilus* Spores in Concentrated Medium Ingredients to *D* Values in Deionized Water

Medium Ingredient (50% Weight/Volume)	Ratio : $\dfrac{D \text{ value in test medium}}{D \text{ value in water}}$
Fructose	0.46
Ammonium sulfate	0.65
Cerelose (glucose)	0.89
Lactose	0.93
Sucrose	1.42
Monosodium glutamate	2.1
Glycerol	3.0
Proline	3.1

Modified from Junker, 2001.

range of environmental conditions and quoted *D* values (at 121°C) of 1.3–5 min (equivalent to values of *k* between 1.77 and 0.46 min^{-1}). Junker (2001) showed that *G. stearothermophilus* spores were more susceptible to steam sterilization when suspended in growth media, as compared with water, probably due to the stimulation of spore germination. Thus, *D* and *k* values determined in water may be considered as the "worst case scenario." However, Junker's comparison of *D* values in concentrated nutrient solutions presents a very different picture, as shown in Table 5.5. While spores in 50% fructose, ammonium sulfate, cerelose, and lactose solutions were all more susceptible to wet heat than in water, spores suspended in 50% glycerol or proline were three times more resistant. Thus, the sterilization of these concentrated feeds must take into account the increased resistance of spores in these solutions.

Commercial preparations of *G. stearothermophilus* spores are available in the form of impregnated filter paper strips and are manufactured such that the *D* value of the spores (in deionized water) are within stated specifications. These spore strips may be used to determine the actual *D* value in the fermentation medium, thus ensuring that sterilization calculations are based on realistic figures.

METHODS OF BATCH STERILIZATION

The batch sterilization of the medium for a fermentation may be achieved either in the fermentation vessel or in a separate mash cooker. Richards (1966) considered the relative merits of in situ medium sterilization and the use of a special vessel. The major advantages of a separate medium sterilization vessel may be summarized as:

1. One cooker may be used to serve several fermenters and the medium may be sterilized as the fermenters are being cleaned and prepared for the next fermentation, thus saving time between fermentations.

2. The medium may be sterilized in a cooker in a more concentrated form than would be used in the fermentation and then diluted in the fermenter with sterile water prior to inoculation. This would allow the construction of smaller cookers.
3. In some fermentations, the medium is at its most viscous during sterilization and the power requirement for agitation is not alleviated by aeration as it would be during the fermentation proper. Thus, if the requirement for agitation during in situ sterilization were removed, the fermenter could be equipped with a less powerful motor. Obviously, the sterilization kettle would have to be equipped with a powerful motor, but this would provide sterile medium for several fermenters.
4. The fermenter would be spared the corrosion that may occur with medium at high temperature.

The major disadvantages of a separate medium sterilization vessel may be summarized as:

1. The cost of constructing a batch medium sterilizer is much the same as that for the fermenter.
2. If a cooker serves a large number of fermenters complex pipework would be necessary to transport the sterile medium, with the inherent dangers of contamination.
3. Mechanical failure in a cooker supplying medium to several fermenters would render all the fermenters temporarily redundant. The provision of contingency equipment may be prohibitively costly.

Overall, the pressure to decrease the "down time" between fermentations has tended to outweigh the perceived disadvantages of using separate sterilization vessels. Thus, sterilization in dedicated vessels is the method of choice for batch sterilization. However, as pointed out by Corbett (1985), the fact that separate batch sterilizers are used, lends further weight to the argument that continuous sterilization should be adopted in preference to batch. The capital cost of a separate batch sterilizer is similar to that of a continuous one and the problems of transfer of sterile media are then the same for both batch and continuous sterilization. Thus, two of the major objections to continuous systems (capital cost and aseptic transfer) may be considered as no longer relevant.

DESIGN OF CONTINUOUS STERILIZATION PROCESSES

The design of continuous sterilization cycles may be approached in exactly the same way as for batch sterilization systems. The continuous system includes a time period during which the medium is heated to the sterilization temperature, a holding time at the desired temperature, and a cooling period to restore the medium to the fermentation temperature. The temperature of the medium is elevated in a continuous heat exchanger and is then maintained in an insulated serpentine holding coil for the holding period. The length of the holding period is dictated by the length of the coil and the flow rate of the medium. The hot medium is then cooled to the fermentation temperature using two sequential heat exchangers—the first utilizing the incoming medium as the cooling source (thus conserving heat by heating-up the incoming medium) and the second using cooling water. As discussed earlier, the major advantage of the continuous process is that a much higher temperature may be utilized, thus reducing the holding time and reducing the degree of nutrient degradation. The required Del factor may be achieved by the combination of temperature and holding time that gives an acceptably small degree of nutrient decay. Richards (1968) quoted the following example to illustrate the range of temperature-time regimes that may be employed to achieve the same probability of obtaining sterility. The Del factor used by Richards for the example sterilization was 45.7 and the following temperature time regimes were calculated such that they all gave the same Del factor value of 45.7:

Temperature (°C)	Holding time
130	2.44 min
135	51.9 s
140	18.9 s
150	2.7 s

Furthermore, because a continuous process involves treating small increments of medium, the heating-up and cooling-down periods are very small compared with those in a batch system. There are two types of continuous sterilizer that may be used for the treatment of fermentation media: the indirect heat exchanger and the direct heat exchanger (steam injector).

The most suitable indirect heat exchangers are of the double-spiral type, which consists of two sheets of high-grade stainless steel, which have been curved around a central axis to form a double spiral, as shown in Fig. 5.11. The ends of the spiral are sealed by covers. A full-scale example is shown in Fig. 5.12. To achieve sterilization

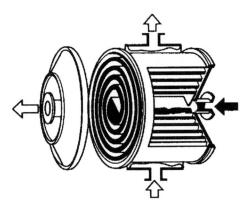

FIGURE 5.11 A Schematic Representation of a Spiral Heat Exchanger

Alfa-Laval Engineering Ltd, Camberley, Surrey, UK

FIGURE 5.12 Industrial Scale Spiral Heat Exchanger

Alfa-Laval Engineering Ltd., Camberley, Surrey, UK

temperatures steam is passed through one spiral from the center of the exchanger and medium through the other spiral from the outer rim of the exchanger, in countercurrent streams. Spiral heat exchangers are also used to cool the medium after passing through the holding coil. Incoming unsterile medium is used as the cooling agent in the first cooler so that the incoming medium is partially heated before it reaches the sterilizer and, thus, heat is conserved.

The major advantages of the spiral heat exchanger are:

1. The two streams of medium and cooling liquid, or medium and steam, are separated by a continuous stainless steel barrier with gasket seals being confined to the joints with the end plates. This makes cross contamination between the two streams unlikely.
2. The spiral route traversed by the medium allows sufficient clearances to be incorporated for the system to cope with suspended solids. The exchanger tends to be self-cleaning which reduces the risk of sedimentation, fouling, and "burning-on."

Indirect plate heat exchangers consist of alternating plates through which the countercurrent streams are circulated. Gaskets separate the plates and failure of these gaskets can cause cross-contamination between the two streams. Also, the clearances between the plates are such that suspended solids in the medium may block the exchanger and, thus, the system is only useful in sterilizing completely soluble media. However, the plate exchanger is more adaptable than the spiral system in that extra plates may be added to increase its capacity.

The continuous steam injector injects steam directly into the unsterile broth. The advantages and disadvantages of the system have been summarized by Banks (1979):

1. Very short (almost instantaneous) heating up times.
2. It may be used for media containing suspended solids.
3. Low capital cost.
4. Easy cleaning and maintenance.
5. High steam utilization efficiency.

However, the disadvantages are:

1. Foaming may occur during heating.
2. Direct contact of the medium with steam requires that allowance be made for condense dilution and requires "clean" steam, free from anticorrosion additives.

In some cases the injection system is combined with flash cooling, where the sterilized medium is cooled by passing it through an expansion valve into a vacuum chamber. Cooling then occurs virtually instantly. A flow chart of a continuous sterilization system using direct steam injection is shown in Fig. 5.13. In some cases a combination of direct and indirect heat exchangers may be used (Svensson, 1988). This is especially true for starch-containing broths when steam injection is used for the preheating step. By raising the temperature virtually instantaneously, the critical gelatinization temperature of the starch is passed through very quickly and the increase in viscosity normally associated with heated starch colloids can be reduced.

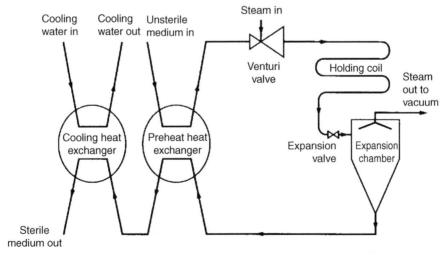

FIGURE 5.13 Flow Diagram of a Typical Continuous Injector-Flash Cooler Sterilizer

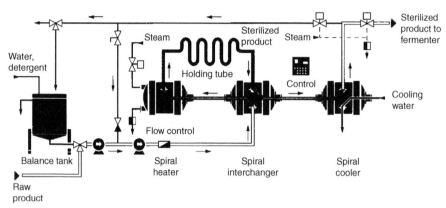

FIGURE 5.14 Flow Diagram of a Typical Continuous Sterilization System Employing Spiral Heat Exchangers

Alfa-Laval Engineering Ltd., Camberley, Surrey, UK.

The most widely used continuous sterilization system is that based on the spiral heat exchangers and a typical layout is shown in Fig. 5.14. Junker et al. (2006) described the replacement of a pilot-plant direct-steam injection process with one based on indirect spiral heat exchangers, thus avoiding the problems of the medium quality being affected by fluctuations in the plant steam quality—particularly the presence of steam anticorrosion additives. The key features of this plant were:

1. A number of feed tanks from 2,000 to 19,000 dm^3 that enabled both a range of medium volumes to be processed and different medium ingredients to be sterilized sequentially and separately.
2. A recycle tank (also referred to as a circulation or surge tank) and recycle facility so that water could be circulated through the system during cleaning and sterilization of the plant.
3. A spiral heat exchanger to raise the temperature of the medium to the sterilization temperature.
4. A "retention loop" to hold the medium at the sterilization temperature for the required time.
5. A "recuperator" spiral heat exchanger to cool the sterilized medium leaving the retention loop against incoming cold, unsterile medium, thus minimizing heat wastage.
6. A cooling spiral heat exchanger supplied with chilled water to cool the sterilized medium leaving the recuperator exchanger to the process temperature.
7. An additional cooling spiral heat exchanger used when the system was used to produce sterile water.
8. A switching station to direct the product stream from the final coolers:
    a. to a distribution system to the fermenters (if medium is sterile),
    b. to the recycle tank for recirculation through the system during sterilization of the system, and
    c. to the sewer during cleaning of the system or should a malfunction occur.

The Del factor to be achieved in a continuous sterilization process has to be increased with an increase in scale, and this is calculated exactly as described in the consideration of the scale up of batch regimes. Thus, if the volume to be sterilized is increased from 1000 to 10,000 dm^3 and the risk of failure is to remain at 1 in 1000 then the Del factor must be increased from 34.5 to 36.8. However, the advantage of the continuous process is that temperature may be used as a variable in scaling up a continuous process so that the increased ∇ factor may be achieved while maintaining the nutrient quality constant. A further advantage of the continuous process is that the distribution system can be devised such that a sterile medium stream can be diverted for use on a small scale. Thus, laboratory or small fermenter scale models of the process can be conducted using medium that has been exposed to exactly the same sterilization regime as the production scale.

Deindoerfer and Humphrey (1961) discussed the application of their Nutrient Quality Criterion (Q) to continuous sterilization scale-up. It will be recalled from our earlier discussion of batch sterilization that:

$$Q = A \cdot t \cdot e^{-(E/RT)}.$$

Therefore, as for the Del factor equation, by taking natural logarithms, and rearranging, the following equation is obtained

$$\ln t = \frac{E}{RT} + \ln \frac{Q}{A}. \tag{5.12}$$

Thus, a plot of the natural logarithms of the time required to achieve a certain Q value against the reciprocal of the absolute temperature will yield a straight line, the slope of which is dependent on the activation energy; that is, a very similar plot to Fig. 5.5 for the Del factor relationship. If both plots were superimposed on the same figure, then a continuous sterilization performance chart is obtained. The example put forward by Deindoerfer and Humphrey (1961) is shown in Fig. 5.15. Thus, in Fig. 5.15, each line of a constant Del factor specifies temperature-time regimes giving the same fractional reduction in spore number and each line of a constant nutrient quality criterion specifies temperature–time regimes giving the same destruction of nutrient. By considering the effect of nutrient destruction on product yield, limits may be imposed on Fig. 5.15 indicating the nutrient quality criterion below which no further increase in yield is achieved (ie, the nutrient is in excess) and the nutrient quality criterion at which the product yield is at its lowest (ie, there is no nutrient remaining). Thus, from such a plot a temperature–time regime may be chosen which gives the required Del factor and does not adversely affect the yield of the process.

The precise adoption of Deindoerfer and Humphrey's approach is possible only if the limiting heat-labile nutrient is identified and the activation energy for its thermal destruction is experimentally determined. As pointed out by Banks (1979), this information may not be available for a complex fermentation medium. However, the compromises proposed by Singh et al. (1989) and Boeck et al. (1989) could allow the application of this approach. As discussed earlier, the Del factor is scale dependent and therefore as the volume to be sterilized is increased so the Del factor should be

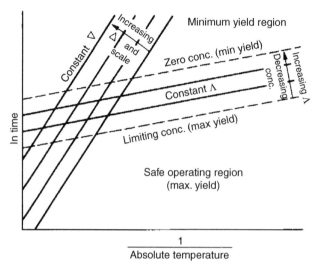

FIGURE 5.15 Continuous Sterilization Performance Chart (Deindoerfer & Humphrey, 1961)

increased if the probability of achieving sterility is to remain the same. However, the nutrient-quality criterion is not scale dependent so that by changing the temperature–time regime to accommodate the attainment of sterility, the nutrient quality may be adversely affected. Examination of Fig. 5.15 indicates that the only way in which the Del factor may be increased without any change in the nutrient quality criterion is to increase the temperature and to decrease the holding time.

When designing a continuous sterilization process based on spiral heat exchangers, it is important to consider the effect of suspended solids on the sterilization process. Microorganisms contained within solid particles are given considerable protection against the sterilization treatment. If the residence time in the sterilizer is insufficient for heat to penetrate the particle then the fermentation medium may not be rendered sterile. The routine solution to this problem is to "over design" the process and expose the medium to a far more severe regime than may be necessary. Armenante and Li (1993) discussed this problem in considerable detail and produced a model to predict the behavior of a continuous system. Their analysis suggested that the temperature of the particle cores is significantly less than that of the bulk liquid. Furthermore, there is a considerable time lag in heat penetrating to the particle cores, resulting in a very different time–temperature profile for the particles as compared with the liquid medium. Thus, the temperature of the particles may not reach the critical point before they leave the sterilizer and heat penetration into the particles will continue downstream of the sterilizer. Armenante and Li's conclusion is that it is the sterilizer and/or the first cooling exchanger that should be "overdesigned" rather than the length of the holding coil. Remember that the first cooling exchanger transfers a significant amount of heat from the sterile medium to the incoming medium and increasing its surface area would give more opportunity for the heat to penetrate the particles. This, coupled with increasing the temperature or residence time in the sterilizer, would ensure that the particle cores are up to temperature before the holding coil is reached. Also, this work suggests that it is unwise to use the direct steam injection method to heat a particulate medium because, again, there will be insufficient time for the heat to penetrate the particles.

An example of the scale up of sterilization regimes is given by the work of Jain and Buckland (1988) on the production of efrotomycin by *Nocardia lactamdurans*. In this case, a beneficial interaction appeared to be occurring between the protein nitrogen source and glucose during sterilization, thus making the protein less available but resulting in a more controlled fermentation. When glucose was sterilized separately, the oxygen demand of the subsequent fermentation was excessive and the fermentation terminated prematurely with very poor product formation. On scaling up the fermentation, it was very difficult to attain the correct sterilization conditions using a batch regime. However, continuous sterilization using direct steam injection allowed the design of a precise process producing sterile medium with the required degree of interaction between the ingredients. The identification of this phenomenon was dependent upon careful monitoring of the small-scale fermentation and consideration being given to sterilization as an important scale-up factor.

When a fermentation is scaled up, it is important to appreciate that the inoculum development process is also increased in scale (see Chapter 6) and a larger seed fermenter may have to be employed to generate sufficient inoculum to start the production scale. Thus, the sterilization regime of the seed fermenter (and its medium) will also have to be scaled up. Therefore, the performance of the seed fermentation should be assessed carefully to ensure that the quality of the inoculum is maintained on the larger scale and that it has not been adversely affected by any increase in the severity of the sterilization regime.

STERILIZATION OF THE FERMENTER

If the medium is sterilized in a separate batch cooker, or is sterilized continuously, then the fermenter has to be sterilized separately before the sterile medium is added to it. This is normally achieved by heating the jacket or coils (see Chapter 7) of the fermenter with steam and sparging steam into the vessel through all entries, apart from the air outlet from which steam is allowed to exit slowly. Steam pressure is held at 15 psi in the vessel for approximately 20 min It is essential that sterile air is sparged into the fermenter after the cycle is complete and a positive pressure is maintained; otherwise a vacuum may develop and unsterile air be drawn into the vessel.

STERILIZATION OF THE FEEDS

A variety of additives may be administered to a fermentation during the process and it is essential that these materials are sterile. The sterilization method depends on the nature of the additive, and the volume and feed rate at which it is administered. If the additive is fed in large quantities then continuous sterilization may be desirable, for example, Aunstrup, Andresen, Falch, and Nielsen (1979) cited the use of continuous heat sterilization for the feed medium of microbial enzyme fermentations. The continuous sterilization system described by Junker et al. (2006) is considered in a previous section and it will be recalled that the system incorporated a range of medium feed tanks of various volumes, thus enabling the sterilization of the different feeds required in the process that are required in different volumes. Feeds could be sterilized sequentially through the sterilizer by separating the different feeds with a plug-flow of water through the system. Filtration is also a commonly used sterilization method for feeds, especially as most feeds will be free of particulates and this is considered in a later section of this Chapter. However, a key consideration in the wet heat sterilization of feeds is the nature and concentration of the feed and how this affects the sterilization process. The water activity (a_w) of a solution is defined as the partial vapor pressure of water (p) in a substance divided by the standard state vapor pressure of water (p_0):

$$a_w = p/p_0$$

Thus, as solute concentration increases, so water activity decreases. Murrell and Scott (1966) demonstrated that the D-value of *G. stearothermophilus* spores increased 20-fold at water activities of between 0.2 and 0.4. Thus, account must be taken of the effect of water activity on the kinetics of spore death in designing the sterilization regime. However, Junker et al. (1999) emphasized that the calculated death rates will only be achieved if sufficient saturated steam is available in the sterilizer. As water activity decreases, so does the vapor pressure which, according to Junker et al, contributed directly to the effectiveness of both air removal and sterilization performance resulting in a slower accumulation rate of F_0 than may be expected, necessitating a longer holding time.

Stowell (1987) described the use of a continuous sterilizer for the addition of oil feeds to industrial scale fermenters, but stressed that each oil has its own characteristics, which makes it impossible to predict the performance of a sterilizer for different oil feeds. Also, oils are sparingly soluble in water resulting in very low moisture content. Steam sterilization is a wet-heat process and the low water content of oils can result in sterilization being effectively based on dry heat rather than wet, requiring significantly higher holding times. Bader et al. (1984) reported that dry spores of *B. macerans* in oil had a D value one hundred times that of wet spores in the same medium. Stowell demonstrated that spores of a process contaminant, *Bacillus circulans*, could survive in an oil suspension for up to 40 min at 150°C.

STERILIZATION OF LIQUID WASTES

Process organisms that have been engineered to produce "foreign" products and therefore contain heterologous genes are subjected to strict containment regulations. Thus, waste biomass of such organisms must be sterilized before disposal.

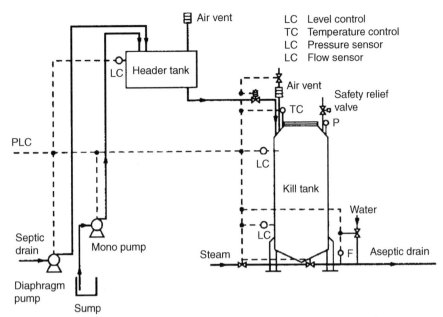

**FIGURE 5.16 A Vessel for the Batch Sterilization of Liquid Waste From a Contained
Fermentation (Jansson, Lovejoy, Simpson, & Kennedy, 1990)**

Sterilization may be achieved by either batch or continuous means but the whole
process must be carried out under contained conditions. Batch sterilization involves
the sparging of steam into holding tanks, whereas continuous processes would employ the type of heat exchangers that have been discussed in the previous section. A
holding vessel for the batch sterilization of waste is shown in Fig. 5.16. Whichever
method, is employed the effluent must be cooled to below 60°C before it is discharged to waste. The sterilization processes have to be validated and are designed
using the Del factor approach considered in the previous sections. However, the
kinetic characteristics used in the calculations would be those of the process organism rather than of *B. stearothermophilus.* Also, the N_t value used in the design calculations would depend on the assessment of the hazard involved should the organism
survive the decontamination process and would thus be smaller than 10^{-3}, used for
medium sterilization. Thus, the sterilization regime used for destruction of the process organism will be different from that used in sterilizing the medium. Junker et al.
(2006) confirmed that the continuous sterilization of waste biomass is the method
of choice for large biotechnology plants and quoted the use of temperatures from
80 to 140°C with holding times ranging from 1 to 5 min. Gregoriades, Luzardo,
Lucquet, and Ryll, 2003 described a continuous process for the inactivation of biological waste arising from mammalian cell cultures. All drains from the cell culture
laboratory, pilot plant, and manufacturing plant were collected in a sump tank that

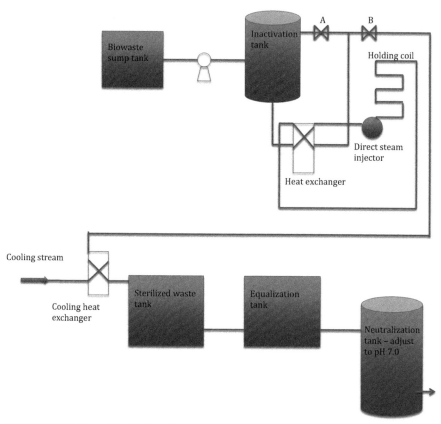

FIGURE 5.17 A Waste Sterilization Process

Waste is pumped from the inactivation tank, via the heat exchanger, to the direct steam injector and through the holding coil and the heat exchanger (thus heating the incoming waste). If the temperature in the coil is less than 80°C then valve A is opened, valve B closed and the flow diverted back to the inactivation tank. When the whole length of the coil has reached temperature then valve A is closed and valve B opened and the sterilized waste is diverted (via the cooling heat exchanger) to the sterilized waste tank and finally discharged to the sewer after pH adjustment.

Redrawn from Gregoriades et al., 2003

acted as a means of controlling the flow of waste to a bioinactivation tank. Waste was then continuously inactivated using a direct steam injector system followed by an insulated holding pipeline such that the waste stream was held at 80°C for 1 min. Cooling of the treated effluent was achieved in two heat exchangers, the first, a heat conservation step, cooling the sterile waste stream against the incoming waste stream and the second using cooling water. The waste was neutralized to pH 7.0 before discharge to the local sewage system. An outline of the plant is shown in Fig. 5.17.

STERILIZATION BY FILTRATION

The range of operations associated with a fermentation process that may involve sterilization by filtration are illustrated in Fig. 5.18 from which it may be seen that both liquid and gas filtration are utilized. The removal of suspended solids from both gas and liquids in these processes may then be described by the following mechanisms:

1. Inertial impaction.
2. Diffusion.
3. Electrostatic attraction.
4. Interception.

1. *Inertial impaction.* Suspended particles in a fluid stream have momentum. The fluid in which the particles are suspended will flow through the filter by the route of least resistance. However, the particles because of their momentum tend to travel in straight lines and may therefore become impacted upon the fibers, where they may then remain. Inertial impaction is more significant in the filtration of gases than in the filtration of liquids.
2. *Diffusion.* Extremely small particles suspended in a fluid are subject to Brownian motion that is random movement due to collisions with fluid molecules. Thus, such small particles tend to deviate from the fluid flow pattern and may become impacted upon the filter fibers. Diffusion is more significant in the filtration of gases than in the filtration of liquids.
3. *Electrostatic attraction.* Charged particles may be attracted by opposite charges on the surface of the filtration medium.
4. *Interception.* The fibers comprising a filter are assembled to define openings of various sizes. Particles that are larger than the filter pores are removed by direct interception. However, a significant number of particles that are smaller than the filter pores are also retained by interception. This may occur by several mechanisms—more than one particle may arrive at a pore simultaneously, an irregularly shaped particle may bridge a pore, once a particle has been trapped by a mechanism other than interception the pore may be partially occluded enabling the entrapment of smaller particles. Interception is equally important mechanism in the filtration of gases and liquids.

Filters have been classified into two types—those in which the pores in the filter are smaller than the particles that are to be removed and those in which the pores are larger than the particles that are to be removed. The former type may be regarded as an absolute filter, so that filters of this type (provided they are not physically damaged) are claimed to be 100% efficient in removing microorganisms. Filters of the latter type are frequently referred to as depth filters, originally composed of felts, woven yarns, and loosely packed fiberglass in packed towers and were widely used for air filtration. However, modern depth filters tend to be manufactured as immobilized fibers (microglass and synthetic polymers) to form a sheet or cartridge that are then accommodated in specialized housings. These systems then avoid the previous

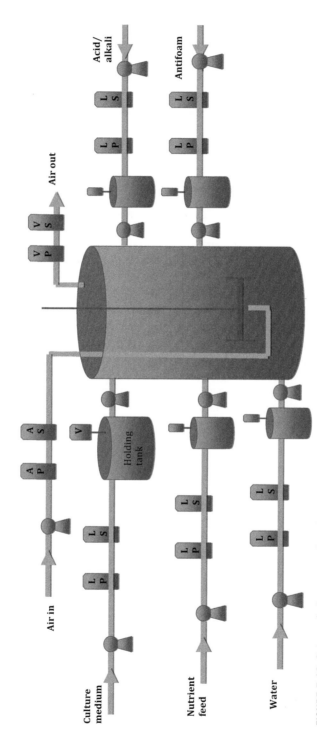

FIGURE 5.18 Schematic Representation of the Processes Streams Associated With a Fermenter That may be Filter-Sterilized

AP, air prefilter; *AS*, air sterilizing filter; *VP*, fermenter vent air prefilter; *VS*, fermenter vent air sterilizing filter; *V*, holding tank vent filter; *LP*, liquid prefilter; *LS*, liquid sterilizing filter.

problem of packed towers in which an increase in the pressure on the filter may result in movement of the material, producing larger channels through the filter, resulting in the loss of integrity. The terms absolute and depth can be misleading as they imply that absolute filtration only occurs at the surface of the filter, whereas absolute filters also have depth and thus filtration occurs within the filter as well as at the surface. Terms that bear more relationship to the construction of filters are "nonfixed pore filters" (corresponding with depth filters) and "fixed pore filters" (corresponding with absolute filters).

Nonfixed pore filters rely on the removal of particles by inertial impaction, diffusion, and electrostatic attraction rather than interception. The packing material contains innumerable tortuous routes through the filter but removal is a statistical phenomenon and, thus, sterility of the product is predicted in terms of the probability of failure (similar to the situation for steam sterilization). Thus, in theory, the removal of microorganisms by a fibrous filter cannot be absolute as there is always the possibility of an organism passing through the filter, regardless of the filter's depth. However, modern nonfixed pore filters can achieve an exceptionally small probability of failure, 10^{-20} or less, such that they may achieve performances equivalent to a fixed pore filter (Wikol et al., 2008) and may be described by a manufacturer as "rated as absolute"—another reason to avoid describing filters as "depth" or "absolute." It is possible that increased pressure applied to a nonfixed pore filter may result in the displacement of previously trapped particles.

Fixed pore filters are constructed so that the filtration medium will not be distorted during operation so that the flow patterns through the filter will not change due to disruption of the material. Pore size is controlled during manufacture so that an absolute rating can be quoted for the filter, that is, the removal of particles above a certain size can be guaranteed. Thus, interception is the major mechanism by which particles are removed. Because fixed pore filters have depth, they are also capable of removing particles that are smaller than the pores by the mechanisms of inertial impaction, diffusion, and attraction and these mechanisms do play significant roles in the filtration of gases. Fixed pore filters are superior for most purposes such that they have absolute ratings, are less susceptible to changes in pressure, and are less likely to release trapped particles. The major disadvantage associated with fixed pore filters was the resistance to flow they presented and, hence, the large pressure drop across the filters which represents a major operating cost.

Filtration companies have developed both fixed pore and nonfixed pore filters based on filter sheets that are pleated and incorporated into a cartridge device in which the filter membrane is supported, and protected from physical damage, by a plastic skeleton. These devices incorporate a large membrane surface area and have minimized the problems associated with early filtration systems. The large surface area reduces pressure-drop across the membrane of fixed pore filters and in nonfixed pore filters the material is immobilized in the cartridge, rather than being loosely packed in a tower, thus preventing its movement during filtration. The cartridges are then accommodated in stainless steel housings and scale-up can then be achieved conveniently by increasing the number of cartridges. The structure of a filter cartridge

(a)

(b)

(c)

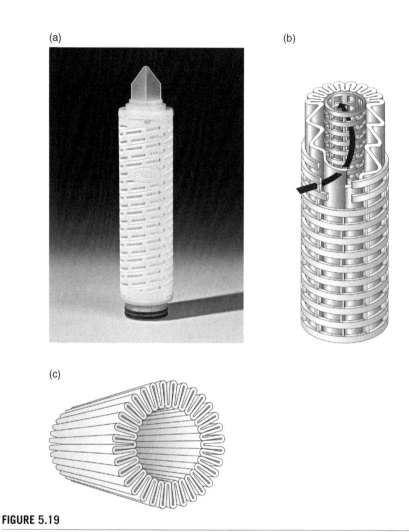

FIGURE 5.19

(a) A filter cartridge. (b) A schematic diagram showing the pleated membrane and supporting structure of a cartridge. (c) A schematic diagram showing the pleated membrane.

Pall Corporation, Portsmouth, UK

and its pleated membrane is shown in Fig. 5.19 and a range of cartridge types in Fig. 5.20a. Examples of stainless steel housings can be seen in Fig. 5.21.

It is important to realize that the filters, themselves, must be sterile and this is achieved in large-scale microbial processes by steam sterilization before and after operation (see Chapter 7). Thus, the materials must be stable at high temperatures and the steam must be free of particulate matter because the filter modules are particularly vulnerable to damage at high temperatures, achieved by filtering the steam

(a)

(b)

FIGURE 5.20

A selection of (a) filter cartridges to be used with stainless steel housings, (b) presterilized plastic capsules containing the filter cartridge.

Parker domnick hunter

itself through stainless steel mesh filters rated at 1 μm. However, filters for both air and liquid filtration are also available as disposable (single-use) capsules in which the pleated filter membrane is welded into a plastic housing with accompanying connections and tubing. Such systems are provided presterilized by gamma irradiation and examples are shown in Fig. 5.20b. The availability of disposable presterilized mixing

FIGURE 5.21 Examples of Stainless-Steel Cartridge Filter Housings

Parker domnick hunter

and holding vessels, along with associated pumps and disposable filters, has enabled the adoption of entire disposable trains from medium preparation through to addition of the sterile medium to a presterilized single-use fermenter, supplied with sterile air via a disposable filter assembly. These single-use systems are predominantly used in animal cell fementations but have also found popularity in small-scale microbial fermentations utilizing disposable reactors, for example, heterologous protein manu-facture, or for the production of inocula for large-scale traditional processes. These systems are discussed in more detail in Chapters 6, and 7.

THEORY OF NONFIXED PORE OR DEPTH FILTERS

Although fermentation companies commonly rely upon the pleated membrane, fixed pore filter systems to ensure sterility, nonfixed pore filters are used as prefilters in conjunction with fixed pore filters and, in certain circumstances, as the primary pro-tective air filter. The use of a prefilter removes larger particles from the fluid stream and thus protects the fixed pore filter and extends its useable life. Thus, it is neces-sary to consider the theory of nonfixed pore filtration. Aiba, Humphrey, & Millis (1973) have given detailed quantitative analysis of the mechanisms associated with depth filtration but this account will be limited to a description of the overall effi-ciency of operation of nonfixed pore filters. Several workers (Ranz & Wong, 1952; Chen, 1955) have put forward equations relating the collection efficiency of a filter bed to various characteristics of the filter and its components. However, a simpler description cited by Richards (1967) may be used to illustrate the basic principles of nonfixed pore filter design.

If it is assumed that if a particle touches a fiber it remains attached to it, and that there is a uniform concentration of particles at any given depth in the filter, then each layer of a unit thickness of the filter should reduce the population entering it by the same proportion; which may be expressed mathematically as:

$$\frac{dN}{dx} = -KN \tag{5.13}$$

where N is the concentration of particles in the air at a depth, x, in the filter and K is a constant.

On integrating Eq. (5.13) over the length of the filter, it becomes:

$$\frac{N}{N_0} = e^{-Kx} \tag{5.14}$$

Where N_0 is the number of particles entering the filter and N is the number of particles leaving the filter.

On taking natural logarithms, Eq. (5.14) becomes:

$$\ln\left(\frac{N}{N_0}\right) = -Kx. \tag{5.15}$$

Eq. (5.15) is termed the log penetration relationship. The efficiency of the filter is given by the ratio of the number of particles removed to the original number pres-ent, thus:

$$E = \frac{(N_0 - N)}{N_0} \tag{5.16}$$

where E is the efficiency of the filter.

But,

$$\frac{(N_0 - N)}{N_0} = 1 - (\frac{N}{N_0}).$$ (5.17)

Substituting

$$N/N_0 = e^{-Kx}$$

Thus:

$$\frac{(N_0 - N)}{N_0} = 1 - e^{-Kx}$$ (5.18)

and

$$E = 1 - e^{-Kx}.$$

The log penetration relationship [Eq. (5.15)] has been used by Humphrey and Gaden (1955) in filter design, by using the concept X_{90}, the depth of filter required to remove 90% of the total number of particles entering the filter; thus:

If N_0 were 10 and x were X_{90}, then N would be 1:

$$\ln(1/10) = -KX_{90}$$

or

$$2.303 \log_{10}(1/10) = -KX_{90}$$
$$2.303(-1) = -KX_{90},$$

therefore,

$$X_{90} = \frac{2.303}{K}.$$ (5.19)

Consideration of Eq. (5.15) indicates that a plot of the natural logarithm of N/N_0 against x, filter length, will yield a straight line of slope K.

The value of K is affected by the nature of the filter material and by the linear velocity of the fluid passing through the filter. Fig. 5.22 is a typical plot for an air filter of K and X_{90} against linear air velocity from which it may be seen that K increases to an optimum with increasing air velocity, after which any further increase in air velocity results in a decrease in K. The increase in K with increasing air velocity is probably due to increased impaction, illustrating the important contribution this mechanism makes to the removal of organisms. The decrease in K values at high air velocities is probably due to disruption of the filter, allowing channels to develop and fibers to vibrate, resulting in the release of previously captured organisms.

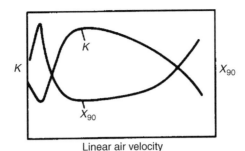

FIGURE 5.22 The Effect of Increasing Linear Air Velocity on *K* and *X*$_{90}$ of a Nonfixed Pore Filtration System (Richards, 1967)

FILTER STERILIZATION OF LIQUIDS

The following liquids are sterilized upstream of the fermenter in a fermentation plant (Fig. 5.18):

- The bulk fermentation medium.
- Nutrient addition feeds, normally comprising one component.
- Water that may be added to a previously sterilized fermenter.
- Acids and alkalis to control pH.
- Buffers—used particularly to control pH during the culture of animal cells. Buffers are also used in the down-stream processing of some parenteral products and have to be sterile.
- Antifoams to control foam formation.

It is worth remembering at this point that fixed pore filtration also occurs in down-stream processing, for the removal of biomass (particularly in animal cell fermentations) and for the sterilization of labile pharmaceutical liquids. These aspects are considered in Chapter 10 but the principles discussed in the following sections are also relevant to down-stream membrane filtration applications.

Sterilization of the fermentation medium

As discussed earlier, most microbial media are steam sterilized. However, media for animal-cell culture cannot be sterilized by steam because they contain heat-labile proteins. Thus, filtration is the method of choice and, as may be appreciated from the previous discussion, fixed pore filtration using hydrophilic membranes is the better system to use. An ideal filtration system for the sterilization of animal cell culture media must fulfill the following criteria:

1. The filtered medium must be free of fungal, bacterial, and mycoplasma contamination.
2. There should be minimal adsorption of medium ingredients to the filter surface.
3. The filtered medium should be free of viruses.

4. The filtered medium should be free of endotoxins.
5. Chemicals from the filter should not leach into the treated medium.

The threat of prion contamination from animal sera used in animal cell media resulted in the replacement of animal proteins with those derived from plant sources. This has resulted in two issues affecting the sterilization process (Jornitz & Meltzer, 2008). Mycoplasma contamination has increased and (due to poorer growth efficiency using plant-derived media) growth promoters such as insulin or insulin-like growth factor have had to be added. Mycoplasmas are smaller than bacteria and, being devoid of cell walls, they are deformable, which means that they may pass through even 0.1 μm fixed pore filters under pressure. Thus, the issues of removing mycoplasma contaminants and the nonspecific adsorption of growth factors have added to the complications of sterilization. Other important aspects that influence the design of the system include the achievable flow rate, total throughput, thermal and mechanical robustness, chemical compatibility, hold-up volume, and disposability (Jornitz & Meltzer, 2008). The total throughput (or filter capacity) is the volume of product that can be passed through a filter before it is blocked (often described as "plugging") and it is this criterion that is key to the choice of the filter system. The use of a prefilter that removes larger particles can increase the throughput of a sterilizing filter, that is, prevent premature blockage. Prefilters are frequently of the non-fixed pore type and protect the vulnerable fixed pore filter. Early cartridge systems for medium sterilization tended to rely on the use of a cascade of up to five filters to ensure a sterile product. For example, a 5 μm prefilter for the removal of coarse precipitates, clot-like material and other gross contaminants; followed by a 0.5 μm for bulk microbial removal, deformable gels, lipid-based materials, and endotoxin reduction and finally two 0.1 μm filters to remove mycoplasmas and endotoxins. To remove viral contamination then a final 0.04-μm filter would be added. Improvements in filter design have resulted in systems composed of a prefilter followed by a 0.1 μm device. Some manufacturers increase the throughput of their filters by incorporating the prefilter and sterilizing filter into one unit by coating the fixed pore membrane with a nonfixed pore layer. The hold-up volume is the volume of medium retained in the filter after the sterilization of a batch of medium and is thus unavailable. This aspect is far more important in down-stream filtration of the product postfermentation as this process stream is much more valuable than the original culture medium and will be discussed in Chapter 10.

The choice of filter must be made on the basis of trials in which candidate filter materials are challenged with the medium to be processed. All manufacturers supply test filters (usually 47 mm in diameter) that can be assessed on a small scale in stainless steel holders. Jornitz and Meltzer (2008) and Lutz, Wilkins, and Carbello (2013) discussed the key aspects to be addressed in these assessments, particularly the issues of filter capacity or throughput (and hence plugging), adsorption and the leaching of chemicals from the filter. However, these initial tests must be followed by a trial of a small-scale version of the pleated membrane device as the structure and geometry of the device will have a considerable impact on the process. Also, trial data from small-scale tests will only be applicable to the scale-up of a process if they are based on diminutive versions of the industrial device. As well as selecting a preferred filter material, these small-scale trials are used to optimize the operating conditions of the filtration process prior to scale-up. Temperature of the medium, differential pressure and pretreatment of the filter have all been shown to impact on filter performance. Lowering the temperature can enhance the total throughput by 30%. Although a lower temperature results in an increase in viscosity, and hence a lower flow rate, it should be remembered that throughput is the key criterion. High differential pressure (ie, flow rate) can result in gel formation on the membrane that drastically reduces throughput. Jornitz and Meltzer suggested that starting a filtration process by flushing with a chilled buffer and using an initial low differential pressure would reduce filter blockage.

Sterilization of fermenter feeds

As discussed earlier, the rationale for sterilizing animal cell culture media by filtration is to avoid the degradation of heat-labile components. Most feeds are heat stable and thus the rationale for their sterilization by filtration is obviously different and peculiar to the feed being sterilized. In animal cell culture, the increasing use of disposable presterilized reactors has resulted in the adoption of disposable equipment for the whole process. Thus, the sterilization of feeds to such animal cell fermentations is achieved using disposable presterilized filter units complete with reservoir, tubing, filter, and couplings. These feeds include nutrients, antifoam, buffers, and alkali for pH control. Sterile buffers are also used extensively in the down-stream processing of biopharmaceuticals and these are routinely filter sterilized. In microbial fermentation steam sterilization has been the standard method for the sterilization of feeds, as well as the bulk medium. However, with the availability of a wide range of filtration systems, the adoption of filter sterilization for some feeds has become more common. As discussed earlier, steam sterilization of oils can be problematic due to the lack of water often resulting in effectively a dry heat process and thus the requirement for a much higher temperature. Thus, filter sterilization of oils and oil-based antifoams is an attractive proposition. Furthermore, the high energy costs associated with steam processes have caused manufacturers to assess whether feed filtration may be a more economic means of sterilization. Thus, the filter sterilization of acids and bases for pH control has become more common.

The choice of filter for feed sterilization is then dependent on the properties of the feed. Acids, alkalis, and buffers for pH control are aggressive liquids but are generally low in contaminants and do not present problems of filter blockage (plugging). Thus, the choice of filter is based on its resistance to the liquid being filtered and the flow rate achievable. The most common filter materials used in these circumstances are polyethersulfone and PTFE (polytetrafluoroethylene). As well as the type of material employed, filter manufacturers have designed the structure of filters specifically for the filtration of low-plugging liquids in which high flow-rate is the key criterion. High flow rates are achieved using thinner membranes that have high "pleatability" and incorporate asymmetric pores that are tapered, that is, wider at the entry side of the filter than at the exit side (Jornitz & Meltzer, 2008). A thinner membrane gives less flow resistance but is more susceptible to damage. The pleatability of the membrane dictates the way in which the membrane can be formatted into a cartridge structure and is a function of both thickness and strength. The edges of the pleated sheet are particularly vulnerable to damage by a high flow rate or sudden changes in pressure. Thus, the design is a compromise giving high permeability with sufficient mechanical strength to withstand the operating conditions. Thus, this reinforces the point made earlier that actual cartridges are assessed in a selection process, not simply a sample of the membrane material. This aspect is discussed in more detail in the next section.

It is easy to overlook the fact that sterile water is also fed to fermentations. This may be done to replace water lost through evaporation or to decrease viscosity of highly viscous fermentations as a means of extending the productive life of the fermentation. Again, filter sterilization is an attractive proposition for the operation as the low microbial contamination should enable a high flow rate to be achieved with very low plugging, provided that a prefilter is incorporated to protect the fixed pore filter from large particles of debris.

Sizing of filters for liquid sterilization

The sizing of filters depends on models that predict the overall throughput of the filter, that is, the volume that can be processed before the filter is blocked or "plugged." The models describe the mechanism of plugging of the filter and described as caking, complete, and gradual. Caking occurs where the particles build up on the surface of the filter rather than in the pores. Complete blocking occurs when a particle arrives at the filter surface and completely blocks the pore. Gradual plugging is caused by particles that are smaller than the pores and the pores become blocked by the gradual build-up of particles. The most commonly applied model for fixed pore medium filtration is gradual plugging and is described as the V_{max} method (Lutz et al., 2013) that involves filter trials performed at a constant pressure across the filter. Blocking of nonfixed pore filters, on the other hand, is best described by complete and caking models and the approach to sizing is based on the P_{max} method that involves filter trials performed at a constant flow rate. Prefilters are usually of the nonfixed pore type and thus the P_{max} method is most widely used for their size prediction.

In the V_{max} method, the equation describing the minimum area (m^2) for filtration is:

$$A_{min} = \frac{V_B}{V_{max}} + \frac{V_B}{(J_i \times t_B)} \tag{5.20}$$

where A_{min} is the minimum area for filtration (m^2); V_B is the batch volume to be filtered (dm^3); V_{max} is the maximum volume of liquid (normalized) that can be filtered at time infinity before the filter becomes blocked (dm^3 m^{-2}), that is, the filter capacity; J_i is the initial volumetric (normalized) flow rate (dm^3 m^{-2} h^{-1}); and t_B is the time of filtration (h).

The sizing parameters, V_{max} and J_i, are both specific terms relating, respectively, the capacity of the filter (the volume that can be filtered before blockage) and the flow rate employed, to the area of the filter. The term V_B/V_{max} describes the filter area required to filter the batch volume before the filter becomes blocked. The term $V_B/(J_i \times t_B)$ describes filter permeability in terms of the area required to accommodate the flow rate being used. Thus, for nonplugging fluids, the V_B/V_{max} value will be very small and primarily J_i will determine the filter area. However, for plugging fluids (meaning that the filter is prone to blockage) then V_B/V_{max} will be significant and will play a major role in the sizing of the filter. Buffers are examples of nonplugging whereas culture media containing materials that bind to the filter surface are plugging fluids.

Rearrangement of Eq. (5.20) gives:

$$\frac{A_{min}}{V_B} = \frac{1}{V_{max}} + \frac{1}{(J_i \times t_B)} \tag{5.21}$$

The term (A_{min}/V_B) is the specific filtration area for the process, that is, m^2 of the filter required per dm^3 of medium to be filtered. The values of V_{max} and J_i are determined experimentally on a laboratory scale using a 47-mm-disc of the chosen filter material and is discussed in detail by Badmington, Wilkins, Payne, and Honig (1995). The test fluid is filtered at a constant pressure and the filtrate collected over a time period. This process can then be described by the equation:

$$\frac{t}{V} = \frac{t}{V_{max}} + \frac{1}{J_i}$$

where t is the filtration time and V is the cumulative volume at time t.

The data are processed by plotting t/V against t. V_{max} is then given by the reciprocal of the slope divided by the area of the filter and J_i is the reciprocal of the intercept divided by the area of the filter. Filter capacity (V_{max}) values for a plugging fluid range from 50 to 3000 dm^3 m^{-2} (Mok, Besnard, Pattnaik, & Raghunath, 2012). For a nonplugging fluid, the filter capacity is well in excess of 5000 dm^3 m^{-2}. The operating pressure and viscosity of the fluid must also be taken into account in the application of Eq. (5.21) in scale-up. Darcy's law describes the flow of a fluid through a porous medium (Mok et al., 2012):

$$J_i = \frac{\Delta P}{\mu \times R}$$

where ΔP is the differential pressure across the membrane (Pa); μ is the viscosity of the liquid (kg m^{-1} s^{-1}) R is the intrinsic resistance of the membrane (m^{-1}).

Darcy's law can be simplified using the term Q, which is $1/R$:

$$J_i = \frac{Q \times \Delta P}{\mu}$$

However, as the intrinsic resistance of the membrane R cannot be measured directly, the term Q' can be introduced which is the membrane permeability for a solution of viscosity μ, that is $Q' = Q/\mu$. The units for Q' are $\text{dm}^3\,\text{m}^{-2}\,\text{s}^{-1}\,\text{Pa}^{-1}$ but are more commonly expressed in this context as $\text{dm}^3\,\text{m}^{-2}\,\text{h}^{-1}\,\text{psi}^{-1}$ (psi being pounds per square inch, meaning that the units of ΔP will also be psi) and thus:

$$J_i = Q' \times \Delta P \text{ or } Q' = \frac{J_i}{\Delta P}$$

The value of Q' for the filtration of a fluid of a certain viscosity can be calculated from the V_{max} experiment in which J_i, it will be remembered, is the flux across the membrane ($\text{dm}^{3-}\,\text{m}^{-2}\,\text{h}^{-1}$). Eq. (5.21) may now be rewritten as:

$$\frac{A_{min}}{V_B} = \frac{1}{V_{max}} + \frac{1}{(Q' \times t_B \times \Delta P)}$$

It should be appreciated that the ΔP referred to in the relationship between J_i and Q' is the specific pressure differential operating during the V_{max} experiment whereas the ΔP referred to in Eqs. (5.20) and (5.21) is a generic term, the value of which may be altered in the optimization of the process.

The term A_{min}/V_B (surface area per unit volume of medium to be filtered) is then used to estimate the filter area required for the large-scale process. This area is then translated into the number of pleated membrane cartridges required to match the calculated area.

The laboratory test for the P_{max} method for sizing nonfixed pore filters (particularly prefilters) takes significantly longer than the V_{max} assessment. The process solution is filtered through a 47 mm disc of the proposed filter at a constant flow rate. The increase in pressure and the volume of filtrate are then measured as well as the turbidity of the filtrate. The differential pressure across the membrane increases as the filter plugs and the maximum capacity of the filter is then indicated when the pressure reaches a predetermined value or the turbidity of the filtrate suddenly increases, indicating that particles have been forced through the filter at the elevated pressure. Flux of the filtrate through the filter ($\text{dm}^3\,\text{m}^{-2}\,\text{h}^{-1}$) is given by:

$$\text{Flux} = \frac{\text{Volume of solution filtered}}{\text{Filtration area} \times \text{Time of filtration}}$$

The resistance of the filter is given by:

$$\text{Resistance} = \frac{\text{Pressure (psi)}}{\text{Flux}}$$

The resistance of the filter is then plotted against its capacity over the time period of the test, as shown in Fig. 5.23. The test should be conducted at several flow rates, including the lowest practical flow rate for the full-scale process.

However, it is important to appreciate the limitations of these approaches and the danger of under-sizing the filter for the scaled-up process. Equally, significantly oversizing the filter system will unnecessarily increase costs. Lutz (2009) presented

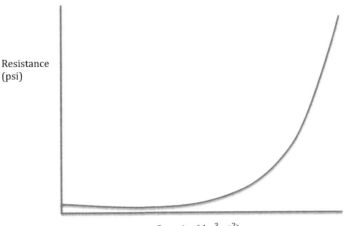

FIGURE 5.23 A Plot of the Data Generated From a P_{max} Determination, Showing the Increase in Resistance Over the Duration of the Experiment

The capacity is the volume of filtrate per unit area of the filter.

a detailed analysis of the major factors contributing to an underestimate of the predicted filter area:

- Direct problems of scale—These have been categorized into differences in the medium, filter construction, operating conditions, and scaling effects, as shown in Table 5.6. A key issue is the difference in performance of the test filter and its

Table 5.6 Factors Contributing to the Differences in Performance of Filter Processes at Different Scales

Factor	Due to	Arising from
Feed solution	Viscosity, temperature, solute concentration, particle concentration, particle size distribution.	Holding time, freeze-thaw issues of storage, mixing, differences in upstream processes.
Filter construction	Filter permeability, solid holding capacity, adsorption/fouling characteristics, mass transfer coefficient.	Filter porosity, filter thickness, pore size distribution, surface coating, feed channel spacers, feed channel height.
Operating conditions	Flux, pressure drop, diafiltration strategy, crossflows, throughput, wetting procedures, cleaning.	Constant pressure versus constant flow, in line resistances, gravitational head pressures, limited pressures, flows or liquid volumes, open versus closed system processing.
Scaling effects	Inaccurate filter areas, pleating effects, in-line flow resistances, flow distribution.	Effective versus installed membrane area, support or drain layer resistances, capsule fittings, different mixing or dead volumes.

Based on Lutz, 2009.

pleated large-scale counterpart—the nature of the pleating can have a major effect on performance, necessitating pilot scale trials to identify potential problems.

- Issues relating to expansion—Process development and optimization is a continual process and thus changes in medium design, flow rate, and volume of medium required may occur. Increasing demand for a product may require increasing the size of the process and thus the filtration plant should be designed to accommodate extra filter cartridges, thus facilitating expansion.
- Problems of variability—Random batch-to-batch variation will occur in the medium to be sterilized (particularly one containing plant or animal-derived materials), filter construction, and operating conditions thus changing the filtration characteristics.

The industry has resolved these issues by applying a "scale-up safety factor"—the ratio of the actual filter area employed divided by the minimum area calculated using the V_{max} approach. Lutz (2009) quoted the safety factor as generally at least a multiple of more than 1.5 and Giglia et al. (2010) and Giglia and Sciola (2011) quoted a commonly used range of 1.3–2.0. This multiple has been arrived at by experience and judgment and is thus an empirical factor. However, the adoption of a safety factor will result in an increase in both capital and running costs and thus over-compensating consequences of scale can result in unnecessary expenditure. Lutz (2009) considered this aspect in detail.

FILTER STERILIZATION OF FERMENTER INLET AIR

Aerobic fermentations require the continuous addition of considerable quantities of sterile air (see Chapter 9). The broadly quoted aeration rate used in a typical fermentation is one volume of air per volume of medium per minute. Thus, a 100 m^3 antibiotic fermentation may need up to 100 m^3 of sterile air per minute. Although it is possible to sterilize air by heat treatment, the most commonly used sterilization process is filtration. The traditional solution to this problem was to use packed filter towers employing fibrous materials such as glass fiber. However, these towers have been replaced by filter membranes, either of fixed pore or nonfixed pore (often described as "depth") construction that are assembled into cartridges, as discussed earlier in the chapter. The validation of nonfixed pore filters used in air filtration enable them to be described as absolute—meaning they have an equivalent probability of failure as a fixed pore system. The nonfixed pore membranes have the advantage of operating at lower pressure and allow higher air flow rates that means that a smaller filter area can be used and operating costs are lower. These systems, like those for the sterilization of liquids, are either contained in stainless steel modules and steam-sterilized or used as disposable capsules (predominantly used for animal cell processes). Air inlet filter systems are designed to remove not only bacterial and fungal propagules but also viruses—particularly bacteriophage in the case of bacterial fermentations. Contamination with phage can result in the lysis of the process organism and protection against such contamination is a major challenge in the industry.

Membrane materials for air filtration must be hydrophobic and thus repel water from the filter surface with polytetrafluoroethylene (PTFE) being the most commonly used material. Expanded PTFE (ePTFE) is a single, continuous structure in which all the fibers and crossover points are connected, thus there are no loose ends or particles, which can be shed into the air stream and contaminate the process (Wikol et al., 2008). PTFE also has the advantage of being resistant to heat and chemical attack. Thus, PTFE filters may be repeatedly steam-sterilized without loss of performance and ammonia may be injected into the air stream, prior to the filter, for pH control. The life expectancy of a filter is a major cost factor in a production-scale fermentation and thus the robust nature of PTFE membranes results in less frequent replacement as well as lower disposal costs associated with used cartridges. As was seen for the filter sterilization of liquids, it is essential that a nonfixed pore prefilter is incorporated up-stream of the sterilizing filter. The prefilter traps large particles such as dust, oil, and carbon (from the compressor), and pipescale and rust (from the pipework) and thus protects the sterilizing filter and increases its lifetime. Common materials used for prefilter construction include glass fiber, polypropylene, and PTFE impregnated borosilicate microfiber. It is also important that the air filters remain dry as wetting significantly reduces their effectiveness. In this context, a coalescing prefilter ensures the removal of water from the air; entrained water is coalesced in the filter (air flow being from the inside of the filter to the outside) and is discharged via an automatic drain. While it is essential to sterilize the sterilizing filter, the prefilter is not subjected to steam sterilization—its purpose is not to produce a sterile product—and, thus, the lifetime of the prefilter is extended. Also, steam sterilizing a prefilter may result in steam condensate carrying previously entrapped phage from the prefilter through the sterilizing filter, which may not be absolute when subjected to a liquid challenge.

The sizing of cartridge air filters is based on the filter information provided by the manufacturer. The manufacturer specifies the flow rates achievable for a cartridge at particular operating conditions (Fig. 5.24) with scale-up being achieved by increasing the number of cartridges.

STERILIZATION OF FERMENTER EXHAUST AIR

In many traditional fermentations, the exhaust gas from the fermenter was vented without sterilization or vented through relatively inefficient depth filters. With the advent of the use of recombinant organisms and a greater awareness of safety and emission levels of allergic compounds the containment of exhaust air is virtually universal (and in the case of recombinant organisms, compulsory). PTFE membrane modules are also used for this application but the system must be able to cope with the sterilization of water-saturated air, at a relatively high temperature and carrying a large contamination level. Also, foam may overflow from the fermenter into the air exhaust line. Thus, some form of pretreatment of the exhaust gas is necessary before

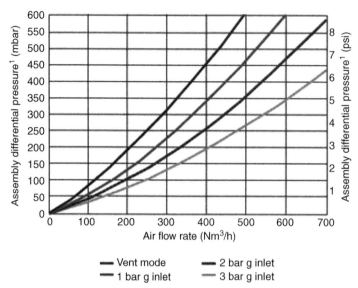

FIGURE 5.24 An Air Filter Performance Chart Showing the Relationship Between Air Flow Rate and Differential Pressure Across the Filter

The term Nm^3 indicates that the volume is that under standard temperature and pressure conditions.

Pall Corporation, Portsmouth, UK

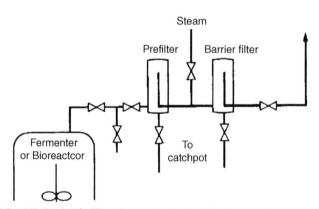

FIGURE 5.25 Dual Hydrophobic Filter System for the Sterilization of Off-Gas From a Fermenter

Pall Corporation, Portsmouth, UK

it enters the filter. This pretreatment may be a hydrophobic prefilter or a mechanical separator to remove water, aerosol particles, and foam. The pretreated air is then fed to a 0.2 μm hydrophobic filter. Again, it is important to appreciate that the filtration system must be steam sterilizable. Fig. 5.25 illustrates the Pall prefilter system and Figs. 5.26 and 5.27 illustrate the mechanical "Turbosep" separator system produced by Parker domnick hunter.

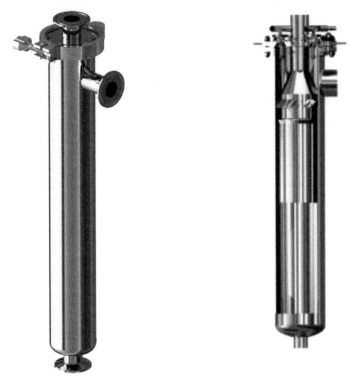

FIGURE 5.26 The Turbosep Mechanical Condensate Separator

Parker domnick hunter

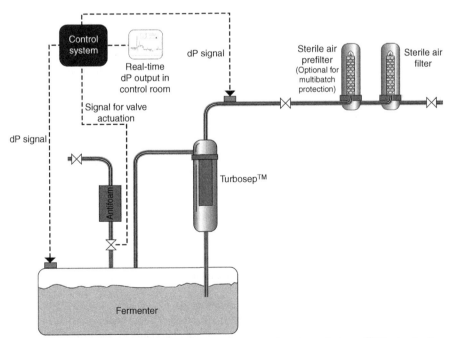

FIGURE 5.27 A Schematic Representation of the Installation of the Turbosep™ Mechanical Condense Separator

Parker domnick hunter

VESSEL VENT FILTERS

Any vessel used as a reservoir must be vented to enable air to leave the vessel when it is being filled and to enter the vessel when it is being emptied. Thus, sterile holding vessels containing liquids such as nutrient feeds, acid, alkali, buffer, antifoam, and water must be vented by a hydrophobic absolute rated, normally, PTFE membrane filter. These filters must be able to cope with the filling and emptying rates employed in the vessel and must obviously operate equally effectively in both forward and reverse directions. Should the filter restrict the entry of air during emptying it is possible for a partial vacuum to develop that, in extreme cases, could lead to implosion of the vessel.

REFERENCES

Aiba, S., Humphrey, A. E., & Millis, N. (1973). *Biochemical engineering* (2nd ed.). New York: Academic Press.

Armenante, P. M., & Li, Y. S. (1993). Complete design analysis of a continuous sterilizer for fermentation media containing suspended solids. *Biotechnology and Bioengineering, 41,* 900–913.

Aunstrup, K., Andresen, O., Falch, E. A., & Nielsen, T. K. (1979). Production of microbial enzymes. In H. J. Peppier, & D. Perlman (Eds.), *Microbial technology* (pp. 282–309). (Vol. 1, 2nd ed.). New York: Academic Press.

Bader, F. G. (1986). Sterilization: prevention of contamination. In A. L. Demain, & N. A. Solomon (Eds.), *Manual of industrial microbiology and biotechnology*. Washington, DC: American Society for Microbiology.

Bader, F. G., Boekeloo, M. K., Graham, H. E., & Cagle, J. W. (1984). Sterilization of oils: data to support the use of a continuous point-of-use sterilizer. *Biotechnology and Bioengineering, 26,* 848–856.

Badmington, F., Wilkins, R., Payne, M., & Honig, E. S. (1995). V_{max} testing for practical microfiltration train scale-up in biopharmaceutical processing. *Pharmaceutical Technology, 19*(9), 64–76.

Banks, G. T. (1979). Scale-up of fermentation processes. In A. Wiseman (Ed.), *Topics in enzyme and fermentation biotechnology* (pp. 170–266). (Vol. 3). Chichester: Ellis Horwood.

Bigelow, W. D. (1921). The logarithmic nature of thermal death time curves. *The Journal of Infectious Diseases, 29*(5), 528–536.

Boeck, L. -V., Wetzel, R. W., Burt, S. C., Huber, F. M., Fowler, G. L., & Joseph, S. A. (1988). Sterilization of bioreactor media on the basis of computer-calculated thermal input designated as F_0. *Journal of Industrial Microbiology, 3,* 305–310.

Boeck, L. V. D., Alford, J. S., Pieper, R. L., & Huber, F. M. (1989). Interaction of media components during bioreactor sterilization: definition and importance of R_0. *Journal of Industrial Microbiology, 4,* 247–252.

Chen, C. Y. (1955). Filtration of aerosols by fibrous media. *Chemical Reviews, 55,* 595–623.

Corbett, K. (1985). Design, preparation and sterilization of fermentation media. In C. L. Cooney, & A. E. Humphrey (Eds.), *Comprehensive biotechnology, the principles of biotechnology: Engineering considerations* (pp. 127–139). Toronto: Pergamon Press.

Deindoerfer, F. H., & Humphrey, A. E. (1959). Analytical method for calculating heat sterilization times. *Applied Microbiology, 7,* 256–264.

Deindoerfer, F. H., & Humphrey, A. E. (1961). Scale-up of heat sterilization operations. *Applied Microbiology*, *9*, 134–145.

Giglia, S., Rautio, K., Kazan, G., Backes, K., Blanchard, M., & Caulmare, J. (2010). Improving the accuracy of scaling from discs to cartridges for dead end microfiltration of biological fluids. *Journal of Membrane Science*, *365*, 437–355.

Giglia, S., & Sciola, L. (2011). Scaling-up normal-flow microfiltration processes. *Bioprocess International*, *9*(9), 58–63.

Gregoriades, N., Luzardo, M., Lucquet, B., & Ryll, T. (2003). Heat inactivation of mammalian cells for biowaste kill system design. *Biotechnology Progress*, *19*, 14–20.

Guizelini, B. P., Vandenberghe, L. P. S., Sella, S. R. B. R., & Soccol, C. R. (2012). Study of the influence of sporulation conditions on heat resistance of *Geobacillus stearothermophilus* used in the development of indicators for steam sterilization. *Archives of Microbiology*, *194*, 991–999.

Helou, C., Mariier, D., Jacalot, P., Abdennebi-Najar, L., Niquet-Leridon, C., Tessier, F. J., & Gadonna-Wideham, P. (2014). Microorganisms and Maillard reaction products: a review of the literature and recent findings. *Amino Acids*, *46*, 267–277.

Humphrey, A. E., & Gaden, E. L. (1955). Air sterilization by fibrous media. *Industrial & Engineering Chemistry*, *47*, 924–930.

Jain, D., & Buckland, B. C. (1988). Scale-up of the efrotomycin fermentation using a computer-controlled pilot plant. *Bioprocess Engineering*, *3*, 31–36.

Jansson, D. E., Lovejoy, P. J., Simpson, M. T., & Kennedy, L. D. (1990). Technical problems in large-scale containment of rDNA organisms. In R. -L. Yu (Ed.), *Fermentation technologies: Industrial applications* (pp. 388–393). London: Elsevier.

Jones, M. C. (1968). The temperature dependence of the lethal rate in sterilization calculations. *Journal of Food Technology*, *3*, 31–38.

Jornitz, M. W., & Meltzer, T. H. (2008). Media and buffer filtration implications. In M. W. Jornitz, & T. H. Meltzer (Eds.), *Filtration and purification in the biopharmaceutical industry* (2nd ed., pp. 439–457). New York: Informa Healthcare.

Junker, B. H. (2001). Technical evaluation of the potential for streamlining of equipment validation for fermentation applications. *Biotechnology and Bioengineering*, *74*(1), 49–61.

Junker, B. H., Beshty, B., & Wilson, J. (1999). Sterilization-in-place of concentrated nutrient solutions. *Biotechnology Bioengineering*, *62*(5), 501–508.

Junker, B. H., Lester, M., Brix, T., Wong, D., & Nuechterlein, J. (2006). A next generation, pilot-scale continuous sterilization system for fermentation media. *Bioprocess Biosystems Engineering*, *28*, 351–378.

Li, L. C., Parasrampuria, J., Bommireddi, A., Pec, E., Dudleston, A., & Mayoral, J. (1998). Moist-heat sterilization and the chemical stability of heat-labile parenteral solutions. *Drug Development and Industrial Pharmacy*, *24*(1), 89–93.

Ling, B., Tang, J., Kong, F., & Mitcham, E. J. W. (2015). Kinetics of food quality changes during thermal processing: a review. *Food Bioprocess Technology*, *8*, 343–358.

Lutz, H. (2009). Rationally defined safety factors for filter sizing. *Journal of Membrane Science*, *341*, 268–278.

Lutz, H., Wilkins, R., & Carbello, C. (2013). Sterile filtration: Principles, best practices and new developments. In P. Kolhe, M. Shah, & N. Rathore (Eds.), *Sterile product development: Formulation, process, quality and regulatory considerations* (pp. 431–459). New York: Springer.

Mann, A., Jiefer, M., & Leuenberger, H. (2001). Thermal sterilization of heat-sensitive products using high-temperature short-time sterilization. *Journal of Pharmaceutical Sciences*, *90*(3), 275–287.

Marshall, C. T., Lilly, M. D., Gbewonyo, K., & Buckland, B. C. (1995). Evaluation of the effect of sterilization conditions on fermentations to produce acidic fibroblast growth factor. *Biotechnology Bioengineering, 47*, 688–695.

Mok, Y., Besnard, L., Pattnaik, P., & Raghunath, B. (2012). Sterilizing-grade filter sizing based on permeability. *Bioprocess International, 10*(6), 58–63.

Murrell, W. G., & Scott, W. J. (1966). The heat resistance of bacterial spores at various water activities. *Journal of General Microbiology, 43*, 411–425.

Nazina, T. N., Tourova, T. P., Poltaraus, A. B., Novikova, E. V., Grigoryan, A. A., Ivanova, A. E., Lysenko, A. M., Petrunyaka, V. V., Osipov, G. A., Belyaev, S. S., & Ivanov, M. V. (2001). Taxonomic study of aerobic thermophilic bacilli: descriptions of *Geobacillus subterraneus* gen. nov., sp. nov. from petroleum reservoirs and transfer of *Bacillus stearothermophilus, Bacillus thermo-catenulatus, Bacillus thermoleovorans, Bacillus kaustophilus, Bacillus thermoglucosidasius* and *Bacillus thermodenitrificans* to *Geobacillus* as the new combinations *G. stearothermophilus, G. thermocatenulatus, G. thermoleovorans, G. kaustophilus, G. thermoglucosidasius* and *G. thermodenitrificans*. *International Journal of Systematic and Evolutionary Microbiology, 51*, 433–446.

Nursten, H. E. (2005). *The maillard reaction—chemistry, biochemistry and implications*. London: Royal Society of Chemistry.

Pistolesi, D., & Mascherpa, V. (2014). F_0 a technical note. http://www.fedegari.com/news/white-papers/discover-our-new-e-book-for-the-use-sterilization.

Pollard, D. (2011). Aseptic operations. In C. Webb (Ed.), *Comprehensive biotechnology, second edition, Vol. 2, engineering fundamentals of biotechnology* (pp. 933–956). Amsterdam: Elsevier.

Ranz, W. E., & Wong, J. B. (1952). Impaction of dusty smoke particles on surface and body collectors. *Industrial & Engineering Chemistry, 44*, 1371–1378.

Richards, J. W. (1966). Fermenter mash sterilization. *Process Biochemistry, 1*(1), 41–46.

Richards, J. W. (1967). Air sterilization with fibrous filters. *Process Biochemistry, 2*(9), 21–25.

Richards, J. W. (1968). *Introduction to industrial sterilization*. London: Academic Press.

Sharma, M.C., & Gurtu, A.K. (1993). Asepsis in bioreactors. In S. Neidleman, & A.I. Laskin (Eds.), *Advances in Applied Microbiology, 39* (pp. 1–27). New York, Academic Press.

Singh, V., Hensler, W., & Fuchs, R. (1989). Optimization of batch fermenter sterilization. *Biotechnology Bioengineering, 33*, 584–591.

Soderberg, A. C. (2014). Fermentation design. In H. C. Vogel, & C. M. Todaro (Eds.), *Fermentation and biochemical engineering handbook, principles, process design and equipment* (pp. 85–108). Oxford: Elsevier.

Stacey, A., & Stacey, G. (2000). Routine quality control testing for cell cultures. In D. Kitchington, & R. Schinazi (Eds.), *Antiviral methods and protocols, methods in molecular medicine* (pp. 27–40). New Jersey: Humana Press.

Stockbridge, P. J., Hinge, R., Lawrence, A., Livingstone, A., Mercier, J. P., Pietrowski, R., & Wiseman, A. (2001). Harvesting and disposition of foreign growth fermenters. *Pharmeuropa, 13*, 8–10.

Stowell, J. D. (1987). Application of oils and fats in antibiotic processes. In J. D. Stowell, A. J. Beardsmore, C. W. Keevil, & J. R. Woodward (Eds.), *Carbon substrates in biotechnology* (pp. 139–160). Oxford: IRL Press.

Svensson, R. (1988). Continuous media sterilization in biotechnical fermentation. *Dechema Monographien, 113*, 225–237.

van Doornmalen, J. P. C. M., & Kopinga, K. (2009). Temperature dependence of F-, D-, and Z-values used in steam sterilization processes. *Journal of Applied Microbiology, 107,* 1054–1060.

Wikol, M., Hartmann, B., Brendle, J., Crane, M., Beuscher, U., Brake, J., & Shickel, T. (2008). Expanded ploytetrafluoroethylene membranes and their applications. In M. W. Jornitz, & T. H. Maltzer (Eds.), *Filtration and purification in the biopharmaceutical industry* (2nd ed., pp. 619–640). New York: Informa Healthcare.

Culture preservation and inoculum development

PRESERVATION OF INDUSTRIALLY IMPORTANT CELL CULTURES AND MICROORGANISMS

The process organism is a company's most precious resource and therefore it is critical that it retains the desirable characteristics that led to its selection. Also, the culture used to initiate an industrial fermentation must be viable and free from contamination. Thus, industrial cultures must be stored in such a way as to eliminate genetic change, protect against contamination, and retain viability. Theoretically, microorganisms and cells in culture may be kept viable by repeated subculture into fresh medium, but, at each cell division, there is a small probability of mutations occurring and because repeated subculture involves many such divisions, there is a high probability that strain degeneration would occur. Also, repeated subculture carries with it the risk of contamination. Thus, preservation techniques have been developed to maintain cultures in a state of "suspended animation" by storing either at reduced temperature or in a dehydrated form. Equally as important as the preservation process itself is the management of the preserved cultures. It is imperative that an effective "culture banking system" is implemented incorporating a validated quality control system to test the performance and purity of the stored strains. The criteria associated with "performance" of the strain must include process productivity as well as viability. This aspect is dealt with in more detail later in this section.

STORAGE AT REDUCED TEMPERATURE

Storage on agar slopes

Microbial cultures grown on agar slopes may be stored in a refrigerator (5°C) or a freezer (−20°C) and subcultured at approximately 6-month intervals. The time of subculture may be extended to one year if the slopes are covered with sterile medicinal grade mineral oil. Although these approaches may be successful in the short term, they cannot be relied upon to maintain master stocks.

Cryopreservation—storage below −135°C

The metabolic activities of both microorganisms and cultured cells may be reduced considerably by storage at very low temperatures (below −135°C and commonly −150 to −196°C). Such low temperatures may be achieved using liquid nitrogen or a mechanical freezer, but the preferred approach is storage in the vapor phase of a liquid nitrogen refrigerator (Smith & Ryan, 2012). The key issue to the effective storage of biological material using cryopreservation is the avoidance of intracellular ice formation. Theoretically, this can be achieved by cooling to below $-137°C$ (the glass transition temperature of water) in milliseconds at which temperature, water becomes an amorphous solid devoid of ice crystals in which molecular movement and interaction has stopped. However, this is not a practical solution and the alternative is to suspend the material in a solution of a cryoprotective agent (CPA) that permeates the cells, deceases the freezing point of water and increases its viscosity, thus reducing the probability of ice crystal formation during cooling. A 10% glycerol solution was commonly used as the cryoprotectant but this has been superseded by dimethyl sulfoxide (DMSO), usually at a concentration of 5–10%. By freezing the cryoprotected cells at a rate of approximately 1°C per min. to below $-135°C$, water is transformed into the desired solid amorphous state (described as vitrified) and the cells are in a state of suspended animation. The stored cells may be "revived" from this state by rapid thawing in a water bath at about 37°C. Although some loss of viability will occur during the freezing and thawing stages the frozen, cryoprotected cultures will be stable provided the storage temperature is maintained below $-135°C$. Thus, viability may be predictable even after a period of many years. However, it is important to appreciate that the cryopreservation process should be optimized for the biological material being preserved and the reader is referred to detailed descriptions of the process provided by Moldenhauer (2008) and Baust, Corwin, and Baust (2011). Also, the reliability of the technique is only as good as that of the storage facilities—liquid nitrogen evaporates and must be replenished and the breakdown of a mechanical freezer could be disastrous. Thus, the equipment must be continuously monitored and an automated appropriate alarm system incorporated.

Cryopreservation is the most universally applicable of all preservation methods and bacteria, fungi, bacteriophage, viruses, algae, yeasts, animal and plant cells, and tissue cultures have all been successfully preserved. However, as discussed in Chapter 2, a key difference between microbial cells in culture and animal cells in culture is that animal cells are susceptible to apoptosis or programed cell death. When microorganisms enter the stationary phase the "general stress response" is induced which protects the cells from damage. However, when animal cells enter the stationary phase then apoptosis may be induced which results in the programed death of the cells. Furthermore, the cooling process can also induce apoptosis (Baust et al., 2011; Bissoyi, Nayak, Pramanik, & Sarangi, 2014). While an understanding of the apoptopic pathway has enabled workers to suppress its expression in many process cell lines (see Chapter 12), it is pertinent to be aware of the phenomenon when generating cultures for preservation. Thus, microbial cultures for cryopreservation are grown into the stationary phase (to induce the protective general stress response) whereas those of cell cultures are harvested at the midexponential phase, to prevent the initiation of apoptosis.

STORAGE IN A DEHYDRATED FORM

Dried cultures

Dried soil cultures have been used widely for culture preservation, particularly for sporulating mycelial organisms. Moist, sterile soil may be inoculated with a culture and incubated for several days for some growth to occur and then allowed to dry at room temperature for approximately 2 weeks. The dry soil may be stored in a dry atmosphere or, preferably, in a refrigerator. The technique has been used extensively for the storage of fungi and actinomycetes and Pridham, Lyons, and Phrompatima (1973) observed that of 1800 actinomycetes dried on soil about 50% were viable after 20-years storage.

Malik (1991) described methods, which extend the approach using substrates other than soil. Silica gel and porcelain beads are suggested alternatives and detailed methods are given for these simple, inexpensive techniques in Malik's discussion.

Lyophilization

Lyophilization, or freeze-drying, is suitable for the preservation of microorganisms but not for animal cell cultures. It involves the freezing of a culture followed by its drying under vacuum, which results in the sublimation of the cell water. The culture is grown to the maximum stationary phase and resuspended in a protective medium such as milk, serum, or sodium glutamate. A few drops of the suspension are transferred to an ampoule, which is then frozen and subjected to a high vacuum until sublimation is complete, after which the ampoule is sealed. The ampoules may be stored in a refrigerator and the cells may remain viable for 10 years or more (Perlman & Kikuchi, 1977).

Lyophilization is very convenient for service culture collections (Snell, 1991) because, once dried, the cultures need no further attention and the storage equipment (a refrigerator) is cheap and reliable. Also, the freeze-dried ampoules may be dispatched as such, still in a state of "suspended animation" whereas liquid nitrogen stored cultures begin to deteriorate in transit. However, freeze-dried cultures are tedious to open and revitalize and several sub-cultures may be needed before the cells regain their typical characteristics. Overall, the technique appears to be second only to liquid nitrogen storage and even when liquid nitrogen is used makes an excellent insurance against the possibility of the breakdown of the nitrogen freezer. The reader is referred to Rudge (1991) and Moldenhauer (2008) for a detailed consideration of lyophilization.

QUALITY CONTROL OF PRESERVED STOCK CULTURES

Whichever technique is used for the preservation of an industrial culture, it is essential to be certain of the quality of the stocks and the security of the storage system. Smith and Ryan (2012) discussed the implementation of the Organisation for Economic Development and Cooperation (OECD) best practice guidelines for the validation of preserved strains. Although these guidelines are intended for national culture collections, they are equally relevant to collections held by research organizations or industry. Key criteria are:

1. Cultures should be preserved by at least two different methods, but if this is not possible then cryopreserved stocks should be maintained in separate locations.
2. The number of transfers or generations of the culture before preservation should be kept to a minimum, thus reducing the risk of degeneration.
3. A bank of identical preserved master cultures must be established as the definitive source of the organism within the organization (described as the "crown-jewel" stock by Gershater, 2010). A bank of "working stock cultures" is sourced from a master culture and thus the master culture bank should contain sufficient ampoules to minimize the likelihood of it being depleted to a level where further master stocks must be generated (Fig. 6.1).
4. The inoculum for each fermentation run is provided by an ampoule from the bank of working stock cultures that must therefore contain sufficient working stocks to minimize the need to revert to a master stock to replenish it. In some microbial processes, the working stock culture is used to generate up to ten agar

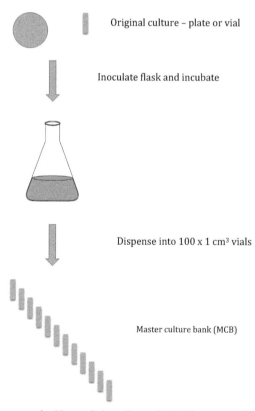

Original culture – plate or vial

Inoculate flask and incubate

Dispense into 100 x 1 cm³ vials

Master culture bank (MCB)

FIGURE 6.1 Development of a Master Culture Bank of 100 Vials, One of Which Will be Used to Generate a "Working Stock Culture Bank", Each of Which Will be Used to Start a Fermentation

slope cultures stored at 4°C, each of which is then used as the starting culture for a fermentation. This approach minimizes the consumption of working stocks but is only suitable if the cultures are stable. Animal cell fermentations should always be initiated using a new working stock culture.

5. Documentation of both the master and working stocks should be such that the ancestry of an individual fermentation can be effectively tracked.

6. The original culture, master, and working preserved stocks should all be evaluated to ensure their quality. This quality assessment should include the purity, the fermentation performance, and variability of the culture.

Gershater (2010) described a procedure to assess productivity and variability that is applicable to the original culture, master stocks, and working stocks. The protocol is shown in Fig. 6.2 and depended on sufficient replication to enable the application

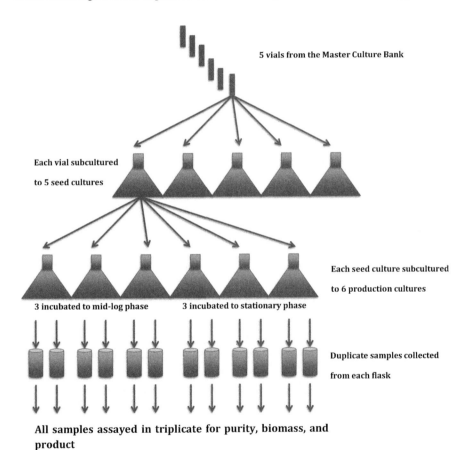

5 vials from the Master Culture Bank

Each vial subcultured to 5 seed cultures

Each seed culture subcultured to 6 production cultures

3 incubated to mid-log phase   3 incubated to stationary phase

Duplicate samples collected from each flask

All samples assayed in triplicate for purity, biomass, and product

FIGURE 6.2 A Variability Test to Ensure the Reproducibility of the Master Culture Bank (MCB)

The number of replicates is such that a statistical analysis of variants can be conducted.

Modified from Gershater, 2010

of analysis of variance procedures. Five stock ampoules are reconstituted and each used to inoculate five seed flasks, each of which (after incubation) is used to inoculate six production flasks, three of which are sampled (in duplicate) at a suitable time point during the fermentation and the remaining three at the end of the process. The samples would be analyzed for biomass and product as well as any other parameters pertinent to the process—for example, substrate utilization. Thus, the five vials give rise to 25 seed cultures that are used to inoculate a total of 150 production flasks generating data at two stages in the final fermentation. As a result, variation can be assessed between replicate ampoules (cultures) and the stages responsible for the variation identified.

The criteria for the verification of the quality of stored cultures are most demanding for those producing heterologous proteins for in vivo or ex vivo human use. For such processes, the manufacturer has to guarantee the origin, performance, stability, and purity of the cell line, using methods described by the Food and Drug Administration (FDA) and the European Medicines Agency (EMEA). The International Conference on Harmonization (ICH) was established in 1990 to facilitate the coordination of the regulatory requirements of Europe, Japan, and the United States of America for approving and authorizing new medicinal products and ICH guidelines have been agreed covering quality, safety, efficacy, and multidisciplinary issues in the industry. ICH Quality Guideline 5 considers the quality of biotechnological products including issues relating to the cell line such as contamination, the expression of the genetic construct, and the stability of the producing system. Because strain degeneration is linked to the number of divisions it undergoes, the guideline introduces the concept of these criteria being demonstrable in cells that have undergone the number of generations produced over the duration of the commercial process. The maximum number of generations (or population doubling level, PDL) is then specified in the process protocol as the "in vitro cell age" which cannot be exceeded in any production run. In vitro cell age is defined as:

> *A measure of the period between thawing of the MCB vial(s) and harvest of the production vessel measured by elapsed chronological time in culture, population doubling level of the cells or passage level of the cells when sub cultured by a defined procedure for dilution of the culture*

O'Callaghan and Racher (2015) described a procedure (Fig. 6.3) to assess cell stability in which a stock culture is sequentially sub-cultured using the process scale conditions until the number of generations lapsed is at least equivalent to the in vitro cell age. Samples are taken throughout this process and cryopreserved for stability assessment. The samples are then reconstituted and used to inoculate simultaneous model production processes and the performance of early subcultures compared with that of sub-cultures beyond the in vitro cell age. The process is illustrated in Fig. 6.3 and the battery of possible tests to which these cultures are then subjected is shown in Table 6.1. This process is designed to eliminate experimental inconsistencies by preserving samples and testing them all under the same experimental conditions using the same batches of media.

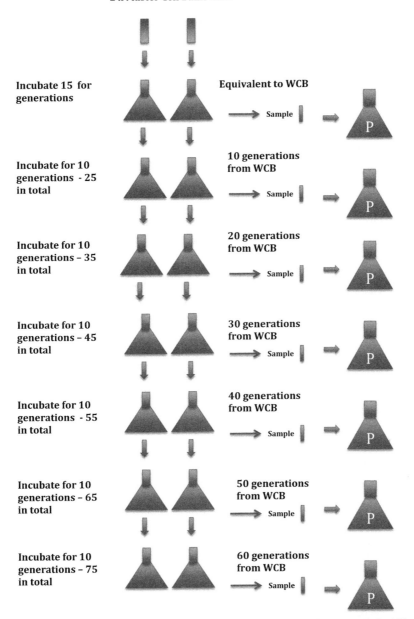

2 x Master Cell Bank vials

Incubate 15 for
generations

Equivalent to WCB

→ Sample

P

Incubate for 10
generations - 25
in total

10 generations
from WCB

→ Sample

P

Incubate for 10
generations – 35
in total

20 generations
from WCB

→ Sample

P

Incubate for 10
generations – 45
in total

30 generations
from WCB

→ Sample

P

Incubate for 10
generations - 55
in total

40 generations
from WCB

→ Sample

P

Incubate for 10
generations – 65
in total

50 generations
from WCB

→ Sample

P

Incubate for 10
generations – 75
in total

60 generations
from WCB

→ Sample

P

FIGURE 6.3 A Study of a Heterologous Protein Production Process to Demonstrate Stability Over the in vitro Cell Age of the Process

An MCB vial is used to initiate a model inoculum program. Samples from each stage are freeze-dried and at the end of the study used to inoculate model production fermentations. *P*, production fermentation.

Modified from O'Callaghan and Rachel, 2015

Table 6.1 Tests Used to Demonstrate the Stability of a Cell Line Used for the Manufacture of a Therapeutic Protein

Integrity of the Protein Product	
Characteristic	**Analytical Method**
Size variation—protein and subunit mass; fragmentation and disulfide shuffling	Reduced and nonreduced SDS electrophoresis
Size variation—Proportions of protein monomers and aggregates	Size exclusion chromatography
Charge variants	Isoelectric focusing and weak cation exchange chromatography
Posttranslational modification	Maldi-ToF-MS analysis of PNGase F-treated product
Amino acid sequence	LC-MS of peptidase treated product
Biological activity	Various assays
DNA and host cell protein contaminants	Various assays
Integrity of Expression Construct	
Characteristic	**Analytical Method**
Gene copy number	qPCR
Structure of expression construct	Southern hybridization analysis
mRNA lengths—changes to transcription unit	Northern hybridization analysis
DNA sequence	Reverse transcription of product mRNA

From O'Callaghan & Racher, 2015.

INOCULUM DEVELOPMENT

It is essential that the culture used to inoculate a fermentation satisfy the following criteria:

1. It must be in a healthy, active state thus minimizing the length of the lag phase in the subsequent fermentation.
2. It must be available in sufficiently large volumes to provide an inoculum of optimum size.
3. It must be in a suitable morphological form.
4. It must be free of contamination.
5. It must retain its product-forming capabilities.

The process adopted to produce an inoculum meeting these criteria is called inoculum development. Hockenhull is credited with the quotation "once a fermentation has been started it can be made worse but not better" (Calam, 1976). Whereas this is an over-statement, it does illustrate the importance of inoculum development. Much of the variation observed in small-scale laboratory fermentations is due to poor inocula being used and, thus, it is essential to appreciate that the establishment of an effective inoculum development program is equally important regardless of the scale of the fermentation. Such a program not only aids consistency on a small scale but also is invaluable in scaling up the fermentation and forms an essential part in progressing a new process.

A critical factor in obtaining a suitable inoculum is the choice of the culture medium. It must be stressed that the suitability of an inoculum medium is determined by the subsequent performance of the inoculum in the production stage. As discussed elsewhere (Chapter 4), the design of a production medium is determined not only by the nutritional requirements of the organism, but also by the requirements for maximum product formation and the limitations of the production process. The formation of product in the seed culture is not an objective during inoculum development so that the seed medium may be of a different composition from the production medium. In an early seminal paper, Lincoln (1960) stated that growing the culture in the "final-type" medium minimizes the lag time in a fermentation. Lincoln's argument is an important one, so the inoculum development medium should be sufficiently similar to the production medium to minimize any period of adaptation of the culture to the production medium, thus reducing the lag phase and the fermentation time. Furthermore, Hockenhull (1980) pointed out the dangers of using very different media in consecutive stages. Major differences in pH, osmotic pressure, and anion composition may result in very sudden changes in uptake rates, which, in turn, may affect viability. Hockenhull also emphasized that for antibiotic fermentations the inoculum medium should contain sufficient carbon and nitrogen to support maximum growth until transfer, so that secondary metabolism remains repressed during the growth of the inoculum. If secondary metabolism is derepressed in the seed fermentation, then selection may enrich the culture with nonproducing variants having a growth advantage over high-producing types. Hockenhull drew attention to Righelato's (1976) work in which it was shown that chemostat culture of *Penicillium chrysogenum* under carbohydrate-limited conditions led to a loss of penicillin synthesizing ability and an increase in the proportion of nonconidiated variants whereas this did not occur in ammonia-, phosphate-, or sulfate-limited conditions. The relevance of this phenomenon is supported by Hockenhull's observation that *P. chrysogenum* inocula produced under nonlimiting conditions are remarkably free from variants whereas variants arise relatively frequently during the carbon-limited production phase. Gershater (2010) also emphasized that secondary metabolites are often growth inhibitory and thus allowing production of a secondary metabolite product in the seed fermentation would result in it being carried over with the inoculum and inhibiting the establishment of the production fermentation. Examples of inoculum and production media are given in Table 6.2, from which it may be seen that inoculum media are, generally, less nutritious than production media and contain a lower level of carbon.

The quantity of inoculum normally used is between 3% and 10% of the medium volume (Lincoln, 1960; Meyrath & Suchanek, 1972; Hunt & Stieber, 1986, Gershater, 2010). A relatively large inoculum volume is used to minimize the length of the lag phase and to generate the maximum biomass in the production fermenter in as short a time as possible, thus increasing vessel productivity. Thus, starting from a working stock culture, the inoculum must be built up in a number of stages to produce

Table 6.2 Inoculum Development and Production Media for a Range of Processes

Process	Inoculum Development Medium	g dm^{-3}	Production Medium	g dm^{-3}	References
Milbemycin	Arkasoy 50	10	Arkasoy 50	10	Warr, Gershater, and Box (1996)
	Casein	2	Casein	2	
	CaCO$_3$	5	CaCO3	5	
	MgSO$_4 \cdot$7H$_2$O	1	MgSO$_4 \cdot$7H$_2$O	1	
	Glucose	10	Fructose	20	
			Starch	50	
Vitamin B$_{12}$	Sugar beet				Spalla et al. (1989)
	Molasses	70		105	
	Sucrose	—		15	
	Betaine	—		3	
	NH$_4$H$_2$PO$_4$	0.8		—	
	(NH$_4$)$_2$SO$_4$	2		2.5	
	MgSO$_4$	0.2		0.2	
	ZnSO$_4$	0.02		0.08	
	5-6 Dimethyl-benzimidazole	0.005		0.025	
Lysine	Cane molasses	5%		20%	Nakayama (1972)
	Corn-steep liquor	1%		—	
	CaCO$_3$	1%		—	
	Soybean meal hydrolysate	—		1.8%	
Clavulanic acid	Soybean flour	1.0%	Soybean flour	1.5%	Butterworth (1984)
	Dextrin	2.0%	Oil	1.0%	
	Pluronic L81 (Antifoam)	0.03%	KH2PO4	0.1%	

sufficient biomass to inoculate the production-stage fermenter. This may involve two or three stages in shake flasks and one to three stages in fermenters, depending on the size of the ultimate vessel. Throughout this procedure, there is a risk of contamination and strain degeneration necessitating stringent quality-control procedures. The greater the number of stages between the working stock culture and the production fermenter the greater the risk of contamination and strain degeneration. Therefore, a compromise may be reached regarding the size of the inoculum to be used and the risk of contamination and strain degeneration. Another factor to be considered in the determination of the inoculum volume is the economics of the process. A seed fermenter 10% of the size of the production fermenter represents a considerable financial investment that must be justified in terms of productivity. A large-scale continuous fermentation for the production of biomass would be expected to operate at steady state in excess of 100 days. Thus, the number of times that the fermenter is inoculated should be very few compared with batch or fed-batch systems. In such circumstances, it may be more economic to compromise on the size of the inoculum and to tolerate a relatively lengthy period of growth up to maximum biomass than to invest a large seed vessel that would be used on very few occasions. This is particularly relevant for biomass continuous processes where one very large fermenter may be used and, thus, any seed vessel would only be servicing the one production vessel.

A key development in animal cell fermentation was the introduction of disposable reactors used for a single fermentation run, as introduced in Chapter 2 and discussed in more detail in Chapter 7. Such reactors can be operated in feedback continuous culture mode (see Chapter 2) in which the biomass is retained within the vessel—so-called perfusion culture when applied to animal cell culture. Thus, the industry has access to single-use vessels designed to produce biomass concentrations far in excess of that obtainable in conventional batch systems and it is not surprising that a number of publications have described their use in inoculum development for animal cell fermentations. They have the advantage of growth rate control (because they can be operated continuously), high sterility assurance and, because very high biomass levels are achievable, can replace much larger traditional seed reactors or reduce the number of stages in the inoculum program. The application of these systems to microbial fermentations has been limited by the mass transfer rates achievable but recent advances in the design of disposable reactors have raised the prospect of single-use vessels being employed in inoculum production for high value microbial fermentations, primarily heterologous proteins, especially in the early stages of inoculum development, perhaps replacing a number of shake flask stages. These developments will be discussed later in this chapter but first a typical inoculum-development program for a microbial fermentation will be described in detail. For a sporulating organism, the process may be modified to facilitate the use of a spore suspension as inoculum and this will be discussed in more detail later in this chapter. A working stock culture (as described earlier in this chapter) is reconstituted and plated on to solidified medium. If the culture has been demonstrated to be sufficiently stable, up to ten colonies of typical morphology of high producers are selected and inoculated on to slopes as the subworking cultures, each sub-working culture being used for a new production run. Alternatively, a more conservative approach would be to utilize one working stock to initiate a fermentation. At this stage, shake flasks may be inoculated to check the productivity of these cultures, the results of such tests being known before the developing inoculum eventually reaches the production plant. Either the working stock or sub-working culture is used to inoculate a shake flask (250 or 500 cm^3 containing 50 or 100 cm^3 medium) that, in turn, is used as inoculum for a larger flask, or a laboratory fermenter, which may then be used to inoculate a pilot-scale fermenter. Culture purity checks are carried out at each stage to detect contamination as early as possible. Although the results of these tests may not be available before the culture has reached the production plant, at least it is known at which stage in the procedure contamination occurred. Junker et al. (2006) discussed the issue of contamination in a range of fermentation processes and reported the use of a below-optimum cultivation temperature in the seed stage fermenter. Naturally, this resulted in a longer incubation time in the seed stage but, as a result, facilitated laboratory-based contamination tests to be completed before the seed, itself, was used as inoculum. Thus, a longer period of inoculum growth was tolerated to ensure the security of the inoculum before it was used to initiate the production stage of the process. The slower growth rate at the lower temperature enabled the seed culture to be in the correct growth phase despite the extended incubation period.

Lincoln (1960) suggested a more elaborate procedure for the development of inoculum for bacterial fermentations that, with minor modifications, is applicable to any type of culture. The procedure involved the use of one working stock culture to develop a bulk inoculum that was subdivided, stored in a frozen state, and used as inocula for several months. A single colony, derived from a sub-master culture, was inoculated into liquid medium and grown to maximum log phase. This culture was then transferred into nineteen times its volume of medium and incubated again to the maximum log phase, at which point it was dispensed in 20 cm^3 volumes, plug frozen and stored at below $-20°C$. At least 3% of the samples were tested for purity and productivity in subsequent fermentation and, provided these were suitable, the remaining samples could be used as initial inocula for subsequent fermentations. To use one of the stored samples as inoculum it was thawed and used as a 5% inoculum for a seed culture, which in turn, was used as a 5% inoculum for the next stage in the program. This procedure ensured that a proven inoculum was used for the penultimate stage in inoculum development. More recently, Junker et al. (2002) reported the successful use of frozen bagged inocula for a range of fungal and bacterial fermentations. In a conventional inoculum development program Junker's group grew the process organism through two shake flask stages and then inoculated a 180 dm^3 (working volume) pilot scale fermenter with pooled shake flasks from the second stage. In the revised program, a culture developed in the pilot fermenter was harvested and dispensed as 2 dm^3 aliquots into sterile plastic bags, with glycerol as a cryoprotectant. The harvest time was assessed by the oxygen consumption rate of the culture. The bags were frozen and stored at $-70°C$ to serve as inocula for future production runs. These stored inocula were sampled to ensure viability, purity, and productivity thus providing a validated bank of inocula. The contents of a defrosted bag were then used directly to inoculate the 180 dm^3 stage in a production fermenter train, thus replacing a seven-day shake flask inoculum stage.

Seth et al. (2013) also described the use of frozen Chinese hamster ovary (CHO) cells inocula for the production of an antibody. In this process, culture produced from a 20 dm^3 perfusion bioreactor was frozen in aliquots and used to inoculate a seed fermenter. This process is discussed later in the chapter.

A 2005 paper by the Merck group, Okonkowski, Kizer-Bentley, Listner, Robinson, and Chartrain (2005), gives an excellent example of the rationale in designing an inoculum train. These workers compared three inoculum development programs for the production of an HIV DNA vaccine using *E. coli* at a production scale of between 30 and 2000 dm^3. The three programs are shown in Fig. 6.4 a, b, and c. Program 1

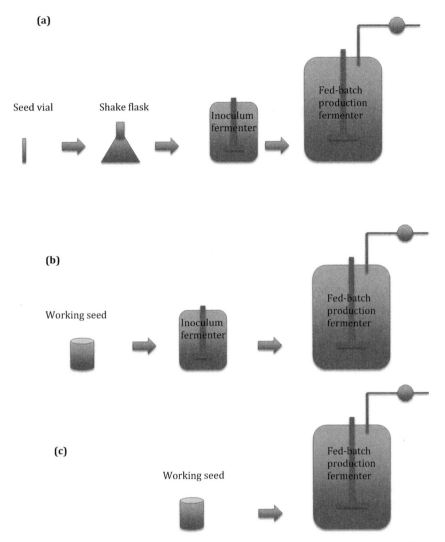

FIGURE 6.4 Three inoculum development programs (a–c) for the production of an HIV DNA vaccine using *E. coli*

Modified from Okonkowski et al., 2005

consisted of a three-stage process with a working master culture as the initial inoculum. Programs 2 and 3 were two and one stage processes each initiated with a large (300 cm^3 or larger), inoculum produced in a 30 dm^3 fermenter and frozen in aliquots with glycerol as a cryoprotectant. The advantage of both these programs over the three-stage variant was the elimination of a shake-flask stage and the concomitant contamination risks during inoculation. The two-stage and one-stage variants were then compared on the basis of eight criteria, weighted according to importance. The

Table 6.3 The Performance of One and Two-Stage Inoculum Development Programs of a DNA Vaccine *E. coli* Fermentation

Process Characteristic	Importance	Program Rating (1, low; 10, high)		Practicality Score (Rating × importance)	
		One-Stage Program	Two-stage Program	One-Stage Program	Two-Stage Program
Minimum cycle time	3	9	4	27	12
High cell growth reproducibility	9	5	7	45	63
Minimum number of seed stages	4	10	5	40	20
High specific plasmid yield	10	10	10	100	100
Robustness to external factors	8	3	7	24	56
Minimum freezer space for inoculum storage	5	2	6	10	30
Minimum over-all cost	7	10	6	70	42
Minimum risk of contamination or mechanical failure	7	9	6	63	42
			Total practi-cality score	379	365

Okonkowski et al., 2005.

performances of the two variants are shown in Table 6.3 from which it can be seen that the one-stage system proved preferable.

CRITERIA FOR THE TRANSFER OF INOCULUM

The physiological condition of the inoculum when it is transferred to the next culture stage can have a major effect on the performance of the fermentation. El-Naggar, El-Bindary, and Nour (2013) demonstrated that inoculum amount and age were amongst the most critical process variables affecting productivity of antibiotics by *Streptomyces anulatus*. The optimum time of transfer must be determined experimentally and then procedures established so that inoculation with an ideal culture may be achieved routinely. These procedures include the standardization of cultural conditions and monitoring the state of an inoculum culture so that it is transferred at the optimum time, that is, in the correct physiological state. A widely used criterion for the transfer of vegetative inocula is biomass and such parameters as packed cell volume, dry weight, wet weight, turbidity, respiration, residual nutrient concentration, and morphological form have been used (Hockenhull, 1980; Hunt & Stieber, 1986; Neves, Vieira, & Menezes, 2001). Ettler (1992) demonstrated that the rheological behavior of *Streptomyces noursei* could be used as the transfer criterion in the nystatin fermentation. The rheology of the seed fermentation changed from Newtonian to non-Newtonian behavior and the optimum inoculum transfer time corresponded with this transformation.

Criteria that may be monitored online are the most convenient parameters to use as indicators of inoculum quality and these would include dissolved oxygen, pH (although pH would normally be controlled in seed fermentations), and oxygen or carbon dioxide in the effluent gas. Parton and Willis (1990) advocated the use of the carbon dioxide production rate (CPR) as a transfer criterion that requires analysis of the fermenter effluent air (see Chapter 8). This approach is suitable only when transfer is being made from a fermenter, but Parton and Willis stressed the importance of adopting this strategy even for the inoculation of laboratory-scale fermentations despite the fact that an adequate inoculum volume could be produced in shake flask. These workers provide an excellent example of the effect of inoculum transfer time on the productivity of a streptomycete secondary metabolite, as shown in Fig. 6.5 a, b, and c. The CPR of the inoculum fermentation and the points at which inoculum was transferred are shown in Fig. 6.5a. The CPRs of the subsequent production fermentations are shown in Fig. 6.5b, from which it may be seen that the three fermentations performed similarly. However, Fig. 6.5c illustrates the very different secondary metabolite production of the three fermentations. Thus, although the time of transfer had only a marginal influence on biomass in the production fermentation, the effect on product formation was critical. It should be emphasized that the amount of biomass transferred was standardized for the three fermentations and, thus, the differences in performance were due to the physiological states of the inocula.

A further example of the value of gas analysis in the validation of inocula is provided by the work of Neves et al. (2001) on the *Streptomyces clavuligerus* clavulanic acid fermentation. A measure of the growth rate at any particular time was given by dividing the carbon dioxide production rate by the total carbon dioxide produced (CPR/Total CO_2). This figure peaked within the 18 and 24 h range and contrasted with the standard protocol of transferring the inoculum after 48 h. By transferring the inoculum at the peak (CPR/Total CO_2) rate, the production fermentation could be terminated after 96 h compared with 139 h for fermentations started with the 48 h culture. As a result, an increase of 22% in the overall process profit was achieved.

In recent years, probes have been developed for online assessment of biomass (see Chapter 8) and these could be invaluable in estimating the time of inoculum transfer (Kiviharju, Salonen, Moilanen, & Eerikainen, 2008). Boulton, Maryan, Loveridge, and Kell (1989) reported the use of a biomass sensor based on dielectric spectroscopy (the Bugmeter) to control the yeast pitching rate (inoculum level) in brewing. The probe measures the dielectric permittivity of viable yeast cells. Using the probe, these workers developed an automatic inoculum dispenser allowing a preset viable yeast mass to be transferred from a yeast storage vessel to the brewery fermentation. Albornoz (2014) investigated the factors that interfered with the dielectric signal and developed methods to deconvolute the output relating to viable biomass from that generated by nonviable

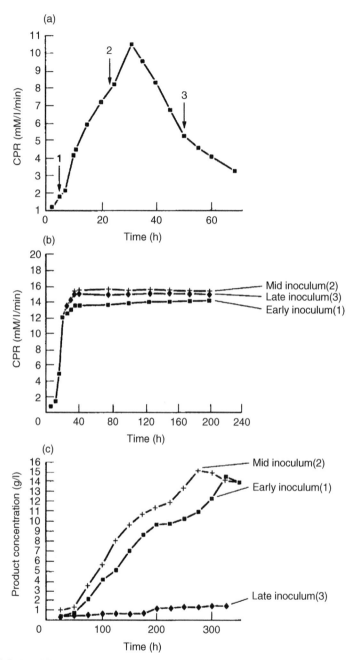

FIGURE 6.5 The Effect of Inoculum Age on Growth and Productivity in a Streptomycete Fermentation

(a) The carbon dioxide production rate (CPR) profile of the inoculum culture showing the points (1, 2 and 3) at which inocula were removed, (b) The effect of inoculum age on the CPR of the production fermentation, (c) The effect of inoculum age on productivity in the production fermentation (Parton & Willis, 1990).

biomass, air bubbles, and media components. Several approaches were successful, particularly the combination of the dielectric probe with an infrared optical probe. Nordon, Littlejohn, Dann, Jenkins, Richardson, and Stimpson (2008) also examined the interference of fermentation ingredients, products, and operating conditions on biomass assessment using near infrared spectroscopy (NIR) and demonstrated the potential of this technique in assessing the transfer time of a streptomycete inoculum. NIR output was shown to correlate with peak CPR and, thus, commensurate with its application to indicate the transfer time. An added advantage of NIR over gas analysis is that the system can be applied (albeit offline) to the analysis of flask cultures.

Alford, Fowler, Higgs, Clapp, and Huber (1992) reported the use of a real-time expert computer system to predict the time of inoculum transfer for industrial-scale fermentations. The system involves the comparison of on-line fermentation data with detailed historical data of the process. A problem with the use of CPR is that the data are not available continuously because the analyzer is not dedicated to any one fermenter, but is analyzing process streams from a large number of vessels via a multiplexer system (see Chapter 8). Thus, a fermentation may have passed a critical stage between monitoring times. Also, occasional false readings may be generated. The expert system enabled the verification of data points as well as prediction of the outcome of the fermentation from early information. Data from seed fermentations were analyzed by the expert system and the transfer time predicted. As a result of this approach operators were able to plan their work more effectively, the need for manual sampling was reduced, and early warning of contamination was provided if the seed-culture profile predicted from early readings was abnormal. Ignova, Montague, Ward, and Glassey (1999) developed an artificial neural network that compared the behavior of the seed stage of a penicillin fermentation with historical data and predicted the likelihood of a poor outcome at the production stage. As a result, it was possible to adopt a more efficient feeding strategy in the production fermentation to ameliorate the potential impact of the "underperforming" inoculum.

Smith and Calam (1980) compared the quality and enzymic profile of differently prepared inocula *Penicillium patulum* (producing griseofulvin) and demonstrated that a low level of glucose 6-phosphate dehydrogenase was indicative of a good quality inoculum. The enzyme profile of good quality inoculum was established early in the growth of the seed culture. Thus, this approach could be used to assess the cultural conditions giving rise to satisfactory inoculum, but would be of less value in determining the time of transfer.

Animal cell, yeast, unicellular bacterial, fungal, and streptomycete fermentations have different requirements for inoculum development and these are dealt with separately.

DEVELOPMENT OF INOCULA FOR ANIMAL CELL PROCESSES

The properties of a successful inoculum of an animal cell fermentation are the same as those for a microbial fermentation—a pure, metabolically active culture that has retained its product-forming ability and is available in a sufficient volume. However, the challenge of producing such an inoculum is greater for an animal cell process. Most microbial fermentations employ an inoculum of between 3% and 10% by volume and many are operated successfully using significantly lower levels. Animal cell fermentations, on the other hand, generally use inocula of between 10% and 50% by volume and thus more stages are required in the inoculum train to build up the desired volume (Kloth, Macisaac, Ghebremariam, & Arunakumari, 2010). Furthermore, the growth rate of animal cells is less than microorganisms so that the incubation periods in the inoculum development process are longer. Thus, the process involves more culture transfers and takes longer, making it more susceptible to contamination. Kloth et al. (2010) reported that it can take between 1 and 6 weeks to produce sufficient inoculum to start a production-scale animal cell fermentation. Furthermore, as animal cell cultures are generally used for the production of heterologous proteins for human use the in vitro cell age (including the number of generations in inoculum build-up) specified for the process must not be exceeded (see earlier discussion).

Kloth et al. (2010) discussed the design of inoculum development programs for animal cell cultures. The initial stage of any process is the reconstitution of the cryo-preserved working stock culture (WSC), normally achieved by heating in a 37°C water bath, thus giving a thaw rate of 20°C per min. As discussed earlier, DMSO is normally used as the cryoprotectant and this must be removed or diluted to facilitate growth. The preserved cells may be separated from the DMSO by centrifugation although this can result in a loss of viability. Dilution of the DMSO is achieved by directly subculturing the defrosted working cell bank culture into fresh medium. However, care must be taken that the degree of dilution is not only sufficient to diminish the inhibitory effect of DMSO but is compatible with the desired ratio of inoculum to fresh medium. Kloth et al. (2010) quote a tolerance range to DMSO of 0.1–0.4%, depending on the cell line, and an inoculum volume between 10% and 50%.

The vast majority of commercial animal cell fermentations use suspension cultures in agitated systems and an example of a typical inoculum development program for such processes is shown in Fig. 6.6 (Kloth et al., 2010). In this example, disposable plastic T-flasks and spinner flasks are used up to a scale of 5 dm^3. T-flasks are disposable plastic flasks providing a large liquid surface area for gas exchange, used for static culture and are available in volumes up to approximately 50 cm^3. Spinner flasks are agitated by a magnetic stirrer and are available as plastic disposable vessels from 100 cm^3 to 3 dm^3 and it has been common practice to use several such vessels at inoculum stages requiring volumes of between 3 and 10 dm^3. Glass resterilizable vessels are available up to 35 dm^3. The inoculum volumes used in this example range between 25% and 30%, apart from the final transfer, which is 14%, reinforcing the fact that animal cell fermentations use larger inoculum volumes, and hence more stages, than microbial processes.

A key development in animal cell fermentation was the introduction of presterilized disposable reactors used for a single fermentation run, as introduced in Chapter 2 and discussed in more detail in Chapter 7. Small-scale disposable vessels are made of rigid plastic but the feasibility of large-scale disposable vessels became a reality in 1999 when Singh developed the first presterilized, flexible, plastic, pillow-like culture bag which could be placed on a rocking platform to generate a wave-like motion within the vessel (Singh, 1999). Presterilized disposable stirred reactors became available in 2006, typically consisting of a flexible plastic bag, acting as a liner retained in a stainless steel

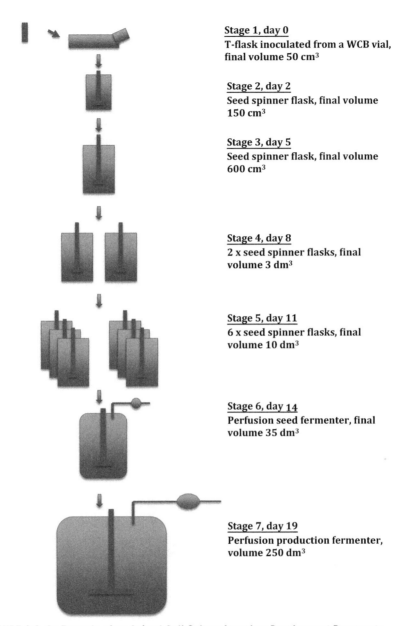

Stage 1, day 0
T-flask inoculated from a WCB vial,
final volume 50 cm^3

Stage 2, day 2
Seed spinner flask, final volume
150 cm^3

Stage 3, day 5
Seed spinner flask, final volume
600 cm^3

Stage 4, day 8
2 x seed spinner flasks, final
volume 3 dm^3

Stage 5, day 11
6 x seed spinner flasks, final
volume 10 dm^3

Stage 6, day 14
Perfusion seed fermenter, final
volume 35 dm^3

Stage 7, day 19
Perfusion production fermenter,
volume 250 dm^3

**FIGURE 6.6 An Example of an Animal Cell Culture Inoculum Development Program to
Generate inoculum for a 250 dm^3 Production Reactor**

Modified from Kloth et al., 2010

supporting vessel. A number of commercial manufacturers now offer these single-use reactors and they can be equipped with disposable pH and dissolved oxygen sensors—for example, the Sartorius wave and stirred reactors shown in Fig. 6.7. Wave reactors are available up to 300 dm^3 and stirred reactors up to 2000 dm^3. Such reactors can be operated in feedback continuous culture mode (see Chapter 2) in which the biomass is retained within the vessel—so-called perfusion culture when applied to animal cell culture. Thus, the industry has access to single-use vessels designed to produce biomass concentrations far in excess of that obtainable in conventional batch systems and it is not surprising that a number of publications have described their use in inoculum

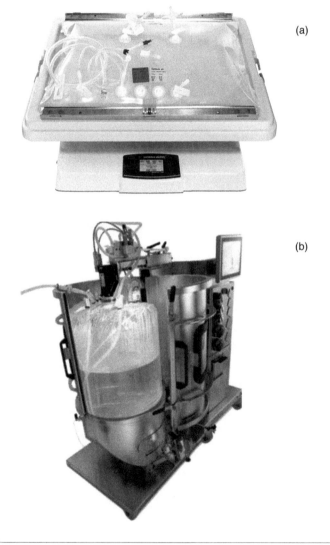

(a)

(b)

FIGURE 6.7

Disposable reactors (a) Rocking bag-type (b) Stirred reactor.

Sartorius Stedim UK Ltd., Epsom, UK

development for animal cell fermentations. They have the advantage of growth rate control (because they can be operated continuously), high sterility assurance and, because very high biomass levels are achievable, can replace much larger traditional seed reactors or reduce the number of stages in the inoculum program. Indeed, such have been the developments in this area that Eibl, Loffelholz, and Eibl (2014) claimed that disposable reactors were becoming standard for animal cell-based seed inoculum production.

A number of approaches have been adopted to exploit the use of disposable reactors in animal cell inoculum development. A single disposable wave reactor can replace the use of multiple shake flasks at a single stage, thus reducing the probability of contamination during transfer. For example, in Fig. 6.6, a 10 dm^3 disposable wave reactor may replace the spinner flasks employed in stage 5. Furthermore, the use of a perfusion disposable reactor would enable the production of a higher cell concentration, thus enabling a smaller reactor to be employed in the inoculum train. The volume-range that can be operated in a single Wave-type reactor is also much greater than that achievable in a stirred vessel, due to the necessity for the medium to completely cover the agitator in the stirred rector. Thus, rather than transferring the culture to the next stage, the initial culture volume in a Wave reactor can be gradually increased, by the addition of medium either as periodic aliquots or as a continuous feed (effectively, fed-batch mode). The CELL-tainer® wave reactor incorporates a segmented system so that the working volume of the reactor can be expanded incrementally from 150 cm^3 to 15 dm^3 by removing plastic supports below the bag. The flexible bag then conforms to the new dimensions dictated by the remaining supports, as shown in Fig. 6.8, and additional medium may be added. Frohlich, Bedard, Gagliardi, and Oosterhuis (2012) demonstrated that such a reactor could replace three stages in an inoculum expansion program (Fig. 6.9).

Cross-sectional view

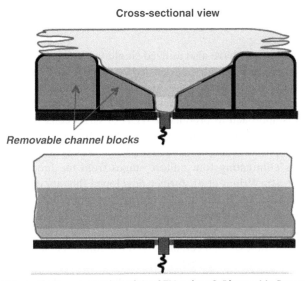

Removable channel blocks

FIGURE 6.8 A Schematic Representation of the CELL-tainer® Disposable Reactor

The reactor bag rests on blocks on the rocking platform incorporates blocks, removing the blocks enables the bag to expand to a greater volume (Celltainer Biotech).

(a)

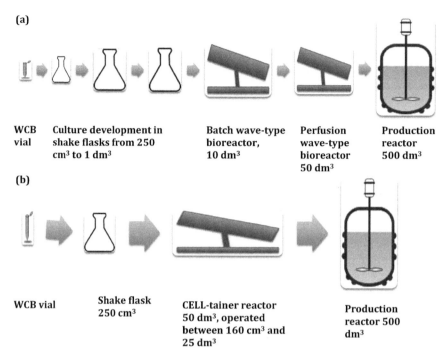

| WCB vial | Culture development in shake flasks from 250 cm³ to 1 dm³ | | Batch wave-type bioreactor, 10 dm³ | Perfusion wave-type bioreactor 50 dm³ | Production reactor 500 dm³ |

(b)

| WCB vial | Shake flask 250 cm³ | CELL-tainer reactor 50 dm³, operated between 160 cm³ and 25 dm³ | Production reactor 500 dm³ |

FIGURE 6.9 The Use of a CELL-tainer Bioreactor to Streamline a Cell Culture Inoculum Development Program

(a) Original inoculum development program. (b) Modified inoculum development program incorporating CELL-tainer reactor.

Modified from Frohlich et al., 2012

A number of workers have also utilized single-use bioreactors, operated in perfusion mode, to produce a high cell density Working Cell Bank (WCB) of vials. Thus, such a WCB can be used to inoculate a larger volume vessel than would be possible in a conventional process. For example, Tao, Shih, Sinacore, and Ryll (2011) used this approach to produce a WCB of CHO cells in 4 cm³ aliquots at a density of 10^8 cells cm^{-3} (compared with 2.5×10^7 cm^{-3} in 1 cm³ aliquots in the conventional process) and inoculated a 20 dm³ Wave reactor directly from the WCB vial, thus eliminating four culture stages from the inoculum program, as shown in Fig. 6.10. Wright et al. further developed this approach by combining the high cell density WCB inoculum with the incorporation of perfusion culture in the penultimate and production stages, as shown in Fig. 6.11. The availability of sterile, single-use, large cryopreservation bags, up to a volume of 20 dm³ has also enabled the storage of high cell density WCB cultures in larger volumes rather than in small vials. Seth et al. (2013) described the use of a single-use perfusion reactor to produce CHO cells at a density of more than 7.0×10^7 cm^{-3} that were then cryopreserved in 150 cm³ bags as a WCB. These bags could then be used as

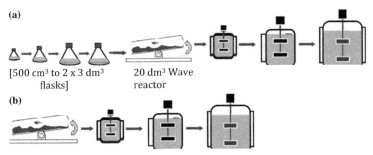

FIGURE 6.10 Comparison of the Use of Standard (a) and High Cell Density (b) Working Cell Banks (WCBs) to Initiate a CHO Inoculum Development Programme

(a) Inoculum program initiated with a 500 cm^3 flask inoculated with a standard WCB vial (1 cm^3 at 2.5 × 10^7 cells cm^{-3}). (b) Inoculum program initiated with a 20 dm^3 Wave reactor inoculated with a high cell density WCB (4.5 cm^3 at 10^8 cells cm^{-3}).

Modified from Tao et al., 2011

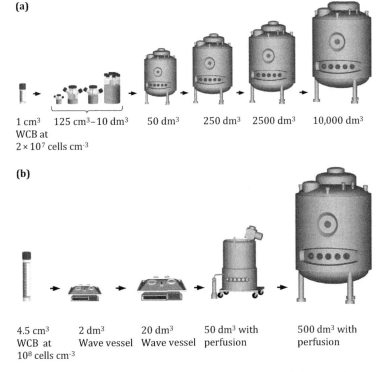

FIGURE 6.11 Alternative Inoculum Development Programs for a CHO Fermentation

(a) Inoculum program initiated with a 125 cm^3 spinner flask inoculated with a standard WCB vial (1 cm^3 at 2.5 × 10^7 cells cm^{-3}) leading to batch reactors. (b) Inoculum program initiated with a 2 dm^3 Wave reactor inoculated with a high cell density WCB (4.5 cm^3 at 10^8 cells cm^{-3}) leading to perfusion reactors.

Modified from Wright et al., 2015

inocula for vessels normally inoculated from a train of cultures. Thus, using these WCBs to inoculate a larger volume could eliminate the initial small-scale cultures of the inoculum development program.

The concept of single-use equipment is not confined to reactors, with a range of presterilized, disposable connectors, tubing, containers, and pumps available to the industry, thus facilitating the secure transfer of culture from one stage to the next (Whitford, 2013).

DEVELOPMENT OF INOCULA FOR YEAST PROCESSES

While the largest industrial fermentations utilizing yeasts are the production of baker's yeast biomass, the brewing of beer, and the production of fuel alcohol, recent processes have also been established for the production of recombinant products.

BAKER'S YEAST

The commercial production of baker's yeast involves the development of an inoculum through a large number of aerobic stages. Although the production stages of the process may not be operated under strictly aseptic conditions, a pure culture is used for the initial inoculum, thereby keeping contamination to a minimum in the early stages of growth. *Saccharomyces cerevisiae* is susceptible to the Crabtree effect (as considered in more detail in Chapter 2) in which aerobic metabolism is repressed by the presence of a high sugar concentration, even under fully aerobic conditions. Under these circumstances the carbon source would be utilized by glycolysis alone and the biomass yield would be poor. Thus, a fed-batch process has been developed in which a low feed rate results in a very low extracellular sugar concentration, thereby circumventing the Crabtree response. Reed and Nagodawithana (1991b) discussed the development of inoculum for the production of bakers' yeast and quoted a process involving eight stages, the first three being aseptic while the remaining stages were carried out in open vessels. Gomez-Pastor, Garre, Matallana, and Perez-Torrado (2011) described a six-stage process in which the first three stages were batch-culture and the last three fed-batch. A summary of a typical inoculum development program for the production of bakers' yeast is given in Fig. 6.12.

BREWING

It has been common practice in the British brewing industry to use the yeast from the previous fermentation to inoculate a fresh batch of wort and there are examples of yeast being transferred from one fermentation to the next *ad infinitum* (Kennedy, Taidi, Aitchison, & Green, 2003). The brewing terms used to describe this process are "crop," referring to the harvested yeast from the previous fermentation, and "pitch," which means to inoculate. One of the major factors that contributed to the continuation of this practice was the wort-based excise laws in the United Kingdom where

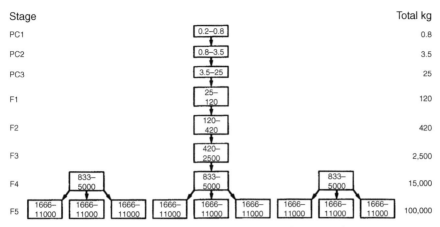

FIGURE 6.12 The Development of Inoculum for the Commercial Production of Bakers' Yeast

PC 1, 2 and 3 are pure culture batch fermentations. F1 and 2 are nonaseptic batch fermentations. F3 and 4 are fed-batch fermentations and F5 is the final fed-batch fermentation (Reed & Nagodawithana, 1991a).

duty was charged on the initial specific gravity of the wort—equivalent to the sugar concentration—rather than the alcohol produced. Thus, dedicated yeast propagation systems were expensive to operate because duty was charged on the sugar consumed by the yeast during growth. It can then be appreciated that the reduced cost of using yeast from a previous fermentation was an attractive proposition (Boulton, 1991). However in 1993, the UK changed its excise beer duty to comply with the European Union Directive to harmonize excise structures necessary for the adoption of the single market. Thus, the basis of the UK excise system changed from initial gravity to the alcohol content of the finished product, removing a major barrier to the widespread adoption of yeast propagation plants. Also, Kennedy et al. (2003) drew attention to the changes in the UK brewing industry in the 1990s resulting in the development of breweries manufacturing particular beers under contract or franchise. Thus, the "subcontracted" brew had to be produced to a highly specified protocol, a major part of which was the integrity of the yeast inoculum.

The dangers inherent in the practice of sequential inoculation were well-recognized as the introduction of contaminants and the degeneration of the strain, the most common degenerations being a change in the degree of flocculence and attenuating abilities of the yeast. In breweries employing top fermentations in open fermenters, these dangers were minimized by collecting yeast to be used for future pitching from "middle skimmings." During the fermentation the yeast cells flocculate and float to the surface, the first cells to do this being the most flocculent and the last cells, the least flocculent. As the head of yeast developed, the surface layer (the most flocculent and highly contaminated yeasts) was removed and discarded and the underlying cells (the "middle skimmings") were harvested and used for subsequent pitching. Therefore, the "middle skimmings" contained cells which had the desired flocculence and which had been protected from contamination by the surface layer of the yeast head. The pitching yeast was treated to reduce the level of contaminating bacteria and remove protein and dead yeast cells by such treatments as reducing the pH of the slurry to 2.5–3, washing with water, washing with ammonium persulfate, and treatment with antibiotics such as polymixin, penicillin, and neomycin (Mandl, Grunewald, & Voerkelius, 1964; Strandskov, 1964; Roessler, 1968; Reed & Nagodawithana, 1991a).

However, traditional open vessels are becoming increasingly rare and the bulk of beer is brewed using cylindro-conical fermenters (see Chapter 7). In these systems, the yeast flocculates and collects in the cone at the bottom of the fermenter where it is subject to the stresses of nutrient starvation, high ethanol concentration, low water activity, high carbon dioxide concentration, and high pressure (Boulton, 1991). Thus, the viability and physiological state of the yeast crop would not be ideal for an inoculum. The viability of the crop may be assessed using a biomass probe of the type described earlier, thus ensuring that at least the correct amount of viable biomass is used to start the next fermentation. However, the physiological state of the biomass will not have been measured by such monitoring procedures. The situation is further complicated by the fact that the harvested yeast is stored before it is used as inoculum. Metabolic activity is minimized during this time by chilling rapidly to about 1°C, suspending in beer and storing in the absence of oxygen. If oxygen is present during the storage period then the yeast cells consume their stored glycogen, which renders them very much less active at the start of the fermentation (Pickerell, Hwang, & Axcell, 1991).

One of the key physiological features of yeast inoculum is the level of glycogen and sterol in the cells. Sterols are required for membrane synthesis but they are only produced in the presence of oxygen. Thus, we have the anomaly of oxygen being required for sterol synthesis, yet anaerobic conditions are required for ethanol production. This anomaly is resolved traditionally by aerating the wort before inoculation. This oxygen allows sufficient sterol synthesis from the stored glycogen early in the fermentation to support growth of the cells throughout the process, that is, after the oxygen is exhausted and the process is anaerobic. Boulton, Jones, and Hinchliffe (1991) developed an alternative approach where the pitching yeast was vigorously aerated prior to inoculation. The yeast was then sterol rich and had no requirement for oxygen during the alcohol fermentation.

The difficulties outlined above and the likelihood of strain degeneration and contamination mean that yeasts are rarely used for more than five to twenty consecutive fermentations (Thorne, 1970; Reed & Nagodawithana, 1991a; Miller, Box, Boulton, & Smart, 2012), which necessitates the periodical production of a pure inoculum. This would involve developing sufficient biomass from a single colony to pitch a fermentation at a level of approximately 2 g of pressed yeast per liter. Hansen (1896) was the first to develop a yeast propagation scheme utilizing a 10% inoculum volume at each stage in the program, employing conditions similar to those used during brewing. However, it is now appreciated that the propagation stage should solely be concerned with the production of biomass and not alcohol. Indeed, the mass propagation of brewer's yeast is usually done in baker's yeast plants (Gomez-Pastor et al., 2011) of the type discussed in the previous section. The low feed-rate fed-batch baker's yeast process produces fully respiratory biomass in an aerobic environment, thus ensuring the best yield. However, brewer's yeast produced in this way may loose its fermentation characteristics and thus Stewart, Hill, and Russell (2013) pointed out the process is adapted to operate at a higher feed-rate, resulting in a higher sugar concentration enabling the yeast to maintain a mixed aerobic and fermentative metabolism. Yeast inoculum produced in this way would also be sterol rich, obviating the need for aerated wort at the start of the brewing process.

However, variability in the performance of propagated yeast has been observed in the first fermentation in which it is pitched. Miller et al. (2012) compared the physiological state of inoculum from freshly propagated yeast biomass with that derived from a previous lager fermentation. The freshly propagated inoculum had the longer lag phase and this was explained by the lack of synchrony in its cell cycle compared with the inoculum that had experienced a brewing fermentation. Thus, the variability associated with the first use of propagated biomass may be reduced if its cell-cycle synchrony could be improved.

FUEL ETHANOL

It is interesting to compare the preparation of inoculum for the *S. cerevisiae* fuel ethanol fermentation with the brewing and baker's yeast processes. The Melle-Boinot process for fuel alcohol production was introduced in Brazil in the 1960s (Zanin, Santana, Bon, & Giordano, 2000) and is an extreme example of using the biomass at the end of a fermentation to inoculate the next. Thus, this process relies on reusing most of the biomass from a previous fermentation to establish a short fed-batch conversion of sugar cane extract to alcohol (Della-Bianca, Basso, Stambuk, Basso, & Gombert, 2013). The biomass is harvested by centrifugation, acid washed and incorporated in its entirety into the next fermentation, which lasts only 6–10 h. The initial biomass is so high that it increases only by 5–10% during the fermentation. Indeed, Della-Bianca quotes this process occurring at least twice daily during a sugarcane season that may last 250 days.

DEVELOPMENT OF INOCULA FOR UNICELLULAR BACTERIAL PROCESSES

The main objective of inoculum development for unicellular bacterial fermentations is to produce an active inoculum, which will give as short a lag phase as possible in subsequent culture. A long lag phase is disadvantageous in that not only time is wasted but also medium constituents are consumed in maintaining a viable culture prior to growth. The length of the lag phase is affected by the size of the inoculum and its physiological condition with mid-exponential cultures tending to reduce the length of the lag phase. As already stated, the inoculum size normally ranges between 3% and 10% of the culture volume but Mulder, Viana, Xavier, and Parachin

(2013) cited a surprisingly wide range of .01–10% being used at a 1000 dm^3 scale for heterologous peptide production by *E. coli*.

The age of the inoculum is particularly important in the growth of sporulating bacteria, for sporulation is induced at the end of the logarithmic phase and the use of an inoculum containing a high percentage of spores would result in a long lag phase in a subsequent fermentation. Hornbaek, Nielsen, Dynesen, and Jakobsen (2004) showed that an early stationary phase inoculum of *Bacillus licheniformis* was more homogeneous, gave a shorter lag phase and higher biomass concentration in the subsequent culture than a late stationary phase inoculum. Although the strain was impaired in spore formation, the late stationary inoculum did produce some spores that contributed to its poorer performance. Sen and Swaminathan (2004) demonstrated that the age and size of the inoculum was a key criterion in the production of surfactin by *Bacillus subtilis* and optimized these variables using a statistical response surface methodology. This optimization approach is discussed in more detail in Chapter 4 and the study showed the importance of inoculum condition in the overall design of a process. Keay, Moseley, Anderson, O'Connor, and Wildi (1972) quote the use of a 5% inoculum of a logarithmically growing culture of a thermophilic *Bacillus* for the production of proteases. Aunstrup (1974) described a two-stage inoculum development program for the production of proteases by *B. subtilis*. Inoculum for a seed fermenter was grown for 1–2 days on a solid or liquid medium and then transferred to a seed vessel where the organism was allowed to grow for a further ten generations before transfering to the production stage. Priest and Sharp (1989) cited the use of a 5% inoculum, still in the exponential phase, for the commercial production of *Bacillus* α-amylase. Underkofler (1976) emphasized that, in the production of bacterial enzymes, the lag phase in plant fermenters could be almost completely eliminated by using inoculum medium of the same composition as used in the production fermenter and employing large inocula of actively growing seed cultures. The inoculum development program for a pilot-plant scale process for the production of vitamin B$_{12}$ from *Pseudomonas denitrificans* is shown in Fig. 6.13 (Spalla, Grein, Garafano, & Ferni, 1989).

The necessity to use an inoculum in an active physiological state is taken to its extreme in the production of vinegar. The acetic-acid bacteria used in the vinegar process are extremely sensitive to oxygen starvation. Therefore, to avoid disturbing the system, the cells at the end of a fermentation are used as inoculum for the next batch by removing approximately 60% of the culture and restoring the original level with fresh medium (Conner & Allgeier, 1976; Xu, Shi, & Jiang, 2011). The advantage of a highly active inoculum apparently outweighs the disadvantages of possible strain degeneration and contaminant accumulation. However, strain stability is a major concern in inoculum development for fermentations employing recombinant bacteria. Sabatie et al. (1991) demonstrated that plasmid stability and productivity in an *E. coli* plasmid-encoded biotin fermentation was greatly improved if stationary, rather then exponential, phase cells were used as inoculum. They postulated that the plasmid copy number may be higher in stationary cells than in exponential ones, thus decreasing the probability of plasmid loss in the subsequent fermentation when a stationary culture is used as inoculum. A stationary phase inoculum would

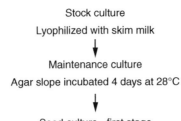

Stock culture

Lyophilized with skim milk

Maintenance culture

Agar slope incubated 4 days at 28°C

Seed culture - first stage

2 dm³ flask containing 0.6 dm³ medium inoculated with culture from one slope; incubated with shaking for 48h at 28°C

Seed culture - second stage

40–80 dm³ fermenter containing 25–50 dm³ medium inoculated with 1–1.2% first stage seed culture. Incubated 25–30h at 32°C

Production culture

500 dm³ fermenter with 300 dm³ medium inoculated with 5% second stage seed culture. Incubated at 32° for 140–160 h

FIGURE 6.13 The Inoculum Development Program for a Vitamin B$_{12}$ Pilot Scale Fermentation Using *Pseudomonas denitrificans* (Spalla et al., 1989)

result in a lag phase, but this disadvantage was more than compensated for by the considerable improvement in plasmid retention and biotin production compared with that obtained using an exponential inoculum. Silva, Queiroz, and Domingues (2012) investigated the effect of cultural conditions on the performance of an *E. coli* batch fermentation and showed that plasmid copy number increased toward the end of almost all fermentation conditions tested. However, the link between plasmid copy number, batch growth phase and plasmid stability is dependent upon the physiological relationship between the bacterial strain and the plasmid and thus it is important that the relationship is quantified for a particular fermentation. Sivashanmugam et al. (2009) described a generic laboratory scale protocol for plasmid-encoded products in which a minimal medium was inoculated with a 1% inoculum of an exponential culture with an OD$_{600}$ between 3 and 5 in LB broth.

In the lactic-acid fermentation the producing organism may be inhibited by lactic acid. Thus, production of lactic acid in the seed fermentation may result in the generation of poor quality inoculum. Yamamoto, Ishizaki, and Stanbury (1993) generated high quality inoculum of *Lactococcus lactis* IO-1 on a laboratory scale using electrodialysis seed culture which reduced the lactate in the inoculum and reduced the length of the lag phase in the production fermentation.

An example of the development of inoculum for an anaerobic bacterial process is provided by the clostridial acetone-butanol fermentation. The process was outcompeted by the petrochemical industry but it has been reestablished in China and

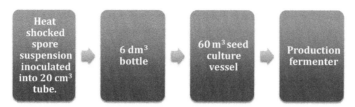

FIGURE 6.14 Outline of the Inoculum Development Program for the Acetone-Butanol Fermentation Operated in China

Modified from Ni and Sun, 2009

Brazil (Ni & Sun, 2009; Green, 2011 and Patakova, Linhova, Rychtera, Paulova, & Melzoch, 2013). Ni and Sun described both batch and continuous processes operating in China. The inoculum development program for the batch process (Fig. 6.14) resembles that described by McNeil and Kristiansen (1986) given in Table 6.4. In both processes the stock culture was heat shocked to stimulate spore germination and to eliminate the weaker spores. Also, inocula of only 6–9 dm³ were used to establish fermenters of 60–90 m³, these being the seed fermenter in the Chinese process and the production fermenter in the other. The use of such small inocula necessitates the achievement of as near perfect conditions as possible to prevent contamination and to avoid an abnormally long lag phase. The continuous Chinese process comprised up to eight 200–500 m³ linked fermenters with inoculum being added periodically, and substrate continuously, to the first two vessels from a 30–60 m³ seed tank. The combined flow from these two vessels then fed into the remaining six vessels in series. This approach enabled the rapid production of biomass in fermenters one and two to be separated from solvent production in the later vessels.

The concept of the use of disposable reactors for the production of inoculum was introduced earlier in this chapter as a technique especially useful for animal cell culture. Although the adoption of disposable culture systems for microbial processes has been hindered by their mass transfer limitations, their application for the

Table 6.4 The Inoculum Development Program for the Clostridial Acetone–Butanol Fermentation (Spivey, 1978)

Stage	Cultural Conditions	Incubation Time (hours)
1.	Heat-shocked spore suspension inoculated into 150 cm³ of potato glucose medium	12
2.	Stage 1 culture used as inoculum for 500 cm³ molasses medium	6
3.	Stage 2 culture used as inoculum for 9 dm³ molasses medium	9
4.	Stage 3 culture used as inoculum for 90,000 dm³ molasses medium	

production of inocula for anaerobic processes would seem an attractive proposition, especially considering the low inoculum volumes used to initiate such fermentations and the susceptibility of these processes to contamination. An interesting example is provided by Jonczyk et al. (2011) who described the use of a rocking motion 10 dm³ bag reactor to culture the anaerobe *Eubacterium ramulus* and achieved biomass levels comparable with a conventional stirred tank reactor and up to five times that achievable in bottle cultivation.

The challenge of using disposable reactors for aerobic microbial cultivation is highlighted by Mahajan, Matthews, Hamilton, and Laird (2010) observation that the oxygen demand of a bacterial process may be 1000 times that of a CHO fermentation. Also, the accompanying heat output of rapidly growing high-density culture may exceed the cooling capacity of the reactor (Mikola, Seto, & Amanullah, 2007). These issues have been addressed by two general approaches, although workers have combined both approaches to achieve optimum results:

1. The control of the oxygen uptake rate of the culture such that it does not exceed the supply capacity of the vessel.
2. The design of the disposable reactor.

The oxygen uptake rate of a culture ($Q_{O_2}x$, mmoles of oxygen $dm^{-3}\ h^{-1}$) is determined by the biomass concentration (x, $g\ dm^3$) and the specific oxygen uptake rate (Q_{O_2}) of the organism (mmoles of oxygen consumed g^{-1} biomass h^{-1}). As discussed in more detail in Chapter 9, Q_{O_2} is directly proportional to the specific growth rate. Thus, controlling the specific growth rate of a culture will control its oxygen uptake rate and this can be achieved by controlling the availability of the limiting substrate in both continuous and fed-batch processes, as discussed in Chapter 2. Fed-batch culture is the most frequently used system for production-scale microbial fermentation and has been widely used to control oxygen consumption by adjusting the feed rate of the limiting substrate (Chapters 2 and 7). Fed-batch culture requires a sophisticated apparatus that enables the substrate to be fed at a predetermined (or changing) flow rate that, in turn, can be feedback controlled by the dissolved oxygen concentration. However, a number of workers have developed methods to achieve the equivalent of fed-batch conditions in simple reactors such as flasks and microplates. These methods are designed to gradually release the limiting substrate into the medium, thus eliminating the need for pumps and complex feedback control. One such approach developed by Panula-Perala et al., (2008) and Krause et al., (2010) has been commercialized by BioSilta as the EnBase® Flo method. The system utilizes starch as the sole carbon source and the activity of added amylase to gradually release glucose to the organism. The growth rate, and hence oxygen demand, is then controlled by the glucose release rate. Thus, this approach is only suitable for strains that are amylase nonproducers and are dependent on the added amylase to generate glucose form starch. Glazyrina et al. (2010) compared the growth of *E. coli* in a Sartorius Biostat Cultibag RM 10 dm³ wave-type single use bioreactor (containing 1 dm³ of medium) in batch culture with the use of the EnBase® glucose-release system. The batch culture became oxygen limited and yielded a biomass OD_{660} of 3–4, while the EnBase®

fed-batch glucose release system supported an OD_{660} of 30, equivalent to 10 g dm^{-3} dry weight. The reactor was also operated with a conventional fed-batch system that required the incorporation of single-use dissolved oxygen and pH probes and this system supported a biomass of OD_{660} of 60 (20 g dm^{-3} dry weight). Thus, these approaches significantly increased the biomass yield to levels feasible for inoculum development. The increased heat output of these *E. coli* fermentations (compared with cell cultures) was controlled by the incorporation of cooling coils into the Biostat reactors. Dreher et al. (2013) achieved an *E. coli* biomass of almost 50 g dm^{-3} dry weight in a fed-batch Sartorius BIOSTAT® Cultibag RM 20/50 (a rocking motion disposable reactor) incorporating a pure oxygen supply and an integral cooling coil. This result was achieved by using a linear fed-batch profile resulting in a decrease in μ over time, thus causing Q_{O_2} to decrease with time. These workers also investigated the performance of a fed-batch disposable stirred tank vessel (BIOSTAT Cultibag STR 50) and achieved a biomass of almost 61 g dm^{-3} dry weight. This result was achieved with an exponentially increasing flow rate such that μ was kept constant at 0.1 h^{-1}. Although the rocking reactor had a much lower capacity for oxygen transfer than the stirred tank, a comparable biomass was attainable by using a feed strategy that kept the oxygen demand below the oxygen transfer capability of the vessel.

Improved reactor design and operation has also enabled the efficient growth of microbial cultures in disposable reactors. Mikola et al. (2007) evaluated a sparged Wave Bioreactor® for the production of *S. cerevisiae* inoculum. The conventional wave type bioreactors depend on air being administered above the liquid surface into the headspace and for oxygen to be incorporated into the medium by the rocking motion of the platform. Although the sparging system proved to be no more effective than the conventional design, the use of air enriched to 60% oxygen produced an effective inoculum culture at a higher volume and under more secure conditions compared with shake-flask cultivation. Mahajan et al. (2010) also used oxygen enrichment in a 5 dm^3 working volume Wave Bioreactor to produce a recombinant *E. coli* biomass of OD_{550} of 15 (approximately 5 g dm^{-3} dry weight). Although such a biomass would not be acceptable for a production fermentation, it was feasible as an inoculum especially when the advantages of disposable reactors of decreased labor costs and lower contamination risks are considered. The performance of production fermentations inoculated with the Wave-generated culture behaved in exactly the same way as the conventional inoculum—both in terms of growth and heterologous protein production. The system was also shown to be economically feasible when compared with a conventional stainless steel stirred tank system.

The CELL-tainer® disposable rocking-platform reactor is a novel device described earlier in this chapter, which enables a wide range of culture volumes to be accommodated in the reactor bag (Fig. 6.8). For microbial culture, the platform of the reactor also combines the conventional rocking motion of a wave reactor with a "to and fro" horizontal movement (Fig. 6.15), thus achieving significantly better oxygen transfer rates. Westbrook, Scharer, Moo-Young, Oosterhuis, and Chou (2014) achieved KLa values of up to 400 h^{-1} in a 3 dm^3 working volume system (see Chapter 9). The system also incorporates an integrated cooling plate in the rocking

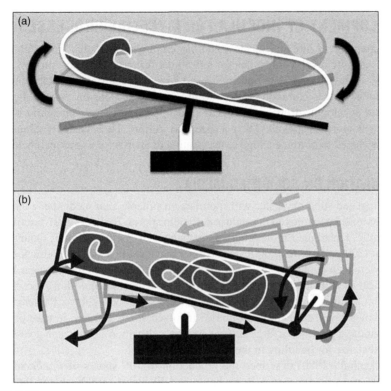

FIGURE 6.15

Diagrammatic representation of a conventional rocking reactor (a) and a CELL-tainer®
reactor (b) The platform of the conventional vessel aerates the culture by rocking up and
down. The platform of the CELL-tainer® combines the rocking motion with a "to and fro"
(horizontal) thus increasing mass transfer (Westbrook et al., 2014).

platform to cope with the increased heat output from a microbial system. CELL-
tainer reactors operated with feedback have been reported to achieve *E. coli* dry
weights of 20 g dm^{-3} at a 10 dm^3 scale (Oosterhuis & van den Berg, 2011) and 45 g
dm^{-3} at a 120 dm^3 scale (Junne, Solymosi, Oosterhuis, & Neubauer, 2013). Thus, the
combination of novel design and fed-batch operation enables a biomass attainment
entirely commensurate with the adoption of such systems for inoculum development. Indeed, Oosterhuis, Neubauer, and Junne (2013) reported the use of the CELL-
tainer system for the production of inoculum of *Corynebacterium glutamicum* in the
production of lysine. As a result, the disposable system replaced the use of pooled
shake-flask cultures, thus producing a more physiologically consistent inoculum and
reducing the risk of contamination in a much simpler infrastructure compared with
that needed for a traditional stainless steel vessel.

DEVELOPMENT OF INOCULA FOR MYCELIAL PROCESSES

The preparation of inocula for fermentations employing mycelial (filamentous) organisms is more involved than that for unicellular bacterial and yeast processes. The majority of industrially important fungi and streptomycetes are capable of asexual sporulation, so it is common practice to use a spore suspension as seed during an inoculum development program. A major advantage of a spore inoculum is that it contains far more "propagules" than a vegetative culture. Three basic techniques have been developed to produce a high concentration of spores for use as an inoculum.

SPORULATION ON SOLIDIFIED MEDIA

Most fungi and streptomycetes will sporulate on suitable agar media but a large surface area must be employed to produce sufficient spores. Parker (1958) described the "roll-bottle" technique for the production of spores of *P. chrysogenum*. Quantities of medium (300 cm^3) containing 3% agar were sterilized in 1 dm^3 cylindrical bottles, which were then cooled to 45°C and rotated on a roller mill so that the agar set as a cylindrical shell inside the bottle. The bottles were inoculated with a spore suspension from a sub-master slope and incubated at 24°C for 6 to 7 days. Parker claimed that although the use of the "roll-bottle" involved some sacrifice in ease of visual examination, it provided a large surface area for cultivation of spores in a vessel of a convenient size for handling in the laboratory.

Hockenhull (1980) described the production of 10^{10} spores of *P. chrysogenum* on a 300 cm^2 agar layer in a Roux bottle and El Sayed (1992) quoted the use of spore suspensions derived from agar media containing between 10^7 and 10^8 cm^{-3}. Butterworth (1984) described the use of a Roux bottle for the production of a spore inoculum of *S. clavuligerus* for the production of clavulanic acid. The spores produced from one bottle containing 200 cm^2 agar surface could be used to inoculate a 75 dm^3 seed fermenter that, in turn, was used to inoculate a 1500 dm^3 fermenter. The clavulanic acid inoculum development program is illustrated in Fig. 6.16. Some representative solidified media for the production of streptomycete and fungal spores are given in Tables 6.5 and 6.6 respectively.

Krasniewski et al. (2006) investigated the bulk production of spores of *Penicillium camamverti* (used in the cheese industry) on solidified medium. They concluded that sporulation was favored by the addition of calcium ions (see later in the following section), a glucose concentration of at least 10 g dm^{-3}, potassium nitrate as the nitrogen source and the avoidance of ammonium sulfate or sodium nitrate.

SPORULATION ON SOLID MEDIA

As discussed in Chapter 1, a solid substrate fermentation is one in which growth occurs on a solid matrix in the absence (or near absence) of free water (Singhania, Patel, Soccol, & Pandey, 2009). Mycelial organisms are well suited to such systems as their penetrating hyphal habit facilitates the colonization of solid substrates. Also,

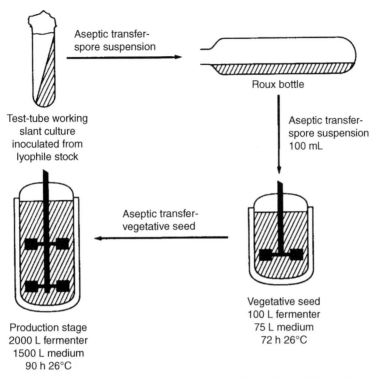

FIGURE 6.16 The Inoculum Development Program for the Production of Clavulanic Acid From *Streptomyces clauuligerus* **(Butterworth, 1984)**

intimate contact between the process organism and a solid substrate enables very high substrate concentrations to be available in a "slow release" format—mimicking a fed-batch process in submerged culture. Thus, the organism grows in its "natural" environment, is controlled by its ability to solubilize the substrate and is effectively substrate-limited. As a result, the complexities of medium design to achieve nutrient limitation (and, hence, sporulation) experienced in submerged sporulation are reduced (see the next section). One of the oldest examples of this technology is the traditional koji fermentation—a Japanese process in which *Aspergillus oryzae* is grown on cooked rice to produce a sporulating culture rich in amylases. The harvested grain and associate fungal biomass is then used as inoculum in the saccharifying stage of the sake brewing process—koji is incubated with cooked rice, water and yeast (*S. cerevisiae*); the *Aspergillus* converts the rice starch to soluble sugars, which the yeast converts to alcohol. This process is at least 5000 years old (Holker, Hofer, & Lenz, 2004) and thus an exceptional successful application of solid substrate fermentation to produce a spore inoculum. Although the koji process is not an aseptic fermentation, there are many examples in the literature of the use of a modified approach to produce spore inocula for submerged fermentations under aseptic

Table 6.5 Solidified Media Suitable for the Sporulation of Some Representative Streptomycetes

Organism	Product	Medium		References
S. aureofaciens	Tetracycline[a]	Malt extract (Difco)	1.0%	Williams, Entwhistle, and Kurylowicz (1974)
		Yeast extract (Difco)	0.4%	
		Glucose	0.4%	
S. erythreus	Erythromycin[a]	Beef extract (Difco)	0.1%	Williams et al. (1974)
		Yeast extract (Difco)	0.1%	
		Casamino acids (Difco)	0.2%	
		Glucose	0.2%	
S. vinaceus	Viomycin[a]	Corn-steep liquor (50% dry matter)	1.0%	Williams et al. (1974)
		Starch	1.0%	
		$(NH_4)_2SO_4$	0.3%	
		NaCl	0.3%	
		$CaCO_3$	0.3%	
S. clavuligerus	Clavulanic[a] acid	Soluble starch	1.0%	Butterworth (1984)
		K_2HPO_4	0.1%	
		$MgSO_4 \cdot 7H_2O$	0.1%	
		NaCl	0.1%	
		$(NH_4)_2SO_4$	0.2%	
		$CaCO_3$	0.2%	
		$FeSO_4 \cdot 7H_2O$	0.001%	
		$MnCl_2 \cdot 4H_2O$	0.001%	
		$ZnSO_4 \cdot 7H_2O$	0.001%	
S. hygroscopicus	Maridomycin	Soluble starch	1.0%	Miyagawa, Suzuki, Higashide, and Uchida (1979)
		Peptone	0.04%	
		Meat extract	0.02%	
		Yeast extract	0.02%	
		N-Z amine (type A)	0.02%	
		Agar	2.0%	

[a]Agar content not quoted.

conditions. Substrates such as barley, hard wheat bran, ground maize, and rice are all suitable for the sporulation of a wide range of fungi. The sporulation of a given fungus is particularly affected by the amount of water added to the cereal before sterilization and the relative humidity of the atmosphere, which should be as high as possible during sporulation (Vezina & Singh, 1975). Singh, Sehgal, and Vezina (1968) described a system for the sporulation of *Aspergillus ochraceus* in which a 2.8 dm^3 Fembach flask containing 200 g of "pot" barley or 100 g of moistened wheat bran produced 5×10^{11} conidia after six days at 28°C and 98% relative humidity. This was

Table 6.6 Solidified Media Suitable for the Sporulation of Some Representative Fungi

Fungus	Medium	(g dm⁻³)	References
		$(g\ dm^{-3})$	
Penicillium chrysogenum	Glycerol	7.5	Booth (1971)
	Cane molasses	7.5	
	Curbay BG	2.5	
	$MgSO_4 \cdot 7H_2O$	0.05	
	KH_2PO_4	0.06	
	Peptone	5.0	
	NaCl	4.0	
	Agar	20	
Aspergillus niger	Molasses	300	Steel et al. (1954)
	KH_2PO_4	0.5	
	Agar	20	

five times the number obtainable from a Roux bottle batched with Sabouraud agar and 50 times the number obtainable from such a vessel batched with Difco Nutrient Agar, incubated for the same time period. Vezina, Sehgal, and Singh (1968) published a list of fungi that are capable of sporulating heavily on cereal grains. El Sayed (1992) quoted the use of cooked rice for the production of spores of *Penicillium* and *Cephalosporium* in penicillin and cephalosporin inoculum development. Sansing and Cieglem (1973) described the mass production of spores of several *Aspergillus* and *Penicillium* species on whole loaves of white bread and Podojil, Blumauerova, Culik, and Vanek (1984) quoted the use of millet for the sporulation of *Streptomyces aureofaciens* in the development of inoculum for the chlortetracycline fermentation (Fig. 6.17).

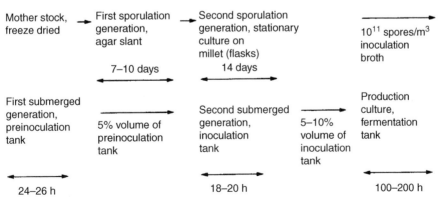

FIGURE 6.17 The Inoculum Development Program for the Production of Chlortetracycline by *Streptomyces aureofaciens* (Podojil, 1984)

These examples illustrate the convenience of simple, very small scale, solid substrate fermentations to produce spore inocula. However, large-scale processes have been developed to produce fungal spores to be used in biotransformations (Larroche, 1996) and as biological control agents. The major disadvantage of such large-scale operations is that they are not aseptic and thus do not lend themselves to the production of inocula for submerged fermentation processes. However, laboratory scale solid substrate fermenters, such as packed-bed reactors, can be operated under aseptic conditions and may prove useful for intensive inoculum generation.

SPORULATION IN SUBMERGED CULTURE

The production of conidia (sporulation) in solidified or solid media by fungi belonging to the Ascomycota is normally induced by the development of hyphae protruding into the atmosphere (Adams, Weiser, & Jae-Hyuk, 1998). However, many such fungi will sporulate in submerged liquid culture (a shake flask or fermenter) provided a suitable nutrient-limited medium is employed (Vezina, Singh, & Sehgal, 1965). This approach is more convenient than the use of solid or solidified media because it is easier to operate aseptically and it may be applied on a large scale. The technique was first adopted by Foster, McDaniel, Woodruff, and Stokes (1945) who induced submerged sporulation in *Penicillium notatum* by including 2.5% calcium chloride in a defined nitrate-sucrose medium. Indeed, Roncal and Ugalde (2003) pointed out that the addition of calcium ions (1–10 mM) to the medium has been reported to induce sporulation in a wide range of *Penicillium* species. An example of the use of submerged sporulation for the production of inoculum for an industrial fermentation is provided by the griseofulvin process. Rhodes, Crosse, Ferguson, and Fletcher (1957) described the conditions necessary for the submerged sporulation of the griseofulvin-producing fungus, *P. patulum,* and the medium utilized is given in Table 6.7. These authors found that for prolific sporulation the nitrogen level had to be limited between 0.05% and 0.1% w/v and that good aeration had to be maintained. Also, an interaction was demonstrated between the nitrogen level and aeration in that the lower the degree of aeration the lower the concentration of nitrogen needed to induce sporulation. Submerged sporulation was induced by inoculating 600 cm^3 of the above medium, in a 2 dm^3 shake flask, with spores from a well-sporulated Czapek-Dox agar culture and incubating at 25°C for 7 days. The resulting suspension of spores was then used as a 10% inoculum for a vegetative seed stage in a stirred fermenter, the seed culture subsequently providing a 10% inoculum for the production fermentation.

Anderson and Smith (1971) used the term "microcycle conidiation" to describe the situation in which the conidia (spores) used to inoculate a submerged culture of *Aspergillus niger* themselves produced both conidiophores (the spore-bearing hyphae) and conidia, without an intervening vegetative stage. This process was induced by incubating the shaken conidial culture in a glutamate medium at 44°C for up to 48 h and then reducing the temperature to 30°C. During the 44°C treatment the conidia swelled, but remained unicellular—referred to as spherical growth (Pazout & Schroder, 1988)—with conidia being produced (by cell division) at the

Table 6.7 Media for the Submerged Sporulation of Selected Fungi

Rhodes et al. (1957): *Penicillium patulum*

Whey powder, to give	Lactose 3.5% Nitrogen 0.05%	
KH$_2$PO$_4$		0.4%
KCl		0.05%
Corn-steep liquor solids to give approx. 0.04% N		0.38%

Foster et al. (1945): *Penicillium notatum*

Sucrose		2.0%
NaNO$_3$		0.6%
KH$_2$PO$_4$		0.15%
MgSO$_4$ · 7H$_2$O		0.05%
CaCl$_2$		2.5%

Vezina et al. (1965): *Aspergillus ochraceus*

Glucose		2.5%
NaCl		2.5%
Corn-steep liquor		0.5%
Molasses		5.0%

lower temperature. Thus, the phenomenon is interpreted as a stress response with sporulation being induced as an aid to survival (Jung, Kim, & Lee, 2014). Pazout and Schroder (1988) succeeded in establishing the spherical growth stage in *Penicillium cyclopium* without exposure to a high temperature stress. This was achieved by careful medium design, the critical conditions being the presence of glucose, glutamic acid, a low phosphate concentration, light, and high oxygen transfer. The low phosphate concentrated enabled the pH to drift from an initial pH of 5.5–4.3 during the first 8–12 h of cultivation and it is this pH drift that has been shown to be a key factor in microcycle conidiation induction. A recent example of the approach is given by the work of Boualem, Gervais, Cavin, and Wache (2014), producing conidia of *P. cambertii* as starter cultures for the production of Camembert cheese. These workers succeeded in inducing microcycle conidiation by a low temperature treatment (18°C) in a synthetic medium developed from Morton (1961) medium for submerged sporulation—an inorganic nitrogen source being essential to the process (Table 6.8). This approach yielded 20 times the spore concentration achieved in solidified medium and 100 times that in conventional submerged sporulation. It is interesting to note that all the media described in this section used to achieve microcycle conidiation were also rich in calcium. Thus, microcycle conidiation can be an exceptionally valuable tool to produce bulk quantities of spores in submerged culture with a minimum of vegetative growth, resulting in a relatively homogeneous liquid inoculum that can be used

Table 6.8 Synthetic Medium (pH 6.0) for the Induction of Microcycle Conidiation in *Penicillium cambertii*

Medium Component	Concentration
Glucose	50 g dm^{-3}
$CaCl_2$	2.94 g dm^{-3}
KH_2PO_4	1 g dm^{-3}
$MgSO_4.7H_2O$	0.5 g dm^{-3}
KNO_3	1 g dm^{-3}
Trace element solution containing:	1 cm^3 dm^{-3}
$ZnSO_4.7H_2O$	225 mg dm^{-3}
$FeSO_4.7H_2O$	200 mg dm^{-3}
$(NH_4)_6O_{24}.4H_2O$	40 mg dm^{-3}
$CuSO_4.5H_2O$	38 mg dm^{-3}
$MnSO_4$	24 mg dm^{-3}

Boualem et al., 2014.

with a low risk of contamination. The problem is that there is no universal approach to induce the phenomenon in fungi and thus significant work is required to persuade a process organism to behave appropriately.

The role of bacterial quorum sensing agents (autoinducers) in the control of morphogenesis and secondary metabolism was introduced in Chapter 2 and is developed in Chapter 3. Table 2.4 (Chapter 2) summarizes the main bacterial quorum sensing molecules and the processes they affect. A number of compounds have also been shown to act as autoinducers in the filamentous fungi, particularly affecting spore germination, the development of the mycelium and sporulation, both sexual and asexual (Ugalde & Rodriguez-Urra, 2014). Hadley and Harrold (1958) demonstrated that *P. notatum* would sporulate in submerged culture in 8 h, rather than 24 h, if the organism were grown in a medium harvested from a previous culture. Furthermore, sterile medium from the culture of 5 different species of penicillia also stimulated the sporulation of *P. notatum*. Forty four years later Roncal, Cordobes, Sterner, and Ugalde (2002) isolated conidiogenone (a diterpene) from the culture broth of *P. cyclopium* that would induce sporulation in submerged culture. The presence of this compound could induce sporulation in the absence of calcium ions and it is claimed that the previous observations of the stimulatory effect of calcium ions was due to their reducing the threshold level of conidiogenone, probably by affecting its binding to a receptor. Schimmel, Coffman, and Parsons (1998) demonstrated that submerged sporulation by *Aspergillus terreus* (the producer of lovastatin) could be enhanced by the addition of butyrolactone 1, a product produced by the fungus and very similar to the quorum sensing molecules of some bacteria, particularly the streptomycetes. Thus, as our understanding of the control of development increases this knowledge can be applied to the development of more efficient production of spore inocula by submerged sporulation.

The differentiation of streptomycetes in submerged culture has also been closely investigated in recent years and is discussed in Chapter 2. It is now clear that the streptomycetes undergo a differentiation process in submerged culture very similar to that seen on solidified media although it appears frequently to stop short of sporulation. Indeed, it has been reported that most streptomycetes do not sporulate in submerged culture (Whitaker, 1992; Rioseras, Lopez-Garcia, Yague, Sanchez, & Manteca, 2014). However, Van Dissel, Claessen, and Van Wezel (2014) suggested that the phenomenon is much more common than was first appreciated and possibly all streptomycetes can undergo such differentiation under specific conditions. Thus, although solid or solidified media tend to be used for the production of streptomycete spore inocula the possibility remains that submerged sporulation processes may be developed in the future.

USE OF THE SPORE INOCULUM

The stage in an inoculum development program at which a large-scale spore inoculum is used varies according to the process; it appears to be common practice that the penultimate stage is so inoculated, but this will depend on the scale of the production fermentation. In the inoculum development program for the early penicillin fermentation described by Parker (1958), the penultimate stage was inoculated with a spore suspension (from a "roll-bottle") and this stage may have produced either a vegetative or a submerged spore inoculum for the final fermentation. For the griseofulvin process, Rhodes et al. (1957) stated that the spore suspension obtained from the submerged sporulation stage could either be used for direct inoculation of the production fermentation or it could be germinated in an inoculum development medium to yield a vegetative inoculum for the final fermentation. The latter course was preferred and an inoculum volume of 7–10% was used. From Figs. 6.16–6.18, it can be seen that in the clavulanic acid process the spore inoculum is used to inoculate the final seed stage, in the chlortetracycline process a vegetative stage is interspersed between the spore inoculated batch and the production fermentation, and in the sagamicin process the spore inoculum is used at a very early stage followed by vegetative growth.

When considering the production of gluconic acid by *A. niger,* Lockwood (1975) discussed the merits of inoculating the final fermentation directly with a spore suspension as compared with germinating the spores in a seed tank to give a vegetative inoculum. Direct spore inoculation would avoid the cost of installation and operation of the seed tanks whereas the use of germinated spores would reduce the fermentation time of the final stage, thus allowing a greater number of fermentations to be carried out per year. However, labor costs for the production of the vegetative inoculum could be almost as high as for the final fermentation although some of these costs may be recovered, in that gluconic acid produced in the penultimate stage would be recoverable from the final fermentation broth and would contribute to the buffering capacity throughout the fermentation. Thus, Lockwood claimed that the choice of inoculum for the production stage depends on the length of the cycle of the fermentation process, plant size and the availability, and cost of labor.

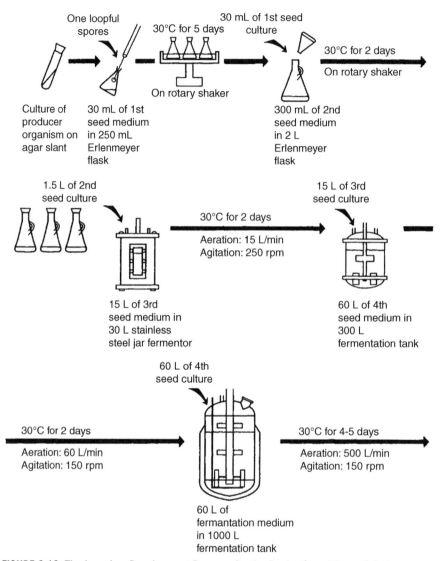

FIGURE 6.18 The Inoculum Development Program for the Production of Sagamicin by
Micromonospora sagamiensis **(Podojil, 1984)**

INOCULUM DEVELOPMENT FOR VEGETATIVE FUNGI

Some fungi will not produce asexual spores and, therefore, an inoculum of vegetative mycelium must be used. *Gibberella fujikuroi* is such a fungus and is used for the commercial production of gibberellin (Borrow et al., 1961). Hansen (1967) described an inoculum development program for the gibberellin fermentation. Cultures were grown on long slants (25×10 mm test tubes) of potato dextrose agar for 1 week at 24°C. Growth from three slants was scraped off and transferred to a 9 dm^3 carboy containing 4 dm^3 of a liquid medium composed of 2% glucose, 0.3% MgSO$_4$ · 7H$_2$O, 0.3% NH$_4$Cl, and 0.3% KH$_2$PO$_4$. The medium was aerated for 75 h at 28°C before transfer to a 100 dm^3 seed fermenter containing the same medium.

The major problem in using vegetative mycelium as initial seed is the difficulty of obtaining a uniform, standard inoculum. The procedure may be improved by fragmenting the mycelium in an homogenizer, such as a Waring blender, prior to use as inoculum. This method provides a large number of mycelial particles and therefore a large number of growing points. Worgan (1968) has given a detailed account of the use of this technique in the preparation of inocula for the submerged culture of the higher fungi.

EFFECT OF THE INOCULUM ON THE MORPHOLOGY OF FILAMENTOUS ORGANISMS IN SUBMERGED CULTURE

As discussed in Chapter 2, when filamentous fungi and bacteria are grown in submerged culture the type of growth varies from the "pellet" form, consisting of compact discrete masses of hyphae, to the filamentous form in which the hyphae form a homogeneous suspension dispersed through the medium—see Fig. 2.2 (Whitaker & Long, 1973; Whitaker, 1992 and Krull et al., 2013). Nielsen (1996) reported three strain-specific mechanisms of pellet generation from a spore inoculum:

1. Spore coagulation—the spores coagulate, germinate and thus form a pellet, for example, *Aspergillus* spp.
2. Noncoagulation—the spores do not coagulate but an individual spore gives rise to a pellet, for example, some streptomycetes.
3. Hyphal agglomeration—the spores germinate, producing hyphae that then agglomerate and form a pellet, for example, *P. chrysogenum*.

The filamentous type of habit gives rise to an extremely viscous non-Newtonian broth which may be very difficult to aerate adequately, whereas the pellet type of habit gives rise to a far less viscous, but also less homogeneous, Newtonian broth (see Chapter 9). In a pelleted culture there is a danger that the mycelium at the center of the pellet may be starved of nutrients and oxygen due to diffusion limitations. In such diffusion-limited pellets the biomass in the center will not grow (or may actually autolyse) with growth being confined to an outer margin, resulting in lower productivity. Thus, to achieve maximum biomass, and hence product yield, most industrial fermentations are designed to operate as homogeneous, dispersed filamentous cultures (Wucherpfennig, Hestler, & Krull, 2011), for example, the penicillin fermentation (Smith & Calam, 1980), the production of streptomycin by *Streptomyces griseus* and turimycin by *S. hygroscopicus* (Whitaker, 1992). The necessity for filamentous growth is taken to the extreme in the Quorn mycoprotein process where *Fusarium graminearium* is produced for human consumption. A highly filamentous morphology is required to produce the desired properties in the product such that it resembles the strength and eating texture of white and soft, red meats (Trinci, 1992). Thus, in this process a median hyphal length of 400 μm is required. However, the pelleted or clumped growth form has been shown to favor the syntheses of some products of mycelial organisms. Examples include citric-acid production from *A. niger* (Al Obaidi & Berry, 1980; Dhillon, Brar, Verma, & Tyagi, 2011), lovastatin from *A. terreus* (Gbewonyo, Hunt, & Buckland, 1992; see also Chapter 9), erythromycin from *Saccharopolyspora erythrea* (Martin & Bushell, 1996, Ghojavand, Bonakdarpour, Heydarian, & Hamedi, 2011), avermectin from *Streptomyces avermitilis* (Yin et al., 2008), and retamycin from *Streptomyces olindensis* (Pamboukian & Facciotti, 2005). The explanation of the beneficial effect of pelleted morphology is controversial (Yague, Lopez-Garcia, Rioseras, Sanchez, & Manteca, 2012) with hypotheses based both on the favorable (low viscosity) rheological properties of particulate cultures and metabolic differentiation associated with the pelleted format.

The relevance of this consideration of mycelial morphology to inoculum development is that the morphology may be influenced considerably by the condition of the spore inoculum, the inoculum development medium, the inclusion of additives in the medium, and the effect of agitation during the fermentation. The effect of agitation is considered in Chapter 9 (Aeration and Agitation) while the other factors are considered here. Posch, Herwig, and Spadiut (2013) emphasized the importance of optimizing the morphology for process performance early in the development of a process and stressed the need to understand the organism's reaction to process parameters.

Effect of the spore inoculum on mycelial morphology

Foster (1949) reported that a high spore inoculum will tend to produce a dispersed form of fungal growth while a low one will favor pellet formation. There is now an extensive literature (reviewed by Wucherpfennig, Kiep, Driouch, Wittmann, & Krull, 2010 and Krull et al., 2013) defining the effect of spore concentration on the mycelial morphology of both fungi and actinomycetes that supports Foster's early observations and representative examples are given in Table 6.9. Thus, the spore inoculum concentration is a key factor in achieving the desired morphological form of industrial fungal and actinomycete fermentations and the general observation can be made that as the spore concentration increases pellet size decreases until eventually a dispersed mycelium is produced. If the production fermentation is to be inoculated with a spore suspension then the spore concentration must be such as to produce the production culture in the desired morphological form; if a vegetative inoculum is to be used to start the production fermentation then, again, the concentration of its spore inoculum must be such as to produce the vegetative inoculum in the desired morphological form. For example, an interesting series of experiments on the effects of inoculum conditions on the morphology of *Penicillium citrinum* were reported by Hosobuchi, Fukui, Matsukawa, Suzuki, and Yoshikawa (1993). This *Penicillium* species synthesizes compound ML-236B, a precursor of pravastatin, a cholesterol-lowering drug. Optimum productivity was achieved when the organism grew as compact pellets in

Table 6.9 The Effect of the Concentration of Spores in a Spore Inoculum on the Morphology of Mycelial Organisms in Submerged Culture

Organism; Product; Desired Morphology	References	Spore Concentration in the Medium at Inoculation (dm^{-3})	Morphology
Penicillium chrysogenum; penicillin; filamentous	Camici, Sermonti, and Chain (1952)	Less than 10^6	Pellets
		More than 10^6	Filamentous
	Calam (1976)	2×10^6	Pellets
		10^7	Fliamentous
	Tucker and Thomas (1992)	Less than 5×10^7	Pellets
		5×10^8	Filamentous
Aspergillus niger, citric acid; pellets	Papagianni and Mattey (2006)	10^7–10^8	Pellets
		10^{11}–10^{12}	Filamentous
Aspergillus niger, heterologous proteins; pellets	Xu et al. (2000)	4×10^9	Pellets
A. terreus; lovastatin; small pellets (<1.5 mm diameter)	Bizukojc and Ledakowicz (2009)	$>2 \times 10^{10}$	Small pellets
Streptomyces coelicolor; actinorhodin; small pellets	Manteca et al. (2008)	10^8	Larger and fewer pellets, antibiotic production delayed
		10^{10}	Smaller and more abundant pellets, earlier onset of antibiotic production.
Streptomyces erythraea; erythromycin; mycelial clumps	Ghojavand et al. (2011)	10^6–10^7	Pellets
		10^8–10^{10}	Mycelial clumps
Saccharopolyspora avermilitis; avermectin; small high-density pellets.	Yao et al. (2009)	Optimum spore concentration: 4×10^9–6×10^9	Small dense pellets gave 10–20% increase in yield.

Modified from Wucherpfenning et al., 2010.

the production fermentation. The vegetative inoculum for the production fermentation had to contain an optimum number of short, filamentous propagules in order to initiate pellet formation in the final culture. This was achieved by using a four-stage inoculum development program (initiated by a spore-inoculated shake flask) with very rich media in the third and fourth cultures. Thus, this system required a dispersed vegetative inoculum to generate a pelleted production fermentation.

Effect of the medium design on mycelial morphology

Although the effects of media on morphological form can be extremely varied, dispersed growth is more likely in rich, complex media, and pelleted growth tends to occur in chemically defined media (Whitaker & Long, 1973). The influence of the medium composition may be explained by its effect on spore germination: a rich, complex medium enabling a greater percentage of spores to germinate compared with a defined medium, thus effectively increasing the number of vegetative propagules. However, the most influential medium property to affect mycelial morphology is the pH, summarized in Table 6.10 (Wucherpfennig et al., 2010). As discussed earlier, the development of pellets may be related to the aggregation of spores or germinated mycelium. As the pH of the medium will influence the ionization of key functional groups on the spore or hyphal surface, and thus the surface charge, it is not surprising that it will have a significant effect on aggregation and, thus, pellet development in some strains. The general conclusion on the effects of pH on mycelium morphology is that an acidic pH favors a dispersed mycelium whereas an alkaline one favors pellet development (Wucherpfennig et al., 2010). For example, Galbraith and Smith (1969) demonstrated that an *A. niger* spore inoculum would produce a dispersed mycelium at a pH below 2.3 and pellets at higher pH. Carlsen, Spohr, Nielsen, and Villadsen (1996) showed a similar pH response in the development of a spore inoculum of *A. oryzae* in a laboratory fermenter—a pH of below 3.5 gave dispersed mycelium, a pH above 5 gave pellets whereas a mixture of both morphologies resulted between pH 4 and 5. In both these examples, it was the aggregation of spores that was affected—in the *A. oryzae* experiments at low pH (2.5–3.5) only freely dispersed spores were found whereas at values above pH 4 aggregates of between 10 and 100 spores were observed. Grimm, Kelly, Krull, and Hempel (2005) demonstrated that there are two separate aggregation steps in the pelleting process

Table 6.10 The Effect of Medium pH on the Morphology of Mycelial Organisms in Submerged Culture

Organism	References	Cultural Conditions	pH Effect
Aspergillus niger	Galbraith and Smith (1969)	Shake flasks; spore inoculum of 8×10^6 spores per flask	Filamentous (dispersed mycelium) at pH < 2.3 Pellets produced at pH > 2.3
Aspergillus oryzae	Carlsen et al. (1996)	Stirred fermenter; spore inoculum of 6–8×10^8 dm^{-3}	Filamentous (dispersed mycelium) at pH 3.0–3.5 Mixture of filamentous and pellets at pH 4.0–5.0 Pellets at pH > 6.0
A. terreus; lovastatin; small pellets (<1.5 mm diameter)	Bizukojc and Ledakowicz (2009)		

Modified from Wucherpfennin et al., 2010.

of *A. niger*—an initial rapid amalgamation of spores followed by a slower process of amalgamation of the germinated spore clumps.

The relative concentrations of medium ingredients can also influence morphological development. For example, nitrogen limitation is claimed to encourage dispersed mycelial development (Formenti et al., 2014). Also, Hunt and Stieber (1986) described the optimization of the inoculum regime of a small-scale streptomycete cephamycin C fermentation. Pellet formation was observed to be detrimental to product formation and the key factor in establishing the correct form appeared to be the concentration of iron in the seed medium, a higher iron concentration giving the optimum inoculum.

Wucherpfennig et al. (2011) examined the effect of medium osmolality (measured by the decrease in freezing point) on the morphology and enzyme productivity of two *A. niger* strains and demonstrated that pellets became smaller as osmolality increased resulting in more mycelial growth. Fructofuranosidase production increased approximately 2.5 times with an increase in osmolality from 0.4 to 3.0 osmole kg^{-1} and that of glucoamylase increased 3 fold with an osmolality shift from 0.2 to 2.4 osmole kg^{-1}.

While medium design may be used to influence morphological form it must be appreciated that such changes may also have a profound (perhaps deleterious effect) on growth and productivity unrelated to the morphological effects. Thus, these observations may have more relevance to the design of an inoculum development medium rather than the production medium, hence producing the inoculum in the correct from to inoculate a production medium designed to support maximum growth and productivity.

Effect of medium additives on mycelial morphology

As discussed earlier, the surface charge of spores will affect their aggregation and, hence, the development of pelleted growth. Thus, a number of workers have utilized polymers, chelators, and surfactants to influence the surface properties of spores, and hence mycelial development. In general, polycationic polymers have been shown to induce aggregation (and hence pelleting) and polyanionic polymers to prevent it by causing electrostatic repulsion between spores, favoring the development of dispersed mycelium (Papagianni, 2004). For example, Elmayergi, Scharer, and Moo-Young (1973) and Trinci (1983) demonstrated that the addition of the polyanions Carbopol and Junlon, respectively, induced spore-inoculated submerged cultures of *A. niger* to develop dispersed mycelia while Hobbs, Frazer, Gardner, Cullum, and Oliver (1989) showed a similar response of *Streptomyces coelicolor* and *Streptomyces lividans* to both these polymers. Unfortunately, while such compounds are suitable for use on a laboratory scale, their application on an industrial scale would be financially prohibitive. However, the use of inorganic particles appears to be a more feasible approach to influence mycelial development at a large scale. Kaup, Ehrich, Pescheck, and Schrader (2008) explored the effect of microparticles (<40 μm diameter) of talc powder ($3MgO.4SiO_2.H_2O$) and aluminum oxide (AlO_3) on the production of the enzyme chloroperoxidase by the filamentous fungus, *Caldariomyces fumago*. This treatment improved enzyme yield by up to ten fold and changed the morphology of the fungus from pellets to single hyphae, the improved yield being attributed to eliminating the problems of oxygen limitation experienced in pelleted cultures. In an extension of their work, Kaup et al. (2008) demonstrated that the technique reduced pellet diameter in a wide range of fungal cultures (including *A. niger* and *Penicilliium chrysogenum*) as well as the streptomycete *S. aureofaciens*.

This approach has now been termed "microparticle enhanced cultivation" (MPEC) and has been refined such that morphology can be precisely modified—from dispersed mycelia to pellets of prescribed size—by modulating the particle material, size, and concentration. Driouch, Roth, Dersch, and Wittmann (2011) controlled the morphology of *A. niger* using different concentrations of magnesium silicate or aluminum oxide particles such that a bespoke fungal form could be produced, ranging from dispersed mycelium to pellets of precise size and number, as shown in Fig. 6.19. Optimization of the production of fructofuranosidase by these

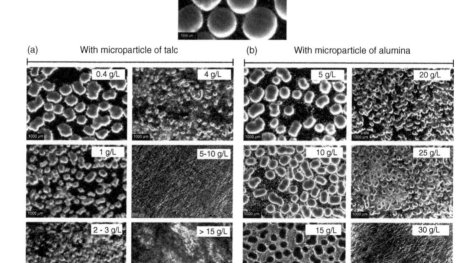

FIGURE 6.19

The effect of microparticles of talc (a) and alumina (b) on the morphology of spore-inoculated submerged cultures of *Aspergillus niger* SKAn1015.

Reproduced with permission from Driouch et al. (2011)

morphologically controlled cultures resulted in a thirty-fold increase in yield. Driouch, Hansch, Wucherpfennig, Krull, and Wittmann (2012) also used 8 μm diameter titanium silicate oxide (TiSiO$_4$) particles to generate bespoke *A. niger* pellets in which the particles were trapped in a mycelium network. Such pellets were not prone to the internal diffusion limitations of conventional pellets and again produced significantly improved fructofuranosidase yields.

Thus, from the earlier discussion the morphology of a mycelial fermentation may be controlled by the addition of microparticles along with the spore inoculum. The concentration of spores in the inoculum could then be optimized in terms of lag time, biomass attainment, and product yield rather than culture morphology, which should be controlled by the microparticles. However, it should be remembered that it is frequently a preculture that is inoculated with a spore suspension and the vegetative mycelium so generated is then used as the inoculum for the production fermentation. Gonciarz and Bizukojc (2014) emphasized this point in their work on the effect of talc microparticles on lovastatin production by *A. terreus*. These workers produced small pellets in precultures of *A. terreus* by optimizing the talc concentration, thus achieving better oxygen penetration into the pellets. The use of such a preculture then significantly improved the lovastatin yield in the production fermentation. The addition of microparticles had no effect on the production fermentation as its morphology was already determined by that of the preculture used as inoculum.

The mechanism underlying the effect of microparticles has yet to be elucidated (Walisko, Krull, Scharder, & Whittmann, 2012). The particles are too small to act as support structures for the developing mycelium but their presence may prevent spore aggregation. Whatever the mechanism, the potential for larger scale fermentation will be decided by economic and practical limitations—for example, the effect on down-stream processing. However, if the approach is beneficial in a preculture stage then these limitations may be lessened. In his discussion of the design of fungal processes, Posch et al. (2013) emphasized the importance of resolving the role of morphology at an early development stage. However, a key difficulty is divorcing the effect of operating conditions on morphology from that on product formation. Microparticle enhanced cultivation may provide the basis of achieving this—the morphology may be determined by the microparticle concentration thereby allowing the investigation of cultural conditions on productivity under a range of morphological types.

ASEPTIC INOCULATION OF PLANT FERMENTERS

The inoculation of plant-scale fermenters may involve the transfer of culture from a laboratory fermenter, a disposable single-use reactor, or spore suspension vessel, to a plant fermenter, or the transfer from one plant fermenter to another. Obviously, it is extremely important that the fermentation is not contaminated during inoculation but if the process is a contained one then it is equally important that the process organism does not escape. Thus, the nature of the inoculation system will be dictated by the containment category of the process (see Chapter 7).

At Containment levels 1 and 2, the addition of inoculum must be carried out in such a way that release of microorganisms is restricted. This should be done by aseptic piercing of membranes, connections with steam locks or the use of disposable systems. At Containment levels 3 and 4, no microorganisms must be released during inoculation or other additions. In order to meet these stringent requirements, all connections must be screwed or clamped and all pipelines must be steam sterilizable (Health and Safety Executive, 1998).

INOCULATION FROM A LABORATORY FERMENTER OR A SPORE SUSPENSION VESSEL

Several systems have been described in the literature that are suitable for inoculating fermentations requiring only Containment levels 1 and 2. To prevent contamination during the transfer process it is essential that both vessels be maintained under a positive pressure and the inoculation port be equipped with a steam supply. Meyrath and Suchanek (1972) described a system for the inoculation of a plant fermenter from a laboratory vessel. The apparatus is shown in Fig. 6.20. The connecting point A is normally covered with the blank plug a and prior to inoculation this plug is slightly loosened, valve E closed and valve F opened to allow steam to exit at A. Valve F is then closed and E opened so that when a is removed sterile air will be released from the vessel. After removal, plug a is placed in strong disinfectant. Blank plug b is then removed and a coupling made between B and A. Valve E is closed and an air line attached to point C, establishing a pressure inside the inoculum fermenter greater than in the plant vessel. Valve E is then opened and the inoculum will be forced into the plant fermenter. After closing valve E the inoculum fermenter may be removed, plug a replaced, and the line steamed out by opening valve F. This system may be modified by using the quick connection devices that are now available (see Chapter 7).

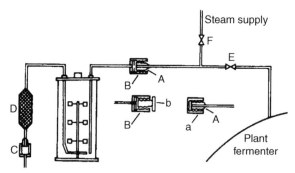

FIGURE 6.20 Inoculation of a Plant Fermenter From a Laboratory Fermenter (Meyrath & Suchanek, 1972)

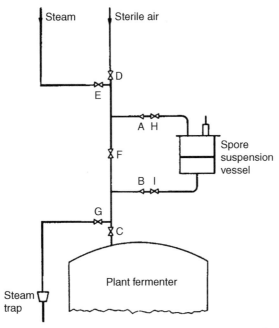

FIGURE 6.21 The Inoculation of a Plant Fermenter From a Spore Suspension Vessel (Parker, 1958)

Parker (1958) described a more complex system for inoculation of a plant fermenter from a spore-suspension vessel. The apparatus is shown in Fig. 6.21. The sterile spore-suspension vessel is batched with the spore suspension in the sterile room. The plant vessel, containing the medium and with blank plugs screwed on at A and B, is sterilized by steam injection. The plugs A and B are slightly loosened to allow steam to emit for 20 min when the whole system is under steam pressure. The blanks are then tightened, the valves E and G shut and sterile air is allowed into the plant fermenter by opening valves D, F, and C. The spore suspension vessel is loosely connected at A and B and valves D, H, I, and C closed and E, F, and G opened. After 20 min steaming A and B are tightened and G and E are closed. D is then opened to establish a positive pressure in the pipework. When the pipework has cooled, the pressure in the plant vessel is reduced to approximately 5 psi, valve F is closed and H, I, and C are opened. This procedure allows the spore suspension to be forced into the fermenter. Valves D, H, I, and C are closed and the suspension vessel replaced by blank plugs at A and B and the pipework steamed by opening valves E and G. A similar system is described by the UK Health and Safety Executive and is shown in Fig. 6.22.

Category 1 and 2 fermentations may also be inoculated by aseptic piercing of a membrane. In this system the inoculum vessel is connected to an inoculating needle assembly (as shown in Fig. 6.23) and the sterile needle pierces a rubber septum set

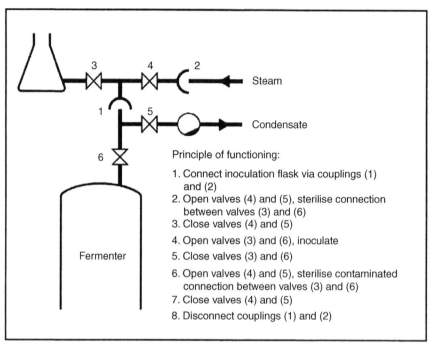

Steam

Condensate

Principle of functioning:

1. Connect inoculation flask via couplings (1) and (2)
2. Open valves (4) and (5), sterilise connection between valves (3) and (6)
3. Close valves (4) and (5)
4. Open valves (3) and (6), inoculate
5. Close valves (3) and (6)
6. Open valves (4) and (5), sterilise contaminated connection between valves (3) and (6)
7. Close valves (4) and (5)
8. Disconnect couplings (1) and (2)

Fermenter

FIGURE 6.22 Inoculation of a Plant Vessel

Health and Safety Executive, 1998

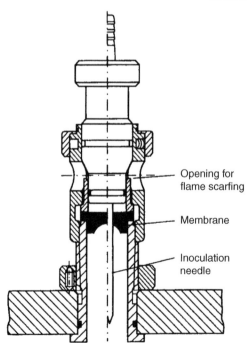

Opening for flame scarfing

Membrane

Inoculation needle

FIGURE 6.23 Needle Inoculation Device (Werner, 1992)

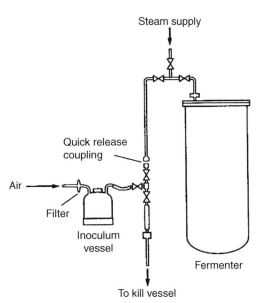

FIGURE 6.24 Inoculum System Suitable for Contained Fermentations (Vranch, 1990)

into a fermenter port. However, the use of a needle presents safety problems and many companies prohibit such systems. Also, aerosols may be created on the removal of a needle.

The inoculation of level 3 or 4 contained fermentations requires the use of modified systems. It must be possible to steam sterilize all the pipework after the inoculation and the condensate from the sterilization must be collected in a kill tank (see Chapters 5 and 7). The inoculation flask is then removed after inoculation and sterilized in an autoclave. One such system is shown in Fig. 6.24 (Vranch, 1990).

INOCULATION OF DISPOSABLE REACTORS AND USE OF DISPOSABLE CONNECTORS

The development of disposable, single-use reactors (particularly for animal cell culture) has been accompanied by the design of single-use connectors, tubing, and harnesses to facilitate aseptic inoculation and sampling. These systems, like their single-use reactor counterparts, are ready-sterilized and are designed to achieve aseptic connections in a nonaseptic environment. A number of commercial manufacturers supply such components, as shown in Fig. 6.25. As discussed earlier in this chapter, disposable single-use reactors are used routinely to develop inocula for animal cell fermentations and are attractive alternatives to conventional systems for the development of microbial inocula. Microbial fermentations (and many animal

(a)

(b)

(c)

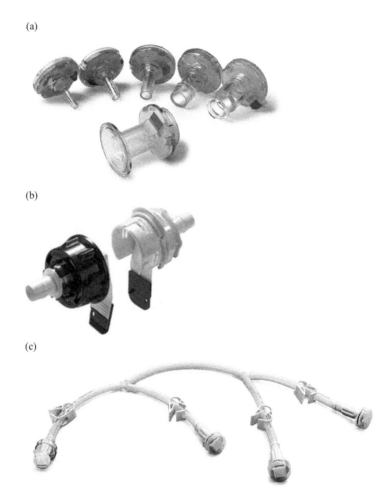

FIGURE 6.25 Disposable Sterile Connectors and Transfer Systems

(a) GE healthcare life sciences Readymate® connectors. (b) Colder products company Aseptiquick® sterile bioprocesing connectors. (c) Sartorius single-use transfer set.

Part a: Courtesy GE Healthcare. Part b: Courtesy Colder Products Company

Part c: Courtesy of Sartorius Stedim

cell processes) tend to use stainless steel reactors at the production stage and, thus, if disposable reactors are to be used as inoculum vessels then means must exist to connect them securely to steam sterilizable couplings on conventional vessels. Several manufacturers produce such systems that enable the connection to be steamed both before, and after, transfer as shown in Fig. 6.26. Provided that the condensate from such a steam sterilization is collected in a kill tank then these systems may also be used in a contained system.

(a)

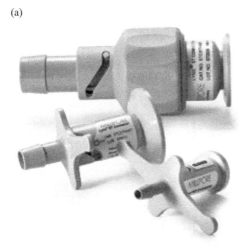

(b)

FIGURE 6.26 Disposable Connectors Enabling Connection Between Single-Use Systems and Conventional Stainless Steel Vessels, Enabling Steam Sterilization of the Connection

(a) Lynx® ST Connectors—enabling connection between disposable systems and a conventional stainless reactor with a steamable connection. (b) Colder Products Company Steam-Thru® Connector.

Part a: Courtesy of EMD Millipore Corporation, Lynx is a trademark of Merck KGaA

Part b: Courtesy of Colder Product Company

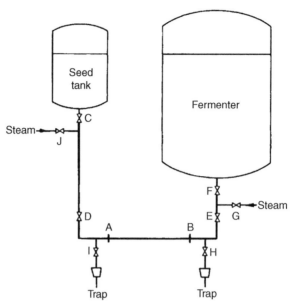

FIGURE 6.27 Inoculation of a Plant Fermenter From Another Plant Fermenter (Parker, 1958)

INOCULATION FROM A PLANT FERMENTER

Fig. 6.27 illustrates the system described by Parker (1958). The two vessels are connected by a flexible pipeline A–B. The batched fermenter is sterilized by steam injection via valves G and J, valves D, I, A, B, H, E, and F being open and valve C closed. Valves H and I lead to steam traps for the removal of condensate. After 20 min at the desired pressure the steam supply is switched off at J and G and the steam-trap valves I and H are closed. F, E, and D are left open so that the connecting pipeline fills with sterile medium. The medium in the fermenter is sparged with sterile air and when it has cooled to the desired temperature the pressure in the seed tank is increased to at least 10 psi while the pressure in the fermenter is reduced to about 2 psi. Valve C is opened and the inoculum is forced into the production vessel. After inoculation is complete the connecting pipeline is resterilized before it is removed. For a contained fermentation the condensate from the two steam traps attached to valves I and H would be directed to a kill tank.

REFERENCES

Adams, T. H., Weiser, J. K., & Jae-Hyuk, Y. (1998). Asexual sporulation in *Aspergillus nidulans*. *Microbiology and Molecular Biology Reviews, 62*(1), 35–54.

Albornoz, M.A.G. (2014). Strategies to overcome interferences during biomass monitoring with dielectric spectroscopy. PhD Thesis Heriot-Watt University.

Alford, J. S., Fowler, G. L., Higgs, R. E., Clapp, D. L., & Huber, F. M. (1992). Development of real-time expert system applications for the on-line analysis of fermentation data. In M. R. Ladisch, & A. Bose (Eds.), *Harnessing biotechnology for the 21st century* (pp. 375–379). Washington, D.C.: American Chemical Society.

Al Obaidi, Z. S., & Berry, D. R. (1980). cAMP concentration, morphological differentiation and citric acid production in *Aspergillus niger*. *Biotechnology Letters, 2*(1), 5–10.

Anderson, J. G., & Smith, J. E. (1971). The production of conidiophores and conidia by newly germinated conidia of *Aspergillus niger* (microcycle conidiation). *Journal of General Microbiology, 69*, 185–197.

Aunstrup, K. (1974). Industrial production of proteolytic enzymes. In B. Spencer (Ed.), *Industrial Aspects of Biochemistry* (pp. 23–46). (Vol. 30). Amsterdam: North Holland, Part 1.

Baust, J. G., Corwin, W. L., & Baust, J. M. (2011). Cell preservation technology. In M. Moo-Young (Ed.), *Comprehensive biotechnology* (pp. 179–190). (Vol. 1). Amsterdam: Elsevier.

Bissoyi, A., Nayak, B., Pramanik, K., & Sarangi, S. K. (2014). Targeting cryopreservation-induced cell death: a review. *Biopreservation and Biobanking, 12*(1), 23–34.

Bizukojc, M., & Ledakowicz, S. (2009). Physiological, morphological and kinetic aspects of lovastatin biosynthesis by *Aspergillus terreus*. *Biotechnology Journal, 4*, 647–664.

Booth, C. (1971). Fungal culture media. In C. Booth (Ed.), *Methods in microbiology* (pp. 49–94). (Vol. 4). London: Academic Press.

Borrow, A., Jefferys, E. G., Kessell, R. H. J., Lloyd, E. C., Lloyd, P. B., & Nixon, I. S. (1961). Metabolism of *Gibberella fujikuroi* in stirred culture. *Canadian Journal of Microbiology, 7*, 227–276.

Boualem, K., Gervais, P., Cavin, J. -F., & Wache, Y. (2014). Production of conidia of *Penicillium camamberti* in liquid medium through microcycles of conidiation. *Biotechnology Letters, 36*, 2239–2243.

Boulton, C. A. (1991). Developments in brewery fermentation. *Biotechnology & Genetic Engineering Reviews, 9*, 127–182.

Boulton, C. A., Maryan, P. S., Loveridge, D. & Kell, D. B. (1989). The application of a novel biomass sensor to the control of yeast pitching rate. *Proceedings of the European Brewing Convention Congress, Zurich* (pp. 653−661).

Boulton, C. A., Jones, A. R., & Hinchliffe, E. (1991). Yeast physiology and fermentation performance. *Proceedings of the Euroepan Brewing Convention Congress, Lisbon*.

Butterworth, D. (1984). Clavulanic acid: Properties, biosynthesis and fermentation. In E. J. Vandamme (Ed.), *Biotechnology of industrial antibiotics* (pp. 225–236). New York: Marcel Dekker.

Calam, C. T. (1976). Starting investigational and production cultures. *Process Biochemistry, 11*(3), 7–12.

Camici, L., Sermonti, G., & Chain, E. B. (1952). Observations on *Penicillium chrysogenum* in submerged culture. Mycelial growth and autolysis. *Bulletin of the World Health Organization, 6*, 265–272.

Carlsen, M., Spohr, A. B., Nielsen, J., & Villadsen, J. (1996). Morphology and physiology of an α-amylase producing strain of *Aspergillus oryzae* during batch cultivations. *Biotechnology Bioengineering, 49*, 266–276.

Conner, H. A., & Allgeier, R. J. (1976). Vinegar: its history and development. *Advances in Applied Microbiology, 20*, 82–127.

Della-Bianca, B. E., Basso, T. O., Stambuk, B. U., Basso, L. C., & Gombert, A. K. (2013). What do we know about the yeast strains from the Brazilian fuel ethanol industry? *Applied Microbiology and Biotechnology, 97*, 979–991.

Dhillon, G. S., Brar, S. K., Verma, M., & Tyagi, R. D. (2011). Recent advances in citric acid bio-production and recovery. *Food Bioprocess Technology, 4,* 505–529.

Dreher, T., Husemann, U., Zahnow, C., de Wilde, D., Adams, T., & Greller, G. (2013). High cell density *Escherichia coli* cultivation in different single-use bioreactor systems. *Chemie Ingenieur Technik, 85*(1/2), 162–171.

Driouch, H., Roth, A., Dersch, P., & Wittmann, C. (2011). Filamentous fungi in good shape: microparticles for tailor-made fungal morphology and enhanced enzyme production. *Bioengineered Bugs, 2*(2), 100–104.

Driouch, H., Hansch, R., Wucherpfennig, T., Krull, R., & Wittmann, C. (2012). Improved enzyme production by bio-pellets of *Aspergillus niger.* Targeted morphology engineering using titanate microparticles. *Biotechnology Bioengineering, 109*(2), 462–471.

Eibl, R. I., Loffelholz, C., & Eibl, D. (2014). Disposable reactors for inoculum production. In R. Portner (Ed.), *Methods in molecular biology: Animal cell biotechnology* (3rd ed., pp. 265–284). New York: Humana Press.

Elmayergi, H., Scharer, J. M., & Moo-Young, M. (1973). Effect of polymer additives on fermentation parameters in a culture of *Aspergillus niger. Biotechnology Bioengineering, 15,* 845–859.

El-Naggar, N. E., El-Bindary, A. A., & Nour, N. S. (2013). Statistical optimization of process variables for antimicrobial metabolites production by *Streptomyces anulatus* NEAE-94 against some multidrug-resistant strains. *International Journal of Pharmacology, 9*(6), 322–334.

El Sayed, A. -H. M. M. (1992). Production of penicillins and cephalosporins by fungi. In D. K. Arora, R. P. Elander, & K. G. Mukerji (Eds.), *Handbook of Applied Mycology, Vol. 4: Fungal Biotechnology* (pp. 517–564). New York: Marcel Dekker.

Ettler, P. (1992). The determination of optimum inoculum quality for submersed fermentation process. *Collection of Czechoslovak Chemical Communications, 57,* 303–308.

Foster, J. W. (1949). *Chemical activities of fungi.* New York: Academic Press, p. 62.

Foster, J. W., McDaniel, L. E., Woodruff, H. B., & Stokes, J. L. (1945). Microbiological aspects of penicillin. Production of conidia in submerged culture of *Penicillium notatum. Journal of Bacteriology, 50,* 365–381.

Formenti, L. R., Norregaard, A., Bolic, A., Hemandez, D. Q., Hagemann, T., Heins, A. -L., Larsson, H., Mauricio-Iglesias, M., Ruhne, U., & Gernaey, K. V. (2014). Challenges in industrial fermentation technology research. *Biotechnology Journal, 9,* 727–738.

Frohlich, B., Bedard, C., Gagliardi, T., & Oosterhuis, N.M.G. (2012). Co-development of a new 2-D rocking single-use bioreactor to streamline cell expansion processes. Presented at: *IBC Bioprocess International Conference.* Providence, RI, USA, 8–12 October, 2012.

Galbraith, J. C., & Smith, J. E. (1969). Filamentous growth of *Aspergillus niger* in submerged shake cultures. *Transactions of the British Mycological Society, 52*(2), 237–246.

Gershater, C. J. L. (2010). Inoculum preparation. In M. C. Flickenger (Ed.), *Encyclopedia of Industrial Biotechnology: Bioprocess, Bioseparation, and Cell Technology* (pp. 1435–1444). New Jersey: John Wiley and Sons.

Gbewonyo, K., Hunt, G., & Buckland, B. (1992). Interactions of cell morphology and transport in the lovastatin fermentation. *Bioprocess Engineering, 8*(1–2), 1–7.

Ghojavand, H., Bonakdarpour, B., Heydarian, S. M., & Hamedi, J. (2011). The inter-relationship between inoculum concentration, morphology, reheology and erythromycin productivity in submerged culture of *Saccaharopolysperma erythraea. Brazilian Journal of Chemical Engineering, 28*(4), 565–574.

Glazyrina, J., Materne, E. -M., Dreher, T., Storm, D., Junne, S., Adams, T., Greller, G., & Neubauer, P. (2010). High cell density cultivation and recombinant protein production with *Escherichia coli* in a rocking-motion-type bioreactor. *Microbial Cell Factories*, *9*, 42.

Gomez-Pastor, R., Garre, E., Matallana, E. & Perez-Torrado, R. (2011). Recent advances in yeast biomass production.In *Biomass–detection, production and usage*. Chapter 11. Matovic (Ed.), Intech. Available from: http://www.intechopen.com/books/biomass-detection-production-and-usage/recent-advances-in-yeast-biomass-production.

Gonciarz, J., & Bizukojc, M. (2014). Adding talc microparticles to *Aspergillus terreus* ATCC 20542 preculture decreases pellet size and improves lovastatin production. *Engineering Life Sciences*, *14*, 190–200.

Green, E. M. (2011). Fermentative production of butanol—the industrial perspective. *Current Opinion in Biotechnology*, *22*, 337–343.

Grimm, L. H., Kelly, S., Krull, R., & Hempel, D. C. (2005). Morphology and productivity of filamentous fungi. *Applied Microbiology and Biotechnology*, *69*, 375–384.

Hadley, G., & Harrold, C. E. (1958). The sporulation of *Penicillium notatum* westling in submerged liquid culture II. The initial sporulation phase. *Journal of Experimental Botany*, *9*(3), 418–425.

Hansen, A. M. (1967). Microbial production of pigments and vitamins. In H. J. Peppier (Ed.), *Microbial Technology* (pp. 222–250). New York: Reinhold.

Hansen, E. C. (1896). Pure culture of systematically selected yeasts in the fermentation industries. In *Practical Studies in Fermentation,* Chapter 1, 1–76 (A. K. Miller, Trans.). Spon, London.

Health and Safety Executive. (1998). *Large-scale contained use of biological agents*. Kew: HSE Books.

Hockenhull, D. J. (1980). Inoculum development with particular reference to *Aspergillus* and *Penicillium*. In J. E. Smith, D. R. Berry, & B. Christiansen (Eds.), *Fungal biotechnology* (pp. 1–24). London: Academic Press.

Hobbs, G., Frazer, C. M., Gardner, D. C. J., Cullum, J. A., & Oliver, S. G. (1989). Dispersed growth of *Streptomyces* in liquid culture. *Applied Microbiology and Biotechnology*, *31*, 272–277.

Holker, U., Hofer, M., & Lenz, J. (2004). Biotechnological advantages of laboratory-sclae solid-state fermentation with fungi. *Allied Microbiology and Biotechnology*, *64*, 175–186.

Hornbaek, T., Nielsen, A. K., Dynesen, J., & Jakobsen, M. (2004). The effect of inoculum age and solid versus liquid propagation on inoculum quality of an industrial *Bacillus licheniformis* strain. *FEMS Microbiology Letters*, *236*(1), 145–151.

Hosobuchi, M., Fukui, F., Matsukawa, H., Suzuki, T., & Yoshikawa, H. (1993). Morphology control of preculture during production of ML-236B, a precursor of pravastatin sodium, by *Penicillium citrinum*. *Journal of Fermentation and Bioengineering*, *76*(6), 476–481.

Hunt, G. R., & Stieber, R. W. (1986). Inoculum development. In A. L. Demain, & N. A. Solomon (Eds.), *Manual of industrial microbiology and biotechnology* (pp. 32–40). Washington, DC: American Society of Microbiology.

Ignova, M., Montague, G. A., Ward, A. C., & Glassey, J. (1999). Fermentation seed quality analysis with self-organising neural networks. *Biotechnology and Bioengineering*, *64*(1), 82–91.

Jonczyk, P., Schmidt, A., Bice, I., Gall, M., Gross, E., Hilmer, J. -M., Bornscheuer, U., & Scheper, T. (2011). Strikt batch-kultivierung von *Eubacterium ramulus* in einem neuartigen eiweg-beutelreaktorsytem. *Chemie Indenieur Technik*, *83*(12), 2147–2152.

Jung, B., Kim, S., & Lee, J. (2014). Microcycle conidiation in filamentous fungi. *Mycobiology*, *42*(1), 1–5.

Junker, B., Seeley, A., Lester, M., Kovatch, M., Schmitt, J., Borysewicz, S., Lynch, J., Zhang, J., & Greasham, R. (2002). Use of frozen bagged seed inoculum for secondary metaboliote and bioconversion processes at the pilot scale. *Biotechnology and Bioengineering*, *79*(6), 628–640.

Junker, B., Lester, M., Leporati, J., Schmitt, J., Kovatch, M., Borysewicz, S., Maciejak, W., Seeley, A., Hesse, M., Connors, N., Brix, T., Creveling, E., & Salmon, P. (2006). Sustainable reduction of bioreactor contamination in an industrial fermentation pilot plant. *Journal of Bioscience and Bioengineering*, *102*(4), 251–268.

Junne, S., Solymosi, T., Oosterhuis, N., & Neubauer, P. (2013). Cultivation of cells and micro-organisms in wave-mixed disposable bag bioreactors at different scales. *Chemie Ingenieur Technik*, *85*(1/2), 57–66.

Kaup, B. -A., Ehrich, K., Pescheck, M., & Schrader, J. (2008). Micro-particle enhanced cultivation of filamentous microorganisms: Increased chloroperoxidase formation by *Caldariomyces fumago* as an example. *Biotechnology and Bioengineering*, *99*(3), 491–498.

Keay, L., Moseley, M. H., Anderson, R. G., O'Connor, R. J., & Wildi, B. S. (1972). Production and isolation of microbial proteases. *Biotechnical and Bioengineering Symposium*, *3*, 63–92.

Kennedy, A. I., Taidi, B., Aitchison, A., & Green, X. (2003). Management of multi-strain, multi-site yeast storage and supply. In K. A. Smart (Ed.), *Brewing yeast fermentation performance* (2nd ed., pp. 131–137). Oxford: Blackwell Science.

Kiviharju, K., Salonen, K., Moilanen, U., & Eerikainen, T. (2008). Biomass measurement on-line: the performance of in situ measurements and software sensors. *Journal of Industrial Microbiology and Biotechnology*, *35*, 657–665.

Kloth, C., Macisaac, G., Ghebremariam, H., & Arunakumari, A. (2010). Inoculum expansion methods, recombinant mammalian cell lines. In M. C. Flickinger (Ed.), *Encyclopedia of Industrial Biotechnology* (pp. 1–30). New Jersey: Wiley.

Krasniewski, I., Molimard, P., Feron, G., Verogoignan, C., Durand, A., Cavin, J. -F., & Cotton, P. (2006). Impact of solid medium composition on the conidiation in *Penicillium camemberti*. *Process Biochemistry*, *41*(6), 1318–1324.

Krause, M., Ukkonen, K., Haataja, T., Ruottinen, M., Glumoff, T., Neubauer, A., Neubauer, P., & Vasala, A. (2010). A novel fed-batch based cultivation method provides high cell-density and improves yield of soluble recombinant proteins in shaken cultures. *Microbial Cell Factories*, *9*, 11.

Krull, R., Wucherpfennig, T., Esfandabadi, M. E., Walisko, R., Melzer, G., Hempel, D. C., Kampen, A., & Whitman, C. (2013). Characterization and control of fungal morphology for improved production performance in biotechnology. *Journal of Biotechnology*, *163*, 112–123.

Larroche, C. (1996). Microbial growth and sporulation behaviour in solid state fermentation. *Journal of Scientific and Industrial Research*, *55*(5/6), 408–423.

Lincoln, R. E. (1960). Control of stock culture preservation and inoculum build-up in bacterial fermentations. *Journal of Biochemical and Microbiological Technology and Engineering*, *2*, 481–500.

Lockwood, L. B. (1975). Organic acid production. In J. E. Smith, & D. P. Berry (Eds.), *The filamentous fungi* (pp. 140–157). (Vol. 1). London: Arnold.

McNeil, B., & Kristiansen, K. (1986). The acetone butanol fermentation. *Advances in Applied Microbiology*, *31*, 61–92.

Mahajan, E., Matthews, T., Hamilton, R., & Laird, M. W. (2010). Use of disposable reacy-ors to generate inoculum cultures for *E. coli* production fermentations. *Biotechnology Progress, 26*(4), 1200–1203.

Malik, K. A. (1991). Maintenance of microorganisms by simple methods. In B. E. Kirsop, & A. Doyle (Eds.), *Maintenance of microorganisms and cultured cells a manual of laboratory methods* (2nd ed., pp. 121–132). London: Academic Press.

Mandl, B., Grunewald, J., & Voerkelius, G. A. (1964). The application of time-saving brewing methods. *Brauwelt, 104*(80), 1541–1543.

Manteca, A., Alvarez, R., Salazar, N., Yague, P., & Sanchez, J. (2008). Mycelium differentia-tion and antibiotic production in submerged cultures of *Streptomyces coelicolor. Applied and Environmental Microbiology, 74*(12), 3877–3886.

Martin, S. M., & Bushell, M. E. (1996). Effect of hyphal morphology on bioreactor per-formance of antibiotic-producing *Sacchsropolyspora erythrea* cultures. *Microbiology, 142*(7), 1783–1788.

Meyrath, J., & Suchanek, G. (1972). Inoculation techniques—effects due to quality and quan-tity of inoculum. In J. R. Norris, & D. W. Ribbons (Eds.), *Methods in microbiology* (pp. 159–209). (Vol. 7B). London: Academic Press.

Mikola, M., Seto, J., & Amanullah, A. (2007). Evaluation of a novel wave bioreactor for aerobic yeast cultivation. *Bioprocess and Biosystems Engineering, 30*, 231–241.

Miller, K. J., Box, W. G., Boulton, C. A., & Smart, K. A. (2012). Cell cycle synchrony of propagated and recycled lager yeast and its impact on lag phase in fermenter. *Journal of the American Society of Brewing Chemists, 70*, 1–9.

Miyagawa, K., Suzuki, M., Higashide, E., & Uchida, M. (1979). Effect of aspartate family amino acids on production of maridomycin. III. *Agricultural and Biological Chemistry, 43*, 1111–1116.

Moldenhauer, J. R. (2008). Preservation of cultures for fermentation processes. In B. McNeil, & L. M. Harvey (Eds.), *Practical fermentation technology* (pp. 125–166). Chichester, UK: Wiley.

Morton, A. G. (1961). The induction of sporulation in mould fungi. *Proceedings of the Royal Society of London B, 153*(953), 548–569.

Mulder, K. C. L., Viana, A. A. B. V., Xavier, M., & Parachin, N. S. (2013). Critical aspects to considered prior to large-scale production of peptides. *Current Protein and Peptide Science, 14*, 556–565.

Nakayama, K. (1972). Lysine and diaminopimelic acid. In K. Yamada, S. Kinoshita, T. Tsunoda, & K. Aida (Eds.), *The microbial production of amino acids* (pp. 369–398). New York: Halsted Press.

Neves, A. A., Vieira, L. M., & Menezes, J. C. (2001). Effects of preculture variability on cla-vulanic acid fermentation. *Biotechnology and Bioengineering, 72*(6), 628–633.

Ni, Y., & Sun, Z. (2009). Recent progress on industrial fermentative production of acetone-butanol-ethanol by *Clostridium acetobutylicum* in China. *Applied Microbiology and Biotechnology, 83*(3), 415–423.

Nielsen, J. (1996). Modelling the morphology of filamentous fungi. *Trends in Biotechnology, 14*, 438–443.

Nordon, A., Littlejohn, D., Dann, A. S., Jenkins, P. A., Richardson, M. D., & Stimpson, S. L. (2008). In situ monitoring of the seed stage of a fermentation process using non-invasive NIR spectroscopy. *Analyst, 133*(5), 660–666.

O'callaghan, P. M., & Racher, A. J. (2015). Building a cell culture process with stable founda-tions: Searching for certainty in an uncertain world. In M. Al-Rubeai (Ed.), *Cell engineering Vol 9–Animal cell culture* (pp. 373–406). Switzerland: Springer International Publishing.

Okonkowski, J., Kizer-Bentley, L., Listner, K., Robinson, D., & Chartrain, M. (2005). Development of a robust, versatile and scalable inoculum ttrain for the production of a DNA vaccine. *Biotechnology Progress*, *21*, 1038–1047.

Oosterhuis, N. M. G., & van den Berg, H. J. (2011). How multipurpose is a disposable reactor? *Biopharmaceutical International*, *24*(3), 1–5.

Oosterhuis, N. M. G., Neubauer, P., & Junne, S. (2013). Single use bioreactors for microbial cultivation. *Pharmaceutical Bioprocessing*, *1*(2), 167–177.

Pamboukian, C. R. D., & Facciotti, M. C. R. (2005). Rheological and morphological characterization of *Streptomyces olindensis* growing in batch and fed-batch fermentations. *Brazilian Journal of Chemical Engineering*, *22*(1), 31–40.

Panula-Perala, J., Siurkus, J., Vasala, A., Wilmanowski, R., Casteieijn, M. G., & Neubauer, P. (2008). Enzyme controlled glucose auto-delivery for high cell density cultivations in microplates and shake flasks. *Microbial Cell Factories*, *7*, 31.

Papagianni, M. (2004). Fungal morphology and metabolite production in submerged mycelial processes. *Biotechnology Advances*, *22*, 189–259.

Papagianni, M., & Mattey, M. (2006). Morphological development of *Aspergillus niger* in submerged citric acid fermentation as a function of the spore inoculum level. Application of neural network and cluster analysis for characterization of mycelial morphology. *Microbial Cell Factories*, *5*(3), 1–12.

Parker, A. (1958). Sterilization of equipment, air and media. In R. Steel (Ed.), *Biochemical engineering—unit processes in fermentation* (pp. 94–121). London: Heywood.

Parton, C., & Willis, P. (1990). Strain preservation, inoculum preparation and inoculum development. In B. McNeil, & L. M. Harvey (Eds.), *Fermentation a practical approach* (pp. 39–64). Oxford: IRL Press.

Patakova, P., Linhova, M., Rychtera, M., Paulova, L., & Melzoch, K. (2013). Novel and neglected issues of acetone-butanol-ethanol (ABE) fermentation by clostridia: *Clostridium* metabolic diversity, tools for process mapping and continuous fermentation systems. *Biotechnology Advances*, *31*(1), 58–67.

Pazout, J., & Schroder, P. (1988). Microcycle conidiation in submerged cultures of *Penicillium cyclopium* attained without temperature changes. *Journal General Microbiology*, *134*, 2685–2692.

Perlman, D., & Kikuchi, M. (1977). Culture maintenance. In D. Perlman (Ed.), *Annual reports on fermentation processes* (pp. 41–48). (Vol. 1). London: Academic Press.

Pickerell, A. T. W., Hwang, A., & Axcell, B. C. (1991). Impact of yeast-handling procedures on beer flavor development during fermentation. *Journal of American Society of Brewing Chemists*, *49*, 87–92.

Podojil, M., Blumauerova, M., Culik, K., & Vanek, Z. (1984). The tetracyclines: Properties, biosynthesis and fermentation. In E. J. Vandamme (Ed.), *Biotechnology of industrial antibiotics* (pp. 259–280). Marcel Dekker: New York.

Posch, A. E., Herwig, C., & Spadiut, O. (2013). Science-based bioprocess design for filamentous fungi. *Trends in Biotechnolgy*, *31*(1), 37–44.

Pridham, T. G., Lyons, A. J., & Phrompatima, B. (1973). Viability of actinomycetes stored in soil. *Applied Microbiology*, *26*, 441–442.

Priest, F. G., & Sharp, R. J. (1989). Fermentation of bacilli. In J. O. Neway (Ed.), *Fermentation process development of industrial organism* (pp. 73–132). New York: Marcel Dekker.

Reed, G., & Nagodawithana, T. W. (1991a). *Yeast technology*. New York: Van Nostrand Reinhold, 112–116.

Reed, G., & Nagodawithana, T. W. (1991b). *Yeast technology*. New York: Van Nostrand Reinhold, 288–290.

Rhodes, A., Crosse, R., Ferguson, T. P., & Fletcher, D. L. (1957). Improvements in or relating to the production of the antibiotic griseofulvin. British Patent 784, 618.

Righelato, R. C. (1976). Selection of strains of *Penicillium chrysogenum* with reduced yeild in continuous culture. *Journal of Applied Chemistry and Biotechnology*, *26*(30), 153–159.

Rioseras, B., Lopez-Garcia, M. T., Yague, P., Sanchez, J., & Manteca, A. (2014). Mycelium differentiation and development of *Streptomyces coelicolor* in lab-scale bioreactors: Programmed cell death, differentiation, and lysis are closely linked to undecylprodigiosib and actinorhodin production. *Bioresources Technology*, *151*, 191–198.

Roessler, J. G. (1968). Yeast management in the brewery, Part II. Yeast handling and treatment. *Brewing Digest*, *43*(9), 94, 96, 98, 102, 115.

Roncal, T., Cordobes, U., Sterner, O., & Ugalde, U. (2002). Conidiation in *Penicilium cyclopium* is induced by conidiogenone, an endogenous direpene. *Eukaryotic Cell*, *1*, 823–829.

Roncal, T., & Ugalde, U. (2003). Conidiation induction in *Penicillium*. *Research in Microbiology*, *154*, 539–546.

Rudge, R. H. (1991). Maintenance of bacteria by freeze-drying. In B. E. Kirsop, & A. Doyle (Eds.), *Maintenance of microorganisms and cultured cells. A manual of laboratory methods* (2nd ed., pp. 31–44). London: Academic Press.

Sabatie, J., Speck, D., Reymund, J., Hebert, C., Caussin, L., Weltin, D., Gloeckler, R., O'Regan, M., Bernard, S., Ledoux, G., Ohsawa, I., Kamogawa, K., Lemoine, Y., & Brown, S. W. (1991). Biotin formation by recombinant strains of *Escherichia coli:* influence of the host physiology. *Journal of Biotechnology*, *20*, 29–50.

Sansing, G. A., & Cieglem, A. (1973). Mass propagation of conidia from several *Aspergillus* and *Penicillium* species. *Applied Microbiology*, *26*, 830–831.

Schimmel, T. G., Coffman, A. D., & Parsons, S. J. (1998). Effect of butyrolactone 1 on the producing fungus, *Aspergillus terreus*. *Applied and Environmental Microbiology*, *64*(10), 3707–3712.

Sen, R., & Swaminathan (2004). Response surface modeling and optimization to elucidate and analyze the effects of inoculum age and size on surfactin production. *Biochemical Engineering Journal*, *21*(2), 141–148.

Seth, G., Hamilton, R. W., Stapp, T. R., Zheng, L., Meier, A., Petty, K., Leung, S., & Chary, S. (2013). Development of a new bioprocess scheme using frozen seed train intermediates to initiate CHO cell culture manufacturing campaigns. *Biotechnology Bioengineering*, *110*(5), 1376–1385.

Silva, F., Queiroz, J. A., & Domingues, C. (2012). Plasmid DNA fermentation strategies: influence on plasmid stability and cell physiology. *Applied Microbiology Biotechnology*, *93*(6), 2571–2580.

Singh, V. (1999). Disposable bioreactor for cell culture using wave-induced motion. *Cytotechnology*, *30*, 149–158.

Singh, K., Sehgal, S. N., & Vezina, C. (1968). Large scale transformation of steroids by fungal spores. *Applied Microbiology*, *16*, 393–400.

Singhania, R. R., Patel, A. K., Soccol, C. R., & Pandey, A. (2009). Recent advances in solid-state fermentations. *Biochemical Engineering Journal*, *44*, 13–18.

Sivashanmugam, A., Murray, V., Cui, C., Zhang, Y., Wang, J., & Li, Q. (2009). Practical protocols for production of very high yields of recombinant proteins using *Escherichi coli*. *Protein Science*, *18*, 936–948.

Smith, G. M., & Calam, C. T. (1980). Variations in inocula and their influence on the productivity of antibiotic fermentations. *Biotechnology Letters*, *2*(6), 261–266.

Smith, D., & Ryan, M. (2012). Implementing best practices and validation of cryopreservation techniques of microorganisms. *The Scientific World Journal*, *2012*, 805659.

Snell, J. J. S. (1991). General introduction to maintenance methods. In B. E. Kirsop, & A. Doyle (Eds.), *Maintenance of microorganisms and cultured cells. A manual of laboratory methods* (2nd ed., pp. 21–30). London: Academic Press.

Spalla, C., Grein, A., Garafano, L., & Ferni, G. (1989). Microbial production of vitamin B_{12}. In E. J. Vandamme (Ed.), *Biotechnology of vitamins, pigments and growth factors* (pp. 257–284). New York: Elsevier.

Spivey, M. J. (1978). The acetone/butanol/ethanol fermentation. *Process of Biochemistry*, *13*(11), 2–3.

Steel, R., Lenz, C., & Martin, S. M. (1954). A standard inoculum for citric acid production in submerged culture. *Canadian Journal of Microbiology*, *1*, 150–157.

Stewart, G. G., Hill, A. E., & Russell, I. (2013). 125th anniversary review: developments in brewing and distilling yeast strains. *Journal of the Institute of Brewing*, *119*(4), 202–220.

Strandskov, F. B. (1964). Yeast handling in a brewery. *Am. Soc. Brewing Chemists Proc*, 76–79.

Thorne, R. S. W. (1970). Pure yeast cultures in brewing. *Process of Biochemistry*, *5*(4), 15–22.

Tao, Y., Shih, J., Sinacore, M., Ryll, T., & Yusuf-Makagiansar, H. (2011). Development and implementation of a perfusion-based high cell density cell banking process. *Biotechnology Progress*, *27*(3), 824–829.

Trinci, A. P. J. (1983). Effect of Junlon on morphology of *Aspergillus niger* and its use in making turbidity measurements of fungal growth. *Transactions of the British Mycological Society*, *58*, 467–473.

Trinci, A. P. J. (1992). Mycoprotein: a twenty year overnight success story. *Mycological Research*, *96*(1), 1–13.

Tucker, K. G., & Thomas, C. (1992). Mycelial morphology: the effect of spore inoculum level. *Biotechnology Letters*, *14*(11), 1071–1074.

Ugalde, U., & Rodriguez-Urra, A. B. (2014). The mycelium blueprint: insights into the cues that shape the filamentous colony. *Applied Microbiology and Biotechnology*, *98*, 8809–8819.

Underkofler, L. A. (1976). Microbial enzymes. In B. M. Miller, & W. Litsky (Eds.), *Industrial microbiology* (pp. 128–164). New York: McGraw-Hill.

Van Dissel, D., Claessen, D., & Van Wezel, G. P. (2014). Morphogenesis of *Streptomyces* in submerged cultures. *Advances in Applied Microbiology*, *89*, 1–46.

Vezina, C., & Singh, K. (1975). Transformation of organic compounds by fungal spores. In J. E. Smith, & D. R. Berry (Eds.), *The filamentous fungi* (pp. 158–192). (Vol. 1). London: Arnold.

Vezina, C., Singh, K., & Sehgal, S. N. (1965). Sporulation of filamentous fungi in submerged culture. *Mycologia*, *57*, 722–736.

Vezina, C., Sehgal, S. N., & Singh, K. (1968). Transformation of organic compounds by fungal spores. *Advances in Applied Microbiology*, *10*, 221–268.

Vranch, S. P. (1990). Containment and regulations for safe biotechnology. In W. G. Hyer (Ed.), *Bioprocessing safety: Worker and community safety and health considerations* (pp. 39–57). Philadelphia, PA: American Society for Testing and Materials.

Walisko, R., Krull, R., Scharder, J., & Whittmann, C (2012). Microparticle based morphology engineering of filamentous microorganisms for industrial bioproduction. *Biotechnology Letters*, *34*(11), 1975–1982.

Warr, S. R. C., Gershater, C. J. L., & Box, S. J. (1996). Seed stage development for improved fermentation performance: increased milbemycin production by *Streptomyces hygroscopicus*. *Journal of Industrial Microbiology*, *16*, 295–300.

Werner, R. G. (1992). Containment in the development and manufacture of recombinant DNA-derived products. In C. H. Collins, & A. J. Beale (Eds.), *Safety in microbiology and biotechnology* (pp. 190–213). Oxford: Butterworth-Heinemann.

Westbrook, A., Scharer, J., Moo-Young, M., Oosterhuis, N., & Chou, C. P. (2014). Application of a two-dimensional disposable rocking bioreactor to bacterial cultivation for recombinant protein production. *Biochemical Engineering Journal*, *88*, 154–161.

Whitaker, A. (1992). Actinomycetes in submerged culture. *Applied Biochemistry and Biotechnology*, *32*, 23–35.

Whitaker, A., & Long, P. A. (1973). Fungal pelleting. *Process Biochemistry*, *8*(11), 27–31.

Whitford, W. G. (2013). Single-use technology supporting the comeback of continuous bioprocessing. *Pharmaceutical Bioprocessing*, *1*(3), 249–253.

Williams, S. T., Entwhistle, S., & Kurylowicz, W. (1974). The morphology of streptomycetes growing in media used for commercial production of antibiotics. *Microbios*, *11*, 47–60.

Worgan, J. T. (1968). Culture of the higher fungi. *Progress in Industrial Microbiology*, *8*, 73–139.

Wright, B., Bruninghaus, M., Vrabel, M., Walther, J., Shah, N., Bae, S. -A., Johnson, T., Yin, J., Zhou, W., Konstantinov, K. (2015). Bioprocess International, *13*(3), 16–25.

Wucherpfennig, T., Kiep, K. A., Driouch, H., Wittmann, C., & Krull, R. (2010). Morphology and rheology in filamentous cultivations. *Advances in Appplied Microbiology*, *72*, 89–136.

Wucherpfennig, T., Hestler, T., & Krull, R. (2011). Morphology engineering—Osmolality and its effect on *Aspergillus niger* morphology and productivity. *Microbial Cell Factories*, *10*(58), 1–15.

Yamamoto, K., Ishizaki, A., & Stanbury, P. F. (1993). Reduction in the length of the lag phase of L-lactate fermentation by the use of inocula from electrodialysis seed cultures. *Journal of Fermentation and Bioengineering*, *76*(2), 151–152.

Yao, X., Chu, X., Huang, Y., Liang, J., Wang, Y., Chu, J., & Zhang, S. (2009). Effect of inoculation methods on avermectin fermentation by *Streptomyces avermitilis*. *Journal of Food Science and Biotechnology*, *2009*(5).

Yague, P., Lopez-Garcia, M. T., Rioseras, B., Sanchez, J., & Manteca, A. (2012). New insights on the development of *Streptomyces* and their relationships with secondary metabolite production. *Current Trends in Microbiology*, *8*, 65–73.

Yin, P., Wang, Y. -H., Zhang, S. -L., Chu, J., Zhuang, Y. -P., Chen, N., Li, X. -F., & Wu, Y. -B. (2008). *Journal of the Chinese Institute of Chemical Engineers*, *39*, 609–615.

Xu, J., Wang, L., Ridgway, D., Gu, T., & Moo-Young, M. (2000). Increased heterologous protein production in *Aspergillus niger* production through extracellular proteases inhibition by pelleted growth. *Biotechnology Progress*, *16*, 222–227.

Xu, Z., Shi, Z., & Jiang, L. (2011). Acetic and propionic acids. In M. Moo-Young (Ed.), *Comprehensive Biotechnology* (pp. 190–199). (Vol. 3).

Zanin, G., Santana, C., Bon, E., Giordano, R., & de Moraes, F., et al. (2000). Brazilian bioethanol program.. *Applied Biochemistry and Biotechnology*, *84–86*(1), 1147–1161.

Design of a fermenter

INTRODUCTION

A research team led by Chaim Weizmann in Great Britain during the First World War (1914–18) developed a process for the production of acetone by a deep liquid fermentation using *Clostridium acetobutylicum*, which led to the eventual use of the first truly large-scale aseptic fermentation vessels (Hastings, 1978). Contamination, particularly with bacteriophages, was often a serious problem, especially during the early part of a large-scale production stage. Initially, no suitable vessels were available and attempts with alcohol fermenters fitted with lids were not satisfactory, as steam sterilization could not be achieved at atmospheric pressure. Large mild-steel cylindrical vessels with hemispherical tops and bottoms were constructed that could be sterilized with steam under pressure. Since the problems of aseptic additions of media or inocula had been recognized, steps were taken to design and construct piping, joints, and valves in which sterile conditions could be achieved and maintained when required. Although the smaller seed vessels were stirred mechanically, the large production vessels were not, and the large volumes of gas produced during the fermentation continually agitated the vessel contents. Thus, considerable expertise was built up in the construction and operation of this aseptic anaerobic process for production of acetone-butanol.

The first true large-scale aerobic fermenters were used in Central Europe in the 1930s for the production of compressed yeast (de Becze & Liebmann, 1944). The fermenters consisted of large cylindrical tanks with air introduced at the base via networks of perforated pipes. In later modifications, mechanical impellers were used to increase the rate of mixing and to break up and disperse the air bubbles. This procedure led to the compressed-air requirements being reduced by a factor of five. Baffles on the walls of the vessels prevented a vortex forming in the liquid. Even at this time, it was recognized that the cost of energy necessary to compress air could be 10–20% of the total production cost. As early as 1932, Strauch and Schmidt patented a system in which the aeration tubes were provided with water and steam for cleaning and sterilizing (Strauch & Schmidt, 1932).

Prior to 1940, the other important fermentation products besides bakers' yeast were ethanol, glycerol, acetic acid, citric acid, other organic acids, enzymes, and sorbose (Johnson, 1971). These processes used highly selective environments such as acidic or anaerobic conditions or the use of an unusual substrate, resulting in contamination being a relatively minor problem compared with the acetone fermentation or the subsequent aerobic antibiotic fermentations.

The decision to use submerged culture techniques for penicillin production, where aseptic conditions, good aeration and agitation were essential, was a very important factor in forcing the development of carefully designed and purpose-built fermentation vessels. In 1943, when the British government decided that surface culture production was inadequate, none of the fermentation plants were immediately suitable for deep fermentation, although the Distillers Company solvent plant at Bromborough only needed aeration equipment to make it suitable for penicillin production (Hastings, 1971). Construction work on the first large-scale plant to produce penicillin by deep fermentation was started on 15th Sep. 1943, at Terre Haute in the United States of America, building steel fermenters with working volumes of 54,000 dm^3 (Callahan, 1944). The plant was operational on 30th Jan. 1944. Unfortunately, no other construction details were quoted for the fermenters.

Initial agitation studies in baffled stirred tanks to identify variables were also reported at this time by Cooper, Fernstrom, and Miller (1944), Foust, Mack, and Rushton (1944) and Miller and Rushton (1944). Cooper's work had a major influence on the design of later fermenters.

BASIC FUNCTIONS OF A FERMENTER

The main function of a fermenter is to provide a controlled environment for the growth of microorganisms or animal cells, to obtain a desired product. In designing and constructing a fermenter, a number of points must be considered:

1. The vessel should be capable of being operated aseptically for a number of days and should be reliable in long-term operation and meet the requirements of containment regulations.
2. Adequate aeration and agitation should be provided to meet the metabolic requirements of the microorganism. However, mixing should not cause damage to the organism nor cause excessive foam generation.
3. Power consumption should be as low as possible.
4. A system of temperature control, both during sterilization and fermentation, should be provided.
5. A system of pH monitoring and control should be provided together with the monitoring and control of other parameters (eg, dissolved oxygen, redox, etc.) as appropriate.
6. Sampling facilities should be provided.
7. Evaporation losses from the fermenter should not be excessive.
8. The vessel should be designed to require the minimal use of labor in operation, harvesting, cleaning, and maintenance.
9. Ideally the vessel should be suitable for a range of processes, but this may be restricted because of containment regulations.

10. The vessel should be constructed to ensure smooth internal surfaces, using welds instead of flange joints whenever possible.
11. The vessel should be of similar geometry to both smaller and larger vessels in the pilot plant or plant to facilitate scale-up (see also Chapter 9).
12. The cheapest materials, which enable satisfactory results to be achieved should be used.
13. There should be adequate service provisions for individual plants (Table 7.1).

The first two points are probably the most critical and are further discussed in Chapters 5 and 9. It is obvious from the aforementioned points that the design of a fermenter will involve cooperation between experts in microbiology, biochemistry, chemical engineering, mechanical engineering, and costing. Although many different types of fermenter have been described in the literature, very few have proved to be satisfactory for industrial aerobic fermentations. The most commonly used ones are based on a stirred upright cylinder with sparger aeration. This type of vessel can be produced in a range of sizes from 1 dm^3 to 1000s of dm^3. Figs. 7.1 and 7.2 are diagrams of typical mechanically agitated and aerated fermenters with one and three multibladed impellers, respectively. Tables 7.2 and 7.3 give geometrical ratios of various of the dimensions which have been quoted in the literature for a variety of sizes of vessel. Moucha, Rejl, Kordac, and Labik (2012) have described the design of multiple-impeller fermenters with particular emphasis on scale-up, oxygen mass transfer, and the limitations of experimental k_La data.

At this stage the discussion will be concerned with stirred, aerated vessels for microbial cell culture. More varied shapes are commonly used for alcohol, biomass production, animal cell culture, and effluent treatment and will be dealt with later in this chapter.

Table 7.1 Service Provisions for a Fermentation Plant

Compressed air
Sterile compressed air (at 1.5–3.0 atm)
Chilled water (12–15°C)
Cold water (4°C)
Hot water
Steam (high pressure)
Steam condensate
Electricity
Stand-by generator
Drainage of effluents
Motors
Storage facilities for media components
Control and monitoring equipment for fermenters
Maintenance facilities
Extraction and recovery equipment
Accessibility for delivery of materials
Appropriate containment facilities

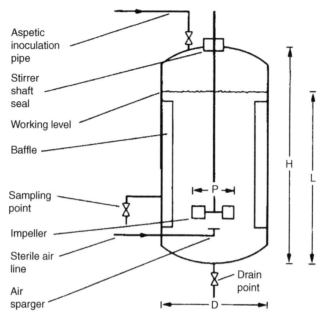

FIGURE 7.1 Diagram of a Fermenter With One Multibladed Impeller

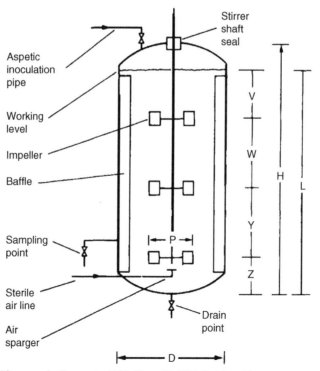

FIGURE 7.2 Diagram of a Fermenter With Three Multibladed Impellers

Table 7.2 Details of Geometrical Ratios of Fermenters With Single Multiblade Impellers (Fig. 7.1)

Dimension	Steel and Maxon (1961)	Wegrich and Shurter (1953)	Blakeborough (1967)
Operating volume	250 dm^3	12 dm^3	—
Liquid height (L)	55 cm	27 cm	—
L/D (tank diameter)	0.72	1.1	1.0–1.5
Impeller diameter (P/D)	0.4	0.5	0.33
Baffle width/D	0.10	0.08	0.08–0.10
Impeller height/D	—	—	0.33

Table 7.3 Details of Geometrical Ratios of Fermenters With Three Multibladed Impellers (Fig. 7.2)

Dimension	Jackson (1958)	Aiba, Humphrey, & Millis (1973)	Paca, Ettler, & Gregr (1976)
Operating volume	—	100,000 dm^3 (total)	170 dm^3
Liquid height (L)	—	—	150 cm
L/D (tank diameter)	—	—	1.7
Impeller diameter	0.34–0.5	0.4	0.33
Baffle width/D	0.08–10	0.095	0.098
Impeller height/D	0.5	0.24	0.37
P/V	0.5–1.0	—	0.74
P/W	0.5–1.0	0.85	0.77
P/Y	0.5–1.0	0.85	0.77
P/Z	—	2.1	0.91
H/D	1.0–1.6	2.2	2.95

ASEPTIC OPERATION AND CONTAINMENT

Aseptic operation involves protection against contamination and it is a well established and understood concept in the fermentation industries, whereas containment involves prevention of escape of viable cells from a fermenter or downstream equipment and is much more recent in origin. Containment guidelines were initiated during the 1970s (East, Stinnett, & Thoma, 1984; Flickinger & Sansone, 1984).

To establish the appropriate degree of containment which will be necessary to grow a microorganism, it, and in fact the entire process, must be carefully assessed for potential hazards that could occur should there be accidental release. Different assessment procedures have been used in the past depending on whether or not the organism contains foreign DNA [genetically engineered/modified (GM)]. Once the hazards are assessed, an organism can be classified into a hazard group for which there is an appropriate level of containment. The procedure, which was adopted within the European Community is outlined in Fig. 7.3. Nongenetically engineered organisms may be placed into a hazard group (1–4) using criteria to assess risk such as those given by Collins (1992):

1. The known pathogenicity of the microorganism.
2. The virulence or level of pathogenicity of the microorganism—are the diseases it causes mild or serious?
3. The number of organisms required to initiate an infection.
4. The routes of infection.
5. The known incidence of infection in the community and the existence locally of vectors and potential reserves.

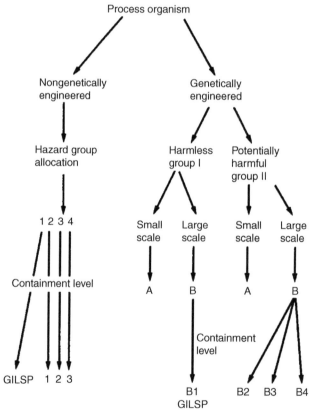

FIGURE 7.3 Categorization of a Process Microorganism and Designation of its Appropriate Level of Containment at Research or Industrial Sites Within the European Federation of Biotechnology

GILSP, good industrial large-scale practice.

6. The amounts or volumes of organisms used in the fermentation process.
7. The techniques or processes used.
8. Ease of prophylaxis and treatment.

It should always be borne in mind that the tenets of Good Laboratory Practice (GLP) and Good Manufacturing Practice (GMP) should always be adhered to.

Once the organism has been allocated to a hazard group, the appropriate containment requirements can be applied (Table 7.4). Hazard group 1 organisms used on a large scale only require Good Industrial Large Scale Practice (GILSP). Processes in this category need to be operated aseptically but no containment steps are necessary, including prevention of escape of organisms. If the organism is placed in Hazard group 4 the stringent requirements of level 3 will have to be met before the process can be operated. Details of hazard categories for a range of organisms can be obtained from Frommer et al. (1989).

Genetically engineered/modified organisms (GMOs) were classified as either harmless (Group I) or potentially harmful (Group II). The process was then classified as either small scale (A: less than 10 dm^3) or large scale (B: more than 10 dm^3) according to the guidelines which can be found in the Health and Safety Executive, 1993. Therefore, large scale processes fell into two categories, IB or IIB. IB processes require containment level B1 and are subject to GILSP, whereas IIB processes were further assessed to determine the most suitable containment level, ranging from B2 to B4 as outlined in Table 7.4. Levels B2 to B4 correspond to levels 1 to 3 for nongenetically engineered organisms.

In Oct. 2014, the UK's Health and Safety Executive issued the "The Genetically Modified Organisms (Contained Use) Regulations 2014," thus adopting EU regulations (HSE, 2014). The most important practical aspect of these regulations is that genetically modified and nongenetically modified organisms are considered under the same terms. Hence the A/B classification for small/large scale culture of GMOs has been removed. In addition these regulations clarified their interaction with the 2002 COSHH (Control of Substances Hazardous to Health) regulations. The 2014 regulations (Schedule 1) describe the classes of contained use:

Class 1. Contained use of no or negligible risk. Containment level 1 is appropriate to protect human health and the environment.
Class 2. Contained use of low risk. Containment level 2 is appropriate to protect human health and the environment.
Class 3. Contained use of moderate risk. Containment level 3 is appropriate to protect human health and the environment.
Class 4. Contained use of high risk. Containment level 4 is appropriate to protect human health and the environment.

Other hazard-assessment systems for classifying organisms have been introduced in many other countries. Production and research workers must abide by appropriate local official hazard lists and be aware of the latest requirements of GLP and GMP.

Table 7.4 Summary of Safety Precautions for Biotechnological Operations in the European Federation for Biotechnology (EFB) (Frommer et al., 1989)

Procedures	GILSP[a]	Containment Category		
		1	2	3
Written instructions and code of practice	+	+	+	+
Biosafety manual	−	+	+	+
Good occupational hygiene	+	+	+	+
Good Microbiological Techniques (GMT)	−	+	+	+
Biohazard sign	−	+	+	+
Restricted access	−	+	+	+
Accident reporting	+	+	+	+
Medical surveillance	−	+	+	+
Primary containment:				
Operation and equipment				
Work with viable microorganisms should take place in closed systems (CS), which minimize (m) or prevent (p) the release of cultivated microorganisms	−	m	P	P
Treatment of exhaust air or gas from CS	−	m	P	P
Sampling from CS	−	m	P	P
Addition of materials to CS, transfer of cultivated cells	−	m	P	P
Removal of material, products and effluents from CS	−	m	P	P
Penetration of CS by agitator shaft and measuring devices	−	m	P	P
Foam-out control	−	m	P	P
Secondary containment:				
Facilities				
Protective clothing appropriate to the risk category	+	+	+	+
Changing/washing facility	+	+	+	+
Disinfection facility	−	+		+
Emergency shower facility	−	−	+	+
Airlock and compulsory shower facilities	−	−	−	+
Effluents decontaminated	−	−	+	+
Controlled negative pressure	−	−	−	+
HEPA filters in air ducts	−	−	+	+
Tank for spilled fluids	−	−	−	+
Area hermetically sealable	−	−	−	+

m, Minimize release. The level of contamination of air, working surface and personnel shall not exceed the level found during microbiological work applying Good Microbiological Techniques.
p, Prevent release. No detectable contamination during work should be found in the air, working surfaces and personnel.
[a]Unless required for product quality, −, not required; +, required.

Problems can occur when different official bodies place the same organism in different hazard categories. In 1989, the European Federation for Biotechnology were aware of this problem with nonrecombinant microorganisms and produced a consensus list (Frommer et al., 1989).

Most microorganisms used in large scale industrial processes are in the lowest hazard group which only require GILSP and GLP/GMP, although some organisms used in bacterial and viral vaccine production and other processes are categorized in higher groups. There is an obvious incentive for industry to use an organism, which poses a low risk as this minimizes regulatory restrictions and reduces the need for expensive equipment and associated containment facilities (Schofield, 1992).

In this chapter, where appropriate, a distinction will be made between equipment, which is used to maintain aseptic conditions for lower containment levels, for example, GILSP and those modifications, which are needed to meet physical containment requirements in order to grow specific cultures, which have been classified in higher hazard groups. Details on design for containment have been given by East et al. (1984), Flickinger and Sansone (1984), Giorgio and Wu (1986), Walker, Narendranathan, Brown, Woodhouse, and Vranch (1987), Turner (1989), Hesselink, Kasteliem, and Logtenberg (1990), Kennedy, Boland, Jannsen, and Frude (1990), Janssen, Lovejoy, Simpson, and Kennedy (1990), Tubito (1991), Hambleton, Griffiths, Cameron, and Melling (1991), Leaver and Hambleton, 1992, Vranch (1992), Van Houten (1992), and Werner (1992).

The fermenter is only one aspect of containment assessment. To meet the standards of the specific level of containment, it will also be necessary to consider the procedures to be used, staff training, the facilities in the laboratory and factory, downstream processing, effluent treatment, work practice, maintenance, etc. (HSE, 2014). It will be necessary to ensure that all these aspects are of a sufficiently high standard to meet the levels of containment deemed necessary for a particular process by a government regulatory body. If these are met, then the process can be operated.

FERMENTER BODY CONSTRUCTION
CONSTRUCTION MATERIALS

In fermentations with strict aseptic requirements, it is important to select materials that can withstand repeated steam sterilization cycles. On a small scale (1–30 dm^3) it is possible to use glass and/or stainless steel. Glass is useful because it gives smooth surfaces, is nontoxic, corrosion proof, and it is usually easy to examine the interior of the vessel. Two basic types of fermenter are used:

1. A glass vessel with a round or flat bottom and a top flanged carrying plate (Fig. 7.4). The large glass containers originally used were borosilicate battery jars (Brown & Peterson, 1950). All vessels of this type have to be sterilized by

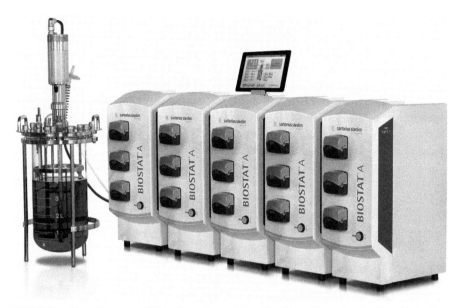

FIGURE 7.4 Glass Fermenter With a Top-Flanged Carrying Plate (Sartorius Stedim, UK, Ltd.)

autoclaving. Cowan and Thomas (1988) state that the largest practical diameter for glass fermenters is 60 cm.

2. A glass cylinder with stainless-steel top and bottom plates (Fig. 7.5). These fermenters may be sterilized in situ, but 30 cm diameter is the upper size limit to safely withstand working pressures (Solomons, 1969). Vessels with two stainless steel plates cost approximately 50% more than those with just a top plate.

At pilot and industrial scale (Figs. 7.6 and 7.7), when all fermenters are sterilized in situ, any materials used will have to be assessed on their ability to withstand pressure sterilization and corrosion and on their potential toxicity and cost. Walker and Holdsworth (1958), Solomons (1969), and Cowan and Thomas (1988) have discussed the suitability of various materials used in the construction of fermenters. Pilot-scale and industrial-scale vessels are normally constructed of stainless steel or at least have a stainless-steel cladding to limit corrosion. The American Iron and Steel Institute (AISI) states that steels containing less than 4% chromium are classified as steel alloys and those containing more than 4% are classified as stainless steels. Mild steel coated with glass or phenolic epoxy materials has occasionally been used (Gordon et al., 1947; Fortune, McCormick, Rhodehamel, & Stefaniak, 1950; Buelow & Johnson, 1952; Irving, 1968). Wood, plastic, and concrete have been used when contamination was not a problem in a process (Steel & Miller, 1970). Galvanic corrosion

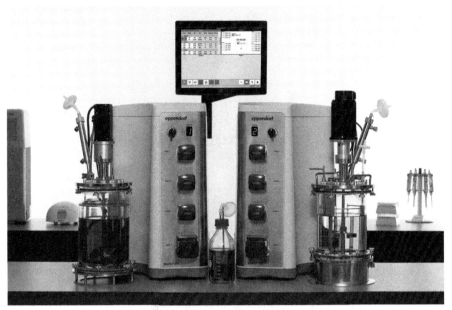

FIGURE 7.5 Three Glass Fermenters With Top and Bottom Plates (Eppendorf AG, Germany)

may also be an issue when dissimilar metals are in direct contact or in contact via an electrolyte solution.

Walker and Holdsworth (1958) stated that the extent of vessel corrosion varied considerably and did not appear to be entirely predictable. Although stainless steel is often quoted as the only satisfactory material, it has been reported that mild-steel vessels were very satisfactory after 12-years use for penicillin fermentations (Walker & Holdsworth, 1958) and mild steel clad with stainless steel has been used for at least 25 years for acetone-butanol production (Spivey, 1978). Pitting to a depth of 7 mm was found in a mild-steel fermenter after 7-years use for streptomycin production (Walker & Holdsworth, 1958).

The corrosion resistance of stainless steel is thought to depend on the existence of a thin hydrous oxide film on the surface of the metal. The composition of this film varies with different steel alloys and different manufacturing process treatments such as rolling, pickling, or heat treatment. The film is stabilized by chromium and is considered to be continuous, nonporous, insoluble, and self-healing. If damaged, the film will repair itself when exposed to air or an oxidizing agent (Cubberly et al., 1980).

The minimum amount of chromium needed to resist corrosion will depend on the corroding agent in a particular environment, such as acids, alkalis, gases, soil, salt, or fresh water. Increasing the chromium content enhances resistance to corrosion, but only grades of steel containing at least 10–13% chromium develop an effective film.

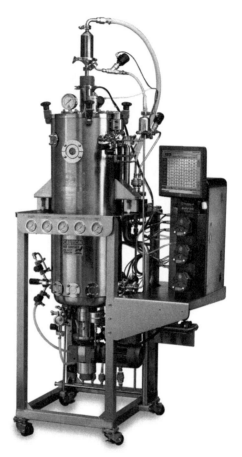

FIGURE 7.6 Stainless Steel Fully Automatic 50-dm³ Fermenter Sterilizable-In-Situ (Eppendorf AG, Germany)

The inclusion of nickel in high percent chromium steels enhances their resistance and improves their engineering properties. The presence of molybdenum improves the resistance of stainless steels to solutions of halogen salts and pitting by chloride ions in brine or sea water. Corrosion resistance can also be improved by tungsten, silicone, and other elements (Cubberly et al., 1980; Duurkoop, 1992).

AISI grade 316 steels, which contain 18% chromium, 10% nickel and 2–2.5% molybdenum are now commonly used in fermenter construction. Table 7.5 (Duurkoop, 1992) gives the classification systems used to identify stainless steels of AISI 316 grades in the United States of America and similar procedures used in Germany, France, the United Kingdom, Japan, and Sweden. In a citric acid fermentation where the pH may be 1–2 it will be necessary to use a stainless steel with 3–4% molybdenum (AISI grade 317) to prevent leaching of heavy metals

FIGURE 7.7 Stainless Steel Pilot Plant Fermenters (Sartorius Stedim, UK, Ltd.)

Table 7.5 Classification Systems Used for Stainless Steels of 316 Grades
(Duurkoop, 1992)

Germany		France	Great Britain	Japan	Sweden	USA	
Werkstoff Nummer	DIN	AFNOR	B.S.	JIS	SS	AISI	UNS
1.4401	X2 CrNiMo 17 12 2	Z 6CND 17.11	316 S 16 316 S 31	SUS 316	2347	316	S31600
1.4404	X2 CrNiMo 17 13 2 G-X CrNiMo 18 10	Z 2CND 18.13 Z 2 CND 17 12 Z 2 CND 19.10 M	316S 11 316 S 12	SUS 316L	2348	316L	S31603
1.4406	X2 CrNiMo 17 12 2	Z 2 CND 17.12 Az	316 S 61	SUS 316LN	—	316LN	S31653
1.4435	X2 CrNiMo 18 14 3	Z 2 CND 17.13	316 S 11 316 S 12	SCS 16 SUS 316L	2353	316L	S31603

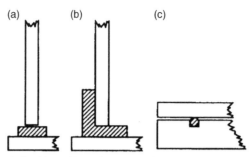

FIGURE 7.8 Joint Seals for Glass–Glass, Glass–Metal and Metal–Metal

(a) Gasket; (b) lip seal; (c) "O" ring in groove.

from the steel which would interfere with the fermentation. AISI Grade 304, which contains 18.5% chromium and 10% nickel, is now used extensively for brewing equipment.

The thickness of the construction material will increase with scale. At 300,000–400,000 dm^3 capacity, 7-mm plate may be used for the side of a vessel and 10-mm plate for the top and bottom, which should be hemispherical to withstand pressures.

At this stage, it is important to consider the ways in which a reliable aseptic seal is made between glass and glass, glass and metal, or metal and metal joints such as between a fermenter vessel and a detachable top or base plate. With glass and metal, a seal can be made with a compressible gasket, a lip seal or an "O" ring (Fig. 7.8). With metal to metal joints only an "O" ring is suitable. This is placed in a groove, machined in either of the end plate, the fermenter body or both. This seal ensures that a good liquid- and/or gas-tight joint is maintained in spite of the glass or metal expanding or contracting at different rates with changes in temperature during a sterilization cycle or an incubation cycle. Nitryl or butyl rubbers are normally used for these seals as they will withstand fermentation process conditions. The properties of different rubbers for seals are discussed by Buchter (1979) and Martini (1984). These rubber seals have a finite life and should be checked regularly for damage or perishing.

A single "O" ring seal is adequate for lower containment levels, a double "O" ring seal is required for higher containment levels and a double "O" ring seal with steam between the seals (steam tracing) is necessary for more stringent containment (Chapman, 1989; Hambleton et al., 1991). In the United States of America, however, simple seals are used at containment levels comparable with higher levels in the United Kingdom (Titchener-Hooker et al., 1993).

Titchener-Hooker et al. (1993) criticized Chapman's proposal to use two seals without a steam trace for a number of reasons including:

a. Double seals are more difficult to assemble correctly.
b. It is difficult to detect failure of one seal of a pair during operation or assembly.
c. Neither of the two seals can be tested independently.
d. Dead spaces between two seals must be considered for contamination.

Leaver (1994) and Titchener-Hooker et al. (1993) consider that correctly fitted single static seals can provide adequate containment for most processes and double static seals with a steam trace should be strictly limited to the small number of processes for which an extreme level of protection may be required.

TEMPERATURE CONTROL

Normally in the design and construction of a fermenter there must be adequate provision for temperature control (see also Chapter 8), which will affect the design of the vessel body. Heat will be produced by microbial activity and mechanical agitation and if the heat generated by these two processes will not be ideal for the particular manufacturing process then heat may have to be added to, or removed from, the system. On a laboratory scale, little heat is normally generated and extra heat has to be provided by placing the fermenter in a thermostatically controlled bath, or by the use of internal heating coils or a heating jacket through which water is circulated.

Once a certain size has been exceeded, the surface area covered by the heating jacket becomes too small to remove the heat produced by the fermentation. When this situation occurs internal coils must be used and cold water is circulated to achieve the correct temperature (Jackson, 1990). Different types of fermentation will influence the maximum size of vessel that can be used with jackets alone.

It is impossible to specify accurately the necessary cooling surface of a fermenter since the temperature of the cooling water, the sterilization process, the cultivation temperature, the type of microorganism, and the energy supplied by stirring can vary considerably in different processes. A cooling area of 50–70 m^2 may be taken as average for a 55,000 dm^3 fermenter and with a coolant temperature of 14°C the fermenter may be cooled from 120 to 30°C in 2.5–4 h without stirring. The consumption of cooling water in this size of vessel during a bacterial fermentation ranges from 500 to 2000 dm^3 h^{-1}, while fungi might need 2000–10,000 dm^3 h^{-1} (Müller & Kieslich, 1966), due to the lower optimum temperature for growth. Such fermentations thus require water recycling with an inline chiller to minimize water usage.

To make an accurate estimate of heating/cooling requirements for a specific process it is important to consider the contributing factors. An overall energy balance for a fermenter during normal operation can be written as:

$$Q_{met} + Q_{ag} + Q_{gas} = Q_{acc} + Q_{exch} + Q_{evap} + Q_{sen} \tag{7.1}$$

where Q_{met}, heat generation rate due to microbial metabolism; Q_{ag}, heat generation rate due to mechanical agitation; Q_{gas}, heat generation rate due to aeration power input; Q_{acc}, heat accumulation rate by the system; Q_{exch}, heat transfer rate to the surroundings and/or heat exchanger; Q_{evap}, heat loss rate by evaporation; Q_{sen}, rate of sensible enthalpy gain by the flow streams (exit—inlet).

Table 7.6 Representative Low and High Values of Calculated Heats (kcal dm^{-3} h^{-1}) of Fermentation for *Bacillus Subtilis* on Molasses at 37°C (Cooney et al., 1969)

	Heats of Fermentation					
	Q_{acc}	Q_{ag}	Q_{evap}	Q_{so}	Q_{exch}	Q_{met}
Low	3.81	3.32	0.023	0.005	0.61	1.12
High	11.3	3.31	0.045	0.010	0.65	8.65

This equation can be rearranged as:

$$Q_{exch} = Q_{met} + Q_{ag} - Q_{gas} - Q_{acc} - Q_{sen} - Q_{evap} \qquad (7.2)$$

Q_{exch} is the heat which will have to be removed by a cooling system.

Atkinson and Mavituna (1991b) have presented data to estimate Q_{met} for a range of substrates, methods to calculate power input for Q_{ag} and Q_{gas}, a formula to calculate the sensible heat loss for flow streams (Q_{sen}) and a method to calculate the heat loss due to evaporation (Q_{evap}). Cooney, Wang, and Mateles (1969) determined representative low and high heats of fermentation values for *Bacillus subtilis* grown on molasses at 37°C (Table 7.6). They concluded that Q_{evap} and Q_{sen} are small contributory factors and $Q_{acc} = 0$ in a steady-state system, Q_{evap} can also be eliminated by using a saturated air stream at the temperature of the broth.

When designing a large fermenter, the operating temperature and flow conditions will determine Q_{evap} and Q_{sen}, the choice of agitator, its speed, and the aeration rate will determine Q_{ag} and the sparger design and aeration rate will determine Q_{gas}.

The cooling requirements (jacket and/or pipes) to remove the excess heat from a fermenter may be determined by the following formula:

$$Q_{exch} = U \cdot A \cdot \Delta T \qquad (7.3)$$

where A, the heat transfer surface available, m^2, Q, the heat transferred, W, U, the overall heat transfer coefficient, W/m^{2}K, ΔT, the temperature difference between the heating or cooling agent and the mass itself, K.

The coefficient U represents the conductivity of the system and it depends on the vessel geometry, fluid properties, flow velocity, wall material, and thickness (Scragg, 1991). $1/U$ is the overall resistance to heat transfer (analogous to $1/K$ for gas-liquid transfer; Chapter 9). It is the reciprocal of the overall heat-transfer coefficient. It is defined as the sum of the individual resistances to heat transfer as heat passes from one fluid to another and can be expressed as:

$$\frac{1}{U} = \frac{1}{h_o} + \frac{1}{h_i} + \frac{1}{h_{of}} + \frac{1}{h_{if}} + \frac{1}{h_w} \qquad (7.4)$$

where h_0, outside film coefficient, W m^{-2} K^{-1}; h_i, inside film coefficient, W m^{-2} K^{-1}; h_{of}, outside fouling film coefficient, W m^{-2} K^{-1}; h_{if}, inside fouling film coefficient, W m^{-2} K^{-1} h_w, wall heat transfer coefficient = k/x, W m^{-2} K^{-1}; k, thermal conductivity of the wall, W m^{-1} K^{-1}; x, wall thickness, m.

There is more detailed discussion of U in Bailey and Ollis (1986), Atkinson and Mavituna (1991b), and Scragg (1991).

Atkinson and Mavituna (1991b) have given three methods to determine ΔT (the temperature driving force) depending on the operating circumstances. If one side of the wall is at a constant temperature, as is often the case in a stirred fermenter, and the coolant temperature rises in the direction of the coolant flow along a cooling coil, an arithmetic mean is appropriate:

$$\Delta T_{am} = \frac{(T_f - T_e) + (T_f - T_i)}{2} \tag{7.5}$$

$$= \frac{T_f - (T_e - T_i)}{2} \tag{7.6}$$

where T_f, the bulk liquid temperature in the vessel; T_e, the temperature of the coolant entering the system; T_i, the temperature of the coolant leaving the system.

If the fluids are in counter- or cocurrent flow and the temperature varies in both fluids then a log mean temperature difference is appropriate:

$$\Delta T_m = \Delta T_e - \Delta T_i / \ln(\Delta T_e / \Delta T_i) \tag{7.7}$$

where ΔT_e, the temperature of the coolant entering; ΔT_i, the temperature of the coolant leaving.

If the flow pattern is more complex than either of the two previous situations then the log mean temperature difference defined in Eq. (7.7) is multiplied by an appropriate dimensionless factor, which has been evaluated for a number of heat-exchanger systems by McAdams (1954).

Appropriate techniques have just been discussed to obtain values for Q_{exch}, U, and ΔT (or ΔT_{am} or ΔT_m). If Eq. (7.3) is now rearranged:

$$A = \frac{Q_{exch}}{U \Delta T} \tag{7.8}$$

Substituting in this equation makes it possible to calculate the heat-transfer surface necessary to obtain adequate temperature control.

AERATION AND AGITATION

Aeration and agitation will be considered in further detail in Chapter 9. In this chapter, it should be stated that the primary purpose of aeration is to provide microorganisms in submerged culture with sufficient oxygen for metabolic requirements, while agitation should ensure that a uniform suspension of microbial cells is achieved in a homogeneous nutrient medium. The type of aeration-agitation system used in a particular fermenter depends on the characteristics of the fermentation process under consideration. Although fine bubble aerators without mechanical agitation have the advantage of lower equipment and power costs, agitation may be dispensed with only when aeration provides sufficient agitation, that is, in processes where broths of low viscosity and low total solids are used (Arnold & Steel, 1958). Thus, mechanical agitation is usually required in fungal and actinomycete fermentations. Nonagitated (tower) fermentations are normally carried out in vessels of a height/diameter ratio of 5:1. In such vessels aeration is sufficient to produce high turbulence, but a tall column of liquid does require greater energy input in the production of the compressed air (Müller & Kieslich, 1966; Solomons, 1980).

The structural components of the fermenter involved in aeration and agitation are:

a. The agitator (impeller).
b. Stirrer glands and bearings.
c. Baffles.
d. The aeration system (sparger).

AGITATOR (IMPELLER)

The agitator is required to achieve a number of mixing objectives, for example, bulk fluid and gas-phase mixing, air dispersion, oxygen transfer, heat transfer, suspension of solid particles, and maintaining a uniform environment throughout the vessel contents. It should be possible to design a fermenter to achieve these conditions; this will require knowledge of the most appropriate agitator, air sparger, baffles, the best positions for nutrient feeds, acid or alkali for pH control, and antifoam addition. There will also be a need to specify agitator size and number, speed, and power input (see also Chapter 9).

Agitators, depending on their type will impart either axial flow (parallel to the impeller shaft) or radial flow (perpendicular to the impeller shaft). Agitators may be classified as disc turbines, vaned discs, open turbines of variable pitch, and propellers together with more recent designs, and are illustrated in Figs. 7.9–7.11. The disc turbine (or Rushton turbine) is the most widely use fermenter agitator and consists of a disc with a series of rectangular vanes set in a vertical plane around the circumference and the vaned disc has a series of rectangular vanes attached vertically to the underside. Air from the sparger hits the underside of the disc and is displaced toward the vanes where the air bubbles are broken up into smaller bubbles. The vanes of a variable pitch open turbine and the blades of a marine propeller are attached directly to a boss on the agitator shaft. In this case, the air bubbles do not initially hit any surface before dispersion by the vanes or blades.

Four other modern agitator developments, the Scaba 6SRGT, the Prochem Maxflo T, the Lightning A315, and the Ekato Intermig (Figs. 7.10 and 7.11), which are

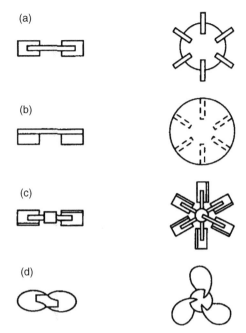

FIGURE 7.9 Types of Agitators

(a) Disc turbine; (b) vaned disc; (c) open turbine, variable pitch; (d) marine propeller (Solomons, 1969).

derived from open turbines, will also be discussed for energy conservation and use in high-viscosity broths.

Since the 1940s a Rushton disc turbine of one-third the fermenter diameter has been considered the optimum design for use in many fermentation processes. It had been established experimentally that the disc turbine was most suitable in a fermenter since it could break up a fast air stream without itself becoming flooded in air bubbles (Finn, 1954). This flooding condition is indicated when the bulk flow pattern in the vessel normally associated with the agitator design (radial with the Rushton turbine) is lost and replaced by a centrally flowing air-broth plume up the middle of the vessel with a liquid flow as an annulus (Nienow, Waroeskerken, Smith, & Konno, 1985; Nienow, 2014; see also Chapter 9). The propeller and the open turbine flood when V_s (superficial velocity, ie, volumetric flow rate/cross-sectional area of fermenter) exceeds 21 m h^{-1}, whereas the flat blade turbine can tolerate a V_s of up to 120 m h^{-1} before being flooded, when two sets are used on the same shaft. Besides being flooded at a lower V_s than the disc turbine, the propeller is also less efficient in breaking up a stream of air bubbles and the flow it produces is axial rather than radial (Cooper et al., 1944). The disc turbine was thought to be essential for forcing the sparged air into the agitator tip zone where bubble break up would occur.

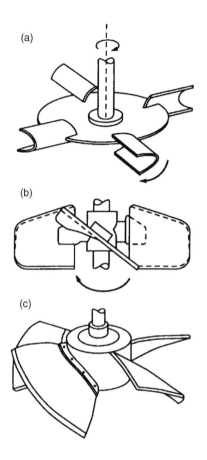

FIGURE 7.10

Diagrams of (a) Scaba agitator; (b) Lightnin' A315 agitator (four blades) and (c) Prochem Maxflo T agitator (four, five or six blades) (Nienow, 1990).

In other studies it has been shown that bubble break up occurs in the trailing vortices associated with all agitator types, which give rise to gas-filled cavities and provided the agitator speed is high enough, good gas dispersion will occur in low-viscosity broths (Smith, 1985). It has been also shown that similar oxygen-transfer efficiencies are obtained at the same power input per unit volume, regardless of the agitator type.

In high-viscosity broths, gas dispersion also occurs from gas filled cavities trailing behind the rotating blades, but this is not sufficient to ensure satisfactory bulk blending of all the vessel contents. When the cavities are of maximum size, the impeller appears to be rotating in a pocket of gas from which little actual dispersion occurs into the rest of the vessel (Nienow & Ulbrecht, 1985).

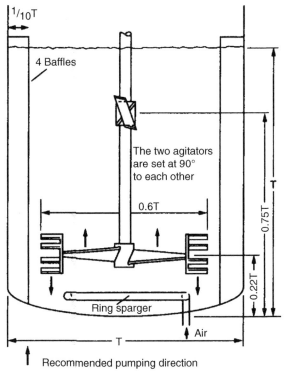

4 Baffles

$^1/_{10}$T

The two agitators
are set at 90°
to each other

0.6T

0.75T

T

0.22T

Ring sparger

Air

T

Recommended pumping direction

FIGURE 7.11 Arrangement for a Pair of Intermig Agitators

Relative dimensions are given as a proportion of the fermenter vessel diameter (T)
(Nienow, 1990).

Recently, a number of agitators have been developed to overcome problems associated with efficient bulk blending (mixing) and oxygen mass transfer in high-viscosity/non-Newtonian fermentation broths. One approach is to combine two classes of impeller—one for mixing, the other for oxygen transfer. The second approach is to use a novel impeller type, which may also be used in combination. This is discussed in detail in Chapter 9.

The Scaba 6SRGT agitator is one, which at a given power input, can handle a high air flow rate before flooding (Nienow, 1992). This radial-flow agitator is also better for bulk blending than a Rushton turbine, but does not give good top to bottom blending in a large fermenter which leads to lower concentrations of oxygen in broth away from the agitators and higher concentrations of nutrients, acid, or alkali or antifoam near to the feed points (Figs. 7.10 and 7.11). John, Bujalski, and Nienow (1998) have described the use of a 3-bladed Scaba axial-flow hydrofoil (the 3SHP1) in a draft tube in combination with an independently driven 6-bladed Rushton turbine. The top impeller, the Scaba impeller, installed to improve bulk flow (mixing), with the lower Rushton turbine used for gas dispersion. Short mixing times, compared with other dual-impeller configurations without a draft tube, were observed. In addition mixing time was controlled by altering the speed of the Scaba impeller.

Another is the Prochem Maxflo agitator. It consists of four, five, or six hydrofoil blades set at a critical angle on a central hollow hub (Gbewonyo, DiMasi, & Buckland 1986; Buckland et al., 1988). A high hydrodynamic thrust is created during rotation, increasing the downward pumping capacity of the blades. This design minimizes the drag forces associated with rotation of the agitator such that the energy losses due to drag are low. This leads to a low power number. The recommended agitator to vessel diameter ratio is greater than 0.4. When the agitator was used with a 800-dm^3 *Streptomyces* fermentation, the maximum power requirement at the most viscous stage was about 66% of that with Rushton turbines. The fall in power was also less in a 14,000-dm^3 fermentation. The oxygen-transfer efficiency was also significantly improved. It was thought that an improvement in bulk mixing was another contributing factor.

Intermig agitators (Fig. 7.11) made by Ekato of Germany are more complex in design. Two units are used (with agitator diameter to vessel diameter ratios of 0.6–0.7) instead of a single Rushton turbine because their power number is so low (Nienow, 1992). A large-diameter air sparger is used to optimize air dispersion (see later section on spargers). The loss in power is less than when aerating with a Rushton turbine. Air dispersion starts from the air cavities which form on the wing tips of the agitator blades. In spite of the downward pumping direction of the wings, the cavities extend horizontally from the back of the agitator blades, reducing the effectiveness of top to bottom mixing in a vessel.

Cooke, Middelton, and Bush (1988) and Nienow (1990, 1992) give comparisons of performance of Rushton turbines with some of the newer designs. Pinelli et al. (2003) have reported the use of Chemineer (Dayton, USA) BT-6 hydrofoil axial impellers, concluding that they are well suited for gas dispersion in fermenters where a broad range of gas flow rates is required. Albaek, Gernaey, & Stocks, 2008) reported improved power consumption profiles in the Hayward Tyler B2 (formerly the APV-B2 or simply B2) axial flow hydrofoil impeller (an alternative to the conventional Rushton turbine) successfully retrofitted to a 550 dm^{-3} pilot- scale fermenter.

These new turbine designs make it possible to replace Rushton turbines by larger low power agitators, which do not lose as much power when aerated, are able to handle higher air volumes without flooding and give better bulk blending and heat/mass transfer characteristics in more viscous media (Nienow, 1990). However, there can be mechanical problems which are mostly of a vibrational nature.

Good mixing and aeration in high viscosity broths may also be achieved by a dual impeller combination, where the lower impeller acts as the gas disperser and the upper impeller acts primarily as a device for aiding circulation of vessel contents. This has been discussed by Nienow and Ulbrecht (1985) and in Chapter 9.

Steel and Maxon (1966) tested a multirod mixing impeller. In a 15,000-dm^3 vessel, the same novobiocin yield and oxygen availability rate were obtained at about half of the power required by a standard turbine-stirred fermenter, but this type of impeller does not appear to have come into general use.

Yang et al. (2012) reported the application of a novel impeller configuration compared to a conventional impeller configuration in cephalosporin C production using *Acremonium chrysogenum* in fermenters up to 12 m^3 working volume. The conventional impeller configuration used four Rushton turbines while the novel configuration utilized two Lightnin A315 axial flow impellers (Fig. 7.9), one Smith turbine ½ pipe radial flow impeller and one Chemineer radial flow impeller with parabolic blades. With the novel impeller configuration, cephalosporin C production was increased by 10%, power consumption decreased by 25% and K_La was enhanced by 15%. Xie et al. (2014) have reported improved mixing and mass transfer using a range of alternative triple-impeller configurations compared with Rushton turbines. Alternatives tested where hollow blade turbines and hydrofoil turbines with axial up-flow and down-flow. Axial up-flow hydrofoil turbines have also shown improved glucoamylase production, lower power consumption, better homogeneity of vessel contents, and improved mass transfer from *Aspergillus niger* compared with Rushton turbines (Tang et al., 2015).

Computational fluid dynamics (CFD) is a branch of fluid mechanics that uses computer based numerical analysis and algorithms to simulate, analyze and solve problems in fluid flow. CFD was originally developed in the 1930s but it has seen rapid growth and application in more recent years due to the widespread availability of computers with increased computing power. There are many examples of CFD use within the fermentation industries. For example, Hristov, Mann, Lossev, and Vlaev (2004) reported the use of CFD in the analysis of mass transfer and mixing in a vessel fitted with triple novel NS impeller types (which had both up-pumping and down-pumping blades) in comparison with Rushton turbines. The key outcome of this analysis was that the novel impellers achieved higher levels of dissolved oxygen that the Rushton turbines under the same conditions. Moilanen, Laakkonen, and Aittamaa (2006) have described the use of CFD to develop a detailed model for aerated fermentations. Gas-liquid mass transfer, bioreaction kinetics, and non-Newtonian behavior, were combined with CFD. The model developed can be used for both fermenter design and scale up.

STIRRER GLANDS AND BEARINGS

The satisfactory sealing of the stirrer shaft assembly has been one of the most difficult problems to overcome in the construction of fermentation equipment which can be operated aseptically for long periods. A number of different designs have been developed to obtain aseptic seals. The stirrer shaft can enter the vessel from the top, side (Richards, 1968) or bottom of the vessel. Top entry is most commonly used, but bottom entry may be advantageous if more space is needed on the top plate for entry ports, and the shorter shaft permits higher stirrer speeds to be used by eliminating the problem of the shaft whipping at high speeds. Originally, bottom entry stirrers were considered undesirable as the bearings would be submerged. Chain, Paladino,

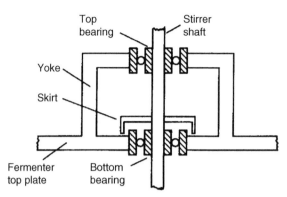

FIGURE 7.12 A Simple Stirrer Seal Based on a Description Given by Rivett et al. (1950)

Ugolini, Callow, and Van Der Sluis (1952) successfully operated vessels of this type, and they have since been used by many other workers. Mechanical seals can be used for bottom entry provided that they are routinely maintained and replaced at recommended intervals (Leaver & Hambleton, 1992).

One of the earliest stirrer seals described was that used by Rivett, Johnson, and Peterson (1950) in a laboratory fermenter (Fig. 7.12). A porous bronze bearing for a 13-mm shaft was fitted in the center of the fermenter top and another in a yoke directly above it. The bearings were pressed into steel housings, which screwed into position in the yoke and the fermenter top. The lower bearing and housing were covered with a skirt-like shield having a 6.5 mm overhang which rotated with the shaft and prevented airborne contaminants from settling on the bearing and working their way through it into the fermenter.

Four basic types of seal assembly have been used: the stuffing box (packed-gland seal), the simple bush seal, the mechanical seal, and the magnetic drive. Most modern fermenter stirrer mechanisms now incorporate mechanical seals instead of stuffing boxes and packed glands. Mechanical seals are more expensive, but are more durable and less likely to be an entry point for contaminants or a leakage point for organisms or products which should be contained. Magnetic drives, which are also quite expensive, have been used in animal cell culture vessels.

Stuffing box (Packed-gland seal)

The stuffing box (Fig. 7.13) has been described by Chain, Paladino, Ugolino, Callow, and Van Der Sluis (1954). The shaft is sealed by several layers of packing rings of asbestos (in the past) or cotton yarn, pressed against the shaft by a gland follower. At high stirrer speeds the packing wears quickly and excessive pressure may be needed to ensure tightness of fit. The packing may be difficult to sterilize properly because of unsatisfactory heat penetration and it is necessary to check and replace the packing rings regularly. Parker (1958) described a split stuffing box with a lantern ring. Steam under pressure was continually fed into it. Chain et al. (1954) used two stuffing boxes

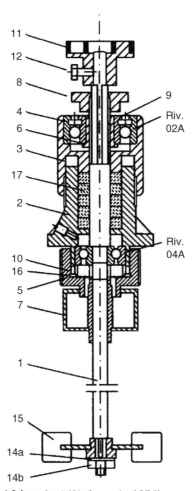

FIGURE 7.13 Packed-Gland Stirrer Seal (Chain et al., 1954)

Components: *1*, agitator shaft; *2*, stuffing box; *3*, upper cap; *4*, lock ring; *5*, lower cap; *6*, chuck; *7*, greasecup; *8*, lock ring; *9*, lock nut; *10*, distance ring; *11*, half coupling; *12*, half coupling; *14a*, washer; *14b*, nut; *15*, impeller; *16*, shim; *17*, packing rings.

on the agitator shaft with a space in between kept filled with steam. Although, at one time, stuffing box-bearings were commonly used in large-scale vessels, operational problems, particularly contamination, have led to their replacement by mechanical seal bearings for most processes. However, these seals are sufficient for the requirements of lower containment requirements (Werner, 1992).

Mechanical seal

The mechanical seal assembly (Figs. 7.14 and 7.15) is now commonly used in both small and large fermenters. The seal is composed of two parts, one part is stationary

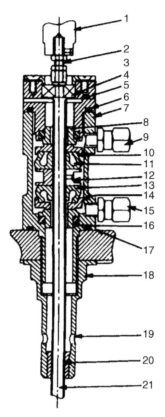

FIGURE 7.14 Mechanical Seal Assembly (Elsworth et al., 1958)

Components: *1*, flexible coupling; *2*, stirrer shaft; *3*, bearing housing; *4*, ball journal fit on mating parts; *5*, two slots for gland leaks, only one shown; *6*, "O"-ring seal; *7*, seal body; *8*, stationary counter-face sealed to body with square-section gasket; *9*, exit port for condensate, fitted with unequal stud coupling; *10*, rotating counter-face; *11*, bellows; *12*, shaft muff; *13*, as 11; *14*, as 10; *15*, entry port for condensate, as 9; *16*, as 8; *17*, as 6; *18*, shaft bush support; *19*, leak holes; *20*, Ferobestos bush; *21*, ground shaft.

in the bearing housing, the other rotates on the shaft, and the two components are pressed together by springs or expanding bellows. The two meeting surfaces have to be precision machined, the moving surface normally consists of a carbon-faced unit while the stationary unit is of stellite-faced stainless steel. Steam condensate can be used to lubricate the seals during operation and serve as a containment barrier. Single mechanical seals are used with a steam barrier in fermenters for primary containment at lower levels (Werner, 1992), whereas double mechanical seals are typically used in vessels with the outer seal as a backup for the inner seal for primary containment at higher levels (Werner, 1992; Leaver & Hambleton, 1992; Fig. 7.15). With

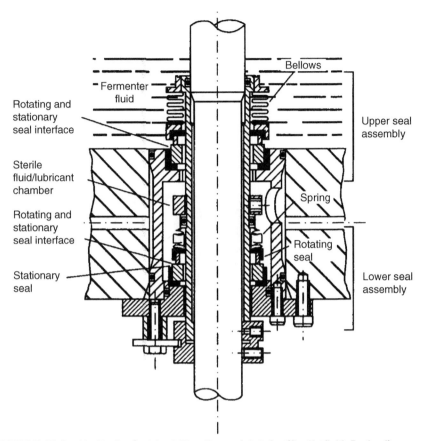

FIGURE 7.15 Double Mechanical Seal (New Brunswick Scientific, Hatfield, England)

such higher hazards the condensate is piped to a kill tank. Monitoring of the steam condensate flowing out of the seal is an effective way for checking for seal failure. Disinfectants are alternatives for flushing the seals (Werner, 1992).

Magnetic drives

The problems of providing a satisfactory seal when the impeller shaft passes through the top or bottom plate of the fermenter may be solved by the use of a magnetic drive in which the impeller shaft does not pierce the vessel (Cameron & Godfrey, 1969). A magnetic drive (Fig. 7.16) consists of two magnets: one driving and one driven. The driving magnet is held in bearings in a housing on the outside of the head plate and connected to a drive shaft. The internal driven magnet is placed on one end of the impeller shaft and held in bearings in a suitable housing on the inner surface of the headplate. When multiple ceramic magnets have been used, it has been possible to transmit power across a gap of 16 mm. Using this drive, a Newtonian fluid can be stirred in baffled vessels of up to 300-dm^3 capacity at speeds of 300–2000 rpm. It

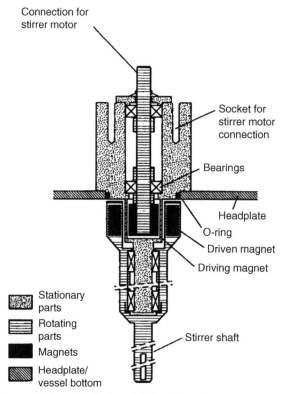

Connection for
stirrer motor

Socket for
stirrer motor
connection

Bearings

Headplate

O-ring

Driven magnet

Driving magnet

Stationary
parts

Rotating
parts

Magnets

Headplate/
vessel bottom

Stirrer shaft

FIGURE 7.16 Diagram of Magnetically Coupled Top Stirrer Assembly (Applikon Dependable Instruments, Schiedam, Netherlands)

would be necessary to establish if adequate power could be transmitted between magnets to stir viscous mould broths or when wanting high oxygen transfer rates in bacterial cultures. Walker et al. (1987) have described the development of a magnetic drive suitable for microbial fermentations up to 1500 dm^3 which could be used when higher containment levels are specified. The stirring mechanism is ideal for animal cell culture to minimize the chances of potential contamination.

BAFFLES

Four baffles are normally incorporated into agitated vessels of all sizes to prevent a vortex and to improve aeration and agitation efficiency and to prevent vortexing. In vessels over 3-dm^3 six or eight baffles may be used (Scragg, 1991). Baffles are metal strips roughly one-tenth of the vessel diameter and attached radially to the wall (Figs. 7.1 and 7.2 and Tables 7.2 and 7.3). The agitation effect is only slightly increased with wider baffles, but drops sharply with narrower baffles (Winkler, 1990). Walker and Holdsworth (1958) recommended that baffles should be installed so that a gap existed between them and the vessel wall, so that there was a scouring action around and behind the baffles, thus minimizing microbial growth on the baffles and the fermenter walls. Extra cooling coils may be attached to baffles to improve the cooling capacity of a fermenter without unduly affecting the geometry.

AERATION SYSTEM (SPARGER)

A sparger may be defined as a device for introducing air into the liquid in a fermenter. Three basic types of sparger have been used and may be described as the porous sparger, the orifice sparger (a perforated pipe), and the nozzle sparger (an open or partially closed pipe). A combined sparger-agitator may be used in laboratory fermenters (Fig. 7.17) and is discussed briefly in a later section.

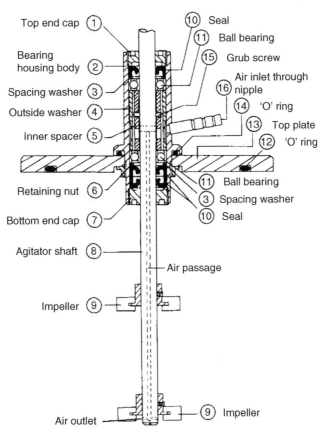

FIGURE 7.17 Diagram of Bearing Housing With Combined Agitator-Sparger (Inceltech L.H. Reading, England)

Porous sparger

The porous sparger of sintered glass, ceramics or metal, has been used primarily on a laboratory scale in nonagitated vessels. The bubble size produced from such spargers is always 10–100 times larger than the pore size of the aerator block (Finn, 1954). The throughput of air is low because of the pressure drop across the sparger and there is also the problem of the fine holes becoming blocked by growth of the microbial culture.

Orifice sparger

Various arrangements of perforated pipes have been tried in different types of fermentation vessel with or without impellers. In small stirred fermenters, the perforated pipes were arranged below the impeller in the form of crosses or rings (ring sparger), approximately three-quarters of the impeller diameter. In most designs the air holes were drilled on the under surfaces of the tubes making up the ring or cross. Walker and Holdsworth (1958) commented that in production vessels, sparger holes should be at least 6 mm (1/4 inch) diameter because of the tendency of smaller holes to block and to minimize the pressure drop.

In low viscosity fermentations sparged at 1 vvm (volume of air^{-1} volume of medium^{-1} minute^{-1}) with a power input of 1 W kg^{-1}, Nienow et al. (1988) found that the power often falls to below 50% of its unaerated value when using a single Rushton disc turbine which is one-third the diameter of the vessel and a ring sparger smaller than the diameter of the agitator. If the ring sparger was placed close to the disc turbine and its diameter was 1.2 times that of the disc turbine, a number of benefits could be obtained (Nienow et al., 1988). A 50% higher aeration rate could be obtained before flooding occurred, the power drawn was 75% of the unaerated value, and a higher K_La could be obtained at the same agitator speed and aeration rate. These advantages were lost at viscosities of about 100 m Pas.

Orifice spargers without agitation have been used to a limited extent in yeast manufacture (Thaysen, 1945), effluent treatment (Abson & Todhunter, 1967) and later in the production of single-cell protein in the air-lift fermenter, which are discussed in a later section of this chapter (Taylor & Senior, 1978; Smith, 1985).

Nozzle sparger

Most modern mechanically stirred fermenter designs from laboratory to industrial scale have a single open or partially closed pipe as a sparger to provide the stream of air bubbles. Ideally the pipe should be positioned centrally below the impeller and as far away as possible from it to ensure that the impeller is not flooded by the air stream (Finn, 1954). The single-nozzle sparger causes a lower pressure loss than any other sparger and normally does not get blocked.

Combined sparger–agitator

On a small scale (1 dm^3), Herbert, Phipps, and Tempest (1965) developed the combined sparger-agitator design, introducing the air via a hollow agitator shaft and emitting it through holes drilled in the disc between the blades and connected to the base of the main shaft. The design gives good aeration in a baffled vessel when the agitator is operated at a range of rpm (Fig. 7.17).

ACHIEVEMENT AND MAINTENANCE OF ASEPTIC CONDITIONS

Once the design problems of aeration and agitation have been solved, it is essential that the design meets the requirements of the degree of asepsis and containment demanded by the particular process being considered. It will be necessary to be able to sterilize, and keep sterile, a fermenter and its contents throughout a complete growth cycle. There may be also a need to protect workers and the environment from exposure to hazardous microorganisms or animal cells (Van Houten, 1992). As mentioned earlier, the containment requirements depend on the size of the fermentation vessel.

The following operations may have to be performed according to certain specifications to achieve and maintain aseptic conditions and containment during a fermentation:

1. Sterilization of the fermenter.
2. Sterilization of the air supply and the exhaust gas.
3. Aeration and agitation.
4. The addition of inoculum, nutrients, and other supplements.
5. Sampling.
6. Foam control.
7. Monitoring and control of various parameters.

On a small scale, below 10 dm^3, the biohazard risk can be controlled by a combination of containment cabinets and work practices (Van Houten, 1992). When the volume of culture exceeds 10 dm^3, GILSP is required for those nonpathogenic nontoxigenic agents, which have an extended history of large scale use. For this category there should be prevention of contamination of the product, control of aerosols, and minimization of the release of microorganisms during sampling, addition of material, transfer of cells, and removal of materials, products, and effluents. It should be appreciated that the majority of fermentations fall into this category.

At higher containment levels, the following points need to be considered when designing a fermenter or other vessel, so that it can operate as a contained system (Tubito, 1991):

1. All vessels containing live organisms should be suitable for steam sterilization and have sterile vent filters. This is discussed in Chapter 5.
2. Exhaust gases from vessels should pass through sterile filters. This is discussed in Chapter 5.
3. Seals on flange joints should be fitted with a single "O"-ring at the lower levels of containment. Flange joints on vessels for high hazard levels need double "O"-rings or double "O"-rings plus a steam barrier. This has been discussed in an earlier section of this chapter.

4. Appropriate seals should be provided for entry ports for sensor probes, inoculum, sampling, medium addition, acid, alkali, and antifoam. This will be discussed later in this chapter.

5. Rotating shafts into a closed system should be sealed with a double acting mechanical seal with steam or condensate between the seals. This has been discussed in an earlier section in this chapter.

6. During operation a steam barrier should be maintained in all fixed piping leading to the "contained" vessels.

7. Provision of appropriate pressure relief facilities will be discussed later in this chapter.

Further details for containment are given by Giorgio and Wu (1986), Hesselink et al. (1990), Kennedy et al. (1990), Janssen et al. (1990), Hambleton et al. (1991), Tubito (1991), Leaver and Hambleton (1992), Van Houten (1992), Vranch (1992), and Werner (1992).

STERILIZATION OF THE FERMENTER

The fermenter should be so designed that it may be steam sterilized under pressure. The medium may be sterilized in the vessel or separately, and subsequently added aseptically. If the medium is sterilized in situ its temperature should be raised prior to the injection of live steam to prevent the formation of large amounts of condensate. This may be achieved by steam being introduced into the fermenter coils or jacket. As every point of entry to and exit from the fermenter is a potential source of contamination, steam should be introduced through all the entry and exit points except the air outlet from which steam should be allowed to leave.

All pipes should be constructed as simply as possible and slope toward drainage points to make sure that steam reaches all parts of the equipment and is not excluded by siphons or pockets of condensate or mash. Each drainage point in the pipework should be fitted with a steam trap. This will be described in the section on valves and steam traps. Parker (1958), Chain et al. (1954), and Müller and Kieslich (1966) and others have all stressed the need to eliminate fine fissures or gaps such as flange seals which might be filled with nutrient solutions and microorganisms. Hambleton et al. (1991) described a high specification pilot scale fermenter with surfaces free of crevices greater than 0.05-mm depth, which is needed if the vessel was to be used for animal cells in suspension culture or on microcarriers. For long-term aseptic operation welded joints should be used wherever possible, even though sections may have to be cut out and rewelded during maintenance and repair (Smith, 1980).

STERILIZATION OF THE AIR SUPPLY

Sterile air will be required in very large volumes in many aerobic fermentation processes. Although there are a number of ways of sterilizing air, only two have found permanent application. These are heat and filtration. Heat is generally too costly for full-scale operation (see also Chapter 5).

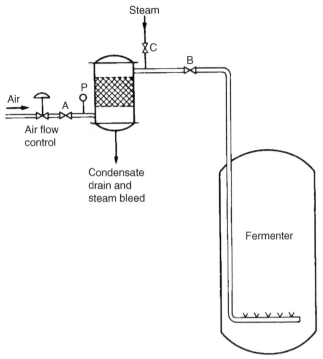

FIGURE 7.18 An Arrangement of Packed Air Filter and Fermenter (Richards, 1968)

Historically, glass wool, glass fiber or mineral slag wool have been used as filter material, but currently most fermenters are fitted with cartridge-type filters as discussed in Chapter 5. However, before the filter may be used it, too, must be sterilized in association with the fermenter. Two procedures are commonly followed depending on the construction of the filter unit.

Fig. 7.18 shows the simple unit described by Richards (1968). During sterilization the main nonsterile air-inlet valve A is shut, and initially the sterile air valve B is closed. Steam is applied at valve C and air is purged downward through the filter to a bleed valve at the base. When the steam is issuing freely through the bleed valve, the valve B is opened to allow steam to pass into the fermenter as well as the filter. It is essential to adjust the bleed valve to ensure that the correct sterilization pressure is maintained in the fermenter and filter for the remainder of the sterilization cycle.

An alternative approach is to use a steam-jacketed air filter (Fig. 7.19). At the beginning of a sterilization cycle the valve A will be closed and steam passed through valves B and C, and bled out of D. Simultaneously steam will be passed into the steam jacket through valve F and out of G. When steam is issuing freely from valve D, valve F, may be opened and steam circulated into the fermenter. The bleed valve D will have to be adjusted to ensure that the correct pressure is maintained. Once the sterilization cycle is complete, valves B and E are closed and A is opened to allow air to

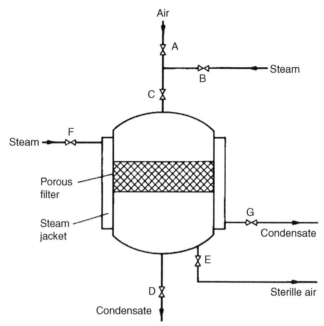

FIGURE 7.19 Design for a Simple Steam-Jacketed Air-Filter

pass through the heated filter and out of valve D to dry the filter. Finally the steam supply to the steam jacket is stopped. Valve D is closed and valve E opened, thus introducing sterile air into the fermenter to achieve a slight positive pressure in the vessel.

STERILIZATION OF THE EXHAUST GAS FROM A FERMENTER

Sterilization of the exhaust gas can be achieved by 0.2-μm filters on the outlet pipe (Fig. 5.21). Under normal operation aerosol formation may occur in the fermenter, and moisture and solid matter may then plug the filter. To ensure satisfactory operation a cyclone separator (for solids) and a coalescer (for liquids) would be included upstream of two filters in series. The filters should be checked regularly to ensure that no viable cells are escaping. A test procedure to ensure integrity has been described by Hesselink et al. (1990).

ADDITION OF INOCULUM, NUTRIENTS, AND OTHER SUPPLEMENTS

To prevent contamination when operating a fermenter requiring GILSP, it is essential that both the addition vessel and the fermenter should be maintained at a positive pressure and that the addition port is equipped with a steam supply.

At low hazard levels, the addition of inoculum, nutrients, etc. must be carried out in such a way that release of microorganisms is restricted. This should be done by aseptic piercing of membranes or connections with steam locks. At higher hazard levels, no microorganisms must be released during inoculation or other additions. In order to meet these stringent requirements, all connections must be screwed or clamped and all pipelines must be steam sterilizable (Werner, 1992).

Further details of the aseptic inoculation of laboratory, pilot-plant, and production fermenters are described in Chapter 6.

SAMPLING

The sampling points fitted to larger fermenters also illustrate the principles for maintaining sterility. A sterile barrier must be maintained between the fermenter contents and the exterior when the sample port is not being used and it must be sterilizable after use. A simple design (Fig. 7.20) is described by Parker (1958). In normal operation valves A, B, and C are closed and a barrier is formed by submerging the end of the sampling port in 40% formalin or a suitable substitute. A sample is obtained by removing the container of formalin and closing valve A. Valves B and C are then opened until the piping has been sterilized by steam. Valves B and C are then partially closed to allow a slow stream of steam and condensate out of the sampling port. Valve A is then opened slightly to cool the piping. The broth is discarded. Valve C is then closed and a sample is collected. Valve A is then closed and the piping is resterilized and left in the out of use arrangement.

An alternative arrangement for a sampling port is illustrated in Fig. 7.21 where the sterile barrier between the sample port and the exterior is a condensed flow of steam between valves C and D, valves A and B being closed. Valve D is connected to a steam trap to avoid condensate accumulation. To sterilize valve B prior to sampling, valve C is partially closed, valve D completely closed and valve B partially opened to allow a slow stream of steam and condensate out of the sampling port. Valve A is

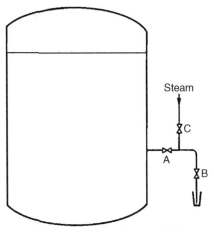

FIGURE 7.20 Simple Design for a Sampling Port (Parker, 1958)

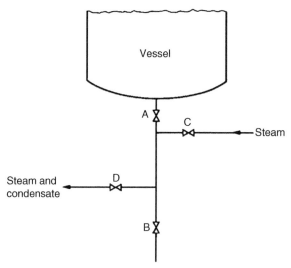

FIGURE 7.21 An Alternative Simple Sample Port

opened briefly to cool the pipe and the broth is discarded. Valve C is then closed and a sample is collected. Valve A is then closed and the piping is resterilized. In between collecting samples valves C and D are left partially open.

A more modern sterilizable system, as illustrated in Fig. 7.22, has been described by Werner (1992). In the closed position steam enters through the entry point 6 and passes through the sampling hole 11 and into the sampling hole 2. When sampling, the handle is pushed so that the hole in the sampling tube is in the fermenter and the vessel contents may be sampled. When the sampler is closed the unit can be resterilized.

Marshall, Webb, Matthews, and Dean (1990) recognized that sampling requires good mechanical design and good operator practice to ensure sterility. Complex valve sequencing can lead to operator error. The automatic MX-3 rotary valve (Fig. 7.23) was developed by Marshall Biotechnology Ltd and is now marketed by New Brunswick Scientific Ltd. In this device, broth is continuously recycled to and from the fermenter and up to 12 samples can be taken automatically in sample bottles according to a preprogrammed sequence. The storage of the bottles in an integral refrigeration block reduces spoilage until they can be removed for analysis. This sampler can be used for GILSP.

In all the sampling devices described aforementioned, steam is used merely to maintain aseptic conditions. If the process organism has to be contained then the sample vessel must be designed accordingly. At lower containment levels, removal of broth should be carried out in such a way that release of microorganisms is minimized. Such a system (Fig. 7.24) is described by Janssen et al. (1990). The sampling bottle assembly is attached to the fermenter by a double-end shut-off quickconnect B and air is vented from the vessel during sampling by a suitable membrane filter. The

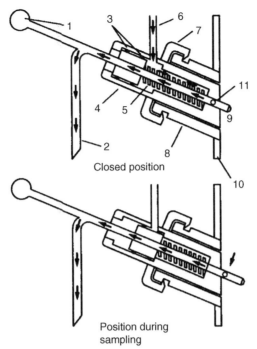

FIGURE 7.22 Sterilizable Sampling System for Category 1 (Werner, 1992)

Key: *1*, handle for opening and closing; *2*, piping for sample or condensate, respectively; *3*, O-ring; *4*, housing; *5*, spring; *6*, steam inlet; *7*, union nut; *8*, welding socket; *9*, product; *10*, wall of the bioreactor; *11*, port of sampling tube.

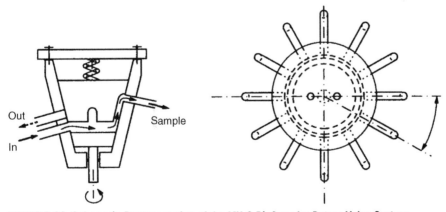

FIGURE 7.23 Schematic Representation of the MX-3 BioSampler Rotary Valve System

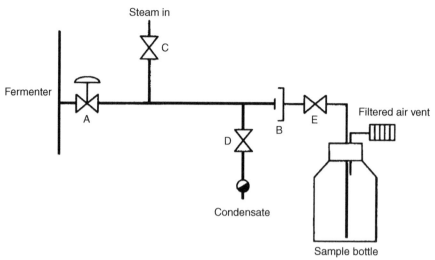

FIGURE 7.24 Sample System for Level 1 (Janssen et al., 1990)

pipe work is steamed for 15 min using 1.3 bar steam with the valve E on the sample bottle closed. With valves C and D closed and A and B open, a sample can be taken, then the pipe is resterilized before removing the sample bottle.

At higher containment levels, sampling should be done in a closed system (Figs. 7.25 and 7.26), and all piping coming into contact with a sample must be sterilizable with steam. The sampling vessel should be closed during transport and samples should be examined under containment conditions corresponding to those specified for the process (Werner, 1992).

FEED PORTS

Additions of nutrients and acid/alkali to small fermenters are normally made via silicone tubes which are autoclaved separately and pumped by a peristaltic pump after aseptic connection. In large fermenter units, the nutrient reservoirs and associated piping are usually an integrated part which can be sterilized with the vessel. However, there may also be ports which are used intermittently (Fig. 7.27). These can be sterilized in situ with steam after connection has been completed and before any additions are made.

SENSOR PROBES

Double "O" ring seals have been used for many years to provide an aseptic seal for glass electrodes in stainless steel housings in fermenters. This system is suitable for lower level hazards, provided that release of microorganisms is minimized and there are adequate disinfection procedures for dealing with leakages (Werner, 1992).

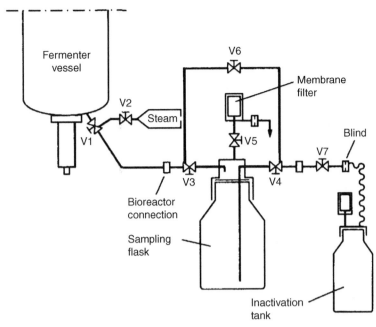

Valve	Sterilization sampling flask	Sampling	Sterilization after sampling of sampling flask	Removal of sampling flask
V1	Closed	Open	Closed	Closed
V2	Open	Closed	Open	Closed
V3	Open	Open	Closed	Closed
V4	Open	Closed	Closed	Closed
V5	Open	Open	Open	Closed
V6	Closed	Closed	Open	Closed
V7	Open	Open	Open	Closed

FIGURE 7.25 Containment Sampling Unit for Chapter 2 (Werner, 1992)

At higher containment levels, probes are fitted with triple "O"-ring seals (Hambleton et al., 1991). Although double "O"-ring seals with steam tracing have been described, they are not normally considered to be feasible (Leaver & Hambleton, 1992). The use of preinserted backup probes is recommended as a means for dealing with probe failure rather than using a retractable electrode housing during a fermentation cycle because of the danger of leakage of broth (Hambleton et al., 1991).

FOAM CONTROL

In any fermentation it is important to minimize foaming. When foaming becomes excessive, there is a danger that filters become wet resulting in contamination. There

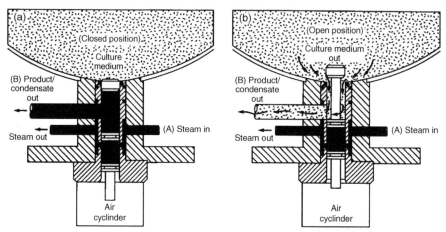

FIGURE 7.26 Resterilizable Harvest-Sampling Valve for Level 2 Containment (New Brunswick, Hatfield, England)

This spool-type valve is connected to the bottom of the fermenter vessel in the closed position (a), pressurized steam circulates throughout the entire valve body and through the product condensate line (B) via a steam inlet line (A). Aseptic withdrawal of samples is achieved with the valve in the open position (b). To prevent possible contamination when the plunger is raised, steam is circulated to the lower valve area. Action of the plunger is controlled by an air cylinder.

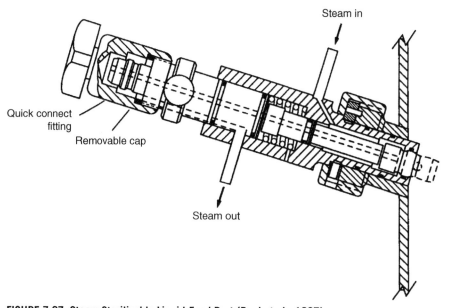

FIGURE 7.27 Steam Sterilizable Liquid-Feed Port (Becl et al., 1987)

is also the possibility that siphoning will develop, leading to the loss of all or part of the contents of the fermenter. Methods for foam control are considered in Chapter 8 and antifoams are discussed in Chapter 4. In certain circumstances antifoams may cause problems with aeration or downstream processing. Foam breakers, which break down foam by an impact mechanism created by some type of rotating mechanism inside the fermenter are manufactured by Bioengineering AG (Switzerland), Chemap AG (Switzerland), Electrolux (Sweden), Frings (Germany), and New Brunswick Scientific Ltd (United States of America). Mechanical foam breakers used in Acetators for vinegar production are described later in this chapter. Domnick Hunter (UK) manufacture the "Turbosep," in which foam is directed over stationary turbine blades in a separator and the liquid fraction is returned to the fermenter.

MONITORING AND CONTROL OF VARIOUS PARAMETERS

These factors are considered in detail in Chapter 8.

VALVES AND STEAM TRAPS

Valves attached to fermenters and ancillary equipment are used for controlling the flow of liquids and gases in a variety of ways. The valves may be:

1. Simple ON/OFF valves which are either fully open or fully closed.
2. Valves which provide coarse control of flow rates.
3. Valves which may be adjusted very precisely so that flow rates may be accurately controlled.
4. Safety valves which are constructed in such a way that liquids or gases will flow in only one direction (nonreturn valves).

When making the decision as to which valves to use in the design and construction of a fermenter, it is essential to consider the following points:

1. Will the valve serve its chosen purpose? Is it suitable for aseptic operation or contained processes, and of the correct dimensions? Is the pressure drop across the valve tolerable?
2. Will the valve withstand the rigors of the process? The materials used to construct the valve should be suited to the process. It is also important to know whether corrosive liquids are used or synthesized during the process. The maximum operating temperature and pressure of the process should be known.
3. Are there welds or flanges in the valve?
4. Is the valve one which can be operated automatically?
5. The cost and availability of suitable valves.
6. Can the valve be used for containment purposes?

A wide range of valves is available, but not all of them are suitable for use in fermenter construction (Solomons, 1969; Kemplay, 1980).

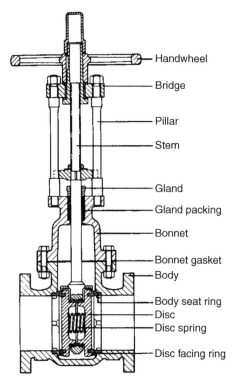

Handwheel
Bridge
Pillar
Stem
Gland
Gland packing
Bonnet
Bonnet gasket
Body
Body seat ring
Disc
Disc spring
Disc facing ring

FIGURE 7.28 Sectional View of a Two-Piece Gate Valve (British Valve Manufacturers Association, 1972)

The valves described in this section open and close by (1) raising or lowering the blocking unit with a screw thread (rising stem), (2) a drilled sphere or plug, or a disc rotating in between two bearings, and (3) a rubber diaphragm or tube which is pinched.

GATE VALVES

In this valve (Fig. 7.28), a sliding disc is moved in or out of the flow path by turning the stem of the valve. It is suitable for general purposes on a steam or a water line for use when fully open or fully closed and therefore should not be used for regulating flow. The flow path is such that the pressure drop is minimal, but unfortunately it is not suitable for aseptic conditions as solids can pack in the groove where the gate slides, and there may be leakage round the stem of the valve which is sealed by a simple stuffing box. This means that the nut around the stem and the packing must be checked regularly.

GLOBE VALVES

In this valve (Fig. 7.29), a horizontal disc or plug is raised or lowered in its seating to control the rate of flow. This type of valve is very commonly used for regulating the flow

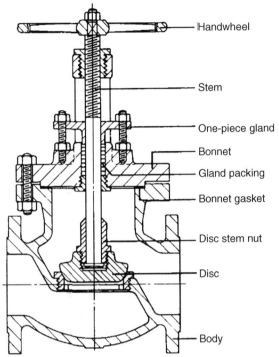

Handwheel

Stem

One-piece gland

Bonnet

Gland packing

Bonnet gasket

Disc stem nut

Disc

Body

FIGURE 7.29 Globe Valve With Outside Screw and Conventional Disc (Kemplay, 1980)

of water or steam since it may be adjusted rapidly. It is not suitable for aseptic operation because of potential leakage round the valve stem, which is similar in design to that of the gate valve. There is a high-pressure drop across the valve because of the flow path.

In both the gate and globe valves it is possible to incorporate a flexible metallic membrane around the stem of the valve, to replace the standard packing. This modified type of valve can be operated aseptically, but is bigger and more expensive. Valves with nonrising stems have been used, but they are still potential sources of contamination.

PISTON VALVES

The piston valve (Fig. 7.30) is similar to a globe valve except that flow is controlled by a piston passing between two packing rings. This design has proved in practice to be very efficient under aseptic operation. It is important to sterilize them partly open so that steam can reach as far as possible into the valve body. There may be blockage problems with mycelial cultures. The pressure drop is similar to that of a globe valve.

NEEDLE VALVES

The needle valve (Fig. 7.31) is similar to the globe valve, except that the disc is replaced by a tapered plug or needle fitting into a tapered valve seat. The valve can be

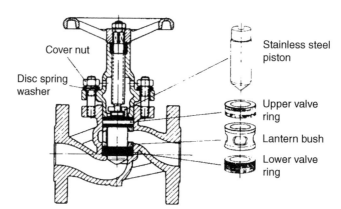

Cover nut

Disc spring washer

Stainless steel piston

Upper valve ring

Lantern bush

Lower valve ring

FIGURE 7.30 Piston Valve (Kemplay, 1980)

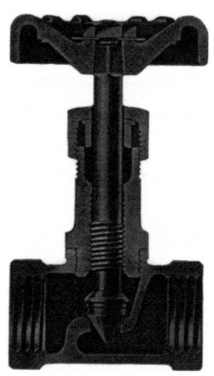

FIGURE 7.31 Needle Valve for Accurate Control of Flow Rate (Kemplay, 1980)

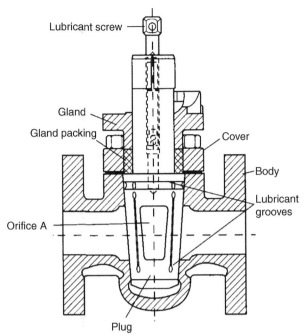

FIGURE 7.32 Sectional View of Lubricated Taper Plug Valve (British Valve Manufacturers Association, 1972)

used to give fine control of steam or liquid flow. Accurate control of flow is possible because of the variable orifice formed between the tapered plug and the tapered seat. The aseptic applications are very limited.

PLUG VALVES

In this valve, there is a parallel or tapered plug sitting in a housing through which an orifice, A, has been machined. The valve shown in Fig. 7.32 is in the closed position. When the plug is turned through 90 degrees, the valve is fully open and the flow path is determined by the cross-sectional area of A, which may not be as large as that of the pipeline. This type of valve has a tendency to leak or seize up, but the use of lubricants and/or sealants may overcome these problems. If suitable packing sleeves are incorporated into the valve it will be suitable for use in a steam line as it is quick to operate, has protected seals, a minimal pressure drop, and a positive closure. It can also provide good flow control.

BALL VALVES

This valve (Fig. 7.33) has been developed from the plug valve. The valve element is a stainless-steel ball through which an orifice is machined. The ball is sealed between

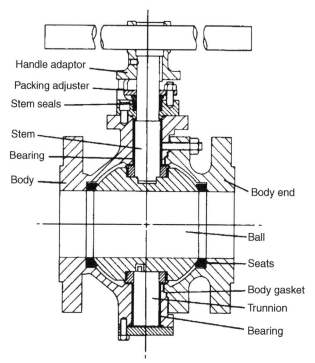

FIGURE 7.33 Sectional View of End-Entry Ball Valve (British Valve Manufacturers Association, 1972)

two wiping surfaces which wipe the surface and prevent deposition of matter at this point. The orifice in the ball can be of the same diameter as the pipeline, giving an excellent flow path. The valve is suitable for aseptic operation, can handle mycelial broths and can be operated under high temperatures and pressures. The pressure and temperature range is normally limited by the PTFE seat and stem seals.

BUTTERFLY VALVES

The butterfly valve (Fig. 7.34) consists of a disc which rotates about a shaft in a housing. The disc closes against a seal to stop the flow of liquid. This type of valve is normally used in large diameter pipes operating under low pressure where absolute closure is not essential. It is not suitable for aseptic operation.

PINCH VALVES

In the pinch valve (Fig. 7.35), a flexible sleeve is closed by a pair of pinch bars or some other mechanism, which can be operated by compressed air remotely or automatically. The flow rate can be controlled from 10% to 95% of rated flow capacity (Pikulik, 1976). The valve is suitable for aseptic operation with fermentation broths,

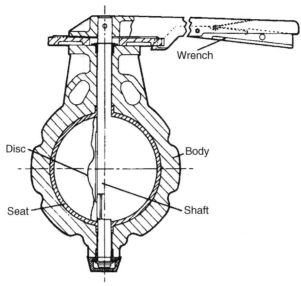

FIGURE 7.34 Sectional View of Wafer-Pattern Butterfly Valve (British Valve Manufacturers Association, 1972)

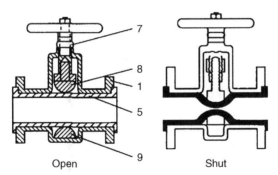

FIGURE 7.35 Sectional View of Pinch Valve in Open and Shut Position

1, body; *5*, flexible tube; *7*, spindle; *8*, top pinch bar; *9*, lower pinch bar (Kemplay, 1980).

even when mycelial, as there are no dead spaces in the valve structure, and the closing mechanism is isolated from the contents of the piping. Obviously, the sleeve of rubber, neoprene, etc., must be checked regularly for signs of wear.

DIAPHRAGM VALVES

Like the pinch valve, the diaphragm valve (Fig. 7.36) makes use of a flexible closure, with or without a weir. It may also be fitted with a quick action lever. This

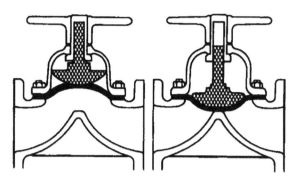

FIGURE 7.36 Sectional Views of Weir-Type Diaphragm Valves in Open and Closed Positions (Thielsch, 1967)

valve is very suitable for aseptic operation provided that the diaphragm is of a material which will withstand repeated sterilization. The valve can be used for ON/OFF, flow regulation, and for steam services within pressure limits. Diaphragm failure, which is often due to excessive handling, is the primary fault of the valve. Ethylene propylene diene modified (EPDM) is now the preferred material. Hambleton et al. (1991) consider that a diaphragm valve with a steam seal on the "clean" side (APV Ltd, Crawley, UK) is a potentially safer valve. However their widescale use would make a fermentation plant much more complex and expensive. Steam barriers on valves have been used by ICI plc for the "Pruteen" air-lift fermenter (Smith, 1985; Sharp, 1989).

MOST SUITABLE VALVE

Among these group of valves which have just been described, globe and butterfly valves are most commonly used for ON/OFF applications, gate valves for crude flow control, needle valves for accurate flow control and ball, pinch or diaphragm valves for all sterile uses.

Ball and diaphragm valves are now the most widely used designs in fermenter and other biotechnology equipment (Leaver & Hambleton, 1992). Although ball valves are more robust, they contain crevices which make sterilization more difficult.

CHECK VALVES

The purpose of the check valve is to prevent accidental reversal of flow of liquid or gas in a pipe due to breakdown in some part of the equipment. There are three basic types of valve: swing check, lift check, and combined stop and check with a number of variants. The swing check valve (Fig. 7.37) is most commonly used in fermenter designs. The functional part is a hinged disc, which closes against a seat ring when the intended direction of flow is accidentally reversed.

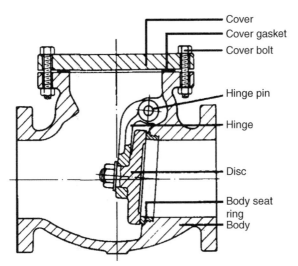

Cover
Cover gasket
Cover bolt

Hinge pin

Hinge

Disc

Body seat
ring
Body

FIGURE 7.37 Sectional View of Swing Check Valve (Kemplay, 1980)

PRESSURE-CONTROL VALVES

When planning the design of a plant for a specific process, the water, steam, and air should be at different, but specified pressures and flow rates in different parts of the equipment. For this reason, it is essential to control pressures precisely and this can be done using reduction or retaining valves.

PRESSURE-REDUCTION VALVES

Pressure-reduction valves are incorporated into pipelines when it is necessary to reduce from a higher to a lower pressure, and be able to maintain the lower pressure in the downstream side within defined limits irrespective of changes in the inlet pressure or changes in demand for gas, steam, or water.

PRESSURE-RETAINING VALVES

A pressure-retaining valve will maintain pressure in the pipeline upstream of itself, and the valve is designed to open with a rising upstream pressure. It is constructed with a reverse action of the pressure-reducing valve.

SAFETY VALVES

Safety valves must be incorporated into every air or steam line and vessel, which is subjected to pressure to ensure that the pressure will never exceed the safe upper limit recommended by the manufacturer or a code of practice and to satisfy government legislation and insurance companies. They must also be of the correct type and

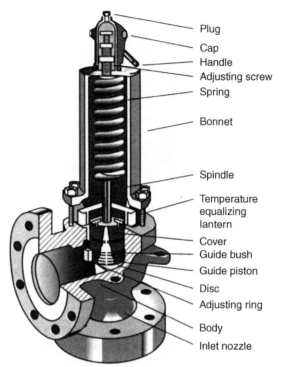

Plug
Cap
Handle
Adjusting screw
Spring

Bonnet

Spindle

Temperature
equalizing
lantern

Cover
Guide bush
Guide piston
Disc
Adjusting ring

Body
Inlet nozzle

FIGURE 7.38 Section of a Safety Valve (Kemplay, 1980)

size to suit the operating conditions and be in sufficient numbers to protect the plant. The reliability of such valves is crucial.

In the simplest valves (Fig. 7.38), a spindle is lifted from its seating against the pressure of gas, steam or liquid. Once the pressure falls below the value set by the tensioned spring, the spindle should return to its original position. However, the valve may stick open if waste material lodges on the valve seat and plant operators may interfere with the release pressure setting. Bursting/rupture discs may be used as an alternative and are of a more hygienic design than some valves (Leaver & Hambleton, 1992).

For GILSP operational categories, venting the escaping gas through the factory roof in emergencies would be satisfactory. At higher containment levels it would be necessary to treat the escaping gas. A kill tank with a HEPA venting filter is one possible solution, but not fully satisfactory.

It is also important to ensure that vessels do not collapse when vacuums occur. Vacuum release valves have been designed to cope with cold rinsing of a hot vessel or absorption of CO_2 when cleaning with caustic solutions (Maule, 1986). Shuttlewood (1984) has discussed design pressure and vacuum pressure in relation to vessel thickness requirements for design codes for American and British standards of safety.

STEAM TRAPS

In all steam lines, it is essential to remove any steam condensate which accumulates in the piping to ensure optimum process conditions. This may be achieved by incorporating steam traps, which will collect and remove automatically any condensate at appropriate points in steam lines. A steam trap has two elements. One is a valve and seat assembly which provides an opening, which may be of variable size, to ensure effective removal of any condensate. This opening may operate on an open/close basis so that the average discharge rate matches the steam condensation rate or the dimensions of this opening may be varied continually to provide a continuous flow of condensate The second element is a device which will open or close the valve by measuring some parameter of the condensate reaching it to determine whether it should be discharged (Armer, 1991).

The steam trap may be designed to operate automatically on the basis of:

1. The density of the fluid by using a float (ball or bucket) which will float in water or sink in steam (Fig. 7.39).
2. By measuring the temperature of the fluid, closing the valve at or near steam temperature and opening it when the fluid has cooled to a temperature, say 8°C below the saturated steam temperature of the steam. The sensing element is often a stainless-steel capsule filled with a water-alcohol mixture, which expands when steam is present and presses a ball into a valve seat or contracts when cooler condensate is present and lifts the ball from the seating. These are used in thermostatic and balanced pressure steam traps (Figs. 7.40 and 7.41).
3. By measuring the kinetic effects of the fluid in motion. At a given pressure drop, low-density steam will move at a much greater velocity than higher density condensate. The conversion of pressure energy into kinetic energy can be used to control the degree of opening of a valve. This type of steam trap is not used very widely.

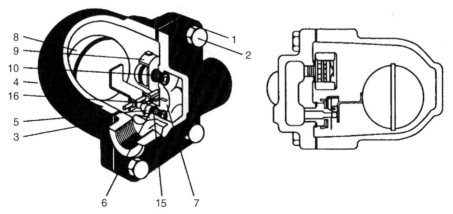

FIGURE 7.39 Ball Float Steam Trap With Thermostatic Air Vent (Spirax/Sarco, Cheltenham, England)

5, Valve seat; *8*, ball float and lever; *9*, air vent.

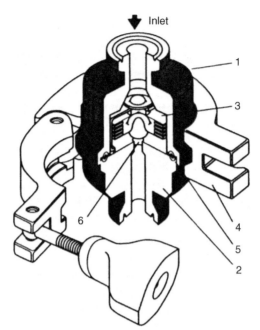

FIGURE 7.40 Balanced Pressure Steam Trap (Spirax/Sarco, Cheltenham, England)

1, Body; *2*, end connection; *3*, element; *4*, tri-clover clamp; *5*, Tri-Clover joint gasket; *6*, ball seat.

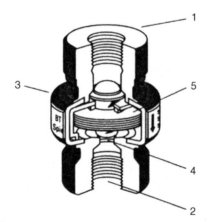

FIGURE 7.41 Stainless Steel Thermostatic Steel Trap (Spirax/Sarco, Cheltenham, England)

1, Inlet; *2*, outlet; *3*, body; *4*, ball seat; *5*, element.

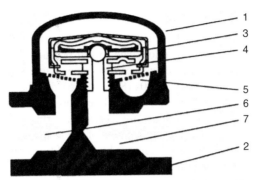

FIGURE 7.42 SBP 30 Sealed Balanced Pressure Thermostatic Steam Trap (Spirax/Sarco, Cheltenham, England)

1, Cover; *2*, body; *3*, element/capsule; *4*, ball seal assembly; *5*, strainer screen; *6*, inlet; *7*, outlet.

Armer (1991) thinks that balanced pressure thermostatic traps are the most suitable for autoclaves and sterilizers. Some can operate close to steam-saturation temperature with little back up of condensate. Sarco Ltd (Cheltenham, UK) make a hermetically sealed balanced pressure steam trap (Fig. 7.42), which has been used with a high containment level pilot-plant fermenter (Hambleton et al., 1991).

Any steam condensate may be (1) returned to the boiler, (2) used in a steam condensate line, (3) vented to waste, and (4) piped to a kill tank in a containment location.

COMPLETE LOSS OF CONTENTS FROM A FERMENTER

Leaver and Hambleton (1992) have suggested building a wall or bund around a fermenter which would retain the entire contents where they to leak out. The escaping fluid would then be pumped into a suitable vessel for sterilization. The floor should be covered with an impervious epoxy based surface and coved up the walls (Kennedy et al., 1990).

TESTING NEW FERMENTERS

When a new fermenter (10 dm^3 or larger) has finally been assembled in a factory it must be hydraulically pressure tested and checked by independent inspectors using nationally approved test procedures. A vessel which has passed the approved tests is given a certificate recognized by insurance companies, which will allow it to be operated in a laboratory or factory. If any subsequent modifications are made to a vessel in this size range, it must be retested and certificated before it can be legally used. Thus a number of fermenter manufacturers cut extra "O" ring grooves and steam traces in head plates or ports to satisfy higher containment-level needs so that the vessels will not have to be modified subsequently and to avoid extra pressure checks.

SCALE-UP OF FERMENTERS

Scale-up is the increase in scale or size of a fermenter, for example, from laboratory scale to pilot scale or from pilot scale to full production scale. Difficulties in the scale-up of fermentation processes were first recognized as an issue for industrial penicillin production in the early 1940s (Schmidt, 2005). The difficulties surrounding scale-up revolve around the fact that different process parameters react differently when fermenter size and volume are increased. Perhaps the major factor to be considered on scale-up is that the surface area of the fermenter broth increases by a squared function while its volume increases by a cubed function and so air flow rate, oxygen and carbon dioxide transfer, foaming, and agitation rate will all be affected. Scale-up (and scale-down) in relation to aeration and agitation is discussed in detail in Chapter 9. In essence compromises must be made when scaling up fermentation as power input, for example, would increase to an unfeasibly large, and economically unviable, extent if other parameters (eg, impeller tip speed) remain constant. Thus it is impossible to maintain the same level of mixing on scale-up as it leads to less favorable mixing behavior (Schmidt, 2005).

Papers by Junker (2004) and Junker et al. (2009) provide much useful guidance on the key parameters in the scale-up of a range of industrial scale fermentations (bacterial and fungal). Garcia-Ochoa and Gomez (2009) have reviewed in detail the relationship between oxygen transfer rate (OTR) and scale-up in microbial fermentations and González-Sáiz, Garrido-Vidal, and Pizarro (2009) describe the application of modeling and simulation to the scale-up of aerated industrial scale vinegar production.

OTHER FERMENTATION VESSELS

Some of the other forms of fermentation vessels will now be considered. These vessels have more limited applications and have been developed for specific purposes or closely related processes. Some are historical developments, such as the earliest forms of packed tower, others were being developed in parallel with the standard mechanically stirred fermenter during the 1940s, while other approaches are more recent including the development and widespread introduction of disposable or single-use fermenters. Aspects of this topic have been discussed by Prokop and Votruba (1976), Katinger (1977), Hamer (1979), Levi, Shennan, and Ebbon (1979), Solomons (1980), Schugerl (1982, 1985), Sittig (1982), Winkler (1990), Atkinson and Mavituna (1991a), Oosterhuis, Hudson, D'Avino, Zijlstra, and Amanullah (2011), Zhong (2011), and Doran (2013).

WALDHOF-TYPE FERMENTER

The investigations on yeast growth in sulfite waste liquor in Germany, Japan, and the United States of America led to the development of the Waldhof-type fermenter (Inskeep, Wiley, Holderby, & Hughes, 1951; Watanabe, 1976). Inskeep et al. (1951)

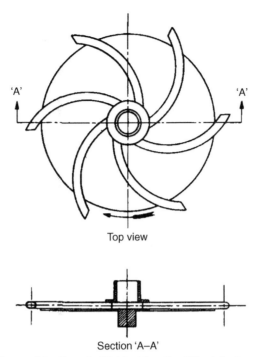

'A'   'A'

Top view

Section 'A–A'

FIGURE 7.43 Top View and Section of a Waldhof Aeration Wheel (Inskeep et al., 1951)

have given a description of a production vessel based on a modification of the original design of Zellstofffabrik Waldhof. The fermenter was of carbon steel, clad in stainless steel, 7.9 m in diameter and 4.3-m high with a center draught tube 1.2 m in diameter. A draught tube was held by tie rods attached to the fermenter walls. The operating volume was 225,000 dm^3 of emulsion (broth and air) or 100,000 dm^3 of broth without air. Nonsterile air was introduced into the fermenter through a rotating pin-wheel type of aerator, composed of open-ended tubes rotating at 300 rpm (Fig. 7.43). The broth passed down the draught tube from the outer compartment and reduced the foaming.

ACETATORS AND CAVITATORS

Fundamental studies by Hromatka and Ebner (1949) on vinegar production showed that if *Acetobacter* cells were to remain active in a stirred aerated fermenter, the distribution of air had to be almost perfect within the entire contents of the vessel. They solved the full-scale problem by the use of a self-aspirating rotor (Ebner, Pohl, & Enekel, 1967). In this design (Fig. 7.44), the turning rotor sucked in air and broth and dispersed the mixture through the rotating stator (d). The aerator also worked without a compressor and was self-priming.

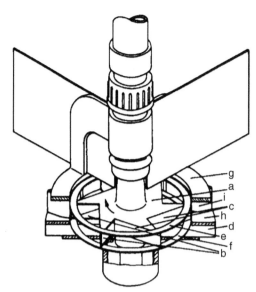

FIGURE 7.44 Axonometrie View of the Self-Priming Aerator Used With the Frings Generator (Ebner et al., 1967)

The turbine is designed as a hollow body (a) with openings, which are arranged radially and open against the direction of rotation (b). The openings are shielded by vertical sheets (c). The turbine sucks liquid from above and below and mixes it with air sucked in through the openings. The suspension is thrown through the stator (d) toward the circumference of the tank. An upper and lower ring on the turbine (e,f) helps to direct and regulate the air–liquid suspension. The stator (d) consists of an upper and lower ring (g,h) which are connected by vertical sheets (i) inclined at about 30 degree toward the radius.

Vinegar fermentations often foam and chemical antifoams were not thought feasible because they would decrease aeration efficiency (Chapter 9) and additives were not desirable in vinegar. A mechanical defoamer therefore had to be incorporated into the vessel and as foam builds up it is forced into a chamber in which a rotor turns at 1000–1450 rpm. The centrifugal force breaks the foam and separates it into gas and liquid. The liquid is pumped back into the fermenter and the gas escapes by a venting mechanism. Descriptions of the design and various sizes of model have been given by Ebner et al. (1967). Fermenters of this design are manufactured by Heindrich Frings, Bonn, Germany. An illustration of the basic components is given in Fig. 7.45. In 1981, 440 acetators were in operation all over the world with a total production of $767 \times 10^6 \, dm^3 \, year^{-1}$. The major vinegar producers were the United States of America ($152 \times 10^6 \, dm^3$), France ($90 \times 10^6 \, dm^3$), and Japan ($46 \times 10^6 \, dm^3$), while the remainder was produced by over 50 countries (Ebner & Follmann, 1983).

Chemap AG of Switzerland manufacture the Vinegator. A self-aspirating stirrer and a central suction tube aerates a good recirculation of liquid. Additional air is provided by a compressor. Foam is broken down by a mechanical defoamer (Ebner & Follman, 1983).

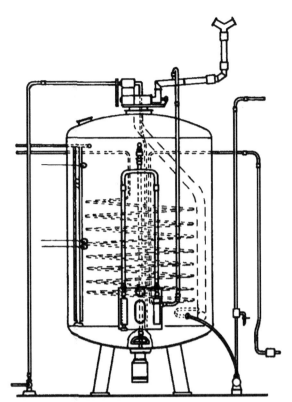

FIGURE 7.45 Diagram of a Section Through a Frings Generator Fermenter Used for the Manufacture of Vinegar

The fermenter, which can be used semicontinuously or continuously, employs vortex stirring (Greenshields, 1978).

At least three other vinegar fermenters are no longer manufactured. The Bourgeois process was sold in Europe between 1955 and 1980 and the Fardon process between 1960 and 1975. The Yeomans cavitator was sold in the United States of America between 1959 and 1970 (Cohee & Steffen, 1959; Mayer, 1961; Ebner & Follman, 1983). The fermenter had an agitator of different design, but similar operating principles to the acetator. Uniform distribution of air bubbles was obtained by means of the circulation pattern created by the centrally located draught (draft) tube. The agitator withdrew liquid from the draught tube and pushed liquid into the main part of the vessel. The outer level rose and overflow occurred back into the top of the draught tube.

TOWER OR BUBBLE COLUMN FERMENTER

Tower or bubble column fermenters (Fig. 7.46) are structurally very simple and are not mechanically agitated. Aeration and agitation are achieved by gas sparging and

Air out

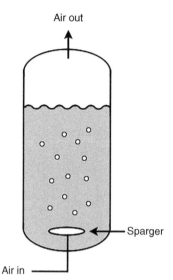

Sparger

Air in

FIGURE 7.46 A Bubble Column Fermenter

consequently they consume less energy than stirred reactors and because of their simple design and lack of moving parts are a low cost option (Doran, 2013). It is difficult to formulate a single definition, which encompasses all the types of tower fermenter. Their main common feature appears to be their height:diameter ratio or aspect ratio. Such a definition has been given by Greenshields and Smith (1971) who described a tower fermenter as an elongated nonmechanically stirred fermenter with an aspect ratio of at least 6:1 for the tubular section or 10:1 overall, through which there is a unidirectional flow of gases. However an aspect ratio of 3:1 is common in baker's yeast production. Several different types of tower fermenter exist and these will be examined in broad groups based on their design.

The simplest types of fermenter are those that consist of a tube which is air sparged at the base (bubble columns). This type of fermenter was first described for citric acid production on a laboratory scale (Snell & Schweiger, 1949). This batch fermenter was in the form of a glass column having a height:diameter ratio of 16:1 with a volume of 3 dm^3. Humid sterile air was supplied through a sinter at the base. Steel, Lentz, and Martin (1955) reported an increase in scale to 36 dm^3 for a fermenter of this type. Pfizer Ltd has always used nonagitated tower vessels for a range of mycelial fermentation processes including citric acid and tetracyclines (Solomons, 1980; Carrington, Dixon, Harrop, & Macaloney, 1992). Pfizer Ltd have since sold their citric acid interests to Arthur Daniels Midland who are operating such vessels up to 23 m high (Burnett, 1993). Volumes of between 200 and 950 m^3 have been reported elsewhere (Rohr, Kubicek, & Kominek, 1983).

In 1965 the brewing industry began to use tower fermenters, which were more complex in design and could be operated continuously. Hall and Howard (1965) described small-scale fermenters that consisted of water jacketed tubes of various dimensions which were inclined at angles of 9–90 degrees to the horizontal. Air and mash were passed in at the base and effluent beer was removed at the top.

A vertical-tower beer fermenter design (Chapter 2) was patented by Shore, Royston, and Watson (1964). Perforated plates were positioned at intervals in the tower to maintain maximum yeast production. The settling zone which could be of various designs, was to provide a zone free of rising gas so that the cells could settle and return to the main body of the tower and the clear beer could be removed. This design must be considered as an intermediate between single- and multistage systems. Towers of up to 20,000 dm^3 capacity and capable of producing up to 90,000 dm^3 day^{-1} have been installed. Greenshields and Smith (1971) commented that it was difficult to predict the upper operating limits for these fermenters. Experiments with particular yeast strains in pilot-size towers were essential to establish optimum full-scale operating conditions.

The next group of tower fermenters are the multistage systems, first described by Owen (1948) and Victorero (1948) for brewing beer, although these systems were not used on an industrial scale. Later work reported using these systems includes continuous cultures of *E. coli* (Kitai, Tone, & Ozaki, 1969), bakers' yeast (Prokop, Erikson, Fernandez, & Humphrey, 1969), and activated sludge (Lee, Erikson, & Fan, 1971; Besik, 1973). The fermenters used by all these workers were basically similar. Each consisted of a column forming the body of the vessel and a number of perforated plates, which were positioned across the fermenter, dividing it into compartments. Approximately 10% of the horizontal plate area was perforated. The possibility of introducing media into individual stages independently was discussed by Lee et al. (1971). Besik (1973) described a down-flow tower in which substrate was fed in at the top and overflowed through down spouts to the next section while air was supplied from the base. Schugerl, Lucke, and Oels (1977) have written a comprehensive general review.

CYLINDRO-CONICAL VESSELS

The use of cylindro-conical vessels in the brewing of lager was first proposed by Nathan (1930), but his ideas were not adopted for the brewing of lagers and beers until the 1960s (Hoggan, 1977). Breweries throughout the world have now adopted this method of brewing. The vessel (Fig. 7.47) consists of a stainless-steel vertical tube with a hemispherical top and a conical base with an included angle of approximately 70 degrees (Boulton, 1991).

Aspect ratios are usually 3:1 and fermenter heights are 10–20 m. Operating volumes are chosen to suit the individual brewery requirements, but are often 150,000–200,000 dm^3. Vessels are not normally agitated unless a particularly flocculant yeast is used, but small impellers may be used to ensure homogeneity when filling with wort (Boulton, 1991). In the vessel, the wort is pitched (inoculated) with yeast and the fermentation proceeds for 40–48 h. Mixing is achieved by the generation of carbon dioxide bubbles that rise rapidly in the vessel. Temperature control is monitored by probes positioned at suitable points within the vessel. A number of cooling jackets

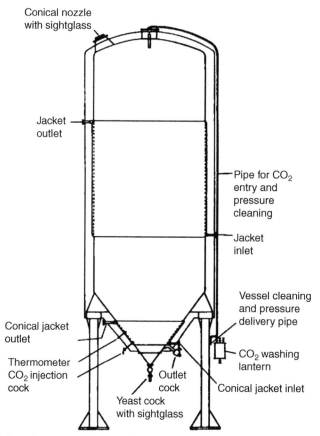

FIGURE 7.47 Cylindro-Conical Fermentation Vessel (Hough, Briggs, & Stevens, 1971)

are fitted to the vessel wall to regulate and cause flocculation and settling of the yeast (Ulenberg, Gerritson, & Huisman, 1972; Maule, 1986; Boulton, 1991). The fermentation is terminated by the circulation of chilled water via the cooling jackets which results in yeast flocculation. Thus, it is necessary to select a yeast strain, which will flocculate readily in the period of chilling. Part of this yeast may be withdrawn and used for repitching another vessel. The partially cleared beer may be left to allow a secondary fermentation and conditioning. Some of the advantages of this vessel in brewing are:

1. Reduced process times may be achieved due to increased movement within the vessel.
2. Primary fermentation and conditioning may be carried out in the same vessel.
3. The sedimented yeast may be easily removed since yeast separation is good.
4. The maturing time may be reduced by gas washing with carbon dioxide.

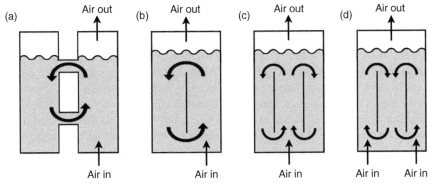

FIGURE 7.48 Schematic Representation of Air-Lift Fermenters

(a) External loop. (b) Split internal loop. (c) Concentric tube, internal riser. (d) Concentric tube, annulus riser.

AIR-LIFT FERMENTERS

An air-lift fermenter (Fig. 7.48) is essentially a riser tube (liquid ascending) where air is introduced connected to a downcomer tube (liquid descending). Liquid circulates between the riser and the downcomer in contrast to a simple bubble column reactor. There are essentially four design variants as shown in Fig. 7.48. These are; (1) the external loop air-lift fermenter where the riser and downcomer are two separate but connected vessels; (2) the split internal loop where the riser and downcomer are located on each side of a single vessel; (3) a concentric tube single vessel with the riser located in the central tube; and (4) a concentric tube single vessel with the riser located in the annulus (outer) tube. Air or gas mixtures are introduced into the base of the riser by a sparger during normal operating conditions. The driving force for circulation of medium in the vessel is produced by the difference in density between the liquid column in the riser (excess air bubbles in the medium) and the liquid column in the downcomer (depleted in air bubbles after release at the top of the loop). Shear stresses on the cultured organisms are low (hence their adoption in fungal, animal, and plant cell culture as well as in bacterial fermentations) and energy consumption for aeration is generally lower than for stirred tank systems (Varley & Birch, 1999; Guieysse, Quijano, & Munoj, 2011). Circulation times in loops of 45-m height may be 120 s. More details on liquid circulation and mixing characteristics are discussed by Chen (1990) and Guieysse et al. (2011). This type of vessel can be used for continuous culture. The first patent for this vessel was obtained by Schüller and Seidel (1940).

It would be uneconomical to use a mechanically stirred fermenter to produce SCP (single-cell protein) from methanol as a carbon substrate, as heat removal would be needed in external cooling loops because of the high rate of aeration and agitation required to operate the process. To overcome these problems, particularly that of cooling the medium when mechanical agitation is used, air-lift fermenters with outer or inner loops (Fig. 7.47) were chosen. Development work for operational processes for SCP has been done by ICI plc in Great Britain (Taylor &

Senior, 1978; Smith, 1980), Hoechst AG-Uhde GmbH in Germany (Faust, Prave, & Sukatsch, 1977) and Mitsubishi Gas Chemical Co. Inc. in Japan (Kuraishi, Teroa, Ohkouchi, Matsuda, & Nagai, 1977). Although ICI plc initially used an outer-loop system in their pilot plant, all three companies preferred an inner-loop design for large-scale operation. Hamer (1979) and Sharp (1989) have reviewed these fermenters. In the ICI plc continuous process, air and gaseous ammonia were introduced at the base of the fermenter. Sterilized methanol, other nutrients and recycled spent medium were also introduced into the downcomer. Heat from this exothermic fermentation was removed by surrounding part of the downcomer with a cooling jacket in the pilot plant, while at full scale (2.3×10^6 dm^3) it was found necessary to insert cooling coils at the base of the riser.

Unfortunately, the production of SCP for animal feed has not proved an economic proposition because of the price of methanol and the competition from animal feeds based on arable protein crops. ICI plc's vessel at Billingham, United Kingdom has now been dismantled.

In 1964, Rank Hovis McDougall decided to develop a protein-rich food primarily for human consumption (Trinci, 1992). They have grown *Fusarium graminearium* on a wheat starch based medium using a modified ICI plc 40 m^3 air-lift fermenter (Fig. 7.49) to produce the mycoprotein 'Quorn'. The use of an air-lift fermenter for culture of a mycelial fungus would seem unusual as lower rates of oxygen transfer occur in a viscous culture which gives rise to lower biomass yields. Because low shear conditions are present in the vessel, long fungal hyphae can be cultured (the preferred product form) even though production yields are only 20 g dm^{-3}. At the present time, a fermenter to produce 10,000 tons per annum of mycoprotein is considered economically feasible.

Okabe, Ohta, and Park (1993) modified a 3-dm^3 air-lift fermenter by putting stainless steel four-mesh sieves at the top and bottom of the draught tube to manipulate the morphology of *Aspergillus terreus* for optimum production of itaconic acid as the culture circulates in the vessel flow path. The fungal morphology was an intermediate state between pellets and pulp. Using this vessel, the itaconic acid production rate (g dm^{-1} h^{-1}) was double that obtained with a stirred fermenter or an air-lift fermenter with a conventional draught tube.

Carrington et al. (1992) used a 20 m^3 pilot-scale bubble column fermenter fitted with an internal helical cooling coil (Fig. 7.50) or a solid draught tube. The fermentation studied was a commercial *Streptomyces* antibiotic fermentation in a complex medium, which produced a viscous non-Newtonian broth. Tracer studies indicated that the vessel fitted with only the cooling coil behaved like an air-lift fermenter with a region of good mixing in the zone above the cooling coil. The coil acted as a leaky draught tube with back mixing taking place between the coils into the riser section. No poorly oxygenated zones were observed. Liquid velocities of 1 m sec^{-1} were measured giving circulation times of 9–12 s and mixing times of 14–18 s. The $K_L a$ at different power inputs and viscosities was found to increase almost linearly with increasing power input and decreased exponentially with increasing viscosity. When a solid draught tube was installed inside the cooling coil the circulation time

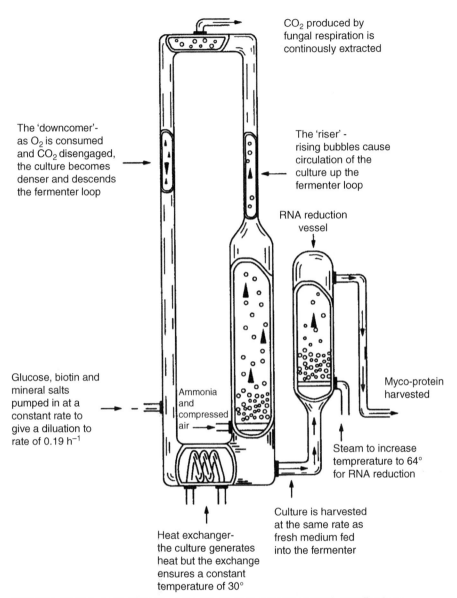

CO$_2$ produced by
fungal respiration is
continuously extracted

The 'downcomer'-
as O$_2$ is consumed
and CO$_2$ disengaged,
the culture becomes
denser and descends
the fermenter loop

The 'riser' -
rising bubbles cause
circulation of the
culture up the
fermenter loop

RNA reduction
vessel

Glucose, biotin and
mineral salts
pumped in at a
constant rate to
give a diluation to
rate of 0.19 h^{-1}

Ammonia
and
compressed
air

Myco-protein
harvested

Steam to increase
tempreature to 64°
for RNA reduction

Heat exchanger-
the culture generates
heat but the exchange
ensures a constant
temperature of 30°

Culture is harvested
at the same rate as
fresh medium fed
into the fermenter

FIGURE 7.49 Schematic Diagram of the Air-Lift Fermenter Used by Marlow Foods at Billingham, England, for the Production of Myco-Protein in Continuous Flow Culture (Trinci, 1992)

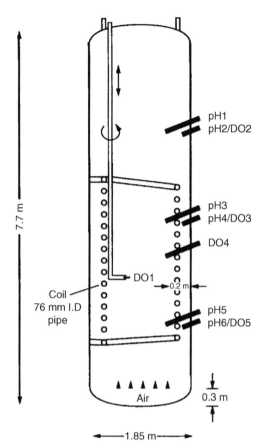

FIGURE 7.50 Schematic Diagram of 20-m³ Bubble Column Fermenter Fitted With Internal Cooling Coil (Carrington et al., 1992)

(pH and DO indicate positions of pH and oxygen electrodes.)

was similar, but the mixing time increased to 18–24 s. The K_La was also determined at different power inputs and viscosities. This gave a reduction of 5%–25% in K_La with the biggest reduction at a high viscosity.

Wu and Wu (1990) compared K_Las in a range of mesh draught tubes and a solid draught tube in an air-lift vessel of 15 dm³ working volume. When a 24-mesh tube was used at high superficial gas velocities, the K_La was double that of the same vessel with a solid draught tube.

Bakker, van Can, Tramper, and Gooijer (1993) have developed a multiple air-lift fermenter in which three air-lift fermenters with internal loops are incorporated into one vessel (Fig. 7.51). Fresh medium is fed into the central compartment, depleted medium overflows into the middle compartment, from here to the outer compartment where medium is eventually discharged. The hydrodynamics and mixing in

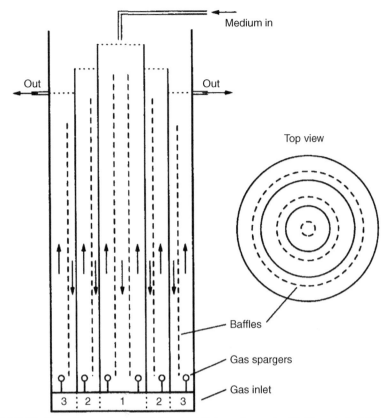

Medium in

Out

Out

Top view

Baffles

Gas spargers

Gas inlet

3 : 2 : 1 : 2 : 3

FIGURE 7.51 Three-Compartment Multiple Air-Lift Loop Fermenter (Bakker et al., 1993)

Cross-sectional side and top view.

the middle compartment were found to be comparable with those obtained with conventional internal loop air-lift vessels, although some tangential liquid flow was observed in this compartment.

De Medeiros Burkert, Maldonado, Filho, and Rodrigues (2005) have compared lipase production by the filamentous fungus *Geotrichum candidum* in a stirred tank and an air-lift fermenter each of 2.2 dm^{-3} working volume. Optimal conditions for the stirred fermenter were an agitation rate of 300 rpm and an aeration rate of 1 v.v.m. For the air-lift fermenter the optimal aeration rate was 2.5 v.v.m. The overall lipase yields were similar but were obtained over a much shorter time period for the air-lift fermenter resulting in ~60% greater productivity for the air-lift fermenter.

DEEP-JET FERMENTER

Some designs of continuous culture fermenter achieve the necessary mechanical power input with a pump to circulate the liquid medium from the fermenter through

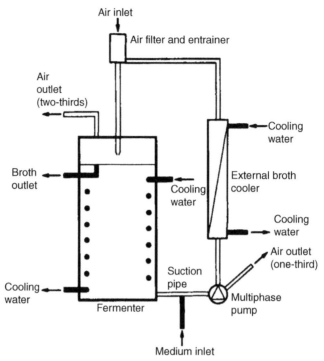

FIGURE 7.52 Diagram of the Vogelbusch Deep-Jet Fermenter System (From Schreier, 1975; Hamer, 1979)

a gas entrainer and back to the fermenter (Fig. 7.52; Hamer, 1979; Meyrath & Bayer, 1979). Two basic construction principles have been used for the gas entrainer nozzles. The injector and the ejector (Fig. 7.53). In an injector a jet of medium is surrounded by a jet of compressed air. The gas from the outlet enters the larger tube with a nozzle velocity of 5–100 m s^{-1} and expands in the tube to form large air bubbles, which are dispersed by the shear of the water jet. In an ejector the liquid jet enters into a larger converging–diverging nozzle and entrains the gas around the jet. The gas, which is sucked into the converging-diverging jet is dispersed in that zone. One of the industrial-scale fermenters using the ejector design is marketed by Vogelbusch (Vogelbusch AG, Vienna, Austria). Partially aerated medium is pumped by a multiphase pump through a broth cooler to an air entrainer above the fermenter. The air–medium mixture falls down a slightly conical shaft at a high velocity and creates turbulence in the fermenter. Two-thirds of the exhaust gas is vented from the fermenter headspace and the remainder via the multiphase pump. Oxygen-transfer rates of 4.5 g dm^{-3} h^{-1} with an energy consumption of 1 kW h^{-1} kg^{-1} have been achieved for industrial-scale yeast production from whey using such a fermentation system.

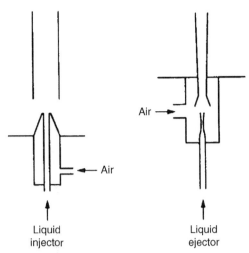

Air →

← Air

Liquid
injector

Liquid
ejector

FIGURE 7.53 Gas Entrainer Nozzles of Deep-Jet Fermenters

CYCLONE COLUMN

Dawson (1974) developed the cyclone column, particularly for the growth of filamentous cultures (Fig. 7.54). The culture liquid was pumped from the bottom to the top of the cyclone column through a closed loop. The descending liquid ran down the walls of the column in a relatively thin film. Nutrients and air were fed in near the base of the column while the exhaust gases left at the top of the column. Good gas exchange, lack of foaming, and limited wall growth have been claimed with this fermenter. Dawson (1974, 1988) has listed a number of potential bacterial, fungal, and yeast applications including the batch production of a vaccine for scours in calves with the vessel being operated as batch, continuous or fed-batch.

PACKED TOWERS, BIOFILTERS, AND OTHER FIXED FILM PROCESSES

The packed tower is a well-established application of immobilized (fixed film) cells. A vertical cylindrical column is packed with pieces of some relatively inert material, for example, wood shavings, twigs, coke, aggregate, or plastic. Initially both medium and cells are fed into the top of the packed bed. Once the cells have adhered to the support and are growing well as a thin film, fresh medium is added at the top of the column and the fermented medium is removed from the bottom of the column. The best known example is the vinegar generator, in which ethanol was oxidized to acetic acid by strains of *Acetobacter* supported on beech shavings; the first recorded use was in 1670 (Mitchell, 1926). More recently, packed towers and many other fixed film reactors, such as trickling filters, fluidized beds, and rotating biological contactors, have been used for both aerobic and anaerobic sewage and effluent treatment. These are discussed in more detail in Chapter 11. They are also used in gas treatment (bioscrubbers) where gaseous material such as volatile organic compounds (VOCs),

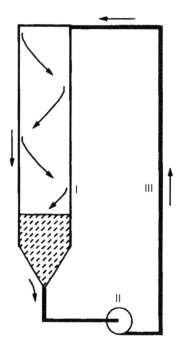

FIGURE 7.54 Schematic Diagram of a Cyclone Column Fermenter

I, Cyclone column; *II*, circulating pump; *III*, recirculating limb (Dawson, 1974).

hydrogen sulfide and ammonia are first entrained into a liquid absorbant in a countercurrent packed column and then degraded microbially by suspended organisms in a bioreactor (Cabrera et al., 2011).

The use of immobilized microorganisms has been widely studied in recent years for pharmaceutical and biological production and (as aforementioned) pollution control. Immobilization may be natural or mediated by a variety of physicochemical processes such as entrapment within a membrane or cross-linking to an inert support. Immobilization allows for short media residence times and high loading rates without "washout" of the organisms. In addition to conventional fixed film reactors, immobilized cells can also be utilized in stirred tank reactors, air-lift reactors, and fluidized bed reactors (Nemati & Webb, 2011).

SOLID-STATE FERMENTERS

Solid-state fermentation (SSF) is defined as fermentation using solids in the absence (or near absence) of free water. There must be sufficient water present however for microbial growth and reaction. Microorganisms, most commonly filamentous fungi, grow on the surface of moist particles (the substrate) in beds where the voidage between particles contains a continuous air phase. SSF has been used to produce traditional fermented food products (eg, soy sauce) for many centuries. From the 1970s there has been increased interest in using SSF to produce products such as fuels, industrial chemicals, pharmaceutical products, enzymes, and many others. More recently, there has been interest in utilizing agricultural wastes as a nutrient source via SSF. In this way, conventional disposal of such wastes would be reduced and a value-added product obtained. In addition, largescale cultivation using SSF will minimize water use and the generation of wastewater when compared to conventional submerged fermentation (Pandey, 2003; Mitchell, De Lima Luz, Krieger, & Berovic, 2011).

Solid-state fermenters can be classified into four groups dependent on their aeration and agitation strategies.

1. The substrate bed is not force aerated and remains static or is agitated infrequently during the fermentation.
2. The substrate bed is force aerated but remains static or is agitated infrequently during the fermentation.
3. The substrate bed is not force aerated but is frequently or continuously agitated.
4. The substrate bed is both force aerated and frequently or continuously agitated.

There are a number of design variants for SSFs. The simplest and most traditional are tray bioreactors in which a tray (or similar) contains the substrate, inoculum and water and is placed in an incubation room or cabinet. Tray bioreactors are an example of the group 1 system above. Packed bed bioreactors consist of a static bed sited on a perforated base through which air is blown. These are classified as group 2 SSFs. Rotating drum and stirred drum reactors (group 3 SSFs) are configured to allow frequent agitation via rotation or stirring along with forced air introduction and exit (Mitchell et al., 2011).

MEMBRANE FERMENTERS

Membrane reactors are designed for the in-situ separation of cells from the growth medium utilizing specialized membranes to retain cells in the fermenter and thus aid production and purification in a single stage, that is, the bioreaction (fermentation) and membrane operation (cell separation/downstream processing) are combined in a single vessel. Their advantages are high cell densities and high productivity but problems with membrane fouling are a serious drawback (Zhong, 2011). The membrane can also act as a reactant/nutrient supplier or as a selective barrier for product removal. They can also act as a support on/in which the bioactive component is immobilized. It may be located (immersed) within the fermenter or located externally to the fermenter (Giorno et al., 2011). Membrane fermenters have found particular application in wastewater treatment (see also Chapter 11).

SINGLE USE AND DISPOSABLE FERMENTERS

Disposable bioreactors are a relatively recent development that are increasingly gaining acceptance for microbial and cell culture applications. They have a number of

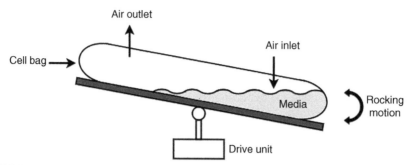

FIGURE 7.55 Schematic Representation of a Wave or Rocking Type Disposable Fermenter

advantages such as ease of use, rapid implementation, flexibility of operation, fewer utility requirements such as steam and water, no cleaning or cleaning validation, and reduced labor costs. Presterilized disposable pH and dissolved oxygen sensors should reduce contamination risk (Hahnel et al., 2011). However, it should be noted that single-use/disposable technologies can be regarded as an additional waste stream in the fermentation process which does not exist in conventional systems (see also Chapter 11).

Disposable reactors may be used both as production scale vessels and seed vessels (to replace shake flask and stirred tank reactors). There are many vessel types available which differ in design, scale, instrumentation, and power requirements with the most common commercially available disposable bioreactors being wave-mixed/rocking (where the rocking motion promotes mixing and oxygen transfer, Fig. 7.55), orbitally shaken and stirred vessels (Eibl, Kaiser, Lombriser, & Eibl, 2010). Further consideration of single-use/disposable fermenter agitation and oxygen transfer is given in Chapter 9.

Oosterhuis et al. (2011) describe the use of presterilized rocking type (wave) reactors and stirred tank disposable vessels with working volumes between 1 and 2000 dm^{-3}. Wave/rocking type reactors are widely used in cell culture but can also find application in aerobic yeast fermentations (Mikola, Seto, & Amanullah, 2007), bacterial fermentations (Westbrook, Scharer, Moo-Young, Oosterhuis, & Chou, 2014; Glazyrina et al., 2010) and fungal fermentations (Jonczyk et al., 2013). Langer and Rader (2014) review the developments in single-use/disposable technologies over the last decade.

ANIMAL CELL CULTURE

Interest in the in vitro cultivation of animal cells has developed because of the need for largescale production of monoclonal antibodies from hybridoma cells, hormones, vaccines, and other products which are difficult or impossible to produce synthetically or by using other culture techniques.

Animal cells are usually more nutritionally demanding than microbial cells. They lack cell walls, which makes them sensitive to shear and extremes of osmolality. The doubling times are normally 12–48 h and cell densities in suspension cultures rarely exceed 10^6–10^7 cells cm^{-3}. Two distinct modes of growth can be recognized:

1. *Anchorage dependent cells.* These cells require a solid support for their replication. They produce cellular protrusions (pseudopodia) which allow them to adhere to positively charged surfaces and often grow as monolayers.
2. *Anchorage independent cells.* These cells do not require a support and can grow as a suspension in submerged culture. Established and transformed cell lines are normally in this category.

There are also intermediate categories of cells, which may be grown as anchorage dependent or suspension cells. It is also possible to grow some anchorage dependent cells in suspension, provided the cells can be grown on suitable microcarriers.

A range of free and immobilized culture systems have therefore been developed for the culture of different lines of animal cells and their products. At the laboratory scale designs are simple with little control or instrumentation and include roller bottles and spinner flasks with volumes ranging from around 100 mL to several liters. Low oxygen transfer rates are the major limitation on scale-up of this type of device. Wave agitation of suspended cultures in disposable plastic bags of up to 500 dm^{-3} have been developed with oxygen transferred through the gas permeable bag walls. Disposable, ready-to-use, stirred tank reactors of 200 dm^{-3} have also been reported (Hahnel et al., 2011; Taya and Kino-oka, 2011; Oosterhuis et al., 2011). Some useful introductions include Griffiths (1986, 1988), Propst, Von Wedel, and Lubiniecki (1989), Lavery (1990), Bliem, Konopitzky, and Katinger (1991), and Taya and Kino-Oka (2011).

Shear is a phenomenon recognized as being critical to the scale up of animal cell culture processes, irrespective of the cell line or reactor configuration (Bliem & Katinger, 1988). Shear may influence the cell culture causing damage, which may result in cell death or metabolic changes. The shear sensitivity of animal cells may vary between cell lines, the phase of growth or with a change to fresh medium.

Mijnbeek (1991) has reviewed research on shear stress of free and immobilized animal cells in stirred fermenters and air-lift fermenters. It was concluded that the predominant damage mechanism in both types of vessel was due to sparging and the breakup of bubbles on the medium surface. This type of damage causing cell death might be reduced by increasing the height to diameter ratio in the vessel, increasing the bubble size, decreasing the gas flow rate, or by adding protective agents. Other damage mechanisms in stirred fermenters are caused by cell–microcarrier eddies and microcarrier–microcarrier interactions. Damage of this type may be reduced by reducing the impeller speed, impeller diameter, microcarrier size, and concentration or by increasing the viscosity of the medium. Hence, many developments in animal cell fermenter design have concentrated on modified low shear impellers and bubble free air sparging via gas permeable and microporous membranes.

The maximum cell densities obtainable in stirred and air-lift fermenters are often only 10^6–10^7 cells cm^{-3}. This is not ideal if secreted proteins are present only in very low concentrations and mixed with other proteins present in the original medium. A number of modified fermenters and reactors have been developed to grow cells at higher concentrations using microcarriers, encapsulation, perfusion, glass beads, or hollow fibers to obtain the required product at higher concentrations.

STIRRED FERMENTERS

Unmodified stirred fermenters have been used for the batch production of some virus vaccines (Propst et al., 1989), but modified vessels are used for most cultures (Propst et al., 1989; Lavery, 1990). The modifications made to fermenters are to reduce the possibility of cell damage due to shear, heat, or contamination. Marine propellers revolving at a slow speed (10–100 rpm) will normally provide adequate mixing. Hemispherical bottoms on the vessels will ensure better mixing of the broth at slow stirrer speeds. Water jacket heating is often preferred since heating probes may give rise to localized zones of high temperature, which might damage some of the cells. Magnetic driven stirrers may be used to reduce the risk of contamination. A novel sparger–impeller design, which improves aeration at slow speeds is incorporated into the Celligen system manufactured by New Brunswick, United States of America (Fig. 7.56; Becl, Stiefel, & Stinnett, 1987). When the impeller rotates, the swept-back ports produce a negative pressure inside the hollow impeller complex. This creates a suction lift that produces highly efficient circulation and gas transfer at low rpm without damaging the cells. Gases are introduced through a ring sparger into the medium as it circulates through a fine stainless steel mesh jacket, which excludes cells and microcarriers. Because the gas sparging is restricted to a relatively small zone, foaming is reduced, and any foam formed is broken up by the mesh in the foam eliminator chamber.

AIR-LIFT FERMENTERS

The operating scale of fermenters for suspended animal cell culture has been increasing to meet the demand for licensed therapeutic recombinant proteins. This is likely to be a continuing trend. Although stirred tanks are generally the reactor of choice, air-lift fermenters are also used (Varley & Birch, 1999). Air-lift fermenters have proved ideal for growth of some cell lines because of the gentle mixing action and reduced shear forces when compared with those in stirred vessels. The absence of a stirrer and associated seals excludes a potential source of contamination (Griffiths, 1988; Propst et al., 1989; Lavery, 1990).

Vessels of 1000–2000 dm^3 are commercially available. Celltech Ltd, UK, have used such vessels to produce monoclonal antibodies from hybridoma cells (Wilkinson, 1987). Varley and Birch (1999) give a detailed review of the design and scale-up of air-lift fermenters for animal cell culture emphasizing the importance of balancing the needs of mixing, oxygen mass transfer, and shear stresses. Air-lift fermenters are considered in more detail earlier in this chapter.

RADIAL FLOW FERMENTERS

A radial flow fermenter for the cultivation of mammalian cells to produce antibodies was originally developed by the Kirin Brewery Co., Japan. The growth medium

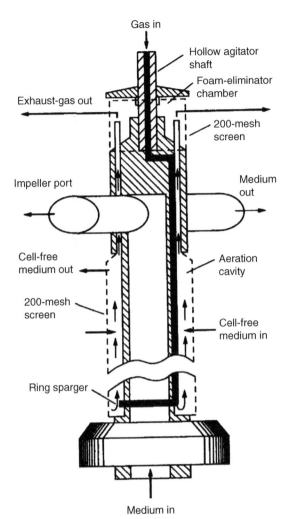

Gas in

Hollow agitator shaft

Foam-eliminator chamber

Exhaust-gas out

200-mesh screen

Impeller port

Medium out

Cell-free medium out

Aeration cavity

200-mesh screen

Cell-free medium in

Ring sparger

Medium in

FIGURE 7.56 Sparger-Impeller in the Celligen Cell-Culture Fermenter (New Brunswick Ltd, Hatfield, England)

flows in a radial pattern across carrier beads and consequently, supply of oxygen and nutrients can be achieved with low shear stresses. The system is compact, simple, and easy to operate. The radial flow fermenter was scaled up to 5 dm^{-3} for the production of interleukin-6 using baby hamster kidney (BHK) cells (Taya & Kino-Oka, 2011).

MICROCARRIERS

Microcarriers may provide a solution to the problem of growing anchorage-dependent cultures in suspension culture in fermenters, by providing the necessary surface for attachment. Animal cells normally have a net negative surface charge and will attach to a positively charged surface by electrostatic forces. Van Wezel (1967) made use of this property and attached anchorage dependent cells to chromatographic grade DEAE Sephadex A-50 resin beads and suspended the coated beads in a slowly stirred liquid medium. The density of the electrostatic charges on the microcarrier surface is critical if cell growth at high bead concentrations is to be achieved. If the net charge density on the bead surface is too low, cell attachment will be restricted. When the charge density is too high, apparent toxic effects will limit cell growth (Fleischaker, 1987). A number of microcarrier beads, manufactured from dextran, cellulose, gelatin, plastic, or glass, are now commercially available (Fleischaker, 1987; Butler, 1988). Dextran microcarriers have been used for larges-cale production of viral vaccines and interferon.

ENCAPSULATION

At least three methods of encapsulation have been developed (Griffiths, 1988; Lavery, 1990). Encapsel, a technique developed by Damon Biotechnology, United States of America, traps the animal cells in sodium alginate spheres, which are then coated with polylysine to form a semipermeable membrane. The enclosed alginate gel is solubilized with sodium citrate to release the cells into free suspension within the capsules, which are usually 50–500-μm diameter. After a few weeks growth it is possible to obtain cell concentrations of 5×10^8 cm^{-3} and product levels 100 times higher than with free cells can be achieved. The high molecular weight products are retained within the capsules. This technique has been used commercially for monoclonal antibody production.

In a second method, the cells are entrapped in calcium alginate which will allow high molecular weight products to diffuse into the medium. Unfortunately, the spheres tend to be 0.5–1.0 mm diameter and slow diffusion into the spheres may cause nutrient limitations. Alternatively, the cells can be entrapped in agarose beads in which the cells are contained in a honeycomb matrix within the gel. These capsules have a wide size distribution and a low mechanical strength compared with alginate ones.

HOLLOW FIBER CHAMBERS

Anchorage dependent cells can be cultured at densities of 10^8 cells cm^{-3} using bundles of hollow fibers held together in cartridge chambers (Hirschel & Gruenberg, 1987; Griffiths, 1988). The cells are grown in the extra capillary spaces (ECS) within the cartridge. Medium and gases diffuse through from the capillary lumea to the ECS. The molecular weight cut-off of the fiber walls may be selected so that the product is retained in the ECS or released into the perfusing medium. Many chambers will be needed for scale-up because of diffusion limitations in larger chambers. These have been used to study production of monoclonal antibodies, viruses, gonadotropin, insulin, and antigens.

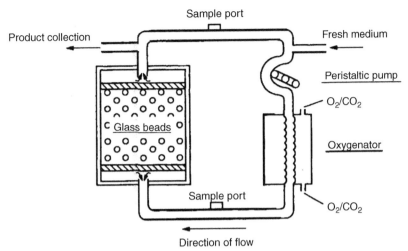

FIGURE 7.57 Schematic Diagram of a Glass Bead Reactor (Browne, Figueroa, Costello, Oakley, & Maclukas, 1988)

PACKED GLASS BEAD REACTORS

Packed glass bead reactors have proved to be useful for long-term culture of attached dependent cell lines. It is possible to obtain cell densities of 10^{10} viable cells in a 1-dm^3 vessel with moderate medium flow rates through the vessel (Fig. 7.57; Propst et al., 1989). Increasing the size of vessels causes problems with mass transfer of oxygen and nutrients and scale up can be achieved by increasing the number of small vessels. This technique is available only as a contract production service from Bioresponse Inc. of California, United States of America.

PERFUSION CULTURES

Perfusion culture is a technique where modified fermenters of up to 100 dm^3 are gently stirred and broth is withdrawn continuously from the vessel and passed through a stainless steel or ceramic filter. This type of culture is sometimes referred to as spin culture, since the filter is spun to prevent blocking with cells. The filtered medium is pumped to a product reservoir and fresh medium is pumped into the culture vessel. The rate of addition and removal can be regulated depending on the cell concentration in the vessel. With this method, it is possible to obtain cell densities 10–30 times higher than the maximum cell density in an unmodified fermenter (Lydersen, 1987; Tolbert, Srigley, & Prior, 1988; Lavery, 1990). This procedure has been used commercially by Invitron Corp. United States of America. This company has patented a gentle "sail" agitator to prevent cell damage in a 100-dm^3 vessel. Attachment dependent cells have been grown on microcarriers using sail agitators rotating at 8–12 rpm.

At Hybridtech Inc. United States of America, animal cells have been immobilized on a ceramic matrix and medium is perfused through this matrix. In this way high cell densities can be maintained. The apparatus is marketed as the "Opticeli" (Lydersen, 1987).

REFERENCES

Abson, J. W., & Todhunter, K. H. (1967). Effluent disposal. In N. Blakeborough (Ed.), *Biochemical and biological engineering science* (pp. 309–343). (Vol. 1). London: Academic Press.

Aiba, S., Humphrey, A. E., & Millis, N. F. (1973). *Biochemical Engineering* (2nd ed.). New York: Academic Press, Chapter 11, pp. 303–316.

Albaek, M. O., Gernaey, K. V., & Stocks, S. M. (2008). Gassed and ungassed power draw in a pilot scale 550 litre fermentor retrofitted with up-pumping hydrofoil B2 impellers in media of different viscosity and with very high power draw. *Chemical Engineering Science, 63,* 5813–5820.

Armer, A. (1991). Steam utilization. In D. A. Snow (Ed.), *Plant Engineer's Reference Book.* London: Butterworth-Heinemann.

Arnold, R. H., & Steel, R. (1958). Oxygen supply and demand in aerobic fermentations. In R. Steel (Ed.), *Biochemical Engineering* (pp. 149–181). London: Heywood.

Atkinson, B., & Mavituna, F. (1991a). Reactors *biochemical engineering and biotechnology handbook* (2nd ed.). London: Macmillan, pp. 607–668.

Atkinson, B., & Mavituna, F. (1991b). Heat transfer *biochemical engineering and bioengineering handbook* (2nd ed.). London: Macmillan, pp. 793–829.

Bailey, J. E., & Ollis, D. F. (1986). *Biochemical Engineering Fundamentals* (2nd ed.). New York: McGraw-Hill, pp. 512–532.

Bakker, W. A. M., van Can, H. J. L., Tramper, J., & de Gooijer, C. D. (1993). Hydrodynamics and mixing in a multiple air-lift loop reactor. *Biotechnology and Bioengineering, 42,* 994–1001.

Becl, G., Stiefel, H., & Stinnett, T. (1987). Cell-culture bioreactors. *Chemical Engineering, 94,* 121–129.

Besik, F. (1973). Multi-stage tower-type activated sludge process for complete treatment of sewage. *Water Sewage Works* (September), pp. 122–127.

Blakeborough, N. (1967). Industrial fermentations. In N. Blakeborough (Ed.), *Biochemical and Biological Engineering Science* (pp. 25–48). (Vol. 1). London: Academic Press.

Bliem, R., & Katinger, H. (1988). Scale-up engineering in animal cell technology: Part II. *Trends in Biotechnology, 6,* 224–230.

Bliem, R., Konopitzky, K., & Katinger, H. (1991). Industrial animal cell reactor systems: aspects of selection and evaluation. *Advances in Biochemical Engineering/Biotechnology, 44,* 1–26.

Boulton, C. A. (1991). Developments in brewery fermentation. *Biotechnology and Genetic Engineering Reviews, 9,* 127–181.

British Valve Manufacturers Association (1972). *Technical Reference Book on Values for the Control of Fluids* (3rd edition). British Valve Manufacturers Association.

Brown, W. E., & Peterson, W. H. (1950). Factors affecting production of penicillin in semi-pilot plant equipment. *Industrial & Engineering Chemistry, 42,* 1769–1774.

Browne, P. G., Figueroa, G., Costello, M. A. G., Oakley, R., & MacIukas, S. M. (1988). Protein production from mammalian cells grown on glass beads. In R. E. Spier, & J. B. Griffiths (Eds.), *Animal Cell Biotechnology* (pp. 251–262). (Vol. 3). London: Academic Press.

Buchter, H. H. (1979). *Industrial Sealing Technology*. New York: Wiley.

Buckland, B. C., Gbewonyo, K., DiMassi, D., Hunt, G., Westerfield, G., & Nienow, A. W. (1988). Improved performance in viscous mycelial fermentations by agitator retrofitting. *Biotechnology and Bioengineering, 31*, 737–742.

Buelow, G. H., & Johnson, M. J. (1952). Effect of separation on citric acid production in 50 gallon tanks. *Industrial and Engineering Chemistry, 44*, 2945–2946.

Burnett, J.M. (1993). The industrial production of citric acid. *124th Meeting of the Society for General Microbiology*. Canterbury, UK.

Butler, M. (1988). A comparative review of microcarriers available for the growth of anchorage dependent animal cells. In R. E. Spiers, & J. B. Griffiths (Eds.), *Animal Cell Biotechnology* (pp. 284–323). (Vol. 3). London: Academic Press.

Cabrera, G., Ramirez, M., & Cantero, D. (2011). Biofilters. In M. Moo-Young (Ed.), *Comprehensive Biotechnology* (pp. 303–319). (Vol. 2). Oxford: Elsevier.

Callahan, J. R. (1944). Large scale production by deep fermentation. *Chem. Metal. Eng., 51*, 94–98.

Cameron, J., & Godfrey, E. I. (1969). The design and operation of high-power magnetic drives. *Biotechnology and Bioengineering, 11*, 967–985.

Carrington, R., Dixon, K., Harrop, A. J., & Macaloney, G. (1992). Oxygen transfer in industrial air agitated fermentations. In M. R. Ladisch, & A. Bose (Eds.), *Harnessing Biotechnology for the 21st Century* (pp. 183–188). Washington, DC: American Chemical Society.

Chain, E. B., Paladino, S., Ugolini, F., Callow, D. S., & Van Der Sluis, J. (1952). Studies on aeration. 1. *Bulletin of the World Health Organization, 6*, 83–98.

Chain, E. B., Paladino, S., Ugolino, F., Callow, D. S., & Van Der Sluis, J. (1954). A laboratory fermenter for vortex and sparger aeration. *Rep. Inst. Sup. Sanita. Roma (Engl. Ed.), 17*, 61–120.

Chapman, G. (1989). Client requirements for supply of contained bioreactors and associated equipment. In T. Salisbury (Ed.), *Proceedings of DTI, HSE, SCI, symposium on large scale bioprocessing safety, laboratory report, LR., 746 (BT)* (pp. 58–62). Stevenage: Warren Spring Laboratories.

Chen, N. Y. (1990). The design of airlift fermenters for use in biotechnology. *Biotechnology and Genetic Engineering Reviews, 8*, 379–396.

Cohee, R. F., & Steffen, G. (1959). Make vinegar continuously. *Food Engineering, 3*, 58–59.

Collins, G. H. (1992). Hazard groups and containment categories in microbiology and biotechnology. In C. H. Collins, & A. J. Beale (Eds.), *Safety in Industrial Microbiology and Biotechnology* (pp. 23–33). London: Butterworth-Heinemann.

Cooke, M., Middelton, J. C., & Bush, J. R. (1988). Mixing and mass transfer in filamentous fermentations. In R. King (Ed.), *Second International Conference on Bioreactor Fluid Dynamics* (pp. 37–64). London: Elsevier.

Cooney, C. L., Wang, D. I. C., & Mateles, R. I. (1969). Measurement of heat evolution and correlation with oxygen consumption during microbial growth. *Biotechnology and Bioengineering, 11*, 269–280.

Cooper, F. M., Fernstrom, G. A., & Miller, S. A. (1944). Performance of gas-liquid contactors. *Industrial and Engineering Chemistry, 36*, 504–509.

Cowan, C. T., & Thomas, C. R. (1988). Materials of construction in the biological process industries. *Process Biochemistry, 23*(1), 5–11.

Cubberly, W. H., Unterweiser, P. M., Benjamin, D., Kirk-patrick, C. W., Knoll, V., & Nieman, K. (1980). Stainless steels in corrosion service. In *Metals Handbook* (9th ed.), (Vol. 3), (pp. 56–93). Ohio: American Society for Metals.

Dawson, P. S. S. (1974). The cyclone column fermenter. *Biotechnology and Bioengineering Symposium, 4,* 809–819.

Dawson, P. S. S. (1988). The cyclone column and continuous phased culture. In J. Gavora, D. F. Gerson, J. Luong, A. Storer, & J. H. Woodley (Eds.), *Biotechnology Research and Applications* (pp. 141–154). London: Elsevier.

De Becze, G., & Liebmann, A. J. (1944). Aeration in the production of compressed yeast. *Industrial and Engineering Chemistry, 36,* 882–890.

De Medeiros Burkert, J. F., Maldonado, R. R., Filho, F. M., & Rodrigues, M. I. (2005). Comparison of lipase production by *Geotrichum candidum* in stirring and airlift fermenters. *Journal of Chemical Technology and Biotechnology, 80,* 61–67.

Doran, P. M. (2013). *Bioprocess Engineering Principles* (2nd ed.). Oxford: Academic Press, pp. 761–852.

Duurkoop, A. (1992). Stainless steel reflects the quality of Applikon's bioreactors. *Bioteknowledge (Applikon, Holland), 12*(1), 4–6.

East, D., Stinnett, T., & Thoma, R. W. (1984). Reduction of biological risk in fermentation processes, by physical containment. *Developments in Industrial Microbiology, 25,* 89–105.

Ebner, H., & Follman, H. (1983). Vinegar. In G. Reed (Ed.), *Biotechnology* (pp. 425–446). (Vol. 5). Weinheim: Verlag Chemie.

Ebner, H., Pohl, K, & Enekel, A. (1967). Self-priming aerator and mechanical defoamer for microbiological process. *Biotechnology and Bioengineering, 9,* 357–364.

Eibl, R., Kaiser, S., Lombriser, R., & Eibl, D. (2010). Disposable bioreactors: the current state-of-the-art and recommended applications in biotechnology. *Applied Microbiology and Biotechnology, 86,* 41–49.

Elsworth, R., Capell, G. A., & Telling, R. C. (1958). Improvements in the design of a laboratory culture vessel. *Journal of Applied Bacteriology, 21,* 80–85.

Faust, U., Prave, P., & Sukatsch, D. A. (1977). Continuous biomass production from methanol by *Methylomonas clara. Journal of Fermentation Technology, 55,* 605–614.

Finn, R. F. (1954). Agitation-aeration in the laboratory and in industry. *Bacteriology Reviews, 18,* 254–274.

Fleischaker, R. (1987). Microcarrier cell culture. In B. K. Lydersen (Ed.), *Large Scale Cell Culture Technology* (pp. 59–79). Munich: Hanser.

Fortune, W. B., McCormick, S. L., Rhodehamel, H. W., & Stefaniak, J. J. (1950). Antibiotics development. *Industrial and Engineering Chemistry, 42,* 191–198.

Foust, H. G., Mack, D. E., & Rushton, J. H. (1944). Gas-liquid contacting by mixers. *Industrial and Engineering Chemistry, 36,* 517–522.

Frommer, W., et al. (1989). Safety biotechnology. III. Safety precautions for handling microorganisms of different risk classes. *Applied Microbiology and Biotechnology, 30,* 541–552.

Flickinger, M. C., & Sansone, E. B. (1984). Pilot-and production-scale containment of cytotoxic and oncogenic fermentation processes. *Biotechnology and Bioengineering, 26,* 860–870.

Garcia-Ochoa, F., & Gomez, E. (2009). Bioreactor scale-up and oxygen transfer rate in microbial processes: an overview. *Biotechnology Advances, 27,* 153–176.

Gbewonyo, K, DiMasi, D., & Buckland, B. C. (1986). The use of hydrofoil impellers to improve oxygen transfer efficiency in viscous mycelial fermentations. In *International Conference on Biorector Fluid Dynamics* (pp. 281–299). UK: BHRA, Cranfield.

Giorgio, R. J., & Wu, J. J. (1986). Design of large scale containment facilities for recombinant DNA fermentations. *Trends in Biotechnology, 4,* 60–65.

Giorno, L., De Bartolo, L., & Drioli, E. (2011). Membrane bioreactors. In M. Moo-Young (Ed. in chief), *Comprehensive Biotechnology, Vol. 2,* 2nd ed., (pp. 263–288). Oxford: Elsevier.

Glazyrina, J., Materne, E.-M., Dreher, T., Storm, D., Junne, S., Adams, T., Greller, G., & Neubauer, P. (2010). High cell density cultivation and recombinant protein production with *Escherichia coli* in a rocking-motion-type reactor. *Microbial Cell Factories, 9,* 42–53.

González-Sáiz, J.-M., Garrido-Vidal, D., & Pizarro, C. (2009). Scale up and design of processes in aerated-stirred fermenters for the industrial production of vinegar. *Journal of Food Engineering, 93,* 89–100.

Gordon, J. J., Grenfell, E., Knowles, E., Legg, B. J., McAllister, R. C. A., & White, T. (1947). Methods for penicillin production in submerged culture on a pilot scale. *Journal of General Microbiology, 1,* 171–186.

Greenshields, R. N. (1978). Acetic acid: vinegar. In A. H. Rose (Ed.), *Economic Microbiology* (pp. 121–186). (Vol. 2). London: Academic Press.

Greenshields, R. N., & Smith, E. L. (1971). Tower fermentation systems and their applications. *Chemical Engineering (London), 249,* 182–190.

Griffiths, J. B. (1986). Scaling up of animal cell cultures. In R. I. Freshney (Ed.), *Animal Cell Culture—A Practical Approach* (pp. 33–69). Oxford: IRL Press.

Griffiths, J. B. (1988). Overview of cell culture systems and their scale-up. In R. E. Spier, & J. B. Griffiths (Eds.), *Animal Cell Biotechnology* (pp. 179–220). (Vol. 3). London: Academic Press.

Guieysse, B., Quijano, G., & Munoz, R. (2011). Airlift bioreactors. In M. Moo-Young (Ed. in chief), *Comprehensive Biotechnology, Vol. 2,* 2nd ed., (pp. 199–212). Oxford: Elsevier.

Hahnel, A., Putz, F. B., Iding, K., Neidiek, T., Gudermann, F., & Lutkemeyer, D. (2011). Evaluation of a disposable stirred tank bioreactor for cultivation of mammalian cells. *BMC Proceedings, 5*(8), 54–55.

Hall, R. D., & Howard, G. A. (1965). Improvements in or relating to brewing of beer. British Patent 979,491.

Hambleton, P., Griffiths, B., Cameron, D. R., & Melling, J. (1991). A high containment polymodal pilot-plant- design concepts. *Journal of Chemical Technology and Biotechnology, 50,* 167–180.

Hamer, G. (1979). Biomass from natural gas. In A. H. Rose (Ed.), *Economic Microbiology* (pp. 315–360). (Vol. 4). London: Academic Press.

Hastings, J. J. H. (1971). Development of the fermentation industries in Great Britain. *Advances in Applied Microbiology, 16,* 1–45.

Hastings, J. J. H. (1978). Acetone–butanol fermentation. In A. H. Rose (Ed.), *Economic Microbiology* (pp. 31–45). (Vol. 2). London: Academic Press.

Health and Safety Executive (1993). A guide to the Genetically Modified Organisms (Contained Use) Regulations, 1992. H.M.S.O., London.

Herbert, D., Phipps, P. J., & Tempest, D. W. (1965). The chemostat: design and instrumentation. *Laboratory Practice, 14,* 1150–1161.

Hesselink, P. G. M., Kasteliem, J., & Logtenberg, M. T. (1990). Biosafety aspects of biotechnological processes: testing and evaluating equipment and components. In P. L. Yu (Ed.), *Fermentation Technology: Industrial Applications* (pp. 378–382). London: Elsevier.

Hirschel, M. D., & Gruenberg, M. L. (1987). An automated hollow fiber system for the large scale manufacture of mammalian cell secreted product. In B. K. Lydersen (Ed.), *Large Scale Cell Culture Technology* (pp. 113–144). Munich: Hanser.

Hoggan, J. (1977). Aspects of fermentation in conical vessels. *Journal of the Institute of Brewing, 83,* 133–138.

Hough, J. S., Briggs, D. E., & Stevens, R. (1971). *Malting and Brewing Science*. London: Chapman and Hall.

Hristov, H. V., Mann, R., Lossev, V., & Vlaev, S. D. (2004). A simplified CFD for three-dimensional analysis of fluid mixing, mass transfer and bioreaction in a fermenter equipped with triple novel geometry impellers. *Transactions of the Institution of Chemical Engineers*, *82*(C1), 21–34.

Hromatka, O., & Ebner, H. (1949). Untersuchungen über sie Essiggerung. 1. Fesselgarung und Durchlufrungsver-fahren. *Enzymologia*, *13*, 369–387.

HSE. (2014). The genetically modified organisms (contained use) regulations 2014. H.M.S.O., London.

Inskeep, G. G., Wiley, A. J., Holderby, J. M., & Hughes, L. P. (1951). Food yeast from sulphite liquor. *Industrial and Engineering Chemistry*, *43*, 1702–1711.

Irving, G. M. (1968). Construction materials for breweries. *Chemical Engineering (N.Y.)*, *75*(14), 102–104100.

Jackson, T. (1958). Development of aerobic fermentation processes: penicillin. In R. Steel (Ed.), *Biochemical Engineering* (pp. 185–221). London: Heywood.

Jackson, A. T. (1990). Basic heat transfer. In *Process Engineering in Biotechnology* (pp. 58–71). Milton Keynes: Open University Press.

Janssen, D. E., Lovejoy, P. J., Simpson, M. T., & Kennedy, L. D. (1990). Technical problems in containment of rDNA organisms. In P. L. Yu (Ed.), *Fermentation Technologies: Industrial Applications* (pp. 388–393). London: Elsevier.

John, A. H., Bujalski, W., & Nienow, A. W. (1998). The performance of a proto-fermenter containing independently driven dual impellersin a draft tube (IDDIDT): mixing times. *Transactions of the Institution of Chemical Engineers*, *76*(C), 199–207.

Johnson, M. J. (1971). Fermentation—yesterday and tomorrow. *Chemical Technology*, *1*, 338–341.

Jonczyk, P., Takenberg, M., Hartwig, S., Beutel, S., Berger, R. G., & Scheper, T. (2013). Cultivation of shear stress sensitive microorganisms in disposable bag reactor systems. *Journal of Biotechnology*, *167*, 370–376.

Junker, B. (2004). Scale-up methodologies for *Escherichia coli* and yeast fermentation processes. *Journal of Bioscience and Bioengineering*, *97*(6), 347–364.

Junker, B., Walker, A., Hesse, M., Lester, M., Vesey, D., Christensen, J., Burgess, B., & Connors, N. (2009). Pilot scale process development and scale-up for antifungal production. *Bioprocess Biosystems Engineering*, *32*, 443–458.

Katinger, H. W. D. (1977). New fermenter configuration. In J. Meyrath, J.D. Bu'Lock (Eds.), *Biotechnology, fungal differentiation, FEMS, symposium*, vol. 4 (pp. 137–155). London: Academic Press.

Kemplay, J. (1980). *Valve Users Manual*. London: Mechanical Engineering Publications.

Kennedy, L. D., Boland, M. J., Jannsen, D. E., & Frude, M. J. (1990). Designing a pilot-scale fermentation facility for use with genetically modified micro-organisms. In P. L. Yu (Ed.), *Fermentation Technologies: Industrial Applications* (pp. 383–387). London: Elsevier.

Kitai, A., Tone, H., & Ozaki, A. (1969). The performance of a perforated plate column as a multistage continuous fermenter. 1. Washout and growth phase differentiation in the column. *Journal of Fermentation Technology*, *47*, 333–339.

Kuraishi, M., Teroa, L., Ohkouchi, H., Matsuda, N., & Nagai, I. (1977). SCP-process development with methanol as substrate. *DECHEMA Monograph (1978)*, *83*(1704–1723), 111–124.

Langer, E. S., & Rader, R. A. (2014). Single-use technologies in biopharmaceutical manufacturing: A 10 year review of trends and the future. *Engineering in Life Sciences*, *14*, 238–243.

Lavery, M. (1990). Animal cell fermentation. In B. McNiel, & L. M. Harvey (Eds.), *Fermentation—A Practical Approach* (pp. 205–220). Oxford: IRL Press.

Leaver, G. (1994). Interpretation of regulatory requirements to large scale biosafety—The role of the industrial biosafety project. In P. Hambleton, T. Salusbury, & J. Melling (Eds.), *International Safety Aspects in Biotechnology*. Amsterdam: Elsevier.

Leaver, G., & Hambleton, P. (1992). Designing bioreactors to minimize or prevent inadvertant release into the workplace and natural environment. *Pharmac. Technol. (April)*, *4*(3), 18–26.

Lee, S. S., Erikson, L. E., & Fan, L. T. (1971). Modeling and optimization of a tower type activated sludge system. *Biotechnology and Bioengineering Symposium*, *2*, 141–173.

Levi, J. D., Shennan, J. L., & Ebbon, G. P. (1979). Biomass from liquid n-alkanes. In A. H. Rose (Ed.), *Economic Microbiology* (pp. 362–419). (Vol. 4). London: Academic Press.

Lydersen, B. K. (1987). Perfusion cell culture system based on ceramic matrices. In B. K. Lydersen (Ed.), *Large Scale Cell Culture Technology* (pp. 169–192). Munich: Hanser.

Marshall, R. D., Webb, G., Matthews, T. M., & Dean, J. F. (1990). Automatic aseptic sampling of fermentation broth. In *Practical Advances in Fermentation Technology* (pp. 6.1–6.11). Institute of Chemical Engineering North Western Branch Symposium Papers No. 3. University of Manchester Institute of Science and Technology.

Martini, L. J. (1984). *Practical Seal Design*. New York: Dekker.

Maule, D. R. (1986). A century of fermenter design. *Journal of the Institute of Brewing*, *92*, 137–147.

Mayer, E. (1961). Vinegar by oxidative fermentation of alcohol. US Patent 2,997,424.

McAdams, W. H. (1954). *Heating and Cooling Inside Tubes. Heat Transmission* (3rd ed.). New York: McGraw-Hill, pp. 202–251.

Meyrath, J., & Beyer, K. (1979). Biomass from whey. In A. H. Rose (Ed.), *Economic Microbiology* (pp. 207–269). (Vol. 4). London: Academic Press.

Mijnbeek, G. (1991). Sheer stress effects on cultured animal cells. *Bioteknowledge (Applikon)*, *1*, 3–7.

Mikola, M., Seto, J., & Amanullah, A. (2007). Evaluation of a novel Wave Bioreactor® cellbag for aerobic yeast cultivation. *Bioprocess Biosystem Engineering*, *30*, 231–241.

Miller, F. D., & Rushton, J. H. (1944). A mass velocity theory for liquid agitation. *Industrial and Engineering Chemistry*, *36*, 499–503.

Mitchell, C. A. (1926). *Vinegar: Its Manufacture and Examination* (2nd ed.). London: Griffin.

Mitchell, D. A., De Lima Luz, L. F., Krieger, N., & Berovic, M. (2011). Bioreactors for solid-state fermentation. In M. Moo-Young (Ed. in chief), *Comprehensive Biotechnology, Vol. 2*, 2nd ed. (pp. 347–360). Oxford: Elsevier.

Moilanen, P., Laakkonen, M., & Aittamaa, J. (2006). Modeling aerated fermenters with computational fluid dynamics. *Industrial & Engineering Chemistry Research*, *45*, 8656–8663.

Moucha, T., Rejl, F. J., Kordac, M., & Labik, L. (2012). Mass transfer characteristics of multiple-impeller fermenters for their design and scale-up. *Biochemical Engineering Journal*, *69*, 17–27.

Müller, R., & Kieslich, K. (1966). Technology of the microbiological preparation of organic substances. *Angewandte Chemie (International Ed. in English)*, *5*, 653–662.

Nathan, L. (1930). Improvements in the fermentation and maturation of beer. *Journal of the Institute of Brewing*, *36*, 538–550.

Nemati, M., & Webb, C., (2011). Immobilized cell bioreactors. In M. Moo-Young (Ed. in chief), *Comprehensive Biotechnology, Vol. 2*, 2nd ed. (pp. 331–346). Oxford: Elsevier.

Nienow, A. W. (1990). Agitators for mycelial fermentations. *Trends in Biotechnology*, *8*, 224–233.

Nienow, A. W. (1992). New agitators versus Rushton turbines: a critical comparison of transport phenomena. In M. R. Ladisch, & A. Bose (Eds.), *Harnessing Biotechnology for the 21st Century* (pp. 193–196). Washington, DC: American Chemical Society.

Nienow, A. W. (2014). Stirring and stirred tank reactors. *Chemie Ingenieur Technik, 86*(12), 2063–2074.

Nienow, A. W., & Ulbrecht, J. J. (1985). Fermentation broths. In J. J. Ulbrecht, & G. E. Patterson (Eds.), *Mixing of Liquids by Mechanical Agitation* (pp. 203–235). New York: Gordon and Breach.

Nienow, A. W., Waroeskerken, M. M. C. G., Smith, J. M., & Konno, M. (1985). On the flooding/loading transition and the complete dispersal conditions in aerated vessels agitated by a Rushton turbine. *Proceedings of fifth European mixing conference* (pp. 143–154). Cranfield: BHRA.

Nienow, A. W., Allsford, K. V., Cronin, D., Huoxing, L., Haozhung, W., & Hudcova, V. (1988). The use of large ring spargers to improve the performance of fermenters agitated by single and multiple standard Rushton turbines. In R. King (Ed.), *2nd International Conference on Bioreactor Fluid Dynamics* (pp. 159–177). London: Elsevier.

Okabe, M., Ohta, N., & Park, Y. S. (1993). Itaconic acid production in an air-lift bioreactor using a modified draft tube. *Journal of Fermentation and Bioengineering, 76,* 117–122.

Oosterhuis, N. M. G., Hudson, T., D'Avino, A., Zijlstra, G. M., & Amanullah, A. (2011). Disposable bioreactors. In M. Moo-Young (Ed. in chief), *Comprehensive Biotechnology, Vol. 2,* 2nd ed. (pp. 249–261). Oxford: Elsevier.

Owen, W. L. (1948). Continuous fermentation. *Sugar, 43,* 36–38.

Paca, J., Ettler, P., & Gregr, V. (1976). Hydrodynamic behaviour and oxygen transfer rate in a pilot plant fermenter. *Journal of Applied Chemistry and Biotechnology, 26,* 310–317.

Pandey, A. (2003). Solid-state fermentation. *Biochemical Engineering Journal, 13,* 81–84.

Parker, A. (1958). Sterilization of equipment, air and media. In R. Steel (Ed.), *Biochemical Engineering—Unit Processes in Fermentation* (pp. 94–121). London: Heywood.

Pikulik, A. (1976). Selecting and specifying valves for new plants. *Chemical Engineering NewYork, 83,* 168–190.

Pinelli, D., Bakker, A., Myers, K. J., Reeder, M. F., Fasano, J., & Magelli, F. (2003). Some features of a novel gas dispersion impeller in a dual-impeller configuration. *Transactions of the Institute of Chemical Engineers, 81*(A), 448–454.

Prokop, A., & Votruba, J. (1976). Bioengineering problems connected with the use of conventional and unconventional raw materials in fermentation. *Folia Microbiologica, 21,* 58–69.

Prokop, A., Erikson, L. E., Fernandez, J., & Humphrey, A. E. (1969). Design and physical characteristics of a multistage continuous tower fermenter. *Biotechnology and Bioengineering, 11,* 945–966.

Propst, C. L., Von Wedel, R. J., & Lubiniecki, A. S. (1989). Using mammalian cells to produce products. In J. O. Neway (Ed.), *Fermentation Process Development of Industrial Organisms* (pp. 221–276). New York: Dekker.

Richards, J.W. (1968). Design and operation of aseptic fermenters. In *Introduction to Industrial Sterilization* (pp. 107–122). London: Academic Press.

Rivett, R. W., Johnson, M. J., & Peterson, W. H. (1950). Laboratory fermenter for aerobic fermentations. *Industrial and Engineering Chemistry, 42,* 188–190.

Rohr, M., Kubicek, C. P., & Kominek, J. (1983). Citric acid. In H. Dallweg (Ed.), *Biotechnology* (pp. 419–454). (Vol. 3). Weinheim: Verlag Chemie.

Schmidt, F. R. (2005). Optimization and scale up of industrial fermentation processes. *Applied Microbiology and Biotechnology, 68,* 425–435.

Schofield, G. M. (1992). Current legislation and regulatory frameworks. In C. H. Collins, & A. J. Beale (Eds.), *Safety in Industrial Microbiology and Biotechnology* (pp. 6–22). London: Buttterworth-Heinemann.

Schreier, K. (1975). High-efficiency fermenter with deep-jet aerators. *Chemiker Zeitung, 99*, 328–331.

Schugerl, K. (1982). New bioreactors for aerobic processes. *International Chemical Engineering, 22*, 591–610.

Schugerl, K. (1985). Nonmechanically agitated bioreactor systems. In C. L. Cooney, & A. E. Humphrey (Eds.), *Comprehensive Biotechnology* (pp. 99–118). (Vol. 2). Oxford: Pergamon Press.

Schugerl, K., Lucre, J., & Oels, U. (1977). Bubble column bioreactors. *Advances in Biochemical Engineering, 7*, 1–84.

Schüller, M., & Seidel, M. (1940). Yeast production and fermentation. US Patent 2,188,192.

Scragg, A. H. (1991). *Bioreactors in Biotechnology—A Practical Approach*. Chichester: Ellis Horwood, pp. 112–125.

Sharp, D. H. (1989). The development of Pruteen by ICI. In *Bioprotein Manufacture* (pp. 53–78). Chichester: Ellis Horwood.

Shore, D. T., Royston, M. G., & Watson, E. G. (1964). Improvements in or relating to the production of beer. British Patent 959,049.

Shuttlewood, J. R. (1984). *Brew Distilling Int.* (August), p. 22.

Sittig, W. (1982). The present state of fermentation reactors. *Journal of Chemical Technology and Biotechnology, 32*, 47–58.

Smith, J. M. (1985). Dispersion of gases in liquids: the hydrodynamics of gas dispersion in low viscosity liquids. In J. J. Ulbrecht, & G. K. Patterson (Eds.), *Mixing of Liquids by Mechanical Agitation* (pp. 139–201). New York: Gordon and Breach.

Smith, S. R. L. (1980). Single cell protein. *Philosophical Transactions of the Royal Society, B290*, 341–354.

Snell, R. L., & Schweiger, L. B. (1949). Production of citric acid by fermentation. US Patent 2,492,667.

Solomons, G. L. (1969). *Materials and Methods in Fermentation*. London: Academic Press.

Solomons, G. L. (1980). Fermenter design and fungal growth. In J. E. Smith, D. R. Berry, & B. Kristiansen (Eds.), *Fungal Biotechnology* (pp. 55–80). London: Academic Press.

Spivey, M. J. (1978). The acetone-butanol-ethanol fermentation. *Process Biochemistry, 13*(11), 2–425.

Steel, R., & Maxon, W. D. (1961). Power requirements of a typical actinomycete fermentation. *Industrial and Engineering Chemistry, 53*, 739–742.

Steel, R., & Maxon, W. D. (1966). Studies with a multiple-rod mixing impeller. *Biotechnology and Bioengineering, 8*, 109–115.

Steel, R., & Miller, T. L. (1970). Fermenter design. *Advances in Applied Microbiology, 12*, 153–188.

Steel, R., Lentz, C. P., & Martin, S. M. (1955). Submerged citric acid production of sugar beet molasses: increase in scale. *Canadian Journal of Microbiology, 1*, 299–311.

Strauch and Schmidt (1932). German Patent 552,241. Cited by de Becze and Liebmann (1944).

Tang, W., Pan, A., Lu, H., Xia, J., Zhuang, Y., Zhang, S., Chu, J., & Noorman, H. (2015). Improvement in glucoamylase production using axial impellers with low power consumption and homogenous mass transfer. *Biochemical Engineering Journal, 99*, 167–176.

Taya, M., & Kino-Oka, M., (2011). Bioreactors for animal cell cultures. In M. Moo-Young (Ed. in chief), *Comprehensive Biotechnology, Vol. 2,* 2nd ed. (pp. 373–381). Oxford: Elsevier.

Taylor, I. J., & Senior, P. J. (1978). Single cell proteins: a new source. *Endeavour, 2*, 31–34.

Thaysen, A. C. (1945). Production of food yeast. *Food, 14,* 116–119.

Thielsch, H. (1967). Manufacture, fabrication and joining of commercial piping. In R. C. King (Ed.), *Piping Handbook* (5th ed., pp. 7.1–7.300). New York: McGraw-Hill.

Titchener-Hooker, N. J., Sinclair, P. A., Hoare, M., Vranch, S.P., Cottam, A., & Turner, M. K. (1993). The specification of static seals for contained operations; an engineering appraisal. *Pharmaceutical Technol. (Europe)* October, 26–30.

Tolbert, W. R., Srigley, W. R., & Prior, C. P. (1988). Perfusion culture systems for large-scale pharmaceutical production. In R. E. Spiers, & J. B. Griffiths (Eds.), *Animal Cell Biotechnology* (pp. 374–393). (Vol. 3). London: Academic Press.

Trinci, A. P. J. (1992). Myco-protein: a twenty year overnight success story. *Mycolological Research, 96,* 1–13.

Tubito, P. J. (1991). Contamination control facilities for the biotechnology industry. In W. Whyte (Ed.), *Cleanroom Design* (pp. 85–120). Chichester: Wiley.

Turner, M. K. (1989). Categories of large-scale containment for manufacturing processes with recombinant organisms. *Biotechnology and Genetic Engineering Reviews, 7,* 1–43.

Ulenberg, G. H., Gerritson, H., & Huisman, J. (1972). Experiences with a giant cylindro-conical tank. *Master Brewers Association of the Americas Technical Quarterly, 9,* 117–122.

Van Houten, J. (1992). Containment of fermentations: comprehensive assessment and integrated control. In M. R. Ladisch, & A. Bose (Eds.), *Harnessing Biotechnology for the 21st Century* (pp. 415–418). Washington, DC: American Chemical Society.

Van Wezel, A. L. (1967). Growth of cell strains and primary cells on microcarriers in homogenous cultures. *Nature (London), 216,* 64–65.

Varley, J., & Birch, J. (1999). Reactor design for large scale suspension animal cell culture. *Cytotechnology, 29,* 177–205.

Victorero, F. A. (1948). Apparatus for continuous fermentation. US Patent 2,450,218.

Vranch, S. P. (1992). Engineering for safe bioprocessing. In C. H. Collins, & A. J. Beale (Eds.), *Safety in Industrial Microbiology and Biotechnology* (pp. 176–189). London: Butterworth-Heinemann.

Walker, J. A. H., & Holdsworth, H. (1958). Equipment design. In R. Steel (Ed.), *Biochemical Engineering* (pp. 223–273). London: Heywood.

Walker, P. D., Narendranathan, T. J., Brown, D. C., Woodhouse, F., & Vranch, S. P. (1987). Containment of micro-organisms during fermentation and downstream processing. In M. S. Verrall, & M. J. Hudson (Eds.), *Separation for Biotechnology* (pp. 469–479). Chichester: Ellis Horwood.

Watanabe, K. (1976). Production of RNA. In K. Ogata, S. Kinoshita, T. Tsonoda, & K. Aida (Eds.), *Microbial Production of Nucleic Acid Related Substances* (pp. 55–65). New York: Kodansha, Tokyo and Halsted.

Wegrich, R. H., & Shurter, R. A. (1953). Development of a typical aerobic fermentation. *Industrial and Engineering Chemistry, 45,* 1153–1160.

Werner, R. G. (1992). Containment in the development and manufacture of recombinant DNA-derived products. In C. H. Collins, & A. J. Beale (Eds.), *Safety in Industrial Microbiology and Biotechnology* (pp. 190–213). London: Butterworth-Heinemann.

Westbrook, A., Scharer, J., Moo-Young, M., Oosterhuis, N., & Chou, C. P. (2014). Application of a two-dimensional disposable rocking bioreactor to bacterial cultivation for recombinant protein production. *Biochemical Engineering Journal, 88,* 154–161.

Wilkinson, P. J. (1987). The development of a large scale production process for tissue culture products. In G. W. Moody, & P. B. Baker (Eds.), *Bioreactors and Biotransformation* (pp. 111–120). London: Elsevier.

Winkler, M. A. (1990). Problems in fermenter design and operation. In M. A. Winkler (Ed.), *Chemical Engineering Problems in Biotechnology* (pp. 215–350). London: SCI/Elsevier.

Wu, W. T., & Wu, J. Y. (1990). Air lift reactor with draught tube. *Journal of Fermentation and Bioengineering*, *70*, 359–361.

Xie, M., Xia, J., Zhou, Z., Chu, J., Zhuang, Y., & Zhang, S. (2014). Flow pattern, mixing, gas hold-up and mass transfer coefficient of triple-impeller configurations in stirred tank bio-reactors. *Industrial and Engineering Chemistry Research*, *53*, 5941–5953.

Yang, Y., Xia, J., Li, J., Chu, J., Li, L., Wang, Y., Zhuang, Y., & Zhang, S. (2012). A novel impeller configuration to improve fungal physiology performance and energy conservation for cephalosporin C production. *Journal of Biotechnology*, *161*, 250–256.

Zhong, J.-J. (2011). Bioreactor engineering. In M. Moo-Young (Ed. in chief), Comprehensive Biotechnology, Vol. 2, 2nd ed. (pp. 165–177). Oxford: Elsevier.

FURTHER READING

Anderson, J. G., & Blain, J. A. (1980). Novel developments in microbial film reactors. In J. E. Smith, D. R. Berry, & B. Kristiansen (Eds.), *Fungal Biotechnology* (pp. 125–152). London: Academic Press.

Instrumentation and control

INTRODUCTION

The success of a fermentation depends upon the existence of defined environmental conditions for biomass and product formation. To achieve this goal it is important to understand what is happening to a fermentation process and how to control it to obtain optimal operating conditions. Thus, conditions such as temperature, pH, degree of agitation, oxygen concentration in the medium, and other factors such as nutrient/substrate concentrations may have to be kept constant during the process. The provision of such conditions requires careful monitoring (data acquisition and analysis) of the fermentation so that any deviation from the specified optimum might be corrected by a control system. Criteria which are monitored frequently are listed in Table 8.1, along with the control processes with which they are associated. As well as aiding the maintenance of constant conditions, the monitoring of a process may provide information on the progress of the fermentation. Such information may indicate the optimum time to harvest or that the fermentation is progressing abnormally which may be indicative of contamination or strain degeneration. Thus, monitoring equipment produces information indicating fermentation progress as well as being linked to a suitable control system.

In initial studies the number of factors which are to be controlled may be restricted in order to gain more knowledge regarding a particular fermentation. Thus, the pH may be measured and recorded but not maintained at a specified pH or the dissolved oxygen concentration may be determined but no attempt will be made to prevent oxygen depletion.

Also, it is important to consider the need for a sensor and its associated control system to interface with a computer (to be discussed in a later section). This chapter will consider the general types of control systems which are available, specific monitoring and control systems, and the role of computers and more recent computational developments such as artificial neural networks and fuzzy control. More information on intrumentation and control has been written by Flynn (1983, 1984), Armiger (1985), Bull (1985), Rolf and Lim (1985), Bailey and Ollis (1986), Kristiansen (1987), Montague, Morris, and Ward (1988), Dusseljee and Feijen (1990), Atkinson and Mavituna (1991), Royce (1993), Alford (2006), Moo-Young (2011), and Doran (2013).

Table 8.1 Process Sensors and Their Possible Control Functions

Category	Sensor	Possible Control Function
Physical	Temperature	Heat/cool
	Pressure	
	Agitator shaft power	
	rpm	
	Foam	Foam control
	Weight	Change flow rate
	Flow rate	Change flow rate
Chemical	pH	Acid or alkali addition, carbon source feed rate
	Redox	Additives to change redox potential
	Oxygen	Change feed rate
	Exit-gas analysis	Change feed rate
	Medium analysis	Change in medium composition

It is apparent from Table 8.1 that a considerable number of process variables may need to be monitored during a fermentation. Methods for measuring these variables, the sensors or other equipment available and possible control procedures are outlined later.

There are three main classes of sensor:

1. Sensors which penetrate into the interior of the fermenter, for example, pH electrodes, dissolved-oxygen electrodes.
2. Sensors which operate on samples which are continuously withdrawn from the fermenter, for example, exhaust-gas analyzers.
3. Sensors which do not come into contact with the fermentation broth or gases, for example, tachometers, load cells.

It is also possible to characterize a sensor in relation to its application for process control:

1. *In line sensor.* The sensor is an integrated part of the fermentation equipment and the measured value obtained from it is used directly for process control.
2. *On line sensor.* Although the sensor is an integral part of the fermentation equipment, the measured value cannot be used directly for control. An operator must enter measured values into the control system if the data is to be used in process control.
3. *Off line sensor.* The sensor is not part of the fermentation equipment. The measured value cannot be used directly for process control. An operator is needed for the actual measurement (eg, medium analysis or dry weight sample) and for entering the measured values into the control system for process control.

When evaluating sensors to use in measurement and control it is important to consider response time, gain, sensitivity, accuracy, ease and speed of calibration, stability, reliability, output signal (continuous or discontinuous), materials of construction, robustness, sterilization, maintenance, availability to purchase, and cost (Flynn, 1983, 1984; Royce, 1993).

METHODS OF MEASURING PROCESS VARIABLES
TEMPERATURE
The temperature in a vessel or pipeline is one of the most important parameters to monitor and control in any process. It may be measured by mercury-in-glass thermometers, bimetallic thermometers, pressure bulb thermometers, thermocouples, metal-resistance thermometers, or thermistors. Metal-resistance thermometers and thermistors are used in most fermentation applications. Accurate mercury-in-glass thermometers are used to check and calibrate the other forms of temperature sensors, while cheaper thermometers are still used with laboratory fermenters. Temperature measurements may also need to be correlated to other analyses in fermentations. For example, refractive index may be used in lieu of density measurements in the control and supervision of alcoholic fermentations. However, refractive index is usually measured at 20°C and hence if the fermentation progresses at temperatures other than 20°C, temperature correction of refractive index measurements will be required (Jimenez-Marquez, Vazquez, Ubeda, & Sanchez-Rojas, 2016).

Mercury-in-glass thermometers
A mercury-in-glass thermometer may be used directly in small laboratory scale fermenters, but its fragility restricts its use. In larger fermenters it would be necessary to insert it into a thermometer pocket in the vessel, which introduces a time lag in registering the vessel temperature. This type of thermometer can be used solely for indication, not for automatic control or recording.

Electrical resistance thermometers
It is well known that the electrical resistance of metals changes with temperature variation. This property has been utilized in the design of resistance thermometers. The bulb of the instrument contains the resistance element, a mica framework (for very accurate measurement) or a ceramic framework (robust, but for less accurate measurement) around which the sensing element is wound. A platinum wire of 100 Ω resistance is normally used. Leads emerging from the bulb are connected to the measuring element. The reading is normally obtained by the use of a Wheatstone bridge circuit and is a measure of the average temperature of the sensing element. This type of thermometer does have a greater accuracy ($\pm$ 0.25%) than some of the other measuring devices and is more sensitive to small temperature changes. There is a fast response to detectable changes (1–10 s), and there is no restriction on distance between the very compact sensing point (30 $\times$ 5 mm) and the display point of

reproducible readings. These thermometers are normally enclosed in stainless-steel sheaths if they are to be used in large vessels and ancillary equipment.

Thermistors

Thermistors are semiconductors made from specific mixtures of pure oxides of iron, nickel, and other metals. Their main characteristic is a large change in resistance with a small temperature change. The change in resistance is a function of absolute temperature. The temperature reading is obtained with a Wheatstone bridge or a simpler or more complex circuit depending on the application. Thermistors are relatively cheap and have proved to be very stable, give reproducible readings, and can be sited remotely from the read-out point. Their main disadvantage is the marked nonlinear temperature versus resistance curve.

Temperature control

The use of water jackets or pipe coils within a fermenter as a means of temperature control has been described in Chapter 7. In many small systems there is a heating element, 300–400 W capacity being adequate for a 10 dm^3 fermenter, and a cooling water supply; these are on or off depending on the need for heating or cooling. The heating element should be as small as possible to reduce the size of the "heat sink" and result in overshoot when heating is no longer required. In some cases it may be better to run the cooling water continuously at a low but steady rate and to have only the heating element connected to the control unit. This can be an expensive mode of operation if the water flows directly to waste. Recirculating thermostatically heated water through fermenters of up to 10 dm^3 capacity can give temperature control of ±0.1°C.

In large fermenters, where heating during the fermentation is not normally required, a regulatory valve at the cooling-water inlet may be sufficient to control the temperature. There may be provision for circulation of refrigerated water or brine if excessive cooling is required. Steam inlets to the coil and jacket must be present if a fermenter is being used for batch sterilization of media.

Low agitation speeds are often essential in animal cell culture vessels to minimize shear damage. In these vessels, heating fingers can create local "hot-spots" which may cause damage to cells very close to them. Heating jackets which have a lower heat output proportional to the surface area (water or silicone rubber covered electrical heating elements—Chapter 7) are used to overcome this problem.

Temperature (and dissolved oxygen) control in fermenters using varying regimes of substrate feeding have been investigated by Vanags, Rychtera, Ferzik, Vishkins, and Viesturs (2007). Restrepo, Gonzalez, and Orduz (2002) have compared the use of on/off control with proportional/integral/derivative (PID) temperature control in *Bacillus thuringiensis* fermentations.

FLOW MEASUREMENT AND CONTROL

Flow measurement and control of both gases and liquids is important in process management.

Gases

One of the simplest methods for measuring gas flow to a fermenter is by means of a variable area meter. The most commonly used example is a rotameter, which consists of a vertically mounted glass tube with an increasing bore and enclosing a free-moving float which may be a ball or a hollow thimble. The position of the float in the graduated glass tube is indicative of flow rate. Different sizes can cater for a wide range of flow rates. The accuracy depends on having the gas at a constant pressure, but errors of up to ±10% of full-scale deflection are quoted (Howe, Kop, Siev, & Liptak, 1969). The errors are greatest at low flow rates. Ideally, rotameters should not be sterilized and are therefore normally placed between a gas inlet and a sterile filter. There is no provision for online data logging with the simple rotameters. Metal tubes can be used in situations where glass is not satisfactory. In these cases the float position is determined by magnetic or electrical techniques, but this provision has not been normally utilized for fermentation work. Rotameters can also be used to measure liquid flow rates, provided that abrasive particles or fibrous matter are not present.

The use of oxygen and carbon dioxide gas analyzers for effluent gas analysis requires the provision of very accurate gas-flow measurement if the analyzers are to be used effectively. For this reason thermal mass flowmeters have been utilized for the range 0–500 dm^3 min^{-1}. These instruments have a ±1% full-scale accuracy and work on the principle of measuring a temperature difference across a heating device placed in the path of the gas flow (Fig. 8.1). Temperature probes such as thermistors are placed upstream and downstream of the heat source, which may be inside or outside the pipeline.

The mass flow rate of the gas, Q, can be calculated from the specific heat equation:

$$H = QC_p(T_2 - T_1)$$

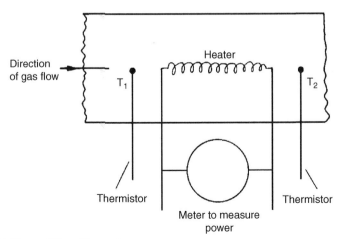

FIGURE 8.1 Thermal Mass Flowmeter

where H = heat transferred, Q = mass flow rate of the gas, C_p = specific heat of the gas, T_1 = temperature of gas before heat is transferred to it, and T_2 = temperature of gas after heat is transferred to it.

This equation can then be rearranged for Q:

$$Q = \frac{H}{C_p(T_2 - T_1).}$$

A voltage signal can be obtained by this method of measurement which can be utilized in data logging.

Control of gas flow is usually by needle valves. Often this method of control is not sufficient, and it is necessary to incorporate a self-acting flow-control valve. At a small scale, such valves as the "flowstat" are available (G.A. Platon, Ltd, Wella Road, Basingstoke, Hampshire, UK.). Fluctuations in pressure in a flow-measuring orifice cause a valve or piston pressing against a spring to gradually open or close so that the original, preselected flow rate is restored. In a gas "flowstat," the orifice should be upstream when the gas supply is at a regulated pressure and downstream when the supply pressure fluctuates and the back pressure is constant. Valves operating by a similar mechanism are available for larger scale applications.

Liquids

The flow of nonsterile liquids can be monitored by a number of techniques (Howe et al., 1969), but measurement of flow rates of sterile liquids presents a number of problems which have to be overcome. On a laboratory scale flow rates may be measured manually using a sterile burette connected to the feed pipe and timing the exit of a measured volume. The possible use of rotameters has already been mentioned in the previous section. A more expensive method is to use an electrical flow transducer (Howe et al., 1969), which can cope with particulate matter in the suspension and measure a range of flow rates from very low to high (50 cm^3 min^{-1} to 500,000 dm^3 min^{-1}) with an accuracy of ±1%. In this flowmeter (Fig. 8.2) there are two windings outside the tube, supplied with an alternating current to create a magnetic field. The voltage induced in the field is proportional to the relative velocity of the fluid and the magnetic field. The potential difference in the fluid can be measured by a pair of electrodes, and is directly proportional to the velocity of the fluid.

In batch and fed-batch fermenters, a cheaper alternative is to measure flow rates indirectly by load cells (see Section "Weight"). The fermenter and all ancillary reservoirs are attached to load cells, which monitor the increases and decreases in weight of the various vessels at regular time intervals. Provided the specific gravities of the liquids are known it is possible to estimate flow rates fairly accurately in different feed pipes. This is another technique which may be used with particulate suspensions.

Another indirect method of measuring flow rates aseptically is to use a metering pump, which pumps liquid continuously at a predetermined and accurate rate.

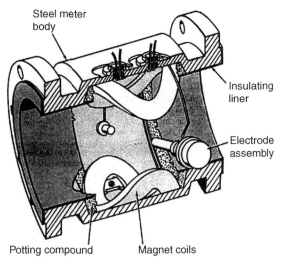

FIGURE 8.2 A Cut-Away View of a Short-Form Magnetic Flowmeter (Howe et al., 1969)

A variety of metering pumps are commercially available including motorized syringes, peristaltic pumps, piston pumps, and diaphragm pumps. Motorized syringes are used only when very small quantities of liquid have to be added slowly to a vessel. In a peristaltic pump, liquid is moved forward gradually by squeezing a compressible tubing (eg, silicone tubing) held in a semicircular housing. Failure (splitting) of the tubing is an obvious drawback. A variety of sizes of tubes can be used in different pumps to produce different known flow rates over a very wide range. Suspensions can be handled since the liquid has no direct contact with moving parts.

A piston pump contains an accurately machined ceramic or stainless-steel piston moving in a cylinder normally fitted with double ball inlet and outlet valves. The piston is driven by a constant-speed motor. Flow rates can be varied within a defined range by changing the stroke rate, the length of the piston stroke, and by using a different piston size. Sizes are available from cubic centimeters per hour to thousands of cubic decameter per hour and all can be operated at relatively high working pressures. Unfortunately, they cannot be used to pump fibrous or particulate suspensions. Piston pumps are more expensive than comparable sized peristaltic pumps but do not suffer from tube failure.

Leakage can occur via the shaft housing of a piston pump. The problem can be prevented by the use of a diaphragm pump. This pump uses a flexible diaphragm to pump fluid through a housing (Fig. 8.3) with ball valves to control the direction of flow. The diaphragm may be made of, for example, Teflon, neoprene, stainless steel, and is actuated by a piston. A range of sizes of pumps is available for flow rates up to thousands of cubic decimeter per hour.

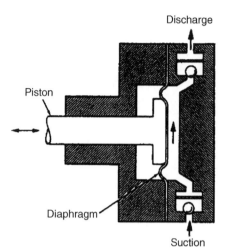

FIGURE 8.3 A Direct-Driven Diaphragm Pump (Howe et al., 1969)

PRESSURE MEASUREMENT

Pressure is one of the crucial measurements that must be made when operating many processes. Pressure measurements may be needed for several reasons, the most important of which is safety. Industrial and laboratory equipment is designed to withstand a specified working pressure plus a factor of safety. It is therefore important to fit the equipment with devices that will sense, indicate, record, and control the pressure. The measurement of pressure is also important in media sterilization. In a fermenter, pressure will influence the solubility of gases and contribute to the maintenance of sterility when a positive pressure is present.

One of the standard pressure measuring sensors is the Bourdon tube pressure gauge (Fig. 8.4), which is used as a direct indicating gauge. The partial coil has an elliptical cross-section (A–A) which tends to become circular with increasing pressure, and because of the difference between the internal and external radii, gradually straightens out. The process pressure is connected to the fixed socket end of the tube while the sealed tip of the other end is connected by a geared sector and pinion movement, which actuates an indicator pointer to show linear rotational response (Liptak, 1969).

When a vessel or pipeline is to be operated under aseptic conditions a diaphragm gauge can be used (Fig. 8.5). Changes in pressure cause movements of the diaphragm capsule which are monitored by a mechanically levered pointer.

Alternatively, the pressure could be measured remotely using pressure bellows connected to the core of a variable transformer. The movement of the core generates a corresponding output. It is also possible to use pressure sensors incorporating strain gauges. If a wire is subject to strain its electrical resistance changes; this is due, in part, to the changed dimensions of the wire and the change in resistivity which occurs due to the stress in the wire. The output can then be measured over

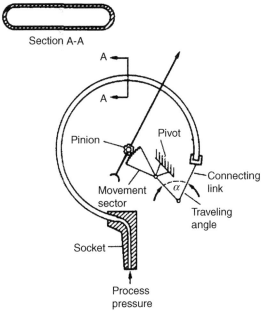

FIGURE 8.4 "C" Bourdon Tube Pressure Gauge (Liptak, 1969)

long distances. Another electrical method is to use a piezoelectric transducer. Certain solid crystals such as quartz have an asymmetrical electrical charge distribution. Any change in shape of the crystal produces equal, external, unlike electric charges on the opposite faces of the crystal. This is the piezoelectric effect. Pressure can therefore be measured by means of electrodes attached to the opposite surfaces of the crystal. Bioengineering AG (Wald, Switzerland) have made a piezoelectrical transducer with integral temperature compensation, to overcome pyroelectric effects, and built into a housing which can be put into a fermenter port.

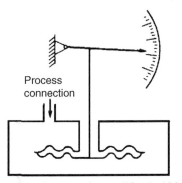

FIGURE 8.5 Nested Diaphragm-Type Pressure Sensor (Liptak, 1969)

It will also be necessary to monitor and record atmospheric pressure if oxygen concentrations in inlet and/or exit gases are to be determined using oxygen gas analyzers (see in later section). Paramagnetic gas analyzers are susceptible to changes in barometric pressure. A change of 1% in pressure may cause a 1% change in oxygen concentration reading. This size of error may be very significant in a vessel where the oxygen consumption rate is very low and there is very little difference between the inlet and exit gas compositions. The pressure changes should be constantly monitored to enable the appropriate corrections to be made.

PRESSURE CONTROL

Different working pressures are required in different parts of a fermentation plant. During normal operation a positive head pressure of 1.2 atmospheres absolute is maintained in a fermenter to assist in the maintenance of aseptic conditions. This pressure will obviously be raised during a steam-sterilization cycle (Chapter 5). The correct pressure in different components should be maintained by regulatory valves (Chapter 7) controlled by associated pressure gauges.

SAFETY VALVES

Safety valves (Chapter 7) should be incorporated at various suitable places in all vessels and pipeline layouts which are likely to be operated under pressure. The valve should be set to release the pressure as soon as it increases markedly above a specified working pressure. Other provisions will be necessary to meet any containment requirements. Bursting disc designed to fail at a given pressure or vacuum may also be incorporated into the fermenter body.

AGITATOR SHAFT POWER

A variety of sensors can be used to measure the power consumption of a fermenter. On a large scale, a Watt meter attached to the agitator motor will give a fairly good indication of power uptake. This measuring technique becomes less accurate as there is a decrease in scale to pilot scale and finally to laboratory fermenters, the main contributing factor being friction in the agitator shaft bearing (Chapter 7). Torsion dynamometers can be used in small-scale applications. Since the dynamometer has to be placed on the shaft outside the fermenter the measurement will once again include the friction in the bearings. For this reason strain gauges mounted on the shaft within the fermenter are the most accurate method of measurement and overcome frictional problems (Aiba, Okamoto, & Satoh, 1965; Brodgesell, 1969). Aiba et al. (1965) mounted four identical strain gauges at 45°C to the axis in a hollow shaft. Lead wires from the gauges passed out of the shaft via an axial hole and electrical signals were then picked up by an electrical slip-ring arrangement. Theoretical treatment of the strain gauge measurements has been covered by Aiba, Humphrey, and Millis (1973).

RATE OF STIRRING

In all fermenters it is important to monitor the rate of rotation (rpm) of the stirrer shaft. The tachometer used for this purpose may employ electromagnetic induction voltage generation, light sensing, or magnetic force as detection mechanisms (Brodgesell, 1969). Obviously, the final choice of tachometer will be determined by the type of signal which is required for recording and/or process control for regulating the motor speed and other ancillary equipment. Provision is often made on small laboratory fermenters to vary the rate of stirring. In most cases it is now standard practice to use an ac slip motor that has an acceptable torque curve that is coupled to a thyristor control. At pilot or full scale, the need to change rates of stirring is normally reduced. When necessary it can be done using gear boxes, modifying the sizes of wheels and drive belts, or by changing the drive motor, which is the most expensive alternative.

FOAM SENSING AND CONTROL

The formation of foam is a difficulty in many types of microbial fermentation which can create serious problems if not controlled. It is common practice to add an antifoam to a fermenter when the culture starts foaming above a certain predetermined level. The methods used for foam sensing and antifoam additions will depend on process and economic considerations. The properties of antifoams have been discussed elsewhere (Chapters 4 and 7), as has their influence on dissolved oxygen concentrations (Chapter 9).

A foam sensing and control unit is shown in Fig. 8.6. A probe is inserted through the top plate of the fermenter. Normally the probe is a stainless-steel rod, which is insulated except at the tip, and set at a defined level above the broth surface. When the foam rises and touches the probe tip, a current is passed through the circuit of the probe, with the foam acting as an electrolyte and the vessel acting as an earth.

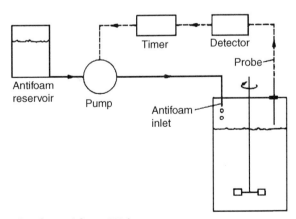

FIGURE 8.6 Foam Sensing and Control Unit

The current actuates a pump or valve and antifoam is released into the fermenter for a few seconds. Process timers are routinely included in the circuit to ensure that the antifoam has time to mix into the medium and break down the foam before the probe is programmed after a preset time interval to sense the foam level again and possibly actuate the pump or valve (feed-on-demand). Alternatively antifoam may be added slowly at a predetermined rate by a small pump so that foaming never occurs and therefore there is no need for a sensing system.

A number of mechanical antifoam devices have been described including discs, propellers, brushes, or hollow cones attached to the agitator shaft above the surface of the broth. The foam is broken down when it is thrown against the walls of the fermenter. Other devices which have been manufactured include horizontal rotating shafts, centrifugal separators, and jets spraying on to deflector plates (Hall, Dickinson, Pritchard, & Evans, 1973; Viesturs, Kristapsons, & Levitans, 1982). Unfortunately most of these devices have to be used in conjunction with an antifoam.

WEIGHT

A load cell offers a convenient method of determining the weight of a fermenter or feed vessel. This is done by placing compression load cells in or at the foot of the vessel supports. When designing the support system for a fermenter or other vessel, the weight of which is to be measured by load cells, the principle of the three-legged stool should be remembered. Three feet will always rest in stable equilibrium even if the supporting surface is uneven. If more feet are provided, each additional feet must be fitted with means of adjustment or precision packing to ensure equal load bearing on all the feet.

A load cell is essentially an elastic body, usually a solid or tubular steel cylinder, the compressive strain of which under axial load may be measured by a series of electrical resistance strain gauges which are cemented to the surface of the cylinder. The load cell is assembled in a suitable housing with electrical cable connecting points. The cell is calibrated by measuring compressive strain over the appropriate range of loading. Changes of resistance with strain which are proportional to load are determined by appropriate electrical apparatus.

It is therefore possible to use appropriately sized load cells to monitor feed rates from medium reservoirs, acid and base utilization for pH control, and the use of antifoam for foam control. The change in weight in a known time interval can be used indirectly as a measure of liquid flow rates.

MICROBIAL BIOMASS

Real-time estimation of microbial biomass in a fermenter is an obvious requirement, yet it has proved very difficult to develop a satisfactory sensor. Most monitoring has been done indirectly by dry weight samples (made quicker with microwave ovens), cell density (spectrophotometers), cell numbers (Coulter counters) or by the use of gateway sensors which will be discussed later in this chapter. Cell density can also be determined in-situ using probes, which measure optical loss by absorption/light scattering caused by cells in the fermentation broth (Luong, Mahmoud, & Male, 2011). This technique uses near infrared wavelength light (see in later section) rather than visible wavelengths conventionally used in offline spectrophotometers. Other alternative approaches are real-time estimation of a cell component which remains at a constant concentration, such as nicotinamide adenine dinucleotide (NAD), by fluorimetry or measurement of a cell property which is proportional to the concentration of viable cells, such as radio frequency capacitance.

The measurement of NAD, provided that it remains at a constant concentration in cells, would be an ideal indirect method for continuous measurement of microbial biomass. In pioneer studies, Harrison and Chance (1970) used a fluorescence technique to determine NAD–NADH levels inside microbial cells growing in continuous culture. Einsele, Ristroph, and Humphrey (1978) mounted a fluorimeter on a fermenter observation port located beneath the culture surface which enabled the measurement of NADH fluorescence in situ, making it possible to determine bulk mixing times in the broth and to follow glucose uptake by monitoring NADH levels. Beyeler, Einsele, and Fiechter (1981) were able to develop a small sterilizable probe for fitting into a fermenter to monitor NADH, which had high specificity, high sensitivity, high stability, and could be calibrated in situ. In batch culture of *Candida tropicalis,* the NAD(P)H-dependent fluorescence signal correlated well with biomass, so that it could be used for online estimation of biomass. Changes in the growth conditions, such as substrate exhaustion or the absence of oxygen, were also very quickly detected.

Schneckenburger, Reuter, and Schoberth (1985) used this technique to study the growth of methanogenic bacteria in anaerobic fermentations. They thought cost was a problem, fluorescence equipment being too expensive for routine biotechnology applications when the minimum price was about US$10,000. Ingold (Switzerland) have developed the Fluorosensor, a probe which can be integrated with a small computer or any data transformation device (Gary, Meier & Ludwig, 1988).

Adenosine 5' triphosphate (ATP) can also be used as an indirect measurement of microbial biomass and microbial activity. Following extraction of ATP from the cells using, for example, dimethylsulfoxide (DMSO), it is then reacted with reduced luciferin catalyzed by firefly luciferase, which then yields a photon of light per ATP molecule. ATP concentration can then be correlated to the biomass present (Luong et al., 2011).

Dielectric spectroscopy can be used online to monitor biomass. Details of the theory and principles of this technique have been described by Kell (1987). This sensor has proved ideal for yeast cells and is now being used by the brewing industry to control yeast pitching rates (Boulton, Maryan, Loveridgev, & Kell, 1989).

MEASUREMENT AND CONTROL OF DISSOLVED OXYGEN

In most aerobic fermentations it is essential to ensure that the dissolved oxygen concentration does not fall below a specified minimal level. Since the 1970s steam sterilizable oxygen electrodes have become available for this monitoring (Fig. 8.7). Details of electrodes are given by Lee and Tsao (1979).

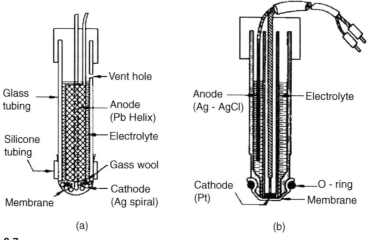

Glass tubing — Vent hole
— Anode (Pb Helix)
Silicone tubing — Electrolyte
— Gass wool
Membrane — Cathode (Ag spiral)

(a)

Anode (Ag - AgCl) — Electrolyte
Cathode (Pt) — O - ring
— Membrane

(b)

FIGURE 8.7

Construction of dissolved-oxygen electrodes: (a) galvanic, (b) Polarographic (Lee & Tsao, 1979).

These electrodes measure the partial pressure of the dissolved oxygen and not the dissolved oxygen concentration. Thus at equilibrium, the probe signal of an electrode will be determined by:

$$P(O_2) = C(O_2) \times P_T$$

where $P(O_2)$ is the partial pressure of dissolved oxygen sensed by the probe, $C(O_2)$ is the volume or mole fraction of oxygen in the gas phase, and P_T is the total pressure.

The actual reading is normally expressed as percentage saturation with air at atmospheric pressure, so that 100% dissolved oxygen means a partial pressure of approximately 160 mmHg.

Pressure changes can have a significant effect on readings. If the total pressure of the gas equilibrating with the fermentation broth varies, the electrode reading will change even though there is no change in the gas composition. Changes in atmospheric pressure can often cause 5% changes and back pressure due to the exit filters can also cause increase in readings. Allowance must also be made for temperature. The output from an electrode increases by approximately 2.5% per °C at a given oxygen tension. This effect is due mainly to increases in permeability in the electrode membrane. Many electrodes have builtin temperature sensors which allow automatic compensation of the output signal. It is also important to remember that the solubility of oxygen in aqueous media is influenced by the composition. Thus, water at 25°C and 760 mmHg pressure saturated with air will contain 8.4 mg O_2 dm^{-3}, while 25% w/w NaCl in identical conditions will have an oxygen solubility of 2.0 mg O_2 dm^{-3}. However, the measured partial pressure outputs for O_2 would be the same even though the oxygen concentrations would be very different. Therefore it is best to calibrate the electrode in percentage oxygen saturation. More details on oxygen electrodes and their calibration has been given by Hailing (1990).

In small fermenters (1 dm^3), the commonest electrodes are galvanic and have a lead anode, silver cathode and employ potassium hydroxide, chloride, bicarbonate or acetate as an electrolyte. The sensing tip of the electrode is a Teflon, polyethylene, or polystyrene membrane, which allows passage of the gas phase so that an equilibrium is established between the gas phases inside and outside the electrode. Because of the relatively slow movement of oxygen across the membrane, this type of electrode has a slow response of the order of 60 s to achieve a 90% reading of true value (Johnson, Borkowski, & Engblom, 1964). Buhler and Ingold (1976) quote 50 s for 98% response for a later version. These electrodes are therefore suitable for monitoring very slow changes in oxygen concentration and are normally chosen because of their compact size and relatively low cost. Unfortunately, this type of electrode is very sensitive to temperature fluctuations, which should be compensated for by using a thermistor circuit. The electrodes also have a limited life because of corrosion of the anode.

Polarographic electrodes, which are bulkier than galvanic electrodes, are more commonly used in pilot and production fermenters, needing instrument ports of 12, 19, or 25 mm diameter. Removable ones need a 25 mm port. They have silver anodes which are negatively polarized with respect to reference cathodes of platinum or gold, using aqueous potassium chloride as the electrolyte. Response times of 0.05–15 s to achieve a 95% reading have been reported (Lee & Tsao, 1979). The electrodes which can be very precise may be both pressure and temperature compensated. Although a polarographic electrode may initially cost six times more than the galvanic equivalent, the maintenance costs are considerably lower as only the membrane should need replacing.

Fast response phase fluorometric sterilizable oxygen sensors have been developed (Bambot, Holavanahali, Lakowicz, Carter, & Rao, 1994). The sensor utilizes the differential quenching of a fluorescence lifetime of a chromophore, tris(4,7-diphenyl-1,10-phenanthroline) ruthenium(II) complex, in response to the partial pressure of oxygen. The fluorescence of this complex is quenched by oxygen molecules resulting in a reduction of fluorescence lifetime. Thus, it is possible to obtain a correlation between fluorescence lifetime and the partial pressure of oxygen. However, at room temperature when a Clark-type oxygen electrode shows a linear calibration, the optical sensor shows a hyperbolic response. The sensitivity of the optical sensor when compared with an oxygen electrode is significantly higher at low oxygen tensions whereas the sensitivity is low at high oxygen tensions. The sensor is autoclavable, free of maintenance requirements, stable over long periods, and gives reliable measurements of low oxygen tensions in dense microbial cultures.

Optical sensors for both dissolved oxygen and dissolved carbon dioxide (see in later sections) are now available commercially. They are more costly than electrochemical sensors but have advantages in terms of response time and no interferences from, for example, pH and ionic strength of the fermentation broth (Luong et al., 2011).

If it is necessary to increase the dissolved oxygen concentration in a medium this may be achieved by increasing the air flow rate or the impeller speed or a combination of both the processes. This may however lead to an increase in foam generation which will require its own control strategy. A further option, which may be employed is to use pure or enhanced oxygen systems such as those used in effluent treatment (Chapter 11).

Control of dissolved oxygen concentration by varying the substrate feeding regime has been described by Vanags et al. (2007) and in phenol degradation by Vojta, Nahlik, Paca, and Komarkova (2002). Li, Chen, and Lun (2002) described the control of oxygen transfer rates using two-stage oxygen supply control in pyruvate production. Model based geometrical control algorithm for dissolved oxygen concentration/control in antibiotic fermentation have been described by Gomes and Menawat (2000). Due to varying biomass concentrations and the rheological properties of the fermentation broth, dissolved oxygen control in fed-batch filamentous fungal fermentations are challenging. Bodizs et al. (2007) report the use of a cascade control system and computation of an appropriate set point for dissolved oxygen concentration to improve control performance. Similarly Vanags, Hrynko, and Viesturs (2010) describe the use of cascade control and a computer algorithm for the control of oxygen concentration in fed batch yeast fermentations.

INLET AND EXIT-GAS ANALYSIS

The measurement and recording of the inlet and/or exit gas composition is important in many fermentation studies. By observing the concentrations of carbon dioxide and oxygen in the entry and exit gases in the fermenter and knowing the gas flow rate it is possible to determine the oxygen uptake rate of the system, the carbon dioxide evolution rate, and the respiration rate of the microbial culture.

The oxygen concentration can be determined by a paramagnetic gas analyzer. Oxygen has a strong affinity for a magnetic field, a property which is shared with only nitrous and nitric oxides. The analyzers may be of a deflection or thermal type (Brown, Kaminski, Blake, & Brodgesell 1969).

In the deflection analyzer, the magnetic force acts on a dumb-bell test body that is free to rotate about an axis (Fig. 8.8). The magnetic force which is created around the test body is proportional to the oxygen concentration. When the test body swings out of the magnetic field a corrective electrostatic force must be applied to return it to the original position. Electrostatic force readings can therefore be used as a measure of oxygen concentrations.

In a thermal analyzer, a flow through "ring" element is the detector component (Fig. 8.9). After entering the ring, the paramagnetic oxygen content of the sample is attracted by the magnetic field to the central glass tube where resistors heat the gases. The resistors are connected into a Wheatstone bridge circuit to detect variations in resistance due to flow-rate changes. The oxygen in the heated sample loses a high proportion of its paramagnetism. Cool oxygen in the incoming gas flow will now be attracted and displace the hot oxygen. This displacement action produces a convection current. The flow rate of the convection current is a function of the oxygen concentration and can be detected by resistors. The resulting gas flow cools the

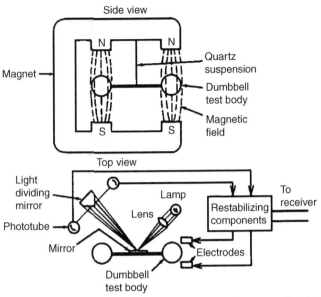

FIGURE 8.8 A Deflection-Type Paramagnetic Oxygen Analyzer (Brown et al., 1969)

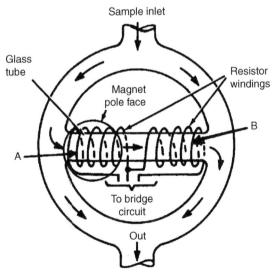

FIGURE 8.9 The Measuring Element in a Thermal-Type Paramagnetic Oxygen Analyzer (Brown et al., 1969)

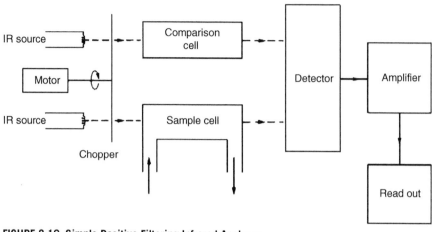

FIGURE 8.10 Simple Positive Filtering Infrared Analyzer

winding A and heats the winding B, and it is the resulting temperature difference that imbalances the Wheatstone bridge.

Carbon dioxide is commonly monitored by infrared analysis using a positive filtering method. The unit (Fig. 8.10) consists of a source of infrared energy, a "chopper" to ensure that energy passes through each side of the optical system, a sample cell, a comparison (or reference) cell, and an infrared detector sensitized at a wavelength at which the gas of interest absorbs infrared energy. In this case the detector will be filled with carbon dioxide. This optical system senses the reduced radiation energy of the measuring beam reaching the detector, which is due to the absorption in the carbon dioxide in the sample cell.

It is expensive to have separate carbon dioxide and oxygen analyzers for each separate fermenter. Therefore it may be possible to couple up a group of fermenters via a multiplexer to a single pair of gas analyzers (Meiners, 1982). Gas analysis readings can then be taken in rotation for each fermenter every 30–60 min. In many cases this will be adequate. Alternatively the gas analyzers can be replaced by a mass spectrometer, which can analyze a number of components as well as oxygen and carbon dioxide (see in later sections).

The importance of off-gas analysis (both O_2 and CO_2) for the optimization of secondary metabolite production from myxobacterial fermentations using a zirconium dioxide O_2 sensor and an infra-red CO_2 sensor has been reported by Huttel and Muller (2012).

pH MEASUREMENT AND CONTROL

In batch culture the pH of an actively growing culture will not remain constant for very long. In most processes there is a need for pH measurement and control during the fermentation if maximum yield of a product is to be obtained. Rapid changes in pH can often be reduced by the careful design of media, particularly in the choice of carbon and nitrogen sources, and also in the incorporation of buffers or by batch feeding (Chapters 2 and 4). The pH may be further controlled by the addition of appropriate quantities of ammonia or sodium hydroxide if too acidic, or sulfuric acid if the change is to an alkaline condition. Normally the pH drift is only in one direction.

pH measurement is now routinely carried out using a combined glass reference electrode that will withstand repeated sterilization at temperatures of 121°C and pressures of 138 kN m^{-2}. The electrodes may be silver/silver chloride with potassium chloride or special formulations (eg, Friscolyt by Ingold) as an electrolyte. Occasionally calomel/mercury electrodes are used. The electrode is connected via leads to a pH meter/controller. If the electrode and its fermenter have to be sterilized in an autoclave then the associated leads and plugs to a pH meter must be able to withstand autoclaving and retain their electrical resistance. Repeated sterilization may gradually change the performance of the electrode. The long culture times associated with animal cell culture or continuous culture of any cells makes a withdrawal option highly desirable to allow for servicing the electrode or troubleshooting without interrupting the fermentation. The housing for this option needs to be carefully designed to ensure that the fermentation does not become contaminated when the electrode is withdrawn, serviced, resterilized, and reinserted into the housing (Gary et al., 1988).

Ingold electrodes contain a ceramic housing in the reference half cell which has pore dimensions capable of preventing fungal or bacterial infections. However, it is often desirable in animal-cell culture to sterilize the reference electrolyte as well as the electrode surfaces and seals. Both liquid and gel filled electrodes are available. The liquid system that gives a faster response, is most stable and more accurate. When the electrode is pressurized above the operating pressure in a fermenter the liquid electrolyte will gradually flow out of the ceramic diaphragm and prevent fouling, particularly from proteins in the fermentation broth precipitating on the membrane after contact with the electrolyte (Gary et al., 1988).

Readers should consult Hailing (1990) for more information on calibration and checking, sterilization, routine maintenance, and problems in use.

Control units, to be discussed later in this chapter, may be simple ON/OFF or more complex. In the case of the ON/OFF controller, the controller is set to a predetermined pH value. When a signal actuates a relay, a pinch valve is opened or a pump started, and acid or alkali is pumped into the fermenter for a short time which is governed by a process timer (0–5 s). The addition cycle is followed by a mixing cycle which is governed by another process timer (0–60 s) during which time no further acid or alkali can be added. At the end of the mixing cycle another pH reading will indicate whether or not there has been adequate correction of the pH drift. In small volumes the likelihood of overshoot is minimal. A recording unit may be wired to the pH meter to monitor the pH pattern throughout a process cycle.

Shinskey (1973) has discussed pH control of batch processes using proportional and proportional plus derivative control (see in later sections) when overshoot of the set point is to be avoided. In the case of proportional action, the controller must be adjusted so that a valve on an acid feed-line is shut when the error is zero. However, overshoot

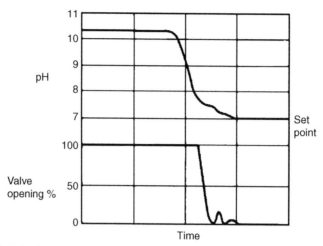

FIGURE 8.11 Valve Opening and pH Changes with Proportional Plus Derivative Control (Shinskey, 1973)

is possible as there may be a delay in closing the valve once the set point is achieved. In some cases, the overshoot cannot be corrected because of lack of alkali nor may it be desirable. Therefore to preclude an overshoot, the valve must be closed before the controlled variable reaches the set point. This may be done using proportional plus derivative control. The derivative action will need careful adjustment. Too little derivative action will cause some overshoot while too much will lead to the premature closure of the valve. This premature closure may be only for a short time before the valve opens again to give a response pattern as shown in Fig. 8.11. Zhu, Cao, Zhang, Zhang, and Zou (2011) describe the successful use of a three stage pH control strategy to maximize the cell growth and exopolysaccharide production in *Tremella fuciformis*. pH was controlled at 4.0, 4.5, and 5.0 for the first 72 h. The pH was then adjusted to 6.0 for the following 24 h and finally further adjusted to pH 6.5 and 7.0 for the final 24 h. This resulted in an increase in dry cell weight and exopolysaccharide production compared with no pH control and pH control at constant values of between pH 5.5 and 7.0. Moon et al. (2015) have shown the importance of pH control (at pH 6.0) in hydrogen production fermentations with significantly reduced hydrogen production at pH deviations of only +/− 0.5 pH units. They also report major population changes from *Clostridium* to lactic acid bacteria as pH deviates from the optimum. Li, Srivastava, Suib, Li, and Parnas (2011) describe the importance of pH control in acetone-butanol-ethanol (A-B-E) fermentations to the ratios of acetone, butanol, and ethanol generated.

REDOX

Aspects of redox potential have been reviewed by Jacob (1970), Kjaergaard (1977) and Hailing, 1990. It is a measure of the oxidation-reduction potential of a biological system and can be determined as a voltage (mV), the value in any system depending on the equilibrium of:

$$\text{Reduced form} \quad \rightleftharpoons \quad \text{Oxidized form} + \text{electron}(s)$$
$$\text{(negative value)} \qquad \qquad \text{(positive value)}$$

The measuring electrode consists of gold, platinum, or iridium which is welded to a copper lead. The interpretation of results presents difficulties and is confusing (Hailing, 1990). The culture is not at redox equilibrium until possibly the end of a growth cycle. During the cycle, although some redox half-reactions may be in equilibrium, they cannot all be in a dynamic system. The microorganisms can also be at a different redox potential from the broth. It has been speculated that the probe signal is indicating something about the relative concentrations of uncertain and probably varying chemical species. This is far from the ideal for a sensor which should only be measuring a specific factor. If the broth contains traces of oxygen this will probably dominate the signal. It is the ability to detect low concentrations of oxygen in media (1 ppm) where redox electrodes may have a good application for determining oxygen availability in anaerobic or microaerophilic processes operated on a small scale (Kjaergaard & Joergensen, 1979).

Hailing (1990) gives further details on routine handling, sterilization, testing, and calibration.

CARBON DIOXIDE ELECTRODES

The measurement of dissolved carbon dioxide is possible with an electrode, since a pH or voltage change can be detected as the gas goes into solution. The first available electrode consisted of a combined pH electrode with a bicarbonate buffer (pH 5) surrounding the bulb and ceramic plug, with the solution being retained by a PTFE membrane held by an O-ring. Unfortunately, this electrode was not steam sterilizable. This basic design has been modified so that dissolved CO_2 from the sample permeates a stainless-steel reinforced silicone bi-layer membrane and dissolves in the internal bicarbonate electrolyte of the electrode. The subsequent pH shift is determined by an internal pH element. This system is extremely sensitive to shifts in the pH element as one pH unit change represents a tenfold increase in pCO_2, but precautions have been taken in the design to ensure minimal drift of the sensor. It is possible to calibrate the electrode on-line using specially formulated buffer solutions. This version can be steam sterilized (Gary et al., 1988).

ON-LINE ANALYSIS OF OTHER CHEMICAL FACTORS

If good control of a fermentation is to be obtained, then all chemical factors which can influence growth and product formation ought to be continuously monitored. The lack of online sensors for fermentation monitoring and control is commonly noted in the literature (Rudnitskaya & Legin, 2008). This ideal situation has not yet been achieved but a number of techniques have been developed.

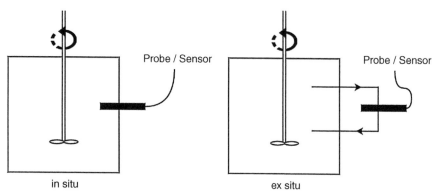

FIGURE 8.12 Location of Probes and Sensors

Within the fermenter (in situ) or in a flow through cell (ex situ).

CHEMICAL AND ION-SPECIFIC SENSORS

Sensors and probes for the online detection and control of most analytes can be described as in situ where the sensor/probe is located within the fermenter or ex situ where it is located in a flow through cell outside the fermenter (Fig. 8.12). Ion-specific sensors have been developed to measure NH_4^+, Ca^{2+}, K^+, Mg^{2+}, PO_4^{3-}, SO^{2-}, etc. The response time of these electrodes varies but is generally a matter of a few seconds.

Chemical sensors for specific analytes or groups of analytes show great promise for the fermentation industries as bioprocesses are described as the most complex in all fields of process engineering (Chen, Nguang, Li, & Chen, 2004). Chemical sensors have many attractive features for fermentation applications such as: low cost; relatively simple instrumentation; easy automation and minimal sample preparation. However, chemical sensors in complex media can show poor selectivity. (Rudnitskaya & Legin, 2008). One of the more recent approaches is to use arrays of multiple sensors (eg, conducting polymer sensors) rather than individual sensors. For liquid samples these have been described as "electronic tongues" and for gaseous samples, "electronic noses" (Rudnitskaya & Legin, 2008 and Ghosh, Bag, Sharma, & Tudu, 2015).

ENZYME AND MICROBIAL ELECTRODES (BIOSENSORS)

Enzyme or microbial cell electrodes (biosensors) can be used in some analyses. A suitable enzyme or microbial cell which produces a change in pH or involves a redox reaction is generally chosen and immobilized on a membrane (or similar substrate) held in close proximity to a transducer generating a signal, which can be correlated to the analyte concentration (Fig. 8.13). Enfors (1981) describes glucose determination by coimmobilizing glucose oxidase and catalase. More recently, it has been possible to use a ferrocene derivative as an artificial redox carrier to shuttle electrons from glucose oxidase to an electrode, thus making the device largely independent of oxygen concentrations (Higgins & Boldot, 1992). Enzyme electrodes are also

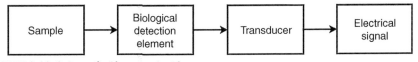

FIGURE 8.13 Schematic Diagram of a Biosensor

commercially available to monitor cholesterol, triglycerides, lactate, acetate, oxalate, methanol, ethanol, creatine, ammonia, urea, amino acids, carbohydrates, and penicillin (Higgins, Swain, & Turner, 1987; Luong, Mulchandani, & Guilbault, 1988; Higgins & Boldot, 1992).

Ideally, an electrode that can be inserted into a fermenter and steam sterilized is required, but none is yet available which can be used in this manner, even when using enzymes which are stable at high temperatures which can be obtained from thermophilic microorganisms. Hewetson, Jong, and Gray (1979) prepared sterilized penicillinase electrodes by assembling sterile components or by standing assembled components in chloroform before placing in a fermenter.

Moeller, Grunberg, Zehnsdorf, Strehlitz, and Bley (2010) report the use of a quasi-online glucose amperometric biosensor (based on glucose dehydrogenase) to control citric acid production from glucose by *Yarrowia lipolyptica*. In repeated fed-batch fermentations, controlling glucose concentration at 20 g/L using a proportional controller, citric acid production was increased 32% compared to simple batch fermentation. pH (6.0), temperature (30°C) and pO_2 (50%) were also controlled.

Backer et al. (2011) have developed a silicon-based biosensor chip utilizing amperometric enzyme sensors. The system was use to monitor both glutamate and glutamine in cell-culture fermentation. Glutamate is determined via glutamate oxidase and glutamine via a combination of glutaminase and glutamate oxidase.

NEAR INFRARED SPECTROSCOPY

Near infrared spectroscopy uses light of wavelengths between the visible and the infrared regions. Important bonds such as aliphatic C—H, aromatic or alkene C—H, N—H, and O—H absorb in the near infrared region and hence are a very valuable tool in the analysis of fermentation broths. In addition it is possible to place near infrared probes within a fermenter (Luong et al., 2011). The use of near infrared spectroscopy has seen large increases in its development and application over the last 25 years across a range of industries, including the fermentation industry, as it is a nondestructive, environmentally friendly, and rapid technique with low running costs. It does however require calibration with mathematical and statistical tools (chemometrics) to extract the required analytical data from the spectra produced (Porep, Kammerer, & Carle, 2015).

Hammond and Brookes (1992) have described the development of near infrared spectroscopy (NIR; 460–1200 nm) for rapid, continuous, and batch analysis of components of fermentation broths. In samples from an antibiotic fermentation, they used NIR absorbance bands to simultaneously estimate fat (in the medium), techoic acid (biomass), and antibiotic (the product). Fat analysis has been made possible with a fiber optic sensor placed in situ through a port in the fermenter wall. The assay time for an antibiotic has been reduced from 2 h to 2 min. A method has also been developed to measure alkaline protease production in broths. Vaccari et al. (1994) have used this technique to measure glucose, lactic acid, and biomass in a lactic acid fermentation. Finn, Harvey, and Mcneil (2006) have described the use of NIR to monitor biomass, glucose, ethanol, and protein content in fed-batch bakers yeast fermentations. Biomass and protein content were analyzed using the whole sample but because of the light scattering effects of the biomass glucose and ethanol were analyzed in filtered samples. Cervera et al. (2009) have reviewed the application of NIR in the monitoring and control of cell culture and fermentations. They describe how various factors can negatively affect the chemometric models developed to interpret NIR spectra. Practical aspects considered include temperature, hydrodynamic conditions in the fermenter and biomass morphology.

NIR spectroscopy has also been applied to anaerobic digestion for the optimization of biogas production (Luck, Buge, Plettenburg, & Hoffmann, 2010). The use of NIR in brewing and wine making and in many other areas of the food and beverage industries has been reviewed by Porep et al. (2015).

MASS SPECTROMETERS

The mass spectrometer can be used for online analysis since it is very versatile and has a response time of less than 5 s for full-scale response and taking about 12 s for a sample stream. It allows for monitoring of gas partial pressures (O_2, CO_2, CH_4, etc.), dissolved gases (O_2, CO_2, CH_4, etc.) and volatiles (methanol, ethanol, acetone, simple organic acids, etc.). Heinzle, Dunn, and Bourne (1981) combined a data processor with a mass spectrometer equipped with a capillary gas inlet to measure gas partial pressures and a membrane inlet to detect dissolved gases and volatiles. It is therefore possible to multiplex this type of system to analyze several fermenters sequentially. Merck and Company Inc. have installed an integrated gas- and liquid-phase analysis system incorporating a mass spectrometer to sample a number of fermenters (Omstead & Greasham, 1989). The availability of low cost mass spectrometers makes their use in a laboratory or pilot plant financially feasible as an alternative to oxygen and carbon dioxide analyzers, as well as analyzing for a range of other gaseous and volatile compounds.

CONTROL SYSTEMS

The process parameters which are measured using probes and sensors described in the previous sections may be controlled using control loops.

A control loop consists of four basic components:

1. A measuring element.
2. A controller.

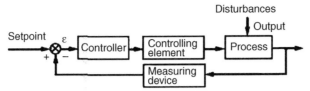

FIGURE 8.14 A Feedback Control Loop (Rolf & Lim, 1985)

3. A final control element.
4. The process to be controlled.

In the simplest type of control loop, known as feedback control (Fig. 8.14), the measuring element senses a process property such as flow, pressure, temperature, etc., and generates a corresponding output signal. The controller compares the measurement signal with a predetermined desired value (set point) and produces an output signal to counteract any differences between the two. The final control element receives the control signal and adjusts the process by changing a valve opening or pump speed and causing the controlled process property to return to the set point.

MANUAL CONTROL

A simple example of control is manual control of a steam valve to regulate the temperature of water flowing through a pipeline (Fig. 8.15). Throughout the time of operation a plant operative is instructed to monitor the temperature in the pipe. Immediately the temperature changes from the set point on the thermometer, the operative will take appropriate action and adjust the steam valve to correct the temperature deviation. Should the temperature not return to the set point within a reasonable time, further action may be necessary? Much depends on the skill of individual

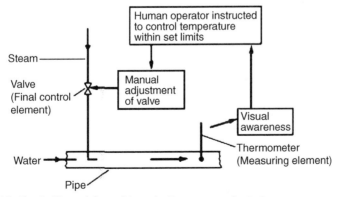

FIGURE 8.15 Simple Manual-Control Loop for Temperature Control

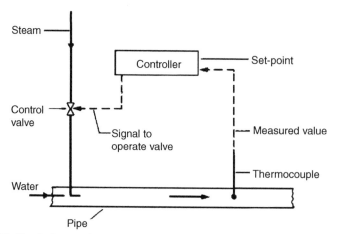

FIGURE 8.16 Simple Automatic Control Loop for Temperature Control

operatives in knowing when and how much adjustment to make. This approach with manual control may be very costly in terms of labor and should always be kept to a strict minimum when automatic control could be used instead. A justifiable use of manual control may be in the adjustment of minor infrequent deviations.

AUTOMATIC CONTROL

When an automatic control loop is used, certain modifications are necessary. The measuring element must generate an output signal which can be monitored by an instrument. In the case of temperature control, the thermometer is replaced by a thermocouple, which is connected to a controller which in turn will produce a signal to operate the steam valve (Fig. 8.16).

Automatic control systems can be classified into four main types:

1. Two-position controllers (ON/OFF).
2. Proportional controllers.
3. Integral controllers.
4. Derivative controllers.

Two-position controllers (ON/OFF)

The two-position controller, which is the simplest automatic controller, has a final control element (valve, switch, etc.) which is either fully open (ON) or fully closed (OFF). The response pattern to such a change will be oscillatory. If there is instant response then the pattern will be as shown in Fig. 8.17.

If one considers the example of the heating of a simple domestic water tank controlled by a thermostat operating with ON/OFF, then there will be a delay in response when the temperature reaches the set point and the temperature will continue rising above this point before the heating source is switched off. At the other extreme, the

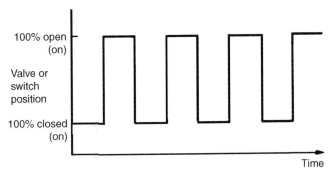

FIGURE 8.17 Oscillatory Pattern of a Simple Two-Position Valve or Switch

water will continue cooling after the heating source has been switched on. With this mode of operation an oscillatory pattern will be obtained with a repeating pattern of maximum and minimum temperature oscillating about the set point, provided that all the other process conditions are maintained at a steady level (Fig. 8.18).

If this type of controller is to be used in process control then it is important to establish that the maximum and minimum values are acceptable for the specific process, and to ensure that the oscillation cycle time does not cause excessive use of valves or switches.

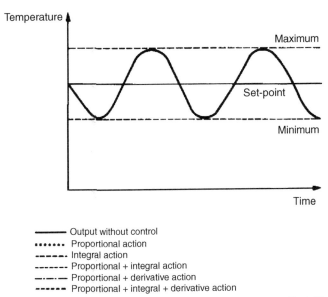

FIGURE 8.18 Oscillatory Pattern of the Temperature of a Domestic Water Tank (No Water Being Drawn Off) Using ON/OFF Control of the Heating Element

ON/OFF control is not satisfactory for controlling any process parameter where there are likely to be large sudden changes from the equilibrium. In these cases alternative forms of automatic control must be used.

In more complex automatic control systems three different methods are commonly used in making error corrections. They are: proportional, integral, and derivative. These control methods may be used singly or in combinations in applying automatic control to a process, depending upon the complexity of the process and the extent of control required. Since many of the controllers used in the chemical industries are pneumatic, the response to an error by the controller will be represented by a change in output pressure. Pneumatic controllers are still widely used because they are robust and reliable. In other cases, when the controller is electronic, the response to an error will be represented as a change in output current or voltage.

Proportional control

Proportional control can be explained as follows: the change in output of the controller is proportional to the input signal produced by the environmental change (commonly referred to as error) which has been detected by a sensor.

Mathematically it can be expressed by the following equation:

$$M = M_0 + K_c \Sigma$$

where, M = output signal, M_0 = controller output signal when there is no error, K_c = controller gain or sensitivity Σ = the error signal.

Hence, the greater the error (environmental change), the larger is the initial corrective action which will be applied. The response to proportional control is shown in Fig. 8.18, from which it may be observed that there is a time of oscillation which is reduced fairly quickly. It should also be noted that the controlled variable attains a new equilibrium value. The difference between the original and the new equilibrium value is termed the offset.

The term K_c (controller gain) is the multiplying factor (which may be dimensioned) which relates a change in input to the change in output.

$$\Delta I \rightarrow \boxed{\text{Controller}} \rightarrow \Delta O$$
$$\text{(Change in input)} \qquad \text{(Change in output)}$$

Then

$$\Delta I = K_c \Delta O.$$

K_c may contain conversion units if there is an electrical input and a pressure output or vice versa.

If the input to the controller is 1 unit of change, then:

1. with a controller gain of 1, the output will be 1 unit,
2. with a controller gain of 2, the output will be 2 units, etc.

On many controllers K_c is graduated in terms of proportional band instead of controller gain.

Now
$$K_c \propto \frac{1}{PB} \text{ (proportional band)}$$

or
$$K_c = c' \frac{1}{PB} \text{ where } c' \text{ is a constant.}$$

This quantity PB is defined as the error required to move the final control element over the whole of its range (eg, from fully open to fully shut) and is expressed as a percentage of the total range of the measured variable (eg, two extremes of temperature).

A fermenter with a heating jacket will be used as an example (Fig. 8.19). A thermocouple is connected to a temperature controller which has a span of 10°C covering the range 25–35°C, with a set point at 30°C. The controller valve which is controlled by a pressure regulator, is fully open at 5 psig (46 kN m^{-2}) and fully closed at 15 psig (138 kN m^{-2}), while the set point of 30°C corresponds to a control pressure on the valve of 10 psig (92 kN m^{-2}). When the controller gain is 1, a change of 10°C will cause a pressure change of 10 psig when the valve will be fully open at 25°C and fully closed at 35°C. Thus, the proportional band is 100%. If the controller gain is 2, a 5°C change will cause the valve to go from fully open to fully closed, that is, 27.5–32.5°C. In this case the proportional band width will be:

$$\frac{\text{Actual band width}}{\text{Total band range}} = \frac{5°C}{10°C} \times 100 = 50\%.$$

Either side of this band the pressure will be constant. In the case of a controller gain of 4, a 2.5°C change will cause a pressure change of 10 psig (92 kNm^{-2}), which will cause the control valve to go from fully open to fully closed (28.75–31.25°C). In this case the proportional band will be

$$= \frac{2.5°C}{10°C} \times 100 = 25\%.$$

These results are summarized in Table 8.2 for controller gains of 1, 2, and 4.

When the proportional band is very small (the controller gain is high), the control mode can be likened to simple ON/OFF, with a high degree of oscillation but no offset. As the proportional band is increased (low controller gain) the oscillations are reduced but the offset is increased. Settings for proportional band width are normally a compromise between degree of oscillation and offset. If the offset is not desirable it can be eliminated by the use of proportional control in association with integral control (see Fig. 8.20 and later sections). The application of a proportional controller (PB = 15%) coupled to an enzyme based glucose sensor to optimize the production of citric acid from glucose has been reported (Moeller et al., 2010).

Table 8.2 The Effect of Controller Gain (K_c) on Bandwidth of a Proportional Temperature Controller

	Measured Temperature (°C)	Pressure Output (psig) at K_c = 1 psig/1°C	Pressure Output (psig) at K_c = 2 psig/1°C	Pressure Output (psig) at K_c = 4 psig/1°C
	35	15	15	15
	34	14	15	15
	33	13	15	15
	32.5	12.5	15	15
	32	12	14	15
	31.25	11.25	12.5	15
	31	11	12	14
Set→	30	10 Prop, band 100%	10 Prop, band 50%	10 Prop, band 25%
point	29	9	8	6
	28.75	8.75	7.5	5
	28	8	6	5
	27.5	7.5	5	5
	27	7	5	5
	26	6	5	5
	25	5	5	5

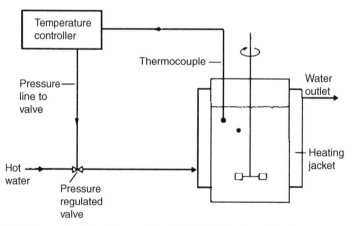

FIGURE 8.19 A Fermenter with a Temperature-Controlled Heating Jacket

Integral control

The output signal of an integral controller is determined by the integral of the error input over the time of operation. Thus:

$$M = M_0 + \frac{1}{T_1} \int \Sigma \, dt$$

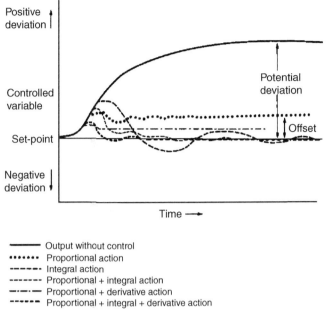

Positive deviation

Controlled variable

Set-point

Negative deviation

Potential deviation

Offset

Time

—————— Output without control
••••••• Proportional action
— — — • Integral action
------ Proportional + integral action
—•—•— Proportional + derivative action
— — — — Proportional + integral + derivative action

FIGURE 8.20 Typical Controlled Plant Responses

where T_i = integral time.

It is important to remember that the controller output signal changes relatively slowly at first as time is required for the controller action to integrate the error. It is evident from Fig. 8.18 that the maximum deviation from the set point is significant when compared with the use of proportional control for control of the chosen parameter, and the system takes longer to reach steady-state. There is, however, no offset which is advantageous in many control processes.

Derivative control

When derivative control is applied, the controller senses the rate of change of the error signal and contributes a component of the output signal that is proportional to a derivative of the error signal. Thus:

$$M = M_0 + T_d \frac{d\Sigma}{dt}$$

where T_d is a time rate constant.

It is important to remember that if the error is constant there is no corrective action with derivative control. In practice, derivative control is never used on its own. The response curve has therefore been deliberately omitted from Fig. 8.20.

Fig. 8.21 demonstrates the response of derivative control to sinusoidal error inputs. The output is always in a direction to oppose changes in error, both away from and toward the set point, which in this example results in a 90 degree phase

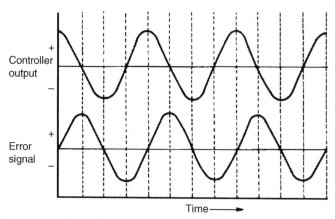

FIGURE 8.21 Response of a Derivative Controller to Sinusoidal Error Inputs

shift. This opposition to a change has a fast damping effect and this property is very useful in combination with other modes of control which will be discussed later.

COMBINATIONS OF METHODS OF CONTROL

Three combinations of control systems are used in practice:

1. Proportional plus integral (PI).
2. Proportional plus derivative (PD).
3. Proportional plus integral plus derivative (PID).

Proportional plus integral control

When proportional plus integral control is used, the output response to an error gives rise to a slightly higher initial deviation in the output signal compared with that which would be obtained with proportional control on its own (Fig. 8.20). This is due to a contribution in the signal from integral control. However, the oscillations are soon reduced and there is finally no offset. This mode of control finds wide applications since the proportional component is ideal in a process where there are moderate changes, whereas the integral component will allow for large load changes and eliminate the offset that would have occurred.

Proportional plus derivative control

If proportional plus derivative control is used, the output response to an error will lead to reduced deviations, faster stabilization and a reduced offset (Fig. 8.20) compared with proportional control alone. Because the derivative component has a rapid stabilizing influence, the controller can cope with rapid load changes.

Proportional plus integral plus derivative control

The combination of proportional plus integral plus derivative normally provides the best control possibilities (Fig. 8.20). The advantages of each system are retained. The maximum deviation and settling time are similar to those for a proportional plus derivative controller while the integral action ensures that there is no offset. This method of control finds the widest applications because of its ability to cope with wide variations of patterns of changes which might be encountered in different processes. There are many literature examples of PID control being used in fermentation applications, for example, Jimenez-Marquez et al. (2016) and Restrepo et al. (2002).

CONTROLLERS

The first primary automatic controllers were electronic control units, which were adjusted manually to set up desired PID response patterns to a disturbance in a control loop. These controllers are relatively expensive and some knowledge of control engineering is necessary to make the correct adjustments to obtain the required control responses.

The availability of cheap computers and suitable computer programs to mimic PID control and handle a number of control loops simultaneously has made it possible to use some very complex control techniques for process optimization. Some aspects of this work on computer applications will be discussed later in this chapter.

MORE COMPLEX CONTROL SYSTEMS

In certain situations PID control is not adequate to control a disturbance. Control may be difficult when there is a long time lag between a change in a manipulated variable and its effect on the measured variable. Consider the example of a heat exchanger where the water temperature is regulated by the flow rate of steam from a steam valve (Hall, Higgins, Kennedy, & Nelson, 1974). By the time the effects of a change in steam flow influence the hot water temperature, a considerable energy change has occurred in the heat exchanger which will continue to drive the hot water temperature away from the set-point after a correction has been made to the steam-flow valve. This lag will lead to cycling of the measured variable about the set point. Cascade control can solve this problem. When cascade control is used (Fig. 8.22), the output of one controller is the set-point for another. Each controller has its own measured variable with only the primary controller having an individual set-point and only the secondary controller providing an output to the process. In more complex cascade systems more loops may be included.

When cascade control is used to control water temperature in a heat exchanger process (Fig. 8.23), there is the primary loop (slow or outer loop), consisting of the temperature sensor and the primary temperature controller of the process water temperature, and the secondary loop (fast or inner loop) consisting of the steam flow sensor, the steam flow controller, its process variable (steam flow), the control valve

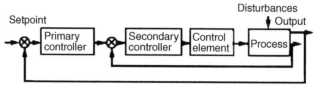

FIGURE 8.22 A Cascade Feedback Control Loop (Rolf & Lim, 1985)

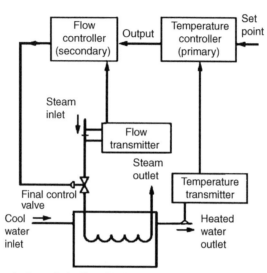

FIGURE 8.23 Cascade Control of a Heat-Exchange Process (Hall et al., 1974)

and the process. The addition of the secondary controller, whose measured variable is steam flow, allows steam flow variations to be corrected immediately before they can affect the hot water temperature. This extra control loop should help to minimize temperature cycling.

Feedforward (anticipatory) control (Fig. 8.24) makes it possible to utilize other disturbances besides the measured values of the process that have to be controlled and enable fast control of a process. The cascade control system (Fig. 8.23) which has just been discussed would be adequate provided that the water inlet pressure and temperature do not fluctuate. When fluctuations occur, feedforward control may be used, but this will require the measurement of both inlet water temperature and flow rate with both variables combined to control the steam flow rate (Fig. 8.25). In the example shown (Hall et al., 1974), a computer is programmed to determine the heat energy required to be added to the cool inlet water to bring it to the appropriate temperature at the outlet and allow the correct amount of steam to enter the heating coils. Inputs for temperature and flow are made from the inlet pipe to the computer. Calculations can then be made to determine the required heat input to raise the water

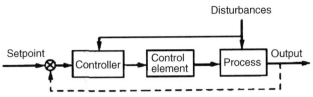

FIGURE 8.24 A Feedforward Control Loop (Rolf & Lim, 1985)

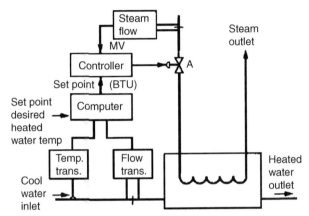

FIGURE 8.25 Feedforward Control of a Heat-Exchange Process (Hall et al., 1974)

temperature in the heat exchanger. The output control signal to the valve A is computed to allow the correct amount of steam to enter the heating coils.

Bodizs et al. (2007) report the use of a cascade control system and computation of an appropriate set point for dissolved oxygen concentration to improve performance of dissolved oxygen control in fed-batch filamentous fungal fermentations. The utilization of cascade control and a computer algorithm is also described for the control of oxygen concentration in fed batch yeast fermentations Vanags et al. (2010).

When one or more of the process variables or characteristics is not known and cannot be measured directly, then adaptive control should be considered. Adaptive control algorithms deal with unknowns and uncertainties by continuous adaptation of the control law to compensate for the unknowns/uncertainties. The main classes of adaptive control are gain scheduling, model reference adaptive control and self-tuning control (Teixeira, Oliveira, Alves, & Carrondo, 2011). In gain scheduling, which is the simplest form of adaptive control, information about the process is used to adapt the parameters of a PI or PID controller (refer to earlier section for information on controller gain/proportional band). In self tuning control, the algorithm learns from experience and self adjusts the parameters to improve control performance after process deviations. Fuzzy systems and artificial neural networks have been used for the adaptation mechanism of controller set points. Model reference

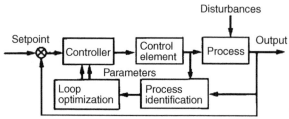

FIGURE 8.26 Adaptive Control (Rolf & Lim, 1985)

adaptive control systems utilize a reference model to adapt control parameters so that the plant response matches the reference model response (Teixeira et al., 2011). The use of adaptive control strategies have been described in fed batch yeast cultures (Renard & Vande Wouwer, 2008), in anaerobic wastewater treatment (Petre, Selisteanu, & Sendrescu, 2013) and in continuous bioreactors (Jang, Chern, & Chou, 2013).

The online identification of process characteristics and the subsequent use of this information to improve the process constitute adaptive control (Hall et al., 1974). The sequences and the interactions in the adaptive control loop are outlined in Fig. 8.26. Adaptive control is useful in circumstances where the process dynamics are not well defined or change with time. This may be most useful in controlling a batch fermentation where considerable and often complex changes may occur (Bull, 1985; Dusseliee and Feijen, 1990). Adaptive control strategies have been used in fed-batch yeast cultivation (Montague et al., 1988), amino acid production (Radjai, Hatch, & Cadman, 1984) and penicillin production (Lorenz et al., 1985).

When a process variable which is difficult, impractical or impossible to measure, inferential control can be used. Here an easily measured variable is used to infer the magnitude of the difficult to measure variable. For example, oxygen uptake rate measurements can be used to estimate (infer) substrate concentration (Teixeira et al., 2011).

Model predictive control (MPC) is a class of control algorithm which applies a model of the process to predict the future response of the process. At each sampling time point, the future process behavior is optimized. The objective of MCP is to always persue the optimal state (Teixeira et al., 2011).

It is also possible to prepare sequential programs using a computer based control system. Consider control of the dissolved oxygen concentration in a fermentation broth. This may be changed by altering the agitation rate, the air or gas flow rate, the partial pressure of oxygen in the inlet gas or the total fermenter pressure. In practice, combinations of these variables may be used either sequentially or simultaneously using suitable computer programs. Initially the agitation rate can be increased to respond to decreases in dissolved oxygen concentration. When a predetermined maximum agitation rate is reached, the air flow can be steadily increased to a preset maximum, followed by the third and subsequent stages.

COMPUTER APPLICATIONS IN FERMENTATION TECHNOLOGY

Since the widespread initial use of computers in the 1960s for modeling fermentation processes (Yamashita & Murao, 1978) and in process control for production of glutamic acid (Yamashita, Hisoshi, & Inagakj, 1969) and penicillin (Grayson, 1969), there have been numerous publications on computer applications in fermentation technology (Rolf & Lim, 1985; Bushell, 1988; Whiteside & Morgan, 1989; Fish, Fox, & Thornhill, 1989). Initially, the use of large computers was restricted because of their cost but reductions in costs and the availability of cheaper small computers has made them virtually ubiquitous within the fermentation industry. The availability of efficient and powerful small computers has led to their use for pilot plants and laboratory systems since the financial investment for online computers amounts to a relatively insignificant part of the whole system. In this section a historical perspective of computer use and applications is given together more recent developments.

Three distinct areas of computer function were recognized by Nyiri (1972):

1. *Logging of process data.* Data logging is performed by the data acquisition system which has both hardware and software components. There is an interface between the sensors and the computer. The software should include the computer program for sequential scanning of the sensor signals and the procedure of data storage.
2. *Data analysis* (*Reduction of logged data*). Data reduction is performed by the data-analysis system, which is a computer program based on a series of selected mathematical equations. The analyzed information may then be put on a print out, fed into a data bank or utilized for process control.
3. *Process control.* Process control is also performed using a computer program. Signals from the computer are fed to pumps, valves, or switches via the interface. In addition the computer program may contain instructions to display devices or teletypes, to indicate alarms, etc.

At this point it is necessary to be aware that there are two distinct fundamental approaches to computer control of fermenters. The first is when the fermenter is under the direct control of the computer software. This is termed Direct Digital Control (DDC) and will be discussed in the next section. The second approach involves the use of independent controllers to manage all control functions of a fermenter and the computer communicates with the controller only to exchange information. This is termed Supervisory Set-Point Control (SSC) and will be discussed in more detail in the Section "Process Control".

It is possible to analyze data, compare it with model systems in a data store, and use control programs, which will lead to process optimization. It is important to be aware of these different applications, since this will influence the size and type of computer system, which will be appropriate for the precise role that it is intended to perform, whether in a laboratory, a pilot plant, or manufacturing plant, or a combination of these three.

COMPONENTS OF A COMPUTER-LINKED SYSTEM

When a computer is linked to a fermenter to operate as a control and recording system, number of factors must be considered to ensure that all the components interact and function satisfactorily for control and data logging. A DDC system will be used as an example to explain computer controlled addition of a liquid from a reservoir to a fermenter. A simple outline of the main components is given in Fig. 8.27. A sensor S in the fermenter produces a signal which may need to be amplified and conditioned in the correct analogue form. At this stage it is necessary to convert the signal to a digital form which can be subsequently transmitted to the computer. An interface is placed in the circuit at this point. This interface serves as the junction point for the inputs from the fermenter sensors to the computer and the output signals from the

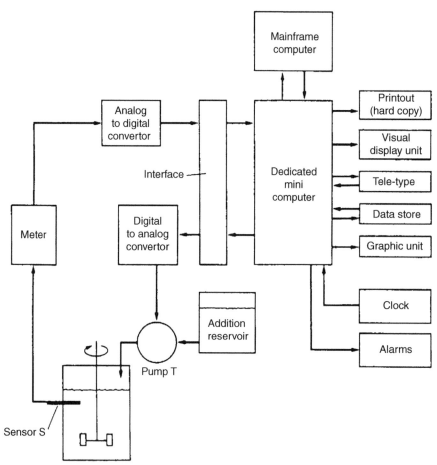

FIGURE 8.27 Simplified Layout of Computer-Controlled Fermenter with only One Control Loop Shown

computer to the fermenter controls such as a pump T attached to an additive reservoir. Digital to analogue conversion is necessary between the interface and the pump T.

A sensor will generate a small voltage proportional to the parameter it is measuring. For example, a temperature probe might generate 1 V at 10°C and 5 V at 50°C. Unfortunately, the signal cannot be understood by the computer and must be converted by an analogue to digital converter (ADC) into a digital form.

The accuracy of an ADC will depend on the number of bits (the unit of binary information) it sends to the computer. An 8-bit converter will work in the range 0–255 and it is therefore able to divide a signal voltage into 256 steps. This will give a maximum accuracy of 100/256, which is approximately 0.4%. However, a 10-bit converter can give 1024 steps with an accuracy of 100/1024, which is approximately 0.1%. Therefore when a parameter is to be monitored very accurately, a converter of the appropriate degree of accuracy will be required. The time taken for an ADC to convert voltage signals to a digital output will vary with accuracy, but improved accuracy leads to slower conversion and hence slower control responses. It is also important to ensure that the voltage ranges of the sensors are matched to the ADC input range. More detailed discussion is given by Whiteside and Morgan (1989).

A digital to analogue converter (DAC) converts a digital signal from the computer into an electrical voltage which can be used to drive electrical equipment, for example, a stirrer motor. Like the ADC, the accuracy of the DAC will be determined by whether it is 8-bit, 10-bit, 12-bit, etc., and will, for example, determine the size of steps in the control of speed of a stirrer motor.

The computer itself is dedicated solely to one or more fermenters. This computer is coupled to a real-time clock, which determines how frequently readings from the sensor(s) should be taken and possibly recorded. The other ancillary equipment may be linked directly to the computer.

The computer may be connected to a large main frame computer for random access, not on a real-time scale, but for long-term data storage and retrieval and for complex data analysis which will not be utilized subsequently in real-time control.

It is also possible to develop programs so that online instruments can be checked regularly and recalibrated when necessary. Swartz and Cooney (1979) were able to routinely recalibrate a paramagnetic oxygen analyzer and an infrared carbon dioxide analyzer every 12 h utilizing a program, which connected a gas of known composition to the analyzers and subsequently monitored the analyzer outputs.

DATA LOGGING

The simplest task for a computer is data logging. Parameters such as those listed in Table 8.1 can be measured by sensors which produce a signal which is compatible with the computer system.

Programs have been developed so that by reference to the real-time clock, the signals from the appropriate sensors will be scanned sequentially in a predetermined pattern and logged in a data store. Typically, this may be 2–60 s intervals, and the data is stored. In preliminary scanning cycles the values are compared with predefined limit values, and deviations from these values result in an error output or if more extreme then an alarm may be activated. In the final cycle of a sequence, say every 5–60 min, the program instructs that the sensor readings are permanently recorded on a print out or in a data store.

Thus, it is now possible to record data continuously for a range of parameters from a number of fermenters simultaneously using minimal manpower, provided that the capital outlay is made for fermenters with suitable instrumentation coupled with adequate computer facilities.

DATA ANALYSIS

A number of the monitoring systems were described as "Gateway Sensors" by Aiba et al. (1973) and are given in Table 8.3. Gateway sensors are so called because the information they yield can be processed to give further information about the fermentation. More details of analysis of direct measurements, indirect measurements, and estimated variables have been discussed by Zabriskie (1985) and Royce (1993).

The respiratory quotient of a culture may be calculated from the metered gas-flow rates and analyzes for oxygen and carbon dioxide leaving a known volume of culture in the fermenter. This procedure was used to monitor growth of *Candida utilis* in a 250 dm^3 fermenter, to follow or forecast events during operation (Nyiri, Toto, & Charles, 1975).

If one defines the fraction of substrate which is converted to product then it is possible to write mass balances for C, H, O, and N with the measurement of only a few quantities (O_2, CO_2, NH_3, etc.). All the other quantities can be calculated, including biomass and yield, if the biomass elemental composition is known (Chapter 2). This procedure was used for the analysis of a bakers' yeast fermentation (Cooney, Wang,

Table 8.3 Gateway Sensors (Aiba et al., 1973)

Sensor	Information that may be Determined from the Sensor Signal
pH	Acid product formation
Dissolved oxygen	Oxygen-transferrate
Oxygen in exit gas	
Gas-flow rate	Oxygen-uptake rate
Carbon dioxide in exit gas	
Gas-flow rate	Carbon dioxide evolution rate
Oxygen-uptake rate	
Carbon dioxide evolution rate	Respiratory quotient
Sugar-level and feed rate	
Carbon dioxide evolution rate	Yield and cell density

& Wang, 1977). Biomass production can be regarded as a stoichiometric relationship in which substrate is converted, in the presence of oxygen and ammonia to biomass, carbon dioxide and water:

Carbon source-energy + oxygen + ammonium → cells + water + carbon dioxide.

Thus, the equation can be written in the form:

$$a C_x H_y O_z + b O_2 + c NH_3 \rightarrow d C_r H_s O_t N_u + H_2O + f CO_2$$

where a, b, c, d, e, and f are moles of the respective reactants and products. $C_x H_y O_z$ is the molecular formula of the substrate where x, y, and z are the specific carbon, hydrogen, and oxygen atom numbers. Biomass is represented by $C_r H_s O_t N_u$ where r, s, t, and u are the corresponding numbers of each element in the cell. This technique was developed to use with bakers' yeast fermentations (Cooney et al., 1977; Wang, Cooney, & Wang, 1977).

PROCESS CONTROL

Armiger and Moran (1979) recognized three levels of process control that might be incorporated into a system. Each higher level involves more complex programs and needs a greater overall understanding of the process. The first level of control, which is routinely used in the fermentation industries, involves sequencing operations, such as manipulating valves or starting or stopping pumps, instrument recalibration, online maintenance and fail safe shutdown procedures. In most of these operations the time base is at least in the order of minutes, so that high-speed manipulations are not vital. Two applications in fermentation processes are sterilization cycles and medium batching.

The next level of computer control involves process control of temperature, pH, foam control, etc. where the sensors are directly interfaced to a computer (DDC; Fig. 8.27). When this is done separate controller units are not needed. The computer program determines the set point values and the control algorithms, such as PID, are part of the computer software package. Better control is possible as the control algorithms are mathematically stored functions rather than electrical functions. This procedure allows for greater flexibility and more precise representation of a process control policy. The system is not very expensive as separate electronic controllers are no longer needed, but computer failure can cause major problems unless there is some manual backup or fail safe facility is incorporated.

The alternative approach is to use a computer in a purely supervisory role. All control functions are performed by an electronic controller using a system illustrated in Fig. 8.28, where the linked computer only logs data from sensors and sends signals to alter set points when instructed by a computer program or manually. This system is known as SSC or DSC. When SSC is used, the modes of control are limited to proportional, integral, and derivative control (or a combination of the three) because the direct control of the fermenter is by an electronic controller.

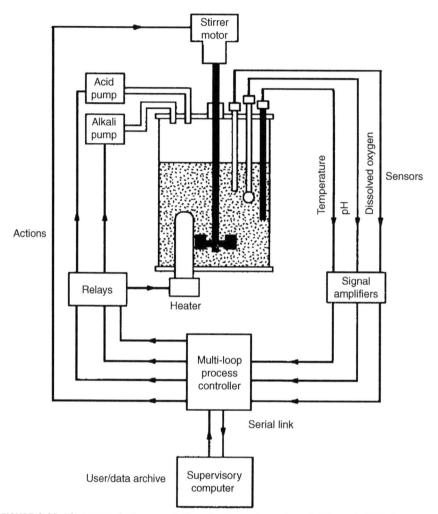

FIGURE 8.28 Diagrammatic Representation of a Supervisory Setpoint Control (SSC) System for Fermenters

This example illustrates a system controlling temperature by means of heating only, dissolved oxygen tension by stirrer speed and pH by the addition of acid and alkali. All control functions are performed by the intelligent process controller and the computer only communicates with this in order to log data and send new setpoints when instructed to do so by the user (Whiteside & Morgan, 1989).

However, in the event of computer failure the process controller can be operated independently.

Whiteside and Morgan (1989) have discussed some of the relative merits of DDC and SSC systems and given case histories of the installation and operation of both systems.

The most advanced level of control is concerned with process optimization. This will involve understanding a process, being able to monitor what is happening and being able to control it to achieve and maintain optimum conditions. Firstly, there is a need for suitable online sensors to monitor the process continuously. Many are now available for dissolved oxygen, dissolved carbon dioxide, pH, temperature, etc. All these sensors have been discussed earlier in this chapter. Second, it is important to develop a mathematical model that adequately describes the dynamic behavior of a process. Shimizu (1993) has stressed the vital role which these models play in optimization and reviewed the use of this approach in batch, fed-batch, and continuous processes for biomass and metabolites. This approach with appropriate online sensors and suitable model programs has been used to optimize bakers' yeast production (Ramirez, Durand, & Blachere, 1981; Shi, Shimizu, Watanabe, & Kobayashi, 1989), an industrial antibiotic process and lactic acid production (Shi et al., 1989).

Although much progress has been made in the ability to control a process, sensors are not always available to monitor online for many metabolites or other parameters in a fermentation broth thus delaying or making a fast response difficult for online control action (see Section "More Complex Control Systems"). Also, it is possible that not all the important parameters in a process have been identified and the mathematical model derived to describe a process may be inadequate. Because of these limitations, an artificial neural network (ANN) may be developed to achieve better control (Karim & Rivera, 1992; Doran, 2013). ANNs are one of the knowledge information processing methods—a mulitilayered network model (Honda & Kobayashi, 2011). These are highly interconnected networks of nonlinear processing units arranged in layers with adjustable connecting strengths (weights). In simpler neural networks there is one input layer, one hidden layer and one output layer (Fig. 8.29). Unlike recognized knowledge-based systems, neural networks do not need information in the form of a series of rules, but learn from process examples from which they derive their own rules. This makes it possible to deal with nonlinear systems and approximate or limited data.

When training a neural network the aim is to adjust the strengths of the interconnections (neurons) so that a set of inputs produces a desired set of outputs. The inputs may be process variables such as temperature, pH, flow rates, pressure, and other direct or indirect measurements which give information about the state of the process. The process outputs obtained (biomass, product formation, etc.) produce the teacher signal(s) which trains the network. The difference between the desired output and the value predicted by the network is the prediction error. Adjustments are made to minimize the total prediction error by modifying the interconnection strengths until no further decrease in error is achieved. Commercial computer packages are now available to help to determine, which of the input variables to use for training and to determine the optimum number of interconnections and hidden layers (Glassey, Montague, Ward, & Kara, 1994). Readers requiring more detail of the theory of neural networks should consult Karim and Rivera (1992).

This method of control has been widely described for fermentation processes. It has been used in a case study on ethanol production by *Zymomonas mobilis* (Karim & Rivera, 1992), in real-time variable estimation and control of a glucoamylase

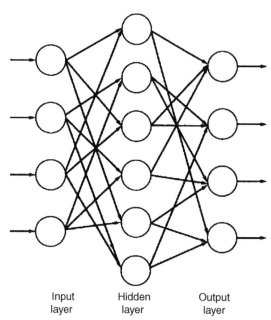

Input   Hidden   Output
layer   layer   layer

FIGURE 8.29 Two-Layer Neural Network (Not all the Possible Interconnections are Shown)

fermentation (Linko & Zhu, 1992) and recombinant *Escherichia coli* fermentations (Glassey et al., 1994). More recently ANNs have been described in controlling glucose feed rate in fed-batch bakers yeast fermentation (Hisbullah, Hussain, & Ramachandran, 2002), in controlling dissolved oxygen and pH in laboratory scale *Saccharomyces* fermentation (Meszaros, Andrasik, Mizsey, Fonyo, & Illeova, 2004), in extracellular protease production by Pseudomonas sp. (Dutta, Dutta, & Banerjee, 2004), in fed-batch yeast fermentations (Gadkar, Mehra, & Gomes, 2005) and in the fed batch production of foreign proteins (Laursen, Webb, & Ramirez, 2007).

In industrial systems where a significant amount of online and offline process data may be available, but there are tight time restraints imposed on process optimization, the potential for developing a relatively accurate neural network model within short time scales becomes very attractive (Glassey et al., 1994).

Fuzzy control is a process control method using "fuzzy reasoning" (also described as "fuzzy inference"). This type of control can transform human reasoning gained from skilled operators into rules suitable for process control and have been widely applied in fermentation and related bioprocesses. Fuzzy control coupled with ANNs (Fuzzy neural networks FNNs) have been studied as a tool for fuzzy modeling (Honda & Kobayashi, 2011). Fonseca, Schmitz, Fileti, and da Silva (2013) describe the use of a fuzzy-proportional/integral controller for fermenter temperature control.

REFERENCES

Aiba, S., Okamoto, R., & Satoh, K. (1965). Two sorts of measurements with a jar type of fermenter-power requirements of agitations and capacity coefficient of mass transfer in bubble aeration. *Journal of Fermentation Technology, 43*, 137–145.

Aiba, S., Humphrey, A. E., & Millis, N. F. (1973). *Biochemical Engineering* (2nd ed.). New York: Academic Press.

Alford, J. S. (2006). Bioprocess control: advances and challenges. *Computers & Chemical Engineering, 30*, 1464–1475.

Armiger, W. B. (1985). Instrumentation for monitoring and controlling reactors. In C. L. Cooney, & A. E. Humphrey (Eds.), *Comprehensive biotechnology* (pp. 133–148). (Vol. 2). Oxford: Pergamon Press.

Armiger, W. B., & Moran, D. M. (1979). Review of alternatives and rationale for computer interfacing and system configuration. *Biotechnology and Bioengineering Symposium, 9*, 215–225.

Atkinson, B., & Mavituna, F. (1991). Measurement and instrumentation. In *Biochemical engineering and biotechnology handbook* (pp. 1023–1057). Basingstoke: Macmillan.

Backer, M., Delle, L., Poghossian, A., Biselli, M., Zang, W., Wagner, P., & Schoning, M. J. (2011). Electrochemical sensor array for bioprocess monitoring. *Electrochimica Acta, 56*, 9673–9678.

Bailey, J. E., & Ollis, D. F. (1986). Instrumentation and control. In *Biochemical engineering fundamentals* (pp. 658–725). New York: McGraw-Hill.

Bambot, S. B., Holavanahali, R., Lakowicz, J. R., Carter, G. M., & Rao, G. (1994). Phase fluorometric sterilizable optical oxygen sensor. *Biotechnology and Bioengineering, 43*, 1139–1145.

Beyeler, W., Einsele, A., & Fiechter, A. (1981). Fluorometric studies in bioreactors: methods and applications. *Abstracts of communications. Second European Congress of Biotechnology, Eastbourne* (pp. 142). London: Society of Chemical Industry.

Bodizs, L., Titica, M., Faria, N., Srinivasan, B., Dochain, D., & Bonvin, D. (2007). Oxygen control for an industrial pilot-scale fed-batch filamentous fungal fermentation. *Journal of Process Control, 17*, 595–606.

Boulton, C. A., Maryan, P. S., Loveridgev, D., & Kell, D. B. (1989). The application of a novel biomass sensor to the control of yeast pitching rate. In: *Proceedings of the 22nd European Brewing Convention, Zurich* (pp. 653–661). European Brewing Convention.

Brodgesell, A. (1969). Miscellaneous sensors. In B. G. Liptak (Ed.), *Instrument Engineer's Handbook* (pp. 914–943). (Vol. 1). Philadelphia, PA: Chilton.

Brown, J., Kaminski, E., K, R., Blake, A. C., & Brodgesell, A. (1969). Analytical measurements. In B. G. Liptak (Ed.), *Instrument Engineer's Handbook* (pp. 713–918). (Vol. 1). Philadelphia: Chilton.

Buhler, H., & Ingold, W. (1976). Measuring pH and oxygen in fermenters. *Process Biochemistry, 11*(3), 19–2224.

Bull, D. N. (1985). Instrumentation for fermentation process control. In C. L. Cooney, & A. E. Humphrey (Eds.), *Comprehensive biotechnology* (pp. 149–163). (Vol. 2). Oxford: Pergamon Press.

Bushell, M. E. (1988). *Computers in fermentation technology. Progress in industrial microbiology* (Vol. 25). Amsterdam: Elsevier.

Cervera, A. E., Petersen, N., Lantz, A. E., Larsen, A., & Gernay, K. V. (2009). Application of near-infrared spectroscopy for monitoring and control of cell culture and fermentation. *Biotechnology Progress, 25*(6), 1561–1581.

Chen, L. Z., Nguang, S. K., Li, X. M., & Chen, X. D. (2004). Soft sensors for on-line biomass measurements. *Bioprocess and Biosystems Engineering, 26*, 191–195.

Cooney, C. L., Wang, H. Y., & Wang, D. I. C. (1977). Computer-aided balancing for prediction of fermentation parameters. *Biotechnology and Bioengineering, 19*, 55–67.

Doran, P. M. (2013). *Bioprocess engineering principles* (2nd ed.) (pp. 778–789). Oxford: Academic Press.

Dusseliee, P. J. B., & Feijen, J. (1990). Instrumentation and control. In B. McNeil, & L. M. Harvey (Eds.), *Fermentation: A practical approach* (pp. 149–172). IRL Press: Oxford.

Dutta, J. J., Dutta, P. K., & Banerjee, R. (2004). Optimisation of culture parameters extracellular protease production from a newly isolated *Pseudomonas* sp. using response surface and artificial neural network models. *Process Biochemistry, 39*(12), 2193–2198.

Einsele, A., Ristroph, D. I., & Humphrey, A. E. (1978). Mixing times and glucose uptake measured with a fluorimeter. *Biotechnology and Bioenginnering, 20*, 1487–1492.

Enfors, S. O. (1981). An enzyme electrode for control of glucose concentration in fermentation broths. In: *Abstracts of communications. Second congress of biotechnology, Eastbourne* (pp. 141). London: Society for Chemical Industry.

Finn, B., Harvey, L. M., & Mcneil, B. (2006). Near-infrared spectroscopic monitoring of biomass, glucose, ethanol and protein content in a high cell density baker's yeast fed-batch bioprocess. *Yeast, 23*, 507–517.

Fish, N. M., Fox, R. I., & Thornhill, N. F. (1989). *Computer applications in fermentation technology—modelling and control of biotechnological processes*. London: Society for Chemical Industry/Elsevier.

Flynn, D. S. (1983). Instrumentation for fermentation processes. In A. Halme (Ed.), *Modelling and control of fermentation processes* (pp. 5–12). Pergamon Press: Oxford.

Flynn, D. S. (1984). Instrumentation and control of fermenters. In J. E. Smith, D. R. Berry, & B. Kristiansen (Eds.), *Filamentous Fungi* (pp. 77–100). (Vol. 4). London: Arnold.

Fonseca, R. R., Schmitz, J. E., Fileti, A. M. F., & da Silva, F. V. (2013). A fuzzy-split range control system applied to a fermentation process. *Bioresource Technology, 142*, 475–482.

Gadkar, K. G., Mehra, S., & Gomes, J. (2005). On-line adaptation of neural networks for bioprocess control. *Computers and Chemical Engineering, 29*, 1047–1057.

Gary, K., Meier, P., & Ludwig, K. (1988). General aspects of the use of sensors in biotechnology with special emphasis on cell cultivation. In J. Gavora, D. F. Gerson, J. Luong, A. Stover, & J. H. Woodley (Eds.), *Biotechnology research and applications* (pp. 155–164). London: Elsevier.

Ghosh, A., Bag, A. K., Sharma, P., & Tudu, B. (2015). Monitoring of the fermentation process and detection of optimum fermentation time of black tea using an electronic tongue. *Sensors Journal, IEEE, 15*(11), 6255–6262.

Glassey, J., Montague, G. A., Ward, A. C., & Kara, B. V. (1994). Enhanced supervision of recombinant *E. coli* fermentations via artifical neural networks. *Process Biochemistry, 29*, 387–398.

Gomes, J., & Menawat, A. S. (2000). Precise control of dissolved oxygen in bioreactors—a model based geometric algorithm. *Chemical Engineering Science, 55*(1), 67–78.

Grayson, P. (1969). Computer control of batch fermentations. *Process Biochemistry, 4*(3), 43–44.

Hailing, P. J. (1990). pH, dissolved oxygen and related sensors. In B. McNeil, & L. M. Harvey (Eds.), *Fermentation—a practical approach* (pp. 131–147). Oxford: IRL.

Hall, M. J., Dickinson, S. D., Pritchard, R., & Evans, J. I. (1973). Foams and foam control in fermentation processes. *Progress in Industrial Microbiology, 12*, 169–231.

Hall, G. A., Higgins, S. P., Kennedy, R. H., & Nelson, J. M. (1974). Principles of automatic control. In D. M. Considine (Ed.), *Process instrumentation and control handbook* (2nd ed.). New York: McGraw-Hill, Section 18.

Hammond, S. V., & Brookes, I. K. (1992). Near infrared spectroscopy—A powerful technique for at-line and on-line analysis of fermentation. In M. R. Ladisch, & A. Bose (Eds.), *Harnessing biotechnology for the 21st century* (pp. 325–333). Washington, DC: American Chemical Society.

Harrison, D. E. F., & Chance, B. (1970). Fluorimetrie technique for monitoring changes in the level of reduced nicotinamide nucleotides in continuous cultures of micro-organisms. *Applied Microbiology*, *19*, 446–450.

Heinzle, E., Dunn, I. J., & Bourne, J. R. (1981). Continuous measurement of gases and volatiles during fermentations using mass spectrometry. In *Abstracts of communications. Second European congress of biotechnology, Eastbourne* (pp. 24). London: Society of Chemical Industry.

Hewetson, J. W., Jong, T. H., & Gray, P. P. (1979). Use of an immobilized penicillinase electrode in the monitoring of the penicillin fermentation. *Biotechnology and Bioengineering Symposium*, *9*, 125–135.

Higgins, I. J., & Boldot, J. (1992). Development of practical electrochemical biological sensing devices. In M. R. Ladisch, & A. Bose (Eds.), *Harnessing biotechnology for the 21st century* (pp. 316–318). Washington, DC: American Chemical Society.

Higgins, I. J., Swain, A., & Turner, A. P. F. (1987). Principles and applications of biosensors in microbiology. *Journal of Applied Bacteriology Symposium Supplement*, *16*, 95S–105S.

Hisbullah, M. A., Hussain, K. B., & Ramachandran (2002). Comparative evaluation of various control schemes for fe-batch fermentation. *Bioprocess Biosystems Engineering*, *24*, 309–318.

Honda, H., & Kobayashi, T. (2011). Fuzzy control of bioprocess. In M. Moo-Young (Ed.), *Comprehensive biotechnology* (pp. 864–872). (Vol. 2). Oxford: Elsevier.

Howe, W. H., Kop, J. G., Siev, R., & Liptak, B. G. (1969). Flow measurement. In B. G. Liptak (Ed.), *Instrument Engineer's Handbook* (pp. 411–567). (Vol. 1). Philadelphia, PA: Chilton.

Huttel, S., & Muller, R. (2012). Methods to optimize myxobacterial fermentations using off-gas analysis. *Microbial Cell Factories*, *11*(59), 1–11.

Jacob, H. E. (1970). Redox potential. In J. R. Morris, & D. W. Ribbons (Eds.), *Methods in microbiology* (pp. 91–123). (Vol. 2). London: Academic Press.

Jang, M. -F., Chern, Y. -J., & Chou, Y. -S. (2013). Robust adaptive controller for continuous bioreactors. *Biochemical Engineering Journal*, *81*, 136–145.

Jimenez-Marquez, F., Vazquez, J., Ubeda, J., & Sanchez-Rojas, J. L. (2016). Temperature dependence of grape must refractive index and its application to winemaking monitoring. *Sensors and Actuators B: Chemical*, *225*, 121–127.

Johnson, M. J., Borkowski, J., & Engblom, C. (1964). Steam sterilisable probes for dissolved oxygen measurement. *Biotechnology and Bioengineering*, *6*, 457–468.

Karim, M. N., & Rivera, S. L. (1992). Artificial neural networks in bioprocess state estimation. *Advances in Biochemical Engineering Biotechnology*, *46*, 1–33.

Kell, D. B. (1987). The principles and potential of electrical admittance spectroscopy; an introduction. In A. P. F. Turner, I. Karube, & G. S. Wilson (Eds.), *Biosensors: Fundamentals and applications* (pp. 427–468). Oxford: University Press.

Kjaergaard, L. (1977). The redox potential: its use and control in biotechnology. *Advances in Biochemical Engineering*, *7*, 131–150.

Kjaergaard, L., & Joergensen, B. B. (1979). Redox potential as a state variable in fermentation systems. *Biotechnology and Bioengineering Symposium, 9,* 85–94.

Kristiansen, B. (1987). Instrumentation. In J. D. Bu'Lock, & B. Kristiansen (Eds.), *Basic biotechnology* (pp. 253–280). London: Academic Press.

Laursen, S. O., Webb, D., & Ramirez, W. F. (2007). Dynamic hybrid neural network model of an industrial fed-batch fermentation process to produce foreign protein. *Computers and Chemical Engineering, 31,* 163–170.

Lee, Y. H., & Tsao, G. T. (1979). Dissolved oxygen electrodes. *Advances in Biochemical Engineering, 13,* 36–86.

Li, Y., Chen, J., & Lun, S. (2002). Oxygen-supply control mode for efficient production of pyruvate in batch process. *Journal of Chemical Industry and Engineering (China), 53*(12), 1232.

Li, S. -Y., Srivastava, R., Suib, S. L., Li, Y., & Parnas, R. S. (2011). Performance of batch, fed-batch and continuous A-B-E fermentation with pH control. *Bioresource Technology, 102,* 4241–4250.

Linko, P., & Zhu, Y. -H. (1992). Neural network modelling for real-time variable estimation and prediction in the control of glucoamylase fermentation. *Process Biochemistry, 27,* 275–283.

Liptak, B. G. (1969). Pressure mesurement. In B. G. Liptak (Ed.), *Instrument Engineer's Handbook* (pp. 171–263). (Vol. 1). Philadelphia, PA: Chilton.

Lorenz, T., Frueh, K., Diekmann, J., Hiddessen, R., Lue-bbert, A., & Schuegerl, K. (1985). On-line measurement and control in a bubble column reactor with an external loop. *Chemical Engineering and Technology, 57,* 116–117.

Luck, S., Buge, G., Plettenburg, H., & Hoffmann, M. (2010). Near-infrared spectroscopy for process control and optimization of biogas plants. *Engineering Life Sciences, 10*(6), 537–543.

Luong, J. H. T., Mulchandani, A., & Guilbault, G. G. (1988). Developments and applications of biosensors. *Trends in Biotechnology, 8,* 310–316.

Luong, J. H. T., Mahmoud, K. A., & Male, K. B. (2011). Instrumentation and analytical methods. In M. Moo-Young (Ed.), *Comprehensive biotechnology* (pp. 829–838). (Vol. 2). Oxford: Elsevier.

Meiners, M. (1982). Computer applications in fermentation. In V. Krumphanzl, B. Sikyta, & Z. Vanek (Eds.), *Overproduction of microbial products* (pp. 637–649). London: Academic Press.

Meszaros, A., Andrasik, A., Mizsey, P., Fonyo, Z., & Illeova, V. (2004). Computer control of pH and DO in a laboratory fermenter using a neural network technique. *Bioprocess and Biosystems Engineering, 26*(5), 331–340.

Moeller, L., Grunberg, M., Zehnsdorf, A., Strehlitz, B., & Bley, T. (2010). Biosensor online control of citric acid production from glucose by *Yarrowia lipolytica* using semicontinuous fermentation. *Engineering in Life Sciences, 10*(4), 311–320.

Moon, C., Jang, S., Yun, Y. -M., Lee, M. -K., Kim, D. -H., Kang, W. -S., Kwak, S. -S., & K.I.M, M. -S. (2015). Effect of the accuracy of pH control on hydrogen fermentation. *Bioresource Technology, 179,* 595–601.

Moo-Young, M. (2011). In M. Moo-Young (Ed.), *Comprehensive biotechnology* (Vol. 2). Oxford: Elsevier.

Montague, G., Morris, A., & Ward, A. (1988). Fermentation monitoring and control: a perspective. *Biotechnology and Genetic Engineering Reviews, 7,* 147–188.

Nyiri, L. K. (1972). A philosophy of data acquisition, analysis and computer control of fermentation processes. *Developments in Industrial Microbiology, 13,* 136–145.

Nyiri, L. K., Toto, G. M., & Charles, M. (1975). Measurement of gas-exchange conditions in fermentation processes. *Biotechnology and Bioengineering, 17*, 1663–1678.

Omstead, D. R., & Greasham, R. H. (1989). Integrated fermenter sampling and analysis. In N. M. Fish, R. I. Fox, & N. F. Thornhill (Eds.), *Computer applications in fermentation technology: Modelling and control of technological processes* (pp. 5–13). London: SCI-Elsevier.

Petre, E., Selisteanu, D., & Sendrescu, D. (2013). Adaptive and robust adaptive control strategies for anaerobic wastewater treatment bioprocesses. *Chemical Engineering Journal, 217*, 363–378.

Porep, J. U., Kammerer, D. R., & Carle, R. (2015). On-line application of near infra-red (NIR) spectroscopy in food production. *Trends in Food Science and Technology, 46*, 211–230.

Radjai, M. K., Hatch, R. T., & Cadman, T. W. (1984). Optimization of amino acid production by automatic self-tuning digital control of redox potential. *Biotechnical and Bioengineering Symposium, 14*, 657–679.

Ramirez, A., Durand, A., & Blachere, H. T. (1981). Optimal baker's yeast production in extended fed-batch culture using a computer coupled pilot-fermentor. In *Abstracts of communications. Second European congress of biotechnology, Eastbourne* (pp. 26). London: Society of Chemical Industry.

Renard, F., & Vande Wouwer, A. (2008). Robust adaptive control of teast fed-batch cultures. *Computers and Chemical Engineering, 32*, 1238–1248.

Restrepo, A., Gonzalez, A., & Orduz, S. (2002). Cost effective control strategy for small applications and pilot plants: on-off valves with temporized PID controller. *Chemical and Engineering Journal, 89*, 101–107.

Rolf, M. J., & Lim, H. C. (1985). Systems for fermentation process control. In C. L. Cooney, & A. E. Humphrey (Eds.), *Comprehensive biotechnology* (pp. 165–174). (Vol. 2). Oxford: Pergamon Press.

Royce, P. N. (1993). A discussion of recent developments in fermentation monitoring from a practical perspective. *Critical Reviews in Biotechnology, 13*, 117–149.

Rudnitskaya, A., & Legin, A. (2008). Sensor systems, electronic tongues and electronic noses for the monitoring of biotechnological processes. *Journal of Industrial Microbiology and Biotechnology, 35*, 443–451.

Schneckenburger, H., Reuter, B. W., & Schoberth, S. M. (1985). Fluorescence techniques in biotechnology. *Trends in Biotechnology, 3*, 257–261.

Shi, Z., Shimizu, K., Watanabe, N., & Kobayashi, T. (1989). Adaptive on-line optimizating control of bioreactor systems. *Biotechnology and Bioengineering, 33*, 999–1009.

Shimizu, K. (1993). An overview on the control system design of bioreactors. *Advances in Biochemistry Engineering Journal of Biotechnology, 50*, 65–84.

Shinskey, F. G. (1973). *pH and pion control in process and waste streams*. New York: Wiley, Chapter 9.

Swartz, J. R., & Cooney, C. L. (1979). Indirect fermentation measurements as a basis for control. *Biotechnology and Bioengineering Symposium, 9*, 95–101.

Teixeira, A. P., Oliveira, R., Alves, P. M., & Carrondo, M. J. T. (2011). Online control strategies. In M. Moo-Young (Ed.), *Comprehensive biotechnology* (pp. 875–881). (Vol. 2). Oxford: Elsevier.

Vaccari, G., Dosi, E., Campsi, A. L., Gonzalez-Varay, R. A., Matleuzzi, D., & Mantovani, C. (1994). A near infrared spectroscopy technique for the control of fermentation processes: an application to lactic acid fermentation. *Biotechnology and Bioengineering, 43*, 913–917.

Vanags, J., Rychtera, M., Ferzik, S., Vishkins, M., & Viesturs, U. (2007). Oxygen and temperature control during the cultivation of microorganisms using substrate feeding. *Engineering in Life Sciences, 7*(3), 247–252.

Vanags, J., Hrynko, V., & Viesturs, U. (2010). Development and application of a flexible controller in yeast fermentations using pO_2 cascade control. *Engineering in Life Sciences*, *10*(4), 321–332.

Viesturs, U. E., Kristapsons, M. Z., & Levitans, E. S. (1982). Foam in microbiological processes. *Advances in Biochemical Engineering*, *21*, 169–224.

Vojta, V., Nahlik, J., Paca, J., & Komarkova, E. (2002). Development and verification of the control system for fed-batch phenol degradation processes. *Chemical and Biochemical Engineering Quarterly*, *16*(2), 59–67.

Wang, H. Y., Cooney, C. L., & Wang, D. I. C. (1977). Computer-aided baker's yeast fermentations. *Biotechnology and Bioengineering*, *19*, 69–86.

Whiteside, M. C., & Morgan, P. (1989). Computers in fermentation. In T. N. Bryant, & J. W. T. Wimpenny (Eds.), *Computers in microbiology—A practical approach* (pp. 175–190). Oxford: IRL Press.

Yamashita, S., & Murao, C. (1978). Fermentation process. *Journal Society of Instrument and Control Engineering*, *6*(10), 735–740.

Yamashita, S., Hisoshi, H., & Inagakj, T. (1969). Automatic control and optimization of fermentation processes—glutamic acid. In D. Perlman (Ed.), *Fermentation advances* (pp. 441–463). New York: Academic Press.

Zabriskie, D. W. (1985). Data analysis. In C. L. Cooney, & A. E. Humphrey (Eds.), *Comprehensive biochemistry* (pp. 175–190). (Vol. 2). Oxford: Pergamon Press.

Zhu, H., Cao, C., Zhang, S., Zhang, Y., & Zou, W. (2011). pH control modes in a 5-L stirred tank bioreactor for cell biomass and exopolysaccharide production by *Tremella fuciformis* spore. *Bioresource Technology*, *102*, 9175–9178.

Aeration and agitation

INTRODUCTION

The majority of fermentation processes are aerobic and, therefore, require the provision of oxygen. If the stoichiometry of respiration is considered, then the oxidation of glucose may be represented as:

$$C_6H_{12}O_6 + 6O_2 = 6H_2O + 6CO_2$$

Thus, 192 g of oxygen are required for the complete oxidation of 180 g of glucose. However, both components must be in solution before they are available to a microorganism and oxygen is approximately 6000 times less soluble in water than is glucose (a fermentation medium saturated with oxygen contains approximately 7.6 mg dm^{-3} of oxygen at 30°C). Thus, it is not possible to provide a microbial culture with all the oxygen it will need for the complete oxidation of the glucose (or any other carbon source) in one addition. Therefore, a microbial culture must be supplied with oxygen during growth at a rate sufficient to satisfy the organisms' demand.

The oxygen demand of an industrial fermentation process is normally satisfied by aerating and agitating the fermentation broth. However, the productivity of many fermentations is limited by oxygen availability and, therefore, it is important to consider the factors which affect a fermenter's efficiency in supplying microbial cells with oxygen. This chapter considers the requirement for oxygen in fermentation processes, the quantification of oxygen transfer, and the factors that will influence the rate of oxygen transfer into solution.

OXYGEN REQUIREMENTS OF INDUSTRIAL FERMENTATIONS

Although a consideration of the stoichiometry of respiration gives an appreciation of the problem of oxygen supply, it gives no indication of an organism's true oxygen demand as it does not take into account the carbon that is converted into biomass and products. A number of workers have considered the overall stoichiometry of the conversion of oxygen, a source of carbon, and a source of nitrogen into biomass and have used such relationships to predict the oxygen demand of a fermentation. A selection of such equations is shown in Table 9.1. Villadsen, Nielsen, and Liden (2011) described these equations as "black box stoichiometries" in which the complexities of the biochemical network are represented as a single equation. The physiology of an organism changes under different cultural conditions and thus care must be taken

Table 9.1 Stoichiometric Equations Describing Oxygen Demand in a Fermentation

Equation	Terms Used	References
$6.67CH_2O + 2.1O_2 = C_{3.92}H_{65} + 2.75CO_2 + 3.42H_2O$	$C_{3.92}H_{65}O_{1.94}$ is 100 g (dry weight) of yeast cells; CH_2O is carbohydrate	Darlington (1964)
$7.14CH_2 + 6.135O_2 = C_{3.92}H_{65}O_{1.94} + 3.22CO_2 + 3.89H_2O$	CH_2 is hydrocarbon	Darlington (1964)
$(A/Y) - B = C$	A = Amount of oxygen for combustion of 1 g of substrate to CO_2, H_2O and NH_3, if nitrogen is present in the substrate	Johnson (1964)
	B = Amount of oxygen required for the combustion of 1 g cells to CO_2, H_2O and NH_3	
	Y = Cell yield (g cells g^{-1} substrate)	
	C = g oxygen consumed for the production of 1 g of cells	
$(dx/dt)/Y + mx + p = (dCO_2/dt) = -(dO_2/dt) RQ$	x = biomass concentration	Righelato, Trinci, Pirt, and Peat (1968)
	t = time	
	Y = g biomass g^{-1} carbon substrate	
	m = maintenance	
	p = allowance for antibiotic production	
$Y_{O/P} = (0.53/Y_{p/G}) - (0.6 X/P) - 0.43$	$Y_{O/P}$ = g oxygen consumed g^{-1} glucose	Cooney (1979)
	$Y_{P/C}$ = g sodium penicillin G produced g^{-1} glucose	
	X = g cells (dry weight) produced	
	P = g sodium penicillin G produced	
$Y_O = \{(32C + 8H - 160)/YM\} - 1.58$	Y_O = g oxygen consumed g^{-1} cells produced	Mateles (1971)
	Y = Cell yield (g cells g^{-1} substrate)	
	M = Molecular weight of the carbon source	
	C, H, and O = Number of atoms of carbon, hydrogen, and oxygen per molecule of carbon source	

in applying these equations. However, from these determinations it may be seen that a culture's demand for oxygen is very much dependent on the source of carbon in the medium. Thus, the more reduced the carbon source, the greater will be the oxygen demand. From Darlington's and Johnson's equations (Table 9.1) it may be seen that the production of 100 g of biomass from hydrocarbon requires approximately three times the amount of oxygen to produce the same amount of biomass from carbohydrate. This point is also illustrated in Table 9.2. However, it must be remembered that the high carbon content of hydrocarbon substrates means that high yield factors (grams biomass per gram substrate consumed) are obtained and the decision to use such substrates is based on the balance between the advantage of high biomass yield

Table 9.2 Oxygen Requirements of a Range of Microorganisms Grown on a Range of Substrates

Substrate	Organism	Oxygen Requirement (g O_2 g^{-1} dry wt.)	References
Glucose	*Escherichia coli*	0.4	Schulze and Lipe (1964)
Methanol	*Pseudomonas C*	1.2	Goldberg, Rock, Ben-Bassat, and Máteles (1976)
Octane	*Pseudomonas* sp.	17	Wodzinski and Johnson (1968)

After Mateles, 1979.

and the disadvantage of high oxygen demand and heat generation. These points are discussed in more detail in Chapter 4.

Darlington's, Johnson's, and Máteles' equations only include biomass production and do not consider product formation, whereas Cooney's and Righelato et al's and Villadsen et al's equations consider product formation. Ryu and Hospodka (1980) used Righelato's approach to calculate that the production of 1 g penicillin consumes 2.2 g of oxygen. However, it is inadequate to base the provision of oxygen for a fermentation simply on an estimation of overall demand, because the metabolism of the culture is affected by the concentration of dissolved oxygen in the broth. The effect of dissolved oxygen concentration on the specific oxygen uptake rate (Q_{O_2}, millimoles of oxygen consumed per gram dry weight of cells per hour) has been shown to be of the Michaelis–Menten type, as shown in Fig. 9.1.

From Fig. 9.1 it may be seen that the specific oxygen uptake rate increases with increase in the dissolved oxygen concentration up to a certain point (referred to as the critical dissolved oxygen concentration, C_{crit}) above which no further increase

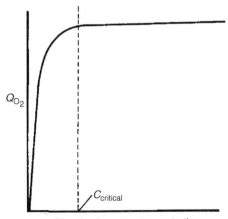

Dissolved oxygen concentration

FIGURE 9.1 The Effect of Dissolved Oxygen Concentration on the Q_{O_2} of a Microorganism

Table 9.3 Critical Dissolved Oxygen Concentrations for a Range of Microorganisms (Riviere, 1977; Doran, 2013)

Organism	Temperature	Critical Dissolved Oxygen Concentration (mmoles dm^{-3})
Azotobacter sp.	30	0.018
Escherichia coli	37	0.008
Saccharomyces sp.	30	0.004
Penicillium chrysogenum	24	0.022
Chinese hamster ovary cells (CHO)	37	0.020

in oxygen uptake rate occurs. Some examples of the critical oxygen levels for a range of microorganisms and CHO cells are given in Table 9.3. Thus, maximum biomass production may be achieved by satisfying the organism's maximum specific oxygen demand by maintaining the dissolved oxygen concentration greater than the critical level. If the dissolved oxygen concentration were to fall below the critical level then the metabolism of the cells may be metabolically disturbed. However, it must be remembered that it is frequently the objective of the fermentation technologist to produce a product of the microorganism rather than the organism itself and that metabolic disturbance of the cell by oxygen starvation may be advantageous to the formation of certain products. Equally, provision of a dissolved oxygen concentration far greater than the critical level may have no influence on biomass production, but may stimulate product formation. Thus, the aeration conditions necessary for the optimum production of a product may be different from those favoring biomass production.

Hirose and Shibai's (1980) investigations of amino acid biosynthesis by *Brevibacterium flavum* provide an excellent example of the effects of the dissolved oxygen concentration on the production of a range of closely related metabolites. *Brevibacterium flavum* is effectively the same organism as *Corynebacterium glutamicum*, the industry's chosen strain for producing amino acids (see Chapter 3). These workers demonstrated the critical dissolved oxygen concentration for *B. flavum* to be 0.01 mg/dm^3 and considered the extent of oxygen supply to the culture in terms of the degree of "oxygen satisfaction," that is the respiratory rate of the culture expressed as a fraction of the maximum respiratory rate. Thus, a value of oxygen satisfaction below unity implied that the dissolved oxygen concentration was below the critical level. The effect of the degree of oxygen satisfaction on the production of a range of amino acids is shown in Fig. 9.2. From Fig. 9.2, it may be seen that the production of members of the glutamate (glutamine, proline, and arginine) and aspartate (lysine, threonine, and isoleucine) families of amino acids was affected detrimentally by levels of oxygen satisfaction below 1.0, whereas optimum production of phenylalanine, valine, and leucine occurred at oxygen satisfaction levels of 0.55, 0.60, and 0.85, respectively. The biosynthetic routes of the amino acids are shown in Fig. 9.3, from which it may be seen that the glutamate and aspartate

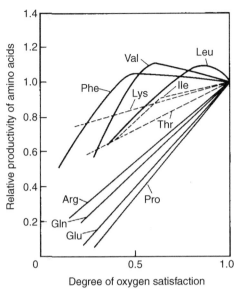

FIGURE 9.2 The Effect of Dissolved Oxygen on the Production of Amino Acids by *Brevibacterium flavum* **(Hirose & Shibai, 1980)**

families are all produced from tricarboxylic acid (TCA) cycle intermediates, whereas phenylalanine, valine and leucine are produced from the glycolysis intermediates, pyruvate and phosphoenol pyruvate. Oxygen excess should give rise to abundant TCA cycle intermediates, whereas oxygen limitation should result in less glucose being oxidized via the TCA cycle, allowing more intermediates to be available for phenylalanine, valine, and leucine biosynthesis. Thus, some degree of metabolic disruption results in greater production of pyruvate derived amino acids. However, the sensitivity to oxygen limitation of the two TCA derived amino acid families differed significantly, with the synthesis of the glutamate family being far more affected than the aspartate. Under biotin limitation, *B. flavum* is deficient in α-ketoglutarate dehydrogenase activity (converting α-ketoglutarate to succinyl co-A) such that the TCA cycle is completed via the glyoxylate cycle (see Chapter 3). Thus, oxaloacetate, the precursor of the aspartate amino acid family, might be expected to be generated via the glyoxylate cycle. However, oxaloacetate may also be produced from the glycolytic intermediate, phophoenolpyruvate (PEP), by the CO_2 fixing reaction catalyzed by (PEP) carboxylase.

$$PEP + HCO_3^- \gg \text{oxaloacetate} + \text{inorganic phosphate}$$

Thus, Hirose and Shibai speculated that the resistance to oxygen limitation displayed by the aspartate amino acid family was due to the synthesis of oxaloacetate via the non-TCA/glyoxylate pathway. Almost 20 years later, Hua, Fu, Yang, and Shimizu (1998) used metabolic flux analysis to explore this relationship. Metabolic

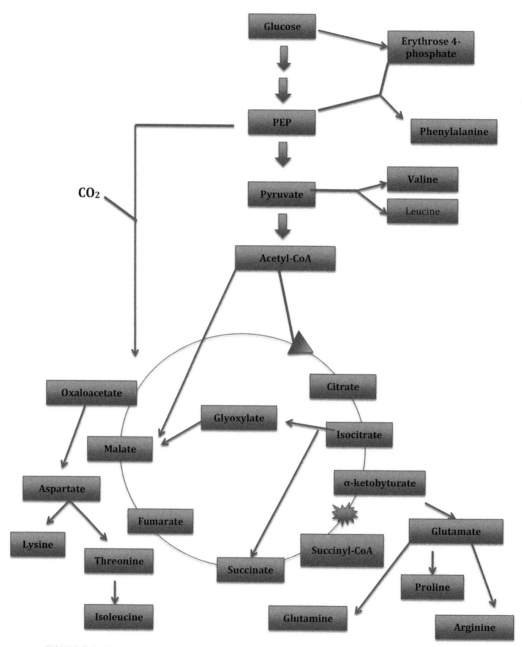

FIGURE 9.3 Glycolysis and the TCA and Glyoxylate Cycles and Associated Amino Acid Biosynthetic Pathways in *Brevibacterium flavum*

PEP, phosphenol pyruvate; ⁂, inactive α-ketobutyrate dehydrogenase.

flux analysis is a stoichiometric model of metabolism enabling the calculation of intracellular fluxes based either upon uptake rates of substrates and output rates of metabolites or on labeling studies (Antoniewicz, 2015). This approach confirmed that the PEP-carboxylase route contributes significantly to lysine synthesis under oxygen limited conditions—a finding which was developed by Becker et al. (2011) in their genetic manipulation of the strain (discussed in Chapter 3).

The TCA cycle intermediate, succinic acid (or succinate), is a key building block for the industrial production of a range of bulk chemicals. Although succinate is manufactured mainly by a chemical process from petrochemical sources, there is considerable interest in the development of a microbiological, and thus more environmental friendly, route to its production (Tsuge, Hasunuma, & Kondo, 2015). Although *C. glutamicum* requires a high oxygen transfer rate to support growth, Inui et al. (2004) demonstrated that the strain *C. glutamicum* R, ATCC13032, could produce succinate when supplemented with bicarbonate and maintained in a nongrowing oxygen-starved condition. Thus, these workers exploited the CO_2 fixing reaction catalyzed by phosphoenolpyruvate carboxylase to generate oxaloacetate which was then converted to succinate by the reductive (or reverse) TCA cycle via malate and fumarate, as can be seen in Fig. 9.4. A two-stage process was developed in which the organism was grown under aerobic conditions and the resulting biomass harvested and resuspended in minimal medium and maintained under anaerobic conditions with bicarbonate supplementation.

Xu et al. (2009) also applied metabolic flux analysis to study the effect of dissolved oxygen supply on the production of the glutamate family member, arginine, by *Corynebacterium crenatum*. These workers confirmed that a high oxygen supply rate was required for both arginine production and cell growth in the first phase of the culture. After 20 h the growth rate declined and growth eventually ceased after 48 h but arginine synthesis continued for a further 48 h. By applying different oxygen transfer rates and analysing the flux through to arginine it was demonstrated that a high transfer rate was beneficial during the growth phase but a lower transfer rate maintained a higher flux to arginine during the final 24 h of the culture. Thus, a two-stage system was developed in which an initial high oxygen transfer rate was used during the growth phase followed by a staged decrease over 6 h to a lower rate for the remainder of the process.

An excellent example of the genetic improvement of a strain to enhance the beneficial effects of oxygen limitation is given by the work of Hasegawa et al. (2012) on L-valine synthesis. As discussed earlier, valine is synthesized from the glycolytic intermediate, pyruvate, and its production is favored by oxygen limitation Fig. 9.3. The production of one mole of valine from one mole of glucose results in the generation of two moles of NADH and the consumption of 2 moles of NADPH, one by acetohydroxyacid isomeroreductase (AHAIR) and one by the amination of oxoglutarate in the regeneration of glutamate consumed in the transamination of ketoisovalerate. However, *C. glutamicum* lacks the enzyme nicotinamide nucleotide transhydrogenase that, in many organisms, catalyzes the reversible interconversion between NADH and NAD. Thus, there is an inherent cofactor imbalance in the valine

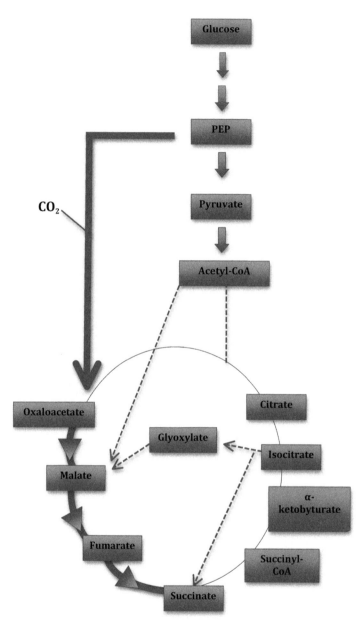

FIGURE 9.4 Pathways of *C. glutamicum* under oxygen deprivation enabling succinate synthesis via the reductive TCA cycle

PEP, Phosphenol pyruvate; ➡, dominant metabolic route.

fermentation. Furthermore, the flux through glycolysis is controlled by the activity of glyceraldehyde-3-phosphate dehydrogenase mediated by the NADH/NAD$^+$ ratio. Thus, an accumulation of NADH under oxygen deprivation reduces glucose consumption and further limits valine synthesis. Hasegawa et al. (2012) addressed these anomalies by engineering the organism and replacing the NADPH-requiring steps with NADH-dependent reactions. Thus, the NADPH-dependent AHAIR was replaced with the NADH-dependent equivalent from *Lysinibacillus sphaericus* and the NADPH-dependent transaminase (utilizing glutamate as the amino donor) replaced by the NADH-dependent *L. sphaericus* isoleucine dehydrogenase (utilizing NH$_3$ as amino donor). The resulting strain increased the valine production from 5.4 to 63% of the theoretical yield from glucose. Again, the process was conducted in two stages—biomass production under oxygen sufficiency followed by valine production under oxygen limitation.

It is frequently assumed that fully aerobic conditions are required for heterologous protein production in microbial hosts. However, Baumann et al. (2008) demonstrated that oxygen-limitation was beneficial for the production of an antibody Fab (fragment antigen-binding) fragment by *Pichia pastoris* cultivated on glucose. Furthermore, this group demonstrated (Baumann, Adelantado, Lang, Mattanovich, & Ferrer, 2011) a link between oxygen limitation, ergosterol synthesis, and production of the heterologous protein. Ergosterol is a component of the cell membrane and requires oxygen for its biosynthesis suggesting that oxygen limitation influences membrane structure and, in turn, protein secretion. In a study of the scale-up of a *S. cerevisiae* heterologous protein Fu et al. (2014) attributed improved protein productivity at the 10,000 dm^3 scale (compared with a 10 dm^3 fermentation) to oxygen limitation and its possible influence on the membrane lipid composition. Interestingly, Ukkonen, Veijola, Vasala, and Neubauer (2013) demonstrated that lower aeration rates also stimulated Fab production by *E. coli* but as bacterial cell membranes do not contain sterols the explanation of the phenomenon cannot have the same basis as that in yeasts.

An example of the effect of dissolved oxygen on secondary metabolism is provided by the work done by Zhou, Holzhauer-Rieger, Dors, and Schugerl (1992) on cephalosporin C synthesis by *Cephalosporium acremonium*. These workers demonstrated that the critical oxygen concentration for cephalosporin C synthesis during the production phase was 20% saturation. At dissolved oxygen concentrations below 20% cephalosporin C concentration declined and penicillin N increased. The biosynthetic pathway to cephalosporin C is shown in Fig. 9.5, from which it may be seen that there are three oxygen-consuming steps in the pathway:

1. Cyclization of the tripeptide, α-amino-adipyl-cysteinyl-valine into isopenicillin N.
2. The ring expansion of penicillin N into deacetoxycephalosporin C (DAOC).
3. The hydroxylation of DAOC to give deacetylcephalosporin C.

DAOC did not accumulate at low oxygen concentrations and, thus, it appears that the most oxygen sensitive step in the pathway is the ring expansion enzyme (expandase) resulting in the accumulation of penicillin N under oxygen limitation.

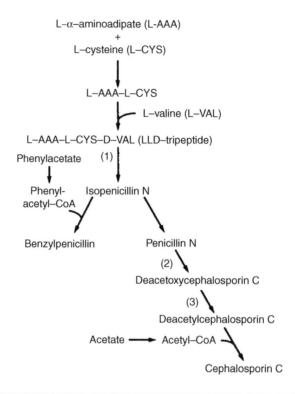

FIGURE 9.5

The biosynthesis of Cephalosporin C, indicating the OXYGEN consuming steps:
1. isopenicillin-N-synthase,
2. deacetoxycephalosporin C synthase (commonly called expandase),
3. deacetyl cephalosporin C synthase (commonly called hydroxylase).

The insecticide, spinosad, is a secondary metabolite produced by *Saccharopolyspora spinosa*. Bai et al. (2015) investigated the effect of dissolved oxygen on the biomass specific ATP content (mg ATP g^{-1} dry weight), the expression of key spinosad genes and spinosad yield. This work resulted in the development of a four-stage process, with the dissolved oxygen concentration (DO) being controlled to optimize either growth or spinosad production over the duration of the fermentation. An initial DO of 40%, increasing to 50%, sustained the growth phase following which optimum secondary metabolite production was achieved by progressively reducing the DO to 30% and 25%.

The requirement for a high dissolved oxygen concentration by many fermentations has resulted in the development of process techniques to ensure that the fermentation does not exceed the oxygen-supply capabilities of the fermentation vessel. The oxygen demand of a fermentation largely depends on the concentration of the biomass and its respiratory activity, which is related to the growth rate. By limiting the initial concentration of the medium, the biomass in the vessel may be kept at a reasonable level and by supplying some nutrient component as a feed, the rate of growth, and hence the respiratory rate, may be controlled. These techniques of medium design and nutrient feed are discussed in Chapters 2 and 4 and later in this chapter.

OXYGEN SUPPLY

Oxygen is normally supplied to microbial cultures in the form of air, this being the cheapest available source of the gas. The method for provision of a culture with a supply of air varies with the scale of the process:

1. Laboratory-scale microbial cultures may be aerated by means of the shake-flask technique where the culture (50–100 cm^3) is grown in a conical flask (250–500 cm^3) shaken on a platform contained in a controlled environment chamber. Animal cell cultures are frequently grown in "T-flasks" which provide a large surface area for oxygen diffusion—the "T" referring to the total surface area available for cell growth; thus, a T-25 flask has a 25 cm^2 growth area. Larger growth areas are provided by "roller bottles"—as the name suggests, bottles that rotated on revolving rollers.

2. Microbial pilot- and industrial-scale fermentations are normally carried out in stirred, aerated vessels, termed fermenters, of the type described in Chapter 7. Laboratory scale experiments using culture volumes upward of approximately 500 cm^3 are also performed in stirred, aerated fermenters as this enables the cultural conditions to be better monitored and controlled, and facilitates the addition of supplements and the removal of samples. Some fermenters are so designed that adequate oxygen transfer is obtained without agitation and the design of these systems (termed bubble columns and air-lift fermenters) is also discussed in Chapter 7. Large-scale animal cell cultures have been grown in a range of fermenter types, including stirred aerated vessels based on microbial systems. The earliest such stirred vessel was termed a spinner flask and consisted of a glass flask with a suspended magnetic stirrer—a system which has been scaled up to tens of dm^3. However, in 1999, Singh revolutionized animal cell culture with the introduction of a presterilized, flexible, plastic, pillow-like culture bag that could be placed on a rocking platform to generate a wave-like motion within the vessel. These systems and their application are discussed in Chapters 6 and 7.

As discussed in detail in Chapter 7, most fermenters are stirred and aerated. The impellers (or agitators) used to stir a vessel are described according to the type of flow caused by their rotation. An axial flow impeller moves liquid up and down the vessel whereas a radial flow impeller moves the liquid out from the impeller to the vessel sides and back again. The traditional impeller that has been widely used in fermenter design is the radial flow Rushton turbine, also described as a disc turbine and consists of a central disc with (normally) six attached vertical blades (Fig. 9.6). Air is introduced under the turbine and as the airflow hits the central disc, it is directed toward the turbine blades, rather than up the stirrer shaft. However, in the last 20 years a number of alternative impellers have been widely adopted and these will be considered in a later section and in Chapter 7.

FIGURE 9.6 A Rushton Turbine Impeller (Chemineer UK)

Bartholomew, Karrow, Sfat, and Wilhelm (1950) represented the transfer of oxygen from air to the cellular site of respiration, during a fermentation, as occurring in a number of steps:

1. Diffusion of the oxygen from the bulk gas phase within the air bubble to the gas-liquid interface.
2. The transfer of oxygen across the gas-liquid interface.
3. The diffusion of the dissolved oxygen through a relatively stagnant "boundary layer" surrounding the air bubble to the bulk fermentation medium.
4. The transfer of the dissolved oxygen from outer edge of the stagnant boundary layer, through the fermentation medium to the suspended organism—be it single bacterial or animal cells, aggregates of cells, dispersed mycelium, mycelial aggregates or pellets.
5. The diffusion of the dissolved oxygen through another relatively stagnant boundary layer to the cell or mycelium surface, or through aggregates or pellets to the cell or mycelium surface.
6. Uptake of the dissolved oxygen by the fermentation organism across its cell membrane.
7. Transfer of the dissolved oxygen within the cell to the enzyme site.

These workers demonstrated that the limiting step in the transfer of oxygen from air to the cell in a *Streptomyces griseus* fermentation was step (3), the transfer of oxygen across the bubble boundary layer into the bulk liquid. These findings have been shown to be correct for nonviscous fermentations but it has been demonstrated that transfer may be limited by either of the stages (4) or (5) in certain highly viscous fermentations. Indeed, in pelleted mycelial fermentations the diffusion of oxygen into the depths of the pellet may be so poor that the cells are deprived of oxygen and die and autolyse resulting in a hollow biomass sphere. The difficulties inherent in such fermentations are discussed later in this chapter.

The rate of oxygen transfer from air bubble to the liquid phase may be described by the equation:

$$dC_L/dt = K_L a(C^* - C_L) \qquad (9.1)$$

where C_L is the concentration of dissolved oxygen in the fermentation broth (mmoles dm^{-3}), t is time (hours), dC_L/dt is the change in oxygen concentration over a time period, that is, the oxygen-transfer rate (mmoles O_2 dm^{-3} h^{-1}), K_L is the mass transfer coefficient (cm h^{-1}), a is the specific gas/liquid interface area per liquid volume (cm^2 cm^{-3}), $C*$ is the saturated dissolved oxygen concentration (mmoles dm^{-3}).

K_L may be considered as the sum of the reciprocals of the resistances to the transfer of oxygen from gas to liquid, and the difference between the saturated dissolved oxygen concentration and the actual concentration in the fermentation broth ($C*-C_L$) may be considered as the "driving force" across the resistances. It is extremely difficult to measure both K_L and "a," the gas–liquid interface area, in a fermentation and, therefore, the two terms are generally combined in the term $K_L a$, the volumetric mass-transfer coefficient, the units of which are reciprocal time (h^{-1}). The volumetric mass-transfer coefficient is used as a measure of the aeration capacity of a fermenter. The larger the $K_L a$, the higher the aeration capacity of the system. The $K_L a$ value will depend upon the design and operating conditions of the fermenter and will be affected by such variables as aeration rate, agitation rate and impeller design. These variables affect "K_L" by reducing the resistances to transfer and affect "a" by changing the number, size, and residence time of air bubbles. It is convenient to use $K_L a$ as a yardstick of fermenter performance because, unlike the oxygen-transfer rate, it is unaffected by dissolved oxygen concentration. However, the oxygen transfer rate is the critical criterion in a fermentation and, as may be seen from Eq. (9.1), it is affected by both $K_L a$ and dissolved oxygen concentration. The dissolved oxygen concentration reflects the balance between the supply of dissolved oxygen by the fermenter and the oxygen demand of the organism. If the $K_L a$ of the fermenter is such that the oxygen demand of the organism cannot be met, the dissolved oxygen concentration will decrease below the critical level (C_{crit}). If the $K_L a$ is such that the oxygen demand of the organism can be easily met the dissolved oxygen concentration will be greater than C_{ait} and may be as high as 70–80% of the saturation level. Thus, the $K_L a$ of the fermenter must be such that the optimum oxygen concentration for product formation can be maintained in solution throughout the fermentation.

DETERMINATION OF $K_L a$ VALUES

The determination of the $K_L a$ of a fermenter is essential in order to establish its aeration efficiency and to quantify the effects of operating variables on the provision of oxygen. This section considers the merits and limitations of the methods available for the determination of $K_L a$ values. It is important to remember at this stage that dissolved oxygen is usually monitored using a dissolved oxygen electrode (see Chapter 8) which records dissolved oxygen activity or dissolved oxygen tension (DOT) while the equations describing oxygen transfer are based on dissolved oxygen concentration. The solubility of oxygen is affected by dissolved solutes so that pure water and a fermentation medium saturated with oxygen would have different dissolved oxygen concentrations yet have the same DOT, that is, an oxygen electrode would record 100% for both. Thus, to translate DOT into concentration the solubility of oxygen in the fermentation medium must be known and this can present difficulties.

SULFITE OXIDATION TECHNIQUE

Cooper, Fernstrom, and Miller (1944) were the first to describe the determination of oxygen-transfer rates in aerated vessels by the oxidation of sodium sulfite solution. This technique does not require the measurement of dissolved oxygen concentrations but relies on the rate of conversion of a 0.5 M solution of sodium sulfite to sodium sulfate in the presence of a copper or cobalt catalyst:

$$Na_2SO_3 + 0.5O_2 = Na_2SO_4$$

The rate of reaction is such that as oxygen enters solution it is immediately consumed in the oxidation of sulfite, so that the sulfite oxidation rate is equivalent to the oxygen-transfer rate. The dissolved oxygen concentration, for all practical purposes, will be zero and the $K_L a$ may then be calculated from the equation:

$$OTR = K_L a \cdot C* \tag{9.2}$$

where OTR is the oxygen transfer rate.

The procedure is carried out as follows: the fermenter is batched with a 0.5 M solution of sodium sulfite containing 10^{-3} M Cu^{2+} ions and aerated and agitated at fixed rates; samples are removed at set time intervals (depending on the aeration and agitation rates) and added to excess iodine solution which reacts with the unconsumed sulfite, the level of which may be determined by a back titration with standard sodium thiosulfate solution. The volumes of the thiosulfate titrations are plotted against sample time and the oxygen transfer rate may be calculated from the slope of the graph.

The sulfite oxidation method has the advantage of simplicity and, also, the technique involves sampling the bulk liquid in the fermenter and, therefore, removes some of the problems of conditions varying through the volume of the vessel. However, the method is time consuming (one determination taking up to 3 h, depending on the aeration and agitation rates) and is notoriously inaccurate. Bell and Gallo (1971) demonstrated that minor amounts of surface-active contaminants (such as amino acids, proteins, fatty acids, esters, lipids, etc.) could have a major effect on the accuracy of the technique and apparent differences in aeration efficiency between vessels could be due to differences in the degree of contamination. Also, the rheology of a sodium sulfite solution is completely different from that of a fermentation broth, especially a mycelial one so that it is impossible to relate the results of sodium sulfite determinations to real fermentations (Garcia-Ochoa & Gomez, 2009). To quote Van't Riet and Tramper (1991) "It can safely be said that the application of this method should be strongly discouraged."

GASSING-OUT TECHNIQUES

The estimation of the $K_L a$ of a fermentation system by gassing-out techniques depends upon monitoring the increase in dissolved oxygen concentration of a solution during aeration and agitation. The oxygen transfer rate will decrease during the period of aeration as C_L approaches $C*$ due to the decline in the driving force ($C*-C_L$). The oxygen

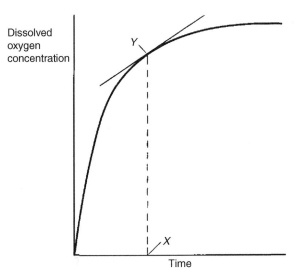

FIGURE 9.7 The Increase in Dissolved Oxygen Concentration of a Solution Over a Period of Aeration

The oxygen transfer rate at time X is equal to the slope of the tangent at point Y.

transfer rate, at any one time, will be equal to the slope of the tangent to the curve of values of dissolved oxygen concentration against time of aeration, as shown in Fig. 9.7.

To monitor the increase in dissolved oxygen over an adequate range it is necessary first to decrease the oxygen level to a low value. Two methods have been employed to achieve this lowering of the dissolved oxygen concentration—the static method and the dynamic method.

Static method of gassing out

In this technique, first described by Wise (1951), the oxygen concentration of the solution is lowered by gassing the liquid out with nitrogen gas, so that the solution is "scrubbed" free of oxygen. The deoxygenated liquid is then aerated and agitated and the increase in dissolved oxygen monitored using either a polarographic dissolved oxygen probe or the more recently developed optical device (see Chapter 8). The increase in dissolved oxygen concentration has already been described by Eq. (9.1), that is:

$$dC_L/dt = K_La(C^*-C_L)$$

and depicted in Fig. 9.7. Integration of Eq. (9.1) yields:

$$\ln(C^*-C_L) = -K_L \, at. \qquad (9.3)$$

Thus, a plot of $\ln(C^*-C_L)$ against time will yield a straight line of slope—K_La, as shown in Fig. 9.8. This technique has the advantage over the sulfite oxidation method in that it is very rapid (normally taking up to 15 min) and may utilize the fermentation medium, to which may be added dead cells or mycelium at a concentration equal to

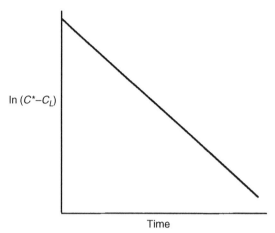

FIGURE 9.8 A Plot of the ln(C^* – C_L) Against Time of Aeration, the Slope of Which Equals $-K_La$

that produced during the fermentation. However, it is important to appreciate that the response time of the oxygen sensor may be inadequate to reflect the true change in the rate of oxygenation over a short period of time, although the use of an optical sensor with a faster response time would minimize this problem. The probe response time (T_p) is defined as the time needed to record 63% of a stepwise change and this should be much smaller than the mass transfer response time of the system (I/K_La). According to Van't Riet (1979), the use of commercially available polarographic electrodes, with a response time of 2–3 s, should enable a K_La of up to 360 h^{-1} to be measured with little loss of accuracy; but for estimations of higher K_La values it would be necessary to incorporate a correction factor into the calculation, as discussed by Taguchi and Humphrey (1966), Heineken (1970, 1971), Wernau and Wilke (1973) and Tribe, Briens, and Margaritas (1995). However, Van't Riet and Tramper (1991) estimated the response time of electrodes more conservatively as up to 100 s, depending on the state of the electrode membrane. Thus, careful electrode maintenance is a prerequisite for the use of this technique. Also, more recent publications quote a factor of ten as being the desirable threshold when calculating the maximum K_La measurable for an electrode response time, that is, the mass transfer response time (I/K_La) should be greater than 10 probe response times (T_p) (Garcia-Ochoa & Gomez, 2009, Gourich, Vial, Azher, Belhaj Soulami, & Ziyad, 2008) suggesting that a probe with a response time of 5 s should only be used to measure K_La values of more than 72 h^{-1} when a correction factor is applied.

While the method is acceptable for small-scale vessels, there are severe limitations to its use on large-scale fermenters that have high gas residence times. When the air supply to such a vessel is resumed after deoxygenation with nitrogen, the oxygen concentration in the gas phase may change with time as the nitrogen is replaced with air. Thus, C^* will no longer be constant. Although correction factors have been derived to compensate for this phenomenon, Van't Riet and Tramper (1991) concluded that the method should not be used for vessels over 1-m-high.

Dynamic method of gassing out

Taguchi and Humphrey's 1966 publication was a major step forward in ease of assessing $K_L a$ values (Rao, Moreira, & Brorson, 2009). These workers utilized the respiratory activity of a growing culture in the fermenter to lower the oxygen level prior to aeration. Therefore, the estimation has the advantage of being carried out during a fermentation and thus should give a more realistic assessment of the fermenter's performance. As for static gassing-out, the method depends upon the use of a dissolved oxygen probe, either polarographic or optical, which may necessitate the use of the response-correction factors referred earlier. The procedure involves stopping the supply of air to the fermentation that results in a linear decline in the dissolved oxygen concentration due to the respiration of the culture, as shown in Fig. 9.9. The slope of the line AB in Fig. 9.9 is a measure of the respiration rate of the culture. At point B the aeration is resumed and the dissolved oxygen concentration increases until it reaches concentration X. Over the period, BC, the observed increase in dissolved oxygen concentration is the difference between the transfer of oxygen into solution and the uptake of oxygen by the respiring culture as expressed by the equation:

$$dC_L/dt = K_L a(C^* - C_L) - xQ_{O_2}$$ (9.4)

where x is the concentration of biomass and Q_{O_2} is the specific respiration rate (mmoles of oxygen g^{-1} biomass h^{-1}).

The term xQ_{O_2} is given by the slope of the line AB in Fig. 9.9. Eq. (9.4) may be rearranged as:

$$C_L = -1/K_L a\{(dC_L/dt) + xQ_{O_2}\} + C^*$$ (9.5)

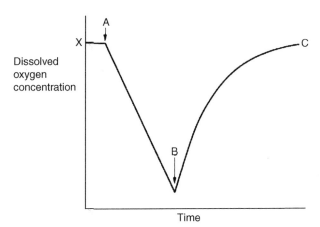

FIGURE 9.9 Dynamic Gassing Out for the Determination of $K_L a$ Values

Aeration was terminated at point A and recommenced at point B.

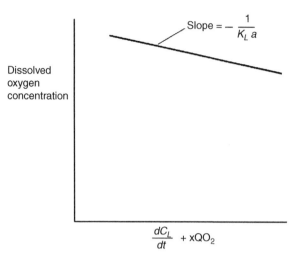

FIGURE 9.10 **The Dynamic Method for Determination of $K_L a$ Values**

The information is gleaned from Fig. 9.9. by taking tangents of the curve, BC, at various values of C_L.

Thus, from Eq. (9.5), a plot of C_L versus $dC_L/dt + xQ_{O_2}$ will yield a straight line, the slope of which will equal to $-1/K_L a$, as shown in Fig. 9.10. This technique is convenient in that the equations may be applied using DOT rather than concentration because it is the rates of transfer and uptake that are being monitored so that the percentage saturation readings generated by the electrode may be used directly.

The dynamic gassing-out method has the advantage over the previous methods of determining the $K_L a$ during an actual fermentation and may be used to determine $K_L a$ values at different stages in the process. The technique is also rapid and only requires the use of a dissolved-oxygen probe, of the membrane type. A major limitation in the operation of the technique is the range over which the increase in dissolved oxygen concentration may be measured. It is important not to allow the oxygen concentration to drop below C_{crit}, during the deoxygenation step as the specific oxygen uptake rate will then be limited and the term xQ_{O_2} would not be constant on resumption of aeration. The occurrence of oxygen-limited conditions during deoxygenation may be detected by the deviation of the decline in oxygen concentration from a linear relationship with time, as shown in Fig. 9.11.

When the oxygen demand of a culture is very high, it may be difficult to maintain the dissolved oxygen concentration significantly above C_{crit} during the fermentation so that the range of measurements which could be used in the $K_L a$ determination would be very small. Thus, it may be difficult to apply the technique during a fermentation that has an oxygen demand close to the supply capacity of the fermenter.

Although the difficulty presented by nitrogen degassing does not arise with the dynamic method, it is also not suitable for use with vessels in excess of 1-m-high.

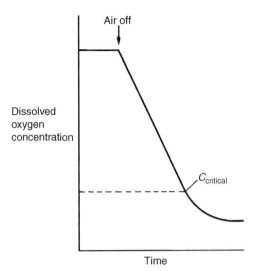

FIGURE 9.11 The Occurrence of Oxygen Limitation During the Dynamic Gassing out of a Fermentation

Van't Riet and Tramper (1991) pointed out that in such vessels the time taken to establish an equilibrium population of air bubbles would be significant and the gas–liquid interface area would change over the aeration period resulting in a considerable underestimate of the K_La value achievable under normal operating conditions. Both the dynamic and static methods are also unsuitable for measuring K_La values in viscous systems. This is due to the very small bubbles (<1 mm diameter) formed in a viscous system that have an extended residence time compared with "normal" sized bubbles. Thus, the gassing out techniques are only useful on a small scale with nonviscous systems.

OXYGEN-BALANCE TECHNIQUE

The K_La of a fermenter may be measured during a fermentation by the oxygen balance technique which determines, directly, the amount of oxygen transferred into solution in a set time interval. The procedure involves measuring the following parameters:

1. The volume of the broth contained in the vessel, V_L (dm^3).
2. The volumetric air flow rates measured at the air inlet and outlet, Q_i and Q_o, respectively (dm^3 min^{-1}).
3. The total pressure measured at the fermenter air inlet and outlet, P_i and P_o, respectively (atm. absolute).
4. The temperature of the gases at the inlet and outlet, T_i and T_o, respectively (K).
5. The mole fraction of oxygen measured at the inlet and outlet, y_i and y_o, respectively.

The oxygen transfer rate may then be determined from the following equation (Wang et al., 1979):

$$OTR = (7.32 \times 10^5 / V_L)(Q_i P_i y_i / T_i - Q_o P_o / T_o) \tag{9.6}$$

Where 7.32×10^5 is the conversion factor equaling (60 min h^{-1}) [mole/22.4 dm^3 (STP)] (273 K/l atm).

These measurements require accurate flow meters, pressure gauges, and temperature-sensing devices as well as gaseous oxygen analyzers (see Chapter 8). The ideal gaseous oxygen analyser is a mass spectrometer analyser that is sufficiently accurate to detect changes of 1–2%.

The $K_L a$ may be determined, provided that C_L and C^* are known, from Eq. (9.1):

$$OTR = K_L a(C^* - C_L) \text{ or } K_L a = OTR/(C^* - C_L)$$

C_L may be determined using a dissolved-oxygen electrode and in this case the slow response time is not an important factor because a rate of change is not being measured, simply the steady-state oxygen concentration. However, it should be remembered that an electrode simply measures the oxygen tension at one point and it is, therefore, advisable to monitor the oxygen tension at a number of points in the vessel with a number of electrodes and to use an average value. Also, the DOT reading must be converted to concentration, which necessitates knowing the oxygen solubility in the fermentation medium. The value of C^* is frequently taken as that value which is in equilibrium with the oxygen concentration of the gas outlet. Wang et al. (1979) claimed that this approach was adequate for small-scale fermenters but on a large scale there may be a considerable difference between the dissolved oxygen concentration in equilibrium with the inlet and outlet gases. Therefore, these workers suggested that the behavior of the gas in transit in the fermenter would approximate to plug flow conditions and a logarithmic mean value for the dissolved oxygen concentration should be used.

The oxygen-balance technique appears to be the simplest method for the assessment of $K_L a$ and has the advantage of measuring aeration efficiency during a fermentation. The sulfite oxidation and static gassing-out techniques have the disadvantage of being carried out using either a salt solution or an uninoculated, sterile fermentation medium. Although, as Banks (1977) suggests, these techniques are adequate for the comparison of equipment or operating variables, it should not be assumed that the values obtained are those actually operating during a fermentation. This may be the case for bacterial or yeast fermentations where the rheology of the suspended cells in the broth is similar to that in a sterile medium or a salt solution, but it is certainly not true for fungal and streptomycete processes where the rheology is quite different.

Tuffile and Pinho (1970) compared a number of methods for the determination of $K_L a$ values in viscous streptomycete fermentations. The techniques used were static gassing-out, dynamic gassing-out, and the oxygen-balance method. Tuffile and Pinho did not make it clear whether nonrespiring mycelium was present during their static gassing-out procedure, but from their results it would appear that it was present

Table 9.4 $K_L a$ Values for a 300-dm^3 Fermenter Containing a 90-h Culture of *S. aureofaciens* (Tuffile & Pinho, 1970)

Method of $K_L a$ Determination	Measured Oxygen Uptake Rate (mmoles dm^{-3}h^{-1})	$K_L a$ (h^{-1})
Static gassing out	—	58.2
Dynamic gassing out	6.6	58,2
Oxygen balance	20.1	108.0

in the vessel. Thus, the rheology of the fermenter contents would appear to have been similar for the different determinations. The $K_L a$ values, determined by the different techniques, for a 300 dm^3 fermenter containing a 90-h culture of *Streptomyces aureofaciens* are shown in Table 9.4.

From Table 9.4 it may be seen that the $K_L a$ values for the two gassing-out techniques were very similar but there was a considerable difference between the oxygen-uptake rates and the $K_L as$ determined by the dynamic method and the balance method. Tuffile and Pinho (1970) claimed that the low oxygen-uptake rate determined by the dynamic method was due to air bubbles remaining in suspension in the mash during the dynamic gassing-out period. Thus, the decline in oxygen concentration after the cessation of aeration was not a measure of the oxygen-uptake rate but the difference between oxygen uptake and the transfer of oxygen from entrapped bubbles. It was demonstrated that a large number of bubbles remained suspended in the medium 15 min after aeration had been stopped. The use of the low oxygen-uptake rate in the calculation of the $K_L a$ would result in an artificially low $K_L a$ being determined. Heijnen, Riet, and Wolthuis (1980) also observed anomalies in determining $K_L a$ values in viscous systems due to the presence of very small bubbles having a much longer residence time than the more abundant large bubbles in the vessel.

Overall, it would appear that the balance method is the most desirable technique to use and the extra cost of the monitoring equipment involved should be a worthwhile investment. A key advantage is that oxygen uptake rates, oxygen transfer rates, and $K_L a$ values can be determined online without any disturbance to the culture. Schaepe et al. (2013) applied this approach to an *E. coli* fermentation and demonstrated that to maintain the dissolved oxygen concentration at 25% of saturation $K_L a$ had to increase from below 100 h^{-1} at the start of the fermentation to 2,500 h^{-1} to cope with the peak volumetric oxygen demand.

Before considering the factors that may affect the $K_L a$ of a fermenter it is necessary to consider the behavior of fluids in agitated systems.

FLUID RHEOLOGY

Fluids may be described as Newtonian or non-Newtonian depending on whether their rheology (flow) characteristics obey Newton's law of viscous flow. Consider a fluid contained between two parallel plates, area A and distance x apart. If the lower plate is moved in one direction at a constant velocity, the fluid adjacent to the moving plate will move in the same direction and impart some of its momentum to the "layer" of liquid directly above it causing it to move in the same direction at a slightly lower velocity. Newton's law of viscous flow states that the viscous force, F, opposing motion at the interface between the two liquid layers, flowing with a velocity gradient of du/dx, is given by the equation:

$$F = \mu A(dv/dx) \tag{9.7}$$

where μ is the fluid viscosity, which may be considered as the resistance of the fluid to flow.

Eq. (9.7) may be written as:

$$F/A = \mu(dv/dx)$$

F/A is termed the shear stress (τ) and is the applied force per unit area, dv/dx is termed the shear rate (γ) and is the velocity gradient. Thus:

$$\tau = \mu\gamma \tag{9.8}$$

Eq. (9.8) conforms to the general relationship:

$$\tau = K\gamma^n \tag{9.9}$$

where K is the consistency coefficient and n is the flow behavior index or power law index.

For a Newtonian fluid n is 1 and the consistency coefficient is the viscosity that is the ratio of shear stress to shear rate. Thus, a plot of shear stress against shear rate, for a Newtonian fluid, would produce a straight line, the slope of which would equal the viscosity. Such a plot is termed a rheogram (as shown in Fig. 9.12).

Thus, a Newtonian liquid has a constant viscosity regardless of shear, so that the viscosity of a Newtonian fermentation broth will not vary with the agitation rate. However, a non-Newtonian liquid does not obey Newton's law of viscous flow and

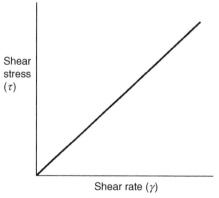

Shear stress (τ)

Shear rate (γ)

FIGURE 9.12 A Rheogram of a Newtonian Fluid

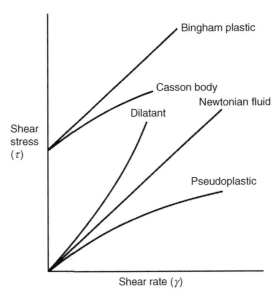

FIGURE 9.13 Rheogram of Fluids of Different Properties

does not have a constant viscosity. The value for n [Eq. (9.9)] of such a fluid deviates from 1 and its behavior is said to follow a power law model. Thus, the viscosity of a non-Newtonian fermentation broth will vary with agitation rate and is described as an apparent viscosity (μ_a). A plot of shear stress against shear rate for a non-Newtonian liquid will deviate from the relationship depicted in Fig. 9.12, depending on the nature of the liquid. Several types of non-Newtonian liquids are recognized and typical rheograms of types important in the study of culture fluids are given in Fig. 9.13, and their characteristics are discussed later.

BINGHAM PLASTIC RHEOLOGY

Bingham plastics are similar to Newtonian liquids apart from the fact that shear rate will not increase until a threshold shear stress is exceeded. The threshold shear stress is termed the yield stress or yield value, τ_0. A linear relationship of shear stress to shear rate is given once the yield stress is exceeded and the slope of this line is termed the coefficient of rigidity or the plastic viscosity. Thus, the flow of a Bingham plastic is described by the equation:

$$\tau = \tau_0 + n\gamma$$

where n is the coefficient of rigidity and τ_0 is the yield stress.

There have been some claims of mycelial fermentation broths displaying Bingham plastic characteristics (Table 9.5). Everyday examples of these fluids include toothpaste and clay.

Table 9.5 Some Examples of the Rheological Nature of Fermentation Broths

Organism	Rheological Type	References
Penicillium chrysogenum	Bingham plastic	Deindoerfer and Gaden (1955)
Streptomyces kanamyceticus	Bingham plastic	Sato (1961)
Penicillium chrysogenum	Pseudoplastic	Deindoerfer and West (1960)
Endomyces sp.	Pseudoplastic	Taguchi, Imanaka, Teramoto, Takatsu, and Sato (1968)
Penicillium chrysogenum	Casson body	Roels et al. (1974)
Xanthomonas campestris	Pseudoplastic	Charles (1978)

PSEUDOPLASTIC RHEOLOGY

The apparent viscosity of a pseudoplastic liquid decreases with increasing shear rate. Most polymer solutions behave as pseudoplastics. The decrease in apparent viscosity is explained by the long chain molecules tending to align with each other at high shear rates resulting in easier flow. The flow of a pseudoplastic liquid may be described by the power law model, Eq. (9.9), that is:

$$\tau = K(\gamma)^n$$

K has the same units as viscosity and may be taken as the apparent viscosity. The flow-behavior index is less than unity for a pseudoplastic liquid, the smaller the value of n, the greater the flow characteristics of the liquid deviate from those of a Newtonian fluid. Eq. (9.9) may be converted to the logarithmic form as:

$$\log \tau = \log K + \text{n} \log \gamma \tag{9.10}$$

Thus, a plot of log shear stress against log shear rate will produce a straight line, the slope of which will equal the flow-behavior index and the intercept on the shear stress axis will be equal to the logarithm of the consistency coefficient.

Many workers have demonstrated that mycelial fermentation broths display pseudoplastic properties as shown in Table 9.5.

DILATANT RHEOLOGY

The apparent viscosity of a dilatant liquid increases with increasing shear rate. The flow of a dilatant liquid may also be described by the Eq. (9.9) but in this case the value of the flow-behavior index is greater than 1, the greater the value the greater the flow characteristics deviate from those of a Newtonian fluid. Thus, the values of K and n may be obtained from a plot of log shear stress against log shear rate. Fortunately, fermentation broths do not exhibit this type of behavior—an everyday example is liquid cement slurry.

CASSON BODY RHEOLOGY

Casson (1959) described a type of non-Newtonian fluid, termed a Casson body, which behaved as a pseudoplastic in that the apparent viscosity decreased with the increasing shear rate but displayed a yield stress and, therefore, also resembled a Bingham plastic. The flow characteristics of a Casson body may be described by the following equation:

$$\sqrt{\tau} = \sqrt{\tau_0} + K_c \sqrt{\gamma} \tag{9.11}$$

where K_c is the Casson viscosity.

A plot of $\sqrt{\tau}$ against $\sqrt{\gamma}$ will give a straight line, the slope of which will equal the Casson viscosity and the intercept of the $\sqrt{\tau}$ axis will equal $\sqrt{\tau_0}$.

Roels, Van Denberg, and Vonken (1974) claimed that the rheology of a penicillin broth could be best described in terms of a Casson body.

Therefore, to determine the rheological nature of a fluid it is necessary to construct a rheogram that requires the use of a viscometer that is accurate over a wide range of shear rates. Furthermore, the testing of mycelial suspensions may present special difficulties. These problems have been considered in detail by Van't Riet and Tramper (1991), whose book should be consulted for methods of assessing the rheological properties of mycelial fluids.

FACTORS AFFECTING $K_L a$ VALUES IN FERMENTATION VESSELS

A number of factors have been demonstrated to affect the $K_L a$ value achieved in a fermentation vessel. Such factors include the airflow rate employed, the degree of agitation, the rheological properties of the culture broth and the presence of antifoam agents. If the scale of operation of a fermentation is increased (so-called "scale-up") it is important that the optimum $K_L a$ found on the small scale is employed in the larger scale fermentation. The same $K_L a$ value may be achieved in different sized vessels by adjusting the operational conditions on the larger scale and measuring the $K_L a$ obtained. However, quantification of the relationship between operating variables and $K_L a$ should enable the prediction of conditions necessary to achieve a particular $K_L a$ value. Thus, such relationships should be of considerable value in scaling-up a fermentation and in fermenter design.

EFFECT OF AIRFLOW RATE ON $K_L a$

Mechanically agitated reactors

The effect of airflow rate on $K_L a$ values in conventional agitated systems is illustrated in Fig. 9.14. The quantitative relationships between aeration and $K_L a$ for agitated vessels are considered in the subsequent section on power consumption.

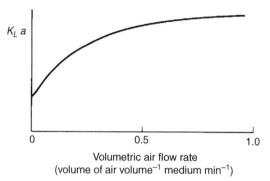

FIGURE 9.14 The Effect of Air-Flow Rate on the K_La of an Agitated, Aerated Vessel

The air-flow rate employed rarely falls outside the range of 0.5–1.5 volumes of air per volume of medium per minute and this tends to be maintained constant on scale-up. Airflow rate primarily affects the specific gas-liquid interface area, "a." A higher flow rate would potentially provide a higher number of bubbles provided the agitator is capable of dispersing the increased airflow into small bubbles. Small bubbles have higher specific surface areas ($cm^2\ cm^{-3}$) than large ones and will also circulate more in suspension, thus giving a higher hold-up time and contributing further to gas exchange. However, if the impeller is unable to disperse the incoming air then extremely low oxygen transfer rates may be achieved due to the impeller becoming "flooded." Flooding is the phenomenon where the air-flow dominates the flow pattern and is due to an inappropriate combination of air flow rate and speed of agitation resulting in the impeller being surrounded by gas such that it cannot pump adequately (see also Chapter 7). Nienow et al. (1977) categorized the different flow patterns produced by a disc turbine that occur under a range of aeration and agitation conditions (Fig. 9.15) and these have been widely discussed in the literature, for example, by Van't Riet and Tramper (1991), Van't Riet (1983),

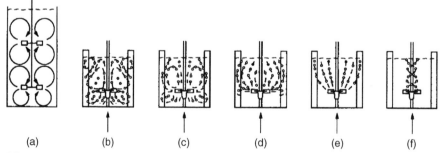

FIGURE 9.15 The Effect of Air-Flow Rate on the Flow Pattern in Stirred Vessels (After Nienow et al., 1977)

(a) Nonaerated, (b–f) increasing air flow rates.

Villadsen et al. (2011), Doran (2013), and Nienow (2014). Fig. 9.15a shows the flow profile of a nonaerated vessel and Fig. 9.15b–f the profiles with increasing air flow rate. As air-flow rate increases the flow profile changes from one dominated by agitation (Fig. 9.15b,c) to one dominated by air flow (Fig. 9.15d–f) until finally the air flow rate is such that the air escapes without being distributed by the agitator (Fig. 9.15f). Nienow et al. (1977) proposed that the onset of flooding is represented by Fig. 9.15d and the desired pattern giving complete gas dispersion is represented by Fig. 9.15c. Using two dimensionless variables, the gas flow number (or aeration number) and the Froude number, Nienow correlated the onset of flooding to the operational conditions of Rushton turbine agitators. The gas flow number is the ratio of the gas flow rate to the impeller pumping capacity:

$$Fl_g = F_g/(ND^3)$$

(9.12)

The Froude number is the ratio of inertial to gravitational forces:

$$Fr = (N^2 D)/g$$

(9.13)

where Fl_g is the gas flow number or aeration number, F_g is the volumetric gas flow rate (m^3 min^{-1}), N is the stirrer speed (rpm), D is the impeller diameter (m), Fr is the Froude number, g is gravitational acceleration (m s^{-2}).

Nienow et al's equation for the transition into flooding (Fig. 9.15f) is:

$$Fl_g = 30(D/D_T)^{3.5} Fr$$

(9.14)

and the equation describing complete gas dispersion (Fig. 9.15c) is:

$$Fl_g = 0.2(D/D_T)^{0.5} Fr^{0.5}$$

(9.15)

where D_T is the diameter of the fermenter (m). If the agitator speed is increased above that predicted in equation (9.15) then the air recirculation becomes significant, at which point:

$$Fl_g = (D/D_T)^5 13 Fr^2$$

(9.16)

Equations (9.14 and 9.15) may be rearranged to predict the critical stirrer speeds for flooding (N_F) and dispersed (N_{Disp}) conditions (Villadsen et al., 2011):

$$N_F = \left(\frac{g}{30} F_g\right)^{1/3} T^{7/6} D^{-2.5}$$

(9.17)

$$N_{Disp} = (5g^{1/2} F_g)^{1/2} T^{1/4} D^{-2}$$

(9.18)

The key practical implication of Eqs. (9.14–9.18) is the influence of the impeller diameter (relative to the diameter of the fermenter) on the gas flow rate that can be accommodated by the fermenter (Nienow, 1998; Doran, 2013; Nienow, 2014). Doran (2013) cites the example of a 10% increase in impeller diameter enabling a

100% increase in the airflow rate before flooding occurs whereas a 10% increase in the impeller rotational speed only increases the flow rate by about 30%.

These equations have been derived for Rushton turbines operating in fluids with a viscosity similar to that of water. The onset of flooding in these impellers is due to the accumulation of air bubbles behind the blades and the development of gas-filled cavities that eventually results in the agitator effectively rotating in a pocket of gas (Nienow & Ulbricht, 1985). However, more recent impellers incorporate half-pipe, concave blades that enable the impeller to cope with a very much higher airflow rate than the Rushton due to the air bubbles leaving the blades of the impeller rather than accumulating behind them. These impellers are also able to cope more effectively with high viscosity broths. Examples include the Scaba SRGT and Chemineer CD6 and BT6, shown in Fig. 9.22 and Fig. 9.23 and are discussed later in this chapter and in Chapter 7.

Nonmechanically agitated reactors

Bubble columns and air-lift reactors are not mechanically agitated and, therefore, rely on the passage of air to both mix and aerate.

1. *Bubble columns*

   The flow pattern of bubbles through a bubble column reactor is dependent on the gas superficial velocity (volumetric air flow rate/cross sectional area of the vessel, cm second^{-1}). At gas velocities of below 1–4 cm s^{-1} the bubbles will rise uniformly through the medium (Van't Riet & Tramper, 1991) and the only mixing will be that created in the bubble wake. This type of flow is referred to as homogeneous. At higher gas velocities bubbles are produced unevenly at the base of the vessel and bubbles coalesce resulting in local differences in fluid density. The differences in fluid density create circulatory currents and flow under these conditions is described as heterogeneous as shown in Fig. 9.16.

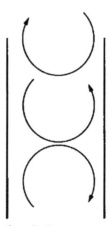

FIGURE 9.16 Schematic Representation of a Heterogeneous Flow Regime in a Bubble Column

Flooding in a bubble column is the situation when the airflow is such that it blows the medium out of the vessel. This requires superficial gas velocities approaching 1 m s^{-1} that are not attainable on commercial scales (Van't Riet & Tramper, 1991).

The K_La in a bubble column is essentially dependent on the superficial gas velocity. Heijnen and Van't Riet (1984) reviewed the subject and demonstrated that the precise mathematical relationship between K_La and superficial gas velocity is dependent on the coalescent properties of the medium, the type of flow, and the bubble size. Unfortunately, these characteristics are rarely known for a commercial process that makes the application of these equations problematical. However, Van't Riet and Tramper (1991) claimed that the relationship derived for noncoalescing, nonviscous, large bubbles (6 mm diameter) will give a reasonably accurate estimation for most nonviscous situations:

$$K_La = 0.32\,(V_s^c)^{0.7} \tag{9.19}$$

Where V_s^c is the superficial air velocity corrected for local pressure.

However, viscosity has an overwhelming influence on K_La in a bubble column that Deckwer, Nguyen-Tien, Schumpe, and Serpe-men (1982) expressed as:

$$K_La = c\pi^{-0.84} \tag{9.20}$$

Where π is the liquid dynamic viscosity (N s m^{-2}).

The practical implication of this equation is that bubble columns cannot be used with highly viscous fluids. Van't Riet and Tramper (1991) suggested that the upper viscosity limit for a bubble column was 100×10^{-3} N s m^{-2} at which point the K_La would have decreased 50-fold compared with a reactor batched with water.

2. *Air-lift reactors*

The structure of air-lift reactors is discussed in Chapter 7. The difference between a bubble column and an air-lift reactor is that liquid circulation is achieved in the air-lift in addition to that caused by the bubble flow. The reactor consists of a vertical loop of two connected compartments, the riser and downcomer. Air is introduced into the base of the riser and escapes at the top. The degassed liquid is denser than the gassed liquid in the riser and flows down the downcomer. Thus, a circulatory pattern is established in the vessel—gassed liquid going up in the riser and degassed liquid coming down the downcomer.

For a given air-lift reactor and medium K_La varies linearly with superficial air velocity on a log–log scale over the normal range of velocities (Chen, 1990). However, it should be remembered that the circulation in an air-lift results in the bubbles being in contact with the liquid for a shorter time than in a corresponding bubble column. Thus, the K_La obtained in an air-lift will be less than that obtained in a bubble column at the same superficial air velocity, that is, less than $0.32\,(V_s^c)^{0.7}$. The advantage of the air-lift lies in the circulation achieved, but this is at the cost of a lower K_La value.

As for a bubble column flooding will not occur within the normal operating superficial air velocities and should not be a problem on a large scale.

EFFECT OF THE DEGREE OF AGITATION ON K_La

The degree of agitation has been demonstrated to have a profound effect on the oxygen-transfer efficiency of an agitated fermenter. Banks (1977) claimed that agitation assisted oxygen transfer by:

1. increasing the area available for oxygen transfer by dispersing the air in the culture fluid in the form of small bubbles,
2. delaying the escape of air bubbles from the liquid,
3. preventing coalescence of air bubbles,
4. decreasing the thickness of the liquid film at the gas–liquid interface by creating turbulence in the culture fluid.

The degree of agitation may be measured by the amount of power consumed in stirring the vessel contents. The power consumption may be assessed by using a dynamometer, by using strain gauges attached to the agitator shaft and by measuring the electrical power consumption of the agitator motor (see Chapter 8). The assessment of electrical consumption is suitable only for use with large-scale vessels.

Relationship between K_La and power consumption

A large number of empirical relationships have been developed between K_La, power consumption and superficial air velocity, which take the form of:

$$K_La = k(P_g/V)^x V_s^y$$

where P_g is the power absorption in an aerated system, V is the liquid volume in the vessel, V_s is the superficial air velocity, k, x, and y are empirical factors specific to the system under investigation.

Cooper et al. (1944) measured the K_Las of a number of agitated and aerated vessels (up to a volume of 66 dm^3) containing one impeller, using the sulfite oxidation technique, and derived the following expression:

$$K_La = k(P_g/V)^{0.95} V_s^{0.67}. \tag{9.21}$$

Thus, it may be seen from Eq. (9.21) that the K_La value was claimed to be almost directly proportional to the gassed power consumption per unit volume. However, Bartholomew (1960) demonstrated that the relationship depended on the size of the vessel and the exponent on the term P_g/V varied with scale as follows:

Scale	Value of exponent on P_g/V
Laboratory	0.95
Pilot plant	0.67
Production plant	0.5

Bartholomew's vessels contained more than one impeller, whereas those of Cooper et al. (1944) contained only one. It is probable that the upper impellers would consume more power relative to their contribution to oxygen transfer than would the lowest impeller, thus affecting the value of the exponent term. Thus, it is important to appreciate that such relationships are scale-dependent when using them in scale-up calculations.

Many workers have produced similar correlations and these have been reviewed by Van't Riet (1983), Winkler (1990), and Garcia-Ochoa and Gomez (2009). Winkler (1990) pointed out that the common feature of these relationships is that the values of x and y are less than unity and it can be seen from Table 9.6, that more recent correlations have confirmed this observation. This means that increasing power input or air

Table 9.6 Value of the Exponents on P_g/V and V_s in the Equation: $K_L a = k(P_g/V)^x V_s^y$

References	Exponent on P_g/V	Exponent on V_s	Reactor Volume and Agitator
Cooper et al., 1944	0.95	0.67	Up to 66 dm³, single FBT
Yagi and Yoshida, 1975	0.8	0.3	12 dm³, single FBT
Figueiredo and Calderbank (1979)	0.6	0.8	600 dm³, single FBT
Van't Riet (1979)	0.4	0.5	2–2600 dm³, any
Nishikawa, Nakamura, Yagi, and Hashimoto (1981)	0.8	0.33	2.7–170 dm⁻³, FBT and FBP
Chandrasekharan and Calderbank (1981)	0.55	$0.55.D^{-0.5}$	50–1430 dm³, FBT
Davies, Gibilaro, Middleton, Cooke, and Lynch (1985)	0.8	0.45	20–180 dm³, FBT
Kawase and Moo-Young (1988)	1.0	0.5	—
Ogut and Hatch (1988)	—	0.7	100 dm³, FBP
Linek, Sinkule, and Benes (1991)	0.65	0.4	20 dm³, FBT
Pedersen, Andersen, Nielsen, and Villadsen (1994)	—	0.5–0.7	15 dm³, two FBT
Gagnon, Lounes, and Thilbault (1998)	0.6–0.8	0.5	22 dm³, FBT
Arjunwadkar, Sarvanan, Kulkarni, and Pandit (1998)	0.68	0.4–0.58	5 dm³, FBT and PBT
Vasconcelos, Orvalho, Rodriguez, and Alves (2000)	0.62	0.49	5 dm³, two FBT
Garcia-Ochoa and Gomez (1998, 2001)	0.6	0.5–0.67	2–25 dm³, CBP, PBP
Puthli, Rathod, and Pandit (2005)	0.57–0.98	0.53	2 dm³, FBT, FBP, PBP

CBT, curved blade turbine; CBP, curved blade paddle; FBP, flat blade paddle; PBP, pitched blade paddle; PBT, pitched blade turbine.
Modified from Garcia-Ochoa and Gomez, 2009.

flow becomes progressively less efficient as a means of increasing K_La as inputs rise. Thus, high oxygen-transfer rates are achieved at considerable expense.

From this discussion it is evident that the K_La of an aerated, agitated vessel is affected significantly by the consumption of power during stirring and, hence, the degree of agitation. Although it is not possible to derive a relationship between K_La and power consumption which is applicable to all situations it is possible to derive a relationship between the two which is operable within certain limits and should be a useful guide in practical design problems. If it is accepted that such relationships between power consumption and K_La are of some practical significance, it is of considerable importance to relate power consumption to the operating variables that may affect it. Quantitative relationships between power consumption and operating variables may be useful in:

1. Estimating the amount of power that an agitation system will consume under certain circumstances, which could assist in fermenter design.
2. Providing similar degrees of power consumption (and, hence, agitation and, therefore, K_Las) in vessels of different size.

Relationship between power consumption and operating variables

Rushton, Costich, and Everett (1950) investigated the relationship between power consumption and operating variables in baffled, agitated vessels using the technique of dimensional analysis. They demonstrated that power absorption during agitation of nongassed Newtonian liquids could be represented by a dimensionless group termed the power number, defined by the expression:

$$N_p = P/(\rho N^3 D^5) \tag{9.22}$$

where N_p is the power number, P is the external power from the agitator, ρ is the liquid density, N is the impeller rotational speed, D is the impeller diameter.

Thus, the power number is the ratio of external force exerted (P) to the inertial force imparted ($\rho N^3 D^5$) to the liquid. The motion of liquids in an agitated vessel may be described by another dimensionless number known as the Reynolds number that is a ratio of inertial to viscous forces:

$$N_{Re} = (\rho D^2 N)/\mu \tag{9.23}$$

where N_{Re} is the Reynolds number and μ is the liquid viscosity.

Yet another dimensionless number, termed the Froude number, relates inertial force to gravitational force and is given the term:

$$N_{Fr} = (N^2 D)/g \tag{9.24}$$

where N_{Fr} is the Froude number and g is the gravitational force.

Rushton et al. (1950) demonstrated that the power number was related to the Reynolds and Froude numbers by the general expression:

$$N_p = c(N_{Re})^x (N_{Fr})^y \tag{9.25}$$

where c is a constant dependent on vessel geometry but independent of vessel size, x and y are exponents.

Examples of the values of c, x, and y are considered later.

However, in a fully baffled agitated vessel the effect of gravity is minimal so that the relationship between the power number and the other dimensionless numbers becomes:

$$N_p = c(N_{Re})^x \tag{9.26}$$

Therefore substituting from Eqs. (9.22) and (9.23):

$$P/(\rho N^3 D^5) = c(\rho D^2 N/\mu)^x \tag{9.27}$$

Values for P at various values of N, D, μ, and ρ may be determined experimentally and the Reynolds and power numbers for each experimental situation may then be calculated. A plot of the logarithm of the power number against the logarithm of the Reynolds number yields a graph termed the power curve. A typical power curve for a baffled vessel agitated by a flat-blade turbine is illustrated in Fig. 9.17 and such a curve would apply to geometrically similar vessels regardless of size.

From Fig. 9.17 it may be seen that a power curve is divisible into three clearly defined zones depicting different types of fluid flow:

1. The laminar or viscous flow zone where the logarithm of the power number decreases linearly with an increase in the logarithm of the Reynolds number. The slope of the graph is equal to x, the exponent in Eq. (9.27) and is obviously equal to -1. The power absorbed in this region is a function of the viscosity of the liquid and the Reynolds number is less than 10.
2. The transient or transition zone, where there is no consistent relationship between the power and Reynolds numbers. The value of x (ie, the slope of the plot) is variable and the value of the Reynolds number is between 10 and 10^4.

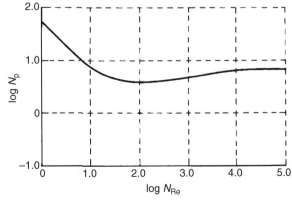

FIGURE 9.17 A Typical Power Curve for a Baffled Vessel Agitated by a Flat-Blade Turbine

3. The turbulent flow zone, where the power number is a constant, independent of the Reynolds number so that the value of x is zero and the value of the Reynolds number is in excess of 10^4.

If the values of the exponent, x are substituted into Eq. (9.27) for the zones of viscous and turbulent flow, then the following terms are given:

$$\text{For viscous flow } P = c\mu N^2 D^3. \tag{9.28}$$

$$\text{For turbulent flow } P = c\rho N^3 D^5. \tag{9.29}$$

From these equations it may be seen that power consumption is dependent on the viscosity of the liquid in the region of viscous flow and that increased speed of agitation, or an increase in the impeller diameter, results in a proportionally greater increase in power transmission to a liquid in turbulent flow than to one in viscous flow. Conditions of viscous flow are rare in fermentation processes, the majority of fermentations exhibiting flow characteristics in either the turbulent or transition zones. If turbulent flow is demonstrated to occur in a fermentation then Eq. (9.29) applies and power consumption is independent of viscosity. The equation may then be used to predict the power requirements of the fermentation and to predict the operating conditions of different sized vessels to achieve the same agitation regime, as outlined by Banks (1979). Power consumption on the small scale may be represented as:

$$P_{sm} = c\rho N_{sm}^3 D_{sm}^5 \tag{9.30}$$

and on the large scale as:

$$P_L = c\rho N_L^3 D_L^5$$

where the subscripts sm and L refer to the small and large scales, respectively. Maintaining the same power input per unit volume:

$$P_{sm}/P_L = V_{sm}/V_L = (c\rho N_{sm}^3 D_{sm}^5)/(c\rho N_L^3 D_L^5) \tag{9.31}$$

where V is the volume.

Assuming the vessels to be geometrically similar then c will be the same regardless of scale and as the same broth would be employed ρ would remain the same for both systems

$$V_{sm}/V_L = (N_{sm}^3 D_{sm}^5)/(N_L^3 D_L^5) \tag{9.32}$$

For geometrically similar vessels

$$D_{sm}/D_L = (V_{sm}/V_L)^{1/3}$$

Therefore, substituting for D_{sm}/D_L in

$$N_L = N_{sm}(V_{sm}/V_L)^{2/9}. \tag{9.33}$$

However, if transient flow conditions occur in a fermentation then it is necessary to construct a complete power curve for such predictions and this is discussed later in the chapter.

The work of Rushton et al. (1950) was carried out using ungassed liquids whereas the vast majority of fermentations are aerated. It is widely accepted that aeration of a liquid decreases the power consumption during agitation because an aerated liquid, containing suspended air bubbles, is less dense than an unaerated one and large gas-filled cavities generated behind the agitator blades decrease the hydrodynamic resistance of the blades. A number of workers have produced correlations of gassed power consumption, ungassed power consumption and operating variables, that of Michell and Miller (1962) being widely used:

$$P_g = k(P^2ND^3/Q^{0.56})^{0.45}$$

Where Q is the volumetric air flow rate.

However, more recent correlations have been elucidated which are applicable over a wider range of operating conditions than that of Michell and Miller. Hughmark (1980) produced the following correlation from 248 sets of published data:

$$P_g/P = 0.1\,(Q/NV)^{-0.25}(N^2D^4/gWV^{0.67})^{-0.2}$$

where Q is the volumetric air flow rate, g is the acceleration due to gravity, and W is the impeller blade width.

Using dimensional analysis:

$$P_g/P = 0.0312 \cdot Fr^{-1.6} \cdot Re^{00.64}\, N_a^{-0.38} \cdot (T/D)^{0.8}$$

where N_a is the aeration number, equals Q/ND, and T is the vessel diameter.

Provided it is remembered that these expressions are not particularly accurate they may be used to predict power consumption in gassed systems where turbulent flow is known to be operating. However, it should be remembered that in nonmycelial fermentations the greatest power demands often occur during agitation when the system is not gassed, that is, during the sterilization of the medium in situ or if the air supply were to fail. Thus, in designing the system care must be taken to ensure that the agitator motor is sufficiently powerful to agitate the ungassed system and for fixed speed motors the operating speed should be specified with respect to the ungassed power draw (Gbewonyo et al., 1986).

From the foregoing account it may be seen that reasonable techniques exist to relate operating variables to power consumption and, hence, to the degree of agitation which may be shown to have a proportional effect on K_La. However, these techniques apply to Newtonian fluids and are not directly applicable to the study of non-Newtonian systems. A non-Newtonian fluid does not have a constant viscosity, which creates difficulties in utilizing relationships that rely on being able to determine the fluid viscosity. These difficulties may be avoided if the agitation system is capable of maintaining turbulent-flow conditions during the fermentation, because under such conditions power consumption is independent of the Reynolds number and, hence, of viscosity. However, the high viscosities of the majority of mycelial fermentation broths make fully turbulent flow conditions impossible, or extremely difficult, to achieve. Such fermentations tend to exhibit transient zone flow conditions that necessitate the construction of complete power curves to correlate power consumption with operating variables. The fact that the viscosity of a non-Newtonian liquid is affected by shear rate means that the viscosity of a non-Newtonian fermentation broth will not be uniform throughout the fermenter because the shear rate will be higher near the agitator than elsewhere in the vessel. Thus, the determination of the impeller Reynolds number is made difficult by not knowing the viscosity of the fermentation broth. Metzner and Otto (1957) proposed a solution to this paradox by introducing the concept of average shear rate (y) related to the agitator shaft speed in the vessel, by the equation:

$$y = kN \tag{9.34}$$

where k is a proportionality constant.

Metzner and Otto determined the value of the proportionality constant to be 13 for pseudoplastic fluids in conventional, baffled reactors agitated by single, flat-blade turbines. Several groups of workers have determined values of k under a wide range of operating variables; the values range from approximately 10–13. Metzner, Feehs, Ramos, Otto, and Toothill (1961) suggested that a compromise value of 11 could be used for calculation purposes, with relatively little loss of accuracy, which would obviate the necessity to determine k for each circumstance. Therefore, provided that the rheological properties of a fermentation broth are known, an apparent average viscosity of the fluid may be calculated using the average shear rate which would enable the calculation of the impeller Reynolds number for each value of the impeller rotational speed, thus enabling a power curve to be constructed. Such a power curve may be used to predict the power requirements of a fermentation and to scale up a fermentation on the basis of power consumption per unit volume. Metzner and Otto's approach was not widely used despite the paper being rated as an "outstanding contribution" to the field of mixing in a review recognizing 21 of the most influential contributions (Calabrese, Kresta, & Liu, 2014). However, there are now examples of its adoption in describing non-Newtonian fermentations. Gallindo and Nienow (1992), Gabelle, Augier, Carvalho, Rousset, and Morchain (2011), Gabelle et al. (2012), and Xie et al. (2015) all utilized Metzner and Otto's numerical value of "k" in their analyses of the mixing of non-Newtonian viscous fluids which is addressed in a later section considering the operation of viscous fermentations.

EFFECT OF MEDIUM AND CULTURE RHEOLOGY ON K_La

As can be seen from the previous section, the rheology of a fermentation broth has a marked influence on the relationship between K_La and the degree of agitation. The objective of this section is to discuss the effects of medium and culture rheology on oxygen transfer during a fermentation. A fermentation broth consists of the liquid medium in which the organism grows, the microbial biomass and any product that is secreted by the organism. Thus, the rheology of the broth is affected by the composition of the original medium and its modification by the growing culture, the concentration and morphology of the biomass and the concentration and rheological properties of the microbial products. Therefore, it should be apparent that fermentation broths vary widely in their rheological properties and significant changes in broth rheology may occur during a fermentation.

Medium rheology

Fermentation media frequently contain starch as a carbon source that may render the medium non-Newtonian and relatively viscous. However, as the organism grows it will degrade the starch and thus modify the rheology of the medium and reduce its viscosity. Such a situation was described by Tuffile and Pinho (1970) in their study of the growth of *Streptomyces aureofaciens* on a starch-containing medium. Before inoculation, the medium displayed Bingham plastic characteristics with a well-defined yield stress and an apparent viscosity of approximately 18 pseudopoise; after 22 h the organism's activity had decreased the medium viscosity to a few pseudopoise and modified its behavior to that of a Newtonian liquid; from 22 h onward the apparent viscosity of the broth gradually increased, due to the development of the mycelium, up to a maximum of approximately 90 pseudopoise and the rheology of the broth became increasingly pseudoplastic in nature. Thus, this example suggests that the rheological problems presented by the medium are minor compared with those presented by a high mycelial biomass, especially when it is considered that the total oxygen demand is relatively low in the early stages of a fermentation. However, it is worth remembering that in nonviscous unicellular fermentations the highest power draw will occur when the medium is sterilized in situ when the vessel is not being aerated and this will correspond with the time when a starch-based medium is at its most viscous.

Effect of microbial biomass on K_La
Agitator design for non-Newtonian fermentations

The biomass concentration and its morphological form in a fermentation has been shown to have a profound effect on oxygen transfer. Most unicellular bacterial and yeast fermentations tend to give rise to relatively nonviscous Newtonian broths in which conditions of turbulent flow may be achieved. Such fermentations present relatively few oxygen-transfer problems. However, mycelial organisms such as the filamentous fungi and streptomycetes can produce highly viscous non-Newtonian broths that present major difficulties in oxygen provision, the productivities of many such fermentations being limited by oxygen availability. Moreover, these filamentous organisms can grow in a variety of morphological forms—ranging from dispersed filaments to compact, suspended pellets depending on the cultural conditions, as shown in Fig. 6.19. As discussed in Chapters 2 and 6, the situation is complicated by the fact that the nature of the mycelial growth (whether filamentous, ie, dispersed filaments, or pelleted) can have a significant impact on both the products produced and the productivity of the process. For example, dispersed filamentous growth is required for penicillin, streptomycin, and turimycin production while pelleted growth is required for citric acid, lovastatin, erythromycin, and avermectin production. The morphology of an organism can be directed toward the optimum form for the process by the design of the medium, the medium pH, the medium osmolality, the concentration of the spore inoculum, and by the addition of inorganic particles to the medium. These aspects are discussed in detail in Chapter 6.

The two extremes of filamentous morphology in submerged culture, dispersed mycelium and discrete pellets, result in non-Newtonian and Newtonian broths respectively and, thus, it is the dispersed mycelium habit that presents the greatest challenge in terms of oxygen transfer. Banks (1977) stressed the difference in the pattern of oxygen uptake between unicellular and mycelial fermentations as illustrated in Fig. 9.18. In both unicellular and mycelial fermentations the pattern of total oxygen uptake is very similar during the exponential growth phase, up to the point of oxygen limitation. However, during oxygen limitation, when arithmetic growth occurs, the oxygen uptake rate remains constant in a unicellular system whereas it decreases in a mycelial one. Banks claimed that the only possible explanation for such a decrease is the increasing viscosity of the culture caused by the increasing mycelial concentration resulting in a decreasing oxygen transfer rate.

Several groups of workers have demonstrated the detrimental effect of the presence of mycelium on oxygen transfer. Fig. 9.19 represents some of the data of Deindoerfer and Gaden (1955) illustrating the effect of *Penicillium chrysogenum* mycelium on $K_L a$. Buckland et al. (1988), using different agitator systems, reported that the $K_L a$ decreased approximately in proportion with the square root of the broth viscosity, that is:

$$K_L a \propto 1/\sqrt{\text{viscosity}}.$$

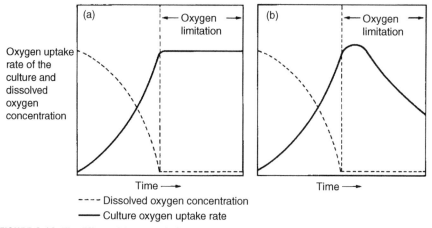

FIGURE 9.18 The Effect of Oxygen Limitation on the Culture Oxygen Uptake Rate

(a) A typical bacterial fermentation, (b) a typical fungal fermentation (Banks, 1977).

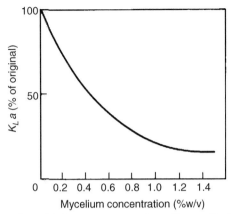

FIGURE 9.19 The Effect of *Penicillium chrysogenum* Mycelium on K_La in a Stirred Fermenter (Deindoerfer & Gaden, 1955)

Steel and Maxon (1966) investigated the problem of oxygen provision to mycelial clumps in the *Streptomyces niveus* novobiocin fermentation and demonstrated that high dissolved oxygen levels (60–80%) occurred in oxygen-limited cultures. It was concluded that, although oxygen was being transferred into solution, the dissolved gas was not reaching a large proportion of the biomass. Thus, as well as K_La being affected adversely by a high viscosity broth, efficient mixing also becomes extremely important in these systems. These workers also demonstrated that, at constant power input, small impellers were superior to large impellers in transferring oxygen from the gas phase to the microbial cells. Wang and Fewkes (1977) confirmed Steel and Maxon's work by demonstrating that the critical dissolved oxygen concentration (C_{crit}) for *S. niveus* in a fermentation varied depending on the degree of agitation and the size of the impeller. Remember that C_{crit} is the dissolved oxygen concentration below which oxygen uptake is limited, that is, it is a physiological characteristic of the organism. It was concluded that the limiting factor was the diffusion of oxygen to the cell surface through a dense mycelial mass. At higher agitation rates, biomass within clumps would be receiving oxygen and would thus contribute to the measured respiration rate whereas at low agitation rates such mycelium would be oxygen limited, that is, the heterogeneity of the system increased at low agitation rates. Wang and Fewkes examined their results in terms of the impeller's ability to produce turbulent shear stress (oxygen transfer into solution) and pumping power (mixing). Turbulent shear stress is proportional to N^2D^2 and impeller pumping power is proportional to ND^3 (where N is the impeller rotational speed and D is the impeller diameter). Thus, the ratio of impeller turbulent shear stress to impeller pumping is proportional to:

$$N^2D^2/ND^3 \text{ or } N/D \text{ (cm sec)}^{-1}.$$

It was demonstrated that the observed critical dissolved-oxygen concentration decreased exponentially as the shear stress to pumping ratio increased, over the range

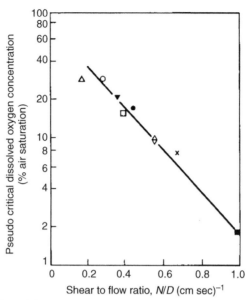

FIGURE 9.20 The Effect of Shear to Flow Ratio on the Observed Critical Oxygen Concentration of *S. niveus* (Wang & Fewkes, 1977)

0.2–1.0 (cm sec) $^{-1}$, as shown in Fig. 9.20. Thus, an increase in the ratio of impeller shear stress to impeller pumping decreases the transport resistance of oxygen to the cell surface resulting in a lower dissolved oxygen concentration maintaining a higher respiration rate. Wang and Fewkes' analysis quantifies Steel and Maxon's observation that smaller impellers gave better oxygen transfer to the cells of *S. niveus,* in that the smaller impeller would have a larger shear stress to impeller pumping power ratio.

Wang and Fewkes' correlations are particularly relevant when it is considered that many mycelial broths are pseudoplastic. The viscosity of a pseudoplastic broth will decrease with increasing shear stress so that viscosity increases with increasing distance from the agitator. Air introduced into the fermenter tends to rise through the vessel by the route of least resistance, that is, through the well-stirred, less viscous central zone. Thus, stagnant zones, receiving little oxygen, may occur in the vessel. Therefore, it is essential that the agitation regime employed creates the correct balance of turbulence (and hence the transfer of oxygen into solution) and pumping power (mixing) to circulate the broth through the region of high shear. It should also be appreciated that ineffective mixing, and thus heterogeneous conditions, make temperature, pH, antifoam, and fed-batch control problematic in these fermentations. The situation is further complicated by the mixing requirements in large-scale vessels necessitating the use of multiple impellers on a common shaft. Due to their radial pumping properties, multiple Rushton turbines tend to give rise to relatively discrete zones around each impeller thus enhancing the heterogeneity of the process.

The quantification of the problem of oxygen transfer and mixing was also considered by Van't Riet and Van Sonsbeek (1992) in the context of the critical time for mass transfer. It is assumed that oxygen transfer into solution in a stirred, aerated reactor takes place only in the stirrer region. If one considers an aliquot of aerated broth leaving the agitator zone, it will be circulated through the vessel and eventually return to the agitator. The dissolved oxygen imparted to the broth should sustain the respiration of the organisms in that aliquot during the circulation. The time it takes for the oxygen in the aliquot to be exhausted will be:

$$t_{cro} = C_L(ag)/OUR$$

where t_{cro} is the time for oxygen to be exhausted, $C_L(ag)$ is the dissolved oxygen concentration in the zone of the agitator, OUR is the oxygen uptake rate.

If the circulation time for the vessel exceeds t_{cro} then oxygen starvation will occur in the aliquot before it returns to the agitator. To prevent this occurring the dissolved oxygen concentration at the agitator should be high, but this would reduce the driving force of oxygen into solution and the oxygen transfer rate would decrease. The alternative approach is to achieve the balance of mass transfer and pumping power (broth circulation) already discussed.

As discussed in Chapter 7 the most widely used fermenter agitator is a disc turbine (Rushton turbine). Van't Riet (1979) and Chapman, Nienow, Cooke, and Middleton (1983) demonstrated that *for nonviscous broths* the K_La is dependent only on the power dissipated in the vessel and is independent of impeller type (at least those impellers included in the study). However, it is obvious from the foregoing discussion that the choice of impeller type is particularly relevant for viscous, non-Newtonian fermentations and this realization has resulted in the development of a range of agitators which address the dual problems of oxygen transfer and mixing in viscous fermentations. Two general approaches were initially employed:

- The use of two different types of impeller in the vessel (one providing mixing and the other gas engagement)
- Replacement of Rushton turbines in existing fermenters with novel impellers capable of achieving the desired levels of both mixing and gas engagement in the one impeller—a process referred to as retrofitting.

The advantages of both approaches have also been combined by adopting different novel impellers in combination, as will be seen from the following discussion.

Legrys and Solomons (1977) and Anderson, LeGrys, and Solomons (1982) gave an excellent analysis of the problem and is an early example of combining adequate pumping power and mass transfer in mycelial fermentations by using two impellers, a bottom-mounted disc turbine and a top-mounted curled-blade (hydrofoil) impeller, combined with a draught tube (Fig. 9.21). A draught tube is a tube running down the center of the vessel that directs the circulation of the liquid. The bottom turbine produced a high degree of turbulence and radial mixing while the top-mounted impeller produced axial mixing with a high flow velocity, resulting in the circulation of one

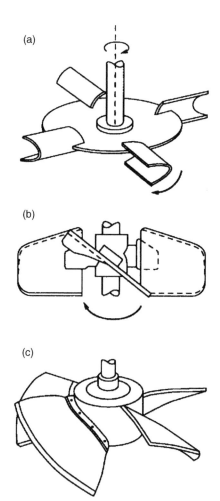

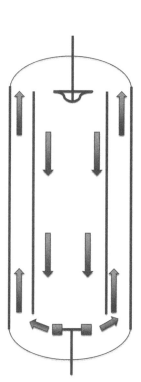

FIGURE 9.21 Representation of Legrys and Solomon's (1977) Fermenter Design Incorporating a Top-Mounted Hydrofoil Impeller and a Bottom-Mounted Disc Turbine Combined with a Draft Tube to Compartmentalize Fluid Circulation

FIGURE 9.22 Agitators Used in Filamentous Fermentations

(a) Scaba agitator; (b) Lightnin' A315; (c) Prochem Maxflo (Nienow, 1990).

tank volume around the draft tube in 20–30 s. Thus, the mycelium was recirculated through the oxygenation zone of the vessel before it became oxygen limited. Cooke et al. (1988) extended Legrys and Solomon's approach using a combination of radial flow and axial flow agitators in a 60-dm^3 fermenter intended for non-Newtonian fermentations. The radial flow agitator was an ICI Gasfoil that is similar to the Scaba SRGT illustrated in Fig. 9.22, being a disc turbine with concave blades. However, in this case the combination was not successful due to minimal fluid movement at the vessel walls that would have created significant cooling problems. (Fig. 9.23)

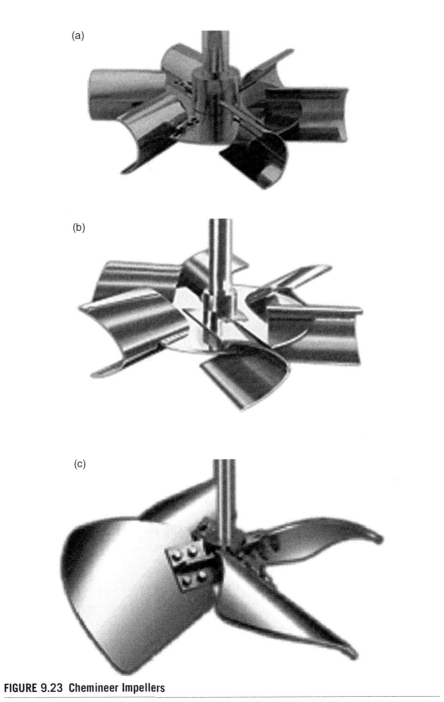

FIGURE 9.23 Chemineer Impellers

(a) CD6 (b) BT6 (c) Maxflo WSE (Chemineer UK Ltd.).

The construction of pilot and industrial scale fermenter vessels is an extremely expensive operation, with major investment committed in the design and choice of motor, gearbox, bearings, and shaft. Thus, it is uneconomic to address the limitations of existing fermenters by incorporating changes to impellers that also require a higher-powered motor and up-grading of the associated power-train. Improved mixing by a Rushton turbine could be achieved by increasing the impeller diameter. The power required for a given stirrer speed depends upon the forces resisting the turning of the impeller and is described by the equation:

$$P = 2\pi NM$$

where P is the power input, N is the impeller rotational speed, and M is the torque, which is the measure of the turning force on an object.

Thus, if the diameter of the impeller is increased then the power input must be increased accordingly, necessitating the installation of a higher power motor and the associated up-graded power train—an uneconomic situation. Eq. (9.28) for power number in the turbulent zone was given previously as:

$$N_p = P/(\rho N^3 D^5)$$

where N_p is the power number, P is the external power from the agitator, ρ is the liquid density, N is the impeller rotational speed, D is the impeller diameter.

Thus,

$$P = N_p \rho N^3 D^5$$

Rushton turbines have a relatively high power number. However, if the Rushton is replaced by an impeller with a lower power number then the power draw will be less or a larger impeller may be used to replace the Rushton at the same power draw. Doran (2013) explained this approach using the following equations: If the new and original impellers consume the same power then:

$$(N_p \rho N^3 D^5)_{new} = (N_p \rho N^3 D^5)_{original}$$

As the same medium is used then liquid density will be the same and the same stirrer speed will be employed then the diameter of the replacement impeller may be calculated from:

$$(D)^5_{new} = \frac{(Np)_{original}}{(Np)_{new}} (D)^5_{original}$$

or:

$$(D)_{new} = \left(\frac{(N_p)_{original}}{(N_p)_{new}} \right)^{1/5} (D)_{original} \qquad (9.35)$$

In the 1980s and 1990s a valuable series of papers by Merck (the pharmaceutical company) scientists and their academic partners reported on the retrofitting of a range of axial-flow impellers to pilot and production scale fermenters. Gbewonyo et al. (1986) evaluated the performance of an axial flow down-pumping hydrofoil impeller, the Prochem Maxflo (Fig. 9.22) in the avermectin fermentation employing *Streptomyces avermitilis* in a 600-dm^3 working volume vessel. The avermectin process is challenging because the broth is extremely viscous, the fermentation requires fairly high oxygen-transfer rates and the situation is complicated by the shear sensitivity of the mycelium. The results of this investigation may be summarized as follows:

1. The impeller pumped the broth axially, that is from the top to the bottom of the fermenter, which is very different from the Rushton turbine that pumped radially, outward from the agitator.
2. The Prochem agitator supported a significantly higher oxygen uptake rate than did the Rushton turbine—probably due to improved mixing reducing the time an aliquot of the culture would spend in oxygen depleted conditions.
3. The power number of the Prochem was 1.1 compared with 6.5 for the Rushton turbine. Thus, a low power number indicates a low power draw and, hence, the Prochem agitator drew significantly less power than did the Rushton turbine, making the former far more economical to operate. This observation is strengthened by Nienow's work on a similar hydrofoil impeller, the Lightnin' A315, which gave a power number of 0.75 compared with 5.2 for a Rushton turbine. Also, as indicated earlier a larger Prochem impeller may be used at the same power draw as the original Rushton.
4. The relationships between K_La and power consumption per unit volume at a viscosity of 700 cp were as follows:

$$\text{Rushton } K_La = 51(P/V)^{0.58},$$
$$\text{Prochem } K_La = 129(P/V)^{0.59}.$$

    These figures reinforce the previous point, demonstrating that the power requirement for the Prochem agitator is approximately 50% of that for the Rushton.
5. Raising the power of the Prochem had a greater effect on oxygen transfer at high viscosity than it did at low viscosity. This points to the key role that bulk mixing plays in a viscous fermentation and suggests that it is at least as important as bubble breakup (at which the Prochem is mediocre).
6. Unlike a Rushton turbine, the Prochem agitator did not generate high shear forces, which is advantageous for a shear sensitive organism.
7. Compared with a Rushton turbine, the power draw of the agitator did not decrease significantly under gassed conditions. A Rushton turbine draws much less power when gassed or importantly, the converse, that power consumption is much greater without aeration. Thus, the motor for a Rushton turbine must be sufficient to cope with unaerated agitation or a variable speed motor must be used but, as pointed out by Nienow (1998), precise safety cut outs must be incorporated to avoid damage to the motor in the event of failure of the air supply.

8. The avermectin yields were slightly better in the Prochem fermenter, but these were achieved with approximately 40% less power consumption.

The same group (Buckland et al., 1988; Buckland, Gbewonyo, Hallada, Kaplan, & Masurekar, 1989, Nienow, Hunt, & Buckland, 1994; Nienow, Hunt, & Buckland, 1996) extended their experiments using 19-m^3 vessels equipped with 3 or 4 hydrofoils containing a variety of rheological fluids–water, viscous fungal biomass and xanthan solutions. Nienow used Metzner and Otto's approach to calculate average viscosity in the non-Newtonian systems, enabling the calculation of Reynolds numbers. This work reinforced the same basic conclusions: bulk mixing is extremely important in viscous fermentations, an axial flow hydrofoil impeller results in lower power costs and the decline in power consumption under gassed conditions is significantly less for the hydrofoil compared with a Rushton turbine. However, the down-pumping hydrofoils caused significant vibration due to the opposing flow of upwardly moving air and the downward flow from the hydrofoils. As a result, the support structures of the fermenters were modified and this was also shown to be successful at the 45-m^3 scale (Nienow et al., 1996). Junker, Mann, and Hunt (2000) confirmed that the axial-flow impellers had been installed in Merck's factory as well as in the pilot plant. It is interesting to note that Legrys and Solomon's earlier combined use of a down-pumping hydrofoil and a Rushton turbine included a draw-tube that segregated the upward and downward flows, thus presumably negating vibration problems. The next stage in the design of alternative agitator systems addressed the problem of vibration by the development of up-pumping hydrofoil impellers which could be used in combination with half-pipe blade turbines.

The superior performance of half-pipe blade turbines was considered in our earlier discussion of the phenomenon of agitator flooding. The first of these impellers was designed by Van't Riet, Boom, and Smith (1976) and the genre became known as "Smith Impellers." The original Smith impeller was a Rushton turbine with the six blades replaced by half-pipes concave in the direction of rotation. The Chemineer CD6 was introduced in the 1980s and is very similar to the original design whist the Chemineer BT6 was introduced in the 1990s and has asymmetric concave blades such that there is a greater overhang at the top of the blade, allowing a higher gas handling capacity (Bakker, 1998). Bakker (2000) summarized the characteristics of these impellers and emphasized their low Power Numbers and resistance to flooding. The CD6 and BT6 have Power Numbers of 2.8–3.2 and 2.3 respectively compared with that of a Rushton turbine of 4.5–6.2. As discussed previously, a lower Power Number means lower power consumption and, hence, more economic operation combined with the attractiveness of the use of larger impellers for the same energy expenditure. Both the CD6 and BT6 have been combined with both down-pumping and up-pumping hydrofoil impellers, the latter being the more successful as the problems of vibration are lessened (Bakker, 2000). An examples of the combination of axial flow and Smith type impellers in a mycelial fermentation are given by Yang et al. (2012). Yang's investigation of a 12 m^3 cephalosporin C fermentation using *Acremonium chrysogenum* compared the performance of four Rushton turbines on the same stirrer shaft with a combination of a BTD6 (at the bottom of the shaft) followed by a conventional Smith impeller and two Lightnin A315 (axial flow) impellers. The novel combination not only gave superior antibiotic production (attributed to the improved homogeneity in the vessel) but did so with a 25% decrease in power consumption. Nienow summarized the progress made in reactor design in his 2014 review and suggested that the most favored agitator system is now a combination of multiple up-pumping impellers, frequently with a Smith-type concave impeller as the lower one.

Manipulation of mycelial morphology

The previous section considered engineering solutions to the problem of oxygen transfer in viscous mycelial fermentations. However, as mentioned in that section, filamentous organisms may also grow in the form of pellets rather than dispersed mycelium that usually results in a nonviscous Newtonian broth. Thus, provided that pellet morphology is commensurate with production of the required metabolite, the cultural conditions may be manipulated to induce the more favorable rheological structure. The key aspect here is the level of understanding of the relationship between morphology and productivity (Posch, Herwig, & Spadiut, 2013; Formenti et al., 2014; Posch & Herwig, 2014). The wide range of cultural conditions that affect morphological form in submerged culture have been reported in an extensive literature and are discussed in detail in Chapter 6. However, these cultural conditions may also affect metabolite productivity independently of their effect on morphology—thus, it is extremely difficult to establish a de facto relationship between morphological form and productivity. Posch et al. (2013) and Posch and Herwig (2014) emphasized the importance of quantifying morphological form using such techniques as image analysis and flow cytometry such that online analysis can give an insight into the development of the culture throughout the process as well as adding to the knowledge base of the fermentation. The novel technique referred to as microparticle enhanced cultivation (MPEC, and discussed in Chapter 6) has provided an extremely powerful tool to not only control morphology but to explore the relationships between morphology, cultural conditions and productivity. MPEC involves the addition to the culture of insoluble inorganic particles (for example, talc and aluminium oxide), the concentration of which determines the morphological form (Walisko, Krull, & Schrader, 2012, 2015). Thus, it is possible to grow a filamentous organism (fungus or actinomycete) in a desired morphological form simply by incorporating the inorganic particles at a particular concentration—independent of medium design. Thus, this very promising technique may be used to induce the most favorable rheological form of a process organism while employing cultural conditions to give optimum product formation.

Examples of the beneficial effect of pelleting on culture rheology include Buckland (1993), Ghojavand, Bonakdarpour, Heydarian, and Hamedi (2011) and Cai et al. (2014). Buckland reported that the $K_L a$ attained in the lovastatin *Aspergillus terreus* fermentation was 20 h^{-1} with a filamentous culture and 80 h^{-1} with a pelleted one at the same power input. Ghojavand et al. (2011) showed that a *Saccharopolyspora erythrea* consisting of small clumps gave a lower viscosity broth than its dispersed mycelium counterpart coupled with improved erythromycin production. Using a genetic approach to the problem, Cai et al. deleted the *AgkipA* gene (shown to affect morphological form) from *Aspergillus glaucus* resulting in the development of a low-viscosity, shear resistant pelleted culture compared with the high viscosity, shear sensitive wild type. The improved strain increased production of its secondary metabolite (asergiolide A, an antitumor agent) by 82%. However, not all pelleted cultures are Newtonian: Metz, Kossen, and Van Suijdam (1979) demonstrated that pellet suspensions could be non-Newtonian but confirmed that they did give rise to low viscosity broths. Also, although the pellet type of growth tends to produce a low viscosity Newtonian broth in which turbulent flow conditions may be achieved, it may also give rise to problems of oxygen availability if the pellets become too large. A large pellet may be so compact that its center may be unaffected by the turbulent forces occurring in the bulk of the fermentation broth so that the passage of oxygen within the pellet is dependent on simple diffusion; this may result in the center of the pellet being oxygen limited. Thus, to maintain the intrapellet oxygen concentration at an adequate level it would be necessary to maintain a high dissolved oxygen concentration to ensure an effective diffusion gradient (Steel & Maxon, 1966, Kobayashi, Van Dedem, & Moo-Young, 1973, and Wang & Fewkes, 1977). If oxygen limitation does occur within a pellet then only its outer layer would contribute to its growth and the center may autolyse. The diffusion of oxygen into the center of a pellet will be influenced by the size of the pellet, and thus it is important to control pellet size. Indeed, Yano, Kodama, and Yamada (1961) developed an equation to predict the maximum radius (R_{crit}) of a pellet that avoids nutrient limitation at its centre and represents the ratio of nutrient supply due to diffusion and the substrate uptake by the organism; in terms of oxygen limitation, the relationship would be:

$$R_{crit} = \sqrt{\frac{6 \times C_L \times D_{eff}}{QO_2 x}} \qquad (9.36)$$

Where R_{crit} is the critical radius of the pellet (m), C_L is the dissolved oxygen concentration at the pellet surface (kg m^{-3}), D_{eff} is the effective diffusion of oxygen in the pellet, $Q_{O_2} x$ is the oxygen volumetric uptake rate (kg oxygen m^{-3}s^{-1}), D_{eff} is the product of the molecular diffusion coefficient and the porosity of the pellet.

A number of workers have applied this analysis combined with the measurement of the oxygen profile in pellets using microelectrodes in an attempt to understand the physiological conditions experienced by the mycelium within a pellet. The range of R_{crit} values reported have been between 50 and 200 μm, depending on the density of the pellet (Nielsen, 1996; Cui, van der Lans, & Luyben, 1998; Hille, Neu, Hempel, & Horn, 2005, and Wucherpfennig, Kiep, Driouch, Wittmann, & Krull, 2010). Schugerl, Wittler, and Lorenz (1988) monitored the dissolved oxygen concentration within pellets of *P. chrysogenum* and demonstrated that, provided they were smaller than 200 μm in radius, the oxygen concentration in the center of the pellet was not limiting. Similarly, Buckland (1993) reported that pellets of *Aspergillus terreus* in the lovastatin fermentation had to be smaller than 90 μm in radius to avoid oxygen limitation in the center of the pellet and the conditions of the fermentation had to be carefully controlled to maintain the optimum pellet size but the cultural conditions used to achieve this end were not revealed. The incorporation of the gene for green fluorescent protein (GFP, originally isolated from the jellyfish *Aequorea victoria*) into an organism is frequently used to report gene expression. The protein will fluoresce when stimulated with light of a specific wavelength and the technique is particularly valuable when used with fluorescent or confocal laser scanning microscopy (CLSM). This approach has been used by a number of workers to visualize gene expression within pellets and mycelial clumps of mycelial organisms as an indication of metabolic activity. For example, Driouch, Roth, Dersch, and Wittman (2010) showed that 1000 µm pellets of *Aspergillus niger* produced GFP only in the outer 200 µm edge whereas talc microparticle-induced dispersed mycelium was active throughout its biomass. This work was extended (Driouch, Hansch, Wucherpfennig, Krull, & Wittmann, 2012) using titanium microparticles and, intriguingly, these particles (depending on concentration) could not only induce small pellets of biomass rather than dispersed mycelium but larger pellets with a more diffuse structure. Such pellets demonstrated high activity to a depth of 500 µm within their more open structure. Gao, Zeng, Yu, Dong, and Chen (2014) also employed microparticles to manipulate the morphology and thus enhance the lipid productivity of *Mortierella isabellina*. Using CLSM to visualize lipid expression, these workers showed that large pellets were only productive at the edge whereas dispersed mycelium was homogeneously productive. Thus, the rheological benefits of the pelleted morphology must be considered in the context of the problems of diffusional limitation within the pellets and a balance struck by manipulating pellet size and structure. As can be seen from these two examples, the application of MPEC has considerable potential to achieve this end.

There is a considerable literature addressing the effects of agitation and aeration conditions on mycelium morphology in bioreactors and, thus, the possible use of shear stress to achieve the optimum morphology for a process (Serrano-Carreon, Galindo, Rocha-Valadez, Holguin-Salas, & Corkidi, 2015). A very early example is given by Dion, Casilli, Sermonti, and Chain (1954) who showed that the morphology of *P. chrysogenum* was influenced by the degree of agitation in that short, branched mycelium was produced at high agitation rates compared with long hyphae produced at low agitation rates. However, while shear stress may influence morphology, and thus reduce viscosity and increase oxygen transfer rate, it may also injure the mycelium and ultimately limit productivity. Thus, any mycelial process has to be optimized to strike a balance between achieving the desired form and damaging the organism. Damage to an organism in submerged culture is caused by eddies—an eddy being the swirling of a fluid caused by agitation. In such a system, kinetic energy is cascaded from large eddies to smaller ones due to inertial forces—the range of eddy sizes being called the inertial sub-range. Below this range the turbulent energy is converted to heat due to the resistance of viscosity (the viscous sub-range). The system is thus elegantly described by Richardson (1922) poem:

> *"Big whorls have little whorls*
> *That feed on their velocity*
> *And little whorls have lesser whorls*
> *And so on to viscosity."*

The diameter of the terminal eddy in the inertial sub-range (the transition between the inertial and viscous sub-ranges) is given by the Kolmogorov equation:

$$\lambda_k = (v^3/\varepsilon_T)^{0.25} \tag{9.37}$$

where λ_k is the Kolmogorov scale, diameter of the terminal eddy in the inertial sub-range (m), v is the kinematic viscosity (m^2 s^{-1}) and $v = \mu/\rho_L$ where μ is dynamic viscosity (Pa s), and ρ is liquid density (kg m^{-3}), ε_T is the mean specific energy dissipation rate (m^2 s^{-3}).

The key issue regarding eddy size is that damage is caused by eddies that are the same size or smaller than the process organism. Large eddies carry the cells while small ones can act in opposite rotational directions on the surface of the cell, causing shear forces. In their review of the effects of hydrodynamics on filamentous fungi Serrano-Carreon et al. (2015) reported typical values of λ_k as ranging between 10 and 50 μm. Thus, bacterial cells are very resistant to shear forces but larger organisms (such as filamentous fungi, animal cells and plant cells and fungal and streptomycete pellets) are far more susceptible. From Eq. (9.37) it can be seem that the size of the terminal eddy is determined by the viscosity and the power supplied to the system—as power increases or viscosity decreases the eddies become smaller and, thus, potentially more damaging. However, ε_T is the mean energy dissipation rate and Cutter (1966) showed that energy dissipation rates close to a Rushton turbine can be approximately seventy times the average. Thus, the degree of damage imparted in the impeller zone is far greater than that in other parts of the fermenter. These observations led workers to consider the effect of shear being due to a combination of the energy dissipation at the impeller and the circulation frequency of the biomass through the impeller zone (van Suijdam & Metz, 1981; Reuss (1988); Oh, Nienow, Al-Rubeai, & Emery, 1989 & Smith, Lilly, & Fox, 1990). Smith et al. (1990) analyzed the penicillin fermentation and adapted van Suijdam and Metz's approach, proposing that the breakup frequency of particles would be dependent on the "energy dissipation/circulation function" (ECDF):

$$ECDF = (P/D^3)(1/t_c) \tag{9.38}$$

where P is the power input, D the impeller diameter, and $1/t_c$ the circulation frequency.

These workers showed that both penicillin production rate and mean hyphal length decreased when the stirrer speed was increased from 800 to 1000 rpm in 10 dm^3 fermenters and further decreased at 1200 rpm. Although the data were insufficient to demonstrate the generality of the relationship with ECDF they did support the concept of a damage-inducing zone around the impellers through which the mycelium is periodically circulated. However, this relationship was specific to vessels employing Rushton turbines and the assay of damage was confined to mean hyphal length—at that time no assay being available to determine the effects of ECDF on mycelial aggregates. Justen, Paul, Nienow, and Thomas (1996) extended the approach to other types of impeller and related ECDF to morphology by using image analysis to measure the mean area of (in their terminology) mycelial clumps, as well as hyphal length. Thus, the definition of ECDF was modified to include the term k, a geometrical factor to accommodate different impeller types (both radial and axial flow):

$$ECDF = (P/(kD^3))(1/t_c) \qquad (9.39)$$

where $k = (\pi/4)(W/D)$ with W/D being the blade height/impeller diameter ratio for axial flow impellers.

In a further refinement of the relationship k was modified to k' to accommodate the different number of trailing vortices accompanying each impeller type. This is relevant as it is in these areas behind the impeller blades that the maximum impact on suspended particles takes place. Thus, the values of k' for different impeller types are as follows:

6 blade pitched blade turbine (6 blades, 6 vortices); $k' = k$
6 blade Rushton turbine (6 blades, 12 vortices); $k' = 2k$
4 blade paddle impeller (4 blades, 8 vortices); $k' = (8/6)\,k$
3 blade axial flow (3 blades, 3 vortices); $k' = k/2$

The key findings may be summarized as:

- At the same power input per unit volume the hyphal damage was dependent on impeller geometry, with damage increasing as the impeller diameter: fermenter diameter ratio decreased. To achieve equal power input per unit volume, such impellers would be operating at a higher speed.
- ECDF correlated with both hyphal and morphological damage regardless of scale used (1.4, 20, and 180 dm^3) whereas correlations of damage with impeller tip speed and power input per unit volume were scale-specific. Thus, only ECDF was deemed suitable to scale-up fermentations where hyphal and mycelial damage were critical.
- The incorporation of the geometric factor k' enabled a very strong correlation to be made between hyphal and mycelial damage and the revised ECDF for the range of impellers employed in the study.

The extensive literature on the morphology of mycelial processes has been addressed by a number of authoritative reviews compiled by Krull and Bley (2015). In their contribution to this text, Serrano-Carreon et al. (2015) summarized the impact of hydrodynamic stress on mycelial morphology as follows:

1. The overall mycelial size is inversely proportional to the global energy supplied to the reactor and, particularly, is an inverse function of the energy dissipated in the vicinity of the impellers, as well as of the time that the biomass is present in that zone.
2. In general, within a range of low dissipated energy, the increase in power improves the production of fungal metabolites; however, this effect could be due to improved dissolved oxygen supply.

3. In the range of high dissipated energy an increase in power leads to saturation or decreasing profiles in terms of metabolite production.
4. The mechanism of mycelial damage has been shown to be associated with the Kolmogorov eddy size; however, very few works have reported estimated values based on experimental power drawn.

Several workers have discussed the possible advantages of reducing the viscosity of a mycelial fermentation, in its later stages, by diluting the broth with either water or fresh medium. Sato (1961) increased the yield of a kanamycin fermentation, displaying Bingham plastic rheology, by 20% by diluting the broth 5% by volume with sterile water. Taguchi (1971) achieved a 50% reduction in the viscosity of an *Endomyces* broth by diluting 10% with water or fresh medium. A scheme has been put forward for the control of viscosity and dissolved oxygen concentration in a hypothetical fermentation. These workers proposed that, as the critical dissolved oxygen concentration is approached, a set volume of broth could be removed from the fermenter and replaced with fresh medium. The process could be repeated in a step-wise manner as the system became oxygen limited, which could be determined by dissolved oxygen concentration or viscosity measurements. Thus, by using such techniques the viscosity may be controlled and maintained below the level that may cause oxygen limitation. Kuenzi (1978) reported an instance where the very slow feeding of medium to a *Cephalosporium* culture resulted in the organism growing in the form of long filaments that produced a highly viscous culture that could not be adequately aerated. The design of fed-batch processes such that efficient control may be achieved over the process is discussed in a subsequent section of this chapter and in Chapter 2.

The production of *Fusarium graminearium* biomass for human food in the ICI-RHM mycoprotein (Quorn®) fermentation (see Chapter 1) presents a very different problem from those of most other fungal fermentations. It is essential that the organism grows as long hyphae so that the biomass can be processed into a textured food product. Long hyphae are susceptible to shear forces, so to maintain the morphological form of the organism an air-lift reactor is used, despite the fact that the viscous broth severely limits the attainable oxygen transfer rate. This limitation of the air-lift fermenter means that only a relatively low biomass concentration may be maintained in the vessel compared with that in a stirred system, but this is an acceptable penalty to pay for the correct morphological form.

Effect of microbial products on oxygen transfer

Generally speaking, the product of a fermentation contributes relatively little to the viscosity of the culture broth. However, the exception is the production of microbial polysaccharides, where the broths tend to be highly viscous (30,000 cp, Sutherland & Ellwood, 1979) and non-Newtonian. Charles (1978) demonstrated that the bacterial cells in a polysaccharide fermentation made a minimal contribution to the high culture viscosity that was due primarily to the polysaccharide product. However, fungal polysaccharide processes (for example, pullulan production by *Aureobasidium pullulans*) is a particularly complex process as both the producing organism and the product contribute significantly to the viscosity (Seviour, McNeil, Fazenda, & Harvey, 2011). Furthermore, the situation is complicated by the shear stress affecting the morphology of the producing fungus, as described in the previous section. Normally, microbial polysaccharides tend to behave as pseudoplastic fluids, although some have also been shown to exhibit a yield stress (Seviour et al., 2011). The yield stress of a polysaccharide can make the fermentation particularly difficult because, beyond a certain distance from the impeller, the broth will be stagnant and productivity in these regions will be practically zero (Gallindo & Nienow, 1992). Thus, bacterial polysaccharide fermentations present problems of oxygen transfer and bulk mixing similar to those presented by mycelial fermentations and mycelial polysaccharide fermentations are doubly cursed. To illustrate these processes the xanthan and pullulan fermentations will be considered in more detail.

1. Xanthan production by *Xanthomonas campestris*

   X. campestris is a strictly aerobic bacterium and thus it has a high oxygen demand that must be satisfied by the fermentation process in the presence of the highly viscous, pseudoplastic product. Thus, similar stirrer configurations to those discussed in the previous section on mycelial fermentations have been used in xanthan fermentations. Such systems combine good gas entrainment and liquid circulation in an attempt to avoid the stagnant zones associated with pseudoplastic fluids. Gallindo and Nienow (1992) investigated the behavior of a hydrofoil impeller, the Lightnin' A315, in a simulated xanthan fermentation. These workers adopted Metzner and Otto's approach to construct power curves. Better agitator performance was achieved when its pumping direction was upward rather than downward (as discussed for mycelial fermentations) resulting in lower power loss on aeration and less torque fluctuations. It was concluded that such agitators may give improved mixing in a xanthan fermentation provided that the polysaccharide concentration is below 25 kg m^{-3}. A novel solution to the problem was proposed by Oosterhuis and Koerts (1987). These workers designed an air-lift loop reactor incorporating a pump to circulate the highly viscous broth. The system was operated on a 4-m^3 scale and proved to be much more efficient than a stirred tank reactor.

   Despite the apparent promise of air-lift systems stirred vessels still appear to be the vessels of choice. A commonly observed phenomenon is that as agitation rate is increased during the fermentation to cope with the increasing oxygen volumetric uptake rate a point is reached where further increase in agitation is counterproductive and polysaccharide yield falls (Casas, Santos, & Garcia-Ochoa, 2000). This observation was attributed to hydrodynamic damage to the producing organism but the mechanism of such damage has been explained by the "Kolmogorov eddy size" being equal to, or smaller than, the size of the organism. Hewitt and Nienow's (2007) analysis of hydrodynamic damage to bacterial cells is discussed later in this chapter (see Section "The Influence of Scale-Down Studies") from which it can be appreciated that hydrodynamic damage to bacterial unicells is unlikely as they are significantly smaller than the eddy size generated under normal fermenter operating conditions. However, *X. campestris* secretes xanthan as a capsule and thus each cell is surrounded by a significant layer of polysaccharide effectively increasing its size which may increase sensitivity to hydrodynamic stress. The other explanation of the adverse effect of high agitation suggested by Seviour et al. (2011) is that the accompanying high dissolved oxygen concentration may cause oxidative stress to the bacteria and oxidative damage to the polysaccharide.

2. Pullulan production by *A. pullulans*

From the discussion of the impact of mycelial morphology on process performance it will be appreciated that the combination of the production of a viscous pseudoplastic product by a filamentous fungus, itself able to produce a viscous pseudoplastic broth, represents a challenging fermentation. It is apparent that pullulan synthesis by *A. pullulans* occurs in swollen unicellular units and chlamydospores while filamentous mycelium synthesizes other polysaccharides (Campbell, Siddique, McDougall, & Seviour, 2004; Orr, Zheng, Campbell, McDougall, & Seviour, 2009). However, the literature on the impact of cultural conditions on the morphology of submerged cultures of *A. pullulans* and the link with pullulan production is confused due to the frequent lack of structural analysis of the synthesized polysaccharide (Seviour et al., 2011). While several reports correlate the unicellular, swollen morphology and pullulan synthesis with high agitation (McNeil & Kristiansen, 1987) there are conflicting reports suggesting the advantages of low dissolved oxygen and low shear stress (Wecker & Onken, 1991). However, Dixit, Mehta, Gahlawat, Prasad, and Choudry (2015) used response surface methodology (see Chapter 4 for a discussion of the method) to optimize the effects of aeration, agitation, and impeller position on $K_L a$, morphology and pullulan production, demonstrating strong correlation between increased agitation and high $K_L a$ with unicellular morphology and pullulan synthesis.

EFFECT OF FOAM AND ANTIFOAMS ON OXYGEN TRANSFER

The high degree of aeration and agitation required in a fermentation frequently gives rise to the undesirable phenomenon of foam formation. In extreme circumstances the foam may overflow from the fermenter via the air outlet or sample line resulting in the loss of medium and product, as well as increasing the risk of contamination. The presence of foam may also have an adverse effect on the oxygen-transfer rate. Hall, Dickinson, Pritchard, and Evans (1973) pointed out that Waldhof and vortex-type fermenters (see Chapter 7) were particularly affected due to the bubbles becoming entrapped in the continuously recirculating foam, resulting in high bubble residence times and, therefore, oxygen-depleted bubbles. The presence of foam in a conventional agitated, baffled fermenter may also increase the residence time of bubbles and therefore result in their being depleted of oxygen. Furthermore, the presence of foam in the region of the impeller may prevent adequate mixing of the fermentation broth. Thus, it is desirable to break down a foam before it causes any process difficulties and, as discussed in Chapter 7, this may be achieved by the use of mechanical foam breakers or chemical anti-foams. However, mechanical foam control consumes considerable energy and is not completely reliable so that chemical antifoams are preferred (Van't Riet & Van Sonsbeek, 1992; Junker, 2007). It is interesting to note that a degree of foaming can be beneficial as higher gas hold-up can decrease power dissipation and actually increase mass transfer rate and reduce power consumption (Berovic, 1992). Thus, low foam levels that can be accommodated in the process without recourse to antifoam control may be advantageous.

All antifoams are surfactants and may, themselves, be expected to have some effect on oxygen transfer. The predominant effect observed by most workers is that antifoams tend to decrease the oxygen-transfer rate, as discussed by Aiba, Humphrey, and Milus (1973), Hall et al. (1973), and Junker (2007). Antifoams cause the collapse of bubbles in foam but they may favor the coalescence of bubbles within the liquid phase, resulting in larger bubbles with reduced surface area to volume ratios and hence a reduced rate of oxygen transfer (Van't Riet and Van Sonsbeek, 1992). Also, antifoam may accumulate at the gas–liquid interface resulting in an increase in the resistance to oxygen diffusion (Pelton, 2002). Thus, a balance must be struck between the necessity for foam control and the deleterious effects of the controlling agent. Foaming increases gas hold-up in the vessel that, in turn, increases broth volume, thus influencing the liquid height in the fermenter at which it is practical to operate. If inadequate space is provided above the liquid level for foam control, then copious amounts of antifoam must be used to prevent loss of broth from the vessel. Van't Riet and Van Sonsbeek (1992) observed that, above a critical liquid height, the $K_L a$ value decreases dramatically due to the excessive use of antifoams. Thus, it may be more productive to operate a vessel at a lower working volume.

Methods for foam control are considered in Chapter 8 and antifoams are discussed in Chapter 4.

BALANCE BETWEEN OXYGEN SUPPLY AND DEMAND

Both the demand for oxygen by a microorganism and the supply to the organism by the fermenter have been considered in this chapter. This section attempts to bring these two aspects together and considers how processes may be designed such that the oxygen uptake rate of the culture does not exceed the oxygen transfer rate of the fermenter.

The volumetric oxygen uptake rate of a culture is described by the term, $Q_{O_2} x$, where Q_{O_2} is the specific oxygen uptake rate (mmoles O_2 g^{-1} biomass h^{-1}) and x is biomass concentration (g dm^{-3}). Thus, the units of $Q_{O_2} x$ are mmoles oxygen $dm^{-3} h^{-1}$.

The volumetric oxygen transfer rate (also measured as mmoles O_2 $dm^{-3} h^{-1}$) of a fermenter is given by Eq. (9.1), that is:

$$dC_L / dt = K_L a(C^* - C_L).$$

It will also be recalled that the dissolved oxygen concentration during the fermentation should not fall below the critical dissolved oxygen concentration (C_{crit}) or the dissolved oxygen concentration that gives optimum product formation. Thus, it is necessary that the oxygen-transfer rate of the fermenter matches the oxygen uptake rate of the culture while maintaining the dissolved oxygen above a particular concentration. A fermenter will have a practical maximum $K_L a$ dictated by the operating conditions of the fermentation and thus, to balance supply and demand it must be the demand that is adjusted to match the supply. This may be achieved by:

1. Controlling biomass concentration.
2. Controlling the specific oxygen uptake rate.
3. A combination of (1) and (2).

CONTROLLING BIOMASS CONCENTRATION

Mavituna and Sinclair (1985a) developed a method to predict the highest biomass concentration (termed the critical biomass or x_{crit}) that can be maintained under fully aerobic conditions in a fermenter of known $K_L a$. Thus, x_{crit} is the biomass concentration that gives a volumetric uptake rate ($Q_{O_2} x_{crit}$) equal to the maximum transfer rate of the fermenter, that is, $K_L a$ ($C^* - C_{crit}$). If C_{crit} is defined as the dissolved oxygen concentration when:

$$Q_{O_2} = 0.99 Q_{O_2} \text{ max}$$

then the volumetric oxygen uptake rate when the dissolved oxygen concentration is C_{crit} will be:

$$0.99 Q_{O_2} \text{ max} \cdot x_{crit}.$$

If the oxygen transfer rate is equal to the uptake rate when the dissolved oxygen concentration equals C_{crit}. Then:

$$K_L a(C^* - C_{crit}) = 0.99 Q_{O_2} \text{ max} \cdot x_{crit}. \tag{9.40}$$

Eq. (9.40) may be used to calculate x_{crit} for a fermenter with a particular $K_L a$ value:

$$x_{crit} = K_L a(C^* - C_{crit})/0.99 \ Q_{O_2} \text{ max} \tag{9.41}$$

Eq. (9.41) may also be modified to calculate the biomass concentration that may be maintained at any fixed dissolved oxygen concentration above C_{crit}:

$$x = K_L a(C^* - C_L)/Q_{O_2} \text{ max}.$$

Mavituna and Sinclair presented this model graphically as shown in Fig. 9.24. The upper graph represents the relationship between the dissolved oxygen concentration and the volumetric oxygen transfer rate achievable in three fermenters (plots 1, 2 and 3 represent fermenters of increasing $K_L a$ values) while the lower graph represents the

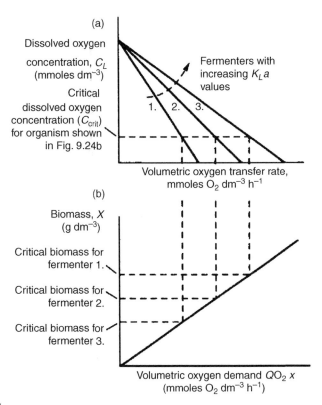

FIGURE 9.24

(a) The relationship between dissolved oxygen concentration and the oxygen transfer rate attainable in 3 fermenters with increasing K_La values, (b) The relationship between biomass concentration and oxygen uptake rate of a process organism. The same scales are used for dC_L/dt and $Q_{O_2}x$ allowing x_{crit} to be determined (Mavituna and Sinclair, 1985a).

relationship between biomass and the volumetric oxygen uptake rate of the culture. The x axes of both graphs are drawn to the same scale. A construction is drawn on the upper graph linking C_{crit} to the oxygen-transfer rates attainable in each of the three fermenters. This construction is extended to the lower graph indicating the oxygen uptake rates equal to the transfer rates attainable at C_{crit}. Finally, from the lower graph, the biomass concentrations (x_{crit}) which would give rise to the uptake rates equal to the transfer rates may be determined. Again, this figure may be used to predict the maximum biomass concentration that may be maintained at any dissolved oxygen concentration above C_{crit}.

It should be appreciated that these authors intended this model to be used only as a method for preliminary design (Mavituna & Sinclair, 1985b). Thus, x_{crit} is interpreted as a target which cannot be exceeded and, in practice, oxygen limitation will probably occur below this value. The mechanism for limiting the biomass concentration will be the concentration of the limiting substrate in the medium that, for batch culture, may be determined from the equation:

$$S_R = x_{crit}/Y$$

where S_R is the initial limiting substrate concentration and Y is the yield factor and it is assumed that the limiting substrate is exhausted on entry into the stationary phase.

The technique may also be applied to continuous and fed-batch culture but it must be appreciated that Q_{O_2} is affected by specific growth rate and the relevant Q_{O_2} value for the growth rate employed would have to be utilized in the calculations. The method should be very useful for the initial design of unicellular bacterial or yeast fermentations where biomass has no effect on $K_L a$. However, in viscous fermentations the biomass concentration influences the $K_L a$ considerably, as discussed in a previous section. Thus, the $K_L a$ will decline with increasing biomass concentration, which makes the application of the technique more problematical.

CONTROLLING THE SPECIFIC OXYGEN UPTAKE RATE

Specific oxygen-uptake rate is directly proportional to specific growth rate so that, as μ increases, so does Q_{O_2}. Thus, Q_{O_2} may be controlled by the dilution rate in continuous culture. Although very few commercial fermentations are operated in continuous culture, fed-batch culture is widely used in industrial fermentations and provides an excellent tool for the control of oxygen demand. The kinetics and applications of fed-batch culture are discussed in Chapter 2. The most common way in which the technique is applied to control oxygen demand is to link the nutrient addition system to a feed-back control loop using a dissolved oxygen electrode as the sensing element (see Chapter 8). If the dissolved oxygen concentration declines below the set point then the feed rate is reduced and when the dissolved oxygen concentration rises above the set point the feed rate may be increased. A pH electrode may also be used as a sensing unit in a fed-batch control loop for the control of oxygen demand—oxygen limitation being detected by the development of acidic conditions. These techniques are particularly important in the growth-stage of a secondary metabolite mycelial fermentation prior to product production when the highest growth rate commensurate with the oxygen transfer rate of the fermenter is required. A full discussion of the operation of fed-batch systems is given in Chapter 2.

SCALE-UP AND SCALE-DOWN

Scale-up means increasing the scale of a fermentation, for example from the laboratory scale to the pilot plant scale or from the pilot plant scale to the production scale. Increase in scale means an increase in volume and the problems of process scale-up are due to the different ways in which process parameters are affected by the size of the unit. It is the task of the fermentation technologist to increase the scale of a fermentation without a decrease in yield or, if a yield reduction occurs, to identify the factor which gives rise to the decrease and to rectify it. The major factors involved in scale-up are:

1. *Inoculum development.* An increase in scale may mean that extra stages have to be incorporated into the inoculum development program. As well as the design of the inoculum train, the stability of the strain must be assessed to ensure that it can retain its productivity for the increased number of generations required. This aspect is considered in Chapter 6.
2. *Sterilization.* Sterilization is a scale dependent factor because the number of contaminating micro-organisms in a fermenter must be reduced to the same absolute number regardless of scale. Thus, when the scale of a process is increased the sterilization regime must be adjusted accordingly, which may result in a change in the quality of the medium after sterilization. This aspect is considered in detail in Chapter 5.
3. *Environmental parameters.* The increase in scale may result in a changed environment for the organism. These environmental parameters may be summarized as follows:
   a. Nutrient availability,
   b. pH,
   c. Temperature,
   d. Oxygen transfer rate and dissolved oxygen concentration,
   e. Shear conditions,
   f. Dissolved carbon dioxide concentration and carbon dioxide removal rate,
   g. Foam production,
   h. Increased heterogeneity.

All the mentioned parameters are affected by agitation and aeration, either in terms of bulk mixing or the provision of oxygen. Points a, b, c, and h are related to bulk mixing while d, e, f, and g are related to air flow, agitation, and oxygen transfer. Thus, agitation and aeration tend to dominate the scale-up literature. However, it should always be remembered that inoculum development and sterilization difficulties may be the reason for a decrease in yield when a process is scaled up and that achieving the correct aeration/agitation regime is not the only problem to be addressed.

INFLUENCE OF SCALE-DOWN STUDIES

From the list of environmental parameters affected by aeration and agitation it will be appreciated that it is extremely unlikely that the conditions of the small-scale fermentation will be replicated precisely on the large scale. The problem of scale-up was extremely well illustrated by Fox (1978) in his description of the "scale-up window." The scale-up window represents the boundaries imposed by the environmental parameters and cost on the aeration/agitation regime and is shown in Fig. 9.25. Suitable conditions of mixing and oxygen transfer can be obtained with a range of

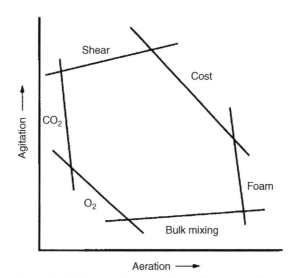

FIGURE 9.25 The "Scale-Up" Window Defining the Operating Boundaries for Aeration and Agitation in the Scale-Up of a Fermentation

After Fox, 1978 reproduced from Lilly, 1983.

aeration/agitation combinations. The two axes of Fig. 9.25 are agitation and aeration and the zone within the hexagon represents suitable aeration/agitation regimes. The boundary of the hexagon is defined by the limits of oxygen supply, carbon dioxide accumulation, shear damage to the cells, cost, foam formation and bulk mixing. For example, the agitation rate must fall between a minimum and maximum value—mixing is inadequate below the minimum level and shear damage to the cells is too great above the maximum value. The limits for aeration are determined at the minimum end by oxygen limitation and carbon dioxide accumulation and at the maximum end by foam formation. The shape of the window will depend on the fermentation—for example, the supply of oxygen would be irrelevant in an anaerobic fermentation, whereas the limitation due to shear may be of major importance in the scale-up of mycelial fermentations. Thus, the most important criteria for a particular fermentation must be established and the scale-up of aeration/agitation based on reproducing those characteristics. However, many processes have been scaled-up based on preconceptions—either of the key limiting factor on which the scale-up was based or the explanation of poor performance in the scaled-up reactor. For example, Hewitt and Nienow (2007) addressed a number of situations where the failure of a scaled-up process to match the lab-scale productivity was attributed to problems of "shear sensitivity." It will be recalled from our previous discussion of the effect of shear on mycelial organisms that damage is caused by Kosmogorov eddies of the same size (or smaller) than the process organism. Hewitt and Nienow cited three pieces of work (Hewitt and Nebe-von-Caron, 2004; Chamsartra, Hewitt, & Nienow, 2005 and Boswell, Nienow, Gill, Kocharunchitt, & Hewitt, 2003) all three of which demonstrated that the Kolmogorov eddies generated by the normal range of agitation rates were significantly lager than the unicellular bacteria or yeasts used in the process. In all these cases any effect of shear forces on organism performance occurred under conditions of agitation far in excess of those employed in the process. Thus, shear damage was not an acceptable explanation for the poor scale-up of these bacterial and yeast processes. Furthermore, it can be said that shear damage is highly unlikely to be an issue for unicellular yeast and bacterial processes in general. However, the large size of mycelial organisms, whether as dispersed filaments or pellets would be susceptible to shear damage, as discussed earlier. Interestingly, the susceptibility of animal cells to shear has been significantly over-stated which, according to Nienow (2006), has held back the development of animal cell culture on a large scale.

The solution to effective scale-up is the acquisition of knowledge of the process and the most significant development to achieve this has been the increasing influence of the philosophy of scale-down on fermentation development. In their discussion of scale-up Villadsen et al. (2011) recommended the advice of Baekeland, the inventor of Bakelite,—"Commit your blunders on the small scale and make your profits on a large scale." This quotation summarizes perfectly the objectives of scale-down—laboratory-scale experiments that mimic the large-scale and generate key information about the physiology, robustness and performance of the process organism (under conditions that are realistic for a production process) before it is plunged into a large-scale fermenter. The small-scale experiments should address the ability of the organism to adapt to the environmental conditions of the large-scale process—particularly the heterogeneity typical in an industrial scale fermenter—and thus reflect the limitations imposed by the scale of manufacture. This approach is very much in line with the United States' Food and Drug Administration (FDA) "Quality by design" initiative in the pharmaceutical industry in which quality should be built into a product based on a thorough understanding of the product and the process by which it is manufactured (Rathore and Winkle, 2009). Neubauer et al. (2013) discussed the interface between product and bioprocess development and pointed out that product development refers to invention (or discovery) of the product whereas process development is the evolution of the manufacturing system. Thus, the objective of the discovery process is to develop an effective compound whereas that of process development is to design the manufacturing process that will bring the invention to the market place. These authors emphasized the complexity of process development in the biotechnology industry that, due to its interdisciplinary nature, can generate severe problems in transforming innovative ideas into viable products. Thus, the scale-down approach has been used to develop the knowledge base to enable the logical scale-up of a new process. Moreover, and very importantly, the approach can also be used to increase the knowledge of an established process by mimicking it on a laboratory scale and thus facilitate its continuous improvement.

Noorman (2011) summarized the five steps of scale-down:

1. Analyzing the industrial fermenter to be used in a new process or the vessel being used in an existing one.
2. Mimicking the large-scale conditions in a lab-scale system.
3. Optimizing the conditions or resolving bottlenecks in the scaled-down system.

4. Translating the findings back to the industrial fermenter.
5. Implementing the improvement on the large scale.

The aspects of scale-down that are simplest to address are:

- *Medium design.* Only media relevant to the industrial situation should be used in development experiments, thus avoiding false hopes being raised by employing media giving high yields that cannot be achieved economically.
- *Medium preparation and sterilization.* If the medium is to be batch sterilized on the large scale its exposure time at a high temperature will be much greater than that experienced in the laboratory or pilot plant. Thus, the sterilization times on the smaller scales should be increased to mimic the industrial situation. Alternatively, medium sterilized in the production fermenter may be used in the laboratory and pilot plant, which ensures that any issues with process water are also encountered in the laboratory. This highlights the advantage of continuous sterilization where little loss of medium quality occurs. Furthermore, medium from the same continuous sterilizer may be used for both full-scale and small-scale vessels.
- *Inoculation procedures.* Due to a range of circumstances, it may not always be possible to inoculate every production fermentation with inoculum in optimum condition. The scale-down approach can be used to predict the consequences of such events by mimicking these situations in the laboratory, for example by storing inoculum or using inocula of different ages, thus preempting problems of variability in the full-scale process.
- *Number of generations.* An industrial-scale fermentation requires a greater number of generations than does a laboratory one; this may place more severe stability criteria on the process strain than may have been appreciated on the small scale. The industrial situation may be modeled in the laboratory by using serial subculture to ensure that the strain is sufficiently stable. This approach is particularly pertinent in the development of recombinant fermentations.

The issues that are more difficult to address at a scaled-down level are those associated with the mode of growth and lack of homogeneity on the large scale. Fed-batch culture is by far the most common production method and thus, ideally, it should be incorporated into laboratory-scale experimentation at an early stage. However, achieving fed-batch on a small scale is a significant challenge. The major difference between laboratory and large-scale culture is that the power input achievable on the small scale cannot be matched at the large scale, giving rise to problems of heterogeneity in the production process. Furthermore, it can be difficult to measure the range of conditions experienced by a process organism in a production reactor so that even full appreciation of the problem itself can be problematic.

Two approaches have been used to achieve effective scale-down:

1. The development of small-volume reactors that enable early experimentation to be conducted in industrially relevant conditions. The conventional small-volume reactor for microbial fermentation is the shake flask—frequently 250 cm^3

containing 50 cm^3 of medium—and similar systems have been developed for animal cell culture. While shake-flasks are cheap to run they are inconvenient to sample and offer very limited opportunity to control culture parameters. Microwell (microtiter) plates are a very attractive means of miniaturizing culture systems and the possibility of having 96 "fermenters" in one plastic plate is obviously tempting. The automation associated with microwell plates also facilitates their use—the plates may be shaken, absorbance can be monitored using plate-readers and robotic devices can be used to make additions and take samples. Krause et al. (2010) and Siurkus et al. (2010) used the EnBase® technology system to achieve fed-batch culture in shake flasks and microwell plates at a very early product development stage of heterologous protein production by *E. coli*. As discussed in Chapter 6, the EnBase® system uses starch as the sole carbon source and the activity of added amylase to gradually release glucose to the organism. The growth rate is then controlled by the glucose release rate. Thus, this approach is only suitable for organisms that are amylase nonproducers and are dependent on the added amylase to generate glucose from starch. These workers demonstrated that growth rate controlled culture could be achieved in microwells and shake flasks and resulted in lower oxygen uptake rates, lower acetate production and less pH drift resulting in improved protein production. Thus, the cultures were controlled such that their metabolic rate was within the control capacity of the systems. Furthermore, the optimum conditions identified using these systems were commensurate with the use of fed-batch culture in larger scale cultures later in the process development program.

A number of academic and commercial fermenter manufacturers have developed bench-scale fermenters that may be run in parallel and have been designed for both microbial and animal cell culture These range in size from a few cm^3 to several hundred cm^3, utilizing either single use, presterilized vessels or conventional re-usable vessels. The incorporation of fluorescence monitoring of pH and dissolved oxygen, the use of robotic systems to make additions (and thus control pH and foam) and the possibility of fed-batch operation makes such systems attractive propositions for scale-down studies. Bareither et al. (2015) reported that the use of a fully instrumented system accommodating 24 disposable 250 cm^3 reactors with robotic addition and sampling could execute a set of experiments up to three to five times faster than a conventional approach (Fig. 9.26). Thus, the significant investment required for such a system may be worthwhile to reduce significantly the development time to get a product to the market place.

2. The development of small-scale simulators that mimic the heterogeneous conditions of a large-scale reactor. The objective of this approach is to investigate the reaction of a process organism to the conditions that may be experienced in the large-scale vessel: primarily the problem of heterogeneity. In a very detailed analysis of the scale-up of *E. coli* and yeast fermentations, Junker (2004) reported mixing times of 2–4 min in a 19 m^3 (19000 dm^3) fermenter while dissolved oxygen was completely utilized within a minute if it were not replenished. Thus, it is inevitable that the process organism will experience a cyclic "feast and famine"

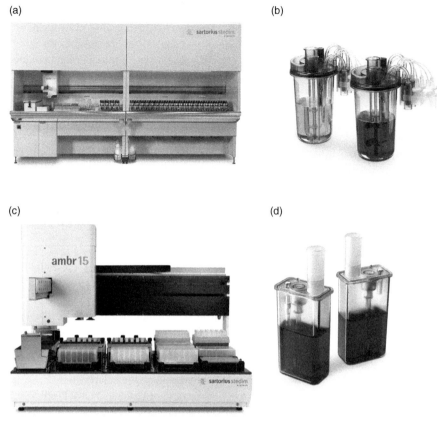

FIGURE 9.26 Single-Use "Scaled-Down" Reactors

(a) Sartorius ambr 250 system incorporating either 12 or 24 disposable reactors.
(b) Disposable 250 cm³ microbial fermenter for the ambr 250. (c) Sartorius ambr 15 system incorporating 24 or 48 disposable reactors (d) Disposable 15 cm³ cell culture reactor for the ambr 15.

as it passes through the gas entrainment zone of the fermenter and out into the more remote parts of the vessel before it returns in an anaerobic state to conditions of oxygen excess. Furthermore, the addition of feeds (substrate addition and pH control) almost invariably occurs at the top of the vessel. When it is considered that glucose may be fed at the top of a fed-batch fermenter at a concentration of 500 g dm⁻³ (Enfors et al., 2001) it is not surprising that a significant concentration gradient exists in the vessel. Enfors et al. showed marked compartmentalization in a 22 m³ fermenter stirred with four Rushton turbines. Much higher glucose concentrations were found in the addition zone compared with the bulk of the medium. It will be recalled that the rationale of a fed-batch process is to control the growth rate by limiting the nutrient available in the feed, yet in a large-scale vessel the organism will again be presented with a huge variation in feed concentration depending on its position in the reactor. These varying conditions will not only result in changes in metabolic flux through pathways and variation in growth rate but also transient changes in gene expression. Thus, it is important not to be misled by the apparent homogeneity witnessed in small-scale vessels and that deviation from the predicted performance of a scaled-up fermentation may be a reflection of the complex physiological periodicity being experienced by the process organism.

Hewitt and Nienow (2007), Neubauer and Junne (2010) and Takors (2012) have all reviewed the use of small-scale simulators of large-scale processes. These simulations of the heterogeneity of production-scale fermenters have been achieved by either making rapid pulse changes to the conditions within a vessel or by designing systems that are inherently heterogeneous—for example, plug flow reactors or multi-stage reactors. In a plug-flow system, the culture is introduced at the bottom of a vertical, unmixed tubular reactor and is pumped through it resulting in the concentration of the substrate decreasing along its height, while the concentration of biomass increases. The use of multi-stage reactors involves the culture being cycled from one vessel to another (or several in series) before being returned to the "mother" vessel. The subsidiary reactor may also be replaced with a plug-flow fermenter. Such systems rely on rapid sampling and sample treatment that can immortalize the cellular state such that the effects of the changes can be assayed. Schematic representations of these scaled-down fermenters are shown in Fig. 9.27.

The following consideration of a small selection of experiments reported in the literature gives an indication of the value of small-scale simulations in unraveling the complex interactions in a large-scale vessel. Enfors et al. (2001) reported the results of a major collaborative European project on the interaction between reactor fluid dynamics and the physiology of the process organism. These workers investigated the performance of an *E. coli* fed-batch culture that yielded 53 g dm^{-3} biomass in a 5 dm^3 fermenter but only 32 g dm^{-3} in a 22 m^3 vessel. Interestingly, the viability at the small scale was 84% compared with >99% on the large scale. The biomass discrepancy was attributed to the poor mixing in the large-scale vessel resulting in significant pH, dissolved oxygen and glucose gradients. Acetate and formate were also produced in the large-scale vessel indicating some anaerobic metabolism. The behavior of the organism was investigated in a simulation reactor of the type shown in Fig. 9.27b—a stirred reactor coupled with a plug flow vessel to which the glucose and pH feeds were administered. Thus, the stirred vessel represented the well-mixed zone of the large fermenter and the plug-flow the poorly mixed addition zone. The simulation gave very similar results to the performance in the large-scale fermenter (lower biomass and higher viability coupled with acetate and formate production) supporting the hypothesis that heterogeneity at the large scale affected the organism's physiology. Furthermore, the transcription of stress-related genes occurred in the plug-flow element after only 14 s but stopped in the well-mixed reactor section. Thus, it is envisaged that such changes in gene expression may occur between the

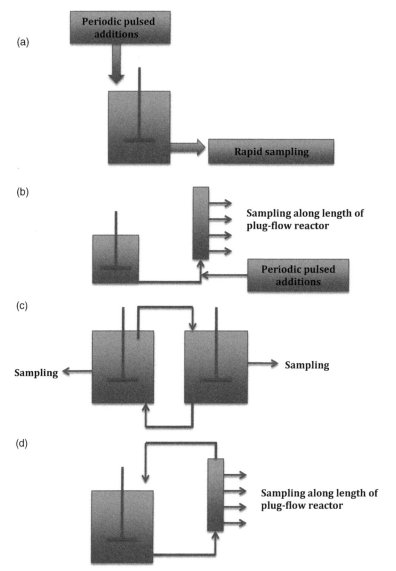

FIGURE 9.27 Scaled-Down Model Systems Simulating Heterogeneity Experienced in the Large-Scale Fermenter

Periodic pulsed feeding to multistage systems: (a) Stirred fermenter. (b) Plug-flow. (c) Stirred fermenters. (d) Stirred + plug-flow.

high glucose/low oxygen zones and low glucose/high oxygen zones in the production vessel. The improved viability under the heterogeneous (both large and simulated) conditions is a fascinating observation that may be explained by the induction of the stress responses (see Chapter 2) that made the cells more robust. Thus, heterogeneity is not necessarily "all a bad thing."

Kaβ et al. (2014) employed a two-stage simulator (a stirred tank reactor coupled with a plug-flow reactor) to investigate susceptibility of *C. glutamicum* to oscillations in dissolved oxygen and substrate concentrations. *C. glutamicum* is used extensively for the commercial production of amino acids. Substrate limitation coupled with fully aerobic conditions were provided in the stirred vessel while substrate excess and oxygen limitation occurred in the plug-flow reactor. The organism reacted remarkably quickly to the changing conditions, organic acids (lactate and succinate) being detected after 20 s of oxygen limitation. The accumulated organic acids were rapidly cometabolized with glucose in the aerobic sector accompanied by a rapid increase in respiratory activity. This accommodation of rapidly changing conditions was not accompanied by any changes in protein concentration or mRNA levels nor was lysine or biomass production affected. Thus, *C. glutamicum* appears to be remarkably tolerant of heterogeneous conditions and was described by the authors as "currently the most robust aerobic industrial microorganism in the field".

The supposed susceptibility of animal cells to hydrodynamic stress has been discussed in depth by Nienow (2006) who showed that shear sensitivity should not be considered a major problem. Nienow et al. (2013) used a scaled-down stirred tank reactor linked to a plug flow reactor to study the effect of fluid dynamic stress on Chinese hamster ovary (CHO) cells producing IgG protein. Stress was applied both in the stirred tank vessel by increased agitation rate and in the plug-flow reactor by pumping through narrow flow constrictions. Regardless of the method used, no effect was observed on cell density, product titre or product quality.

APPROACHES TO SCALE-UP OF AERATION AND AGITATION

It can be appreciated from the previous discussion that scale-down experiments can give an insight into the factors limiting a full-scale fermentation and indicate the approach to be adopted in the scale-up. Furthermore, employing conditions in both product and process development which are commensurate with those feasible on the production scale will facilitate the progression of the fermentation to full-scale. A key issue in this respect is the conflict between maintaining both oxygen transfer and mixing times constant at both scales. Doran (2013) derived the following equation describing the relationship between the power input required at different scales to maintain a constant mixing time:

$$P_L = P_S \left(\frac{V_L}{V_S} \right)^{5/3}$$

where P_L and P_S are the power inputs for the large and small scales respectively; V_L and V_S are the fermenter working volumes at the large and small scales respectively.

It follows that scaling up from a 1 m³ vessel to a 100 m³ vessel would require the power input of the larger vessel to be approximately 2000 times that of the pilot fermenter—an objective which is neither economically nor technically possible. Thus, it is important that the concept of compromise is fully appreciated before embarking on a scale-up exercise.

A prerequisite of the methods discussed here is that the small and large vessels are geometrically similar. However, this goal is rarely achieved and a range of acceptable ratios were proposed by Einsele (1978):

$D/D_T = 0.3 - 0.45$ (where D is the impeller diameter and D_T the fermenter diameter)
$H/D_T = $ or < 2.0 (where H is the height of the liquid in the fermenter)

The number of impellers is 2 or 3.

Details of industrial installations are rare in the literature but two excellent papers by Junker (2004 and 2007) discuss scale-up systems at Merck. Junker (2004) suggested that the parameter relating liquid height to fermenter diameter is better expressed as the "tangent to tangent" tank height (H_{t-t}) to fermenter diameter. The values of these ratios for vessels employed at Merck (volumes ranging from 30 to 19000 dm³) were:

$D/D_T = 0.33$ for vessels with Rushton turbines
$D/D_T = 0.46$–0.5 for vessels with Maxflo T axial flow impellers
$D/D_T = 0.43$–0.5 for vessels with A315 Lightnin axial flow impellers
$H_{t-t}/D_T = 1.7$–2-1 for vessels ranging from 30 to 1900 dm³ and 2.9 for a 19000 dm³ vessel

The approach to the scale-up problem is threefold:

1. The identification of the principal environmental domain affected by aeration and agitation in the fermentation, for example, oxygen concentration, shear, bulk mixing.
2. The identification of a process variable (or variables) which controls the identified environmental domain.
3. The calculation of the value of the process variable to be used on the large scale that will result in the replication of the same environmental conditions on both scales.

The process variables, which affect mixing and mass transfer, are summarized in Table 9.7 (Oldshue, 1985; Scragg, 1991; Junker, 2004). Thus, if dissolved oxygen concentration is perceived as the over-riding environmental condition then power consumption per unit volume and volumetric air flow rate per unit volume should be maintained constant on scale-up. However, as a result, the other parameters will not be the same in the larger scale and, therefore, neither will the environmental factors that they influence. This phenomenon is well illustrated by Oldshue's much-quoted example summarized in Table 9.8 where a 125-fold increase in scale is represented. If power consumption per unit volume is kept constant then impeller tip speed (ie, shear) increases and flow min⁻¹ vol⁻¹ (ie, mixing) decreases. If mixing is kept constant,

Table 9.7 The Effect of Process Variables on Mass Transfer or Mixing Characteristics

Process Variable	Mass Transfer or Mixing Characteristic Affected
Power consumption per unit volume	Oxygen transfer rate
Volumetric air flow rate	Oxygen transfer rate
Superficial air velocity	Oxygen transfer rate
Constant dissolved oxygen concentration	Oxygen transfer rate
Aeration number	Oxygen transfer rate
Impeller tip speed	Shear rate
Pumping rate	Mixing time
Reynolds number (see previous section)	Heat transfer

Table 9.8 The effect of the choice of scale-up criteria on operating conditions in the scaled-up vessel. Based on scale-up from 80 dm^3 to 10^4 dm^3

Criterion Used in Scale-Up	Effect on the Operating Conditions on the Large Scale (Large Scale Value/Small Scale Value)			
	P	P/V	Flow min^{-1} vol^{-1}	ND_i
P/V	125.0	1.0	0,34	1.7
Flow min^{-1} vol^{-1}	3125.0	25.0	1.0	5.0
ND_i (Impeller tip speed)	25.0	0.2	0.2	1.0
Reynolds number	0.2	0.0016	0.04	0.2

Based on Oldshue, 1985.

an enormous (and totally uneconomic) increase in power is required and shear increases fivefold. If impeller tip speed (shear) is kept constant then power consumption (hence, K_La) and mixing decrease. As shown earlier, by Doran's analysis, these figures also illustrate that it is economically impossible to maintain the same degree of mixing on scale-up and, therefore, a decrease in yield may be due to mixing anomalies.

The most important environmental domains affected by aeration and agitation for the majority of fermentations are oxygen concentration and shear. Thus, the most widely used scale-up criteria are the maintenance of a constant K_La or constant shear conditions.

As discussed earlier, shear is very important for mycelial fermentations as it has a significant effect on morphology but plays little role in unicellular bacterial and yeast processes. Understanding of the effect of shear forces on mycelial morphology has increased significantly in recent years (Serrano-Carreon et al., 2015). However, if there is a poor comprehension of the relationship between shear and morphology for the fermentation being scaled-up, then constant impeller tip speed can be used as "a rule of thumb" (Junker, 2004) and a good example is provided by the scale-up of the vancomycin fermentation (Junker, 2007). Impeller tip speed (ITS in m s^{-1}) is given by the equation:

$$ITS = \pi ND \qquad (9.42)$$

where N is impeller rotational speed (revolutions per second).

Junker (2004) reported impeller tip speeds of 4.7–7.6 m s^{-1} in fermenters ranging in size from 30 to 19000 dm^3, values in broad agreement for industrial fermenters cited by Einsele (1978).

Constant K_La may be achieved on the basis of constant power consumption per unit volume and constant volumetric air-flow rate. The operating variable dictating constant power consumption in geometrically similar vessels is the agitator speed. The agitator speed on the large scale is then calculated from the correlations between K_La and power consumption and between power consumption and operating variables. An example of this approach is given in the previous section describing the effects of operating variables on power consumption.

Hubbard (1987) and Hubbard, Harris, and Wierenga (1988) summarized their procedure for scaling up both Newtonian and non-Newtonian fermentations and proposed two methods to determine the large-scale conditions:

Method 1—Based on a constant K_La

1. Determine the volumetric air flow rate (Q) on the large scale based on maintaining a constant Q/V (volume of air volume^{-1} medium minute^{-1}, where V = the working volume of the fermenter)
2. Calculate the agitator speed that will give the same K_La on the large scale; this is achieved using the correlations between power consumption and N and between K_La and power consumption.

Method 2—Based on constant impeller tip speed

1. Calculate the agitator speed keeping the impeller tip speed constant, πND_i.
2. Calculate Q from power correlations and K_La correlations.

The accuracy of these scale-up techniques is only as good as the power and K_La correlations, so it is worth expending some considerable time to test the validity of potential correlations for the fermentation in question. Examples of these correlations are given in Table 9.6. Interestingly, Junker (2004) calculated the values of the exponent of P_g/V for a range of *E. coli* and yeast fermentations and showed them to range from 0.96–0.98. She emphasized that many of the earlier correlations were based on antibiotic (mycelial) fermentations whereas more recent scale-up activity was related to heterologous protein production utilizing unicells, thus the choice of exponent is very much dependent on the type of fermentation.

When the parameter for scale-up has been chosen, and the large-scale conditions predicted, the next step is to run test the prediction at the increased scale.

Table 9.9 Effect of Fermenter Volume on Specific Volumetric Air Flow Rate (Q/V) and Superficial Air Velocity

Fermenter volume (dm³)	Q/V (volume of air volume⁻¹ medium min⁻¹)	Superficial air velocity (cm s⁻¹)
800	1.0	2.0
1900	1.0	2.8
19000	1.0	8.1

Data from Junker, 2004.

Other conditions may then have to be adjusted to account for the issues raised in Table 9.8. In particular, scale-up of the air flow rate based on Q/V (volume of air volume^{-1} medium min^{-1}) should also be reviewed because Q/V is related to the cube route, whereas the escape of air from the fermentation fluid is related to the square route. Thus, superficial air velocity (Q/ cross sectional area of the fermenter) takes into account the surface area from which the air escapes. As can be seen from Table 9.9 while Q/V is kept constant with increasing scale so superficial air velocity increases. As a result, both foaming and flooding of the impeller may occur at the larger scale—thus, specific air flow rate (Q/V) tends to be reduced with increasing scale.

Junker (2004) reported the scale-up of both *E. coli* and yeast fermentations based on maintaining similar minimum dissolved oxygen levels. Thus, cascade control of agitation, pressure and air flow rate may be used at the large scale to maintain the required C_L value.

SCALE-UP OF AIR-LIFT REACTORS

Bubble columns and air-lift vessels tend to be scaled-up on the basis of geometric similarity and superficial gas velocity (Scragg, 1991). The relationship between $K_L a$, superficial gas velocity and viscosity for a bubble column is given by the equation:

$$K_L a = C \cdot V_S^a \cdot \mu_a^b$$

where V_s is the superficial air velocity (m s^{-1}), μ_a is the apparent viscosity (Pa s).

For air-lift reactors the equation can be modified to take into account the ratio of the cross-sectional areas of the down-comer to riser (A_D/A_R):

$$K_L a = C \cdot V_S^a \cdot \mu_a^b \cdot \left(1 + {A_D}\middle/{A_R}\right)^c \tag{9.43}$$

The constant C depends on geometric similarity and values for the exponents a, b and c are given in Table 9.10 (Garcia-Ochoa & Gomez, 2009). It can be seen that the exponents for both μ_a and $(1 + A_D/A_R)$ are negative and thus a decrease in $K_L a$ is predicted with an increase in viscosity and increase in the ratio of the cross sectional areas of downcomer to riser. The riser is the zone of aeration whereas the downcomer

Table 9.10 Exponent Values for V_S; μ_a and $(1 + A_D/A_R)$ (Garcia-Ochoa & Gomez, 2009)

References	System	V_S	μ_a	$1 + A_D/A_R$
Deckwer, Furckhart and Zoll (1974); Deckwer et al. (1982)	Water and water +electrolyte	0.7–1.3		
Jackson and Shen (1978)	Water	1.2		
Bello Robinson and Moo-Young (1984)	Water and water +electrolyte	0.9		−1.0
Godbole Shumpe, Shah and Carr (1984)	CMC	0.44–0.59	−0.8 to −1.0	
Al-Masry and Abasaeed (1998)	Newtonian	0.76	−0.76	−2.41
Sanchez et al. (2000)	Tap and sea water	0.94–1.17		

is unaerated. Also, the narrower the down-comer the higher the descending velocity of the culture resulting in the fluid spending less time in the nonaerated section. Thus, the organism, as well as the $K_L a$, will be affected by this ratio as it is exposed to extremes of oxygen levels in the riser and downcomer. It is critical that the response of the organism to these variations is tested in the laboratory in a scaled-down version of the fermentation.

The major difference between scales of in these systems will be the height of the vessels resulting in increased pressure at the base of the larger vessel. This would result in higher oxygen and carbon dioxide solubility that would give a higher $K_L a$ but might result in carbon dioxide inhibition.

REFERENCES

Aiba, S., Humphrey, A. E., & Milus, N. (1973). *Biochemical Engineering*. London: Academic Press.

Al-Masry, W. A., & Abasaeed, A. E. (1998). On the scale-up of external loop airlift reactors: Newtonian systems. *Chemical Engineering Science*, *53*, 4085–4094.

Anderson, C., LeGrys, G. A., & Solomons, G. L. (1982). Concepts in the design of large-scale fermenters for viscous culture broths. *Chemical Engineering*, *377*, 43–49.

Antoniewicz, M. R. (2015). Methods and advances in metabolic flux analysis: a mini-review. *Journal of Industrial Microbiology and Biotechnology*, *42*, 317–325.

Arjunwadkar, S. J., Sarvanan, A. B., Kulkarni, P. R., & Pandit, A. B. (1998). Gas liquid mass transfer in dual impeller bioreactor. *Biochemical Engineering Journal*, *1*, 99–106.

Bai, Y., Zhou, P. -P., Fan, P., Tong, Y., Wang, H., & Yu, L. -J. (2015). Four-stage dissolved oxygen strategy based on multi-scale analysis for improving spinosad yield by *Saccharopolyspora spinosa ATCC48460*. *Microbial Biotechnology*, *8*(3), 561–568.

Bakker, M.H.I. (1998). Impeller assembly with asymmetric concave blades. US Patent 5791780.

Bakker, A. (2000). A new gas dispersion impeller with vertically asymmetric blades. In *The Online CFM Book*. Available from: http://www.bakker.org/cfm.

Banks, G. T. (1977). Aeration of moulds and streptomycete culture fluids. In A. Wiseman (Ed.), *Topics in enzyme and fermentation biotechnology* (pp. 72–110). (Vol. 1). Chichester: Ellis Horwood.

Banks, G. T. (1979). Scale-up of fermentation processes. In A. Wiseman (Ed.), *Topics in enzyme and fermentation biotechnology* (pp. 170–267). (Vol. 3). Chichester: Ellis Horwood.

Bareither, R., Goldfeld, M., Kistler, C., Tait, A., Bargh, N., Oakeshott, R., O'Neill, K., Hoshan, L., & Pollard, D. (2015). Automated disposable small-scale bioreactor for high-throughput process development: implementation of the 24 bioreactor array. *Pharmaceutical Bioprocessing*, *3*(3), 185–197.

Bartholomew, W. H. (1960). Scale-up of submerged fermentations. *Advanced Applications of Microcomputer*, *2*, 289–300.

Bartholomew, W. H., Karrow, E. O., Sfat, M. R., & Wilhelm, R. H. (1950). Oxygen transfer and agitation in submerged fermentations. Mass transfer of oxygen in submerged fermentations of Streptomyces griseus. *Industrial and Engineering Chemistry*, *42*(9), 1801–1809.

Baumann, K., Maurer, M., Dragosits, M., Cos, O., Ferrer, P., & Mattanovich, D. (2008). Hypoxic fed-batch cultivation of *Pichia pastoris* increases specific and volumetric productivity of recombinant proteins. *Biotechnology Bioengineering*, *100*(1), 177–183.

Baumann, K., Adelantado, N., Lang, C., Mattanovich, D., & Ferrer, P. (2011). Protein trafficking, ergosterol biosynthesis and membrane physics impact recombinant protein secretion in *Pichia pastoris*. *Microbial Cell Factories*, *10*(93), 1–15.

Becker, J., Zelder, O., Hafner, S., Schroder, H., & Wittmann, C. (2011). From zero to hero—design-based systems metabolic engineering of *Corynebacterium glutamicum* for L-lysine production. *Metabolic Engineering.*, *13*(2), 159–168.

Bell, G. H., & Gallo, M. (1971). Effect of impurities on oxygen transfer. *Process Biochemistry*, *6*(4), 33–35.

Bello, R. A., Robinson, C. W., & Moo-Young, M. (1984). Liquid circulation and mixing characteristics of air-lift contactors. *Canadian Journal of Chemical Engineering*, *62*, 573–577.

Berovic, M. (1992). Foam problems in fermentation processes. In R. N. Mukherjee (Ed.), *Downstream process biotechnology* (pp. 248–261). New Delhi: Tata-McGraw-Hill.

Boswell, C. D., Nienow, A. W., Gill, N. K., Kocharunchitt, S., & Hewitt, C. J. (2003). The impact of fluid mechanical stress on *Saccharomyces cerevisiae* cells during continuous cultivation in an agitated, aerated bioreactor; its implication for mixing in the brewing process and aerobic fermentations. *Transactions of the Institute of Chemical Engineering C.*, *81*, 23–31.

Buckland, B.C. (1993). Mevinolin production. Paper presented at the *Soc. Gen. Microbiol. 124th Meeting*, University of Kent, Canterbury, January 1993.

Buckland, B.C., Gbewonyo, K., Jain, D., Glazomitsky, K., Hunt, G. and Drew, S.W. (1988). Oxygen transfer efficiency of hydrofoil impellers in both 800L and 1900L fermenters. In *Proceedings of the second international conference on bioreactor fluid dynamics*, pp. 1–16 R. King (Ed.). London: Elsevier.

Buckland, B. C., Gbewonyo, K., Hallada, T., Kaplan, L., & Masurekar, P. (1989). In A. L. Demain (Ed.), *Novel microbial products in medicine and agriculture* (pp. 161–169). Amsterdam: Elsevier.

Cai, M., Zhang, Y., Hu, W., Yu, Z., Zhou, W., Jiang, T., Zhou, X., & Zhang, Y. (2014). Genetically shaping morphology of the filamentous fungus *Aspergillus glaucus* for production of the antitumour polyketide aspergiolide A. *Microbial Cell Factories*, *13*, 73–83.

Calabrese, R. V., Kresta, S. M., & Liu, M. (2014). Recongnizing the 21 most influential contributions to mixing research. *Chemical Engineering Progress*, *January*, 20–29.

Campbell, B. S., Siddique, A-B. M., McDougall, B. M., & Seviour, R. J. (2004). Which morphological forms of the fungus Aureobasidium pullulans are responsible for pullulan production? *FEMS Microbiology Letters*, *232*(2), 225–228.

Casas, J. A., Santos, V. E., & Garcia-Ochoa, F. (2000). Xanthan gum production under several operational conditions: molecular structure and rheological properties. *Enzyme and Microbial Biotechnology*, *26*, 282–291.

Casson, N. (1959). *Rheology of disperse systems*. Oxford: Pergamon Press.

Chamsartra, S., Hewitt, C. J., & Nienow, A. W. (2005). The impact of fluid mechanical stress on *Corynebacterium glutamicum* during continuous cultivation in an agitated bioreactor. *Biotechnology Letters*, *27*, 693–700.

Chandrasekharan, k., & Calderbank, P. H. (1981). Further observations on the scale-up of aerated mixing vessels. *Chemical Engineering Science*, *36*(8), 819–823.

Chapman, C. M., Nienow, A. W., Cooke, M., & Middleton, J. C. (1983). Particle–gas–liquid mixing in stirred vessels. *Chemical Engineering Research and Design*, *61*(3), 167–181.

Charles, M. (1978). Technical aspects of the rheological properties of microbial cultures. *Advances in Biochemical Engineering*, *8*, 1–62.

Chen, N. Y. (1990). The design of airlift fermenters for use in biotechnology. *Biotechnology and Genetic Engineering Review*, *8*, 379–396.

Cooke, M., Middleton, J.C., & Bush, J.R. (1988). Mixing and mass transfer in filamentous fermentations. In *Proceedings of the second international conference on bioreactor fluid dynamics* (pp. 37–64). R. King (Ed.). London: Elsevier.

Cooney, C. L. (1979). Conversion yields in penicillin production: theory versus practice. *Process Biochemistry*, *14*(5), 31–33.

Cooper, C. M., Fernstrom, G. A., & Miller, S. A. (1944). Performance of agitated gas-liquid contacters. *Industrial Engineering and Chemistry*, *36*, 504–509.

Cui, Y. Q., van der Lans, R. G. J. M., & Luyben, K. C. M. (1998). Effects of dissolved oxygen tension and mechanical forces on fungal morphology in submerged fermentation. *Biotechnology Bioengineering*, *54*(8), 400–419.

Cutter, L. A. (1966). Flow and turbulence in a stirred tank. *American Institute of Chemical Engineers Journal.*, *12*, 34–44.

Darlington, W. A. (1964). Aerobic hydrocarbon fermentation—A practical evaluation. *Biotechical Bioengineering*, *6*(2), 241–242.

Davies, S. M., Gibilaro, L. G., Middleton, J. C., Cooke, M., & Lynch, P. M. (1985). The application of two novel techniques for mass transfer coefficient determination to the scale up of gas sparged vessels. *Proceedings of the European conference on mixing*, *5*, 27–34.

Deckwer, W. D., Furckhart, R., & Zoll, R. (1974). Mixing and mass transfer in a tall bubble column. *Chemical Engineering Science*, *29*, 2177–2188.

Deckwer, W. D., Nguyen-Tien, K., Schumpe, A., & Serpe-men, Y. (1982). Oxygen mass transfer into aerated CMC solutions in a bubble column. *Biotechnical Bioengineering*, *24*, 461–481.

Deindoerfer, F. H., & Gaden, E. L. (1955). Effects of liquid physical properties on oxygen transfer in penicillin fermentation. *Applied Microbiology*, *3*, 253–257.

Deindoerfer, F. H., & West, J. M. (1960). Rheological examination of some fermentation broths. *Journal of Microbiology Biochemical and Technological Engineering*, *2*, 165–175.

Dion, W. M., Casilli, A., Sermonti, G., & Chain, E. B. (1954). The effect of mechanical agitation on the morphology of *Penicillium chrysogenum* Thom in stirred fermenters. *Rend. 1st Super Sanita*, *17*, 187–205.

Dixit, P., Mehta, A., Gahlawat, G., Prasad, G. S., & Choudry, A. R. (2015). Understanding the effect of interaction among aeration, agitation and impeller positions on mass transfer during pullulan fermentation by *Aureobasidium pullulans*. *RSC Advances*, *5*, 38984–38994.

Doran, P. M. (2013). *Bioprocess engineering principles* (2nd ed.). Oxford: Academic Press.

Driouch, H., Roth, A., Dersch, P., & Wittman, C. (2010). Optimized bioprocess for production of fructofuranosidase by recombinant *Aspergillus niger*. *Applied Microbiology and Biotechnology*, *87*, 2011–2024.

Driouch, H., Hansch, R., Wucherpfennig, W., Krull, R., & Wittmann, C. (2012). Improved enzyme production by bio-pellets of *Aspergillus niger*. Targeted morphology engineering using titanate microparticles. *Biotechnology and Bioengineering*, *109*(2), 462–471.

Einsele, A. (1978). Scaling-up bioreactors. *Process Biochemistry*, *7*, 13–14.

Enfors, S. O., Jahic, M., Rozkov, A., Xu, B., Hecker, M., Jurgen, B., Kruger, E., Schweder, T., Hamer, G., O'Beirne, D., et al. (2001). Physiological responses to mixing in large scale bioreactors. *Journal of Biotechnology*, *85*, 175–185.

Formenti, L. R., Norregaard, A., Bolic, A., Hernandez, D. Q., Hagemann, T., Heins, A. -L., Larsson, H., Mears, L., Mauricio-Iglesias, M., Kruhne, U., & Gernaey, K. V. (2014). Challenges in industrial fermentation technology research. *Biotechnology Journal*, *9*, 727–738.

Figueiredo, L. M., & Calderbank, P. H. (1979). The scale-up of aerated mixing vessels for specified oxygen dissolution rates. *Chemical Engineering Science*, *34*, 1333–1338.

Fox, R.I. (1978). The applicability of published scale-up criteria to commercial fermentation processes. *Proceedings of first European Congress Biotechnology*, Part 1, pp. 80–83.

Fu, Z., Verderame, T. D., Leighton, J. M., Sampey, B. P., Appelbaum, E. R., Patel, P. S., & Aon, J. C. (2014). Exometabolome analysis reveals hypoxia at the up-scaling of a *Saccharomyces cerevisiae* high-cell density fed-batch biopharmaceutical process. *Microbial Cell Factories*, *13*(32), 1–22.

Gabelle, J. -C., Augier, F., Carvalho, A., Rousset, R., & Morchain, J. (2011). Effect of tank size on K_La and mixing time in aerated stirred reactors with non-Newtonian fluids. *Canadian Journal of Chemical Engineering*, *999*, 1–12.

Gabelle, J. -C., Jourdier, E., Licht, R. B., Ben Chaabane, F., Henaut, I., Morchain, J., & Augier, F. (2012). Impact of rheology on the mass transfer coefficient during the growth phase of *Trichoderma reesei* in stirred bioreactors. *Chemical Engineering Science*, *75*, 408–417.

Gagnon, H., Lounes, M., & Thilbault, J. (1998). Power consumption and mass transfer in agitated gas-liquid columns: A comparative study. *Canadian Journal of Chemical Engineering*, *76*, 379–389.

Gallindo, E., & Nienow, A. W. (1992). Mixing of highly viscous simulated fermentation broths with the Lightnin' A-315 impeller. *Biotechnology Progress*, *8*, 233–239.

Garcia-Ochoa, F., & Gomez, E. (1998). Mass transfer in stirred tank reactors for xanthan solutions. *Biochemical Engineering Journal*, *1*, 1–10.

Garcia-Ochoa, F., & Gomez, E. (2001). Estimation of oxygen mass transfer coefficient in stirred tank reactors using artificial neuronal networks. *Enzyme and Microbial Technology*, *28*, 560–569.

Garcia-Ochoa, F., & Gomez, E. (2009). Bioreactor scale-up and oxygen transfer rate in microbial processes: An overview. *Biotechnology Advances*, *27*, 153–176.

Gao, D., Zeng, J., Yu, X., Dong, T., & Chen, S. (2014). Improved lipid accumulation by morphology engineering of oleaginous fungus *Mortierella isabellina*. *Biotechnology and Bioengineering, 111*(9), 1758–1766.

Gbewonyo, K, Dimasi, D., & Buckland, B.C. (1986). The use of hydrofoil impellers to improve oxygen transfer efficiency in viscous mycelial fermentations. In *International Conference on Bioreactor Fluid Dynamics* (pp. 281–299). Cranfield, UK: BHRA.

Ghojavand, H., Bonakdarpour, B., Heydarian, S. M., & Hamedi, J. (2011). The inter-relationships between inoculum concentration, morphology, rheology and erythromycin productivity in submerged culture of *Saccharopolyspora erythraea*. *Brazilian Journal of Chemical Engineering, 28*(4), 565–574.

Godbole, S. P., Shumpe, A., Shah, Y. T., & Carr, N. L. (1984). Hydrodynamics and mass transfer in non-Newtonian solutions in a bubble column. *American Institute of Chemical Engineering Journal, 30*, 213–220.

Goldberg, I., Rock, J. S., Ben-Bassat, A., & Máteles, R. I. (1976). Bacterial yields on methanol, methylamine, formaldehyde and formate. *Biotechnical and Bioengineering, 18*, 1657–1668.

Gourich, B., Vial, Ch., Azher, N. El., Belhaj Soulami, M., & Ziyad, M. (2008). Influence of hydrodynamics and probe response on oxygen mass transfer measurements in a high aspect ratio bubble column reactor: Effect of the coalescence behaviour of the liquid phase. *Biochemical Engineering Journal, 39*, 1–14.

Hall, M. J., Dickinson, S. D., Pritchard, R., & Evans, J. I. (1973). Foams and foam control in fermentation processes. *Progress in Industrial Microbiology, 12*, 171–234.

Hasegawa, S., Uematsu, K., Natsuma, Y., Suda, M., Hiraga, K., Tojima, T., Inui, M., & Yukawa, H. (2012). Improvement of the redox balance increases L-valine production by *Corynebacterium glutamicum* under oxygen deprivation conditions. *Applied and Environmental Microbiology, 78*(3), 691–699.

Heijnen, J. J., Riet, K. W., & Wolthuis, A. J. (1980). Influence of very small bubbles on the dynamic $K_L a$ measurement in viscous gas-liquid systems. *Biotechnical Bioengineering, 22*, 1945–1956.

Heijnen, J. J., & Van't Riet, K. (1984). Mass transfer, mixing and heat transfer phenomena in low viscosity bubble column reactors. *Chemical Engineering Journal, 28*, B21.

Heineken, F. G. (1970). Use of fast-response dissolved oxygen probes for oxygen transfer studies. *Biotechnical and Bioengineering, 12*, 145–154.

Heineken, F. G. (1971). Oxygen mass transfer and oxygen respiration rate measurements utilising fast response oxygen electrodes. *Biotechnical and Bioengineering, 13*, 599–618.

Hewitt, C. J., & Nebe-von-Caron, G. (2004). The application of multi-parameter flow cytometry to monitor individual microbial cell physiological state. In S. O. Enfors (Ed.), *Advances in Biochemical Engineering/Biotechnology: Physiological Stress Responses in Bioprocesses* (pp. 197–223). (Vol. 89). New York: Springer.

Hewitt, C. J., & Nienow, A. W. (2007). The scale-up of microbial batch and fed-batch fermentation processes. *Advances in Applied Microbiology, 62*, 105–136.

Hille, A., Neu, T. R., Hempel, D. C., & Horn, H. (2005). Oxygen profiles and biomass distribution in biopellets of *Aspergillus niger*. *Biotechnology and Bioengineering, 92*(5), 614–623.

Hirose, Y., & Shibai, H. (1980). Effect of oxygen on amino acids fermentation. In M. Moo-Young, C. W. Robinson, & C. Vezina (Eds.), *Advances in Biotechnology* (pp. 329–333). (Vol. 1). Toronto: Pergamon Press.

Hua, Q., Fu, P. C., Yang, C., & Shimizu, K. (1998). Microaerobic lysine fermentations and metabolic flux analysis. *Biochemical Engineering Journal, 2*, 89–100.

Hubbard, D. W. (1987). Scale-up strategies for bioreactors containing non-Newtonian broths. *Annals of New York Academy of Sciences*, *506*, 600–607.

Hubbard, D. W., Harris, L. R., & Wierenga, M. K. (1988). Scale-up for polysaccharide fermentation. *Chemical Engineering Progress*, *84*(8), 55–61.

Hughmark, G. A. (1980). Power requirements and interfacial area in gas-liquid turbine agitated systems. *Industrial & Engineering Chemistry Process Design and Development*, *19*, 638–645.

Inui, M., Murakami, S., Okino, S., Kawaguchi, H., Vertes, A. A., & Yukawa, H. (2004). Metabolic analysis of *Corynebacterium glutamicum* during lactate and succinate productions under oxygen deprivation conditions. *Journal of Molecular Microbiology and Biotechnology*, *7*, 182–196.

Jackson, M. L., & Shen, C. C. (1978). Aeration and mixing in deep tank fermentation systems. *American Institute of Chemical Engineering Journal*, *24*, 63–71.

Johnson, M. J. (1964). Utilisation of hydrocarbons by microorganisms. *Chemical Industry*, *36*, 1532–1537.

Junker, B. H., Mann, Z., & Hunt, G. (2000). Retrofit of CD-6 (Smith) impeller in fermentation vessels. *Applied Biochemistry and Biotechnology*, *89*, 67–83.

Junker, B. (2004). Scale-up methodologies for *Escherichia coli* and yeast fermentation processes. *Journal of Bioscience and Bioengineering*, *97*(6), 347–364.

Junker, B. (2007). Foam and its mitigation in fermentation systems. *Biotechnology Progress*, *23*, 767–784.

Justen, P., Paul, G. C., Nienow, A. W., & Thomas, C. R. (1996). Dependence of mycelial morphology on impeller type and agitation intensity. *Biotechnology and Bioengineering*, *52*, 672–684.

Kaβ, F., Hariskos, I., Michel, A., Brandt, H. -J., Spann, R., Junne, S., Wiechert, W., Neubauer, P., & Oldiges, M. (2014). Assessment of robustness against dissolved oxygen/substrate oscillations for *C, glutamicum* DM1933 in two –compartment bioreactor. *Bioprocess and Biosystems Engineering*, *73*, 1151–1162.

Kawase, Y., & Moo-Young, M. (1988). Volumetric mass transfer coefficients in aerated stirred tank reactors with Newtonian and non-Newtonian media. *Chemical Engineering Research and Development*, *66*, 284–288.

Kobayashi, T., Van Dedem, G., & Moo-Young, M. (1973). Oxygen transfer into mycelial pellets. *Biotechnical Bioengineering*, *15*, 27–45.

Krause, M., Ukkonen, K., Haataja, T., Ruottinen, M., Glumoff, T., Neubauer, A., Neubauer, P., & Vasala, A. (2010). A novel fed-batch based cultivation method provides high cell-density and improves yield of soluble recombinant proteins in shake culture. *Microbial Cell Factories*, *9*(11), 1–11.

Krull, R. and Bley, T. (2015). Filaments in Bioprocesses. Advances in Biochemical Engineering/Biotechnology (Vol. 149). Springer, Cham, Switzerland.

Kuenzi, M. T. (1978). Process design and control in antibiotic fermentation. In R. Hutter, T. Leisinger, Neusch, & W. Wehrli (Eds.), *Antibiotics and other secondary metabolites, biosynthesis and production, FEMS symp* (pp. 39–56). (Vol. 5). London: Academic Press.

Legrys, G.A. and Solomons, G.L. (1977). US patent application 23128.

Lilly, M. D. (1983). Problems in process scale-up. In L. J. Nisbet, & D. J. Winstanley (Eds.), *Bioactive microbial products 2. development and production* (pp. 79–90). London: Academic Press.

Linek, V., Sinkule, J., & Benes, P. (1991). Critical assessment of gassing-in method measuring $k_L a$ in fermenters. *Biotechnology Bioengineering*, *38*, 323–330.

Mateles, R. I. (1971). Calculation of the oxygen required for cell production. *Biotechnology Bioengineering, 13*(4), 581–582.

Mateles, R. I. (1979). The physiology of single cell protein (SCP) production. In A. T. Bull, D. C. Ellwood, & C. Ratledge (Eds.), *Society for General Microbiology Symposium 29, Microbial Technology: Current State and Future Prospects* (pp. 29–52). Cambridge: Cambridge University Press.

Mavituna, F., & Sinclair, C. G. (1985a). A graphical method for the determination of critical biomass concentration for non-oxygen limited growth. *Biotechnological Letters, 7*, 69–74.

Mavituna, F., & Sinclair, G. (1985b). Reply to a comment on 'A graphical method for the determination of critical biomass concentration for non-oxygen limited growth'. *Biotechnological Letters, 7*, 813–814.

McNeil, B., & Kristiansen, B. (1987). Influence of impeller speed upon the pullulan fermentation. *Biotechnology Letters, 9*, 101–104.

Metz, B., Kossen, N. W. F., & Van Suijdam, J. C. (1979). Rheology of mould suspensions. *Advance Biochemical Engineering, 11*, 103–156.

Metzner, A. B., & Otto, R. E. (1957). Agitation of non-Newtonian fluids. *American Institute of Chemical Engineering Journal, 3*(1), 3–10.

Metzner, A. B., Feehs, R. H., Ramos, H. L., Otto, R. E., & Toothill, J. D. (1961). Agitation of viscous Newtonian and non-Newtonian fluids. *American Institute of Chemical Engineering Journal, 7*, 3–9.

Michell, B. J., & Miller, S. A. (1962). Power requirements of gas-liquid agitated systems. *American Institute of Chemical Engineering Journal, 8*, 262–266.

Neubauer, P., Cruz, N., Glauche, F., Junne, S., Knepper, A., & Raven, M. (2013). Consistent development of bioprocesses from microliter cultures to the industrial scale. *Engineering in Life Sciences, 13*, 224–238.

Neubauer, P., & Junne, S. (2010). Scale-down simulators for metabolic analysis of large-scale bioprocesses. *Current Opinion in Biotechnology, 21*, 114–121.

Nielsen, J. (1996). Modelling the morphology of filamentous microorganisms. *Trends in Biotechnology, 14*, 438–443.

Nienow, A. W. (1990). Agitators for mycelial fermentations. *Trends in Biotechnology, 8*, 224–233.

Nienow, A. W. (1998). Hydrodynamics of stirred bioreactors. *Applied Mechanics Reviews, 51*(1), 3–32.

Nienow, A. W. (2006). Reactor engineering in large scale animal cell culture. *Cytotechnology, 50*, 9–33.

Nienow, A. W. (2014). Stirring and stirred-tank reactors. *Chemie-Ingenieur-Technik., 86*(12), 2063–2074.

Nienow, A. W., Wisdom, D. J., & Middleton, J. C. (1977). The effect of scale and geometry on flooding, recirculation and power in gassed stirred vessels. Paper Fl, Second European conference on mixing, March 1977, Cambridge.

Nienow, A. W., & Ulbrecht, J. J. (1985). Gas-liquid mixing in high viscosity systems. In J. J. Ulbrecht, & G. K. Patterson (Eds.), *Mixing of liquids by mechanical agitation* (pp. 203–235). New York: Gordons and Beach.

Nienow, A. W., Hunt, G., & Buckland, B. C. (1994). A fluid dynamic study of the retrofitting of large bioreactors: Turbulent flow. *Biotechnology Bioengineering, 44*, 1177–1186.

Nienow, A. W., Hunt, G., & Buckland, B. C. (1996). A fluid dynamic study using a simulated viscous, shear thinning broth of the retrofitting of large agitated bioreactors. *Biotechnology Bioengineering, 49*, 15–19.

Nienow, A. W., Rielly, C. D., Brosnan, K., Bargh, N., Lee, K., Coopman, K., & Hewitt, C. J. (2013). The physical characterisation of a microscale parallel bioreactor platform with an industrial CHO cell line expressing an IgG4. *Biochemical Engineerng Journal, 76,* 25–36.

Nishikawa, M., Nakamura, M., Yagi, H., & Hashimoto, K. (1981). Gas absorption in in aerated mixing vessels. *Journal of Chemical Engineering Japan, 14*(2), 219–226.

Noorman, H. (2011). An industrial perspective on bioreactor scale-down: What we can learn from combined large-scale bioprocess and model fluid studies. *Biotechnology Journal, 6,* 934–943.

Ogut, A., & Hatch, R. T. (1988). Oxygen transfer, Newtonian and non-Newtonian fluids in mechanically agitated vessels. *Canadian Journal of Chemical Engineering, 66,* 79–85.

Oh, S. K. W., Nienow, A. W., Al-Rubeai, M., & Emery, A. N. (1989). The effect of agitation intensity with and without continuous sparging on the growth and antibody production of hybridoma cells. *Journal of Biotechnology, 12,* 45–62.

Oldshue, J. Y. (1985). Current trends in mixer scale-up techniques. In J. Ulbrecht, & G. K. Patterson (Eds.), *Mixing of liquids by mechanical agitation* (pp. 309–341). New York: Gordon and Breach.

Oosterhuis, N. M. G., & Koerts, K. (1987). Method and reactor vessel for the fermentation production of polysaccharides in particular, xanthan. *European Patent Application, EP 249,* 288.

Orr, D., Zheng, W., Campbell, B. S., McDougall, B. M., & Seviour, R. J. (2009). Culture conditions affect the chemical composition of the exopolysaccharide synthesized by the fungus iAureobasidium pullulans. *Journal of Applied Microbiology, 107,* 691–698.

Pedersen, A. G., Andersen, H., Nielsen, J., & Villadsen, J. (1994). A novel technique based on Kr-85 for quantification of gas-liquid mass transfer in bioreactors. *Chemical Engineering Science, 6,* 803–810.

Pelton, R. (2002). A review of antifoam mechanisms in fermentation. *Journal of Industrial Microbiology and Biotechnology, 29,* 149–154.

Posch, A., Herwig, C., & Spadiut, O. (2013). Science-based bioprocess design for filamentous fungi. *Trends in Biotechnology, 31*(1), 37–44.

Posch, A., & Herwig, C. (2014). Physiological description of multivariate interdependencies between process parameters, morphology and physiology during fed-batch penicillin production. (2014). *Biotechnology Progress, 30*(3), 689–699.

Puthli, M. S., Rathod, V. K., & Pandit, A. B. (2005). Gas-liquid mass transfer studies with triple impeller system on a laboratory scale bioreactor. *Biochemical Engineering Journal, 23,* 25–30.

Rao, G., Moreira, A., & Brorson, K. (2009). Disposable bioprocessing: the future has arrived. *Biotechnology and Bioengineering, 102*(2), 348–356.

Rathore, A. S., & Winkle, H. (2009). Quality by design for biopharmaceuticals. *Nature Biotechnology, 27,* 26–34.

Reuss, M. (1988). Influence of mechanical stress on the growth of *Rhizopus nigricans* in stirred bioreactors. *Chemical Engineering Technology, 11,* 789–800.

Richardson, L. F. (1922). *Weather prediction by numerical process.* Cambridge: Cambridge University Press.

Righelato, R. C., Trinci, A. P. J., Pirt, S. J., & Peat, A. (1968). Influence of maintenance energy and growth rate on the metabolic activity, morphology and conidiation of *Penicillium chrysogenum. Journal of Genetics and Microbiology, 50*(1), 394–412.

Riviere, J. (1977). *Industrial Application of Microbiology* (translated and edited by Moss, M.O. and Smith, J.E.). Surrey University Press, Guildford.

Roels, J. A., Van Denberg, J., & Vonken, R. M. (1974). The rheology of mycelial broths. *Biotechnology and Bioengineering, 16*, 181–208.

Rushton, J. H., Costich, E. W., & Everett, H. J. (1950). Power characteristics of mixing impellers. *Chemical Engineering Progress, 46*, 395–404.

Ryu, D. D. Y., & Hospodka, J. (1980). Quantitative Physiology of *Penicillium chrysogenum* in penicillin fermentation. *Biotechnology and Bioengineering, 22*(2), 289–298.

Sanchez, A., Garcia, F., Contreras, A., Molina, E., & Chisti, Y. (2000). Bubble-column and airlift photobioreactors for algal culture. *American Institute of Chemical Engineering Journal, 46*, 1872–1887.

Sato, K. (1961). Rheological studies on some fermentation broths. (IV) Effect of dilution rate on rheological properties of fermentation broth. *Journal of Fermentation and Technology, 39*, 517–520.

Schaepe, S., Kuprijanov, A., Sieblist, C., Jenzesch, M., Simutis, R., & Lubbert, A. (2013). K_La of stirred tank bioreactors revisited. *Journal of Biotechnology, 168*, 576–583.

Schugerl, K., Wittler, R., & Lorenz, T. (1988). The use of moulds in pellet form. *Trends in Biotechnology, 1*(4), 120–122.

Schulze, K. L., & Lipe, R. S. (1964). Relationship between substrate concentration, growth rate and respiration rate of *E. coli* in continuous culture. *Archive fur Mikrobiologie, 48*, 1–20.

Scragg, A. H. (1991). *Bioreactors in biotechnology. A practical approach.* Chichester: Ellis Horwood.

Serrano-Carreon, L., Galindo, E., Rocha-Valadez, J. A., Holguin-Salas, A., & Corkidi, G. (2015). Hydrodynamics, fungal physiology and morphology. In R. Krull, & T. Bley (Eds.), *Filaments in bioprocesses. Advances in biochemical engineering/biotechnology* (pp. 142–223). (Vol. 149). Cham, Switzerland: Springer.

Seviour, R. J., McNeil, B., Fazenda, M. L., & Harvey, L. M. (2011). Operating bioreactors for microbial exopolysaccharide production. *Critical Reviews in Biotechnology, 31*(2), 170–185.

Siurkus, J., Panula-Perala, J., Horn, U., Kraft, M., Rimseliene, R., & Neubauer, P. (2010). Novel approach of high cell density recombinant bioprocess development: Optimisation and scale-up from microtitre to pilot scales while maintaining fed-batch cultivation mode of *E. coli* cultures. *Microbial Cell Factories, 9*, 35.

Smith, J. J., Lilly, M. D., & Fox, R. I. (1990). The effect of agitation on the morphology and penicillin production of *Penicillium chrysogenum. Biotechnology and Bioengineering., 35*, 1011–1023.

Steel, R., & Maxon, W. D. (1966). Studies with a multiple-rod mixing impeller. *Biotechnology and Bioengineering., 8*, 109–116.

Sutherland, I.W., & Ellwood, D.C. (1979). Microbial exopolysaccharides—industrial polymers of current and future potential. In A. T. Bull, D. C. Ellwood & C. Ratledge (Eds.), Soc. Gen. Microbiology Symposium, **29**, *Microbial Technology: Current State, Future Prospects* (pp. 107–150). Cambridge University Press, Cambridge.

Taguchi, H. (1971). The nature of fermentation fluids. *Advances in Biochemistry and Engineering, 1*, 1–30.

Taguchi, H., & Humphrey, A. E. (1966). Dynamic measurement of the volumetric oxygen transfer coefficient in fermentation systems. *Journal of Fermentation Technology, 44*(12), 881–889.

Taguchi, H., Imanaka, T., Teramoto, S., Takatsu, M., & Sato, M. (1968). Scale-up of glucamylase fermentation by *Endomyces* sp. *Journal of Fermentation Technology, 46*(10), 823–828.

Takors, R. (2012). Scale-up of microbial processes: Impacts, tools and open questions. *Journal of Biotechnology, 160,* 3–9.

Tribe, L. A., Briens, C. L., & Margaritas, A. (1995). Determination of the volumetric mass transfer coefficient ($k_L a$) using the dynamic "gas out-gas in" method: Analysis of errors caused by dissolved oxygen probes. *Biotechnology Bioengineering, 46,* 388–392.

Tsuge, Y., Hasunuma, T., & Kondo, A. (2015). Recent advances in the metabolic engineering of *Corynebacterium glutamicum* for the production of lactate and succinate from renewable resources. *Journal of Industrial Microbiology and Biotechnology, 42,* 375–389.

Tuffile, C. M., & Pinho, F. (1970). Determination of oxygen transfer coefficients in viscous streptomycete fermentations. *Biotechnology and Bioengineering, 12,* 849–871.

Ukkonen, K., Veijola, J., Vasala, A., & Neubauer, P. (2013). Effect of culture medium, host strain and oxygen transfer on recombinant Fab antibody fragment yield and leakage to medium in shaken *E. coli* cultures. *Microbial Cell Factories, 12*(73), 1–14.

Van Suijdam, J. C., & Metz, B. (1981). Influence of engineering variables upon morphology of filamentous moulds. *Biotechnology and Bioengineering, 23,* 111–148.

Van't Riet, K. (1979). Review of measuring methods and results in non-viscous gas-liquid mass transfer in stirred vessels. *Industrial & Engineering Chemistry Process Design and Development, 18*(3), 357–360.

Van't Riet, K. (1983). Mass transfer in fermentation. *Trends in Biotechnology, 1*(4), 113–116.

Van't Riet, K., Boom, J. M., & Smith, J. M. (1976). Power consumption, impeller coalescence and recirculation in aerated vessels. *Transactions of the Institute of Chemical Engineers, 54,* 124–131.

Van't Riet, K., & Tramper, J. (1991). *Basic bioreactor design.* New York: Marcel Dekker.

Van't Riet, K., & Van Sonsbeek, H. M. (1992). Foaming, mass transfer and mixing: Interrelations in large scale fermentations. In M. R. Ladisch, & A. Bose (Eds.), *Harnessing biotechnology for the 21st century* (pp. 189–192). Washington, DC: American Chemical Society.

Vasconcelos, J. M. T., Orvalho, S. C. P., Rodriguez, A. M., & Alves, S. S. (2000). Effect of blade shape on the performance of six blade disk turbine impellers. *Industrial Engineering Chemistry Research, 39,* 203–208.

Villadsen, J., Nielsen, J., & Liden, G. (2011). *Bioreaction engineering principles.* New York: Springer.

Walisko, R., Krull, R., & Schrader, J. (2012). Microparticle based morphology engineering of filamentous microorganisms for industrial bio-production. *Biotechnology Letters, 34,* 1975–1982.

Walisko, R., Moench-Tegeder, J. B., Wucherpfennig, T., & Krull, R. (2015). The taming of the shrew—Controlling the morphology of filamentous eukaryotic and prokaryotic microorganisms. In R. Krull, & T. Bley (Eds.), *Filaments in bioprocesses, advances in biochemical engineering* (pp. 15–82). (149).

Wang, D. I. C., & Fewkes, R. C. J. (1977). Effect of operating and geometric parameters on the behaviour of non-Newtonian mycelial antibiotic fermentations. *Developmental and Industrial Microbiology, 18,* 39–57.

Wang, D. I. C., Cooney, C. L., Demain, A. L., Dunnill, P., Humphrey, A. E., & Lilly, M. D. (1979). *Fermentation and enzyme technology.* New York: Wiley.

Wecker, A., & Onken, U. (1991). Influence of dissolved oxygen concentration and shear arte on the production of pullulan by *Aureobasidium pullulans. Biotechnology Letters, 13,* 155–160.

Wernau, W. C., & Wilke, C. R. (1973). New method for evaluation of dissolved oxygen probe response for K_La determination. *Biotechnology and Bioengineering, 15,* 571–578.

Winkler, M. A. (1990). Problems in fermenter design and operation. In M. A. Winkler (Ed.), *Chemical engineering problems in biotechnology* (pp. 215–350). London: SCI/Elsevier.

Wise, W. S. (1951). The measurement of the aeration of culture media. *Journal of Genetics and Microbiology, 5,* 167–177.

Wodzinski, R. S., & Johnson, M. J. (1968). Yield of bacterial cells form hydrocarbons. *Applied Microbiology, 16,* 1886–1891.

Wucherpfennig, K. A., Kiep, K. A., Driouch, H., Wittmann, C., & Krull, R. (2010). Morphology and rheology in filamentous cultivations. *Advances in Applied Microbiology, 72,* 89–136.

Xie, M-h., Xia, J-y., Zhou, Z., Zhou, G-z., Chu, J., Zhuang, Y-p., Zhang, S-l, & Noorman, H. (2015). Power consumption, local and average volumetric mass transfer coefficient in multiple-impeller stirred bioreactors for xanthan gum solutions. *Chemical Engineering Science, 106,* 144–156.

Xu, H., Dou, W., Xu, H., Zhang, X., Roa, Z., Shi, Z., & Xu, Z. (2009). A two-stage oxygen supply strategy for enhanced L-arginine production by *Corynebacterium crenatum* based on metabolic flux analysis. *Biochemical Engineering Journal, 43,* 41–45.

Yagi, H., & Yoshida, F. (1975). Gas absorption by Newtonian and non-Newtonian fluids in sparged agitated vessels. *Industrial Engineering Chemistry Process Design and Development, 14,* 488–493.

Yang, Y., Xia, J., Li, J., Chu, J., Li, L., Wang, Y., Zhuang, Y., & Zhang, S. (2012). A novel impeller configuration to improve fungal physiology performance and energy conservation for cephalosporin C production. *Journal of Biotechnology, 161,* 250–256.

Yano, T., Kodama, T., & Yamada, K. (1961). Fundamental studies on the aerobic fermentation Part VIII. Oxygen transfer within a mold pellet. *Agricultural and Biological Chemistry, 25*(7), 580–584.

Zhou, W., Holzhauer-Rieger, K., Dors, M., & Schugerl, K. (1992). Influence of dissolved oxygen concentration on the biosynthesis of cephalosporin C. *Enzyme Microbiology and Technology, 14,* 848–854.

The recovery and purification of fermentation products

10

INTRODUCTION

The extraction and purification of fermentation products may be difficult and costly. Ideally, one is trying to obtain a high-quality product as quickly as possible at an efficient recovery rate using minimum plant investment operated at minimal costs. Unfortunately, recovery costs of microbial products may vary from as low as 15% to as high as 70% of the total manufacturing costs (Aiba, Humphrey, & Millis, 1973; Swartz, 1979; Pace & Smith, 1981; Atkinson & Sainter, 1982; Datar, 1986). Obviously, the chosen process, and therefore its relative cost, will depend on the specific product. Atkinson and Mavituna (1991) indicate percentage of total costs being 15% for industrial ethanol, 20–30% for bulk penicillin G and up to 70% for enzymes. The extraction and purification of products such as recombinant proteins and monoclonal antibodies can account for 80–90% of the total processing costs (Doran, 2013). The high (and sometimes dominant) cost of downstream processing will affect the overall objective in some fermentations. A useful overview of relative downstream processing costs is given by Straathof (2011).

If a fermentation broth is analyzed at the time of harvesting, it will be discovered that the specific product may be present at a low concentration (typically $0.1–5\,\mathrm{g\,dm^{-3}}$) in an aqueous solution that contains intact microorganisms, cell fragments, soluble and insoluble medium components, and other metabolic products. The product may also be intracellular, heat labile, and easily broken down by contaminating microorganisms. All these factors tend to increase the difficulties of product recovery. To ensure good recovery or purification, speed of operation may be the overriding factor because of the labile nature of a product. The processing equipment therefore must be of the correct type and also the correct size to ensure that the harvested broth can be processed within a satisfactory time limit. It should also be noted that each step or unit operation in downstream processing will involve the loss of some product as each operation will not be 100% efficient and product degradation may have occurred. Even if the percentage recovery for each step is very high, say for example, 90%, after five steps only around 60% of the initial product will be obtained. Hence, it is also important that the minimum number of unit operations possible are used to maximize product recovery.

The choice of recovery process is based on the following criteria:

1. The intracellular or extracellular location of the product.
2. The concentration of the product in the fermentation broth.
3. The physical and chemical properties of the desired product (as an aid to select separation procedures).
4. The intended use of the product.
5. The minimal acceptable standard of purity.
6. The magnitude of biohazard of the product or broth.
7. The impurities in the fermenter broth.
8. The marketable price for the product.

The main objective of the first stage for the recovery of an extracellular product is the removal of large solid particles and microbial cells usually by centrifugation or filtration (Fig. 10.1). In the next stage, the broth is fractionated or extracted into major fractions using ultrafiltration, reverse osmosis, adsorption/ion-exchange/gel filtration or affinity chromatography, liquid–liquid extraction, two phase aqueous extraction, supercritical fluid extraction, or precipitation. Afterward, the product-containing fraction is purified by fractional precipitation, further more precise chromatographic techniques and crystallization to obtain a product, which is highly concentrated and essentially free from impurities. Other products are isolated using modifications of this flow-stream. Finally, the finished product may require drying.

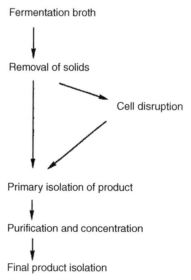

FIGURE 10.1 Stages in the Recovery of Product From a Harvested Fermentation Broth

It may be possible to modify the handling characteristics of the broth so that it can be handled faster with simpler equipment making use of a number of techniques:

1. Selection of a microorganism which does not produce pigments or undesirable metabolites.
2. Modification of the fermentation conditions to reduce the production of undesirable metabolites.
3. Precise timing of harvesting.
4. pH control after harvesting.
5. Temperature treatment after harvesting.
6. Addition of flocculating agents.
7. Use of enzymes to attack cell walls.

It must be remembered that the fermentation and product recovery are integral parts of an overall process. Because of the interactions between the two, neither stage should be developed independently, as this might result in problems and unnecessary expense. Darbyshire (1981) has considered this problem with reference to enzyme recovery. The parameters to consider include time of harvest, pigment production, ionic strength, and culture medium constituents. Large volumes of supernatants containing extracellular enzymes need immediate processing while harvesting times and enzyme yields might not be predictable. This can make recovery programs difficult to plan. Changes in fermentation conditions may reduce pigment formation. Corsano, Iribarren, Montagna, Aguirre, and Suarez (2006) discussed the integration and economic implications of downstream processing in batch and semicontinuous ethanol fermentations.

Certain antifoams remain in the supernatant and may affect centrifugation, ultrafiltration or ion-exchange resins used in recovery stages. Trials may be needed to find the most suitable antifoam (see also Chapter 4). The ionic strength of the production medium may be too high, resulting in the harvested supernatant needing dilution with demineralized water before it can be processed. Such a negative procedure should be avoided if possible by unified research and development programs. Media formulation is dominated by production requirements, but the protein content of complex media should be critically examined in view of subsequent enzyme recovery. This view is also shared by Topiwala and Khosrovi (1978), when considering water recycle in biomass production. They stated that the interaction between the different unit operations in a recycle process made it imperative that commercial plant design and operation should be viewed in an integrated fashion.

Flow sheets for recovery of penicillin, cephamycin C, citric acid, and micrococcal nuclease are given in Figs 10.2–10.5, to illustrate the range of techniques used in microbiological recovery processes. A series of comprehensive flow sheets for alcohols, organic acids, antibiotics, carotenoids, polysaccharides, intra- and extracellular enzymes, single-cell proteins, and vitamins have been produced by Atkinson and Mavituna (1991). Other reviews on separation and purification are available for penicillin (Swartz, 1979), amino acids (Samejima, 1972), enzymes (Aunstrup, 1979; Darbyshire, 1981), single-cell protein (Hamer, 1979) and polysaccharides (Pace &

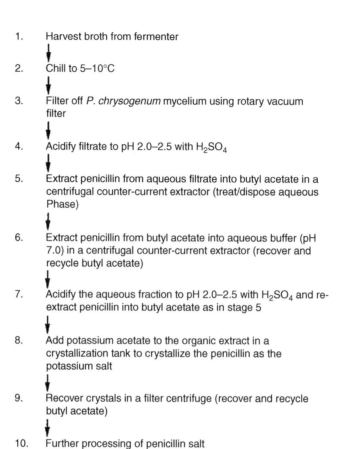

1. Harvest broth from fermenter

2. Chill to 5–10°C

3. Filter off *P. chrysogenum* mycelium using rotary vacuum filter

4. Acidify filtrate to pH 2.0–2.5 with H_2SO_4

5. Extract penicillin from aqueous filtrate into butyl acetate in a centrifugal counter-current extractor (treat/dispose aqueous Phase)

6. Extract penicillin from butyl acetate into aqueous buffer (pH 7.0) in a centrifugal counter-current extractor (recover and recycle butyl acetate)

7. Acidify the aqueous fraction to pH 2.0–2.5 with H_2SO_4 and re-extract penicillin into butyl acetate as in stage 5

8. Add potassium acetate to the organic extract in a crystallization tank to crystallize the penicillin as the potassium salt

9. Recover crystals in a filter centrifuge (recover and recycle butyl acetate)

10. Further processing of penicillin salt

FIGURE 10.2 Recovery and Partial Purification of Penicillin G

Righelato, 1980; Smith & Pace, 1982). In the selection of processes for the recovery of biological products, it should always be understood that recovery and production are interlinked, and that good recovery starts in the fermentation by the selection of, among other factors, the correct media and time of harvesting.

The recovery and purification of many compounds may be achieved by a number of alternative routes. The decision to follow a particular route involves comparing the following factors to determine the most appropriate under a given set of circumstances:

Capital costs.
Processing costs.
Throughput requirements.
Yield potential.
Product quality.
Technical expertise available.
Conformance to regulatory requirements.

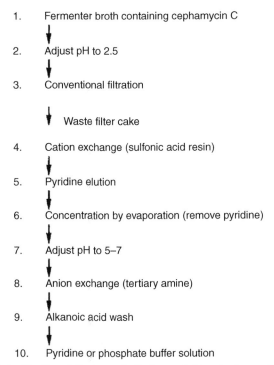

1. Fermenter broth containing cephamycin C

2. Adjust pH to 2.5

3. Conventional filtration

    ↓ Waste filter cake

4. Cation exchange (sulfonic acid resin)

5. Pyridine elution

6. Concentration by evaporation (remove pyridine)

7. Adjust pH to 5–7

8. Anion exchange (tertiary amine)

9. Alkanoic acid wash

10. Pyridine or phosphate buffer solution

FIGURE 10.3 Purification of Cephamycin C: Sequential Ion Exchange Process (Omstead, Hunt, & Buckland, 1985)

Waste treatment needs.

Continuous or batch processing.

Automation.

Personnel health and safety (Wildfeuer, 1985).

The major problem currently faced in product recovery is the large-scale purification of biologically active molecules. For a process to be economically viable, large-scale production is required, and therefore large-scale separation, recovery, and purification. This then requires the transfer of small-scale preparative/analytical technologies (eg, chromatographic techniques) to the production scale while maintaining efficiency of the process, bioactivity of the product and purity of the product so that it conforms with the safety legislation and regulatory requirements. Developments in this field and remaining areas for development are documented by Pyle (1990).

REMOVAL OF MICROBIAL CELLS AND OTHER SOLID MATTER

Microbial cells and other insoluble materials are normally separated from the harvested broth by filtration or centrifugation. Because of the small size of many microbial cells, it will be necessary to consider the use of filter aids to improve

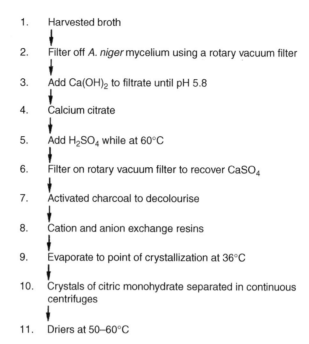

1. Harvested broth

2. Filter off *A. niger* mycelium using a rotary vacuum filter

3. Add Ca(OH)$_2$ to filtrate until pH 5.8

4. Calcium citrate

5. Add H$_2$SO$_4$ while at 60°C

6. Filter on rotary vacuum filter to recover CaSO$_4$

7. Activated charcoal to decolourise

8. Cation and anion exchange resins

9. Evaporate to point of crystallization at 36°C

10. Crystals of citric monohydrate separated in continuous centrifuges

11. Driers at 50–60°C

FIGURE 10.4 Recovery and Purification of Citric Acid (Sodeck et al., 1981)

filtration rates, while heat and flocculation treatments are employed as techniques for increasing the sedimentation rates in centrifugation. Flocculation can also be utilized in other downstream processing operations to aid product recovery. Hao, Xu, Liu, and Liu (2006) report the use of the flocculants chitosan and polyacrylamide on cell debris and soluble protein in the fermentation broth, to enhance the recovery of 1,3-propanediol by reactive extraction and distillation. The methods of cell and cell debris separation described in the following sections have been practiced for many years. Bowden, Leaver, Melling, Norton, and Whittington (1987) review some potential developments in cell recovery. These include the use of electrophoresis and dielectrophoresis to exploit the charged properties of microbial cells, ultrasonic treatment to improve flocculation characteristics and magnetic separations. Although not necessarily for the removal of cells, other downstream operations, which involve the application of an electrical field are also showing potential. One such process is electrodialysis that involves the transfer of ions from a dilute solution to a concentrated one through a semipermeable membrane by applying an electrical field (Moresi & Sappino, 2000). Lopez and Hestekin (2013) report the use of electrodialysis in the separation of organic acids in aqueous solution where the product, sodium butyrate, was successfully transferred from the aqueous phase into an ionic liquid phase through electrodialysis. A recovery rate of 99% was obtained with reduced energy input compared to traditional processing.

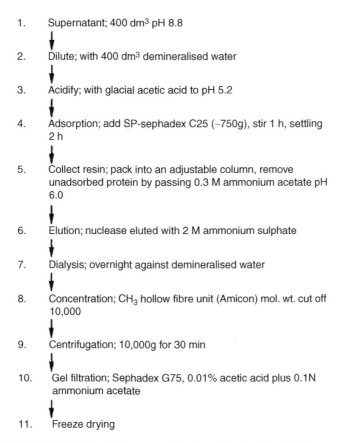

1.   Supernatant; 400 dm^3 pH 8.8

2.   Dilute; with 400 dm^3 demineralised water

3.   Acidify; with glacial acetic acid to pH 5.2

4.   Adsorption; add SP-sephadex C25 (~750g), stir 1 h, settling 2 h

5.   Collect resin; pack into an adjustable column, remove unadsorbed protein by passing 0.3 M ammonium acetate pH 6.0

6.   Elution; nuclease eluted with 2 M ammonium sulphate

7.   Dialysis; overnight against demineralised water

8.   Concentration; CH$_3$ hollow fibre unit (Amicon) mol. wt. cut off 10,000

9.   Centrifugation; 10,000g for 30 min

10.   Gel filtration; Sephadex G75, 0.01% acetic acid plus 0.1N ammonium acetate

11.   Freeze drying

FIGURE 10.5 Purification of Micrococcal Nuclease (Darbyshire, 1981)

FOAM SEPARATION (FLOATATION)

Foam separation depends on using methods, which exploit differences in surface activity of materials. The material may be whole cells or molecular such as a protein or colloid, which is selectively adsorbed or attached to the surface of gas bubbles rising through a liquid, to be concentrated or separated and finally removed by skimming (Fig. 10.6). It may be possible to make some materials surface active by the application of surfactants such as long-chain fatty acids, amines, and quaternary ammonium compounds. Materials made surface active and collected are termed colligends whereas the surfactants are termed collectors. When developing this method of separation, the important variables, which may need experimental investigation are pH, air-flow rates, surfactants, and colligend-collector ratios. The recovery of surface active products is clearly an important potential application of this technique. Davis, Lynch, and Varley (2001) report the use of foam separation in the recovery

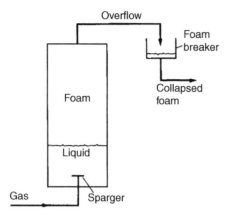

FIGURE 10.6 Schematic Flow Diagram for Foam Fractionation (Wang & Sinskey, 1970)

of the lipopeptide biosurfactant surfactin from *B. subtilis* cultures. They report that improved surfactin recovery can be achieved when foaming was simultaneous with the fermentation stage rather than as a nonintegrated semibatch process.

Rubin, Cassel, Henderson, Johnson, and Lamb (1966) investigated foam separation of *E. coli* starting with an initial cell concentration of 7.2×10^8 cells cm^{-3}. Using lauric acid, stearyl amine, or *t*-octyl amine as surfactants, it was shown that up to 90% of the cells were removed in 1 min and 99% in 10 min. The technique also proved successful with *Chlorella* sp. and *Chlamydomonas* sp. In other work with *E. coli*, Grieves and Wang (1966) were able to achieve cell enrichment ratios of between 10 and 1×10^6 using ethyl-hexadecyl-dimethyl ammonium bromide. DeSousa, Laluce, and Jafelicci (2006) examined the effects of a range of both organic and inorganic additives on floatation recovery of *Saccharomyces cerevisiae*. They report that compounds associated with cellular metabolism such as acetate and ethanol can improve floatation recovery of yeast cells.

PRECIPITATION

Precipitation may be conducted at various stages of the product recovery process. It is a particularly useful process as it allows enrichment and concentration in one step, thereby reducing the volume of material for further processing.

It is possible to obtain some products (or to remove certain impurities) directly from the broth by precipitation, or to use the technique after a crude cell lysate has been obtained.

Typical agents used in precipitation render the compound of interest insoluble, and these include:

1. Acids and bases to change the pH of a solution until the isoelectric point of the compound is reached and pH equals pI, when there is then no overall charge on the molecule and its solubility is decreased.

2. Salts such as ammonium and sodium sulfate are used for the recovery and fractionation of proteins. The salt removes water from the surface of the protein revealing hydrophobic patches, which come together causing the protein to precipitate. The most hydrophobic proteins will precipitate first, thus allowing fractionation to take place. This technique is also termed "salting out."

3. Organic solvents. Dextrans can be precipitated out of a broth by the addition of methanol. Chilled ethanol and acetone can be used in the precipitation of proteins mainly due to changes in the dielectric properties of the solution.

4. Nonionic polymers such as polyethylene glycol (PEG) can be used in the precipitation of proteins and are similar in behavior to organic solvents.

5. Polyelectrolytes can be used in the precipitation of a range of compounds, in addition to their use in cell aggregation.

6. Protein binding dyes (triazine dyes) bind to and precipitate certain classes of protein (Lowe & Stead, 1985).

7. Affinity precipitants are an area of much current interest in that they are able to bind to, and precipitate, compounds selectively (Niederauer & Glatz, 1992).

8. Heat treatment as a selective precipitation and purification step for various thermostable products and in the deactivation of cell proteases (Ng, Tan, Abdullah, Ling, & Tey, 2006).

FILTRATION

Filtration is one of the most common processes used at all scales of operation to separate suspended particles from a liquid or gas, using a porous medium which retains the particles but allows the liquid or gas to pass through. Gas filtration has been discussed in detail elsewhere (Chapters 5 and 7). It is possible to carry out filtration under a variety of conditions, but a number of factors will obviously influence the choice of the most suitable type of equipment to meet the specified requirements at minimum overall cost, including:

1. The properties of the filtrate, particularly its viscosity and density.
2. The nature of the solid particles, particularly their size and shape, the size distribution and packing characteristics.
3. The solids:liquid ratio.
4. The need for recovery of the solid or liquid fraction or both.
5. The scale of operation.
6. The need for batch or continuous operation.
7. The need for aseptic conditions.
8. The need for pressure or vacuum suction to ensure an adequate flow rate of the liquid.

THEORY OF FILTRATION

A simple filtration apparatus is illustrated in Fig. 10.7, which consists of a support covered with a porous filter cloth. A filter cake gradually builds up as filtrate passes

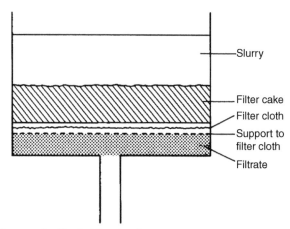

FIGURE 10.7 Diagram of a Simple Filtration Apparatus

through the filter cloth. As the filter cake increases in thickness, the resistance to flow will gradually increase.

Thus, if the pressure applied to the surface of the slurry is kept constant the rate of flow will gradually diminish. Alternatively, if the flow rate is to be kept constant the pressure will gradually have to be increased. The flow rate may also be reduced by blocking of holes in the filter cloth and closure of voids between particles, if the particles are soft and compressible. When particles are compressible, it may not be feasible to apply increased pressure.

Flow through a uniform and constant depth porous bed can be represented by the Darcy equation:

$$\text{Rate of flow} = \frac{dV}{dt} = \frac{KA\,\Delta P}{\mu L} \tag{10.1}$$

where μ, liquid viscosity; L, depth of the filter bed; ΔP, pressure differential across the filter bed; A, area of the filter exposed to the liquid; K, constant for the system.

K itself is a term which depends on the specific surface area s (surface area/unit volume) of the particles making up the filter bed and the voidage Σ when they are packed together. The voidage is the amount of filter-bed area, which is free for the filtrate to pass through. It is normally 0.3–0.6 of the cross-sectional area of the filter bed. Thus K (Kozeny's constant) can be expressed as

$$K = \frac{\Sigma^2}{5(1-\Sigma)^2 s^2}$$

Unfortunately, s and Σ are not easily determined.
In most practical cases L is not readily measured but can be defined in terms of:

V = volume of filtrate passed in time t and
v = volume of cake deposited per unit volume of filtrate.

Then

$$L = \frac{vV}{A}$$

Substituting in Eq. (10.1):

$$\frac{dV}{dt} = \frac{KA^2 \Delta P}{\mu v V} \qquad (10.2)$$

This is a general equation relating the rate of filtration to pressure drop, cross-sectional area of the filter and filtrate retained. Eq. 10.2 can be integrated for filtration at constant pressure.

$$V dV = \frac{KA^2 \Delta P \, dt}{\mu v V} \qquad (10.3)$$

Integrating Eq. (10.3):

$$V^2 = \frac{2 KA^2 \Delta P t}{\mu v} \qquad (10.4)$$

Now in Eq. (10.4), ΔP is constant, μ is generally equal to 1, v can be determined by laboratory investigation and A^2 remains approximately constant. Thus, there is a linear relationship between V^2 and t. By carrying out small-scale filtration trials, it is therefore possible to obtain a value for K. It is then possible to reapply the equation for large-scale filtration calculations.

Although it is also possible to derive the equation for the pressure necessary to maintain a constant filtration rate, it has little practical application. The pressure is made up of two components. First, the pressure needed to pass the constant volume through the filter resistance and, second, an increasing pressure component, which is proportional to the resistance from the increasing cake depth. This filtration procedure would be complex to perform practically, and other methods of filtration are used to achieve constant flow rates, for example, vacuum drum filters.

USE OF FILTER AIDS

It is common practice to use filter aids when filtering bacteria or other fine or gelatinous suspensions which prove slow to filter or partially block a filter. Kieselguhr (diatomaceous earth) is the most widely used material. It has a voidage of approximately 0.85, and, when it is mixed with the initial cell suspension, improves the porosity of a resulting filter cake leading to a faster flow rate. Alternatively, it may be used as an initial bridging agent in the wider pores of a filter to prevent or reduce blinding. The term "blinding" means the wedging of particles which are not quite large enough to pass through the pores, so that an appreciable fraction of the filter surface becomes inactive.

The minimum quantity of filter aid to be used in filtration of a broth should be established experimentally. Kieselguhr is not cheap, and it will also absorb some of the filtrate, which will be lost when the filter cake is disposed. The main methods of using the filter aid are:

1. A thin layer of Kieselguhr is applied to the filter to form a precoat prior to broth filtration.
2. The appropriate quantity of filter aid is mixed with the harvested broth. Filtration is started, to build up a satisfactory filter bed. The initial raffinate is returned to the remaining broth prior to starting the true filtration.
3. When vacuum drum filters are to be used which are fitted with advancing knife blades, a thick precoat filter is initially built up on the drum (later section in this chapter).

In some processes such as microbial biomass production, filter aids cannot be used and cell pretreatment by flocculation or heating must be considered (see later sections in this chapter). In addition it is not normally practical to use filter aids when the product is intracellular and its removal would present a further stage of purification.

Plate and frame filters

A plate and frame filter is a pressure filter in which the simplest form consists of plates and frames arranged alternately. The plates are covered with filter cloths (Fig. 10.8) or filter pads. The plates and frames are assembled on a horizontal framework and held together by means of a hand screw or hydraulic ram so that there is no leakage between the plates and frames, which form a series of liquid-tight compartments. The slurry is fed to the filter frame through the continuous channel formed by the holes in the corners of the plates and frames. The filtrate passes through the filter cloth or pad, runs down grooves in the filter plates and is then discharged through outlet taps to a channel. Sometimes, if aseptic conditions are required, the outlets may lead directly into a pipe. The solids are retained within the frame and filtration is stopped when the frames are completely filled or when the flow of filtrate becomes uneconomically low.

On an industrial scale, the plate and frame filter is one of the cheapest filters per unit of filtering space and requires the least floor space, but it is intermittent in operation (a batch process) and there may be considerable wear of filter cloths as a result of frequent dismantling. This type of filter is most suitable for fermentation broths with a low solids content and low resistance to filtration. It is widely used as a "polishing" device in breweries to filter out residual yeast cells following initial clarification by centrifugation or rotary vacuum filtration. It may also be used for collecting high value solids that would not justify the use of a continuous filter. Because of high labor costs and the time involved in dismantling, cleaning, and reassembly, these filters should not be used when removing large quantities of worthless solids from a broth.

Pressure leaf filters

There are a number of intermittent batch filters usually called by their trade names. These filters incorporate a number of leaves, each consisting of a metal framework of

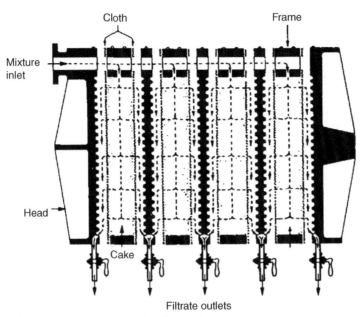

FIGURE 10.8 Flush Plate and Frame Filter Assembly

The cloth is shown away from the plates to indicate flow of filtrate in the grooves between pyramids (Purchas, 1971).

grooved plates, which is covered with a fine wire mesh, or occasionally a filter cloth and often precoated with a layer of cellulose fibers. The process slurry is fed into the filter, which is operated under pressure or by suction with a vacuum pump. Because the filters are totally enclosed it is possible to sterilize them with steam. This type of filter is particularly suitable for "polishing" large volumes of liquids with low solids content or small batch filtrations of valuable solids.

Vertical metal-leaf filter

This filter consists of a number of vertical porous metal leaves mounted on a hollow shaft in a cylindrical pressure vessel. The solids from the slurry gradually build up on the surface of the leaves and the filtrate is removed from the plates via the horizontal hollow shaft. In some designs the hollow shaft can be slowly rotated during filtration. Solids are normally removed at the end of a cycle by blowing air through the shaft and into the filter leaves.

Horizontal metal-leaf filter

In this filter, the metal leaves are mounted on a vertical hollow shaft within a pressure vessel. Often, only the upper surfaces of the leaves are porous. Filtration is continued until the cake fills the space between the disc-shaped leaves or when the operational pressure has become excessive. At the end of a process cycle, the solid cake can be discharged by releasing the pressure and spinning the shaft with a drive motor.

Stacked-disc filter

One kind of filter of this type is the Metafilter. This is a very robust device and because there is no filter cloth and the bed is easily replaced, labor costs are low. It consists of a number of precision-made rings, which are stacked on a fluted rod (Fig. 10.9). The rings (22 mm external diameter, 16 mm internal diameter, and 0.8 mm-thick) are normally made from stainless steel and precision stamped so that there are a number of shoulders on one side. This ensures that there will be clearances of 0.025–0.25 mm when the rings are assembled on the rods. The assembled stacks are placed in a pressure vessel, which can be sterilized if necessary. The packs are normally coated with a thin layer of Kieselguhr, which is used as a filter aid. During use, the filtrate passes between the discs and is removed through the grooves of the fluted rods, while solids are deposited on the filter coating. Operation is continued until the resistance becomes too high and the solids are removed from the rings by applying back pressure via the fluted rods. Metafilters are primarily used for "polishing" liquids such as beer.

CONTINUOUS FILTERS

Rotary vacuum filters

Large rotary vacuum filters are commonly used by industries, which produce large volumes of liquid which need continuous processing. The filter consists of a rotating, hollow, segmented drum covered with a fabric or metal filter, which is partially immersed in a trough containing the broth to be filtered (Fig. 10.10). The slurry is fed on to the outside of the revolving drum and vacuum pressure is applied internally so that the filtrate is drawn through the filter, into the drum and finally to a collecting vessel. The interior of the drum is divided into a series of compartments, to which the vacuum pressure is normally applied for most of each revolution as the drum slowly revolves (~1 rpm). However, just before discharge of the filter cake, air pressure may be applied internally to help ease the filter cake off the drum. A number of spray jets may be carefully positioned so that water can be applied to rinse the cake. This washing is carefully controlled so that dilution of the filtrate is minimal.

It should be noted that the driving force for filtration (pressure differential across the filter) is limited to 1 atmosphere (100 kN m^{-2}) and in practice it is significantly less than this. In contrast, pressure filters can be operated at many atmospheres pressure. A number of rotary vacuum drum filters are manufactured, which differ in the mechanism of cake discharge from the drum:

1. String discharge.
2. Scraper discharge.
3. Scraper discharge with precoating of the drum.

String discharge

Fungal mycelia produce a fibrous filter cake, which can easily be separated from the drum by string discharge (Fig. 10.11). Long lengths of string 1.5 cm apart are threaded over the drum and round two rollers. The cake is lifted free from the upper

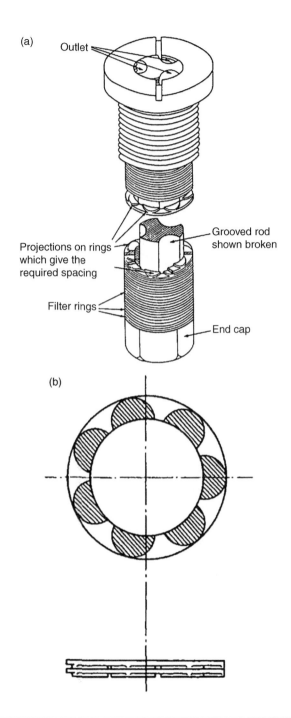

(a) Outlet

Projections on rings
which give the
required spacing

Grooved rod
shown broken

Filter rings

End cap

(b)

FIGURE 10.9

(a) Metafilter pack (Coulson & Richardson, 1991). (b) Rings for metafilter (Coulson & Richardson, 1991).

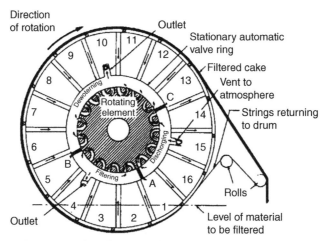

FIGURE 10.10 Diagram of String-Discharge Filter Operation

Sections 1 to 4 are filtering; sections 5 to 12 are dewatering; and section 13 is discharging the cake with the string discharge. Sections 14, 15 and 16 are ready to start a new cycle. A, B and C represent dividing members in the annular ring (Miller et al., 1973).

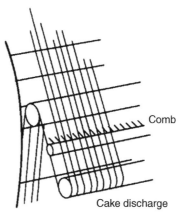

FIGURE 10.11 Cake Discharge on a Drum Filter Using Strings (Talcott, Willus, & Freeman, 1980)

part of the drum when the vacuum pressure is released and carried to the small rollers where it falls free.

Scraper discharge

Yeast cells can be collected on a filter drum with a knife blade for scraper discharge (Fig. 10.12). The filter cake which builds up on the drum is removed by an accurately positioned knife blade. Because the knife is close to the drum, there may be gradual wearing of the filter cloth on the drum.

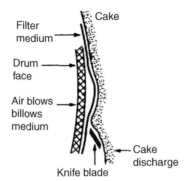

FIGURE 10.12 Cake Discharge on a Drum Using a Scraper (Talcott et al., 1980)

Scraper discharge with precoating of the drum

The filter cloth on the drum can be blocked by bacterial cells or mycelia of actinomycetes. This problem is overcome by precoating the drum with a layer of filter-aid 2–10 cm thick. The cake which builds up on the drum during operation is cut away by the knife blade (Fig. 10.13), which mechanically advances toward the drum at a controlled slow rate. Alternatively, the blade may be operated manually when there is an indication of "blinding" which may be apparent from a reduction in the filtration rate. In either case the cake is removed together with a very thin layer of precoat. A study of precoat drum filtration has been made by Bell and Hutto (1958). The operating variables studied include drum speed, extent of drum submergence, knife advance speed, and applied vacuum. The work indicated that optimization for a new process might require prolonged trials. Although primarily used for the separation of microorganisms from broth, studies have indicated (Gray, Dunnill, & Lilly, 1973) that rotary vacuum filters can be effective in the processing of disrupted cells.

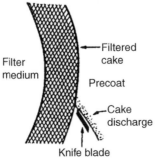

FIGURE 10.13 Cake Discharge on a Precoated Drum Filter (Talcott et al., 1980)

CROSS-FLOW FILTRATION (TANGENTIAL FILTRATION)

In the filtration processes previously described, the flow of broth was perpendicular to the filtration membrane. Consequently, blockage of the membrane led to the lower rates of productivity and/or the need for filter aids to be added, and these were serious disadvantages.

In contrast, an alternative which is rapidly gaining prominence both in the processing of whole fermentation broths (Tanny, Mirelman, & Pistole, 1980; Brown & Kavanagh, 1987; Warren, MacDonald, & Hill, 1991) and cell lysates (Gabler & Ryan, 1985; Le & Atkinson, 1985) is cross-flow filtration. Here, the flow of medium to be filtered is tangential to the membrane (Fig. 10.14a), and no filter cake builds up on the membrane.

The benefits of cross-flow filtration are:

1. Efficient separation, >99.9% cell retention.
2. Closed system; for the containment of organisms with no aerosol formation (see also Chapter 7).
3. Separation is independent of cell and media densities, in contrast to centrifugation.
4. No addition of filter aid (Zahka & Leahy, 1985).

The major components of a cross-flow filtration system are a media storage tank (or the fermenter), a pump, and a membrane pack (Fig. 10.14b). The membrane is usually in a cassette pack of hollow fibers or flat sheets in a plate and frame type

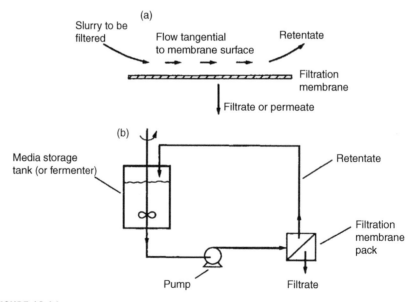

FIGURE 10.14

(a) Schematic diagram of cross-flow filtration. (b) Major components of a cross-flow filtration system.

stack or a spiral cartridge (Strathmann, 1985). In this way, and by the introduction of a much convoluted surface, large filtration areas can be attained in compact devices. Two types of membrane may be used; microporous membranes (microfiltration) with a specific pore size (0.45, 0.22 μm etc.) or an ultrafiltration membrane (see later section) with a specified molecular weight cut-off (MWCO). The type of membrane chosen is carefully matched to the product being harvested, with microporous and 100,000 MWCO membranes being used in cell separations.

The output from the pump is forced across the membrane surface; most of this flow sweeps the membrane, returning retained species back to the storage tank and generally less than 10% of the flow passes through the membrane (permeate). As this process is continued the cells, or other retained species are concentrated to between 5% and 10% of their initial volume. More complex variants of the process can allow in situ washing of the retentate and enclosed systems for containment and sterilization (Mourot, LaFrance, & Oliver, 1989). Russotti et al. (1995) report on a pilot scale system utilizing crossflow-microfiltration (pore sizes 0.22–0.65 μm) to harvest recombinant yeast cells to recover intracellular products.

Many factors influence filtration rate. Increased pressure drop will, up to a point increase flow across the membrane, but it should be remembered that the system is based on a swept clean membrane. Therefore, if the pressure drop is too great the membrane may become blocked. The filtration rate is therefore influenced by the rate of tangential flow across the membrane; by increasing the shear forces at the membrane's surface retained species are more effectively removed, thereby increasing the filtration rate. Adikane, Singh, and Nene (1999) report that a 2.9-fold increase in cross-flow velocity resulted in an average increase in flux across the membrane of 1.8-fold and a 41% reduction in processing time. Higher temperatures will increase filtration rate by lowering the viscosity of the media, though this is clearly of limited application in biological systems. Filtration rate is inversely proportional to concentration, and media constituents can influence filtration rate in three ways. Low molecular weight compounds increase media viscosity and high molecular weight compounds decrease shear at the membrane surface, both leading to a reduction in filtration rate. Finally, broth constituents can "foul" the membrane, primarily by adsorption onto the membrane's surface, causing a rapid loss in efficiency. This can be controlled by increasing the pore size in microporous membranes, modification of the membrane chemistry or media formulation in particular by reducing the use of antifoaming agents (Russotti et al., 1995). Lee, Chang, and Ju (1993) have shown that the pulses of air injected into the flow to a cross-flow filter increase the shear rate at the membrane surface reducing the effects of membrane fouling.

CENTRIFUGATION

Microorganisms and other similar sized particles can be removed from a broth by using a centrifuge when filtration is not a satisfactory separation method. Although a centrifuge may be expensive when compared with a filter it may be essential when:

1. Filtration is slow and difficult.
2. The cells or other suspended matter must be obtained free of filter aids.
3. Continuous separation to a high standard of hygiene is required.

Noncontinuous centrifuges are of extremely limited capacity and therefore not suitable for large-scale separation. The centrifuges used in harvesting fermentation broths are all operated on a continuous or semicontinuous basis. Some centrifuges can be used for separating two immiscible liquids yielding a heavy phase and light phase liquid, as well as a solids fraction. They may also be used for the breaking of emulsions.

According to Stoke's law, the rate of sedimentation of spherical particles suspended in a fluid of Newtonian viscosity characteristics is proportional to the square of the diameter of the particles, thus the rate of sedimentation of a particle under gravitational force is:

$$V_g = \frac{d^2 g (\rho_P - \rho_L)}{18 \mu} \qquad (10.5)$$

where V_g, rate of sedimentation (m s^{-1}); d, particle diameter (m); g, gravitational constant (m s^{-2}); ρ_P, particle density (kg m^{-3}); ρ_L, liquid density (kg m^{-3}); μ, viscosity (kg m^{-1} s^{-1}).

This equation can then be modified for sedimentation in a centrifuge:

$$V_c = \frac{d \omega^2 r (\rho_P - \rho_L)}{18 \mu} \qquad (10.6)$$

where V_c, rate of sedimentation in the centrifuge (ms^{-1}); ω, angular velocity of the rotor (s^{-1}); r, radial position of the particle (m).

Dividing Eq. (10.6) by Eq. (10.5) yields

$$\frac{\omega^2 r}{g}$$

This is a measure of the separating power of a centrifuge compared with gravity settling. It is often referred to as the relative centrifugal force and given the symbol "Z."

It is evident from this formula that factors influencing the rate of sedimentation over which one has little or no control are the difference in density between the cells and the liquid (increased temperature would lower media density but is of little practical use with fermentation broths), the diameter of the cells (could be increased by coagulation/flocculation), and the viscosity of the liquid. Ideally, the cells should have a large diameter, there should be a large density difference between cell and liquid and the liquid should have a low viscosity. In practice, the cells are usually very small, of low density and are often suspended in viscous media. Thus it can be seen that the angular velocity and diameter of the centrifuge are the major factors to be considered when attempting to maximize the rate of sedimentation (and therefore throughput) of fermentation broths.

CELL AGGREGATION AND FLOCCULATION

Following an industrial fermentation, it is quite common to add flocculating agents to the broth to aid dewatering (Wang, 1987). The use of flocculating agents is widely practiced in the effluent-treatment industries for the removal of microbial cells and suspended colloidal matter (Delaine, 1983).

It is well known that aggregates of microbial cells, although they have the same density as the individual cells, will sediment faster because of the increased diameter of the particles (Stokes law). This sedimentation process may be achieved naturally with selected strains of brewing yeasts, particularly if the wort is chilled at the end of fermentation, and leads to a natural clearing of the beer.

Microorganisms in solution are usually held as discrete units in three ways. First, their surfaces are negatively charged and therefore repulse each other. Second, because of their generally hydrophilic cell walls a shell of bound water is associated with the cell which acts as a thermodynamic barrier to aggregation. Finally, due to the irregular shapes of cell walls (at the macromolecular level) steric hindrance will also play a part.

During flocculation, one or more mechanisms besides temperature can induce cell flocculation:

1. Neutralization of anionic charges, primarily carboxyl and phosphate groups, on the surfaces of the microbial cells, thus allowing the cells to aggregate. These include changes in the pH and the presence of a range of compounds, which alter the ionic environment.
2. Reduction in surface hydrophilicity.
3. The use of high molecular weight polymer bridges. Anionic, nonionic, and cationic polymers can be used, though the former two also require the addition of a multivalent cation.

Flocculation usually involves the mixing of a process fluid with the flocculating agent under conditions of high shear in a stirred tank, although more compact and efficient devices have been proposed (Ashley, 1990). This stage is known as coagulation, and is usually followed by a period of gentle agitation when flocs developed initially are allowed to grow in size. The underlying theoretical principles of cell flocculation have been discussed by Atkinson and Daoud (1976).

Nakamura (1961) described the use of various compounds for flocculating bacteria, yeasts and algae, including alum, calcium salts, and ferric salts. Other agents which are now used include tannic acid, titanium tetrachloride and cationic agents such as quaternary ammonium compounds, alkyl amines, and alkyl pyridinium salts. Gasner and Wang (1970) reported a many 100-fold increase in the sedimentation rate of *Candida intermedia* when recoveries of over 99% were readily obtained. They found that flocculation was very dependent on the choice of additive, dosage, and conditions of floc formation, with the most effective agents being mineral colloids and polyelectrolytes. Nucleic acids, polysaccharides, and proteins released from partly lysed cells may also bring about agglomeration. In SCP processes, phosphoric acid has been used as a flocculating agent since it can be used as a nutrient in medium recycle with considerable savings in water usage (Hamer, 1979).

The majority of flocculating agents currently in use are polyelectrolytes, which act by charge neutralization and hydrophobic interactions to link cells to each other. In processes where the addition of some toxic chemicals is to be avoided, alternative techniques have been adopted. One method is to coagulate microbial protein, which has been released from the cells by heating for short periods. Kurane (1990) reports the use of bioflocculants obtained from *Rhodococcus erythropolis*. They are suggested as being safer alternatives to conventional flocculants. Warne and Bowden (1987) suggest the use of genetic manipulation to alter cell surface properties to aid aggregation. Flocculating agents such as crosslinked cationic polymers may also be used in the processing of cell lysates and extracts prior to further downstream processing (Fletcher et al., 1990). Bentham, Bonnerjea, Orsborn, Ward, and Hoare (1990) utilized borax as a flocculating agent for yeast cell debris prior to decanter centrifugation.

RANGE OF CENTRIFUGES

A number of centrifuges will be described which vary in their manner of liquid and solid discharge, their unloading speed and their relative maximum capacities. When choosing a centrifuge for a specific process, it is important to ensure that the centrifuge will be able to perform the separation at the planned production rate, and operate reliably with minimum manpower. Largescale tests may therefore be necessary with fermentation broths or other materials to check that the correct centrifuge is chosen.

Basket centrifuge (perforated-bowl basket centrifuge)

Basket centrifuges are useful for separating mould mycelia or crystalline compounds. The centrifuge is most commonly used with a perforated bowl lined with a filter bag of nylon, cotton, etc. (Fig. 10.15). A continuous feed is used, and when the basket is filled with the filter cake, it is possible to wash the cake before removing it. The bowl may suffer from blinding with soft biological materials so that high centrifugal forces cannot be used. These centrifuges are normally operated at speeds of up to 4000 rpm for feed rates of 50–300 dm^3 min^{-1} and have a solids holding capacity of 30–500 dm^3. The basket centrifuge may be considered to be a centrifugal filter.

Tubular-bowl centrifuge

This is a centrifuge to consider using for particle size ranges of 0.1–200 μm and up to 10% solids in the in-going slurry. Fig. 10.16a shows an arrangement used in a Sharples Super-Centrifuge. The main component of the centrifuge is a cylindrical bowl (or rotor) (A in Fig. 10.16), which may be of a variable design depending on application, suspended by a flexible shaft (B), driven by an overhead motor or air

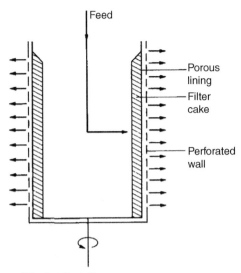

Feed

Porous lining

Filter cake

Perforated wall

FIGURE 10.15 Diagram of Basket Centrifuge

turbine (C). The inlet to the bowl is via a nozzle attached to the bottom bearing (D). The feed which may consist of solids and light and heavy liquid phases is introduced by the nozzle (E). During operation solids sediment on the bowl wall while the liquids separate into the heavy phase in zone (G) and the light phase in the central zone (H). The two liquid phases are kept separate in their exit from the bowl by an adjustable ring, with the heavy phase flowing over the lip of the ring. Rings of various sizes may be fitted for the separation of liquids of various relative densities. Thus the centrifuge may be altered to use for:

1. Light-phase/heavy-phase liquid separation.
2. Solids/light-liquid phase/heavy-liquid phase separation.
3. Solids/liquid separation (using a different rotor, Fig. 10.16b).

The Sharpies laboratory centrifuge with a bowl radius of approximately 2.25 cm can be operated with an air turbine at 50,000 rpm to produce a centrifugal force of approximately 62,000g, but has a bowl capacity of only 200 cm^3 with a throughput of 6–25 dm^3 h^{-1}. The largest size rotor is the Sharpies AS 26, which has a bowl radius of 5.5 cm and a capacity of 9 dm^3, a solids capacity of 5 dm^3 and a throughput of 390–2400 dm^3 h^{-1}.

The advantages of this design of centrifuge are the high centrifugal force, good dewatering, and ease of cleaning. The disadvantages are limited solids capacity, difficulties in the recovery of collected solids, gradual loss in efficiency as the bowl fills, solids being dislodged from the walls as the bowl is slowing down, and foaming. Plastic liners can be used in the bowls to help improve batch cycle time. Alternatively a spare bowl can be changed over in about 5 min.

(a)

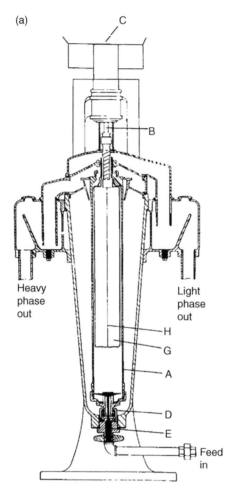

Heavy phase out

Light phase out

Feed in

(b)

FIGURE 10.16

(a) Section of a Sharples Stiper-Centrifuge (Alfa Laval Sharpies, Camberley, UK).
(b) A Sharples Super-Centrifuge assembled for discharge of one liquid phase (Alfa Laval Sharpies, Camberley, UK)

The solid-bowl scroll centrifuge (decanter centrifuge)

This type of centrifuge is used for continuous handling of fermentation broths, cell lysates and coarse materials such as sewage sludge (Fig. 10.17). The slurry is fed through the spindle of an archimedean screw within the horizontal rotating solids bowl. Typically the speed differential between the bowl and the screw is in the range 0.5–100 rpm (Coulson & Richardson, 1991). The solids settling on the walls of

(a)

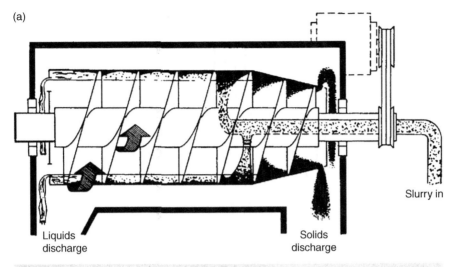

Slurry in

Liquids
discharge

Solids
discharge

(b)

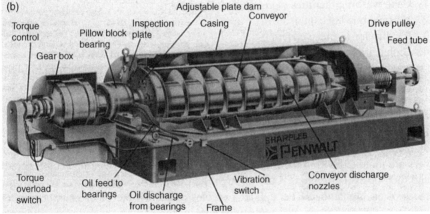

Adjustable plate dam

Conveyor

Torque
control

Inspection
plate

Casing

Drive pulley

Pillow block
bearing

Feed tube

Gear box

Torque
overload
switch

Oil feed to
bearings

Oil discharge
from bearings

Vibration
switch

Conveyor discharge
nozzles

Frame

FIGURE 10.17

(a) Diagram of a solid-bowl scroll centrifuge (Alfa Laval Sharpies Ltd, Camberley, UK).
(b) Cutaway view of a Sharpies Super-D-Canter continuous solid-bowl centrifuge, Model
P-5400 (Alfa Laval Sharpies Ltd, Camberley, UK).

the bowl are scraped to the conical end of the bowl. The slope of the cone helps
to remove excess liquid from the solids before discharge. The liquid phase is discharged from the opposite end of the bowl. The speed of this type of centrifuge is
limited to around 5000 rpm in larger models because of the lack of balance within
the bowl, with smaller models having bowl speeds of up to 10000 rpm. Bowl diam-
eters are normally between 0.2 and 1.5 m, with the length being up to 5 times the
diameter. Feed rates range from around 200 dm^3 h^{-1} to 200 m^3 h^{-1} depending on scale
of operation and material being processed. A number of variants on the basic design
are available:

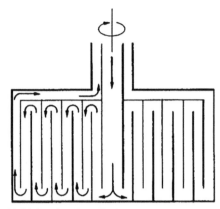

FIGURE 10.18 L.S. of a Multichamber Centrifuge

1. Cake washing facilities (screen bowl decanters).
2. Vertical bowl decanters.
3. Facility for in-place cleaning.
4. Biohazard containment features; steam sterilization in situ, two or three stage mechanical seals, control of aerosols, containment casings, and the use of high pressure sterile gas in seals to prevent the release of microorganisms.

Multichamber centrifuge

Ideally, this is a centrifuge for a slurry of up to 5% solids of particle size 0.1–200 μm diameter. In the multichamber centrifuge (Fig. 10.18), a series of concentric chambers are mounted within the rotor chamber. The broth enters via the central spindle and then takes a circuitous route through the chambers. Solids collect on the outer faces of each chamber. The smaller particles collect in the outer chambers where they are subjected to greater centrifugal forces (the greater the radial position of a particle, the greater the rate of sedimentation).

Although these vessels can have a greater solids capacity than tubular bowls and there is no loss of efficiency as the chamber fills with solids, their mechanical strength and design limits their speed to a maximum of 6500 rpm for a rotor 46-cm diameter with a holding capacity of up to 76 dm^3. Because of the time needed to dismantle and recover the solids fraction, the size and number of vessels must be of the correct volume for the solids of a batch run.

Disc-bowl centrifuge

This centrifuge relies for its efficiency on the presence of discs in the rotor or bowl (Fig. 10.19). A central inlet pipe is surrounded by a stack of stainless-steel conical discs. Each disc has spacers so that a stack can be built up. The broth to be separated flows outward from the central feed pipe, then upward and inward between the discs at an angle of 45 degrees to the axis of rotation. The close packing of the discs assists

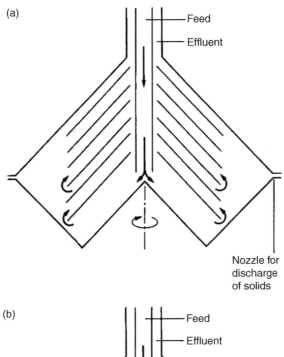

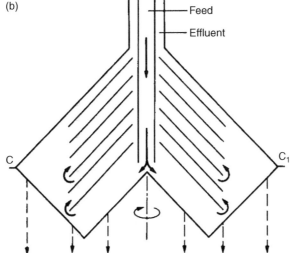

FIGURE 10.19

(a) L.S. of disc-bowl centrifuge with nozzle discharge. (b) L.S. of disc-bowl centrifuge with intermittent discharge. (Solids discharged when rotor opens intermittently along the section C–Cj.)

FIGURE 10.20 Alfa Laval BTUX 510 Disc Stack Centrifuge (Alfa Laval Sharples Ltd, Camberley, UK)

rapid sedimentation and the solids then slide to the edge of the bowl, provided that there are no gums or fats in the slurry, and eventually accumulate on the inner wall of the bowl. Ideally, the sediment should form a sludge which flows, rather than a hard particulate or lumpy sediment. The main advantages of these centrifuges are their small size compared with a bowl without discs for a given throughput. Some designs also have the facility for continuous solids removal through a series of nozzles in the circumference of the bowl or intermittent solids removal by automatic opening of the solids collection bowl. The arrangement of the discs makes this type of centrifuge laborious to clean. However, recent models such as the Alfa Laval BTUX 510 (Alfa Laval Sharpies Ltd, Camberley, Surrey, U.K.) system (Fig. 10.20) are designed to allow for cleaning in situ. In addition, this and similar plant have the facility for in situ steam sterilization and total containment, incorporating double seals to comply with containment regulations (see also Chapter 7). Feed rates range from 45 to 1800 dm^3 min^{-1}, with rotational speeds typically between 5000 and 10,000 rpm. The Westfalia CSA 19–47–476 is also steam sterilizable and has been used for the sterile collection of organisms (Walker, Narendranathan, Brown, Woolhouse, & Vranch, 1987). Similarly, the Westfalia CSA 8 can be modified for contained operation and steam sterilization (Frude & Simpson, 1993).

CELL DISRUPTION

Microorganisms are protected by extremely tough cell walls. In order to release their cellular contents a number of methods for cell disintegration have been developed (Wimpenny, 1967; Hughes, Wimpenny, & Lloyd, 1971; Harrison, 2011). Any potential method of disruption must ensure that labile materials are not denatured by the process or hydrolyzed by enzymes present in the cell. Huang, Andrews, and Asenjo (1991) report the use of a combination of different techniques to release products from specific locations within yeast cells. In this way the desired product can be obtained with minimum contamination. Although many techniques are available which are satisfactory at laboratory scale, only a limited number have been proved to be suitable for large-scale applications, particularly for intracellular enzyme extraction (Wang et al., 1979; Darbyshire, 1981). Containment of cells can be difficult or costly to achieve in many of the methods described later and thus containment requirements will strongly influence process choice. Methods available fall into two major categories:

Physicomechanical methods

1. Liquid shear.
2. Solid shear.
3. Agitation with abrasives.
4. Freeze–thawing.
5. Ultrasonication.
6. Hydrodynamic cavitation.

Chemical and biological methods

1. Detergents.
2. Osmotic shock.
3. Alkali treatment.
4. Enzyme treatment.
5. Solvents.

PHYSICOMECHANICAL METHODS

Liquid shear (high-pressure homogenizers)

Liquid shear is the method which has been most widely used in large scale enzyme purification procedures (Scawen, Atkinson, & Darbyshire, 1980). High-pressure homogenizers used in the processing of milk and other products in the food industry have proved to be very effective for microbial cell disruption. One machine, the

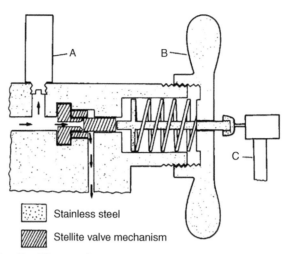

Stainless steel

Stellite valve mechanism

FIGURE 10.21 Details of Homogonizer Valve Assembly (Brookman, 1974)

A, 0–50,000 psi pressure transducer; *B,* pressure-control handwheel; *C,* linear variable displacement transformer; *(→)* direction of flow.

APV-Manton Gaulin-homogenizer (The APV Co. Ltd, Crawley, Surrey, UK), which is a high-pressure positive displacement pump, incorporates an adjustable valve with a restricted orifice (Fig. 10.21). The smallest model has one plunger, while there are several in larger models. During use, the microbial slurry passes through a nonreturn valve and impinges against the operative valve set at the selected operating pressure. The cells then pass through a narrow channel between the valve and an impact ring followed by a sudden pressure drop at the exit to the narrow orifice. There are various discharge valve designs (Fig. 10.22) with sharp edged orifices being preferred for cell disruption. The large pressure drop across the valve is believed to cause cavitation in the slurry and the shock waves so produced disrupt the cells. Brookman (1974) considered the size of the pressure drop to be very important in achieving effective disruption, and as with all mechanical methods, cell size and shape influence ease of disruption (Wase & Patel, 1985). The working pressures are extremely high. Hetherington, Follows, Dunhill, and Lilly (1971) used a pressure of 550 kg cm^{-2} for a 60% yeast suspension. A throughput of 6.4 kg soluble protein h^{-1} with 90% disruption could be achieved with a small industrial machine. In larger models, flow rates of up to 600 dm^3 h^{-1} are now possible and operating pressures of 1200 bar are utilized in some processes (Asenjo, 1990). Darbyshire (1981) has stressed the need for cooling the slurry to between 0 and 4°C to minimize loss in enzyme activity because of heat generation during the process. The increase in slurry temperature is approximately proportional to the pressure drop across the valve. Because of problems caused by heat generation and because cell suspensions can be surprisingly abrasive, it is common practice to operate such homogenizers in a multipass mode but at a lower pressure. The degree of disruption and consequently the amount of protein released will

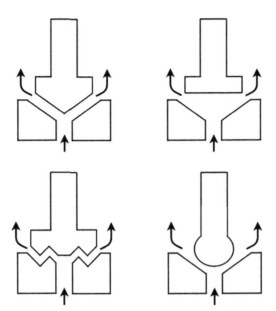

FIGURE 10.22 Discharge Valve Designs in High-Pressure Homogenizers

influence the ease of subsequent separation of the product from the cell debris in high-pressure homogenizers and bead mills (Agerkvist & Enfors, 1990). A careful balance must therefore be made between percentage release of product and the difficulty and cost of further product purification.

Solid shear

Pressure extrusion of frozen microorganisms at around $-25°C$ through a small orifice is a well established technique at a laboratory scale using a Hughes press or an X-press to obtain small samples of enzymes or microbial cell walls. Disruption is due to a combination of liquid shear through a narrow orifice and the presence of ice crystals. Magnusson and Edebo (1976) developed a semicontinuous X-press operating with a sample temperature of $-35°C$ and an X-press temperature of $-20°C$. It was possible to obtain 90% disruption with a single passage of *S. cereuisiae* using a throughput of 10 kg yeast cell paste h^{-1}. This technique might be ideal for microbial products which are very temperature labile.

Agitation with abrasives (high speed bead mills)

Mechanical cell disruption can also be achieved in a disintegrator containing a series of rotating discs/impellers on a central drive shaft and a charge of small beads. Beads are typically $0.1-3$ mm diameter depending on the type of microorganism and impeller tip speeds are in the order of 15 m s^{-1}. The beads are made of mechanically resistant materials such as glass, alumina ceramics and some titanium compounds

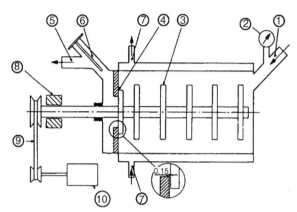

FIGURE 10.23 Simplified Drawing of the Dyno-Muhle KD5 (Mogren et al., 1974)

1, Inlet of suspension; *2,* manometer; *3,* rotating disc; *4,* slit for separation of glass beads from the suspension; *5,* outlet of suspension; *6,* thermometer; *7,* cooling water, inlet and outlet; *8,* bearings; *9,* variable V-belt drive; *10,* drive motor. Cylinder dimensions: inside length 33 cm; inside diameter 14 cm.

(Fig. 10.23). Disruption is achieved through interparticle collision and solid shear (Harrison, 2011). In a small disintegrator, the Dyno-Muhle KD5 (Wiley A. Bachofen, Basle, Switzerland), using a flow rate of 180 dm^3 h^{-1}, 85% disintegration of an 11% w/v suspension of *S. cereuisiae* was achieved with a single pass (Mogren, Lindblom, & Hedenskog, 1974). Although temperatures of up to 35°C were recorded in the disintegrator, the specific enzyme activities were not considered to be very different from values obtained by other techniques. Dissipation of heat generated in the mill is one of the major problems in scale up, though this can generally be overcome with the provision of a cooling jacket. In another disintegrator, the Netzsch LM20 mill (Netzsch GmbH, Selb, Germany), the agitator blades were alternately mounted vertically and obliquely on the horizontal shaft (Fig. 10.24). A flow rate of up to 400 dm^3 h^{-1} was claimed for a vessel with a nominal capacity of 20 dm^3 (Rehacek & Schaefer, 1977).

Freezing–thawing

Freezing and thawing of a microbial cell paste will inevitably cause ice crystals to form and their expansion followed by thawing will lead to some subsequent disruption of cells. It is slow, with limited release of cellular materials, and has not often been used as a technique on its own, although it is often used in combination with other techniques. β-Glucosidase has been obtained from *S. cerevisiae* by this method (Honig & Kula, 1976). A sample of 360 g of frozen yeast paste was thawed at 5°C for 10 h. This cycle was repeated twice before further processing.

Ultrasonication (ultrasonic cavitation)

High frequency vibration (~20 kHz) at the tip of an ultrasonication probe leads to cavitation (the formation of vapor cavities in low pressure regions), and shock waves

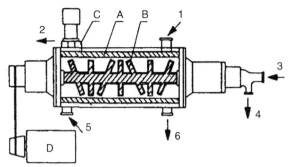

FIGURE 10.24 Simplified Drawing of the Netzsch Model LM-20 Mill (Rehacek & Schaefer, 1977)

A, cylindrical grinding vessel with cooling jacket; *B*, agitator with cooled shaft and discs; *C*, annular vibrating slot operator; *D*, variable-speed-drive motor; *1 and 2*, product inlet and outlet; *3 and 4*, agitator cooling inlet and outlet; *5 and 6*, vessel-cooling inlet and outlet.

generated when the cavities collapse cause cell disruption. The method can be very effective on a small scale (5–500 mL), but a number of serious drawbacks make it unsuitable for large-scale operations. Power requirements are high, there is a large heating effect so cooling is needed, the probes have a short working life and are only effective over a short range. Continuous laboratory sonicators with hold-up volumes of around 10 cm^3 have been shown to be effective (James, Coakley, & Hughes, 1972).

Hydrodynamic cavitation
Cavitation similar to that generated by ultrasonication probes can also be generated by fluid flow. When fluid flows through an orifice an increase in velocity is accompanied by a decrease in pressure of the fluid. When the pressure falls to the vapor pressure of the fluid cavitation occurs resulting in cell damage/disruption (Harrison, 2011).

CHEMICAL AND BIOLOGICAL METHODS
Detergents
A number of detergents will damage the lipoproteins of the microbial cell membrane and lead to the release of intracellular components. The compounds which can be used for this purpose include quaternary ammonium compounds, sodium lauryl sulfate, sodium dodecyl sulfate (SDS) and Triton X-100. Anionic detergents such as SDS disorganize the cell membrane while cationic detergents are believed to act on lipopolysaccharides and phospholipids of the membrane. Nonionic detergents such as Triton X-100 cause partial solubilization of membrane proteins (Harrison, 2011). Unfortunately, the detergents may cause some protein denaturation and may need to be removed before further purification stages can be undertaken. The stability of the desired product must be determined when using any detergent system. Pullulanase is an enzyme which is bound to the outer membrane of *Klebsiella pneumoniae*.

The cells were suspended in pH 7.8 buffer and 1% sodium cholate was added. The mixture was stirred for 1 h to solubilize most of the enzyme (Kroner, Hustedt, Granda, & Kula, 1978). The use of Triton X-100 in combination with guanidine-HCl is widely and effectively used for the release of cellular protein (Naglak & Wang, 1992; Hettwer & Wang, 1989), Hettwer and Wang (1989) obtaining greater than 75% protein release in less than 1 h from *Escherichia coli* under fermentation conditions.

Osmotic shock

Osmotic shock caused by a sudden change in salt concentration will cause disruption of a number of cell types. Cells are equilibrated to high osmotic pressure (typically 1 M salt solutions). Rapid exposure to low osmotic pressure causes water to quickly enter the cell. This increases the internal pressure of the cell resulting in cell lysis. Osmotic shock is of limited application except where the cell wall is weakened or absent. Application on a large scale is limited by the cost of chemicals, increased water use, and possible product dilution (Harrison, 2011).

Alkali treatment

Alkali treatment might be used for hydrolysis of microbial cell wall material provided that the desired product will tolerate a pH of 10.5–12.5 for up to 30 min. Chemical costs can be high both in terms of alkali required and neutralization of the resulting lysate. Darbyshire (1981) has reported the use of this technique in the extraction of L-asparaginase.

Enzyme treatment

There are a number of enzymes which hydrolyze specific bonds in cell walls of a limited number of microorganisms. Enzymes shown to have this activity include lysozyme, produced from hen egg whites and other natural sources, and other enzyme extracts from leucocytes, *Streptomyces* spp., *Staphylococcus* spp., *Micromonospora* spp. *Penicillium* spp., *Trichoderma* spp., and snails. Lysozyme hydrolyses β-1-4 glucosidic bonds in the polysaccharide chains of peptidoglycan causing cell lysis. Although this is probably one of the most gentle methods available, unfortunately it is relatively expensive and the presence of the enzyme(s) may complicate further downstream purification processes. Enzyme lysis in large scale operations is limited by the availability and cost of appropriate enzymes. The use of immobilized lysozyme has been investigated by a number of workers and may provide the solution to such problems (Crapisi, Lante, Pasini, & Spettoli, 1993).

Solvents

Solvents extract the lipid components of the cell membrane causing the release of intracellular components and are applicable across a wide range of microorganisms. Solvents used include alcohols, dimethyl sulfoxide, methyl ethyl ketone, and toluene. However, their toxicity, flammability, and ability to cause protein denaturation requires careful consideration (Harrison, 2011).

Chemical and enzymatic methods for the release of intracellular products have not been used widely on a large scale, with the exception of lysozyme. However, their potential for the selective release of product and that they often yield a cleaner lysate mean that they are potentially invaluable tools in the recovery of fermentation products (Andrews & Asenjo, 1987; Andrews, Huang, & Asenjo, 1990; Harrison, 2011). Enzymes may also be used as a pretreatment to partially hydrolyze cell walls prior to cell disruption by mechanical methods.

LIQUID–LIQUID EXTRACTION

The separation of a component from a liquid mixture by treatment with a solvent in which the desired component is preferentially soluble is known as liquid–liquid extraction. The specific requirement is that a high percentage extraction of product must be obtained but concentrated in a smaller volume of solvent.

Prior to starting a large-scale extraction, it is important to find out on a small scale the solubility characteristics of the product using a wide range of solvents. A simple rule to remember is that "like dissolves like." The important "likeness" as far as solubility relations are concerned is in the polarities of molecules. Polar liquids mix with each other and dissolve salts and other polar solids. The solvents for nonpolar compounds are liquids of low or nil polarity.

The dielectric constant is a measure of the degree of molar polarization of a compound. If this value is known it is then possible to predict whether a compound will be polar or nonpolar, with a high value indicating a highly polar compound. The dielectric constant D of a substance can be measured by determining the electrostatic capacity C of a condenser containing the substance between the plates. If C_0 is the value for the same condenser when completely evacuated then

$$D = \frac{C}{C_0}$$

Experimentally, dielectric constants are obtained by comparing the capacity of the condenser when filled with a given liquid with the capacity of the same condenser containing a standard liquid whose dielectric constant is known very accurately. If D_1 and D_2 are the dielectric constants of the experimental and standard liquids and C_1 and C_2 are the electrostatic capacities of a condenser when filled with each of the liquids, then

$$\frac{D_1}{D_2} = \frac{C_1}{C_2}$$

The value of D_1 can be calculated since C_1 and C_2 can be measured and D_2 is known. The dielectric constants for a number of solvents are given in Table 10.1.

The final choice of solvent will be influenced by the distribution or partition coefficient K where

$$K = \frac{\text{Concentration of solute in extract}}{\text{Concentration of solute in raffinate}}$$

Table 10.1 Dielectric Constants of Solvents at 25°C (Arranged in Order of Increasing Polarity)

Solvent	Dielectric Constant
Hexane	1.90 (least polar)
Cyclohexane	2.02
Carbon tetrachloride	2.24
Benzene	2.28
Di-ethyl ether	4.34
Chloroform	4.87
Ethyl acetate	6.02
Butan-2-ol	15.8
Butan-l-ol	17.8
Propan-l-ol	20.1
Acetone	20.7
Ethanol	24.3
Methanol	32.6
Water	78.5 (most polar)

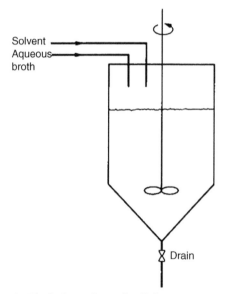

FIGURE 10.25 Diagram of a Single-Stage Extraction Unit

The value of K defines the ease of extraction. When there is a relatively high K value, good stability of product and good separation of the aqueous and solvent phases, then it may be possible to use a single-stage extraction system (Fig. 10.25). A value of 50 indicates that the extraction should be straightforward whereas a value of 0.1 shows that the extraction will be difficult and that a multistage process will be

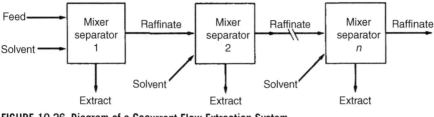

FIGURE 10.26 Diagram of a Cocurrent Flow Extraction System

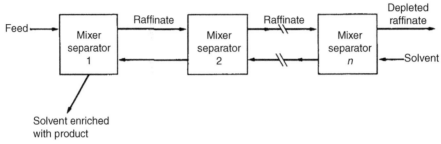

FIGURE 10.27 Diagram of a Countercurrent Extraction System

necessary. Unfortunately, in a number of systems the value of K is low and cocurrent or countercurrent multistage systems have to be utilized. The cocurrent system is illustrated in Fig. 10.26. There are n mixer/separator vessels in line and the raffinate goes from vessel 1 to vessel n. Fresh solvent is added to each stage, the feed and extracting solvent pass through the cascade in the *same* direction. Extract is recovered from each stage. Although a relatively large amount of solvent is used, a high degree of extraction is achieved.

A countercurrent system is illustrated in Fig. 10.27. There are a number of mixer/separators connected in series. The extracted raffinate passes from vessel 1 to vessel n while the product-enriched solvent is flowing from vessel n to vessel 1. The feed and extracting solvent pass through the cascade in *opposite* directions. The most efficient system for solvent utilization is countercurrent operation, showing a considerable advantage over batch and cocurrent systems. Unless there are special reasons the counter-current system should be used. In practice, the series of countercurrent extractions are conducted in a single continuous extractor using centrifugal forces to separate the two liquid phases. The two liquid streams are forced to flow countercurrent to each other through a long spiral of channels within the rotor.

The Podbielniak centrifugal extractor (Fig. 10.28) consists of a horizontal cylindrical drum revolving at up to 5000 rpm about a shaft passing through its axis. The liquids to be run countercurrent are introduced into the shaft, with the heavy liquid entering the

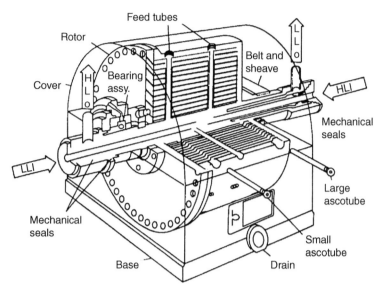

FIGURE 10.28 Diagram of the Podbielniak Extractor (Queener & Swartz, 1979)

HLI, LLI, HLO, and LLO indicate heavy and light liquid in and out.

drum at the shaft while the light liquid is led by an internal route to the periphery of the drum. As the drum rotates, the heavy liquid is forced to the periphery of the drum by centrifugal action where it contacts the light liquid. The solute is transferred between the liquids and the light liquid is displaced back toward the axis of the drum. The heavy liquid is returned to the drum's axis via internal channels. The two liquid streams are then discharged via the shaft. Flow rates in excess of $100,000$ dm^3 h^{-1} are possible in the largest models. Probably the most useful property of this type of extractor is the low hold-up volume of liquid in the machine compared with the throughput.

Penicillin G is an antibiotic which is recovered from fermentation broths by centrifugal countercurrent solvent extraction. At neutral pHs in water penicillin is ionized, while in acid conditions this ionization is suppressed and the penicillin is more soluble in organic solvents. At pH 2 to 3, the distribution ratio of total acid will be

$$K = \frac{(\text{RCOOH})\text{org}}{(\text{RCOOH})\text{aq} + (\text{RCOO}^-)\text{aq}}$$

For penicillin this value may be as high as 40 in a suitable solvent (Podbielniak, Kaiser, & Ziegenhorn, 1970). The penicillin extraction process may involve the four following stages:

1. Extraction of the penicillin G from the filtered broth into an organic solvent (amyl or butyl acetate or methyl iso-butyl ketone).
2. Extraction from the organic solvent into an aqueous buffer.
3. Extraction from aqueous buffer into organic solvent.
4. Extraction of the solvent to obtain the penicillin salt.

At each extraction stage progressively smaller volumes of extradant are used to achieve concentration of the penicillin (Fig. 10.2). Unfortunately, penicillin G has a half-life of 15 min at pH 2.0 at 20°C. The harvested broth is therefore initially cooled to 0–3°C. The cooled broth is then acidified to pH 2–3 with sulfuric or phosphoric acid immediately before extraction. This acidified broth is quickly passed through a Podbielniak centrifugal countercurrent extractor using about 20% by volume of the solvent in the counter flow. Ideally, the hold-up time should be about 60–90 s. The penicillin-rich solvent then passes through a second Podbielniak extractor counter-current to an aqueous NaOH or KOH solution (again about 20% by volume) so that the penicillin is removed to the aqueous phase (pH 7.0 to 8.0) as the salt.

$$RCOOH(org) + NaOH(aq) \rightarrow RCOO^-Na^+ + H_2O$$

These two stages may be sufficient to concentrate the penicillin adequately from a broth with a high titer. Penicillin will crystallize out of aqueous solution at a concentration of approximately 1.5×10^6 units cm^{-3}. If the broth harvested initially contains 60,000 units cm^{-3}, and two fivefold concentrations are achieved in the two extraction stages, then the penicillin liquor should crystallize. If the initial broth titer is lower than 60,000 units cm^{-3} or the extractions are not so effective, the solvent and buffer extractions will have to be repeated. At each stage the spent liquids should be checked for residual penicillin and solvent usage carefully monitored. Since the solvents are expensive and their disposal is environmentally sensitive they are recovered for recirculation through the extraction process. The success of a process may depend on efficient solvent recovery and reuse.

SOLVENT RECOVERY

A major item of equipment in an extraction process is the solvent-recovery plant which is usually a distillation unit. It is not normally essential to remove all the raffinate from the solvent as this will be recycled through the system. In some processes the more difficult problem will be to remove all the solvent from the raffinate because of the value of the solvent and problems which might arise from contamination of the product.

Distillation may be achieved in three stages:

1. Evaporation, the removal of solvent as a vapor from a solution.
2. Vapor–liquid separation in a column, to separate the lower boiling more volatile component from other less volatile components.
3. Condensation of the vapor, to recover the more volatile solvent fraction.

Evaporation is the removal of solvent from a solution by the application of heat to the solution. A wide range of evaporators is available. Some are operated on a batch basis and others continuously. Most industrial evaporators employ tubular heating surfaces. Circulation of the liquid past the heating surfaces may be induced by boiling or by mechanical agitation. In batch distillation (Fig. 10.29), the vapor from the

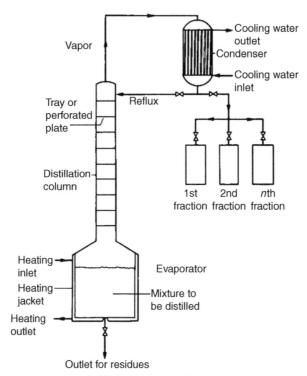

FIGURE 10.29 Diagram of a Batch Distillation Plant With a Tray or Perforated-Plate Column

boiler passes up the column and is condensed. Part of the condensate will be returned as the reflux for countercurrent contact with the rising vapor in the column. The distillation is continued until a satisfactory recovery of the lower-boiling (more volatile) component(s) has been accomplished. The ratio of condensate returned to the column as reflux to that withdrawn as product is, along with the number of plates or stages in the column, the major method of controlling the product purity. A continuous distillation (Fig. 10.30) is initially begun in a similar way as with a batch distillation, but no condensate is withdrawn initially. There is total reflux of the condensate until ideal operating conditions have been established throughout the column. At this stage the liquid feed is fed into the column at an intermediate level. The more volatile components move upward as vapor and are condensed, followed by partial reflux of the condensate. Meanwhile, the less volatile fractions move down the column to the evaporator (reboiler). At this stage part of the bottoms fraction is continuously withdrawn and part is reboiled and returned to the column.

Countercurrent contacting of the vapor and liquid streams is achieved by causing:

1. Vapor to be dispersed in the liquid phase (plate or tray column),
2. Liquid to be dispersed in a continuous vapor phase (packed column).

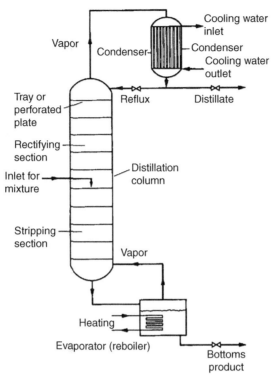

FIGURE 10.30 Diagram of a Continuous Distillation Plant With a Tray or Perforated-Plate Column

The plate or tray column consists of a number of distinct chambers separated by perforated plates or trays. The rising vapor bubbles through the liquid which is flowing across each plate, and is dispersed into the liquid from perforations (sieve plates) or bubble caps. The liquid flows across the plates and reaches the reboiler by a series of overflow wires and down pipes.

A packed tower is filled with a randomly packed material such as rings, saddles, helices, spheres or beads. Their dimensions are approximately one-tenth to one-fiftieth of the diameter of the column and are designed to provide a large surface area for liquid-vapor contacting and high voidage to allow high throughput of liquid and vapor.

The heat input to a distillation column can be considerable. The simplest ways of conserving heat are to preheat the initial feed by a heat exchanger using heat from:

1. The hot vapors at the top of the column,
2. Heat from the bottoms fraction when it is being removed in a continuous process,
3. A combination of both.

Since it is beyond the scope of this text to consider the distillation process more fully the reader is therefore directed to Coulson and Richardson (1991).

TWO-PHASE AQUEOUS EXTRACTION

Liquid–liquid extraction is a well established technology in chemical processing and in certain sectors of biochemical processing. However, the use of organic solvents has limited application in the processing of sensitive biologicals. Aqueous two-phase systems, on the other hand, have a high water content and low interfacial surface tension and are regarded as being biocompatible (Mattiasson & Ling, 1987).

Two-phase aqueous systems have been known since the late 19th century, and a large variety of natural and synthetic hydrophilic polymers are used today to create two (or more) aqueous phases. Phase separation occurs when hydrophilic polymers are added to an aqueous solution, and when the concentrations exceed a certain value two immiscible aqueous phases are formed. Settling time for the two phases can be prolonged, depending on the components used and vessel geometry. Phase separation can be improved by using centrifugal separators (Huddlestone et al., 1991), or novel techniques such as magnetic separators (Wikstrom, Flygare, Grondalen, & Larsson, 1987).

Many systems are available:

1. Nonionic polymer/nonionic polymer/water, for example, polyethylene glycol/dextran.
2. Polyelectrolyte/nonionic polymer/water, for example, sodium carboxymethyl cellulose/polyethylene glycol.
3. Polyelectrolyte/polyelectrolyte/water, for example, sodium dextran sulfate/sodium carboxymethyl cellulose.
4. Polymer/low molecular weight component/water, for example, dextran/propyl alcohol.

The distribution of a solute species between the phases is characterized by the partition coefficient, and is influenced by a number of factors such as temperature, polymer (type and molecular weight), salt concentration, ionic strength, pH, and properties (eg, molecular weight) of the solute. As the goal of any extraction process is to selectively recover and concentrate a solute, affinity techniques such as those applied in chromatographic processes can be used to improve selectivity. Examples include the use of PEG–NADH derivatives in the extraction of dehydrogenases, p-aminobenzamidine in the extraction of trypsin and cibacron blue in the extraction of phosphofructokinase. It is possible to use different ligands in the two phases leading to an increase in selectivity or the simultaneous recovery and separation of several species (Cabral & Aires-Barros, 1993). Wohlgemuth (2011) reports that a comparison of aqueous two-phase separation and ion-exchange chromatography shows that the process yield and costs are lower for the aqueous two-phase process. Two phase aqueous systems have found application in the purification of many solutes; proteins, enzymes (Gonzalez, Pencs, & Casas, 1990; Guan, Wu, Treffry, & Lilley, 1992), recombinant proteins using a PEG/salt system (Gu & Glatz, 2006), β-carotenene and lutein from cyanobacterial fermentations utilizing a PEG/salt process, (Chavez-Santoscoy, 2010), antibiotics (Guan, Zhu, & Mei, 1996), cells and subcellular particles, and in extractive bioconversions. The cost of phase forming polymers and chemicals have limited the use of aqueous two-phase processes in industrial applications (Wohlgemuth, 2011), however some aqueous two-phase systems for handling large-scale protein separation have emerged, the majority of which use PEG as the upper phase forming polymer with either dextran, concentrated salt solution or hydroxypropyl starch as the lower phase forming material (Mattiasson & Kaul, 1986). Hustedt, Kroner, and Papamichael (1988) demonstrated the application of continuous cross-current extraction of enzymes (fumarase and penicillin acylase) by aqueous two-phase systems at production scale.

REVERSED MICELLE EXTRACTION

Reversed micelle extraction is potentially an attractive alternative to conventional solvent extraction for the recovery of bioproducts as the solute of interest remains in an aqueous environment at all times and hence can be considered "biocompatible." A reversed micelle is a nanometer scale droplet of aqueous solution stabilized in a nonpolar environment by a surfactant at the interface between the two liquids (Fig. 10.31). The minimum concentration of surfactant required for micelles is known as the critical micelle concentration (CMC) and is highly system specific. There are a number of limitations to reversed micelle extraction including costs and low rates of mass transfer (Krijgsman, 1992).

SUPERCRITICAL FLUID EXTRACTION

The technique of supercritical fluid extraction utilizes the dissolution power of supercritical fluids, that is, fluids above their critical temperature and pressure. Its advantages include the use of moderate temperatures, good solvent, and transport properties (high diffusivity and low viscosity), and that cheap and nontoxic fluids are available.

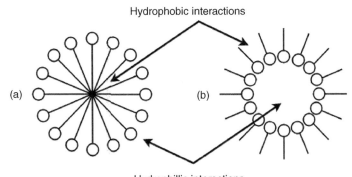

Hydrophobic interactions

(a)   (b)

Hydrophillic interactions

FIGURE 10.31 Micelles

(a) Normal micelle. (b) Reversed micelle.

Supercritical fluids are used in the extraction of hop oils, caffeine, vanilla, vegetable oils, and β-carotene. It has also been shown experimentally that the extraction of certain steroids and chemotherapeutic drugs can be achieved using supercritical fluids. Other current and potential uses include the removal of undesirable substances such as pesticide residues, removal of bacteriostatic agents from fermentation broths, the recovery of organic solvents from aqueous solutions, cell disintegration, destruction and treatment of industrial wastes, and liposome preparation. There are, however, a number of significant disadvantages in the utilization of this technology:

1. Phase equilibria of the solvent/solute system is complex, making design of extraction conditions difficult.
2. The most popular solvent (carbon dioxide) is nonpolar and is therefore most useful in the extraction of nonpolar solutes. Though cosolvents can be added for the extraction of polar compounds, they will complicate further downstream processing.
3. The use of high pressures leads to high capital costs for plant, and operating costs may also be high.

Thus, the number of commercial processes utilizing supercritical fluid extraction is relatively small, due mainly to the existence of more economical processes. However, its use is increasing, for example, the recovery of high value biologicals, when conventional extractions are inappropriate, and in the treatment of toxic wastes (Bruno, Nieto De Castro, Hamel, & Palavra, 1993). Super critical CO_2 extraction has been described by Fabre, Condoret, and Marty (1999) for the extraction of 2-phenylethyl alcohol (rose aroma) from cell free extracts of *Kluyveromyces marxianus* fermentations at pressures of 200 bar and temperatures between 35 and 45°C. The distribution coefficient was found to be twice that of conventional solvent extraction using *n*-hexane with greater than 90% product extraction.

ADSORPTION

Adsorption can be a useful technique for the separation of a product from a dilute aqueous phase and the use of polymer absorbers for the recovery of small molecules is well established. A range of polymers (eg, ion-exchangers) are available on a large scale. After extraction of the product onto the absorber, the product can then be recovered by solvent elution/extraction and the absorber can then be recycled (Wohlgemuth, 2011). Kujawska, Kujawski, Bryjak, and Kujawski (2015) report the application of absorptive polymers and zeolites for product recovery in acetone-butanol-ethanol fermentations. The use of ion-exchange resins for the recovery of lactic acid have been described by Dave, Patil, and Suresh (1997), Rincon, Fuertes, Rodriguez, Rodriguez, and Monteagudo (1997), and Moldes, Alonso, and Parajo (2003), and recombinant hepatitis B antigen by Ng et al. (2006). Further details of adsorption, ionexchange etc. are further covered in the later section on chromatography.

REMOVAL OF VOLATILE PRODUCTS

Distillation (evaporation) is a well established process, which can be used for the separation of volatile products from less volatile materials. Examples of products include ethanol (both alcoholic beverages and biofuel), flavors, and fragrances. Batch and continuous fractional distillation has been addressed earlier in this chapter (solvent recovery).

A relatively new emerging membrane based alternative to distillation for the recovery of volatile products is pervaporation (Vane, 2005). The term is derived from *Per*meation and E*vaporation*. In the process a liquid stream containing two or more miscible components is in contact with one side of a polymeric or inorganic membrane. Components from the liquid stream permeate through the membrane and the "permeate" evaporates into the vapor phase. This vapor can then be condensed. Different chemical entities will have differing affinity for the membrane and different permeation rates and thus with the correct choice of membrane, the desired product can be concentrated into the vapor phase. For example, if the membrane is hydrophobic in nature it will concentrate hydrophobic molecules. A schematic representation of a pervaporation process is shown in Fig. 10.32. There are numerous reported examples of pervaporation being used for the recovery of ethanol (Levandowicz et al., 2011; Staniszewski, Kujawski, & Lewandowska, 2007; Vane, 2005; Ikegami et al., 2003) and acetone-butanol-ethanol (Liu, Liu, & Feng, 2005; Cai et al., 2016; Kujawska et al., 2015).

Other techniques for the recovery of volatile products include flash extraction in acetone-butanol fermentations (Shi, Zhang, Chen, & Mao, 2005) and gas stripping in acetone-butanol-ethanol fermentations (Qureshi & Blaschek, 2001; Kujawska et al., 2015).

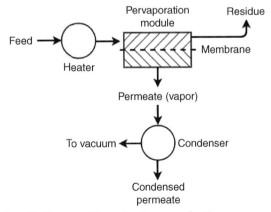

FIGURE 10.32 Schematic Representation of a Pervaporation Process

CHROMATOGRAPHY

In many fermentation processes, chromatographic techniques are used to isolate and purify relatively low concentrations of metabolic products. Chromatographic methods separate solutes based on charge, polarity, size, and affinity. In this context, chromatography will be concerned with the passage and separation of different solutes as liquid (the mobile phase) is passed through a column, that is, *liquid chromatography.* Gas chromatography, when the mobile phase is a gas, is a widely used analytical technique but has little application in the recovery of fermentation products. Depending on the mechanism by which the solutes may be differentially held in a column, the techniques can be grouped as follows:

1. Adsorption chromatography.
2. Ion-exchange chromatography.
3. Gel permeation chromatography.
4. Affinity chromatography.
5. Reverse phase chromatography.
6. High performance liquid chromatography.

Chromatographic techniques are also used in the final stages of purification of a number of products. The scale-up of chromatographic processes can prove difficult, mainly as a result of the pressures used causing compaction of the column packing materials, and there is much current interest in the use of mathematical models and computer programs to translate data obtained from small-scale processes into operating conditions for larger scale applications (Cowan, Gosling, & Laws, 1986; Cowan, Gosling, & Sweetenham, 1987).

ADSORPTION CHROMATOGRAPHY

Adsorption chromatography involves binding of the solute to the solid phase primarily by weak Van de Waals forces. The materials used for this purpose to pack columns include inorganic adsorbents (active carbon, aluminum oxide, aluminum hydroxide, magnesium oxide, silica gel) and organic macroporous resins. Adsorption and affinity chromatography are mechanistically identical, but are strategically different. In affinity systems selectivity is designed rationally while in adsorption selectivity must be determined empirically.

Dihydrostreptomycin can be extracted from filtrates using activated charcoal columns. It is then eluted with methanolic hydrochloric acid and purified in further stages (Nakazawa, Shibata, Tanabe, & Yamamoto, 1960). Some other applications for small-scale antibiotic purification are quoted by Weinstein and Wagman (1978). Active carbon may be used to remove pigments to clarify broths. Penicillin-containing solvents may be treated with 0.25–0.5% active carbon to remove pigments and other impurities (Sylvester & Coghill, 1954).

Macro-porous adsorbents have also been tested. The first synthetic organic macro-porous adsorbents, the Amberlite XAD resins, were produced by Rohm and

Haas in 1965. These resins have surface polarities, which vary from nonpolar to highly polar and do not possess any ionic functional groups. Voser (1982) considers their most interesting application to be in the isolation of hydrophilic fermentation products. He stated that these resins would be used at Ciba-Geigy in recovery of cephalosporin C (acidic amino acid), cefotiam (basic amino acid), desferrioxamine B (basic hydroxamic acid) and paramethasone (neutral steroid).

ION EXCHANGE

Ion exchange can be defined as the reversible exchange of ions between a liquid phase and a solid phase (ion-exchange resin) which is not accompanied by any radical change in the solid structure. Cationic ion-exchange resins normally contain a sulfonic acid, carboxylic acid, or phosphonic acid active group. Carboxy-methyl cellulose is a common cation exchange resin. Positively charged solutes (eg, certain proteins) will bind to the resin, the strength of attachment depending on the net charge of the solute at the pH of the column feed. After deposition solutes are sequentially washed off by the passage of buffers of increasing ionic strength or pH. Anionic ion-exchange resins normally contain a secondary amine, quaternary amine, or quaternary ammonium active group. A common anion exchange resin, DEAE (diethylaminoethyl) cellulose is used in a similar manner to that described earlier for the separation of negatively charged solutes. Other functional groups may also be attached to the resin skeleton to provide more selective behavior similar to that of affinity chromatography. The appropriate resin for a particular purpose will depend on various factors such as bead size, pore size, diffusion rate, resin capacity, range of reactive groups, and the life of the resin before replacement is necessary. Weak-acid cation ion-exchange resins can be used in the isolation and purification of streptomycin, neomycin, and similar antibiotics.

In the recovery of streptomycin, the harvested filtrate is fed on to a column of a weak-acid cationic resin such as Amberlite IRC 50, which is in the sodium form. The streptomycin is adsorbed on to the column and the sodium ions are displaced.

$$RCOO^- Na^+_{(resin)} + streptomycin$$
$$\rightarrow RCOO^- streptomycin^+_{(resin)} + NaOH$$

Flow rates of between 10 and 30 bed volumes per hour have been used. The resin bed is now rinsed with water and eluted with dilute hydrochloric acid to release the bound streptomycin.

$$RCOO^- streptomycin^+_{(resin)} + HCl$$
$$\rightarrow RCOOH_{(resin)} + streptomycin^+ Cl^-$$

A slow flow is used to ensure the highest recovery of streptomycin using the smallest volume of eluent. In one step, the antibiotic has been both purified and concentrated, may be more than 100-fold. The resin column is regenerated to the sodium form by passing an adequate volume of NaOH slowly through the column and rinsing with distilled water to remove excess sodium ions.

$$RCOOH_{(resin)} + NaOH \rightarrow RCOO^-Na^+_{(resin)} + H_2O$$

The resin can have a capacity of 1 g of streptomycin g^{-1} resin. Commercially, it is not economic to regenerate the resin completely, therefore the capacity will be reduced. In practice, the filtered broth is taken through two columns in series while a third is being eluted and regenerated. When the first column is saturated, it is isolated for elution and regeneration while the third column is brought into operation.

Details for isolation of some other antibiotics are given in Weinstein and Wagman (1978). Ion-exchange chromatography may be combined with HPLC in, for example, the purification of somatotropin using DEAE cellulose columns and β-urogastrone in multigram quantities using a cation exchange column (Brewer & Larsen, 1987).

GEL PERMEATION

This technique is also known as gel exclusion and gel filtration. Gel permeation separates molecules on the basis of their size. The smaller molecules diffuse into the gel more rapidly than the larger ones, and penetrate the pores of the gel to a greater degree. This means that once elution is started, the larger molecules which are still in the voids in the gel will be eluted first. A wide range of gels are available, including crosslinked dextrans (Sephadex and Sephacryl) and crosslinked agarose (Sepharose) with various pore sizes depending on the fractionation range required.

One early industrial application, although on a relatively small scale, was the purification of vaccines (Latham, Michelsen, & Edsall, 1967). Tetanus and diphtheria broths for batches of up to 100,000 human doses are passed through a 13 dm^3 column of G 100 followed by a 13 dm^3 column of G 200. This technique yields a fairly pure fraction which is then concentrated 10-fold by pressure dialysis to remove the eluent buffer (Na$_2$HPO$_4$).

AFFINITY CHROMATOGRAPHY

Affinity chromatography is a separation technique with many applications since it is possible to use it for separation and purification of most biological molecules on the basis of their function or chemical structure. This technique depends on the highly specific interactions between pairs of biological materials such as enzyme–substrate, enzyme–inhibitor, antigen–antibody, etc. The molecule to be purified is specifically adsorbed from, for example, a cell lysate applied to the affinity column by a binding substance (ligand) which is immobilized on an insoluble support (matrix). Eluent is then passed through the column to release the highly purified and concentrated molecule. The ligand is attached to the matrix by physical absorption or chemically by a covalent bond. The pore size and ligand location must be carefully matched to the size of the product for effective separation. The latter method is preferred whenever possible. Porath (1974) and Yang and Tsao (1982) have reviewed methods and coupling procedures.

Coupling procedures have been developed using cyanogen bromide, bisoxiranes, disaziridines, and perio-dates, for matrixes of gels and beads. Four polymers, which are often used for matrix materials are agarose, cellulose, dextrose, and polyacrylamide. Agarose activated with cyanogen bromide is one of the most commonly used supports for the coupling of amino ligands. Silica based solid phases have been shown to be an effective alternative to gel supports in affinity chromatography (Mohan & Lyddiatt, 1992).

Purification may be several 1000-fold with good recovery of active material. The method can however be quite costly and time consuming, and alternative affinity methods such as affinity cross-flow filtration, affinity precipitation, and affinity partitioning may offer some advantages (Janson, 1984; Luong, Nguyen, & Male, 1987). Affinity chromatography was used initially in protein isolation and purification, particularly enzymes. Since then many other large-scale applications have been developed for enzyme inhibitors, antibodies, interferon, and recombinant proteins (Janson & Hedman, 1982; Ostlund, 1986; Folena-Wasserman, Inacker, & Rosenbloom, 1987; Nachman, Azad, & Bailón, 1992), and on a smaller scale for nucleic acids, cell organelles, and whole cells (Yang & Tsao, 1982). In the scale-up of affinity chromatographic processes (Katoh, 1987) bed height limits the superficial velocity of the liquid, thus scale-up requires an increase in bed diameter or adsorption capacity.

REVERSE PHASE CHROMATOGRAPHY (RPC)

When the stationary phase has greater polarity than the mobile phase it is termed "normal phase chromatography." When the opposite is the case, it is termed "reverse phase chromatography." RPC utilizes a solid phase (eg, silica) which is modified so as to replace hydrophilic groups with hydrophobic alkyl chains. This allows the separation of proteins according to their hydrophobicity. More-hydrophobic proteins bind most strongly to the stationary phase and are therefore eluted later than less-hydrophobic proteins. The alkyl groupings are normally eight or eighteen carbons in length (C_8 and C_{18}). RPC can also be combined with affinity techniques in the separation of, for example, proteins and peptides (Davankov, Kurganov, & Unger, 1990).

HIGH PERFORMANCE LIQUID CHROMATOGRAPHY (HPLC)

HPLC is a high resolution column chromatographic technique. Improvements in the nature of column packing materials for a range of chromatographic techniques (eg, gel permeation and ion-exchange) yield smaller, more rigid, and more uniform beads. This allows packing in columns with minimum spaces between the beads, thus minimizing peak broadening of eluted species. It was originally known as high *pressure* liquid chromatography because of the high pressures required to drive solvents through silica based packed beds. Improvements in the performance led to the name change and its widespread use in the separation and purification of a wide range of solute species, including biomolecules. HPLC is distinguished from liquid chromatography by the use of improved media (in terms of their selectivity and physical properties) for the solid (stationary) phase through which the mobile (fluid) phase passes.

The stationary phase must have high surface area/unit volume, even size and shape and be resistant to mechanical and chemical damage. However, it is factors such as these which lead to high pressure requirements and cost. This may be acceptable for analytical work, but not for preparative separations. Thus, in preparative HPLC some resolution is often sacrificed (by the use of larger stationary-phase particles) to reduce operating and capital costs. For very high value products, large-scale HPLC columns containing analytical media have been used. Fast protein liquid chromatography (FPLC) is a variant of HPLC which is more suited to large scale purification processes (Doran, 2013).

Affinity techniques can be merged with HPLC to combine the selectivity of the former with the speed and resolving power of the latter (Forstecher, Hammadi, Bouzerna, & Dautre-vaux, 1986; Shojaosadaty & Lyddiatt, 1987).

CONTINUOUS CHROMATOGRAPHY

Although the concept of continuous enzyme isolation is well established (Dunnill & Lilly, 1972), the stage of least development is continuous chromatography. Jungbauer (2013) described the possible continuous chromatography alternatives to batch chromatography. These include annular chromatography, carrousel chromatography, and various configurations of moving bed chromatography. However as far back as 1969, Fox, Calhoun, and Eglinton (1969) developed a continuous-fed annular column for this purpose (Fig. 10.33). It consisted of two concentric cylindrical sections clamped

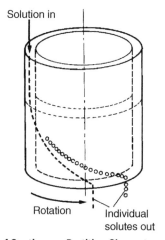

FIGURE 10.33 The Principle of Continuous-Partition Chromatography

---, faster-moving component; ○ ○, slower-moving component (Fox et al., 1969).

to a base plate. The space (1 cm wide) between the two sections was packed with the appropriate resin or gel giving a total column capacity of 2.58 dm^3. A series of orifices in the circumference of the base plate below the column space led to collecting vessels. The column assembly was rotated in a slow-moving turntable (0.4–2.0 rpm). The mixture for separation was fed to the apparatus by an applicator rotating at the same speed as the column, thus allowing application at a fixed point, while the eluent was fed evenly to the whole circumference of the column. The components of a mixture separated as a series of helical pathways, which varied with the retention properties of the constituent components. This method gave a satisfactory separation and recovery but the consumption of eluent and the unreliable throughput rate were not considered to be satisfactory for a large-scale method (Nicholas & Fox, 1969; Dunnill & Lilly, 1972).

MEMBRANE PROCESSES
ULTRAFILTRATION AND REVERSE OSMOSIS

Both processes utilize semipermeable membranes to separate molecules of different sizes and therefore act in a similar manner to conventional filters.

ULTRAFILTRATION

Ultrafiltration can be described as a process in which solutes of high molecular weight are retained when the solvent and low molecular weight solutes are forced under hydraulic pressure (between 2 and 10 atmospheres) through a membrane of a very fine pore size, typically between 0.001 and 0.1 μm. It is therefore used for product concentration and purification. A range of membranes made from a variety of polymeric materials, with different molecular weight cut-offs (500–500,000), are available which makes possible the separation of macromolecules such as proteins, enzymes, hormones, and viruses. It is practical only to separate molecules whose molecular weights are a factor of ten different due to variability in pore size (Heath & Belfort, 1992). Because the flux through such a membrane is inversely proportional to its thickness, asymmetric membranes are used where the membrane (~0.3 μm thick) is supported by a mesh around 0.3 mm thick.

When considering the feasibility of ultrafiltration, it is important to remember that factors other than the molecular weight of the solute affect the passage of molecules through the membranes (Melling & Westmacott, 1972). There may be concentration polarization caused by accumulation of solute at the membrane surface, which can be reduced by increasing the shear forces at the membrane surface either by conventional agitation or by the use of a cross-flow system (see previous section). Second, a slurry of protein may accumulate on the membrane surface forming a gel layer which is not easily removed by agitation. Formation of the gel layer may be partially controlled by careful choice of conditions such as pH (Bailey & Ollis, 1986). Finally, equipment and energy costs may be considerable because of the high pressures necessary; this also limits the life of ultrafiltration membranes.

There are numerous examples of the use of ultrafiltration for the concentration and recovery of biomolecules: viruses (Weiss, 1980), enzymes (Atkinson & Mavituna, 1991), antibiotics (Pandey et al., 1985), xanthan gum (Lo, Yang, & Min, 1997), and surfactin (Chen, Chen, & Juang, 2007). Tessier, Bouchard, and Rahni (2005) described the application of ultrafiltration coupled to nanofiltration in the purification of benzylpenicillin. The nanofiltration membrane was used to concentrate the permeates produced by ultrafiltration with recoveries of ~90% being obtained. Details of large scale applications are given by Lacey and Loeb (1972) and by Ricketts, Lebherz, Klein, Gustafson, and Flickinger (1985). Affinity ultrafiltration (Luong et al., 1987; Luong & Nguyen, 1992) is a novel separation process developed to circumvent difficulties in affinity chromatography. It offers high selectivity, yield, and concentration, but it is an expensive batch process and scale up is difficult.

REVERSE OSMOSIS

Reverse osmosis (also described as hyperfiltration) is a separation process where the solvent molecules are forced by an applied pressure to flow through a semipermeable membrane in the opposite direction to that dictated by osmotic forces, and hence is termed reverse osmosis. It is used for the concentration of smaller molecules than is possible by ultrafiltration as the pores are 1–10 angstroms diameter. Concentration polarization is again a problem and must be controlled by increased turbulence at the membrane surface. Nanofiltration is a modified form of reverse osmosis, which utilizes charged membranes to separate small solutes and charged species based on both charge and size effects (Doran, 2013).

LIQUID MEMBRANES

Liquid membranes are insoluble liquids (eg, an organic solvent) which are selective for a given solute and separate two other liquid phases. Extraction takes place by the transport of solute from one liquid to the other. They are of great interest in the extraction and purification of biologicals for the following reasons:

1. Large area for extraction.
2. Separation and concentration are achieved in one step.
3. Scale-up is relatively easy.

Their use has been reported in the extraction of lactic acid (Chaudhuri & Pyle, 1990) and citric acid using a supported liquid membrane (Sirman, Pyle, & Grandison, 1990). The utilization of selective carriers to transport specific components across the liquid membrane at relatively high rates has increased interest in recent years (Strathmann, 1991). Liquid membranes may also be used in cell and enzyme immobilization, and thus provide the opportunity for combined production and isolation/extraction in a single unit (Mohan & Li, 1974, 1975). The potential use of liquid membranes has also been described for the production of alcohol reduced beer as having little effect on flavor or the physicochemical properties of the product (Etuk & Murray, 1990).

DRYING

The drying of any product (including biological products) is often the last stage of a manufacturing process (McCabe, Smith, & Harriot, 1984; Coulson & Richardson, 1991). It involves the final removal of water or other solvents from a product, while ensuring that there is minimum loss in viability, activity, or nutritional value. Drying is undertaken because:

1. The cost of transport can be reduced.
2. The material is easier to handle and package.
3. The material can be stored more conveniently in the dry state.

A detailed review of the theory and practice of drying can be found in Perry and Green (1984). It is important that as much water as possible is removed initially by centrifugation or in a filter press to minimize heating costs in the drying process. Driers can be classified by the method of heat transfer to the product and the degree of agitation of the product. For some products simple tray driers, where the product is placed on trays over which air is passed in a heated oven may be sufficient. A vacuum may be applied to aid evaporation at lower temperatures. In contact driers, the product is contacted with a heated surface. An example of this type is the drum drier (Fig. 10.34), which may be used for more temperature stable bioproducts. A slurry is run onto a slowly rotating steam heated drum, evaporation takes place and the dry product is removed by a scraper blade in a similar manner as for rotary vacuum

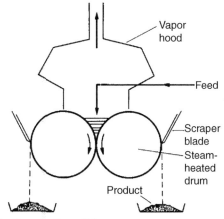

FIGURE 10.34 Cross-Section of a Drum Drier

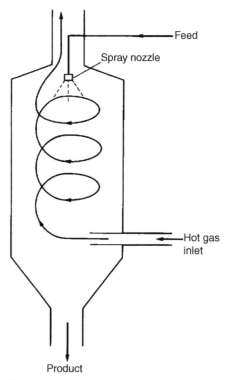

FIGURE 10.35 Counter-Current Spray Drier

filtration. The solid is in contact with the heating surface for 6–15 s and heat transfer coefficients are generally between 1 and 2 kW m^{-2} K^{-1}. Vacuum drum driers can be used to lower the temperature of drying.

A spray drier (Fig. 10.35) is most widely used for drying of biological materials when the starting material is in the form of a liquid or paste. The material to be dried does not come into contact with the heating surfaces, instead, it is atomized into small droplets through, for example, a nozzle or by contact with a rotating disc. The wide range of atomizers available is described in Coulson and Richardson (1991). The droplets then fall into a spiral stream of hot gas at 150–250°C. The high surface area:volume ratio of the droplets results in a rapid rate of evaporation and complete drying in a few seconds, with drying rate and product size being directly related to droplet size produced by the atomizer. The evaporative cooling effect prevents the material from becoming overheated and damaged. The gas-flow rate must be carefully regulated so that the gas has the capacity to contain the required moisture content at the cool-air exhaust temperature (75–100°C). In most processes, the recovery of very small particles from the exit gas must be conducted using cyclones or filters. This is especially important for containment of biologically active compounds. The jet spray drier is particularly suited to handling heat sensitive materials. Operating at a temperature of around 350°C, residence times are approximately 0.01 s because of the very fine droplets produced in the atomizing nozzle.

Spray driers are the most economical available for handling large volumes, and it is only at feed rates below 6 kg min^{-1} that drum driers become more economic.

Freeze drying (also known as lyophilization or cryodesiccation) is an important operation in the production of many biologicals and pharmaceuticals. The material is first frozen and then dried by sublimation in a high vacuum followed by secondary drying to remove any residual moisture. The great benefit of this technique is that it does not harm heat sensitive materials. Freeze drying is generally more energy intensive than other forms of drying.

Fluidized bed driers are used increasingly in the pharmaceutical industry. Heated air is fed into a chamber of fluidized solids, to which wet material is continuously added and dry material continuously removed. Very high heat and mass-transfer rates are achieved, giving rapid evaporation and allowing the whole bed to be maintained in a dry condition.

CRYSTALLIZATION

Crystallization is an established method used in the initial recovery of organic acids and amino acids, and more widely used for final purification of a diverse range of compounds. Crystallization is a two stage process, the formation of nuclei in a supersaturated solution and crystal growth, which proceed simultaneously and can be independently controlled to some extent. Industrial crystallizers may be batch or continuous processes with supersaturation being achieved by cooling or by removal of solvent (evaporative crystallization).

In citric acid production, the filtered broth is treated with $Ca(OH)_2$ so that the relatively insoluble calcium citrate crystals will be precipitated from solution. Checks are made to ensure that the $Ca(OH)_2$ has a low magnesium content, since magnesium citrate is more soluble and would remain in solution. The calcium citrate is filtered off and treated with sulfuric acid to precipitate the calcium as the insoluble sulfate and release the citric acid. After clarification with active carbon, the aqueous citric acid is evaporated to the point of crystallization (Lockwood & Irwin, 1964; Sodeck, Modl, Kominek, & Salzbrunn, 1981; Atkinson & Mavituna, 1991). Crystallization is also used in the recovery of amino acids; Samejima (1972) has reviewed methods for glutamic acid, lysine, and other amino acids. The recovery of cephalosporin C as its sodium or potassium salt by crystallization has been described by Wildfeuer (1985). In 1,3-propanediol fermentations salt by-products (sodium succinate and sodium sulfate) of the fermentation need to be removed before recovery of 1,3-propanediol. Wu, Ren, Xu, and Liu (2010) describe recovery, rather than simply removal, of these salts at high yield and purity by batch crystallization. Buque-Taboada, Straathof, Heijnen, and van der Wielen (2006) review the application of in situ crystallization in by-product recover from fermentation broths as soon as the product is formed when such products have an inhibitory or degrading effect on further product formation.

WHOLE BROTH PROCESSING

The concept of recovering a metabolite directly from an unfiltered fermentation broth is of considerable interest because of its simplicity, the reduction in process stages, and the potential cost savings. It may also be possible to remove the desired fermentation product continuously from a broth during fermentation so that inhibitory effects due to product formation and product degradation can be minimized throughout the production phase (Roffler, Blanch, & Wilke, 1984; Diaz, 1988). It can also be used to continuously remove undesirable byproducts from a fermentation broth which might otherwise inhibit cell growth or degrade a desired extracellular product (Agrawal & Burns, 1997; Demirci & Pongtharangku, 2007).

The continuous downstream processing of biopharmaceuticals via centrifugation, filtration, extraction, precipitation, crystallization, and chromatography has been reviewed by Jungbauer (2013). Ó Meadhra (2005) describes potential systems for the continuous crystallization of products from fermentation.

Bartels, Kleiman, Korzun, and Irish (1958) developed a process for adsorption of streptomycin on to a series of cationic ion-exchange resin columns directly from the fermentation broth, which had only been screened to remove large particles so that the columns would not become blocked. This procedure could only be used as a batch process. Belter, Cunningham, and Chen (1973) developed a similar process for the recovery of novobiocin. The harvested broth was first filtered through a vibrating screen to remove large particles. The broth was then fed into a continuous series of well-mixed resin columns fitted with screens to retain the resin particles, plus the absorbed novobiocin, but allow the streptomycete filaments plus other small particulate matter to pass through. The first resin column was removed from the extraction line after a predetermined time and eluted with methanolic ammonium chloride to recover the novobiocin.

Karr, Gebert, and Wang (1980) developed a reciprocating plate extraction column (Fig. 10.36) to use for whole broth processing of a broth containing 1.4 g dm^{-3} of a slightly soluble organic compound and 4% undissolved solids provided that chloroform or methylene chloride were used for extraction. Methyl-iso-butyl ketone, diethyl ketone, and iso-propyl acetate were shown to be more efficient solvents than chloroform for extracting the active compound, but they presented problems since they also extracted impurities from the mycelia, making it necessary to filter the broth before beginning the solvent extraction. Considerable economies were claimed in a comparison with a process using a Podbielniak extractor, in investment, maintenance costs, solvent usage, and power costs but there was no significant difference in operating labor costs.

An alternative approach is to remove the metabolite continuously from the broth during the fermentation. Cycloheximide production by *Streptomyces griseus* has been shown to be affected by its own feedback regulation (Kominek, 1975). Wang, Kominek, and Jost (1981) have tested two techniques at laboratory scale

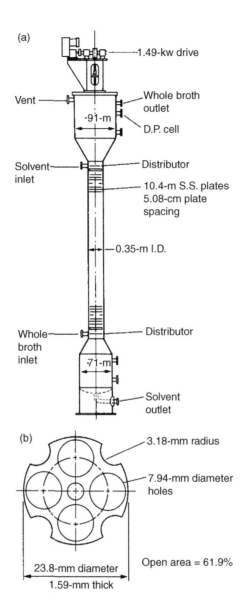

(a)

1.49-kw drive

Vent

Whole broth
outlet

·91-m

D.P. cell

Solvent
inlet

Distributor

10.4-m S.S. plates
5.08-cm plate
spacing

0.35-m I.D.

Distributor

Whole
broth
inlet

·71-m

Solvent
outlet

(b)

3.18-mm radius

7.94-mm diameter
holes

23.8-mm diameter

Open area = 61.9%

1.59-mm thick

FIGURE 10.36

(a) Diagram of a 0.35-m internal diameter reciprocating plate column (Karr et al., 1980).
(b) Plan of a 23.8-m stainless-steel plate for a 25-mm diameter reciprocating plate test
column (Karr et al., 1980).

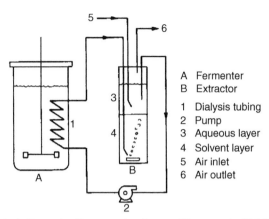

A Fermenter
B Extractor

1 Dialysis tubing
2 Pump
3 Aqueous layer
4 Solvent layer
5 Air inlet
6 Air outlet

FIGURE 10.37 Dialysis-Extraction Fermentation System (Wang et al., 1981)

for improving the production of cycloheximide. In a dialysis method (Fig. 10.37), methylene chloride was circulated in a dialysis tubing loop which passed through a $10 \, dm^{-3}$ fermenter. Cycloheximide in the fermentation broth was extracted into the methylene chloride. It was shown that the product yield could be almost doubled by this dialysis-solvent extraction method to over $1200 \, \mu g \, cm^{-3}$ as compared with a control yield of approximately $700 \, pg \, cm^{-3}$. In a resin method, sterile beads of XAD-7, an acrylic resin, as dispersed beads or beads wrapped in an ultrafiltration membrane, were put in fermenters 48 h after inoculation. Some of the cyclohexi-mide formed in the broth is absorbed by the resin. Recovery of the antibiotic from the resin is achieved by solvents or by changing the temperature or pH. When assayed after harvesting, the control (without resin) had a bioactivity of $750 \, \mu g$ cm^{-3}. Readings of total bioactivity (from beads and broth) for the bead treatment and the membrane-wrapped bead treatments were $1420 \, \mu g \, cm^{-3}$ and $1790 \, \mu g \, cm^{-3}$ respectively. Agrawal and Burns (1997) described the use of a membrane based system for whole broth processing to separate lysozyme from feed mixture con-taining lysozyme, myoglobin, and yeast cells for in situ product removal.

Roffler et al. (1984) reviewed the use of a number of techniques for the in situ recovery of fermentation products:

1. Vacuum and flash fermentations for the direct recovery of ethanol from fermentation broths.
2. Extractive fermentation (liquid–liquid and two-phase aqueous) for the recovery of ethanol, organic acids, and toxin produced by *Clostridium tetani*.
3. Adsorption for the recovery of ethanol and cycloheximide.
4. Ion-exchange in the extraction of salicylic acid and antibiotics.
5. Dialysis fermentation in the selective recovery of lactic acid, salicylic acid, and cycloheximide.

Hansson, Stahl, Hjorth, Uhlen, and Moks (1994) have used an expanded ad-sorption bed for the recovery of a recombinant protein produced by *E. coli* directly from the fermentation broth. The protein was produced in high yields (550 mg dm^{-3}) and >90% recovery together with concentration (volume reduction) and removal of cells was achieved on the expanded bed. Affinity chromatography was used for further purification, and again an overall yield of >90% obtained.

REFERENCES

Adikane, H. V., Singh, R. K., & Nene, S. N. (1999). Recovery of penicillin G from fermentation broth by microfiltration. *Journal of Membrane Science, 162*, 119–123.

Agerkvist, I., & Enfors, S. -O. (1990). Characterisation of *E. coli* cell disintegrates from a bead mill and high pressure homogenizers. *Biotechnology and Bioengineering, 36*(11), 1083–1089.

Agrawal, A., & Burns, M. A. (1997). Application of membrane based preferential transport to whole broth processing. *Biotechnology and Bioengineering, 55*(4), 581–591.

Aiba, S., Humphrey, A. E., & Millis, N. F. (1973). Recovery of fermentation products. In *Biochemical engineering* (2nd ed., pp. 346–392). New York: Academic Press.

Andrews, B. A., & Asenjo, J. A. (1987). Enzymatic lysis and disruption of microbial cells. *Trends in Biotechnology, 5*, 273–277.

Andrews, B. A., Huang, R. B., & Asenjo, J. A. (1990). Differential product release from yeast cells by selective enzymatic lysis. In D. L. Pyle (Ed.), *Separations for Biotechnology 2* (pp. 21–28). London: Elsevier.

Asenjo, J. A. (1990). Cell disruption and removal of insolubles. In D. L. Pyle (Ed.), *Separations for Biotechnology 2* (pp. 11–20). London: Elsevier.

Ashley, M. H. J. (1990). Conceptual design of a novel flocculation device. In D. L. Pyle (Ed.), *Separations for Biotechnology 2* (pp. 29–37). London: Elsevier.

Atkinson, B., & Daoud, I. S. (1976). Microbial flocs and flocculation in fermentation process engineering. *Advances in Biochemical Engineering, 4*, 41–124.

Atkinson, B., & Mavituna, F. (1991). *Biochemical Engineering and Biotechnology Handbook* (2nd ed.). London: Macmillan.

Atkinson, B., & Sainter, P. (1982). Development of downstream processing. *Journal of Chemical Technology and Biotechnology, 32*, 100–108.

Aunstrup, K. (1979). Production, isolation and economics of extracellular enzymes. In L. B. Wingard, & E. Katzir-Katchalski (Eds.), *Applied Biochemistry and Bioengineering* (pp. 27–69). (Vol. 2). New York: Academic Press.

Bailey, J. E., & Ollis, D. F. (1986). *Biochemical Engineering Fundamentals* (2nd ed.). New York: McGraw-Hill, pp. 768–769.

Bartels, C. R., Kleiman, G., Korzun, J. N., & Irish, D. B. (1958). A novel ion-exchange method for the isolation of streptomycin. *Chemical Engineering Progress, 54*, 49–51.

Bell, G. B., & Hutto, F. B. (1958). Analysis of rotary pre-coat filter operation. 1. New concepts. *Chemical Engineering Progress, 54*, 69–76.

Belter, P. A., Cunningham, F. L., & Chen, J. W. (1973). Development of a recovery process for novobiocin. *Biotechnology and Bioengineering, 15*, 533–549.

Bentham, C. C., Bonnerjea, J., Orsborn, G. B., Ward, P. N., & Hoare, M. (1990). The separation of affinity flocculated yeast cell debris using a pilot plant scroll decanter centrifuge. *Biotechnology and Bioengineering, 36*(4), 397–401.

Bowden, C. P., Leaver, G., Melling, J., Norton, M. G., & Whittington, P. N. (1987). Recent and novel developments in the recovery of cells from fermentation broths. In M. S. Verrall, & M. J. Hudson (Eds.), *Separations for Biotechnology* (pp. 49–61). Chichester: Ellis Horwood.

Brewer, S. J., & Larsen, B. R. (1987). Isolation and purification of proteins using preparative HPLC. In M. S. Verrall, & M. J. Hudson (Eds.), *Separations for Biotechnology* (pp. 113–126). Chichester: Ellis Horwood.

Brookman, J. S. G. (1974). Mechanism of cell disintegration in a high pressure homogenizer. *Biotechnology and Bioengineering, 16*, 371–383.

Brown, D. E., & Kavanagh, P. R. (1987). Cross-flow separation of cells. *Process Biochemistry, 22*(4), 96–101.

Bruno, T. J., Nieto De Castro, C. A., Hamel, J. -F. P., & Palavra, A. M. F. (1993). Supercritical fluid extraction of biological products. In J. F. Kennedy, & J. M. S. Cabrai (Eds.), *Recovery Processes for Biological Materials* (pp. 303–354). Chichester: Wiley.

Buque-Taboada, E. M., Straathof, A. J. J., Heijnen, J. J., & van der Wielen, L. A. M. (2006). In situ product recovery (ISPR) by crystallization: basic principles, design, and potential applications in whole-cell biocatalysis. *Applied Microbiology and Biotechnology, 71*, 1–12.

Cabral, J. M. S., & Aires-Barros, M. R. (1993). Liquid-liquid extraction of biomolecules using aqueous two-phase systems. In J. F. Kennedy, & J. M. S. Cabral (Eds.), *Recovery Processes for Biological Materials* (pp. 273–302). Chichester: Wiley.

Cai, D., Chen, H., Chen, C., Hu, S., Wang, Y., Chang, Z., Miao, Q., Qin, P., Wang, Z., Wang, J., & Tan, T. (2016). Gas stripping-pervaporation process for energy saving product recovery from acetone-butanol-ethanol (ABE) fermentation broth. *Chemical Engineering Journal, 287*, 1–10.

Chaudhuri, J. B., & Pyle, D. L. (1990). A model for emulsion liquid membrane extraction of organic acids. In D. L. Pyle (Ed.), *Separations for Biotechnology 2* (pp. 112–121). London: Elsevier Applied Science.

Chavez-Santoscoy, A. (2010). Application of aqueous two-phase systems for potential extractive fermentation of cyanobacterial products. *Chemical Engineering and Technology, 33*(1), 177–182.

Chen, H. -L., Chen, Y. -S., & Juang, R. -S. (2007). Separation of surfactin from fermentation broths by acid precipitation and two stage dead-end ultrafiltration processes. *Journal of Membrane Science, 299*, 114–121.

Corsano, G., Iribarren, O., Montagna, J. M., Aguirre, P. A., & Suarez, E. G. (2006). Economic tradeoffs involved in the design of fermentation processes with environmental constraints. *Chemical Engineering Research and Design, 84*(A10), 932–942.

Coulson, J. M., & Richardson, J. F. (1991) *Chemical Engineering* (Vol. 2). (4th ed.). Oxford: Pergamon Press.

Cowan, G. H., Gosling, I. S., & Laws, J. F. (1986). Physical and mathematical modelling to aid scale-up of liquid chromatography. *Journal of Chromatography A, 363*, 31–36.

Cowan, G. H., Gosling, I. S., & Sweetenham, W. P. (1987). Modelling for scale-up and optimisation of packed-bed columns in adsorption and chromatograpy. In M. S. Verrall, & M. J. Hudson (Eds.), *Separations for Biotechnology* (pp. 152–175). Chichester: Ellis Horwood.

Crapisi, A., Lante, A., Pasini, G., & Spettoli, P. (1993). Enhanced microbial cell lysis by the use of lysozyme immobilised on different carriers. *Process Biochemistry, 28*(1), 17–21.

Darbyshire, J. (1981). Large scale enzyme extraction and recovery. In A. Wiseman (Ed.), *Topics in Enzyme and Fermentation Biotechnology* (pp. 147–186). (Vol. 5). Chichester: Ellis Horwood.

Datar, R. (1986). Economics of primary separation steps in relation to fermentation and genetic engineering. *Process Biochemistry, 21*(1), 19–26.

Davankov, V. A., Kurganov, A. A., & Unger, K. (1990). Reversed phase high performance liquid chromatography of proteins and peptides on polystyrene coated silica supports. *Journal of Chromatography, 500,* 519–530.

Dave, S. M., Patil, S. S., & Suresh, A. K. (1997). Ion exchange for product recovery in lactic acid fermentation. *Separation Science and Technology, 32*(7), 1273–1294.

Davis, D. A., Lynch, H. C., & Varley, J. (2001). The application of foaming for the recovery of Surfactin from *B. subtilis* ATCC21332 cultures. *Enzyme and Microbial Technology, 28,* 346–354.

Delaine, J. (1983). Physico-chemical pretreatment of process effluents. *Institution of Chemical Engineers Symposium Series, 77,* 183–203.

Demirci, A., & Pongtharangku, T. (2007). Online recovery of nisin during fermentation and its effect on nisin production in a biofilm reactor. *Applied Microbiology and. Biotechnology, 74,* 555–562.

DeSousa, S. R., Laluce, C., & Jafelicci, A., Jr. (2006). Effects of organic and inorganic additives on floatation recovery of washed cells of *Saccaromyces cerevisiae* resuspended in water. *Colloids and Surfaces B: Biointerfaces, 48,* 77–83.

Diaz, M. (1988). Three-phase extractive fermentation. *Trends in Biotechnology, 6*(6), 126–130.

Doran, P. M. (2013). *Bioprocess Engineering Principles* (pp. 445–578) (2nd ed.). Oxford: Academic Press.

Dunnill, P., & Lilly, M. D. (1972). Continuous enzyme isolation. *Biotechnology and Bioengineering Symposium, 3,* 97–113.

Etuk, B. R., & Murray, K. R. (1990). Potential use of liquid membranes for alcohol reduced beer production. *Process Biochemistry, 25*(1), 24–32.

Fabre, C. E., Condoret, J. -S., & Marty, A. (1999). Extractive fermentation or aroma with supercritical CO_2. *Biotechnology and Bioengineering, 64*(4), 392–400.

Fletcher, K., Deley, S., Fleischaker, R. J., Forrester, I. T., Grabski, A. C., & Strickland, W. N. (1990). Clarification of tissue culture fluid and cell lysates using Biocryl bioprocessing aids. In D. L. Pyle (Ed.), *Separations for Biotechnology 2* (pp. 142–151). London: Elsevier.

Folena-Wasserman, G., Inacker, R., & Rosenbloom, J. (1987). Assay, purification and characterisation of a recombinant malaria circumsporozoite fusion protein by high performance liquid chromatography. *Journal of Chromatography, 411,* 345–354.

Forstecher, P., Hammadi, H., Bouzerna, N., & Dautre-vaux, M. (1986). Rapid purification of antisteriod anti-bodies by high performance liquid affinity chromatography. *Journal of Chromatography A, 369,* 379–390.

Fox, J. B., Calhoun, R. C., & Eglinton, W. J. (1969). Continuous chromatography apparatus. 1. Construction. *Journal of Chromatography A, 43,* 48–54.

Frude, M. J., & Simpson, M. T. (1993). Steam sterilisation of a Westfalia CSA 8 centrifuge. *Process Biochemistry, 28,* 297–303.

Gabler, R., & Ryan, M. (1985). Processing cell lysate with tangential flow filtration. In D. LeRoith, J. Shiloach, & T. J. Leahy (Eds.), *Purification of Fermentation Products,* (pp. 1–20), ACS, Symposium Series, 271, A.C.S, Washington, D.C.

Gasner, C. C., & Wang, D. I. C. (1970). Microbial recovery enhancement through flocculation. *Biotechnology and Bioengineering, 12,* 873–887.

Gonzalez, M., Pencs, C., & Casas, L. T. (1990). Partial purification of β-galactosidase from yeast by an aqueous two phase system method. *Process Biochemistry*, *25*(5), 157–161.

Gray, P. P., Dunnill, P., & Lilly, M. D. (1973). The clarification of mechanically disrupted yeast suspensions by rotary vacuum precoat filtration. *Biotechnology and Bioengineering*, *15*, 309–320.

Grieves, R. B., & Wang, S. L. (1966). Foam separation of *Escherichia coli* with a cationic surfactant. *Biotechnology and Bioengineering*, *8*, 323–336.

Gu, Z., & Glatz, C. E. (2006). Aqueous two-phase extraction for protein recovery from corn extracts. *Journal of Chromatography B*, *845*, 38–50.

Guan, Y., Wu, X. -Y., Treffry, T. E., & Lilley, T. H. (1992). Studies on the isolation of penicillin acylase from *Escherichia coli* by aqueous two-phase partitioning. *Biotechnology and Bioengineering*, *40*(4), 517–524.

Guan, Y., Zhu, Z., & Mei, L. (1996). Technical aspects of extractive purification of penicillin fermentation broth by aqueous two-phase partitioning. *Separation Science and Technology*, *31*(18), 2589–2597.

Hamer, G. (1979). Biomass from natural gas. In A. H. Rose (Ed.), *Economic Microbiology* (pp. 315–360). (Vol. 4). London: Academic Press.

Hansson, M., Stahl, S., Hjorth, R., Uhlen, M., & Moks, T. (1994). Single step recovery of a secreted recombinant protein by expanded bed adsorption. *Biotechnology*, *12*(3), 285–288.

Hao, J., Xu, F., Liu, H., & Liu, D. (2006). Downstream processing of 1,3-propanediol fermentation broth. *Journal of Chemical Technology and Biotechnology*, *81*, 102–108.

Harrison, S.T. L. (2011) Cell distruption. In: M. Moo-Young (Ed. in chief), *Comprehensive Biotechnology, Volume 2,* (2nd ed.) (pp. 619–639). Oxford: Elsevier.

Heath, C. A., & Belfort, G. (1992). Synthetic membranes in biotechnology. *Advances in Biochemical Engineering/Biotechnology*, *47*, 45–88.

Hetherington, P. J., Follows, M., Dunhill, P., & Lilly, M. D. (1971). Release of protein from baker's yeast (*Saccharomyces cerevisiae*) by disruption in an industrial homogenizer. *Transactions of the Institution of Chemical Engineers*, *49*, 142–148.

Hettwer, D. J., & Wang, H. Y. (1989). Protein release from *Escherichia coli* cells permeabilized with guanidine-HCl and Triton X-100. *Biotechnology and Bioengineering*, *33*, 886–895.

Honig, W., & Kula, M. R. (1976). Selectivity of protein precipitation with polyethylene glycol fractions of various molecular weights. *Analytical Biochemistry*, *72*, 502–512.

Huang, R. -B., Andrews, B. A., & Asenjo, J. A. (1991). Differential product release (DPR) of proteins from yeast: a new technique for selective product recovery from microbial cells. *Biotechnology and Bioengineering*, *38*(9), 977–985.

Huddlestone, J., Veide, A., Kohler, K., Flanagan, J., En-fors, S. -O., & Lyddiatt, A. (1991). The molecular basis of partitioning in aqueous two-phase systems. *Trends in Biotechnology*, *9*, 381–388.

Hughes, D. E., Wimpenny, J. W. T., & Lloyd, D. (1971). The disintegration of microorganisms. In J. R. Norris, & D. W. Ribbons (Eds.), *Methods in Microbiology* (pp. 1–54). (Vol. 5B). London: Academic Press.

Hustedt, H., Kroner, K. -H., & Papamichael, N. (1988). Continuous cross-current aqueous two-phase extraction of enzymes from biomass. Automated recovery in production scale. *Process Biochemistry*, *23*(5), 129–137.

Ikegami, T., Kitamoto, D., Negishi, H., Haraya, K., Matsuda, H., Nitanai, Y., Koura, N., Sano, T., & Yanagishita, H. (2003). Drastic improvement of bioethanol recovery using pervaporation separation technique employing a silicone rubber coated silicate membrane. *Journal of Chemical Technology and Biotechnology*, *78*, 1006–1010.

James, C. J., Coakley, W. T., & Hughes, D. E. (1972). Kinetics of protein release from yeast sonicated in batch and flow systems at 20 kHz. *Biotechnology and Bioengineering, 14*(1), 33–42.

Janson, J. -C. (1984). Large scale purification—state of the art and future prospects. *Trends in Biotechnology, 2*(2), 31–38.

Janson, J. -C., & Hedman, P. (1982). Large-scale chromatography of proteins. *Advances in Biochemical Engineering, 25*, 43–99.

Jungbauer, A. (2013). Continuous downstream processing of biopharmaceuticals. *Trends in Biotechnology, 31*(8), 479–492.

Karr, A. E., Gebert, W., & Wang, M. (1980). Extraction of whole fermentation broth with a Karr reciprocating plate extraction column. *Canadian Journal of Chemical Engineering, 58*, 249–252.

Katoh, S. (1987). Scaling-up affinity chromatography. *Trends in Biotechnology, 5*, 328–331.

Kominek, L. A. (1975). Cycloheximide production by *Streptomyces griseus;* alleviation of end-product inhibition by dialysis extraction fermentation. *Antimicrobial Agents and Chemotherapy, 7*, 856–860.

Krijgsman, J. (1992). Product recovery in bioprocess technology. In R. O. Jenkins (Ed.), *BIOTOL Series* (pp. 247–250). Oxford: Butterworth-Heinemann.

Kroner, K. H., Hustedt, S., Granda, S., & Kula, M. R. (1978). Technical aspects of separation using aqueous two-phase systems in enzyme isolation processes. *Biotechnology and Bioengineering, 20*, 1967–1988.

Kujawska, A., Kujawski, J., Bryjak, M., & Kujawski, W. (2015). ABE fermentation products recovery methods—a review. *Renewable and Sustainable Energy Reviews, 48*, 648–661.

Kurane, R. (1990). Separation of biopolymers: separation of suspended solids by microbial flocculant. In D. L. Pyle (Ed.), *Separations for Biotechnology 2* (pp. 48–54). London: Elsevier.

Lacey, R. E., & Loeb, S. (1972). *Industrial Processing with Membranes.* New York: Wiley-Interscience.

Latham, W. C., Michelsen, C. B., & Edsall, G. (1967). Preparative procedure for the purification of toxoids by gel filtration. *Applied Microbiology, 15*, 616–621.

Le, M. S., & Atkinson, T. (1985). Crossflow microfiltration for recovery of intracellular products. *Process Biochemistry, 20*(1), 26–31.

Lee, C. -K., Chang, W. -G., & Ju, Y. -H. (1993). Air slugs entrapped cross-flow filtration of bacterial suspensions. *Biotechnology and Bioengineering, 41*(5), 525–530.

Liu, F., Liu, L., & Feng, X. (2005). Separation of acetone-butanol-ethanol (ABE) from dilute aqueous solutions by pervaporation. *Separation Purification Technology, 42*, 273–282.

Lo, Y. -M., Yang, S. -T., & Min, D. B. (1997). Ultrafiltration of xanthan gum fermentation broth: process and economic analyses. *Journal of Food Engineering, 31*, 219–236.

Lockwood, L. B., & Irwin, W. E. (1964). Citric acid. In *Kirk-Othmer Encyclopedia of Chemical Technology* (pp. 524–541). (Vol. 5). New York: Wiley.

Lopez, A. M., & Hestekin, J. A. (2013). Separation of organic acids from water using ionic liquid assisted electrodialysis. *Separation and Purification Technology, 116*, 162–169.

Lowe, C. R., & Stead, C. V. (1985). The use of reactive dyestuffs in the isolation of proteins. In M. S. Verrall (Ed.), *Discovery and Isolation of Microbial Products* (pp. 148–158). Chichester: Ellis Horwood.

Luong, J. H. T., & Nguyen, A. -L. (1992). Novel separations based on affinity interactions. *Advances in Biochemical Engineering and Biotechnology, 47*, 137–158.

Luong, J. H. T., Nguyen, A. L., & Male, K. B. (1987). Recent developments in downstream processing based on affinity interactions. *Trends in Biotechnology, 5*, 281–286.

Magnusson, K. E., & Edebo, L. (1976). Large-scale disintegration of micro-organisms by freeze-pressing. *Biotechnology and Bioengineering, 18*, 975–986.

Mattiasson, B., & Kaul, R. (1986). Use of aqueous two-phase systems for recovery and purification in biotechnology. In J.A. Asenjo & J. Hong (Eds.), *Separation, Recovery, Purification in Biotechnology* (pp. 78–92). ACS., Symposium Series 314, A.C.S., Washington, D.C.

Mattiasson, B., & Ling, T. G. I. (1987). Extraction in aqueous two-phase systems for biotechnology. In M. S. Verrall, & M. J. Hudson (Eds.), *Separations for Biotechnology* (pp. 270–292). Chichester: Ellis Horwood.

McCabe, W. L., Smith, J. C., & Harriot, P. (1984). *Unit Operations in Chemical Engineering* (4th ed.). New York: McGraw-Hill.

Melling, J., & Westmacott, D. (1972). The influence of pH value and ionic strength on the ultrafiltration characteristics of a penicillinase produced by *Escherichia coli* strain W3310. *Journal of Applied Chemistry, 22*, 951–958.

Miller, S. A. (1973). Filtration. In R. H. Perry, & C. H. Chilton (Eds.), *Chemical Engineers' Handbook*. New York: McGraw-Hill.

Mogren, H., Lindblom, M., & Hedenskog, G. (1974). Mechanical disintegration of micro-organisms in an industrial homogenizer. *Biotechnology and Bioengineering, 16*, 261–274.

Mohan, R. R., & Li, N. N. (1974). Reduction and separation of nitrate and nitrite by liquid membrane encapsulated enzymes. *Biotechnology and Bioengineering, 16*(4), 513–523.

Mohan, R. R., & Li, N. N. (1975). Nitrate and nitrite reduction by liquid membrane-encapsulated whole cells. *Biotechnology and Bioengineering, 17*(8), 1137–1156.

Mohan, S. B., & Lyddiatt, A. (1992). Silica based solid phases for affinity chromatography. Effect of pore size and ligand location upon biochemical productivity. *Biotechnology and Bioengineering, 40*(5), 549–563.

Moldes, A. B., Alonso, J. L., & Parajo, J. C. (2003). Recovery of lactic acid from simultaneous saccharification and fermentation using anion exchange resins. *Bioprocess and Biosystems Engineering, 25*, 357–363.

Moresi, M., & Sappino, F. (2000). Electrodialytic recovery of some fermentation products from model solutions: techno-economic feasibility study. *Journal of Membrane Science, 164*, 129–140.

Mourot, P., LaFrance, M., & Oliver, M. (1989). Aseptic concentration of microbial cells by cross-flow filtration. *Process Biochemistry, 24*(1), 3–8.

Nachman, M., Azad, A. R. M., & Bailón, P. (1992). Efficient recovery of recombinant proteins using membrane based immunoaffinity chromatography (MIC). *Biotechnology and Bioengineering, 40*(5), 564–571.

Naglak, T. J., & Wang, H. Y. (1992). Rapid protein release from *Escherichia coli* by chemical permeabilization under fermentation conditions. *Biotechnology and Bioengineering, 39*(7), 733–740.

Nakamura, H. (1961). Chemical separation methods for common microbes. *Journal of Biochemical and Microbiological Technology and Engineering, 3*, 395–403.

Nakazawa, K., Shibata, M., Tanabe, K., & Yamamoto, H. (1960). Di-hydrostreptomycin. US patent 2,931,756.

Ng, M. Y. T., Tan, W. S., Abdullah, N., Ling, T. C., & Tey, B. T. (2006). Heat treatment of unclarified *Escherichia coli* homogenate improved the recovery efficiency of recombinant hepatitis B core antigen. *Journal of Virological Methods, 137*, 134–139.

Nicholas, R. A., & Fox, J. B. (1969). Continuous chromatography apparatus. II Application. *Journal of Chromatography, 43*, 61–65.

Niederauer, M. Q., & Glatz, G. E. (1992). Selective precipitation. *Advances in Biochemical Engineering and Biotechnology, 47*, 159–188.

Ó Meadhra, R. (2005). Continuous crystallization strategy for the recovery of fermentation products. *Transactions of the Institute of Chemical Engineers, Part A, 83*(A8), 1000–1008.

Omstead, D. R., Hunt, G. R., & Buckland, B. G. (1985). Commercial production of Cephamycin antibiotics. In H. W. Blanch, S. Drew, & D. I. G. Wang (Eds.), *Comprehensive Biotechnology* (pp. 187–210). (Vol. 3). New York: Pergamon.

Ostlund, C. (1986). Large-scale purification of monoclonal antibodies. *Trends in Biotechnology, 4*(11), 288–293.

Pace, G. W., & Righelato, R. H. (1980). Production of extracellular polysaccharides. *Advances in Biochemical Engineering, 15*, 41–70.

Pace, G. W., & Smith, I. H. (1981). Recovery of microbial polysaccharides. In *Abstracts of Communications. Second European Congress of Biotechnology, Eastbourne* (pp. 36). London: Society of Chemical Industry.

Pandey, R. C., Kalita, C. C., Gustafson, M. E., Kline, M. C., Leidhecker, M. E., & Ross, J. T. (1985). Process developments in the isolation of Largomycin F-II, a chemoprotein antitumor antibiotic. In D. LeRoith, J. Shiloach, & T. J. Leahy (Eds.), *Purification of Fermentation Products* (pp 133–153). ACS Symposium Series 271. ACS, Washington, DC.

Perry, R. H., & Green, D. W. (1984). *Perry's Chemical Engineers' Handbook* (6th ed.). New York: McGraw-Hill.

Podbielniak, W. J., Kaiser, H. R., & Ziegenhorn, G. J. (1970). Centrifugal solvent extraction. In *The History of Penicillin Production. Chemical Engineering Progress Symposium, 66*(100), 44–50.

Porath, j (1974). Affinity techniques—general methods and coupling procedures. In W. B. Jakoby, & M. Wilchek (Eds.), *Methods in Enzymology* (pp. 13–30). (Vol. 34). New York: Academic Press.

Purchas, D. B. (1971). *Industrial Filtration of Liquids* (2nd ed.). London: Leonard Hill, pp. 209.

Pyle, D.L. (Ed.), (1990). *Separations for Biotechnology 2*, SCI/Elsevier Applied Science, London.

Queener, S., & Swartz, R. W. (1979). Penicillins: biosynthetic and semisynthetic. In A. H. Rose (Ed.), *Secondary Products of Metabolism, Economic Microbiology* (pp. 35–122). (Vol. 3). London: Academic Press.

Qureshi, N., & Blaschek, H. P. (2001). Recovery of butanol from fermentation broth by gas stripping. *Renewable Energy, 22*, 557–564.

Rehacek, J., & Schaefer, J. (1977). Disintegration of microorganisms in an industrial horizontal mill of novel design. *Biotechnology and Bioengineering, 19*, 1523–1534.

Ricketts, R. T., Lebherz, W. B., Klein, F., Gustafson, M. E., & Flickinger, M. C. (1985). Application, sterilisation, and decontamination of ultrafiltration systems for large scale production of biologicals. In D. LeRoith, J. Shiloach, T. J. Leahy (Eds.), *Purification of Fermentation Products* (pp. 21–49). ACS., Symposium Series, 271, ACS, Washington, D.C.

Rincon, J., Fuertes, J., Rodriguez, J. F., Rodriguez, L., & Monteagudo, J. M. (1997). Selection of a cation exchange resin to produce lactic acid solutions from whey fermentation broths. *Solvent Extraction and Ion Exchange, 15*(2), 329–345.

Roffler, S. R., Blanch, H. W., & Wilke, C. R. (1984). *In situ* recovery of fermentation products. *Trends in Biotechnology, 2*(5), 129–136.

Rubin, A. J., Cassel, E. A., Henderson, O., Johnson, J. D., & Lamb, J. C. (1966). Microflotation: new low gas-flow rate foam separation technique for bacteria and algae. *Biotechnology and Bioengineering, 8*, 135–151.

Samejima, H. (1972). Methods for extraction and purification. In K. Yamada, S. Kinoshita, T. Tsunoda, & K. Aida (Eds.), *The Microbial Production of Amino Acids* (pp. 227–259). Tokyo: Kodanska.

Scawen, M. D., Atkinson, A., & Darbyshire, J. (1980). Large scale enzyme purification. In R. A. Grant (Ed.), *Applied Protein Chemistry* (pp. 281–324). London: Applied Science Publishers.

Shi, Z., Zhang, C., Chen, J., & Mao, Z. (2005). Performance evaluation of acetone-butanol continuous flash extractive fermentation process. *Bioprocess and Biosystems Engineering, 27*, 175–183.

Shojaosadaty, S. A., & Lyddiatt, A. (1987). Application of affinity HPLC to the recovery and monitoring operations in biotechnology. In M. S. Verrall, & M. J. Hudson (Eds.), *Separations for Biotechnology* (pp. 436–444). Chichester: Ellis Horwood.

Sirman, T., Pyle, D. L., & Grandison, A. S. (1990). Extraction of citric acid using a supported liquid membrane. In D. L. Pyle (Ed.), *Separations for Biotechnology 2* (pp. 245–254). London: Elsevier Applied Science.

Smith, I. H., & Pace, G. W. (1982). Recovery of microbial polysaccharides. *Journal of Chemical Technology and Biotechnology, 32*, 119–129.

Sodeck, G., Modl, J., Kominek, J., & Salzbrunn, G. (1981). Production of citric acid according to the submerged fermentation. *Process Biochemistry, 16*(6), 9–11.

Staniszewski, M., Kujawski, W., & Lewandowska, M. (2007). Ethanol production from whey in bioreactor with co-immobilized enzyme and yeast cells followed by pervaporative recovery of product—kinetic model predictions. *Journal of Food Engineering, 82*, 618–625.

Straathof, A.J. J. (2011). The proportion of downstream costs in fermentative production processes. In M. Moo-Young (Ed. in chief), *Comprehensive Biotechnology, Vol. 2,* (2nd ed.) (pp. 811–814). Oxford: Elsevier.

Strathmann, H. (1985). Membranes and membrane processes in biotechnology. *Trends in Biotechnology, 3*(5), 112–118.

Strathmann, H. (1991). Fundamentals of membrane separation processes. In C.A. Costa, & J.S. Cabrai (Eds.), *Chromatographic and Membrane Processes in Biotechnology* (pp. 153–175). NATO ASI Series, Dordrecht, The Netherlands: Kluwer Academic Publishers.

Swartz, R. R. (1979). The use of economic analyses of penicillin G manufacturing costs in establishing priorities for fermentation process improvement. *Annual Report on Fermentation Processes, 3*, 75–110.

Sylvester, J. C., & Coghill, R. D. (1954). The penicillin fermentation. In L. A. Underkofler, & R. J. Hickey (Eds.), *Industrial Fermentations* (pp. 219–263). (Vol. 2). New York: Chemical Publishing Co.

Talcott, R. M., Willus, C., & Freeman, M. P. (1980). Filtration. In *Kirk-Othmer Encyclopedia of Chemical Technology* (pp. 284–337). (Vol. 10). New York: Wiley.

Tanny, G. B., Mirelman, D., & Pistole, T. (1980). Improved filtration technique for concentrating and harvesting bacteria. *Applied and Environmental Microbiology, 40*(2), 269–273.

Tessier, L., Bouchard, P., & Rahni, M. (2005). Separation and purification of benzylpenicillin produced by fermentation using coupled ultrafiltration and nanofiltration technologies. *Journal of Biotechnology, 116*, 79–89.

Topiwala, H. H., & Khosrovi, B. (1978). Water recycle in biomass production processes. *Biotechnology and Bioengineering, 20*, 73–85.

Vane, L. M. (2005). A review of pervaporation for product recovery from biomass fermentation processes. *Journal of Chemical Technology and Biotechnology, 80*, 603–629.

Voser, W. (1982). Isolation of hydrophylic fermentation products by adsorption chromatography. *Journal of Chemical Technology and Biotechnology, 32*, 109–118.

Walker, P. D., Narendranathan, T. J., Brown, D. C., Woolhouse, F., & Vranch, S. P. (1987). Containment of micro-organisms during fermentation and downstream processing. In M. S. Verall, & M. J. Hudson (Eds.), *Separations for Biotechnology* (pp. 467–482). Chichester: Ellis Horwood.

Wang, D. I. C. (1987). Separation for biotechnology. In M. S. Verall, & M. J. Hudson (Eds.), *Separations for Biotechnology* (pp. 30–48). Chichester: Ellis Horwood.

Wang, D. I. C., & Sinskey, A. J. (1970). Collection of microbial cells. *Advances in Applied Microbiology, 12*, 121–152.

Wang, D. I. C., Cooney, C. L., Demain, A. L., Dunnill, P., Humphrey, A. E., & Lilly, M. D. (1979). Enzyme isolation. In *Fermentation and Enzyme Technology* (pp. 238–310). New York: Wiley.

Wang, H. Y., Kominek, L. A., & Jost, J. L. (1981). On-line extraction fermentation processes. In M. Moo-Young, C. W. Robinson, & C. Vezina (Eds.), *Advances in Biotechnology* (pp. 601–607). (Vol. 1). Toronto: Pergamon Press.

Warne, S. R., & Bowden, C. P. (1987). Biosurface properties and their significance to primary separation: a genetic engineering approach to the utilisation of bacterial auto-aggregation. In M. S. Verrall, & M. J. Hudson (Eds.), *Separations for Biotechnology* (pp. 90–104). Chichester: Ellis Horwood.

Warren, R. K., MacDonald, D. G., & Hill, G. A. (1991). Cross-flow filtration of *Saccharomyces cerevisae*. *Process Biochemistry, 26*(6), 337–342.

Wase, D. A. J., & Patel, Y. R. (1985). Effect of cell volume on disintegration by ultrasonics. *Journal of Chemical Technology and Biotechnology, 35*, 165–173.

Weinstein, M. J., & Wagman, G. H. (1978). *Antibiotics, isolation, separation and purification*. Amsterdam: Elsevier.

Weiss, S. A. (1980). Concentration of baboon endogenous virus in large scale production by use of hollow fibre ultrafiltration technology. *Biotechnology and Bioengineering, 22*(1), 19–31.

Wikstrom, P., Flygare, S., Grondalen, A., & Larsson, P. O. (1987). Magnetic aqueous two phase separation: A new technique to increase rate of phase separation using dextran-ferrofluid or larger iron particles. *Analytical Biochemistry, 167*, 331–339.

Wildfeuer, M. E. (1985). Approaches to cephalosporin C purification from fermentation broth. In D. LeRoith, J. Shiloach, & T.J. Leahy (Eds.), *Purification of fermentation products* (pp. 155–174). ACS Symposium Series 271. ACS, Washington, DC.

Wimpenny, J. W. T. (1967). Breakage of micro-organisms. *Process Biochemistry, 2*(7), 41–44.

Wohlgemuth, R. (2011). Product recovery. In M. Moo-Young (Ed. in chief), *Comprehensive biotechnology, Vol. 2,* 2nd ed. (pp. 591–601). Oxford: Elsevier.

Wu, R. C., Ren, H. J., Xu, Y. Z., & Liu, D. (2010). The final recovery of salt from 1,3-propanadiol fermentation broth. *Separation and Purification Technology, 73*, 122–125.

Yang, C. M., & Tsao, G. T. (1982). Affinity chromatography. *Advances in Biochemical Engineering, 15*, 19–42.

Zahka, J., & Leahy, T.J. (1985). Practical aspects of tangential flow filtration in cell separations. In D. LeRoith, J. Shiloach, & T.J. Leahy (Eds.), Purification of fermentation products (pp. 51–69). ACS Symposium Series 271. ACS, Washington, DC.

FURTHER READING

Levandowicz, G., Bialas, W., Marcewski, B., & Szymanowska, D. (2011). Application of membrane distillation for ethanol recovery during fuel ethanol production. *Journal of Membrane Science, 375*, 212–219.

Perry, R. H., & Chilton, G. H. (1974). *Chemical engineers' handbook* (5th ed.). New York: McGraw-Hill.

Pritchard, M., Scott, J. A., & Howell, J. A. (1990). The concentration of yeast suspensions by crossflow filtration. In D. L. Pyle (Ed.), *Separations for biotechnology 2* (pp. 65–73). London: Elsevier.

Russotti, G., Osawa, A. E., Sitrin, R. D., Buckland, B. C., Adams, W. R., & Lee, S. S. (1995). Pilot-scale harvest of recombinant yeast employing microfiltration: a case study. *Journal of Biotechnology, 42*, 235–246.

Effluent treatment

11

INTRODUCTION

Every fermentation plant utilizes raw materials, which are converted to a variety of products. Depending on the individual process, varying amounts of a range of waste materials are produced. Typical wastes might include unconsumed inorganic and organic media components, microbial cells and other suspended solids, filter aids, waste wash water from cleansing operations, cooling water, water containing traces of solvents, acids, alkalis, human sewage, etc. Historically, it was possible to dispose of wastes directly to a convenient area of land or into a nearby watercourse. This cheap and simple method of disposal is now very rarely possible, nor is it environmentally desirable. With increasing density of population and industrial expansion, together with greater awareness of the damage caused by pollution, the need for treatment and controlled disposal of waste has, and will, continue to grow. Water authorities and similar bodies have become more active in combating pollution caused by domestic and industrial wastes. Legislation in all developed countries now regulates the discharge of wastes, be they gas, liquid, or solid (Fisher, 1977; Hill, 1980; Masters, 1991; Brown, 1992). In the UK, much of the legislation pertaining to waste disposal and pollution is embraced by the Environmental Pollution Act 1990. (HMSO, 1990) The EPA 1990 has been amended, superseded and expanded upon by more recent legislation such as the Water Industry Act 1991, the Water Resources Act 1991, the Environment Act 1995, the Pollution Prevention and Control Act 1995, the Water Act 2003, the Water Framework Directive 2000, and the Environmental Permitting Regulations 2010. All the above Acts, plus others not specifically mentioned but still of potential relevance to the fermentation industry, are published through HMSO and are available to access via legislation.gov.uk. Further information on environmental legislation cab be found in the following text; Bell, McGillivray and Pedersen (2013).

With liquid wastes, it may be possible to dispose of untreated effluents to a municipal sewage treatment works (STW). Permission for disposal to sewer is enforced by the local water company responsible for the local sewerage system and its STWs. Obviously, much will depend on the composition, strength, and volumetric flow rate of the effluent. STWs are planned to operate with an effluent of a reasonably constant composition at a steady flow rate. Thus, if the discharge from an industrial process is large in volume and intermittently produced it may be necessary to install storage tanks on site to regulate the effluent flow. In some locations, municipal sewers are not available or the effluent may be of such a composition

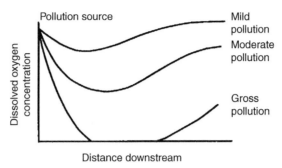

FIGURE 11.1 The Oxygen Sag Curve

that the wastewater treatment company requires some form of pretreatment before discharge to its sewers. In these cases, an effluent-treatment plant will have to be installed at the factory. Whatever the pollutant load of the liquid effluent, its discharge to a sewer will be a cost centered activity, and will incur charges from the treatment company.

Normally, fermentation effluents do not contain toxic materials, which directly affect the aquatic flora or fauna. Unfortunately, most of the effluents do contain high levels of organic matter, which are readily oxidized by microbial attack and so drastically deplete the dissolved oxygen concentration in the receiving water unless there is a large dilution factor. This can be shown by the oxygen sag curve in Fig. 11.1. Different aquatic species have varying tolerances to depleted oxygen levels, and as a consequence, some species will die off in specific stretches of the receiving water, and in other regions a different population capable of growth at lower oxygen levels will develop.

Effluents may be treated in a variety of ways, as will be outlined later in this chapter. In a number of processes, it may be possible to recover waste organic material as a solid and sell it as a byproduct which may be an animal feed supplement or a nutrient to use in fermentation media (Chapter 4). The marketable by-product helps to offset the cost of the treatment process. It is now recognized that water is no longer a cheap raw material (Chapter 4), hence there are considerable advantages in reducing the quantities used and in recycling whenever it is feasible (Ashley, 1982). Obviously the introduction of good "housekeeping" will lead to reductions in the volume of water used and the volume of effluent for the treatment and final discharge. Recycling and reuse of materials, waste minimization, waste reduction at source, and integrated pollution control are now very important factors to consider in the design and operation of any manufacturing facility. (Laing, 1992; Donaldson, 1993; McLeod & O'Hara, 1993). The Waste (England and Wales) Regulations 2011 (HMSO, 2011) in summary state that, "*you have a legal duty of care to take all reasonable steps to keep your waste safe.*" This "Duty of Care" was included within the original EPA of 1990 but the 2011 legislation goes further in stating (in summary) "*you need to take all such measures as are reasonable in the circumstances to apply the waste hierarchy to prevent waste.*" The waste hierarchy states that the most favored option is waste prevention, running through minimization, reuse, recycling, and energy recovery, to the least favored option of waste disposal.

SUSTAINABILITY

Raw material use, energy use, and waste (gas, liquid, and solid) generation are of increasing concern to companies, the government and the public and as a consequence are now integral to the design of fermentation processes. These processes have to some extent an advantage over many industrial processes in that they use relatively mild conditions (eg, pH). However, they generate wastes such as biomass, off-gases, and potentially toxic liquid effluents and require substantial quantities of water and energy. Much of sustainable design is focused on minimizing such wastes together with raw material and energy use. Waste minimization should also improve efficiency, productivity, and profits. Other aspects of sustainability go further and include the assessment of environmental impacts over the life cycle of products, processes, and equipment—life cycle analysis/assessment (LCA). LCA is a tool used to evaluate the environmental impacts associated with a product and the process, which produces it. It is not confined to the manufacture of the product but also extends to the procurement of all materials used in manufacturing and the impact of the product disposal, that is, it is a "cradle to grave" approach (Doran, 2013; Laca, Herrero, & Diaz, 2011). Extended producer responsibility (EPR) places responsibility on the producer for impacts their products may have at the end of the products life cycle (Gavrilescu, 2011).

The main effluent streams produced in fermentation processes are cell biomass (before or after disruption), spent aqueous broth and materials such as solvents used in downstream processing and off-gases, mainly CO_2 but can also include organic vapors. Large quantities of wastewater and dilute cleaning products are produced during cleaning operations and in steam sterilization. Additional solid waste is generated by the plastics and polymers used in gas and liquid filtration and from analytical/quality control facilities. With the growth in the availability, and use, of single-use disposable fermenters and ancillary equipment is generating a new solid waste stream, which may adversely affect environmental sustainability of a fermentation process. However, this is offset as they do not need the cleaning and sterilization facilities that a traditional fermentation process requires (Doran, 2013).

DISSOLVED OXYGEN CONCENTRATION AS AN INDICATOR OF WATER QUALITY

Since oxygen is essential for the survival of most macroorganisms, it is important to ensure that there are adequate levels of dissolved oxygen in rivers, lakes, reservoirs, etc., if they are to be managed satisfactorily. Ideally, the oxygen concentration should be at least 90% of the saturation concentration at the ambient temperature and salinity of the water. It is therefore important to know how effluents containing soluble and particulate organic matter can influence the dissolved oxygen concentration. One widely used method of assessment is the "biochemical oxygen demand" (BOD), which is a measure of the quantity of oxygen required for the oxidation of organic matter in water, by microorganisms present, in a given time interval at a given temperature. The oxygen concentration of the effluent, or a dilution of it, is determined before and after incubation in the dark at 20°C for 5 days. The oxygen decrease can then be determined titrimetrically and the results presented as milligrams of oxygen consumed per cubic decimeter of sample. Mineral nutrients and a suitable bacterial inoculum are usually added to the initial sample to ensure optimal growth conditions. This test is only an estimate of biodegradable material, hence recalcitrant or inhibitory compounds might be overlooked (SCA, 1989).

BOD can also be determined using electrolysis cell respirometric methods. Within the cell, oxygen pressure over the sample is maintained by replacing that used by the microorganisms using an electrolysis reaction. BOD can then be correlated to the amount of oxygen required, that is, which has been generated (Metcalf and Eddy Inc., 2003).

Because the standard BOD test takes 5 days it may be necessary to resort to the "chemical oxygen demand" (COD), a chemical test which only takes a few hours to complete. The test is based on treating the sample with a known amount of boiling acidic potassium dichromate solution for 2.5–4 h and then titrating the excess dichromate with ferrous sulfate or ferrous ammonium sulfate (HMSO, 1972). The oxidized organic matter is taken as being proportional to the potassium dichromate utilized. Most compounds are oxidized virtually to the completion in this test, including those, which are not biodegradable. In circumstances where substances are toxic to microorganisms, the COD test may be the only suitable method available for assessing the degree of treatment required. The BOD:COD ratios for sewage are normally between 0.2:1 and 0.5:1. The ratio values for domestic sewage may be fairly steady. When industrial effluents of variable composition and loading are discharged, the ratio may fluctuate considerably. Very low BOD:COD ratios will indicate high concentrations of nonbiodegradable organic matter and consequently biological effluent treatment processes may be ineffective (Ballinger & Lishka, 1962; Davis, 1971). A number of alternative tests are available to indicate the "oxygen demand" of a wastewater, including total organic carbon (TOC) and permanganate value (HMSO, 1972; American Public Health Association, 1992). Typical BOD:TOC ratios for untreated effluent are in the range 1.2:1–2.0:1.

SITE SURVEYS

A complete survey of industrial operations is essential for any individual site before an economical waste-treatment program can be planned. It is desirable to divide the facility into as many units as possible, as knowledge of the various material streams may show unexpected losses of finished product, solvent wastage, excessive use of

Table 11.1 Factors to Investigate in a Site Survey

Daily flow rate
Fluctuations in daily, weekly, and seasonal flow
BOD/COD
Suspended solids
Turbidity
pH range
Temperature range
Odors and tastes
Color
Hardness
Detergents
Radioactivity
Presence of specific toxins or inhibitors (eg, heavy metals, phenolics etc.)

water, or unnecessary contamination of water, which might be recycled, recovered, or reused within the site. The factors and concentrations where appropriate, listed in Table 11.1 ought to be known at all production rates under which an individual unit may operate in a representative time-period.

The survey may indicate a need for better control of water usage and should identify sources of uncontaminated and contaminated water that might be reused in the factory. Concentrated waste streams should be kept separate if they contain materials that can be profitably recovered. It is also often more economical to treat a concentrate rather than a large volume of a dilute effluent because of the saving on pumps and settling tank capacities, provided that concentrations do not reach toxic or inhibitory levels in biological treatment processes.

The various wastes may be tested in a laboratory and on a pilot scale to assess the best potential methods of chemical and biological treatment. Once the pHs of the effluents are known, samples may be mixed to see if a neutral pH is reached. A variety of tests may be used to establish methods for reducing salt concentrations, coagulating suspended particles and colloidal materials, and for breaking emulsions.

The commonly used biological tests include respirometry, aeration-flask tests (Otto, Barker, Schwarz, & Tjarksen, 1962), and continuous-culture experiments. Small flask respirometers (Warburg or Gilson) and oxygen electrodes are used initially to establish the conditions to use in biooxidation of the effluent, and to test for the presence of toxic materials. Large respirometers (Simpson & Anderson, 1967) are useful for predicting effluent treatment rates and oxygen requirements. The residues in the flasks can be analyzed to see if there are any recalcitrant materials. The use of laboratory continuous-culture vessels fitted with sludge-return pumps and settling tanks can provide detailed information (Ramathan & Gaudy, 1969). Proposed large-scale operating conditions for feed and aeration rates can be tested and their effectiveness assessed. The results from all these experiments may help in the design of a full-scale plant.

If the survey is comprehensive, it should be possible to plan an overall treatment program for a site and to establish:

1. Water sources, which can be combined or reused.
2. Concentrated waste streams, which contain valuable wastes to be recovered as food, animal feed, fertilizer, or fuel.
3. Toxic effluents needing special treatment, or acids or alkalis needing neutralization.
4. The effluent loading expected under maximum production conditions.
5. The effluent(s), which might be discharged directly, without treatment, on to land or to a watercourse and not cause any pollution.
6. The effluent(s), which might be discharged into municipal sewers.

When all the relevant information has been obtained one can predict the size and type of effluent treatment plant required, and thus its capital and operational costs. This can then be compared with water company charges to treat the waste at an STW with and without onsite treatment. It should be remembered that the water company may insist on the onsite treatment before a waste is discharged to the sewer, and will in most cases set consent limits for maximum flow rates and concentrations of specific analytes.

STRENGTHS OF FERMENTATION EFFLUENTS

It is already evident from earlier sections of this chapter that the presence of high levels of particulate or soluble organic matter in water will result in potential high BODs. This is precisely what is being achieved in all large-scale fermentation processes. An initial medium rich in organic matter is converted to biomass and primary and secondary metabolites. Unfortunately, the product often represents a small proportion of the initial raw material, even in an efficiently operated fermentation. The spent wastes remaining after the distillation of whisky may account for 90% of the initial raw organic materials, while in an antibiotic fermentation the effluent may represent in excess of 95%.

Data for a variety of fermentation effluents are summarized in Table 11.2. The BODs of many of these samples are much higher than that of domestic sewage and some may be comparable with strong effluents such as sulfite paper mill liquor. It is evident from these data that fermentation effluents may present serious potential pollution problems and may be expensive to dispose of unless well planned processes are used. A number of steps may be taken to reduce BODs in a process. Some of these will be discussed in this chapter. Careful selection of raw materials may have a significant effect on the type and quantities of effluent being produced. The cheapest raw material which meets the nutrient requirements of the microorganisms may not be ideal if product yield, recovery cost, effluent disposal cost, and possible

Table 11.2 BOD Strengths of Effluents (mg dm^{-3})

Effluent	BOD	Reference
Domestic sewage	350	Boruff (1953)
Sulfite liquor from paper mill	20,000–40,000	
Beer:		
(a) spent grain press	15,000	Abson and Todhunter (1967)
(b) hop-press liquor	7430	
(c) yeast wash water	7400	
(d) spoil beer	up to 100,000	
(e) bottle washings	550	
Maltings:		
(a) suspended solids	1240	Koziorowski and Kucharski (1972)
(b) wastes	20–204	
(c) grain washings	1500	
Brewery effluent	1,400–1,800	Fang et al. (1990)
Industrial alcohol stillage	10,000–25,000	Blaine (1965)
Distillery stillage	10,000–25,000	Jackson (1960)
Yeast production	3,000–14,000	Boruff (1953)
Antibiotic waste	5,000–30,000	Jackson (1960)
Penicillin:		
(a) wet mycelium from filter	40,000–70,000	Boruff (1953)
(b) filtrate	2,150–10,000	
(c) wash water	210–13,800	
Streptomycin spent liquor	2,450–5,900	Koziorowski and Kucharski (1972)
Aureomycin spent liquor	4,000–7,000	
Solvents	up to 2,000,000	

by-product value are considered together. The high BOD value of fungal mycelium (40,000–70,000 mg dm^{-3}) would indicate that any biomass should normally be kept separate from the remainder of an effluent and some of it may be sold as a by-product. It may also be worthwhile to concentrate liquid fractions, for example, industrial alcohol and distillery stillages (10,000–25,000 mg dm^{-3}) will both produce dried solubles fractions, which can be sold.

Metabolites or components of some fermentation effluents may be extremely toxic and polluting and will require complete destruction, for example, by chemical or thermal methods, before disposal. The need for such a treatment strategy will therefore make a significant contribution to the overall cost of the process. One such metabolite is avermectin produced by *Streptomyces avermitilis* fermentations. Here all effluent streams from the process are captured and any avermectin present chemically degraded (Omstead, Kaplan, & Buckland, 1989). Recent studies (Liu et al., 2012) have highlighted the issue of antibiotic resistance genes becoming common in the effluent of wastewater treatment systems treating effluents from antibiotic production facilities.

Sampling and analysis of fermentation waste streams are undertaken for a number of reasons such as (1) to determine the concentration of a range of constituents (eg, BOD, COD, nutrients, toxins, recalcitrant materials, etc.) and so devise appropriate treatment strategies, (2) provide data for regulatory compliance, (3) as an indicator of plant performance. The data collected must be representative of the wastewater being sampled, reproducible by others using the same sampling, and analytical procedures and documented (Metcalf and Eddy Inc., 2003).

TREATMENT AND DISPOSAL OF EFFLUENTS

The effluent disposal procedure, which is finally adopted by a particular manufacturer is obviously determined by a number of factors, of which the most important is the control exercized by the relevant authorities in many countries on the quantity and quality of the waste discharge and the way in which it might be done (Fisher, 1977). The range of effluent-disposal methods, which can be considered are as follows:

1. The effluent is discharged to land, river, or sea in an untreated state.
2. The effluent is removed and disposed of in a landfill site or is incinerated.
3. The effluent is partially treated on site (eg, by lagooning) prior to the further treatment or disposal by one of the other routes indicated.
4. Part of the effluent is untreated and discharged as in (1) or (2); the remainder is treated at a sewage works or at the site before discharge.
5. All of the effluent is sent to the sewage works for treatment, although there might be reluctance by the sewage works to accept it, possibly resulting in some preliminary onsite treatment being required, and discharge rates and effluent composition defined.
6. All the effluent is treated at the factory before discharge.

DISPOSAL
SEAS AND RIVERS

The simplest way of disposal will be on a sea coast or in a large estuary where the effluent is discharged through a pipeline (installed by the factory or local authorities) extending below the low-water mark. In such a case, there may be little preliminary treatment and one relies solely on the degree of dilution in the sea water. This solution to waste disposal is highly unlikely today as a result of tighter regulation noted previously and the fact that most companies and organizations take corporate social responsibility (CSR) more seriously in the 21st century.

If effluents are to be discharged into a river, they must meet the requirements of the local river or drainage authorities. In Britain there is a Royal Commission standard requiring a maximum BOD (5 days) of 20 mg dm^{-3} and 30 mg dm^{-3} of suspended solid matter (the 20:30 standard). Stricter standards are often applied, depending on the use of the receiving water, such as a 10:10 standard; in addition, levels of ammoniacal nitrogen, nitrate, nitrite, and phosphate may be stipulated. There are, as well, often stringent upper limits for toxic metals and chemicals which might kill the fauna (particularly fish and shellfish) and flora, for example, sulfites, cyanides, phenols, copper, zinc, cadmium, arsenic, etc. It is highly unlikely that one would be able to discharge an industrial waste today without some form of pretreatment.

LAGOONS (OXIDATION PONDS)

Lagoons, holding ponds, oxidation ponds, etc., may be used by a number of industries if land is available at a reasonable cost. It is a method often used in seasonal industries, where capital investment in effluent plant is difficult to justify. The lagoon normally consists of a volume of shallow water enclosed by watertight embankments. Shallow oxidation ponds are typically 1–2 m deep while deep oxidation ponds are typically between 2 and 5 m in depth. They can be designed to maintain aerobic conditions throughout, but more commonly decomposition at the surface is aerobic and that nearer the bottom is anaerobic and they are then known as facultative ponds. Oxygen for aerobic degradation is provided both from the surface of the pond, and from algal photosynthesis. Deeper ponds can be mechanically agitated to provide aeration. In some cases sedimented material from the base of the lagoon (sludge) may be recycled. In such cases the process essentially becomes an activated sludge process (Metcalf and Eddy Inc., 2003). Lagoons are simple to build and operate, but are expensive in terms of land requirements as large reactor volumes are required for retention times of perhaps up to 30 days. They may be used as the sole method of treatment, incorporating both physical (sedimentation) and biological processes, but the effluent produced may not reach locally acceptable standards. However, BOD and suspended solids removals can range from approximately 40–80%. Alternatively, they can provide an initial pretreatment or can be used to "polish" effluent from secondary treatment processes. In addition, lagoons can provide a useful "heat sink" if any hot wastewater streams are produced.

SPRAY IRRIGATION

Liquid wastes can be applied directly to land as irrigation water and fertilizer when they are claimed to have a number of beneficial effects on the soil and plants. If this method of disposal is to be used, then it is necessary to have a large area of land near a manufacturing plant in an area of low to medium rainfall. Pipeline costs will often restrict the use of this technique. Colovos and Tinklenberg (1962) described the disposal of antibiotic and steroid wastes with BODs of 5,000–20,000 mg dm^{-3}. These wastes were initially chlorinated to lower the BOD and reduce unpleasant odors and then sprayed on to land until the equivalent of 38 mm of rainfall was reached. This process was repeated at monthly intervals and improved plant growth. It is unlikely that such a technique could be used today for the reasons previously outlined.

When appropriate, solid organic wastes may be spread onto land as a fertilizer and soil conditioner. This practice is common with sewage sludges. The site(s) chosen for land spreading of solid wastes requires careful evaluation and selection for the correct soil characteristics, topography, and location of both groundwater and surface watercourses (Metcalf and Eddy Inc., 2003). Mbagwu and Ekwealor (1990) report the use of spent brewers' grains to improve the productivity of fragile soils. Irrespective of whether the waste is liquid or solid, the concentration of heavy metals and certain organic components will require careful monitoring and control to safeguard the environment and public health as well as to meet legislative requirements.

LANDFILLING

Landfilling is a disposal method for municipal solid waste (MSW) and industrial waste. It utilizes natural or man made voids (eg, disused clay pits) into which the waste is deposited. Both solid and liquid wastes can in theory be deposited depending on the restrictions imposed by the site license (issued by the national environmental protection agency) and by current legislation. Disposal of liquid wastes in this manner is becoming ever more restrictive. Strict controls exist on the amount of liquid and toxic materials which can be accepted because of the threat of groundwater pollution if leachate (a liquid having BOD levels up to 30,000 mg dm^{-3}) escapes from the site through the geotextile and/or clay lining of the landfill. Leachate is generated from liquid deposited in the site, water entering the site naturally via precipitation or surface run-off and by anaerobic microbial action as organic matter in the landfill is degraded. Microbial action similar to that in anaerobic digesters leads to the production of landfill gas (LFG) which, being 50–60% methane can, if collected efficiently, provide a useful source of energy and used in the generation of electricity for use onsite or for sale to the national grid (Freestone, Phillips, & Hall, 1994). In Oct. 1996 the UK government introduced the landfill tax (HMRC, 2015). Currently this tax stands at approximately £80 ton^{-1} for most wastes significantly increasing the cost of this disposal method. Inert wastes carry a tax burden of approximately £2.50 ton^{-1}.

INCINERATION

A number of designs exist for the incineration of solid and/or liquid wastes either onsite or at a commercial incinerator, including rotary kilns, fluidized beds, and multiple hearth furnaces. Similar to landfills, useful energy can be derived from the incineration of wastes in the form of electricity or both electricity and heat in combined heat and power (CHP) plants. Again this energy can be used onsite or to generate revenue by sale off-site (DEFRA, 2013). Combustion temperatures need to be carefully controlled to destroy and prevent the formation of dioxins and furans, formation of which occurs at between 300 and 800°C, and total destruction is effected at temperatures above 1000°C with a retention time of 1 s. Flue gases from the incinerator require cleaning to remove particulates, acid vapors, etc. using electrostatic precipitators, cyclones, and wet scrubbers to comply with local environmental protection standards. Waste disposal by incineration has traditionally been significantly more expensive than landfilling, though the gap has narrowed with the introduction of the landfill tax. In 2012 "gate fees" for incinerators ranged between £32 and £101 ton^{-1} (WRAP, 2012).

DISPOSAL OF EFFLUENTS TO SEWERS

Municipal authorities and water treatment companies that accept trade effluents into their sewage systems will want to be sure that:

1. The sewage works has the capacity to cope with the estimated volume of effluent.
2. The effluent will not interfere with the treatment processes used at the sewage works.
3. There are no compounds present in the effluent, which will pass through the sewage works unchanged and then cause problems when discharged into a watercourse.

It is common practice for water companies to demand preliminary onsite pretreatment before discharge into sewers to minimize the effects of industrial wastes. The actual pretreatment required will depend on the precise nature of the waste and may range from simple sedimentation or neutralization to complex physical, chemical, and biological processes. Large volumes of intermittently produced wastewaters may require storage prior to metered disposal to the sewer.

TREATMENT PROCESSES

Fermentation wastes may be treated onsite or at an STW by any or all of the three following methods:

1. Physical treatment.
2. Chemical treatment.
3. Biological treatment.

The final choice of treatment and disposal processes used in each individual factory will depend on local circumstances.

Treatment processes may also be described in the following manner:

1. Primary treatment; physical and chemical methods, for example, sedimentation, floatation, coagulation, etc.
2. Secondary treatment; biological methods (eg, activated sludge) conducted after primary treatment.

3. Tertiary treatment; physical, chemical, or biological methods (eg, microstrainers, sand filters, and grass plot irrigation) used to improve the quality of liquor from previous stages (Forster, 1985).
4. Sludge conditioning and disposal; physical, chemical, and biological methods. Anaerobic digestion is often used to condition (make it more amenable to dewatering) the sludge produced in previous stages in particular sludge from primary and secondary sedimentation. Following dewatering (eg, by centrifugation using a decanter centrifuge) the sludge can then be disposed of by land spreading, incineration, landfilling, etc.

PHYSICAL TREATMENT

The removal of suspended solids by physical methods before subsequent biological treatment will considerably reduce the BOD of the resulting effluent. In nearly all fermentation processes the cells are separated from the liquid fraction in recovery processes (Chapter 10). Obviously, biomass processes need not be considered. Yeast cells from other processes may be a marketable product, but microbial cells may not always be marketable, particularly when contaminated with filter aid. In these instances, when the cells and filter aid are a waste, the recovered material may be dealt with in two basic ways:

1. The waste is disposed of without any further treatment.
2. The waste bulk is reduced by mechanical dewatering with a filter press, centrifuge, rotary vacuum filter, or belt press. The compressed waste is then incinerated (Grieve, 1978) or disposed of in a landfill site.

Solid wastes are produced in some processes before inoculation. In breweries, where malted grain is still used, coarse screens or "whirlpool" centrifuges may be used to remove spent grain from the wort after it is mashed. About 5 kg (wet weight) of grain are produced per barrel (180 dm³) of beer. If hops are used, rather than hop extracts, they will also be recovered on screens in a "hop back." This residue may amount to 250 g per barrel. Both the spent grain and hop waste may then be mechanically dewatered before being sold or dumped.

The stillage (after distillation) in whisky distilleries may be passed through screens (1 mm openings). These screenings are then removed, mechanically dewatered, and dried in rotary driers to yield a potentially marketable residue known as Distillers' grains. According to a survey in Scotland, about half the whisky distilleries were evaporating the spent waste to a syrup containing 45% solids, mixing with spent grain, drying and selling the final product, "Distillers" Dark Grains', as a low-grade cattle food (Mackel, 1976).

The production of yeast and ethanol through the molasses fermentation usually generates large quantities of colored acidic wastewater with a high organic load. The main colored components are called malanoidins, which are not effectively removed by aerobic or anaerobic biological processes. Liu et al. (2013) report the use of submerged nanofiltration membranes (MWCO ~ 580 Da) for the effective removal of malanoidins from biologically treated molasses fermentation wastewater.

Physical processes installed for primary effluent treatment may include the following stages:

1. Course screens (eg, a bar rack), to remove larger suspended and floating matter are unlikely to be required for fermentation effluents. However, fine screens (for the removal of fine particles) and microscreens may be employed.
2. Comminutors, to reduce particle size to either increase the rate of BOD removal or prevent blockage in subsequent unit operations.
3. Grit removal, to prevent damage to plant in later processes, utilizing one of a number of units such as horizontal flow chambers, constant velocity channels, aerated chambers, and vortex chambers. It is highly unlikely this will be required for fermentation effluents.
4. Flow equalization, for temporal storage of effluent streams to maintain constant flow or BOD loadings to later operations.
5. Sedimentation tanks for the removal of finer suspended matter. These are generally large circular (usually around 10 m in diameter) or rectangular (possibly up to 50 m in length) continuous flow tanks operating at retention times of 1.5–2.5 h (typically 2 h), with facility for the continuous removal of settled sludge and clarified liquor. Particle sedimentation obeys Stoke's Law (Chapter 10) hence coagulation and/or flocculation may be conducted prior to sedimentation to increase particle diameter/density. Sedimentation tanks can remove 65–70% of the incoming suspended solids and, depending on the nature of the waste, up to 40–45% of its BOD load depending on strength of the wastewater (Forster, 1985; Metcalf and Eddy Inc., 2003). They can be operated with or without prior chemical coagulation/flocculation. Similar settlement processes are also conducted after secondary (biological) treatment.
6. Floatation, used to separate solid particles from the liquid phase by the subsurface introduction of fine air bubbles. The bubbles adhere to the solid particles and rise to the surface, where they can be removed.
7. Aeration (air stripping), used in the removal of volatile organic compounds (VOCs). The "off-gases" produced here may require their own treatment strategies, for example, by using granular activated carbon or thermal oxidation (Metcalf and Eddy Inc., 2003).
8. Thermal (combustion) technology has been reported for the treatment of hazardous waste from antibiotic (spiramycin, lincomycin, and kitasamycin) fermentations (Yang et al., 2015).

Physical processes used in tertiary treatment to produce an effluent of better quality than the 30:20 standard include microstrainers, slow sand filters, upflow sand filters, and rapid gravity sand filters. Throughputs vary between around 3 m^3 m^{-2} day^{-1} for slow sand filters and 700 m^3 m^{-2} day^{-1} for microstrainers. Suspended solids removal is generally 50–70% and BOD removal around 30–50%, depending on the technique used. A detailed description of tertiary/advanced treatment is given by Truesdale (1979), Viessman and Hammer (1993), and Metcalf and Eddy Inc., 2003.

CHEMICAL TREATMENT

Fine suspended and colloidal particles in an effluent may be aggregated together to form larger particles which are easier to separate from the liquid phase by co-agulation and/or flocculation (Cooper, 1975; see also Chapter 10). Coagulation is essentially instantaneous whereas flocculation requires some more time and gentle agitation to achieve "aggregation" of the particles. Ferrous or ferric sulfate, ferric chloride, aluminum sulfate (alum), aluminum chloride, calcium hydroxide (lime), and polyelectrolytes are often used as chemical coagulants. A solution of coagulant of the appropriate strength for effective treatment is added to the effluent in a vigor-ously mixed tank, a precipitate or floc forms almost immediately and carries down the suspended solids to form a sludge. This sludge may be drawn off, mechanically dewatered, and subjected to further treatment. Flocs may also be separated from the liquid by filtration. The flocs formed on coagulation may still be small, and will therefore require an extended period to settle, and as a consequence, for a given throughput of effluent a large sedimentation tank will be needed. Increasing the par-ticle diameter by encouraging small flocs to coalesce (flocculation) increases the rate of sedimentation and thus, for a given throughput, a smaller vessel can be operated. Polyelectrolytes, both natural and synthetic, are commonly used as flocculants. Floc-culants act in one of three ways; charge neutralization, polymer bridge formations, or a combination of the two. Following addition, the effluent is gently mixed (turbulent mixing would break up the flocs) by passage through sinuous flocculation channels, hydrodynamic flocculators, or mechanically mixed flocculators (Smethurst, 1988).

The use of Fenton's reagent [Fe(II)/Fe(III)/H_2O_2] as an oxidant has been widely reported in the treatment of fermentation wastewaters either alone or in combina-tion with other treatment processes. Rivas, Beltran, Gimeno, and Alvarez (2003) reported its use on the pretreatment of fermentation brines from the manufacture of table olives obtaining COD reductions of between 0.3 and 1.6 mol of COD per mol of H_2O_2 consumed. Xu, Hamid, Wen, Zhang, and Yang (2014) reported the use of Fenton's reagent in treating highly toxic avermectin fermentation wastewaters using a complex treatment system. After treatment in an upflow anaerobic sludge blanket (UASB) reactor the effluent undergoes Fenton oxidation followed by a membrane bioreactor (MBR). COD reductions of 84.3% were obtained. Xing and Sun (2009) describe the use of Fenton's reagent coupled with coagulation and sedimentation in the treatment of antibiotic fermentation wastewater obtaining 66.6% color removal and 72.4% COD removal.

Phosphate in the effluent stream may be removed chemically by reaction with calcium hydroxide (at pH > 10) to precipitate hydroxyapatite. Phosphate may also be removed in a similar manner using ferric and aluminum salts. Heavy metals can also be removed from effluents by precipitation usually as their hydroxide or sulfide.

Chemical disinfection normally takes place (if at all) following biological treatment. There are numerous chemicals, which have been used historically as disinfectants but the most common in current use are:

1. Chlorine and related compounds such as sodium hypochlorite, calcium hypochlorite, and chlorine dioxide. Reaction products of chlorine in water (hypochlorous acid and monochloroamine) also act as disinfectants. Many potable water treatment plants have moved away from using chlorine gas in recent years to using sodium hypochlorite because of safety concerns over the use of chlorine gas. Disinfection by-products, for example, trihalomethanes (THMs) and haloacetic acids (HAAs) when using oxidants such as chlorine (and ozone) are also a cause for concern. Low concentration chlorine residues have the potential to be toxic if released into the aquatic environment. If necessary the final effluent may be dechlorinated using sulfur dioxide, sodium sulfite, sodium thiosulfate, or activated carbon (Metcalf and Eddy Inc., 2003).

2. Ozone, a strong oxidant, also acts as a disinfectant. It is highly unstable and must be generated onsite by corona discharge. Like chlorine, ozone is highly toxic and consequently stringent safety features need to be in place if it is used. By-product formation is also an issue in ozone disinfection.

3. Although it is not a chemical treatment, it is appropriate to briefly discuss UV disinfection at this point, as it is considered a more environmentally friendly option. The portion of the UV range, which is effective for disinfection is at wavelengths between 220 and 320 nm. Low-pressure mercury vapor lamps in a quartz sleeve produce UV irradiation at 254 nm. UV radiation is effective as a disinfectant as it causes damage to the organisms DNA and hence inhibits replication. For UV disinfection to be effective, the liquid being treated must have very low suspended solids concentrations as particulate matter will effectively "shield" microorganisms from the UV light and hence reduce its effectiveness (Parker & Darby, 1995).

BIOLOGICAL TREATMENT

Most organic-waste materials may be degraded biologically. This process may be achieved aerobically or anaerobically in a number of ways. The most widely used aerobic processes are trickling filters, rotating disc contactors, activated sludge processes, and their modifications. The anaerobic processes are used both in the treatment of specific wastewaters and in sludge conditioning.

AEROBIC PROCESSES

Trickling filters

The term filter in this unit operation is a misnomer, as the action of a trickling filter is not one of filtration, but rather it is a fixed film bioreactor. Settled effluent to be treated is passed down through a packed bed countercurrent to a flow of air. Microorganisms adhering to the packing matrix adsorb oxygen from the upflowing air and

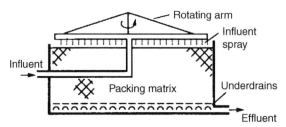

FIGURE 11.2 Schematic Diagram of a Trickling Filter

organic matter from the downflowing effluent; the latter is then metabolized and the effluent stream's BOD reduced. Trickling filters with a stone or rock packing have been used since the early 20th century as a cheap and simple treatment option.

A conventional trickling filter (Fig. 11.2) usually consists of a cylindrical concrete tank 2–3 m in depth and 8–16 m in diameter. Trickling filters utilizing plastic packing may be around 10–12 m in depth and upto 50 m in diameter. Some filters are rectangular in shape, but a rotary system allows more uniform hydraulic loading and are mechanically simpler (Bruce & Hawkes, 1983; Viessman & Hammer, 1993). The tank is packed with a bed of stone (usually granite) or special plastic packings, the bed being underlaid with drains. Plastic packing has the advantage that taller vessels can be used (biotowers) and higher loading rates can be applied. The packing material diameter should be 50–100 mm to give a specific surface of around $100 \text{ m}^2 \text{ m}^{-3}$, and the material should be packed to give a voidage (percentage air space of total bed volume) of 45–55%, which should minimize the risk of the spaces between the packing materials becoming blocked by the microbial film. Synthetic packing material, although more expensive, has a higher surface area and voidage, allowing higher treatment rates per unit volume of bed and reducing the likelihood of blockages. The trickling filter is always followed by a secondary sedimentation tank or humus tank to remove suspended matter (eg, biofilm sloughed off the packing) from the treated effluent. It is generally claimed that the sludge from trickling filters has better thickening properties to that from activated sludge. In conventionally loaded or low-rate filters, the effluent from which the suspended solids have been removed by primary sedimentation is fed on to the upper surface of the bed by spray nozzles or mechanical distributor arms (McKinney, 1962; Higgins, 1968). The effluent trickles gradually through the bed and a slime layer of biologically active material (bacteria, fungi, algae, protozoa, and nematodes) forms on the surface of the support material. The large surface area created in the bed permits close contact between air flowing upwards through the bed, the descending effluent and the biologically active growth. The bacteria in the biological film remove the majority of the organic loading. Complex organic materials are broken down and utilized, nitrogenous matter, and ammonia are oxidized to nitrates and sulfides, and other compounds are similarly oxidized. The higher organisms (protozoa etc.) control the accumulation of the biological film (which prevents the filter from blocking) and improve the settling characteristics of the solids (humus) discharged with the filter effluent. In low-rate filters, the scouring action of the hydraulic load normally has a minor role in removal of any loose micro-bial film. The active slime takes time to develop and can be poisoned by the addition of toxic chemicals. The simple filter is inefficient when operated at abnormally high organic loading rates. Initially, there is a very rapid build up of bacteria, fungi, and algae at the top of the filter, which cannot be controlled by the resident population of worms and larvae. The voids, therefore, block up, resulting in ponding (untreated effluent accumulating on the surface of the filter bed). Film growth can be limited by reducing the dosing frequency, which gives better liquid distribution deeper into the bed and by recycling treated effluent from the filter to increase the hydraulic loading rate providing the retention time is still sufficient for BOD removal.

A trickling filter bed should remove 75–95% of the BOD and 90–95% of the suspended solids at organic loading rates of 0.06–0.12 kg BOD m^{-3} day^{-1} for con-ventional trickling filtration. When part of the treated effluent is being recirculated to dilute the feed and increase the hydraulic load placed on the unit the organic loading can be increased to 0.9–0.15 kg BOD m^{-3} day^{-1}. The increase in hydraulic load thus applied causes greater hydraulic scouring of the bed (preventing blockage), but does not reduce treatment efficiency due to improved wetting of the packing surface and, thus, more efficient use of the biofilm (Forster, 1985). To achieve the Royal Com-mission (20:30) Standard together with a high degree of nitrification, filters being supplied with domestic sewage should receive organic loading rates of 0.07–0.1 kg BOD m^{-3} day^{-1} and hydraulic loading rates of 0.12–0.6 m^{3} m^{-3} day^{-1} (Gray, 1989).

It is possible to modify the trickling filter to increase the capacity for organic loading by the use of two sets of filters and settling tanks in series; this is known as alternating double filtration (ADF). Effluent is applied to the first filter at a high hydraulic and organic loading rates, it passes from this filter through the first settling tank and then on to a second filter and settler. After a period of one to two weeks, the sequence of the filters is reversed and the second filter receives the higher loading. In this way heavy film growth is promoted in the first filter to receive the effluent, but when the filter sequence is reversed it becomes nutrient limited, encouraging ex-cess film removal. Loading rates of 0.32–0.47 kg BOD m^{-3} day^{-1} have been claimed (Forster, 1977), but recommended rates for design purposes are 0.15–0.26 kg BOD m^{-3} day^{-1} (Forster, 1985).

Trickling filters can also be operated in combined aerobic treatment processes. Typically, they are combined with the activated sludge process with a tricking filter preceding an activated sludge reactor. Such combined processes have often been the result of upgrading a treatment plant with the addition of a trickling filter or activated sludge plant. An intermediate sedimentation tank between the two processes may or may not be incorporated (Metcalf and Eddy Inc., 2003).

Cook (1978) has stressed the need to consider the possible intermittency of a factory wastewater treatment process. It was shown that starving of a laboratory-scale trickling filter beyond 48 h resulted in near failure of the filter. This indicated the need to supplement or artificially load the filter to maintain a viable biomass in the system. Alternatively effluent storage and flow balancing can be used to maintain a constant flow to the filter.

Towers

Because trickling filters do not have both a high specific area and a high voidage, they are less suitable for the treatment of large volumes of strong industrial effluents (Table 11.2). Large areas of land would be required for the expensive and extensive traditional filter beds. Towers, up to around 10 m in height, packed with lightweight (30–95 kg m^{-3}) plastic multifaced modules or small random packing units have proved to be a space saving and relatively inexpensive solution to the problem. By comparison traditional stone packings have unit weight of approximately 800–1450 kg m^{-3}. These packings have a relatively open structure for oxygen transfer (specific surface of 90–150 m^2 m^{-3}) and high voidage (70–95%), but are expensive compared with the conventional filter packings. They are capable of coping with high BOD loadings. At a loading of 3.2 kg BOD m^{-3} day^{-1}, a 50% BOD removal may be achieved and at 1.5 kg m^{-3} day^{-1}, 70% removal is possible (Ripley, 1979). The biological film is similar to that formed on the conventional packing and scouring is due to the hydraulic load applied rather than the predation of higher organisms.

Trickling filters with plastic packing can be used in tertiary treatment (ie, after secondary treatment) for nitrification of ammoniacal nitrogen. Influent BOD levels are very low ($\sim$10 mg dm^{-3}).

Biologically aerated filters (BAFs)

BAFs (Fig. 11.3) are a relatively recent development based on the trickling filter. They consist of a packed bed, which provides sites for microbial growth through which air is passed but, unlike trickling filters, the reactor volume is flooded with the effluent to be treated which is passed upward or downward through the reactor (ie, co- or counter-currently to the air supply) depending on the design. The packing matrix may be natural (eg, pumice) or synthetic (eg, polyethylene), and may be either a fixed structure or randomly packed.

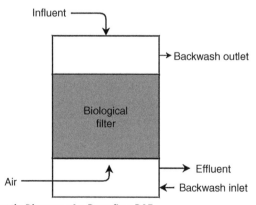

FIGURE 11.3 Schematic Diagram of a Downflow BAF

The combination of aeration and filtration allows high rates of BOD and ammonia removal together with solids capture, so that sedimentation tanks may not be required. However, regular backwashing is essential to remove filtered solids and excess biomass. Organic loading rates for 90% BOD removal are significantly greater than those obtained for trickling filters, being in the range 0.7–2.8 kg BOD m^{-3} day^{-1} (Stephenson, Allan, & Upton, 1993). They are versatile treatment systems, and of those currently in operation design capacities vary between 600 and 70,000 m^3 day^{-1}. In addition to their use as a secondary treatment process, they can also be utilized for tertiary treatment or modified to allow denitrification in a manner similar to that of activated sludge systems and trickling filters.

Rotating biological contactors (RBCs)

RBCs were first introduced in Germany in the 1960s and many thousands of units installed worldwide since then. In this treatment method (Fig. 11.4), a unit composed of closely spaced plastic discs (2–3 m diameter with 1–2 cm spacing between discs), on a central drive shaft are rotated slowly (0.5–1.5 rpm) through the effluent, which is flowing through a vessel constructed of reinforced concrete, steel or glass fiber depending on scale, so that 40–50% of the disc surfaces are submerged (Borchardt, 1970; Pretorious, 1973). The discs, usually made from synthetic material (eg, polystyrene, PVC), are arranged in stages or groups separated by baffles to minimize short circuiting or surging (Forster, 1985) and to enhance specific treatment requirements such as nitrification. Multiple RBC units in series can also be used to the same effect. The discs may be flat or corrugated to increase surface area. A microbial film forms on the discs; this is aerated during the exposed part of the cycle and absorbs nutrients during the submerged part. Shear forces produced as the discs rotate through the liquid control the thickness of the biofilm, with excess biofilm being sloughed from the discs. A sedimentation tank following the biological stage is therefore required to remove biological solids. Like trickling filters no sludge recycle from the settling tank is required. Loading rates of 13 g BOD m^{-2} day^{-1} for domestic sewage and partial treatment of loads of 400 g BOD m^{-2} day^{-1} have been used. To achieve the 20:30 standard the loading rate should not exceed 6 g BOD m^{-2} day^{-1}. RBC systems can also be designed to allow for nitrification to take place both in

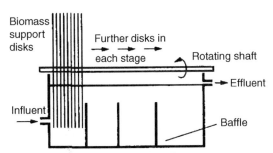

FIGURE 11.4 Schematic Diagram of a Rotating Biological Contactor

secondary and tertiary treatment. Rotating biological contactors are compact, easily covered for health and aesthetic reasons, available as packaged units, simpler to operate under varying loads than trickling filters (the biofilm being wetted at all times) and are easily added onto existing treatment processes. As such they can provide a cost effective method of onsite treatment.

Ware and Pescod (1989) describe the use of full scale anaerobic/aerobic rotating biological contactors for treating brewery wastewaters. Greater than 85% COD removal was obtained in the aerobic stage, but difficulties were experienced in maintaining anaerobic populations.

Fluidized-bed systems

Fluidized-bed reactors in wastewater treatment are relatively recent innovations. The support matrix (sand, anthracite, reticulated foam and activated carbon) has a large surface area (typical particle sizes are 0.4–0.5 mm and the specific surface area is about 1000 m^2 m^{-3}) on which the biofilm adheres and thus they are able to operate at high biomass concentrations with high rates of treatment. This allows strong wastewaters to be treated in small reactors. They are also useful for the treatment of industrial wastewaters when variable loadings are encountered (Cooper & Wheeldon, 1980, 1982). The support matrix is fluidized by the upflow of effluent through the reactor, and the degree of bed expansion is controlled by the flow rate of wastewater. The treated effluent can thus be decanted off without loss of the support matrix and with careful operation a secondary sedimentation tank may not be needed. The support matrix is regularly withdrawn to remove excess biomass. Fluidized-bed systems can be operated aerobically, anaerobically (see later section) or anoxically for denitrification.

Activated sludge processes

The basic activated-sludge process (Fig. 11.5) consists of aerating and agitating the effluent in the presence of a flocculated suspension of microorganisms on particulate organic matter—the activated sludge sometimes referred to as "mixed liquor suspended solids" (MLSS). This process was first reported by Arden & Lockert (1914) and is now the most widely used biological treatment process for both domestic and industrial wastewaters. The raw effluent enters a primary sedimentation tank where

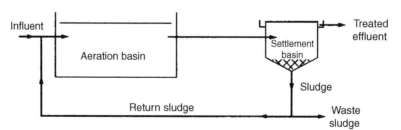

FIGURE 11.5 Simplified Cross-Section of an Activated Sludge Process

coarse solids are removed. The partially clarified effluent passes to a second vessel, which can be of a variety of designs, into which air or oxygen is injected by bubble diffusers, paddles, stirrers, surface aerators, etc. Vigorous agitation ensure that the effluent and oxygen are in contact with the activated sludge. Dissolved oxygen concentrations should be maintained at about 1.5–2.0 mg dm^{-3}. Higher dissolved oxygen concentrations may be used to improve nitrification in activated sludge plants receiving high BOD loadings. It should be noted that higher dissolved oxygen concentrations will increase aeration costs (Metcalf and Eddy Inc., 2003). After a predetermined residence time of several hours, the effluent passes to a second sedimentation tank to remove the flocculated solids. Part of the sludge from the settlement tank is recycled to the aeration tank to maintain the biological activity, that is, high concentrations of MLSS. The overflow obtained from the settlement tank should be of a 20:30 standard or better and be suitable for discharge into most inland waters depending on local discharge consents. The excess sludge is dewatered, to be sold as a soil conditioner, incinerated, or landfilled.

In conventional activated sludge processes, organic loading rates are 0.5–1.5 kg BOD m^{-3} day^{-1} with hydraulic retention times of 5–14 h depending on the nature of the wastewater, giving BOD reductions of 90–95%. High-rate activated-sludge processes can be used as a partial treatment for strong wastes prior to further treatment or discharge to a sewer and are widely used in the food processing and dairy industries. The organic loading rate is 1.5–3.5 kg BOD m^{-3} day^{-1}, and with hydraulic retention times of only 1–2 h, BOD reductions of 60–70% are possible (Gray, 1989).

A number of modifications of the basic process can be used to improve treatment efficiency, or for a more specific purpose such as nitrification and denitrification (Winkler & Thomas, 1978; Gray, 1989). Tapered aeration and stepped feed aeration are used to balance oxygen supply and demand (which is greatest at the point of wastewater entry to the aeration basin) with the amount of oxygen supplied. Contact stabilization exploits biosorption processes and thereby allows considerable reduction in basin capacity (~ 50%) for a given wastewater throughput. Denitrification (the biological reduction of nitrate to nitrite and on to nitrogen gas under anoxic conditions) can be accomplished in an activated-sludge plant when the first part of the basin is not aerated and as a result the returned activated sludge or an internal recycle from the activated sludge tank is denitrified. Denitrification can also be conducted in an anoxic reactor after a conventional aerobic activated sludge reactor. However, this will require a secondary carbon source to be added to the anoxic vessel as denitrification requires a carbon source.

Biological phosphorous removal can also be achieved in modified activated sludge reactors. All include an anaerobic zone followed by an aerobic zone as there must be anaerobic contacting between the activated sludge and the influent wastewater for biological phosphorus removal to occur.

A variation on the activated sludge process is the sequencing batch reactor (SBR). This process uses a "fill and draw" reactor where aeration (reaction) and sedimentation occur in the same vessel. The sequence of operation being: fill with effluent; reaction/aeration; sedimentation and draw (remove treated effluent and sludge). For

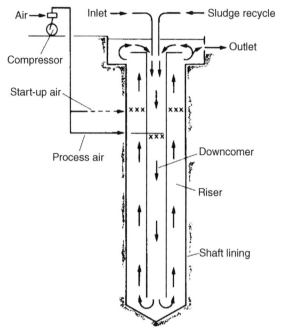

FIGURE 11.6 Deep-Shaft Effluent Treatment Plant (Hemming et al., 1977)

continuous operation at least two SRB tanks must be provided (Metcalf and Eddy Inc., 2003).

In advanced activated-sludge systems, the amount of dissolved oxygen available for biological activity is increased to improve the treatment rate. One vessel of this type is the "Deep Shaft" (Hemming, Ousby, Plowright, & Walker, 1977), which is quite distinctive from the other aeration tanks and has been developed from the ICI SCP process (Taylor & Senior, 1978; Chapter 7). The "Deep Shaft" (Fig. 11.6) consists of a shaft 50–150 m deep, separated into a down-flow section (down-comer) and an up-flow section (riser). The shaft may be 0.5–10 m diameter, depending on capacity. Fresh effluent is fed in at the top of the "Deep Shaft" and air is injected into the down-flow section at a sufficient depth to make the liquid circulate at 1–2 m s^{-1}. The driving force for circulation is created by the difference in density (due to air bubble volume) between the riser and down-flow sections. For starting up, circulation of liquid is stimulated by injecting air at the same depth in the riser. Air injection is then gradually all transferred to the air injection point in the down-comer. Because of the pressure created in the down-comer, oxygen-transfer rates of 10 kg O$_2$ m^{-3} h^{-1} can be achieved and bubble contact times of 3–5 min are possible instead of 15 s in diffused air systems. BOD removal rates of 90% are achievable at organic loadings of 3.7–6.6 kg BOD m^{-3} day^{-1} at hydraulic retention times of 1.17–1.75 h (Gray, 1989). Sludge production was found to be much less than that for conventional sewage-treatment processes.

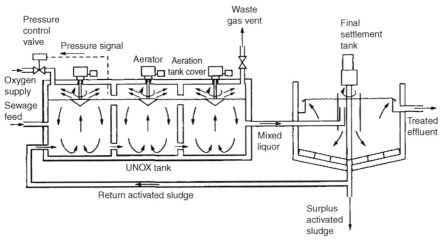

FIGURE 11.7 Schematic Diagram of a Multistage UNOX System (Fuggle, 1983)

Two types of pure oxygen systems have also been developed to increase the rate of oxygen transfer:

1. Closed systems which operate in oxygen-rich atmospheres and,
2. Open systems employing fine bubble diffusers.

An example of the closed system is the UNOX Process developed by the Union Carbide Corporation in the United States of America, and marketed in the United Kingdom by Wimpey Unox (Fig. 11.7). The enclosed oxygenation tank is compartmentalized by baffles. Settled wastewater and returned sludge are fed into the first stage and oxygen ($\sim$ 90% purity) is pumped into the headspace. The oxygen and wastewater move sequentially through the compartments, and the oxygen concentration in the gas phase decreases in each stage as it is consumed by the microorganisms. As the nutrient concentration also falls stage by stage, oxygen supply and demand is balanced. Organic loading rates (when treating municipal wastewater) are 3–4 times higher than those of aerated systems at 2.5–4.0 kg BOD m^{-3} day^{-1}. Pure oxygen systems also have shorter residence times and produce less sludge with better settling qualities than conventional systems. The UNOX process has been successfully used for the treatment of brewery wastewaters containing 2000 mg BOD dm^{-3} at flow rates of 2269 m^3 day^{-1} with a 6.6 h retention time (Brooking et al., 1990). There are a number of drawbacks to using this type of pure oxygen system—safety provision to deal with the high O_2 concentrations, the cost of generating/purchasing oxygen, and poor CO_2 stripping from the effluent.

The VITOX aeration system (Fig. 11.8) developed by the British Oxygen Company is an example of an open tank oxygenation system. Its main advantage is that it can be used in existing aeration tanks either to replace or upgrade conventional aeration without the need to replace or modify existing units. Oxygenation is achieved

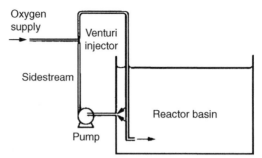

FIGURE 11.8 Simplified VITOX Sidestream Aeration System

by pumping settled wastewater at high pressure through a venturi where oxygen is injected. Turbulence and high pressures thus created ensure high levels of oxygen dissolution. The flexibility of this system means that it is particularly useful for the treatment of high strength intermittently produced wastewaters such as those generated by food processing industries (Gostick et al., 1990).

Membrane biological reactors

Membrane biological reactors (MBRs) have been used for the treatment of both domestic and industrial wastewater and consist of a biological reactor together with microfiltration membranes (pore size 0.1–0.4 µm) to separate the suspended solids and biomass from the clarified effluent (Fig. 11.9). MBRs can be operated both aerobically and anaerobically and the membrane unit may be either internal or external to the biological reactor. MBRs operate at higher MLSS concentrations allowing higher loading rates, shorter retention times, and lower space requirements. However, disadvantages include higher capital cost, the high cost of membrane replacement, higher energy costs, and problems with membrane fouling (Metcalf and Eddy Inc., 2003).

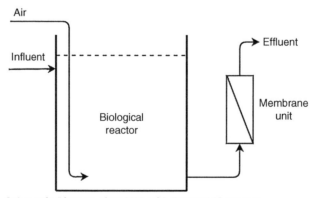

FIGURE 11.9 Schematic Diagram of an MBR with External Membrane

ANAEROBIC TREATMENT

Anaerobic treatment of waste organic materials originated with the use of septic tanks and Imhoff tanks, which have now been replaced by a variety of high-rate digesters (Pohland & Ghosh, 1971). Examples of wastewaters treated by anaerobic processes include domestic wastewater, chemical manufacturing, breweries, distilleries, dairy wastes, fish and shellfish processing, agricultural wastes, landfill leachates, pharmaceuticals, pulp and paper wastes, and a range of food/food processing wastes. Loehr (1968) has listed the following reasons for using anaerobic processes for waste treatment:

1. Higher loading rates can be achieved than are possible for aerobic treatment techniques.
2. Lower power requirements may be needed per unit of BOD treated.
3. Useful end-products such as digested sludge and/or combustible gases may be produced.
4. Organic matter is metabolized to a stable form.
5. There is an alteration of water-binding characteristics to permit rapid sludge dewatering.
6. The reduced amount of microbial biomass leads to easier handling of sludge.
7. Low levels of microbial growth will decrease the possible need for supplementary nutrients with nutritionally unbalanced wastes.

   Disadvantages include the following (Metcalf and Eddy Inc., 2003):

1. Longer start-up times to develop the necessary biomass inventory.
2. May require addition of alkalinity.
3. Biological nitrogen and phosphorous removal is not possible.
4. More sensitive to temperature.
5. Possible more sensitive to toxic/inhibitory materials.
6. Potential for the production of odors and corrosive gases.

Anaerobic digestion

Anaerobic digestion can be used in the treatment of "whole" wastewaters and in the treatment of sludges produced by other treatment processes. Large volumes of wet sludge, which are produced in primary and secondary sedimentation tanks may have to be reduced in volume before disposal. This volume of sludge can be reduced by anaerobic digestion. In sludges containing $20,000–60,000$ mg dm^{-3} solid matter, 80% of the degradable matter may be digested, which will reduce the solids content by 50%. During anaerobic digestion, hydrolytic and fermentative bacteria break down complex organic molecules to volatile fatty acids, organic acids, alcohols, and aldehydes. Hydrogen producing acetogens convert the fermentation products to acetate, H_2, CO_2, and a few one carbon compounds and homoacetogens convert H_2, CO_2, and 1-carbon compounds at acetate. Acetoclastic methanogens then convert acetate to methane and CO_2 and hydrogen utilizing methanogens combine hydrogen and carbon dioxide to yield methane and water. The composition of the biogas produced

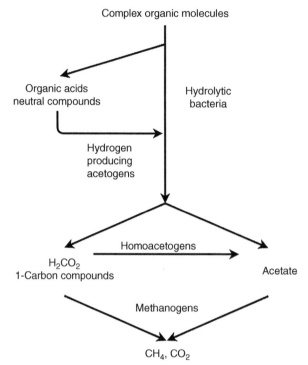

FIGURE 11.10 Overview of the Microbiology and Biochemistry of Anaerobic Digestion

is typically methane (~ 60%) and carbon dioxide (~ 40%). An overview of the bacterial groups and biochemistry of anaerobic digestion is shown in Fig. 11.10. The gas produced (biogas) is a very useful byproduct, and can be burnt as a heating fuel, fed to gas engines to generate electricity or used as a vehicle fuel. As well as being used in sludge digestion and conditioning, anaerobic digesters are also used directly in the treatment of many high strength wastewaters, for example, from the food and agricultural industries (Gray, 1989; Borja, 2011).

A number of anaerobic processes have been developed; completely mixed reactors (often these are simply described as anaerobic digesters) with or without sludge recycle (Fig. 11.11), anaerobic filters, up-flow anaerobic sludge blankets (UASB), anaerobic sequencing batch reactors (ASBR), and anaerobic fluidized beds are the most common. Two stage anaerobic treatment systems utilizing a short retention time completely mixed acidification reactor followed by a UASB methanogenic reactor have been reported (Burgess & Morris, 1984; Ghosh, Ombregt, & Pipyn, 1985). Benefits claimed include improved reliability, higher BOD/COD reductions and greater methane productivity. The use of membrane separation techniques similar to those used in the activated sludge process have also been described (Choo & Lee, 1998).

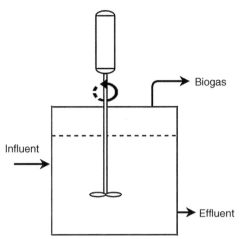

FIGURE 11.11 Schematic of a Completely Mixed Anaerobic Digester Operating Without Sludge Recycle

It has been suggested that the effluent from anaerobic digesters can be processed and supplemented/modified to provide useful chemicals via fermentation with *Clostridium butyricum* (Zacharof, Vouzelaud, Mandale, & Lovitt, 2015). Elbeshbishy, Dhar, Hafez, and Lee (2015) report the application of a number of Clostridia in Acetone-Butanol-Ethanol (ABE) fermentations utilizing renewable carbohydrates.

Bio-hydrogen production has received much attention in recent years. Hydrogen is a clean, effective and renewable fuel, which can be produced by biological methods. For example, anaerobic bacterial fermentative hydrogen production using aquatic algal biomass as the carbon source has been reported by Xia et al. (2015). Trchounian and Trchounian (2015) report the use of *Escherichia coli* in hydrogen production from glycerol as a waste product of biodiesel production.

Anaerobic digesters

The digester tanks used for this process may be up to 12,000 m^3 in volume and equipped with heating coils for accurate temperature control normally in the mesophilic range (25–38°C, usually 28–32°C in anaerobic digestion) to increase the rate of digestion, which is also improved by mechanical agitation (Loll, 1977). Some digesters are operated thermophilically (50–70°C, usually 55–63°C in anaerobic digestion) to increase the rate of degradation and biogas production, though much of the extra gas produced will be consumed in maintaining the digester temperature. The use of thermophilic anaerobic digestion has been investigated for the treatment of high strength (16–25 g COD dm^{-3}), low pH ($\sim$ 3.8) and high temperature winery effluents (Romero, Sales, Cantero, & Galan, 1988). Wiegant, CLaassen, and Lettinga (1985) reported the successful application of thermophilic anaerobic digestion of vinasse in completely mixed and UASB reactors, although methane production was found to decrease at high loading rates due to the presence of inhibitory compounds. Retention times are generally between 10 and 30 days, although with solids recycle retention times may be as low as 0.5 day. Most digesters operate on a semicontinuous "draw and fill" basis with the appropriate volume of digested material being removed and replaced with "fresh" material on a daily basis.

Anaerobic filters

In anaerobic filters, as with the aerobic filters, there is a microbial film growing on an inert support (Chian & DeWalle, 1977). Anaerobic filters may be operated in an upflow or downflow mode, and a wide variety of packings are used; both natural and synthetic (Young, 1983). The specific surface area of the packing is on average 100 m^2 m^{-3} (Song & Young, 1986). The first full-scale anaerobic filters were constructed in the 1970s and in 1982 the Bacardi Corporation brought into operation a 100,000 m^3 plant to treat distillery effluents (Szendrey, 1983). Many other effluents are found to be amenable to the treatment using anaerobic filter systems; antibiotic fermentation wastes, citric acid fermentation wastes, yeast production wastewater and brewery and winery wastewaters (Szendrey, 1983), molasses distillery slops (Silvero et al., 1986), fermentation and pharmaceutical wastes (Bonastre and Paris, 1989), guar gum, chemical processing waste, domestic effluent, landfill leachate and food canning and soft drinks waste (Young, 1991).

Up-flow anaerobic sludge blankets (UASB)

In this system (originally developed in the Netherlands) high levels of active biomass are retained in the reactor by flocculation (Lettinga, van Velsen, Hobma, de Zeeuw, & Klapwiik, 1980; Lettinga et al., 1983). No support media is added to the reactor; instead the flocculated sludge develops in the reactor and acts as a fluidized bed. Feed is pumped through the bed (the sludge blanket), above which fine particles flocculate and settle back to the blanket as sludge, thus preventing washout of organisms. Anaerobic sludge blankets have been found to be effective in the treatment of many wastewaters, including sugar-beet wastes, domestic sewage, slaughterhouse wastes, agricultural wastes (Lettinga et al., 1983), brewery wastes (Fang, Jinfu, & Guohua, 1989; Fang, Guohua, Jinfu, Bute, & Guowei, 1990), winery wastes (Cheng et al., 1990), and distillery wastes (Burgess & Morris, 1984; Ghosh et al., 1985). The use of UASBs has been reported in the treatment of citric acid waste where citric acid wastewater is anaerobically treated together with electrodialysis for methane production (Xu et al., 2015).

ADVANCED OR TERTIARY TREATMENT

Advanced or tertiary wastewater treatment is the further treatment of effluent from secondary processes to remove the suspended, colloidal, and dissolved materials which remain. These constituents may be simple inorganics such as nitrates and metal ions or complex organic molecules, for example, pharmaceutical products

Table 11.3 Advanced Wastewater Treatments

Treatment Process	Example(s) of Wastewater Constituents Removed
Filtration (depth, surface, micro- and ultra-)	Suspended and colloidal solids; microorganisms
Reverse osmosis	Suspended and colloidal solids; dissolved organic matter; dissolved inorganic matter; microorganisms
Electrodialysis	Suspended and colloidal solids; dissolved organic matter; dissolved inorganic matter; microorganisms
Adsorption	Suspended and colloidal solids; dissolved organic matter; microorganisms
Air stripping	Volatile organic compounds; ammonia
Ion exchange	Suspended and colloidal solids; dissolved organic matter; dissolved inorganic matter; microorganisms
Distillation	Suspended and colloidal solids; dissolved organic matter; dissolved inorganic matter; microorganisms
Precipitation	Suspended and colloidal solids; TOC; Phosphorous
Chemical oxidation	Particulate organic matter; TOC
Advanced oxidation processes	Dissolved organic matter

which act as endocrine disrupters in aquatic species (Metcalf and Eddy Inc., 2003). Processes used in advanced treatment and types of constituent removed are shown in Table 11.3.

CONSTRUCTED WETLANDS

Constructed wetlands are becoming increasingly popular for the treatment of both domestic and industrial wastewaters mostly as a "polishing step" following conventional secondary treatment. They have low capital and operating costs, are easy to maintain and resistant to shock loads. Constructed wetlands treat wastewaters by a number of routes; filtration, adsorption, precipitation, ion exchange, plant uptake, and microbial degradation (both aerobic and anaerobic). There are two types of constructed wetland. Free water surface (FWS) wetlands are similar to natural wetlands having a soil bottom, emergent vegetation with the water surface exposed to the atmosphere. The water column depth is generally less that 0.4 m and water flow is low velocity. Subsurface flow (SSF) wetlands are filled with a porous media, such as gravel and soil, and the water level is below the surface. Constructed wetlands are also classified by the types of plants growing within them such that floating (eg, water hyacinth), submerged (eg, *Lobelia* dortmanna) and rooted emergent (eg, common reed, *Phragmites australis*) macrophyte systems are classified (Naja & Volensky, 2011). The treatment of both industrial and domestic/municipal wastewater in FWS and SSF constructed wetlands are designed with an area of around 5–10 m^2 per person equivalent and a length to width ratio of approximately 15:1(Pell & Worman, 2011).

BY-PRODUCTS

The marketing of wastes from fermentation processes has been established for at least 200 years. By the 1700s, brewers' grains, spent hops, and surplus yeast from larger breweries were accumulating in sufficient quantities for specialized trades to develop (Mathias, 1959). Around London, cattle and pigs were fattened on "wash" and brewers' grains. Excess yeast was supplied to bakers and gin distillers.

In any fermentation recovery process there is a need to recognize whether there is a potential marketable waste and, if necessary, to develop a market. Obviously the marketability and cost of reclamation of by-products will be very important in deciding upon a policy of waste recovery. Under favorable market conditions, it has been claimed that the profit on animal feed byproducts from a completely integrated distillery may almost pay the cost of the grain (Blaine, 1965). This claim would now be more difficult to substantiate, due to fuel costs and capital outlay on plant (Quinn & Marchant, 1980; Sheenan & Greenfield, 1980).

Corbin et al. (2015) report the use of grape marc (the remains of grapes that have been pressed for wine fermentation) could generate up to 13 Mt yr^{-1} of waste biomass which could be used as a source of carbohydrate for bioethanol production in the range 270–400 dm^3 t^{-1}.

DISTILLERIES

In grain-based distilleries it has been common practice to recover spent grain and stillage, the waste liquor after the alcohol has been distilled off (Boruff, 1953; Blaine, 1965). The stillage is first passed through screens with 1 mm openings. The screenings are then dewatered with mechanical filter presses and dried in rotary driers (Chapter 10). This product is termed Distillers' Dried Grains (light grains). The screened stillage is concentrated in evaporators to give 25–35% solids in a thick syrup. Stillages, which have been prefiltered may be concentrated to 35–50% solids. This syrup can then be mixed with the pressed screenings and dried to give Distillers' Dried Grains with Solubles (dark grains). Alternatively the evaporated stillage may be dried completely in drum driers to produce Distillers' Dried Solubles. Dark and light grains and dried solubles have all been used as animal-feed supplements. Flachowsky, Baldeweg, Tiroke, Konig, and Schneider (1990) report their use following chemical treatment as a replacement feed for sheep. Dried solubles have also been used as a medium adjunct in the preparation of antibiotics.

In distilleries using cane molasses as a feedstock, evaporated spent wash has been used as a fuel for boilers (Sheenan & Greenfield, 1980). It has proved worth while to recover potassium salts from sugar-beet stillage. The market for the evaporated product must be considered within a range of 50 km of the evaporation plant (Lewicki, 1978).

Boruff (1953) cited an unpublished 1949 survey of American distilleries, which showed that 85% of the stillage solids were being recovered as dried feeds, 14% solids as wet grains, and only 1% was waste. In recent years the traditional distillers' by-products from stillage have become increasingly uneconomic. A number of processes to produce SCP from stillage using *Geotrichum candidum, Candida utilis,* or *C. tropicalis* have been evaluated (Quinn & Marchant, 1980; Sheenan & Greenfield, 1980).

BREWERIES

The three marketable wastes from breweries are spent grain, spent hops, and yeast. The spent grain is recovered from the mash tun and is then sold as animal feed either after pressing in a wet state or after drying in rotary driers. Alternatively, the wet grain may be used in the preparation of silage for cattle. The possible markets for hops are restricted. Some are used as a fertilizer or as a low-grade fuel.

Yeast can be separated from beer by filtration or centrifuging. The yeast slurry is then dried in drum driers. Some of the yeast may be mixed with brewers' spent grains to produce a feed material with a slightly higher protein content than normal brewers' grains. The yeast may also be used directly as a source of vitamins. If it is to be used as a human food it must be debittered to remove the hop bitter substances absorbed on to the yeast cells. The cells are then washed in an alkaline solution, washed with water, and drum dried. Although bakers' yeast was originally obtained as a brewery by-product, this market has diminished considerably. Most bakers' yeast is now produced directly by a distinct production process. Dewatered sludge from brewery wastewater treatment operations has been reported to increase agricultural yields when used as fertilizer (Naylor & Severson, 1984). Lyons (1983) reports on the potential use of brewery effluents as a feedstock for fuel and industrial ethanol production.

Schneider et al. (2013) report the use of brewery wastewaters, containing high concentrations of maltose and glucose, as a carbon source for the production of high value carotenoids and microbial lipids for biodiesel production using *Rhodotorula glutinis*.

AMINO ACID WASTES

The main wastes from glutamic acid or lysine fermentations are cells, a liquor with a high amino-acid content, which can be used as an animal-feed supplement, and the salts removed from the liquor by crystallization, which is a good fertilizer (Renaud, 1980).

FUEL ALCOHOL WASTES

The stillage from ethanol production and wastes from starchy fermentations are, following concentration, saleable as an animal-feed supplement. Wastes from sugar fermentations and distillation can be digested anaerobically and the methane generated used as an energy source (Essien & Pyle, 1983; Faust, Prave, & Schlingmann, 1983; Singh, Hsu, Chen, & Tzeng, 1983). Faust et al. (1983) also suggest the use of CO_2 rich off gases in the food and beverage industries.

REFERENCES

Abson, J. W., & Todhunter, K. H. (1967). Effluent treatment. In N. Blakebrough (Ed.), *Biochemical and biological engineering science* (pp. 309–343). (Vol. 1). London: Academic Press.

American Public Health Association (1992). In A. E. Greenberg, L. S. Clesceri, & A. D. Eaton (Eds.), *Standard methods for the examination of water and wastewater* (18th ed.). Washington, DC: American Public Health Association.

Arden, E., & Lockett, W. T. (1914). Experiments on the oxidation of sewage without the aid of filters. *Surveyor, 45*, 610–620.

Ashley, M. H. J. (1982). The efficient use of water in fermentation processes. *Institute of Chemical Engineers Symposium Series, 78*, 355–368.

Ballinger, D. G., & Lishka, R. J. (1962). Reliability and precision of BOD and COD determinations. *Journal Water Pollution Control Federation, 34*, 470–474.

Bell, S., Mcgillivray, D., & Pedersen, O. (2013). *Environmental law* (8th ed.). Oxford, UK: Oxford University Press.

Blaine, R. K. (1965). Fermentation products. In C. F. Gumham (Ed.), *Industrial waste water control* (pp. 147–166). New York: Academic Press.

Bonastre, N., & Paris, J. M. (1989). Survey of laboratory, pilot and industrial anaerobic filter operations. *Process Biochemistry, 24*(1), 15–20.

Borchardt, J.A. (1970). Biological waste treatment using rotating discs. In *Biological Waste Treatment Biotechnological and Bioengeering Symposium 2* (pp. 131–140).

Borja, R. Biogas Production. In *Comprehensive biotechnology*, Vol. 2, 2nd ed. (pp. 785–797). M. Moo-Young (Ed.), Elsevier: Oxford.

Boruff, G. S. (1953). The fermentation industries. In W. Rudolfs (Ed.), *Industrial wastes. Their disposal and treatment* (pp. 99–131). New York: Reinhold.

Brooking, J., Buckingham, G., & Fuggle, R. (1990). Constraints on effluent plant design for a brewery. In *Effluent Treatment and Waste Disposal* (pp. 109–126). *I. Chem. E. Symp. Ser. 116,*1. Chem. E., Rugby.

Brown, A. (1992). In A. Brown (Ed.), *The UK environment* (pp. 75–104). London: HMSO (Department of the Environment).

Bruce, A. M., & Hawkes, H. A. (1983). Biological filters. In C. R. Curds, & H. A. Hawkes (Eds.), *Ecological aspects of used water treatment, Vol. 3, the processes and their ecology* (pp. 1–111). London: Academic Press.

Burgess, S., & Morris, G. G. (1984). Two-phase anaerobic digestion of distillery effluents. *Process Biochemistry, 19*(5), iv–iv10.

Cheng, S. S., Lay, J. J., Wei, Y. T., Wiu, M. H., Roam, G. D., & Chang, T. C. (1990). A modified UASB process treating winery wastewater. *Water Science and Technology, 22*(9), 167–174.

Chian, E. S. K., & DeWalle, F. B. (1977). Treatment of high strength acidic wastewater with a completely mixed anaerobic filter. *Water Research, 11*, 295–304.

Choo, K. H., & Lee, C. H. (1998). Hydrodynamic behavior of anaerobic biosolids during crossflow filtration in the membrane anaerobic bioreactor. *Water Research, 32*, 3387–3397.

Colovos, G. C., & Tinklenberg, N. (1962). Land disposal of pharmaceutical manufacturing wastes. *Biotechnology and Bioengineering, 4*, 153–160.

Cook, E. E. (1978). Effects of long term endogenous respiration (fasting) on the organic removal capacity of a trickling filter. *Biotechnology and Bioengineering, 20*, 293–296.

Cooper, P. F. (1975). Physical and chemical methods of sewage treatment. Review of present state of technology. *Water Pollution Control, 74*, 303–311.

Cooper, P. F., & Wheeldon, D. H. V. (1980). Fluidised and expanded bed reactors for wastewater treatment. *Water Pollution Control, 79*, 286–306.

Cooper, P. F., & Wheeldon, D. H. V. (1982). Complete treatment of sewage in a two-stage fluidised bed system. *Water Pollution Control, 81*, 447–464.

Corbin, K. R., Hseih, Y. S. Y., Betts, N. S., Byrt, C. S., Henderson, M., Stork, J., DeBolt, S., Fincher, G. B., & Burton, R. A. (2015). Grape marc as a source of carbohydrates for bioethanol: chemical composition, pre-treatment and saccharification. *Bioresource Technology, 193*, 76–83.

Davis, E. M. (1971). BOD vs COD vs TOC vs TOD. *Water and Wastes Engineering, 8*(2), 32–34.

DEFRA (2013). Incineration of Municipal Solid Waste. Available from: www.gov.uk/government/uploads/system/uploads/attachment_data/file/221036/pb13889-incineration-municiple.

Donaldson, J. (1993). Identifying opportunities for chemical waste recycling. *Wastes Management*, 16–18.

Doran, P. M. (2013). *Bioprocess engineering principles* (2nd ed.). Oxford: Academic Press.

Elbeshbishy, E., Dhar, B. R., Hafez, H., & Lee, H. -S. (2015). Acetone-butanol-ethanol production in a novel continuous flow system. *Bioresource Technology, 190*, 315–320.

Essiemn, D., & Pyle, D. L. (1983). Energy conservation in ethanol production by fermentation. *Process Biochemistry, 18*(4), 31–37.

Fang, H. H. P., Jinfu, Z., & Guohua, L. (1989). Anaerobic treatment of brewery effluent. *Biotechnol. Letters, 11*(9), 673–678.

Fang, H. H. P., Guohua, L., Jinfu, Z., Bute, C., & Guowei, G. (1990). Treatment of brewery effluent by UASB process. *Journal of Environmental Engineering, 116*(3), 454–460.

Faust, U., Prave, P., & Schlingmann, M. (1983). An integral approach to power alcohol. *Process Biochemistry, 18*(3), 31–37.

Fisher, N. S. (1977). Legal aspects of pollution. In A. G. Callely, C. F. Forster, & D. A. Stafford (Eds.), *Treatment of industrial effluents* (pp. 18–29). London: Hodder and Stoughton.

Flachowsky, G., Baldeweg, P., Tiroke, K., Konig, H., & Schneider, A. (1990). Feed value and feeding of waste-lage made from distillers solubles, pig slurry solids and ground straw treated with urea and NaOH. *Biological Wastes, 34*(4), 271–280.

Forster, C. F. (1977). Bio-oxidation. In A. G. Callely, C. F. Forster, & D. A. Stafford (Eds.), *Treatment of industrial effluents* (pp. 65–87). London: Hodder and Stoughton.

Forster, C. F. (1985). *Biotechnology and wastewater treatment*. Cambridge: Cambridge University Press.

Freestone, N. P., Phillips, P. S., & Hall, R. (1994). Having the last gas. *Chemistry in Britain*, 48–50.

Fuggle, R.W. (1983) The application of the UNOX activated sludge process to the treatment of coal processing wastewaters. In *Effluent Treatment in the Process Industries*, pp. 105–124. *J. Chem. E. Symp. Ser.* **77,** I. Chem. E., Rugby.

Gavrilescu, M. (2011). Sustainability. In M. Moo-Young (Ed.), *Comprehensive Biotechnology* (pp. 839–851). (Vol. 2). Oxford: Elsevier.

Ghosh, S., Ombregt, J. P., & Pipyn, P. (1985). Methane production from industrial wastes by two-phase anaerobic digestion. *Water Research, 19*(9), 1083–1088.

Gostick, N. A., Wheatley, A.D., Bruce, B.M. and Newton, P.E. (1990) Pure oxygen activated sludge treatment of a vegetable processing wastewater. In *Effluent Treatment and Waste Disposal*, 69-84. I. Chem. E. Symp. Ser. No. 116,1. Chem. E., Rugby.

Gray, N. F. (1989). *Biology of wastewater treatment*. Oxford: Oxford Science Publications.

Grieve, A. (1978). Sludge incineration with particular reference to the Coleshill plant. *Water Pollution Control, 77*, 314–321.

HMRC (2015). Guidance landfill tax rates. Available from: www.gov.uk/government/publi-cations/rates-and-allowances-landfill-tax/landfill-tax-rates-from-1-april-2013.

HMSO. (1972). *Analysis of raw, potable and waste waters*. London: HMSO, pp. 121–122.

HMSO. (1990). *Environmental Protection Act 1990*. London: HMSO.

HMSO. (2011). *Waste (England and Wales) Regulations 2011*. London: HMSO.

Hemming, M. L., Ousby, J. C., Plowright, D. R., & Walker, J. (1977). Deep Shaft—Latest position. *Water Pollution Control, 76*, 441–451.

Higgins, P. M. (1968). Waste treatment by aerobic techniques. *Developments in Industrial Microbiology, 9*, 146–159.

Hill, F. (1980). Effluent treatment in the Bakers' yeast industry. In *Efffluent treatment in the biochemical industries, process biochemistry's third international conference.*

Jackson, J. (1960). The treatment of distillery and antibiotics wastes. In P. C. G. Isaac (Ed.), *Waste treatment* (pp. 226–239). Oxford: Pergamon Press.

Koziorowski, B., & Kucharski, J. (1972). *Industrial waste disposal*. Oxford: Pergamon Press.

Laca, A., Herrero, M., & Diaz, M. (2011). Life cycle assessment in biotechnology. In M. Moo-Young (Ed.), *Comprehensive Biotechnology* (pp. 839–851). (Volume 2). Oxford: Elsevier.

Laing, I. G. (1992). Waste minimisation: the role of process development. *Chemical Industry, 1992*, 682–686.

Lettinga, G., van Velsen, A. F. M., Hobma, S. W., de Zeeuw, W., & Klapwiik, A. (1980). Use of the upflow sludge blanket (USB) reactor concept for biological waste treatment, espe-cially for anaerobic treatment. *Biotechnology and Bioengineering, 22*, 699–734.

Lettinga, G., Hulshoff Pol, L.W., Wiegant, W., de Zeeuw, W., Hobma, S.W., Grin, P., Roersma, R., Sayad, S. & Van Velsen, A.M. F. (1983). Upflow sludge blanket processes. In *Proceedings of the, third international, symposium on anaerobic digestion* (pp. 139–158). Aug. 14–19, Boston, M.A.

Lewicki, W. (1978). Production, application and marketing of concentrated molasses-fermen-tation-effluent (vinasses). *Process Biochemistry, 14*(6), 12–13.

Liu, M., Zhu, H., Dong, B., Zheng, Y., Yu, B., & Gao, C. (2013). Submerged nanofiltration of biologically treated molasses fermentation wastewater for the removal of melanoidins. *Chemical Engineering Journal, 223*, 388–394.

Liu, M., Zhang, Y., Yang, M., Tian, Z., Ren, L., & Zhang, S (2012). Abundance and distribu-tion of tetracycline resistance genes and mobile elements in oxytetracycline production wastewater treatment systems. *Environmental Science & Technology, 46*, 7551–7557.

Loehr, R. C. (1968). Anaerobic treatment of wastes. *Developments in Industrial Microbiol-ogy, 9*, 160–174.

Loll, U. (1977). Engineering, operation and economics of biodigestion. In H. G. Schlegel, & J. Barnes (Eds.), *Microbial energy conversion* (pp. 361–378). Oxford: Pergamon Press.

Lyons, T. P. (1983). Ethanol production in developing countries. *Process Biochemistry, 18*(2), 18–25.

Mackel, C.J. (1976). A study of the availability of distillery by-products to the agricultural industry, 46 pp. Report of the School of Agriculture, University of Aberdeen, Scotland.

McKinney, R. L. (1962). Complete mixing activated sludge treatment of antibiotic wastes. *Biotechnology and Bioengineering, 4*, 181–195.

McLeod, G., & O'Hara, J. (1993). EC proposals for integrated pollution prevention and control. *Chemical Industry, 1993*, 849–851.

Mbagwu, J. S. C., & Ekwealor, G. C. (1990). Agronomic potential of brewers spent grains. *Biological Wastes, 34*(4), 335–347.

Masters, G. M. (1991). *Introduction to environmental engineering and science* (pp. 101–179). New Jersey: Prentice Hall.

Mathias, P. (1959). *The brewing industry in England,* (pp. 1700–1830). Cambridge: University Press.

Metcalf and Eddy Inc. (2003). *Wastewater engineering, treatment and reuse.* New York: McGraw Hill.

Naja, G.M. & Volensky, B. (2011). Constructed wetlands for water treatment. In *Comprehensive biotechnology*, Vol. 6, 2nd ed. (pp. 353–368). M. Moo-Young (Ed.), Elsevier: Oxford.

Naylor, L. M., & Severson, K. Y. (1984). Brewery sludge as a fertilizer. *Biocycle, 25*(3), 48–51.

Omstead, M. N., Kaplan, L., & Buckland, B. G. (1989). Fermentation development and process improvement. In W. G. Campbell (Ed.), *Ivermectin and abamectin* (pp. 33–54). New York: Springer.

Otto, R., Barker, W., Schwarz, D., & Tjarksen, B. (1962). Laboratory testing of pharmaceutical wastes for biological control. *Biotechnology and Bioengineering, 4*, 139–145.

Parker, J. A., & Darby, J. L. (1995). Particle associated coliform in secondary effluents: shielding from ultraviolet light disinfection. *Water Environment Research, 67*, 1065–1072.

Pell, M. and Worman, A. Biological wastewater treatment systems. In *Comprehensive Biotechnology,* Vol. 6, 2nd ed. (pp. 284–286). M. Moo-Young (Ed.), Elsevier: Oxford.

Pohland, F.G. & Ghosh, S. (1971). Developments in anaerobic treatment processes. In *Biological Waste Treatment. Biotech. Bioeng. Symp. 2*, 85–106.

Pretorious, W. A. (1973). The rotating disc unit. A waste treatment system for small communities. *Water Pollution and Control, 72*, 721–724.

Quinn, J. P., & Marchant, R. (1980). The treatment of malt whisky distillery waste using the fungus *Geotrichum candidum. Water Research, 14*, 545–551.

Ramathan, M., & Gaudy, A. F. (1969). Effect of high substrate concentration and cell feedback on kinetic behaviour of heterogeneous populations in completely mixed systems. *Biotechnology and Bioengineering, 11*, 207–237.

Renaud, G (1980). Treatment of effluent from glutamic acid and lysine fermentation. In *Effluent treatment in the biochemical industries, process biochemistry's third international conference.*

Ripley, P. (1979). Process engineering aspects of the treatment and disposal of distillery effluent. *Process Biochemistry, 14*(1), 8–10.

Rivas, F. J., Beltran, F. J., Gimeno, O., & Alvarez, P. (2003). Optimisation of Fenton's reagent usage as a pre-treatment for fermentation brines. *Journal of Hazardous Materials, B96*, 277–290.

Romero, L. I., Sales, D., Cantero, D., & Galan, M. A. (1988). Thermophilic anaerobic digestion of winery waste (vinasses): Kinetics and process optimisation. *Process Biochemistry, 23*(4), 119–125.

SCA (Standing Committee of Analysts) (1989). Five day biochemical oxygen demand (BOD_5). MEWAM, HMSO, London.

Sheenan, G. J., & Greenfield, P. F. (1980). Utilisation, treatment and disposal of distillery waste water. *Water Research, 14*, 257–277.

Schneider, T., Graeff-Honninger, S., French, W. T., Hernandez, R., Merkt, N., Claupein, W., Hetrick, M., & Pham, P. (2013). Lipid and carotenoid production by oleaginous red yeast *Rhodotorula glutinis* cultivated on brewery effluents. *Energy, 61,* 34–43.

Silvero, G. M., Anglo, P. G., Montero, G. V., Pacheco, M. V., Alamis, M. L., & Luis, V. S., Jr. (1986). Anaerobic treatment of distillery slops using an upflow anaerobic filter reactor. *Process Biochemistry, 21*(6), 192–195.

Simpson, J. R., & Anderson, G. K. (1967). Large-volume respirometers with particular reference to waste-treatment. *Progress in Industrial Microbiology, 6,* 141–167.

Singh, V., Hsu, G. G., Chen, G., & Tzeng, G. H. (1983). Fermentation processes for dilute food and dairy wastes. *Process Biochemistry, 18*(2), 13–25.

Smethurst, G. (1988). *Basic water treatment for application world-wide* (2nd ed.). London: Thomas Telford.

Song, K. H., & Young, J. C. (1986). Media design factors for fixed-bed anaerobic filters. *Journal Water Pollution Control Federation, 58,* 115–121.

Stephenson, T., Allan, M., & Upton, J. (1993). The small footprint wastewater treatment process. *Chemistry and Industry, 15,* 533–536.

Szendrey, L.M. (1983). Startup and operation of the Bacardi Corporation anaerobic filter. In *Proceedings of the third international symposium on anaerobic digestion* (pp. 365–377). 14–19 Aug. Boston: MA.

Taylor, I. J., & Senior, P. J. (1978). Single cell proteins: a new source of animal feeds. *Endeavour (N.S.), 2,* 31–34.

Trchouian, K., & Trchouian, A. (2015). Hydrogen production from glycerol by *Escherichia coli* and other bacteria: An overview and perspectives. *Applied Energy, 156,* 174–184.

Truesdale, G. A. (1979). Tertiary treatment. In *Water Pollution Control Technology* (pp. 84–91). London: H.M.S.O.

Viessman, W., & Hammer, M. J. (1993). *Water supply and pollution control* (5th ed.). New York: HarperCollins, pp. 741–748.

Ware, A. J., & Pescod, M. B. (1989). Full scale studies with an anaerobic/aerobic RBC unit treating brewery wastewater. *Water Science & Technology, 21*(4–5), 197–208.

Wiegant, W. M., Claassen, J. A., & Lettinga, G. (1985). Thermophilic anaerobic digestion of high strength wastewaters. *Biotechnology and Bioengineering, 27*(9), 1374–1381.

Winkler, W. A., & Thomas, A. (1978). Biological treatment of aqueous wastes. In A. Wiseman (Ed.), *Topics in enzyme and fermentation biotechnology* (pp. 200–279). (Vol. 2). Chichester: Ellis Horwood.

WRAP (2012). Gate Fees Report 2012 Comparing the cost of alternative treatment options. WRAP, UK.

Xia, A., Cheng, J., Song, W., Su, H., Ding, L., Lin, R., Lu, H., Liu, J., Zhou, J., & Cen, K. (2015). Fermentative hydrogen production using algal biomass as feedstock. *Renewable and Sustainable Energy Reviews, 51,* 209–230.

Xing, Z. -P., & Sun, D. -Z. (2009). Treatment of antibiotic fermentation wastewater by combined polyferric sulfate coagulation, Fenton and sedimentation process. *Journal of Hazardous Materials, 168,* 1264–1268.

Xu, J., Su, X. -F., Bao, J. -W., Chen, Y. -Q., Zhang, H. -J., Tang, L., Wang, K., Zhang, J. -H., Chen, X. -S., & Mao, Z. -G. (2015). Cleaner production of citric acid by recycling its extraction wastewater treated with anaerobic digestion and electrodialysis in an integrated citric acid-methane production process. *Bioresource Technology, 189,* 186–194.

Xu, Q., Hamid, A., Wen, X., Zhang, B., & Yang, N. (2014). Fenton-Anoxic-Oxic/MBR process as a promising process for avermectin fermentation wastewater reclamation. *Separation and Purification Technology*, *134*, 82–89.

Yang, S., Zhu, X., Wang, J., Jin, X., Liu, Y., Qjan, F., Zhang, S., & Chen, J. (2015). Combustion of hazardous biological waste from the fermentation of antibiotics using TG-FTIR and Py-GC/MS techniques. *Bioresource Technology*, *193*, 156–163.

Young, J.C. (1983). The anaerobic filter—past, present and future. In *Proceedings of the third international symposium on anaerobic digestion* (pp. 91–106). 14–19 Aug. Boston: MA.

Young, J. C. (1991). Factors affecting the design and performance of upflow anaerobic filters. *Water Science and Technology*, *24*(8), 133–155.

Zacharof, M. -P., Vouzelaud, C., Mandale, S., & Lovitt, R. W. (2015). Valorization of spent anaerobic digester effluents through the production of platform chemicals using *Clostridium butyricum*. *Biomass and Bioenergy*, *81*, 294–303.

FURTHER READING

Haigh, N. (1990). *EEC Environmental Policy and Britain* (2nd ed.). Harlow: Longman.

Hughes, D. (1992). *Environmental Law* (2nd ed.). London: Butterworths.

Smith, C. (1993). Waste, incineration and the environment. *Waste Planning*, *9*, 7–17.

Tromans, S. (1991). *The Environmental Protection Act 1990, text and commentary*. London: Sweet & Maxwell.

Water Pollution Research Centre (1972). *Water pollution research: Report of the director of water pollution research, 1971*. London: HMSO.

The production of heterologous proteins

12

INTRODUCTION

Cohen, Chang, Boyer, and Helling (1973) published their seminal paper entitled *"Construction of Biologically Functional Bacterial Plasmids In Vitro."* They described the in vitro use of restriction endonucleases and DNA ligase to construct a recombinant plasmid and demonstrated it to be "biologically functional when inserted into *E. coli* by transformation." In their discussion they referred to the procedure as "potentially useful for insertion of specific sequences from prokaryotic or eukaryotic chromosomes or extrachromosomal DNA into independently replicating bacterial plasmids." The application of this technique, and the many others to which it gave rise, changed biology irrevocably and gave birth to what has been termed "new biotechnology" and the production of "biopharmaceuticals." A number of different terms are used in the literature to describe genes that have been introduced into a host from a different organism, including foreign gene, heterologous gene and, when describing a particular gene being expressed, "the gene of interest." Similarly the terms heterologous protein, foreign protein, and recombinant protein are used to describe the protein that is coded by such genes. A biopharmaceutical is a heterologous protein that is used as a therapeutic agent. Somatostatin was the first demonstration of the expression of a synthetic foreign gene in *E. coli* (Itakura et al., 1977) and in 1982 recombinant human insulin became the first heterologous protein (biopharmaceutical) to be approved for medical use. A further eight approvals of heterologous proteins were made in the 1980s comprising two for human growth hormone, two for interferons, one monoclonal antibody, one recombinant vaccine for hepatitis B, one for tissue plasminogen activator, and one for erythropoietin (Walsh, 2012). In the early 1990s the sales of heterologous products were still relatively small compared with "conventional" fermentation products but 20 years later biopharmaceuticals generated sales in excess of $100 billion—equivalent to one-third of the global pharmaceutical market (Nielsen, 2013). Over 400 biopharmaceuticals were on the market in 2015 and another 1300 were in development with approximately 50% undergoing clinical trials (Sanchez-Garcia et al., 2016). As discussed in Chapter 1, many of the new products are developments of existing products that have been enhanced by postsynthesis modifications to enhance drug performance. Table 12.1 (also included in Chapter 1 as Table 1.5) lists a selection of the recombinant proteins licensed for therapeutic use and the production systems used.

Table 12.1 The Major Groups of Recombinant Proteins Developed as Therapeutic Agents

Therapeutic Group	Recombinant Protein	Production System	Clinical Use, Treatment of
Blood factors	Factor VIII	Mammalian cells	Anemia
Thrombolytic	Tissue plasminogen activator	Mammalian cells, E. coli	Clot lysis
Anticoagulants	Hirudin	Saccharomyces cerevisiae	Anticoagulant
Hormones	Insulin	E. coli, S. cerevisiae	Diabetes
	Human growth hormone	E. coli, S. cerevisiae	Hypopituitary dwarfism
	Follicle stimulating hormone	Mammalian cells	Infertility
	Glucagon	E. coli, S. cerevisiae	Type 2 diabetes
Growth factors	Erythropoietin	Mammalian cells	Anemia
	Granulocyte-macrophage colony stimulating factor	E. coli	Bone marrow transplantation
	Granulocyte colony stimulating factor	E. coli, mammalian cells	Cancer
Cytokines	Interferon-α	E. coli	Cancers, hepatitis B leukemia
	Interferon-β	E. coli	Cancers, amyotrophic lateral sclerosis, genital warts

Modified from Dykes (1993) and Walsh (2010).

Not only has the number of products developed significantly since 1982 but so have the range of host cells and expression systems used for production, the technology of the production process, and the methods for purifying and characterizing the products. As is the case for any fermentation process, the production of a heterologous protein is far more than simply the expression of the product. The process chain has to be considered as a whole, from inoculum development to the production fermentation, downstream processing, and validation. McDonald (2011) summarized the following important considerations in the design of a heterologous protein process:

- The choice of host including bacteria, yeast, fungi, mammalian cells, insect cells, and plant cells.
- The need to accomplish various posttranslational modifications (PTMs), for example, protein folding, glycosylation, proteolytic processing, disulfide bridge formation, phosphorylation, and hydroxylation. This aspect strongly influences the choice of host.

- The design of the expression system such that it is commensurate with the host cell including the choice of promoter, ribosome-binding site, and start and stop codons.
- The incorporation of the genetic material—whether it is randomly inserted into the genome, targeted to a specific site, stably maintained extra-chromosomally, or transiently expressed.
- The fermentation process itself—the choice of stainless steel reactors or single-use vessels and their mode of operation, batch, fed-batch or continuous, and the adoption of perfusion (feedback) systems.
- The control of the production process—whether the synthesis of the protein is constitutive or inducible and the impact this has on process operation.
- The optimization of the fermentation process in terms of product concentration (mg product dm^{-3}), volumetric production rate (mg product dm^{-3} h^{-1}), and specific production rate (mg product g^{-1} biomass h^{-1}).
- The downstream processing operations, which will depend upon whether the heterologous protein is "tagged" to facilitate recovery and where it is located—intracellular (soluble or misfolded in an inclusion body), excreted (in the culture broth), or targeted to an organelle.
- The whole process is commensurate with the stringent control required by the regulatory authorities.

The choice of production host is a complex one and the features of each system are summarized in Table 12.2. As discussed in Chapter 1, *E. coli* was the predominant choice in the 1980s, and it still accounts for the production platform of 31% of products (Sanchez-Garcia et al., 2016; Berlec & Strukelj, 2013; Reader, 2013), followed by yeast-based processes (15%) with mammalian cells now being the most common system (predominantly Chinese Hamster Ovarian, CHO, cell lines) at 43%; the remaining 11% are produced by hybridoma cells, insect cells, plant cells, and transgenic plants and animals. The following discussion is limited to the major hosts used for commercial production—bacteria, yeasts, and mammalian cells.

HETEROLOGOUS PROTEIN PRODUCTION BY BACTERIA

Unsurprisingly, *E. coli* has been the organism of choice as a bacterial production platform. This is due to the extensive knowledge of its genetics, biochemistry, and physiology, which is the basis of the biochemist's classification of the bacteria into "*E. coli* and the rest"! *E. coli* also has the advantage, which it shares with many of the "rest," of growing rapidly on simple, cheap medium ingredients and can be grown to a very high biomass concentration making it amenable to relatively easy and economic "scale-up." Furthermore, the host strains used for commercial production are classified as "generally regarded as safe" (GRAS) and thus are amenable for large-scale processes. The detailed knowledge of the genetics of *E. coli* and the availability of plasmids and promoters also meant that a well-stocked molecular biology toolbox was accessible for the manipulation of the organism.

Table 12.2 The Advantages and Disadvantages of Hosts Used for the Production of Heterologous Proteins

Host	Advantages	Disadvantages
E. coli	Sequenced genome, well characterized genetics, and molecular biology	Heterologous proteins poorly secreted
	Reliable expression vectors available	Heterologous protein may accumulate as insoluble inclusion bodies
	High growth rate, high cell densities, simple medium requirements, and straightforward scale-up.	Unable to catalyze PTMs
	GRAS strains available	Presence of an additional methionine molecule at the N-terminal of the heterologous protein
Yeasts	Sequenced genome, well characterized genetics, and molecular biology	Heterologous proteins expressed at relatively low levels compared with *E. coli*
	Reliable expression vectors available	Exported proteins often retained in the periplasmic space, although *Pichia pastoris* is a good secretor of heterologous proteins
	Relatively high growth rate, high cell densities, and simple medium requirements	PTMs may differ from those in animal cells
	GRAS strains available and well-established as a commercial fermentation organisms	
	Able to catalyze some PTMs	
Cultured animal cells	Full PTM machinery present enabling the production of human proteins designed for clinical use	Low growth rates and more difficult to grow, complex medium requirements
		Low cell densities achieved compared with yeast and bacterial systems
		Complex and time lengthy cell line selection systems
		Scale-up more complex than microbial systems
		Lower levels of heterologous proteins produced compared with microbial systems

PTM, posttranslational modification.

However, there are some serious issues that limit the use of *E. coli*—it is unable to perform any PTMs, thus excluding proteins requiring glycosylation and it is unsuitable for the synthesis of hydrophobic proteins or those containing more than 500 amino acids (Harbron, 2015). There are also a number of differences between the structure and mechanisms of expression of prokaryotic and eukaryotic genes that have to be accommodated in the production strategy:

- Eukaryotic genes contain introns, noncoding regions of DNA, that are resolved in the eukaryote by the removal of the corresponding sections of the mRNA by RNA splicing. This RNA processing system does not exist in prokaryotes but the problem is overcome by transforming the prospective gene into a "prokaryotic format" by producing cDNA (double stranded DNA copies) of the processed eukaryotic mRNA using the enzyme reverse transcriptase.
- The eukaryotic gene will not be transcribed unless there is an appropriate promoter sequence upstream of its coding region to which RNA polymerase may bind. Thus, any expression system must incorporate a suitable promoter.
- Prokaryotic ribosomes will not bind to mRNA unless it contains a ribosome-binding site (a specific purine rich sequence) located upstream of the coding region, thus the corresponding DNA sequence must be present upstream of the foreign gene.
- In order to be effectively translated the gene must have a start and stop sequence.

At this stage it is important to distinguish between cloning vectors and expression vectors. A cloning vector is designed to amplify the number of copies of a DNA sequence, that is, the sequence is replicated along with the plasmid that contains it. An expression vector is designed to express the DNA sequence, as well as to amplify it, that is the sequence is transcribed into mRNA and translated into protein. Thus, an expression vector will contain the features described previously that enable expression—the inclusion of a promoter, a ribosome-binding site, and translation start and stop codons. The sequence of events for the isolation of a gene and expressing it in a heterologous host are given in Fig. 12.1, which is based on Walsh's (2014b) summary of the process. The identification and production of the desired DNA sequence is achieved using polymerase chain reaction (PCR) and bioinformatics technology, which together facilitate the amplification of the sequence. The PCR product can then be inserted into a cloning vector, which is introduced into a host by transformation (Fig. 3.27, Chapter 3). Growth of the host amplifies the vector that can then be isolated and the sequence excised and validated. The sequence is then inserted into an expression vector that, in turn, is introduced into the host. It is perfectly possible to introduce the original PCR product directly into the expression vector and circumvent the cloning amplification stage. However, expression vectors tend to be large and are maintained at low copy numbers, which makes them inconvenient to use as a source of the isolated sequence. Thus, the advantage of the inclusion of the cloning vector step is that the insert can be conveniently stored in the cloning vector, which can be readily amplified in its host (Walsh, 2014b). The reader is referred to

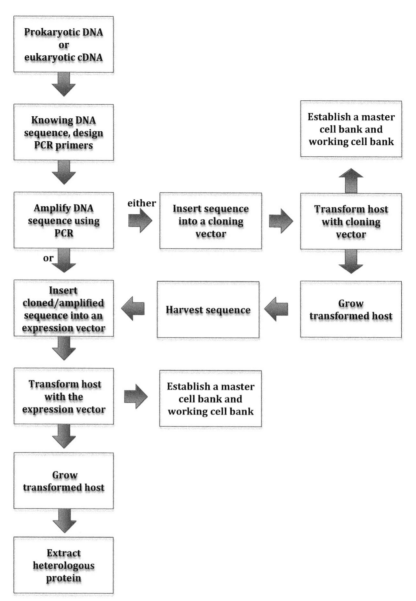

FIGURE 12.1 Outline Protocol for the Production of a Heterologous Protein

Modified from Walsh (2014)

an exceptionally helpful discussion of the overall process by Graslund et al. (2008) who discuss the approaches that may be used in the early experimental stages of expressing a protein.

There are a number of strategies that can be adopted to produce the heterologous protein in the host cell depending on whether the protein is soluble, insoluble, and accumulated as an intracellular aggregate, or secreted. In order to be biologically active, the amino acid chain of a protein generated by translation has to be folded into a three-dimensional structure. It is this stage in the production of a heterologous protein that presents the greatest challenge. Very high concentrations of foreign proteins can be achieved in *E. coli* but in most cases the protein accumulates in the bacterium as insoluble aggregates termed inclusion bodies. Walsh (2014b) cited the following reasons for the formation of inclusion bodies:

- very high local concentrations of the protein result in nonspecific precipitation;
- insufficient chaperones or folding enzymes resulting in the aggregation of partially folded intermediates;
- the inability of the organism to form disulfide bonds due to the reducing environment of the cytoplasm;
- the lack of eukaryotic posttranslational modifying enzymes.

A number of approaches, discussed later in this chapter, have been used to circumvent the development of inclusion bodies and comprise both molecular biological and physiological techniques. However, the development of inclusion bodies may not always be "bad news" as it is possible, in some situations, to isolate the aggregates in a simple downstream processing operation and recover the fully folded, functional protein. Rather than accumulating within the cell, the process may also be designed such that the protein is secreted either into the periplasm (the space between the cytoplasmic membrane of *E. coli* and its lipopolysaccharide (LPS) outer membrane) or into the culture medium. These approaches are summarized in Fig. 12.2.

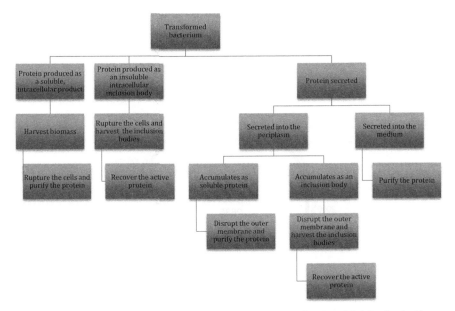

FIGURE 12.2 A Flow Chart Illustrating the Range of Processes Employed for the Production of a Heterologous Protein by *E. coli*

CLONING VECTORS

Figs. 12.3 and 12.4 give examples of the plasmid-based cloning vectors, pBR322 and pUC19, respectively. Both plasmids share the following characteristics:

- They are smaller than natural plasmids, enabling efficient uptake by bacteria, faster replication with low energy consumption, and are robust (less fragile) and thus easier to purify.

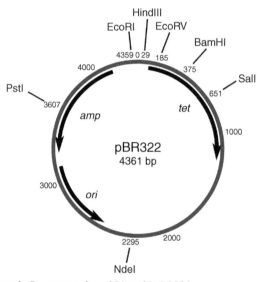

FIGURE 12.3 Schematic Representation of Plasmid pBR322

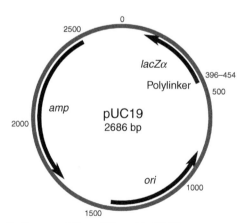

FIGURE 12.4 Schematic Representation of Plasmid pUC19

- Their origins of replication enable the plasmids to replicate independently of the bacterial cell cycle, which means that a large number of copies are produced per cell. This type of replication origin is termed "relaxed" as compared with "stringent" that links plasmid replication with chromosomal replication, resulting in a low copy number per cell.
- They contain single recognition sites for a number of restriction enzymes. It is important that there is only one recognition site for each endonuclease as more than one would result in the plasmid being "cut to pieces" rather than "opened."
- They bear two selectable markers such that colonies containing the manipulated plasmid (and hence the foreign gene) can be distinguished from both those that have not been transformed and those that have been transformed by a plasmid not containing the desired gene.

pUC18/19 were developed from pBR322. The two selectable markers for pBR322 are resistant to the antibiotics ampicillin and tetracycline and each contains several endonuclease recognition sites. Thus, depending on the sequence of the foreign gene, it may be incorporated within either resistance gene. As a result, the interrupted gene will not be expressed and the bacterium will be susceptible to that antibiotic. However, the other selectable marker will still give resistance. Thus, the transformed strain containing the recombinant vector will be resistant to one antibiotic but susceptible to the other and can be isolated by replica plating. The selectable markers of pUC18/19 are ampicillin resistance and β-galactosidase production (the *lacZ* gene). All 13 of the restriction enzyme recognition sites of pUC18/19 are concentrated at a location termed the multiple cloning site (MCS), or polylinker site, that is within *lacZ*. Thus, introduction of a foreign gene at any of the cloning sites will interrupt *lacZ* and prevent its expression. The production of β-galactosidase can be switched on by adding the inducer IPTG (isopropyl-β-D-galactopyranoside) and the activity of the enzyme can then be detected by its hydrolysis of the colorless compound X-gal (5-bromo-4-chloro-3-indolyl-β-galactopyranoside) to a blue derivative. Thus, the desired colonies generated from a transformation with a recombinant vector will be resistant to ampicillin and will remain white in the presence of both IPTG and X-gal, whereas colonies containing the un-engineered vector will be blue, as *lacZ* will be expressed.

EXPRESSION VECTORS

As previously indicated, expression vectors incorporate a number of features in addition to those included in cloning vectors. A schematic representation of a typical *E. coli* expression vector is shown in Fig. 12.5. A relaxed origin of replication is generally employed, giving a relatively high copy number. The most commonly used origins are wild-type ColE1 (enabling 15–20 copies of the plasmid per cell), pMB1, a modified ColE1 (15–60 copies), and a mutant version of pMB1 used in pUC vectors (500–700 copies). Rosano and Ceccarelli (2014) cautioned that a high copy number does not necessarily equate to a high protein yield, as the resulting high metabolic burden can be problematic.

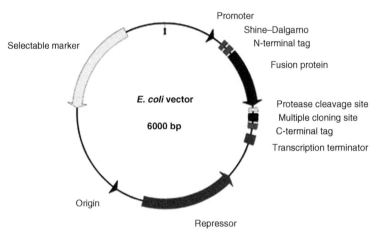

FIGURE 12.5 A Schematic Representation of a Generalized *E. coli* Expression Vector

(Modified from European Molecular Biology Laboratory, 2016)

Selectable markers

A selectable marker must be included as it enables the isolation of plasmid-containing transformants, as discussed for cloning vectors. However, the generation of plasmid-free cells from the isolated transformant can also be a major problem and thus a selecting environment is also important when the transformant is cultured (Rosano & Ceccarelli, 2014). The most common solution is to use antibiotic resistance as the selective marker and to grow the organism in the presence of the antibiotic. Ampicillin resistance is often used but resistance (endowed by the gene *bla*) is due to the production of the enzyme β-lactamase that degrades the antibiotic. However, the enzyme is secreted from the cells and degrades ampicillin in the culture medium, such that the selective force declines after several hours of incubation in a liquid medium, which would be a severe disadvantage in the development of inocula or during a production process. A better solution is the use of tetracycline resistance, based on the efflux of the antibiotic from the cells rather than its destruction, and thus the concentration in the medium is unaffected (Korpimaki, Kurittu, & Karp, 2003). However, the use of antibiotics in large-scale culture is undesirable both in terms of cost and the risk of spreading antibiotic resistance. Thus, "plasmid addiction" systems have been developed in which the bacterium is dependent on the plasmid for an essential gene. An example is given in Chapter 3, of such a system being used in the production of tryptophan where the gene *serA* was deleted from the chromosome of *Corynebacterium glutamicum* and inserted into a recombinant plasmid, making the bacterium dependent on the plasmid for the synthesis of serine.

Promoters

A promoter is essential for the expression of the foreign gene and a wide range of prokaryotic promoters is available, making its choice an important part of the production strategy. The promoter must be sufficiently strong to produce a very high concentration of the heterologous product that can account for up to 50% of the total cellular protein. It should also be inducible so that the synthesis of the protein can be switched on by the addition of an inducer. Production of 50% of its protein as an "unwanted" product is an enormous metabolic burden for a cell to bear and a mutant that has lost the ability to produce the protein will be at a significant growth advantage. Also, some heterologous proteins can be growth inhibitory or even toxic to the host bacterium. Thus, being able to induce product synthesis limits the risk of selection of nonproducers or growth inhibition to the production phase—all other operations, strain storage, inoculum development, and the preproduct synthesis growth phase of the production fermentation, being done in the absence of the inducer. Examples of inducible systems are shown in Table 12.3. The repressor gene shown in Fig. 12.5 codes for the repressor protein that ultimately controls the interaction between the promoter and RNA polymerase, thus controlling gene expression. The mode of action of the repressor depends on the system. In the lac operon, RNA polymerase is prevented from binding to the promoter by the *lac* repressor protein produced by the *lacI* gene. The natural inducer of the *lac* operon, allolactose (derived from lactose), binds to the repressor and neutralizes its activity thus enabling expression of the operon. Thus, if the *lac* promoter is employed, the repressor gene will be *lacI*. A modified repressor gene, *lacI^Q*, gives higher levels of the repressor and prevents basal expression in the absence of the inducer (Calos, 1978). Lactose can be used in the fermentation or its nonmetabolized analog, isopropyl β-D-1-thiogalactopyranoside (IPTG). The advantage of IPTG over lactose is that its concentration will not change whereas lactose will be utilized by the strain and its concentration decreases. However, the use of fed-batch culture can alleviate this situation as lactose can be fed during

Table 12.3 Promoters and Their Inducers Used in the Production of Heterologous Proteins by *E. coli*

Promoter	Inducer	Properties
ara	Arabinose	Rapid induction, strong regulation, repressed by glucose
Cad	Acidic pH	High expression levels, simple induction process but the pH may adversely affect the producer
Lac	Isopropyl-β-D-thiogalactoside (IPTG)	Relatively low-level expression, leaky expression under noninducing conditions, repressed by glucose
Tac	IPTG	Well characterized, high expression but leaky under noninducing conditions
Trp	Indoleacrylic acid	Well characterized, high expression but leaky under noninducing conditions

Modified from Walsh (2014).

the production phase at a slow rate, thus inducing protein production at the end of the growth phase and maintaining a slow growth rate during the production phase.

The *araBAD* operon coordinates the control of three structural genes (*araB*, *araA*, and *araD*) coding for enzymes that enable *E. coli* to utilize arabinose as a carbon source. The regulator gene, *araC*, codes for a protein that can act as both repressor and activator. In the absence of the inducer, arabinose, a dimer of the AraC protein binds to two DNA sites, creating a loop that prevents RNA polymerase binding to the promoter. In the presence of arabinose the binding of one member of the dimer changes, such that the loop is released, and it binds at another site that stimulates expression. This system results in very low background expression. Thus, the repressor gene associated with the *araBAD* promoter would be *araC*. Both the *lac* and *araBAD* operons are catabolite repressed by glucose so an excess of glucose in the growth phase of the fermentation will further secure the repression of the system.

The *trp* operon codes for the tryptophan synthesizing genes that are subject to repression by the end product, tryptophan. The *trp* repressor, coded by the *trpR* gene, is inactive in the absence of tryptophan. However, above a threshold concentration, tryptophan activates the repressor enabling it to bind to the operator and prevent transcription by RNA polymerase. Thus, the *trp* operon is expressed (de-repressed) when tryptophan is exhausted. The *trp* promoter is very strong but is difficult to apply in a fermentation system as its induction relies on depletion of tryptophan. However, the strength of the *trp* system and the convenience of the *lac* system have been combined in the hybrid promoter, *tac*. This promoter combines the -35 region of the *trp* promoter with the -10 region of the *lac* promoter resulting in a strong promoter inducible by lactose or IPTG.

The T7 promoter is recognized by the RNA polymerase of the phage, T7, and has the advantage of supporting very strong expression—the heterologous product accounting for up to 50% of the total cell protein (Graumann & Premstaller, 2006). The T7 promoter is incorporated into the expression vector such that it controls the transcription of the heterologous gene; the T7 promoter is only recognized by T7 RNA polymerase and, thus, the gene coding for T7 RNA polymerase is introduced into the bacterial chromosome under the control of an inducible promoter, normally *lac*. As a result, expression is switched on by the addition of lactose or IPTG, which enables the expression of the RNA polymerase that, in turn, recognizes the T7 promoter and enables expression of the heterologous protein. Expression vectors using this technology are termed pET plasmids.

All the aforementioned promoter systems utilize a chemical induction system, the most common approach being the addition of IPTG, the nonmetabolized analog of lactose. However, this compound must be removed from the product, adding cost to the downstream processing of the product. Both pH and temperature-induced systems have been developed to address this problem but the former approach is available in very few vectors and pH induction will adversely affect the physiological conditions (Valdez-Cruz, Caspeta, Perez, Ramirez, & Trujillo-Roldan, 2010). Temperature-induced systems are based on the λ bacteriophage promoters, λpL and λpR, and their repressor λcI. The promoters pL and pR control the expression of leftward and rightward transcribed genes, respectively. Infection of *E. coli* by λ phage can result in a lysogenic state in which the phage is inserted into the host chromosome and is replicated along with it. The expression of the λ genes is repressed by the binding of the cl repressor to the promoter operators. The shift to the lytic phase is initiated by the autodigestion of cl in response to physiological changes in the host, thus enabling the binding of RNA polymerase to the promoters resulting, ultimately, in the expression of the phage genome. Lieb (1966) isolated a range of λ temperature sensitive (T_s) mutants, including mutant cl[857] in which threonine was substituted for alanine in position 66 of the amino-terminal region of the cl repressor. This temperature sensitive repressor now forms the basis of many thermally induced expression plasmids that incorporate the λpL, λpR, or both promoters and the cl[857] repressor is either integrated into the plasmid or the host chromosome. The fermentation is operated at a 28–30°C to produce the biomass and expression is induced by raising the temperature to 40–42°C. A large-scale process has to be cooled, due to the exothermic nature of growth producing more heat than is lost naturally from the fermenter; thus, increasing the temperature can be achieved by simply reducing cooling. Caspeta, Flores, Perez, Bolivar, and Ramirez (2009) investigated the effect of heating rate on expression and demonstrated that a slow temperature transition favored protein yield, suggesting that reducing the rate of cooling, rather than attempting a rapid temperature rise by the input of heat, would be the method of choice. However, *E. coli* responds to a temperature rise by inducing a variety of complex stress responses that can adversely affect the production of the desired protein. Valdez-Cruz, Ramirez, and Trujillo-Roldan (2011) addressed this aspect and emphasized that thermal induction is not alone in inducing stress responses in *E. coli* —chemical induction also results in similar changes.

The λ system can also be manipulated such that it is controlled by chemical induction. Mieschendahl, Petri, and Hanggi (1986) developed a plasmid/host construction in which the λPL promoter controlled the heterologous gene on an expression plasmid and the λcl repressor was integrated into the bacterial chromosome under the control of the *trp* promoter. It will be recalled that the *trp* operon is subject to feedback repression by tryptophan and is thus switched off (repressed) in the presence of the amino acid. Thus, the *trp* promoter will always be "on" in cells grown in the absence of tryptophan resulting in the synthesis of the λ repressor and repression of heterologous protein expression. The addition of tryptophan to the culture will switch off the *trp* promoter, repress the synthesis of the λ repressor, and thus lift the repression of the plasmid-encoded λPL-controlled heterologous gene.

Ribosome-binding site, start and stop codons

The further inclusions downstream of the promoter to enable expression are the ribosome-binding site (the Shine–Dalgarno sequence), the start codon that initiates translation, the MCS, and the stop codon that terminates translation (Fig. 12.6). The most commonly used start codon in *E. coli* is ATG, with GTG used in a small number of vectors and TTG and TAA hardly ever used. The optimal spacing between the ribosome-binding site and the start codon is 7 ± 2 nucleotides. As is the case for a cloning vector, the multiple cloning site contains a series of unique restriction sites

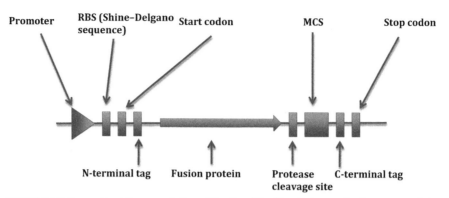

FIGURE 12.6 Generalized Representation of the Possible Sites Between the Promoter and the Stop Codon of an *E. coli* Expression Vector

that facilitate the introduction of the heterologous gene. All organisms use the stop codons TAA, TAG, and TGA but TAA is the preferred sequence for *E. coli*. The termination efficiency may be increased by including two or three stop codons in series.

Tags, fusion proteins, and associated sites

It can be seen from Figs. 12.5 and 12.6 that the generalized structure of an expression vector also includes a number of sites between the start codon and the MCS site and between the MCS and the stop codon. These sites enable a peptide (termed a peptide tag) or a polypeptide (termed a fusion protein) to be added to either the N-terminal or C-terminal end of the heterologous protein. The vector may also cater for the removal of the fusion protein in downstream processing. A protease cleavage site is included in the vector such that an amino acid sequence, recognized by a protease, is incorporated between the fusion protein and the heterologous product that can then be cleaved easily from the fusion protein. These additions can assist in a number of issues that are frequently problematic in heterologous protein production, in particular, detection, purification, expression, solubility, and degradation, and secretion. However, it should be remembered that the expression of proteins in heterologous hosts is used to further a range of outcomes, not just large-scale manufacture. Heterologous production is exceptionally valuable in fundamental science in enabling structural and functional activity studies to be performed on proteins that may be inaccessible, or produced in very small quantities, in their natural biological location. Indeed, in the postgenomic era knowledge of a gene can come before that of its coded protein so that heterologous expression can reveal a novel protein. Thus, tags and protein fusions have been used in a variety of situations and it is important to appreciate that some approaches that are invaluable in the laboratory are too expensive or present downstream processing problems on a large scale. Also, techniques may be applicable in the early stages of the development of a biotechnology product that would not find their way into the industrial large-scale process. Bell, Engleka, Malik, and Strickler (2013) have very helpfully reviewed the use of tags and fusion proteins in this context and broader reviews have been provided by Costa, Almeida, Castro, and Domingues (2014), Rosano and Ceccarelli (2014), and Young, Britton, and Robinson (2012).

1. *Improved detection.* Detection of the heterologous protein is particularly important in the early laboratory stages of expressing a protein as it may be produced in very small amounts. Fusing the protein to a peptide tag that is recognized by an antibody or binding protein then enables detection using Western blotting (a facility that is invaluable when the level is undetectable by SDS-PAGE), immunoprecipitation, and immunofluorescence. Common detection tags include *c*-myc, hemaglutinin antigen, and FLAG (Young et al., 2012). *c*-myc was one of the first detection tags developed and is a 10 amino acid segment of the c-myc protein, a human regulatory gene coding for a transcription factor involved in control of the cell cycle and apoptosis. HA is a nine amino acid sequence derived from the human influenza hemaglutinin protein and it can be positioned at either the N- or C-terminal end of the protein. However, it may influence folding and the function of the protein. FLAG is an eight amino acid hydrophilic peptide (Asp-Tyr-Lys-Asp-Asp-Asp-Asp-Lys) and includes the binding sites for several highly specific monoclonal antibodies. An added advantage is an in-built cleavage site for enterokinase to facilitate tag removal. Because of its hydrophilic nature the tag is likely to be located on the surface of a fusion protein making it accessible to both the detection antibodies and enterokinase.

   A different approach to detection is the use of green fluorescent protein (GFP) as a fusion protein that can be detected both in vivo using fluorescence microscopy and in vitro. Newstead, Kim, von Heijne, Iwata, and Drew (2007) and Young et al. used GFP fusion in high-throughput screens to identify superior-producing clones.

2. *Improved purification.* Extraction of a heterologous protein from the myriad of proteins produced by the host organism, such that its purity is commensurate with its use as a therapeutic agent, is a significant challenge. Thus, the purification rationale has been integrating into the construction of the expression vector. This has been achieved by the use of tags or fusion proteins that will bind to affinity resins in affinity chromatography. Thus, the purification processes are highly specific and, effectively, generic such that a particular system may be used for a wide variety of proteins. The most commonly used purification tag is poly-His, consisting of at least six histidine residues that can be attached to either the N- or C-terminus of the heterologous protein. The purification process is based on immobilized metal-affinity chromatography in which the negatively charged histidine moieties bind with positively charged transition metals (Ni^{2+}, Co^{2+}, Cu^{2+}, and Zn^{2+}) immobilized on a matrix. The cellular extract is added to the column and the bound his-tagged protein eluted using changes in pH or a solution of the metal ion used in the column. The FLAG system can be used for purification as well as detection and anti-FLAG

affinity gels (Sigma-Aldrich) have been developed based on a monoclonal antibody.

The most well-known fusion tag used in the large-scale manufacture of heterologous proteins is the Fc-tag—a 250 amino acid domain of the heavy chain of immunoglobulin furthest from the antigen-binding site (Czajkowsky, Hu, Shao, & Pleass, 2012; Bell et al., 2013). The peptide was first developed in 1989 as a fusion peptide combined with a heterologous protein as an AIDS therapeutic (Capon et al., 1989). The tag enables purification of the fused protein in an affinity chromatography system as it has high affinity to the *Staphylococcus aureus* surface protein, Protein A, and can be eluted simply and economically using a pH gradient. However, this is a secondary advantage of the tag—its primary feature is pharmacological and therapeutic as it increases the serum half-life of the protein in the patient, and thus effectively increases its activity, as well as binding with Fc receptors on immune cells—a key property for vaccines and cancer therapies. In addition, the Fc domain can improve the solubility of the protein product in *E. coli* as it folds independently of the heterologous protein. The final asset in the Fc repertoire is that, as it plays a key role in the therapeutic activity of the product, it is not removed from the protein in downstream processing.

The fusion proteins, glutathione S-transferase (GST) and maltose binding protein (MBP) that are used to enhance expression (discussed in the next section) can also be used in purification. GST binds to a glutathione residue and MBP binds to amylose resin.

3. *Improved expression and solubility.* The early attempts to produce heterologous proteins in bacteria met with the unexpected problem of proteins accumulating in the host as large, insoluble deposits termed inclusion bodies. As discussed earlier, a number of reasons may contribute to this lack of solubility such as very high local protein concentration resulting in precipitation, insufficient chaperones or folding enzymes to cope with the increased protein concentration, the lack of disulfide bridge formation in *E. coli* due to the reducing environment of the cytoplasm, and the absence of PTMs of the foreign protein (Walsh, 2014b). Regardless of the cause, the production of inclusion bodies has generally been considered as an unwelcome event although, as discussed in the next section, this interpretation is not always the case. The most common means of increasing the solubility of the expressed protein, and avoiding inclusion body formation, is its fusion to a soluble protein. The most widely used fusion proteins are also highly expressed and thus both problems of solubility and expression can be addressed by the fusion protein technique. The most commonly used fusion proteins include GST, MBP, N-utilization substance A (Nus A), thioredoxin A (TrxA) and small ubiquitin related modifier (SUMO).

GST is a 26 kDa protein from *Schistosoma japonicum* and compares least favorably with the other fusion proteins as a solubilizer in *E. coli*. It has the advantage of also acting as a purification tag as it will bind to immobilized glutathione that can be used in an affinity chromatography system. Commercial expression vectors also include specific protease cleavage sites to facilitate removal of the fusion protein post purification. MBP is a 42 kDa *E. coli* K12 protein that has been widely used to enhance solubility and expression of eukaryotic proteins expressed in bacteria (Harbron, 2015) and, along with NusA, is claimed to be one of the most effective solubility enhancers (Sun, Tropea, & Waugh, 2010). It shares with GST the advantage of also being a purification tag that will bind to amylose resin. However, amylose affinity chromatography has presented difficulties due to its relatively low binding capacity for some MBP fusion proteins, often necessitating additional affinity steps to achieve purity. Thus, Pryor and Leiting (1997) developed an expression vector that encoded a His_6-tag at the 3′end of the MBP-encoding sequence resulting in high expression of soluble MBP fusion proteins that could be effectively purified using metal affinity chromatography.

NusA is a 55 kDa hydrophilic protein that has a natural role in *E. coli* in both the promotion and prevention of the pausing of RNA polymerase. Translation is slowed down when transcription is paused and this is thought to allow more time for protein folding resulting in a soluble protein (De Marco, Stier, Blandin, & De Marco, 2004). This protein does not have a natural affinity property and thus an affinity tag sequence, such as His_6, has to be included in the expression vector. TrxA is an *E. coli* thioredoxin, the natural role of which is its oxidoreductase activity resulting in the reduction of disulfide bonds through disulfide exchange. It can be effective as a solubilizing agent by attachment at either the N- or C-terminal ends but does not facilitate purification. Thus, like NusA, an affinity tag must be used. SUMO is a small (11 kDa) *S. cerevisiae* protein that is an effective N-terminal solubility enhancer. Its mode of action may be due to a chaperoning effect on the heterologous protein. It does not possess a natural means of purification and thus, like NuSA and TrxA, a facility for the attachment of a purification tag, such as His_6, must be incorporated into the expression vector. A major advantage of SUMO for prokaryotic application is that *S. cerevisiae* produces a specific protease that will cleave the SUMO tag at a Gly–Gly motif. Bell et al. (2013) reported that the SUMO-His_6 system is economically feasible for large-scale protein production due to the simple purification and tag removal processes.

A recent addition to the portfolio of fusion proteins is Fh8, a small antigen (8 kDa) belonging to the calmodulin-like calcium binding proteins, excreted by the liver fluke, *Fasciola hepatica* (Costa et al., 2014). The antigen was produced in *E. coli* as a heterologous protein and proved to be highly soluble and thermo-stable. Costa et al. demonstrated its potential as a solubilizing fusion protein and it has the advantage of its calcium-binding properties providing an economic means of purifying the fused product. The fusion system also increased production as well as solubility and its small size places a lower stress on the producing *E. coli* than do the larger fusion proteins.

4. *Decreased solubility and degradation.* As mentioned in the previous section, the production of inclusion bodies can be advantageous—for example, if the protein

is toxic to the producing cell or is particularly susceptible to degradation. Also, the downstream processing of a process based on inclusion bodies can be a relatively simple process. Once harvested, the inclusion bodies have to be solubilized and, thus, the approach is only suitable for small, easily refolded, proteins and peptides as the detergents and chaotropic agents used may affect the biological activity of more recalcitrant molecules (Young et al., 2012; Ramon, Senorale-Pose, & Marin, 2014). Thus, in certain circumstances it may be advantageous to use a fusion protein that will insolubilize the heterologous product and direct the formation of inclusion bodies. Two such fusion proteins are available— ketosteroid isomerase, a very insoluble 13 kDa protein, and the 27 kDa product of the shortened *TrpE* gene of *E. coli* (*Trp-ΔLE*).

5. *Secretion*. The secretion of a heterologous protein during a production process can have significant potential advantages over its accumulation in the cytoplasm (Yoon, Kim, & Kim, 2010; Overton, 2014). A secreted protein is no longer vulnerable to degradation by intracellular proteases and downstream processing is simplified by the lack of contaminating cellular components. *E. coli*, being a Gram-negative bacterium, has a cell envelope consisting of the cytoplasmic membrane, the cell wall, and an outer, LPS, membrane. The zone between the two membranes, including the cell wall, is termed the periplasmic space, or the periplasm and has an oxidizing, rather than a reducing environment, and is rich in the enzyme disulfide bond oxidoreductase, that catalyzes disulfide bridge formation. Thus, secretion of the heterologous protein into the periplasmic space enables disulfide bridge construction resulting in the correct folding of the protein giving an active, soluble product.

Secretion of proteins to the outside world requires passage through the two membranes. Five different secretory systems have been identified in *E. coli* but the predominant ones are types I and II. Type I secretion is a one-step process in which the transporter spans both membranes and recognizes an exportable protein by the presence of a C-terminal signal peptide. The type II system is a two-step process in which the protein is transported firstly through the cytoplasmic membrane into the periplasm and then from the periplasm through the outer membrane to the culture medium (reviewed by Korotkov, Sandkvist, & Hol, 2012). The first stage of type II passage is usually catalyzed by the Sec-B dependent pathway in which the cytosolic protein chaperone, Sec-B recognizes an exportable preprotein by its N-terminal signal peptide, thus enabling its passage through the cytoplasmic membrane by the Sec translocation system. Sec facilitates the passage of unfolded protein, one amino acid at a time, so the protein will not be transported if it folds before combining with Sec. However, an alternative secretion system, the twin-arginine translocation system (Tat), will transport a folded preprotein (Natale, Bruser, & Driessen, 2008). A further protein complex facilitates the extracellular export of proteins from the periplasm via the outer membrane.

Thus, by exploiting the secretion systems of *E. coli*, the industrial process may be designed to accumulate the desired protein in either the culture medium or the periplasm. However, secretion into the periplasm (the type II pathway) appears to be the more popular approach for industrial secretory production (Yoon et al., 2010) as the outer membrane may be selectively disrupted to release the folded protein without liberating the contents of the bacterium. To facilitate secretion into the periplasm, the heterologous protein must be recognized as a secretory protein and this is normally achieved by appending a signal sequence to the N terminus of the heterologous protein (Natale et al., 2008). Malik (2016) reported that the signal sequence, alone, may not be sufficient for effective transport and more success may be achieved by appending a full-length fusion protein that is stable, soluble, and normally transported to the periplasm. Successful periplasmic expression has been achieved using MBP, a fusion protein also used for cytoplasmic expression. As discussed earlier, MBP also has the advantage of acting as a purification tag due to its affinity for maltose and, thus, can play a dual role in the process.

Some heterologous proteins will partially fold in the cytoplasm and, thus, cannot be transported into the periplasm by the sec system. However, as discussed earlier, the tat system naturally transports such proteins and Matos et al. (2012) exploited this mechanism by expressing the model protein GFP in the periplasm of *E. coli* by tagging it with the signal peptide from the Tat-secreted protein, TorA. Table 12.4 gives some examples of fusion proteins that have been used to facilitate secretion.

It is important to appreciate that signal sequences recognized by a transport system show significant variability and no one fusion tag will be a panacea for the secretion of any heterologous protein. Thus, despite the significant knowledge of these transport systems, the choice of fusion protein is based on performance trials and it is difficult to rationalize why a particular fusion works for one protein and not another. In fact, the same can be said for the choice of promoter and other characteristics of expression vectors. Thus, high-throughput screening, of the types discussed in Chapter 3 for isolating secondary metabolite overproducers, can play a vital role in optimizing these combinations. Han, Kim, and Kim (2014) published an analysis of the periplasmic proteome of *E. coli* (strains K12 and B), newly identifying 30 K12 proteins and 53 B

Table 12.4 Examples of Fusion Proteins Used to Facilitate the Secretion of Heterologous Proteins Into the Periplasmic Space of *E. coli*

Fusion Protein	Translocation System	Heterologous Protein	Reference
Ecotin	Sec	Pepsinogen	Malik, Rudolph, and Sohling (2006)
Ecotin	Sec	Proinsulin	Malik et al. (2007)
Maltose binding protein	Sec	Nanobodies (single domain antibody)	Salema and Fernandez (2013)
TorA signal peptide	Tat	GFP	Matos et al. (2012)

GFP, green fluorescent protein.

proteins as periplasmic. Such a database provides a valuable resource for the analysis of the properties of secreted proteins and the possible identification of candidate fusion proteins.

THE MANIPULATION OF THE HOST BACTERIUM

The discussion so far has concentrated on the design of the expression vector to facilitate the production of a biologically active heterologous protein by the bacterial host. However, the host itself can be manipulated both physiologically and genetically to optimize productivity. This aspect has been reviewed by Liu et al. (2015). The most obvious starting point in this discussion is the choice of host strain. The most common *E. coli* strains used in both laboratory and biotechnology projects are based on *E. coli* K12 (isolated in 1922) and *E. coli* B (isolated in 1918). The reader is referred to Daegelen, Studier, Lenski, Cure, and Kim's (2009) fascinating account of the genealogy of the B strains and their use by the scientific community. Descendants of the B strain appear to be preferred over those of K12 for the production of heterologous proteins due to a number of key characteristics (Yoon, Jeong, Kwon, & Kim, 2009). B strains do not produce flagella resulting in a significant saving of resources and their higher growth rate compared with K12 strains has been attributed to this difference (Posfai et al., 2006). Also, the B strains are superior secreters of periplasmic proteins. They also tend to produce lower levels of acetate, give higher heterologous protein expression, and less protein degradation during downstream processing. Yoon et al. (2012) completed a "multiomics" comparison of the B and K12 strains and related their phenotypes to their genomes, transcriptomes, and proteomes. This confirmed the suitability of the B strains to heterologous protein production due to a greater capacity for amino acid biosynthesis, fewer proteases, lack of flagella, lower acetate accumulation, an additional type II secretion system and a different cell wall and outer membrane structure commensurate with superior secretion. However, K-12 showed a higher expression of heat-shock genes and was more resistant to stress conditions. The transcriptome and proteome data enabled an understanding of the mechanisms underlying the performance of both strains and thus forms the basis for the development of an optimized host.

A number of important features have been added to B strains. Studier and Moffatt (1986) created strain BL21(DE3) by inserting the gene coding for phage T7 RNA polymerase into the chromosome of strain BL21 and placing it under the control of the *lacUV5* promoter. As discussed previously, this is the basis of the pET expression plasmids in which the control of the transcription of the heterologous gene is under the control of the T7 promoter. Induction of the production of T7 RNA polymerase with lactose or IPTG enables the transcription of the heterologous gene. The second feature addressed the problem of amino acids being encoded by more than one codon and each organism having a different bias in their use (Nakamura, Gojobori, & Ikemura, 2000). For example, codons rarely used in *E. coli* include AGG, AGA, and CGA for arginine; AUA for isoleucine; CUA for leucine; GGA for glycine; and CCC for proline. However, the codons of heterologous genes will reflect the codon usage bias of the "donor" resulting in a mismatch between these codons and the tRNA pool in the host. Such a shortage can result in ribosome stalling, leading to shortened proteins and premature mRNA turnover. This problem has been addressed by the development of the Novagen strain BL21(DE3) Rosetta that contains the plasmid, pRARE, facilitating the synthesis of key mammalian-biased codons that are rare in *E. coli*. An alternative approach eradicates the problem at the heterologous gene level by making use of the chemical synthesis of DNA such that the heterologous gene is synthesized containing *E. coli* codons (Overton, 2014).

A problem common to all *E. coli* strains is the presence of LPS in the outer membrane. LPS (or endotoxin) can induce a pyrogenic response in mammals and thus it should be removed from heterologous proteins produced in *E. coli*. However, the purification processes add to the cost of the product and do not result in complete endotoxin removal (Schwarz, Schmittner, Duschl, & Horejs-Hoeck, 2014). The active component of LPS that induces the reaction in mammals is lipid A and it was considered that this element was essential for outer membrane integrity in *E. coli*. However, (Meredith, Aggarwal, Mamat, Lindner, & Woodward, 2006) demonstrated that viability of a strain of *E. coli* K12 that incorporated the nonendotoxic lipid-A precursor, lipid IV_A, in place of lipid-A. Mamat et al. (2015) used this knowledge to develop a strain of *E. coli* BL21(DE3) that was essentially free of endotoxin.

The issue of the production of heterologous proteins as insoluble inclusion bodies was discussed earlier. The difficulty arises because the reducing environment of the *E. coli* cytoplasm does not allow the formation of disulfide bridges in the foreign protein. Periplasmic expression is one solution to this problem but another approach involves the manipulation of the host *E. coli*. The lack of disulfide bonds in cytoplasmic proteins is due to two NADPH consuming systems that maintain free thiol groups in a reduced state—the thioredoxin and glutaredoxin pathways (Prinz, Aslund, Holmgren, & Beckwith, 1997). Double mutants with lesions in both the *trxB* and *gor* genes can form protein disulfide bonds in the cytoplasm, but such mutants grow slowly. Bessette, Aslund, Beckwith, and Georgiou (1999) isolated spontaneous mutants that suppressed the slow growing phenotype of these variants yet maintained the ability to form disulfide bridges. Novagen adopted this technology to produce both K12 and BL21 strains (named Origami) that were capable of cytoplasmic sulfide bridge formation and thus produce soluble folded proteins intracellularly.

E. coli is well known to excrete acetate into the culture medium, referred to as "overflow metabolism," when grown on glucose at high growth rates—a situation that is wasteful of substrate and likely to cause physiological anomalies due to the inhibitory effects of acetate. The phenomenon can be exacerbated when the organism is used for heterologous protein production and is a cause for concern in the design of such processes (Eiteman & Altman, 2006). The role of metabolic flux and throughput analysis in understanding the physiology of industrial strains was discussed in Chapter 3 and the reader is referred to that account for the background to this section. Holms' (1997) throughput analysis of glucose metabolism by *E. coli* ML308 is an excellent model to explore the basis of acetate excretion. Fig. 3.34 from Chapter 3 is repeated here as Fig. 12.7, and represents the throughput of carbon from

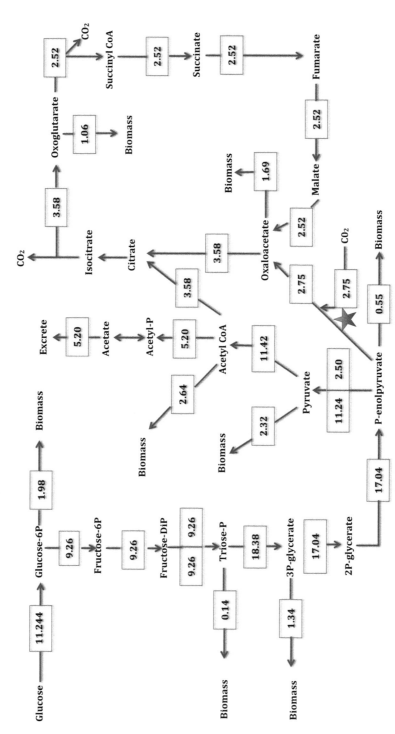

FIGURE 12.7 Throughput Diagram for *E. coli* ML308 Grown on Glucose as Sole Carbon Source in Batch Culture at µmax (Units, Moles kg^{-1} Dry Weight Biomass). ⭐ Represents PEP Carboxylase

Modified from Holms (1997)

glucose in *E. coli* ML308 under aerobic, batch culture conditions growing at the maximum specific growth rate, μ_{max}, of 0.94 h^{-1}. Under these conditions (μ_{max}), the organism secretes significant amounts of acetate, due to the flux of carbon through the pathways being in excess of the requirements of the growth rate. As can be seen in Fig. 12.7, only 2.75 moles of phosphoenolpyruvate (PEP) and 3.58 moles of acetyl CoA are required to feed the TCA cycle, leaving an excess of 5.2 moles of acetate. This overprovision cannot be allowed to accumulate within the organism as pathway intermediates because it would increase the osmotic pressure to an unacceptable level and thus it is exported to the culture medium.

Fig. 12.8 shows the throughput of carbon from glucose in a chemostat culture at the lower specific growth rate of 0.72 h^{-1}. Growth is more efficient at the lower growth rate with the consumption of glucose decreasing from 11.2 moles kg^{-1} dry weight to 8.64, and acetate is not excreted. The inefficient glucose consumption at the high growth rate is entirely accounted for by overflow metabolism excreting acetate—the throughputs of PEP and acetyl CoA to the TCA cycle are identical for both growth rates. Thus, carbon flux leading to PEP and acetyl CoA is in excess of requirements at the high growth rate but in balance with requirements at the lower rate. The carboxylation of PEP with CO_2, catalyzed by PEP carboxylase, is the key anaplerotic reaction that replaces carbon removed from the TCA cycle for biosynthesis. Thus, for both growth rates, 2.75 moles of CO_2 must be combined with 2.75 moles of PEP to replace 1.69 moles of oxaloacetate and 1.06 moles of oxoglutarate (together equivalent to 2.75 moles of PEP) that are converted to amino acids for biomass production. These data suggest that anaplerosis and the TCA cycle operate at maximum throughput at both growth rates and cannot accommodate any further input from glucose, resulting in the "excess" carbon provided at high growth rates being excreted as acetate. Holms suggested that it is the anaplerotic provision to the TCA cycle that limits biosynthesis and thus, in turn, the utilization of precursors.

A number of approaches may be used to reduce acetate excretion. The most obvious is the control of the growth rate, such that it is below the threshold of acetate excretion. The threshold of 0.72 h^{-1} reported for *E. coli* ML308 is quite high compared with other reports in the literature, with values of between 0.14 and 0.48 h^{-1} quoted in Eiteman and Altman's review (2006). Such control may be achieved in either continuous (chemostat) or fed-batch culture, with the latter being by far the more common in large-scale processes. However, it is important to appreciate that in both culture systems the limiting nutrient must be the carbon source (usually glucose) as limitation with another nutrient (eg, the nitrogen source) would control the growth rate but the carbon concentration would be in excess resulting in very high acetate excretion. In fed-batch systems a feedback control system would be used to maintain the glucose concentration in the medium at virtually zero. Such systems are described in Chapter 2.

Reduction of acetate excretion can also be achieved by manipulating the genome but the explanation of this approach requires a closer investigation of the relationships between acetyl CoA and acetate. Fig. 12.9 shows more detail of this relationship than is shown in the overview (Figs. 12.7 and 12.8). Pyruvate is converted to acetate by two routes: via acetyl-CoA and acetyl phosphate, catalyzed by pyruvate dehydrogenase

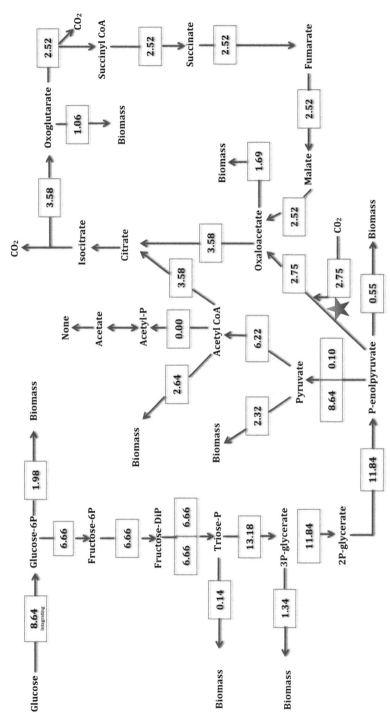

FIGURE 12.8 Throughput Diagram for *E. coli* ML308 Grown on Glucose as Sole Carbon Source in Chemostat Culture at a Specific Growth Rate of 0.72 h^{-1} (Units, Moles kg^{-1} Dry Weight Biomass). ★ **Represents PEP Carboxylase**

Modified from Holms (1997)

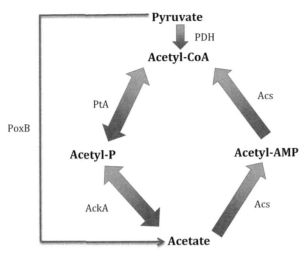

FIGURE 12.9 The Pathways Relating Acetyl CoA and Acetate in *E. coli*

PtA, phosphotransacetylase; *AckA*, acetate kinase; *Acs*, acetyl CoA synthetase; *PDH*, pyruvate deydrogenase; *PoxB*, pyruvate oxidase.

(PDH), phosphotransacetylase (PtA), and acetate kinase (AckA); and directly by pyruvate oxidase. Acetate is recycled back to acetyl-CoA in a two-step process cata-lyzed by acetyl-CoA synthetase (Acs). An obvious genetic engineering approach to reducing acetate accumulation is to remove one or both acetate generating pathways. Several workers have achieved this and whilst disruption of either the PtA–AckA pathway or PoxB reduces acetate excretion they both result in a significant loss of growth efficiency.

A new interpretation of the acetate phenomenon has been proposed that acetate is synthesized at all growth rates but its accumulation is prevented at low growth rates by being recycled to acetyl CoA and its excretion at high growth rates caused by a catabolite repression-mediated downregulation of Acs initiated by the high glucose concentration commensurate with the high specific growth rate (Valgepea et al., 2010; Peebo et al., 2014). Peebo et al. (2014) then approached the problem by addressing the control of Acs by deleting the gene responsible for its inactivation (*pka*), combined with the deletion of the TCA regulator gene (*arcA*), thus facilitating acetate recycling and greater TCA throughput at high growth rates. Thus, acetate lev-els were reduced by enabling its conversion to acetyl CoA that was then drawn into an accelerated TCA cycle. The increased CO_2 production by this strain effectively represents the dispersal of acetate via acetyl CoA and the TCA cycle.

The production of large amounts of recombinant protein that is of no benefit to the producing organism is obviously a huge metabolic burden for the organism to tol-erate. The increased production of acetate in recombinant processes has already been mentioned but another major issue is the induction of the general stress response initiated by the "alarmone," ppGpp. The term "alarmone" describes a molecule that seems hormonal in nature and induces wide-ranging responses to an adverse condition. This response was discussed in Chapter 3 (Fig. 3.41) and is induced, specifically, by amino acid starvation and more generally by nutrient starvation. Heterologous protein production obviously significantly increases the demand for amino acids—a situation that can result in amino acid limitation. Amino acid limitation is detected by the binding of uncharged rRNA molecules to the ribosomes that, in turn, induces the synthesis of the ppGpp synthetase 1 (the ribosomal protein Rel-A) that, in turn, produces ppGpp. The alarmone acts by binding to RNA polymerase and both inhibits the synthesis of stable RNA moieties (rRNA and tRNA) and stimulates the expression of a large number of stress-related genes. Carneiro, Villas-Boas, Ferreira, and Rocha (2011) investigated the performance of a *relA* deletion mutant that did not display the stringent response. Although there were some disadvantages in the metabolic behavior of the strain, it showed enhanced heterologous protein production.

Transcriptomic and proteomic technology enables the observation of changes in gene expression during heterologous protein production and these approaches have resulted in guided metabolic engineering to improve productivity. For example, Choi, Lee, Lee, and Lee (2003) examined transcriptomes of *E. coli* producing a human insulin-like growth factor 1 fusion protein and identified the downregulation of two genes involved in amino acid and nucleotide synthesis—*prsA* and *glpF*. Coexpression of these genes in the producer strain more than doubled the heterologous protein. However, the processing of the vast amount of data generated by these approaches is a major task and the way forward lies in combining this information with C^{13} metabolic flux analysis to develop a stoichiometric model of a process (Carneiro, Ferreira, & Rocha, 2013). These approaches have been successful for the production of simple products, such as, amino acids (see Chapter 3) but their application for heterologous proteins is much more challenging.

HETEROLOGOUS PROTEIN PRODUCTION BY YEAST

S. cerevisiae was the first eukaryotic host to be approved for the production of heterologous proteins. It shares with *E. coli* the advantages of relatively fast growth on simple media and the ability to grow to high cell densities. It is a model eukaryotic organism and thus a wealth of information is available on its physiology, biochemistry, genetics, and molecular biology. Its prominence as a model organism has also ensured that it was in the vanguard of organisms whose biology was explored using "omics" technologies—thus adding to the encyclopedic knowledge of this simple eukaryote. Also, its long history in the fermentation industry as a safe organism makes it an attractive proposition to the regulatory authorities; like many *E. coli* strains, *S. cerevisiae* is classified as GRAS. A frequently quoted advantage of the organism, being a eukaryote, is its ability to perform PTMs of foreign proteins. However, its glycosylation abilities are limited and this makes it unsuitable for the production of glycosylated human therapeutic proteins (Walsh, 2014b). Whilst the fact that *S. cerevisiae* secretes few proteins into the culture medium can make downstream processing easier it also means that the organism is not a good secretor of heterologous proteins and secretion is often limited to the periplasmic space. Compared with *E. coli*, the expression levels can also be considerably lower and are typically about 5% of total protein (Walsh, 2014b).

Yeast species other than *S. cerevisiae* have also been used as host cells, in particular, *P. pastoris*—an organism capable of utilizing methanol as its sole carbon source. *P. pastoris* has a number of advantages as a heterologous host (Potvin, Ahmad, & Zhang, 2012; Ahmad et al., 2014):

- Its metabolism is predominantly aerobic and thus its growth is not limited by the accumulation of ethanol.
- It is well studied, is relatively simple to manipulate and commercial expression systems are available.
- Recombinant proteins can be secreted into the medium.
- It has a strong, tightly regulated, methanol inducible promoter (pAOX1) that can be used to control heterologous protein production.
- Its ability to undertake PTMs is greater than that of *S. cerevisiae*.
- Granted GRAS status.

Other yeasts used for protein production include *Hansenula polymorpha, Klyveromyces lactis*, and *Schizosaccharomyces pombe*. A list of heterologous proteins produced by yeasts is given in Table 12.5. The remainder of this account concentrates on the two most common hosts, *S. cerevisiae* and *P. pastoris*.

The process of heterologous protein production in yeast is very similar to that described in Fig. 12.1. cDNA is still used as the source of the desired gene as only 283 of the 6000 genes of *S. cerevisiae* contain introns (Parenteau et al., 2008). Thus, the mRNA processing machinery of yeasts is very different from that of "higher" eukaryotes. Also, as explained in the next section, yeast expression vectors are "shuttle vectors," meaning that they can replicate in both *E. coli* and yeast cells. Thus, the early stages of cloning are also done in a bacterial system.

YEAST EXPRESSION VECTORS

The design of yeast expression vectors follows the same basic principles described for bacterial expression vectors. However, it is easier to engineer the vector in *E. coli* and thus they are generally shuttle vectors, that is, they can replicate in both the yeast and *E. coli*. As a result, the vectors are based on a DNA backbone derived from a bacterial plasmid (most frequently pBR322) including an *E. coli* origin of replication and a bacterial selectable marker (commonly ampicillin resistance). A generalized representation of a yeast expression vector is shown in Fig. 12.10. The yeast components of the two main types of expression vectors for *S. cerevisiae* are based on yeast centromeric plasmids (YCp) or yeast episomal plasmids (YEp). YCp vectors contain a centromere sequence from the *S. cerevisiae* genome and the vector replication is linked to cell division. Although this gives good stability the copy number of YCp vectors is only 1–2 per cell. YEp vectors contain the ARS (autonomously replicating

Table 12.5 Heterologous Proteins Produced (At least at Research Level) in Yeast (Walsh, 2014)

Heterologous Protein	Host Organism
Hirudin	*S. cerevisiae*; *H. polymorpha*
Hepatitis B surface antigen	*S. cerevisiae, P. pastoris, H. polymorpha*
Human factor XIII	*S. cerevisiae, S. pombe, P. pastoris*
Human plasmin	*P. pastoris*
Human serum albumin	*S. cerevisiae, H. polymorpha, K. lactis*
Tissue plasminogen activator	*S. cerevisiae, K. lactis*
Insulin	*S. cerevisiae*
Interferon-α	*S. cerevisiae*
Interleukin-2	*S. cerevisiae*
Human epidermal growth factor	*S. cerevisiae*
Human growth hormone	*S. cerevisiae*
Human nerve growth factor	*S. cerevisiae*
Influenza viral hemagglutinin	*S. cerevisiae*
Polio viral protein	*S. cerevisiae*
Antibodies/antibody fragments	*S. cerevisiae*
α_1-Antitrypsin	*S. cerevisiae*
Streptokinase	*P. pastoris*
Tumor necrosis factor	*P. pastoris*
Tetanus toxin fragment C	*P. pastoris*

sequence) from the natural yeast 2 μm plasmid. They are able to replicate autonomously and are maintained at an average of 40 copies per cell but must be maintained under selective conditions as otherwise they are easily lost. Thus, the YEp vectors are the episomal vectors of choice and are available from commercial suppliers (Zhang & An, 2010).

Stable episomal vectors are not available for *P. pastoris* and foreign protein production in this organism relies on vectors that do not have an autonomous replication site but their integration into the host genome ensures their stability. As they are integrating vectors they must be designed such that they can be linearized before transformation of the yeast host. Although integration limits the copy number, the stability is such that no selective force is required to maintain the construction, once completed. This is a significant advantage in the production process as complex media may be used and antibiotics are not required (see later discussion). Multiple copies of the foreign gene can be achieved by including multiple copies of it in the plasmid or by using techniques to facilitate multiple integrations of the plasmid (Cereghino & Cregg, 2000). Such vectors are also of the shuttle type and thus include both bacterial sequences to enable their replication and selection in *E. coli*.

Yeast selection markers are required to isolate clones containing the vector and, in the case of episomal vectors, to maintain stability during growth. Both auxotrophic

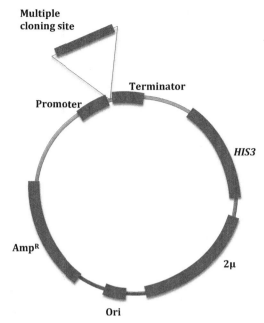

FIGURE 12.10 A Schematic Representation of a Yeast Expression Vector

and resistance markers can be used (Table 12.6) but this requires the host strain to be auxotrophic for the specified compound and precludes the use of complex medium ingredients for episomal vectors. However, the use of a synthetic medium is preferable to employing an antibiotic in the production process. As mentioned earlier, selective force does not have to be used for the maintenance of *P. pastoris* integrating plasmids.

Table 12.6 Selectable Markers for Isolation and Maintenance of *S. cerevisiae* Vector-Containing Cells

Auxotrophic Marker	Enables the Synthesis of:
HIS3	Histidine
LEU2	Leucine
TRP1	Tryptophan
URA3	Uracil

Resistance markers	Phenotype
CUP1	Copper resistance
G418^R	Kanamycin resistance
Zeo^r	Zeocin resistance
TUN^R	Tunicamycin resistance

A variety of promoters, both constitutive and inducible, are available to control the expression of heterologous proteins in *S. cerevisiae* and *P. pastoris* (Table 12.7). An inducible promoter means that protein production can be switched on after growth is completed, thus minimizing the risk of selecting nonproducing mutants. Such a risk is due to the metabolic burden of synthesizing an unnecessary, or possibly toxic, protein. A constitutive promoter tends to be used to simplify the process but inducer addition is a relatively straightforward process for a fed-batch fermentation, which is the favored strategy for most products. The alcohol dehydrogenase promoter of *P. pastoris* is particularly suitable for a tightly controlled fermentation. The organism can be grown on glucose, which strongly represses the heterologous gene production, and then switched to methanol as the carbon source for the production phase.

Other sequences that must be included downstream of the promoter are the ribosome recognition site, the multiple cloning site, and the transcription termination site. The ribosome recognition site in eukaryotes is called the Kozak sequence, the most conserved element of which is the AUG sequence. The translation initiation sites of 95% of yeast mRNA molecules correspond to the first AUG codon at the 5′ end of the message. Thus, it is advisable to remove upstream AUG codons in the heterologous mRNA leader sequence to ensure efficient translation initiation (Curran & Bugeja, 2015). The multiple cloning site of the expression vector enables any of a

Table 12.7 Yeast Promoters That May Be Incorporated Into Expression Vectors (Curran & Bugeja, 2015; Mattanovich et al., 2012; Zhang & An, 2010)

Promoter	Naturally Controls the Expression of:	Constitutive or Inducible
Promoters for *S. cerevisiae*		
CYC1	Cytochrome C oxidase	Constitutive
TEF	Translation elongation factor	Constitutive
GAPDH	Glyceraldehyde-3-phosphate dehydrogenase	Constitutive
PGK	Phosphoglycerate kinase	Constitutive
ADH2	Alcohol dehydrogenase	Inducible by ethanol
GAL1-10	Galactokinase/galactose epimerase 1	Inducible by galactose
Promoters for *P. pastoris*		
GAP	Glyceraldehyde-3-phosphate dehydrogenase	Constitutive
TEF	Translation elongation factor	Constitutive
PGK	Phosphoglycerate kinase	Constitutive
AOX1	Alcohol oxidase	Inducible by methanol
FLD1	Formaldehyde dehydrogenase	Inducible by methanol

number of restriction enzymes to be used to introduce the heterologous DNA, as discussed in our earlier consideration of bacterial expression vectors. Finally, it is critical to include the relevant yeast termination of transcription site at the end of the foreign gene as the yeast may not recognize its native termination sequence. Should the RNA polymerase read through the termination site then very long, unstable mRNA molecules would be produced and can result in a catastrophic reduction in yield.

The construction of vector can also produce the protein in a form to facilitate its purification, as discussed in the earlier consideration of *E. coli* expression vectors. This is achieved by the use of tags, such as, poly histidine or FLAG but rather than these sequences being incorporated into the vector downstream of the promoter as in bacterial vectors (Fig. 12.6) they are incorporated in-frame with the gene of interest (Celik & Calik, 2012).

PROTEIN SECRETION

Retention of the synthesized heterologous protein within the cytoplasm of the yeast can result in very high expression levels. However, some proteins may be degraded in the cell and others may not fold completely resulting in the production of inclusion bodies, as found in *E. coli* systems. Intracellular accumulation of the protein also has a significant impact on its recovery at the end of the process. The product has to be liberated from the yeast biomass by cell breakage—not a trivial task on a large scale—and then separated from the myriad of native intracellular proteins that are freed at the same time. Furthermore, the heterologous protein is then vulnerable to degradation by proteases released from the cells. Although the host strain may be manipulated to remove proteases, the remaining problems associated with intracellular accumulation can only be avoided if the protein is exported from the cell into the culture medium.

Naturally produced proteins that are destined to be exported from the cell are labeled with a hydrophobic signal sequence at the N-terminus. The signal sequence enables the protein to be transported to the endoplasmic reticulum (ER) lumen and then to the Golgi apparatus. Thus, to achieve secretion of a heterologous protein it must also carry the appropriate signal sequence. Because *S. cerevisiae* and *P. pastoris* are both eukaryotes they may recognize the native secretion signal of the foreign protein so it is always worthwhile to include the signal sequence in the initial assessment of expression sequences (Zhang & An, 2010). Alternatively, a specific yeast signal sequence can be incorporated into the recombinant construct. The most common sequences for secretion to the periplasm are those of *S. cerevisiae* invertase (SUC2, 19 amino acids) and acid phosphatase (PHO1, 17 amino acids) and for secretion to the culture medium is that of the α-factor pheromone (α-MF, 20 amino acids). The most frequently used signals for secretion in *P. pastoris* are the *S. cerevisiae* invertase and α-factor as well as its native acid phosphatase (Curran & Bugeja, 2015; Ahmad et al., 2014).

Besides the previously cited advantages of secretion of the protein, secretion also gives the protein access to the organism's PTM machinery and this is considered in the next section.

THE MANIPULATION OF THE HOST

As was the case for *E. coli*, yeasts have also been manipulated both physiologically and genetically to optimize productivity. Fed-batch culture tends to be the production method of choice as it gives good process control without the disadvantages of continuous culture (see Chapter 2). The application of the fed-batch strategy to the culture of *S. cerevisiae* goes back to the development of the bakers' yeast fermentation (Reed & Nagodawithana, 1991). *S. cerevisiae* is very sensitive to the Crabtree effect (Crabtree, 1929) in which respiratory activity is repressed by the presence of free glucose. Thus, the bakers' yeast process employed a slow molasses feed rate such that the metabolism was fully respiratory and growth efficiency was at its maximum. The works of Biener, Steinkamper, and Horn (2012) and Anane, van Ensburg, and Gorgens (2013) are examples of heterologous protein processes employing the same rationale as the bakers' yeast process to give maximum biomass yield. However, Baumann, Adelantado, Lang, Mattanovich, and Ferrer (2011) demonstrated that production of a heterologous fragment antigen-binding (Fab) fragment by *P. pastoris* (a non-Crabtree effect yeast) was favored by oxygen limitation and linked this phenomenon with ergosterol synthesis. Ergosterol is a component of the cell membrane and requires oxygen for its biosynthesis suggesting that oxygen limitation influences membrane structure and, in turn, protein secretion.

P. pastoris is a respiratory yeast and, thus, the Crabtree effect is not a concern. However, the control of its growth rate using fed-batch culture is still a key aspect in its use as an industrial producer of recombinant proteins. The organism has been shown to achieve cell densities of up to 150 g dm^{-3} in fed-batch culture, a biomass yield that contributes significantly to the high levels of heterologous proteins reported (Ahmad et al., 2014). Furthermore, the complexity of the relationship between specific growth rate (μ) and heterologous protein production is illustrated by Looser et al.'s (2015) review that cited examples of negative, positive, and bell-shaped correlations. This underlines the importance of growth rate control (by fed-batch culture) such that the optimum growth rate for a process can be achieved.

Apart from the generation of auxotrophs to complement the vector selection system, genetic modification of the host strains has focused on the secretion and glycosylation pathways. On entry into the ER, nascent proteins are bound to chaperone proteins that enable disulfide bond formation, folding and glycosylation and only properly processed proteins will progress to the Golgi apparatus. Misfolded or insoluble proteins are recognized by the cell quality control system and transferred to the cytosol for degradation (endoplasmic reticulum associated protein degradation, ERAD). The limitation to secretion appears to lie in the complexity of the folding and quality control system. Thus, these have been the targets to improve the secretion of heterologous hosts, primarily overexpression of chaperones, protein disulfide isomerases, and folding helpers (Idiris, Tohda, Kumagai, & Takegawa, 2010). The other aspect of host engineering related to secretion is the development of protease-deficient strains. Degradation of heterologous proteins not only reduces the yield of product but also complicates downstream processing as the degradation products still containing affinity tags may interfere with affinity chromatography systems (Ahmad et al., 2014). Several enzymes have been implicated in protein degradation and, thus, it is difficult to predict the effect of eliminating particular proteases on the production of specific heterologous proteins (Ahmad et al., 2014). However, the consensus is that the key proteases are those coded by *PEP4* (an aspartyl protease) and *PRB1* (a serine protease). Deletion mutants for both of these genes for *S. cerevisiae* and *P. pastoris* have been developed (Tomimoto et al., 2013; Gleeson, White, Meininger, & Komives, 1998, respectively) and given promising results.

Approximately 70% of approved therapeutic proteins are glycoproteins meaning that accurate PTM is essential for the production of such products (Mattanovich et al., 2012). However, whilst the process of N-glycosylation in the ER of *S. cerevisiae* is very similar to that of mammals, the processes in the Golgi apparatus differ. The N-glycans in yeast are highly mannose-rich whereas mammalian proteins are more complex and contain monosaccharides rarely present in yeast, including galactose, fucose, and sialic acid (Stanley, Schachter, & Taniguchi, 2009). Thus, the yeast N-glycosylation pathways have to be modified to produce human-like proteins in these organisms. This approach has been particularly successful with *P. pastoris*. The structures of glycosyltransferases and mannosidases located in the Golgi body are highly conserved between eukaryotic species and consist of three regions—an N-terminal transmembrane (of the Golgi) domain with an extension into the cytoplasm, a stem domain, and a C-terminal catalytic domain. To enable mammalian glycosylation patterns to be achieved in *P. pastoris* then C-terminal catalytic mammalian domains have been fused with the yeast transmembrane and stem domains (Nett et al., 2011). The difficulty in this approach was that it was not possible to predict whether a particular heterologous domain would be effective in *Pichia*. Nett et al. resolved this problem by developing a combinatorial genetic library of different constructs screened using a high-throughput protocol.

HETEROLOGOUS PROTEIN PRODUCTION BY MAMMALIAN CELL CULTURES

The key advantage of mammalian cell cultures over microbial systems for the production of heterologous proteins is their ability to carry out human-like PTMs of their protein products. Up to July 2014, 56% of biopharmaceuticals (recombinant biological products) approved in the United States and the European Union were produced in mammalian systems compared with 35.5% in microbial systems (*E. coli* and yeast) (Walsh, 2014a). The use of Chinese hamster ovary cells (CHO) accounted for 64% of the products produced in mammalian systems, with the annual global revenue from CHO products exceeding US$100 billion (Jadhav et al., 2013). The major advantage of CHO cells is their capacity to grow in suspension culture in serum free medium at high densities and, as discussed in Chapter 4, the manufacture of therapeutic proteins cannot employ animal-derived products in the production process due to the possibility of prion contamination. Furthermore, many important human viral pathogens are unable to replicate in CHO cells (Bandaranayake & Almo, 2014).

Yields of heterologous proteins in the order of $g\ dm^{-3}$ have been obtained from CHO cells, a 100-fold increase over the 1980 processes, and they are envisaged to continue to be key biopharmaceutical manufacturing platforms (Lai, Yang, & Ng, 2013). However, human cell lines are predicted to be increasingly important as hosts for heterologous protein production, particularly for the production of more complex human proteins (Bandaranayake & Almo, 2014). Human cell lines include human embryonic kidney cells (HEK) and the PER.C6 cell line derived from human embryonic retina cells transformed by adenovirus E1. The PER.C6 cell line has the advantage of producing very high cell densities (up to 10^8 cells cm^{-3}) and secreted protein titers of $8\ g\ dm^{-3}$ have been achieved using fed-batch and $15\ g\ dm^{-3}$ using perfusion culture (Kuczewski, Schirmer, Lain, & Zarbis-Papastoitsis, 2011).

TRANSIENT GENE EXPRESSION

Our consideration of the synthesis of heterologous proteins in microbial systems emphasized the development of stable recombinant strains that could be relied upon to manufacture the desired protein over many generations. Similarly, the ideal situation for the production of a heterologous protein in mammalian cells is to transform a cell line such that its ability to produce the product is stably inherited by its progeny. However, the selection of such mammalian cell lines can be a very lengthy process (see later discussion) that may not be commensurate with the aim of the exercise. For example, a fast route to production would be desirable in the following situations: production of sufficient protein in the preclinical development of a product; the synthesis of variants of a protein to be assessed as a potential biopharmaceutical; the production of a protein receptor for use in a screening assay or to determine its structure. Such a route is termed transient gene expression (TGE) in which the gene for the heterologous protein is not incorporated into the host genome, nor stably replicated along with cell division but is expressed in the host cell for, typically, 10 days posttransformation (Baldi, Hacker, Adam, & Wurm, 2007). The rationale of this approach is to grow up a large volume of cells (between one and several hundred dm^3 have been quoted, (Geisse, 2009; Bandaranayake & Almo, 2014) and then transform it with the expression vector (plasmid)—thus, the transformation is done on a bulk scale after growth and the transformed culture will synthesize the protein for a limited time, as shown in Fig. 12.11. Such a process requires bulk preparation of the expression vector in an *E. coli* fermentation and a simple, cheap transformation process that will work on a large scale. At this stage it is pertinent to point out that different terms tend to be used in the description of the manipulation of animal cells compared with that of microorganisms. Whereas "transformation" is used to describe the uptake of DNA by microorganisms, "transfection" is the term commonly employed to describe the same process in cultured animal cells. This is because "transformation" is used in animal cell biology to describe the genetic change that renders a cell able to proliferate in culture, similar to a cancerous cell. Thus, in the rest of this section, the term, "transfection," will be used to describe the uptake of DNA by a mammalian cell.

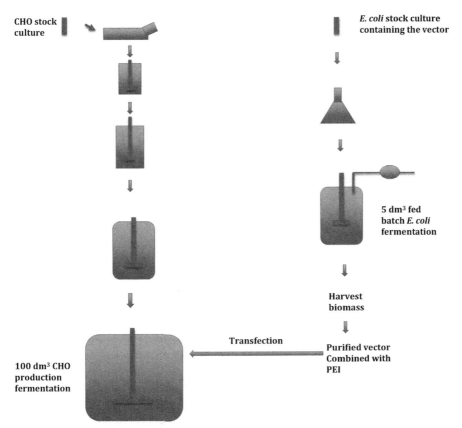

CHO stock culture

E. coli stock culture containing the vector

5 dm³ fed batch *E. coli* fermentation

Harvest biomass

Transfection

Purified vector Combined with PEI

100 dm³ CHO production fermentation

FIGURE 12.11 Diagrammatic Representation of a Transient Gene Expression Process

PEI, polyethylenimine.

To achieve effective transfection, between 1 and 1.25 mg of the expression vector are required dm^{-3} of the mammalian culture. The production of plasmid in a fed-batch *E. coli* fermentation is a relatively routine process and Zhu (2012) reported that between 500 and 2600 mg dm^{-3} plasmid can be produced on a 5 dm^3 scale. Thus, it should not be a problem to produce sufficient expression vector to infect a process of several hundred dm^3. However, it is important that the purified vector is free of contaminating *E. coli* LPS. The expression vector for a TGE process must contain the desired eukaryotic gene (normally as cDNA) controlled by a strong eukaryotic promoter, a transcription terminator, and a bacterial cassette including an origin of replication and a selection marker to enable the proliferation of the vector in the *E. coli* host. Smaller plasmids that can be maintained in *E. coli* at a high copy number are preferred as they are more easily transfected and can be propagated relatively easily in the bacterium (Jager, Bussow, & Schirrmann, 2015). The transfection technique most applicable to the large scale is the use of polyethylenimine (PEI), a cheap, stable cationic polymer that condenses DNA into positively charged particles that bind to the negatively charged cell surface. The DNA-PEI particles are then taken into the cell by endocytosis (Longo, Kavran, Kim, & Leahy, 2013). TGE processes have been developed predominantly using CHO and HEK293 cell lines as they have been shown to be amenable to large-scale transfection (De Jesus & Wurm, 2011). Medium development has been an important factor in the development of successful TGE processes as many use different media for growth and transfection, requiring a complex cell harvesting and resuspension step before transfection (Zhu, 2012). However, it has been reported that if the transfection is performed at a high cell density (20×10^6 cells cm^{-3}) then specific transfection media may not be necessary, thus eliminating the resuspension step (Backliwal et al., 2008).

Commonly reported difficulties with TGE systems include poor expression of the foreign gene and the onset of posttransfection apoptosis. Improved expression in TGE systems has been achieved by the addition of histone deacetylase inhibitors, such as, valproic acid (Wulhfard et al., 2010). Acetylation and deacetylation of histones play a key role in the regulation of gene expression in eukaryotes. The addition of an acetyl group to the lysine monomers in the N-terminal portion of histone proteins removes their positive charge, and therefore their interaction with the negatively charged phosphate groups of DNA, thus enabling more effective transcription. Deacetylation results in the inhibition of transcription, thus explaining the beneficial effect of deacetylase inhibitors in TGE processes. The onset of apoptosis in TGE systems is related to the high cell densities desirable for effective transfection and high productivity (apoptosis is discussed in Chapter 2). This problem may be addressed by utilizing a cell line that has been stably transfected with an antiapoptosis gene (Majors, Betenbaug, Pederson, & Chiang, 2008).

At the time of writing no TGE processes had been approved for the manufacture of a pharmaceutical heterologous protein but it is possible that this approach may be feasible on a large scale provided that the product quality can be shown to be of an acceptable standard that can be replicated effectively from batch to batch.

STABLE GENE EXPRESSION

Although TGE is an attractive proposition as a "fast track" route for the production of heterologous proteins, stable gene expression is considered a key criterion for the commercial production of biopharmaceutical heterologous proteins. Stable gene expression can be achieved when the transfected DNA is integrated into the genome. However, due to the random nature of integration, the cell clones that are produced are highly variable and have to be screened to find a suitable stable cell line. The screening process is very time consuming and can add a year to the development period of a biopharmaceutical.

The cell line selection systems used by most pharmaceutical companies are based on a complementation strategy in which the host CHO cells are deficient in a selectable and amplifiable gene and the expression vector contains both the gene of interest (coding for the desired heterologous protein) and a functioning copy of the gene for which the host is deficient. Although it is also possible to administer the gene of interest and the selectable gene in separate plasmids (vectors), as they tend to integrate at the same site (Wurm, 2004), this approach can be unreliable. Transfection of a cell and integration of the vector enable the cell to grow in minimal medium as the vector is replicated along with cell division. However, one gene copy may not produce the desired yield and thus the next step is to increase (amplify) the copy number of the integrated marker, thus resulting in an associated increase in copy number of the heterologous protein gene. A selective force giving an advantage to a cell line containing several integrated vectors is selected by culturing in the presence of an inhibitor of the enzyme coded by the selectable gene. Thus, growth will be inhibited depending on the number of integrated copies—more gene copies equating to greater resistance to the inhibiter and cell growth. Also, the gene of interest may be under the control of a strong promoter and the selective marker gene under the control of a weak promoter (Fig. 12.12) Thus, the selectable gene requires many copies to facilitate resistance whereas the strong promoter controlling the coamplified gene of interest ensures high expression resulting in improved protein production. Two main systems are used with CHO cells—one based on dihydrofolate reductase (DHFR) (Kaufman & Sharp, 1982) and the other on glutamine synthetase (GS) (Cockett, Bebbington, & Yarranton, 1990). Both systems have been approved by the regulatory authorities for the production of biopharmaceuticals and are the "workhorses" of the industry. DHFR deficiency results in a requirement for thymidine and thus the transfection is done in medium lacking thymidine. Methotrexate (MTX) is an inhibitor of DHFR and selects for cell lines in which the gene for DHFR is amplified, accompanied by the amplification of the gene of interest. In their original paper, Kaufman and Sharp (1982) demonstrated that selected cells contained up to 1000 copies of the transfecting DNA. GS deficiency results in a requirement for glutamine and the enzyme is inhibited by methionine sulfoximine (MSX), thus MSX is used to select for transfected cell lines containing amplified gene copies. The cell lines can be subjected to serial selection in which the concentration of the inhibitor is increased so that the growth of the cells will be dependent on their producing higher levels of the enzyme subject to inhibition such that the inhibitor is quenched by the enzyme concentration—enzyme production, in turn, is dependent upon the number of copies of the gene and their insertion site.

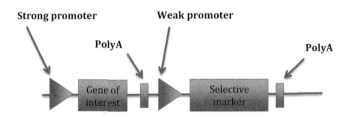

FIGURE 12.12 Locations of the Gene of Interest and the Selective Markers, Controlled by Strong and Weak Promoters Respectively, in a CHO Expression Vector

Modified from Lai et al. (2013)

The mechanics of the selection process is highly labor intensive, requiring the screening of, possibly, thousands of clones generated during the amplification process (Kuystermans & Al-rubeai, 2015). An overview of the process is given in Fig. 12.13. The most common method is limiting dilution using microtiter (multiwell) plates in which the cell suspension is diluted such that an individual well

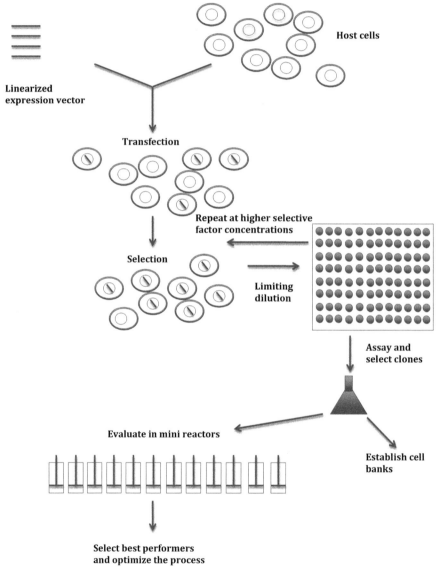

FIGURE 12.13 Diagrammatic Representation of the Selection and Appraisal of Mammalian Cell Lines for the Production of a Heterologous Protein

Modified from Lai et al. (2013)

contains a single cell. Kuystermans and Al-rubeai. (2015) have reviewed the streamlining and automation of the screening process using, for example, cell sorting. Each cell line has to be assayed and promising candidates are examined in flask culture and eventually in reactors. Robotic-assisted scaled-down reactor systems are invaluable in this process so that a large number of clones can be assessed under conditions that mimic the large-scale production process. Such approaches are discussed in more detail in Chapter 9 under "Influence of scale-down studies." The use of systems that enable potential cell lines to be assessed using the culture strategy of the commercial process (eg, fed-batch culture) is a costly option but is likely to shorten significantly the development time of a new product and make scale-up more predictable. The chosen cell line must then be shown to be stable for the in vitro cell age, that is, the number of generations that the cells will undergo in the commercial production process from the thawing of the master cell bank to the harvesting of the production vessel (O'Callaghan & Racher, 2015). This procedure is outlined in Chapter 6 and Fig. 6.3.

The costs of development are particularly relevant when considering the establishment of a manufacturing process for a "biosimilar." Once the patent on a drug has expired then any manufacturer can produce a generic version of the drug provided the requirements of the regulatory authorities are met. Whereas small-molecule pharmaceuticals can be directly copied and approved by the regulatory authorities with relative ease, a heterologous protein is a complex molecule whose structure (particularly PTMs) is process-dependent, necessitating a much more complex regulatory process. Thus, the term "biosimilar" was introduced to describe a generic biopharmaceutical that might be similar to the original drug but not necessarily identical. The development of a biosimilar process has to go through the same procedure as the original patented protein of selecting a stable cell line and gaining regulatory approval of the manufacturing process that may take 5–8 years, whereas that for a small-molecule generic may be achieved in 1–2 years (Kuystermans & Al-Rubeai, 2015). The cost associated with the development of a biosimilar has been reported as between 60 and 200 million euros compared with 1.6–2.4 million euros for a conventional generic medicine (Nefarma, 2013). Furthermore, because the returns will be smaller than for the original protein drug, it is critical that the biosimilar gets to the market as quickly as possible. The economic reality of biosimilar manufacture may be the reason for the dramatic decrease in the rate of biosimilars reaching the market place (Walsh, 2014a), despite earlier predictions that patent expiry would open the floodgates to biosimilars. That is not to underestimate the importance of haste in the development of a patented protein. A patent is normally granted for 20 years but 10–12 years are consumed with the development and clinical approval stages leaving between 8 and 10 years of protection for the approved manufactured product. Thus, any saving in development time enables the producer to make the most of the patent duration. A number of approaches have been developed that may accelerate the development of a biopharmaceutical process and these will be addressed in the following sections.

Expression vector technology

The main objective in designing mammalian cell expression vectors is to improve the efficiency of isolating stable cell clones producing high levels of the heterologous protein. The following strategies have been adopted:

- Controlling the coexpression of the gene of interest and the selective marker gene
- Increasing the rigor (or stringency) of the selective marker
- Incorporating DNA regulatory elements into the vector
- Targeting the integration of the vector to a particular site in the host genome
- An artificial chromosome expression system.

1. *Controlling the coexpression of the gene of interest and the selective marker gene*

    As mentioned earlier, the gene of interest and the selective marker gene are normally incorporated in the one vector and are controlled by different promoters. However, this can result in the transcription of the one gene being suppressed. Furthermore, genetic rearrangements can result in the deletion of the gene of interest, leaving only the selective marker in situ, thus also divorcing selective pressure from the amplification of the gene of interest. This may be resolved by controlling both genes with the one promoter and introducing an internal ribosome entry site (IRES) sequence between them—Fig. 12.14. An IRES is a nucleotide sequence that enables the initiation of translation within a messenger RNA sequence, thus enabling the translation of two different proteins from a single, dicistronic (from two genes) mRNA. By incorporating the gene of interest as the upstream (first) gene in the vector, the expression of the (downstream) selective marker gene becomes dependent on the transcription of the heterologous gene, virtually eliminating the chance of losing the heterologous gene and retaining the selective marker (Rees et al., 1996; Lai et al., 2013).

FIGURE 12.14 Incorporation of an Internal Ribosome Entry Site *(IRES)* into an Expression Vector Enabling the Control of Both the Gene of Interest and the Selective Marker by the Same Promoter

Modified from Lai et al. (2013)

2. *The rigor (or stringency) of the selective marker*

The logic of this approach is to strengthen the selective force used in clone selection by weakening (frequently termed attenuating) the ability of the system to produce the product of the selective gene. Thus, the marker gene of a resistant clone has to be expressed at a high level (eg, due to amplification of the gene) to tolerate the selective pressure, and in so doing elevates the expression of the gene of interest; for example, resistance to MTX in the DHFR system that coselects for the amplification of the linked selective gene and the gene of interest. Because of the greater susceptibility of the strain to the selective inhibitor, this tactic also has the advantage that selective pressure can be exerted by a lower concentration of the inhibitor. As a result, undesirable side effects of the inhibitor, such as low growth rate, can be avoided thus making the selection process shorter. An example of this strategy was given earlier in the employment of a weak promoter for the selective gene and a strong promoter for the gene of interest (Fig. 12.12). Another approach is to lower the translation efficiency of the selective gene by incorporating little-used codons into its sequence (termed codon de-optimization), thereby exploiting the shortage of rare tRNA moieties in the cell. Weakening the productivity of the selective gene and thus strengthening the selective force can also be achieved by the incorporation of signal sequences that destabilize either the mRNA or the expressed protein. Adenylate-uridylate-rich elements (AREs), located in the 3′untranslated region of mRNA, label the mRNA for rapid degradation. Similarly, a peptide sequence rich in proline (P), glutamate (E), serine (S), and threonine (T), termed PEST, indicates a protein with a short intracellular half-life. Thus, the addition of commensurate DNA sequences for these elements to the selective gene reduces the stability of both the mRNA and the protein product. This approach is illustrated in Fig. 12.15 for vectors with either one or two promoters. Finally, modified IRES sequences can result in the decreased expression of the selective gene. Chin, Chin, Chin, Lai, & Ng (2015) investigated the use of multiple strategies and designed a CHO vector that incorporated a common promoter and the gene of interest with combinations of a weakened IRES, codon deoptimization of the DHFR selective gene, a PEST sequence, and an ARE sequence. MTX selection enabled the isolation of cell clones producing up to 1.15 g dm^{-3} heterologous protein within 3 months. The most effective combination of elements was the mutated IRES and the PEST sequence.

3. *Incorporating DNA regulatory elements into the vector*

DNA in eukaryotic chromosomes is combined with protein and RNA in a complex of molecules termed chromatin that packages the DNA into a smaller volume, prevents damage, and also controls gene expression. Histone proteins play a key role in the packaging of the DNA by enabling it to wrap around the proteins to form nucleosomes—often referred to as a "beads on a string." This gives rise to a "loose" chromatin structure, termed euchromatin, which enables access to the DNA by RNA polymerase and thus represents a zone of active transcription. Further packaging results in a much "tighter" structure,

(a)

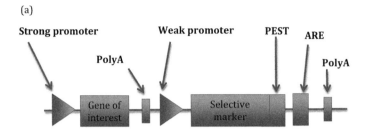

(b)

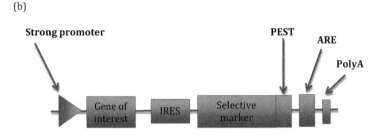

FIGURE 12.15

(a) Locations of the PEST and ARE sequences designed to lower the amount of protein produced by the selective marker gene, in a CHO expression vector in which the gene of interest and the selective marker gene are controlled by different promoters. (b) Locations of the IRES, PEST, and ARE sequences designed to lower the amount of protein produced by the selective marker gene, in a CHO expression vector in which the gene of interest and the selective marker gene are controlled by the same promoter.

Modified from Lai et al. (2013)

termed heterochromatin, which is inaccessible to RNA polymerase and is therefore not transcribed. Heterochromatin may be constitutive (it remains in that form) or facultative (it may be converted to euchromatin). The DNA within constitutive heterochromatin will be poorly transcribed whereas the expression of DNA within facultative heterochromatin may be activated or silenced by the conversion of one form of chromatin into the other. Conversion is mediated by the acetylation and deacetylation of the lysine molecules in the N-terminal tail of histone proteins. Lysine is a di-amino amino acid and acetylation of the side chain amino group removes its positive charge, thus reducing the interaction between the histone and the negatively charged phosphate groups of the DNA. As a result, acetylation results in a loose association between DNA and protein (euchromatin) whereas deacetylation causes the tighter association (heterochromatin). The methylation of the histone lysine residues also plays a major role in the control of gene expression but methylation of the amino group does not affect its charge. Therefore, the role of methylation appears to be the attraction of key enzymes to the histone surface, including histone acetylases and deacetylases that contain methyl-lysine binding sites. Thus, from this discussion it is clear that the interaction of heterologous genes with chromatin is a key aspect of their transcription and, although integrated in the chromosome, the form of chromatin with which they are associated may silence them. This is particularly pertinent when it is remembered that integration into the chromosome is a random event and the nature of the chromatin at the insertion site may dictate whether the gene is expressed. The following paragraph describes DNA sequences that may prevent the chromatin silencing of inserted heterologous genes.

The nuclear matrix is a network of fibers, similar to the cytoskeleton but inside the cell nucleus. Matrix attachment regions (MARs) and scaffold attachment regions (SARs) are DNA sequences that facilitate the attachment of the DNA to the nuclear matrix, thus maintaining a euchromatin structure and enabling transcription. A number of workers have shown that incorporation of MAR regions either side of the heterologous gene increased both the occurrence of high producing clones and gene expression, as shown in Fig. 12.16 (Girod, Zahn-Zabal & Mermod, 2005; Girod et al., 2007). This approach also reduced

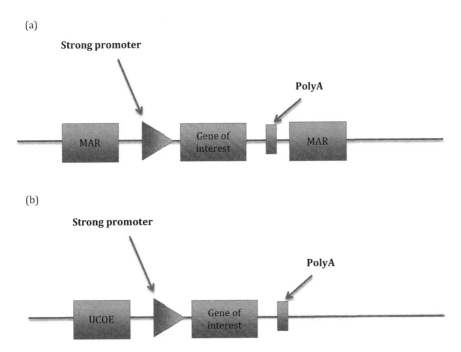

FIGURE 12.16 Expression Vector Constructs Designed to Prevent Chromatin Silencing of the Gene of Interest

(a) Incorporation of MAR sites into an expression vector enabling the maintenance of transcriptionally active chromatin. (b) Incorporation of an UCOE into an expression vector.

Modified from Lai et al. (2013)

the requirement for clone selection and thus hastened the process development. An insulator DNA sequence is able to protect a gene from inappropriate signals produced in its vicinity. Ubiquitous chromatin opening elements (UCOE) are insulators that protect genes from chromatin related silencing by preventing the advance of nearby condensed chromatin (West, Gazner, & Felsenfeld, 2015). Merck Millipore (2014) have marketed a UCOE-based cloning system that incorporates the element upstream of the promoter controlling the heterologous gene, as shown in Fig. 12.14. The claims for this system include heterologous gene expression regardless of its integration site, the prevention of chromatin gene silencing, and high protein yield.

4. *Targeting the integration of the vector to a particular site in the host genome* Only a very small proportion (0.1%) of a cell's genome is actively transcribed, which explains why random integration of transfected cells requires selection and amplification. However, if insertion were directed to occur at a highly transcribed site (a so-called "hotspot") then just a single gene copy may give a yield comparable with amplified insertions at a weak one. This approach is termed site-specific recombination or recombinase mediated cassette exchange (RMCE) and requires the generation of an engineered host cell that is receptive to the introduction of the gene of interest at a specific site. This is achieved by the integration of a reporter gene flanked by short DNA target sequences that are recognized by a specific recombinase enzyme. The chance of the reporter gene integrating at a hotspot will be approximately 0.1% and requires an extensive search. However, once identified, the cell line is used as a master host into which a vector containing the gene of interest, flanked by the same DNA target sequences, can be exchanged for the reporter gene by the recombinase (cotransfected on a separate expression vector) (Qiao, Oumard, Wegloehner, & Bode, 2009). Bandaranayake and Almo (2014) claimed that, once the master cell line had been developed, a RMCE system could be used to generate a suitable production strain in 2–3 months compared with 12 months for the conventional amplification approach. Crawford et al. (2013) reported yields of up to 1 g dm^{-3} antibody using the RMCE technology.

OTHER CONTRIBUTING FACTORS TO THE SUCCESS OF MAMMALIAN CELL CULTURE PROCESSES

The increased yields of heterologous protein achieved by mammalian cell cultures are not simply due to the progress in expression technology. Major contributions have also been made in cell line development, medium design, fed-batch and perfusion culture, and disposable reactors. Medium design, particularly serum-free medium, is addressed in Chapter 4 and the influence of disposable reactors is considered in Chapters 6 and 7. In order to fully exploit the potential of a fed-batch or perfusion process the fermentation should be capable of operating for a long time period at a high cell density. As discussed in Chapter 2, the stationary phase in microbial batch culture and low growth rates in fed-batch and chemostat culture are accompanied by a number of distinct physiological changes enabling the organism to adapt to the stressful conditions associated with the lack of nutrients and the development of an adverse environment. The response of mammalian cells to cell density, nutrient depletion, and toxic products in culture is related to regulatory development in the whole organism in which apoptosis, or programmed cell death, plays a key role—effectively preventing the unregulated proliferation of cells. Thus, the death of such cells in culture due to apoptosis is an active, genetically controlled process thus is a target for genetic manipulation (Kim, Kim, & Lee, 2012).

The most promising target for the control of apoptosis is the *BCL-2* (B-cell lymphoma 2) gene family. This family of genes are antiapoptotic, that is, they prevent apoptosis and are involved in the development of cancers. However, overexpression of BCL-2 genes in a number of cell lines has resulted in increased productivity without concomitant loss in quality (Butler & Meneses-Acosta, 2012).

REFERENCES

Anane, E., van Ensburg, E., & Gorgens, J. F. (2013). Optimisation and scale-up of alpha-glucosidase production by recombinant *Saccharomyces cerevisiae* in aerobic fed-batch culture with constant growth rate. *Biochemical Engineering Journal*, *81*, 1–7.

Ahmad, M., Hirz, M., Pichler, H., & Schwab, H. (2014). Protein expression in *Pichia pastoris*: recent achievements and perspectives for heterologous protein production. *Applied Microbiology and Biotechnology*, *98*, 5301–5317.

Backliwal, G., Hildinger, M., Chenuet, S., Wulhfard, S., De Jesus, M., & Wurm, F. M. (2008). Rational vector design and multi-pathway modulation of HEK 293E cells yield recombinant antibody titers exceeding 1g/l by transient transfection under serum-free conditions. *Nucleic Acids Research*, *36*(15), e96.

Baldi, L., Hacker, D. L., Adam, M., & Wurm, F. M. (2007). Recombinant protein production by large-scale transient gene expression in mammalian cells: state of the art and future perspectives. *Biotechnology Letters*, *29*(5), 677–684.

Bandaranayake, A. D., & Almo, S. C. (2014). Recent advances in mammalian protein production. *FEBS Letters*, *588*(2), 253–260.

Baumann, K., Adelantado, N., Lang, C., Mattanovich, D., & Ferrer, P. (2011). Protein trafficking, ergosterol biosynthesis and membrane physics impact recombinant protein secretion in Pichia pastoris. *Microbial Cell Factories*, *10*(93), 1–15.

Bell, M. R., Engleka, M. J., Malik, A., & Strickler, J. E. (2013). To fuse or not to fuse: What is your purpose? *Protein Science*, *22*(11), 1466–1477.

Berlec, A., & Strukelj, B. (2013). Current state and recent advances in biopharmaceutical production in *Escherichia coli*, yeasts and mammalian cells. *Journal of Industrial Microbiology and Biotechnology*, *40*, 257–274.

Bessette, P. H., Aslund, F., Beckwith, J., & Georgiou, G. (1999). Efficient folding with multiple disulfide bonds in the *Escherichia coli* cytoplasm. *Proceedings of the National Academy of Sciences of the USA*, *96*(24), 13703–13708.

Biener, R., Steinkamper, A., & Horn, T. (2012). Calorimetric control of the specific growth rate during fed-batch cultures of *Saccharomyces cerevisiae*. *Journal of Biotechnology*, *160*, 195–201.

Butler, M., & Meneses-Acosta, A. (2012). Recent advances in technology supporting biopharmaceutical production from mammalian cells. *Applied Microbiology and Biotechnology*, 96(4), 885–894.

Calos, M. P. (1978). DNA sequence for a low-level promoter of the lac repressor gene and an "up" promoter mutation. *Nature*, 274, 762–765.

Capon, D. J., Chamow, S. M., Mordenti, J., Marsters, S. A., Gregory, T., Mitsuya, H., et al. (1989). Designing CD4 immunoadhesins for AIDS therapy. *Nature*, 337, 525–531.

Carneiro, S., Ferreira, E., & Rocha, I. (2013). Metabolic responses to recombinant bioprocesses in *Escherichia coli*. *Journal of Biotechnology*, 164, 396–408.

Carneiro, S., Villas-Boas, S. G., Ferreira, E. C. K., & Rocha, I. (2011). Metabolic footprint analysis of recombinant *Escherichia coli* strains during fed-batch culture. *Molecular Biosystems*, 7, 899–910.

Caspeta, L., Flores, N., Perez, N. O., Bolivar, F., & Ramirez, O. T. (2009). The effect of heating rate on *Escherichia coli* metabolism, physiological stress, transcriptional response, and production of temperature-induced recombinant prote a scale-down study. *Biotechnology and Bioengineering*, 102(2), 468–482.

Celik, E., & Calik, P. (2012). Production of recombinant proteins by yeast cells. *Biotechnology Advances*, 30, 1108–1118.

Cereghino, J. L., & Cregg, J. M. (2000). Protein expression in the methylotrophic yeast *Pichia pastoris*. *FEMS Microbiology Reviews*, 24, 45–66.

Chin, L. C., Chin, H. K., Chin, C. S. H., Lai, E. T., & Ng, S. K. (2015). Engineering selection stringency on expression vector for the production of recombinant human alpha 1-antrypsin using Chinese hamster ovary cells. *BMC Biotechnology*, 15, 44.

Choi, J. H., Lee, S. J., Lee, S. J., & Lee, S. Y. (2003). Enhanced production of insulin-like growth factor 1 fusion protein in *Escherichia coli* by co-expression of the down-regulated genes identified by transcriptome profiling. *Applied and Environmental Microbiology*, 69(8), 4737–4742.

Cockett, M. I., Bebbington, C. R., & Yarranton, G. T. (1990). High level expression of tissue inhibitor of metalloproteases in Chinese hamster ovary cells using glutamine synthetase gene amplification. *Nature Biotechnology*, 8(7), 662–667.

Cohen, S. N., Chang, A. C. Y., Boyer, H. W., & Helling, R. B. (1973). Construction of biologically functional bacterial plasmids in vitro. *Proceedings of the National Academy of Sciences USA*, 70(11), 3240–3244.

Costa, S., Almeida, A., Castro, A., & Domingues, L. (2014). Fusion tags for protein solubility, purification and immunogenicity in *Escherichia coli*: the novel Fh8 system. *Frontiers in Microbiology*, 5, article 63.

Crabtree, H. G. (1929). Observations on the carbohydrate metabolism of tumors. *Biochemical Journal*, 23, 536–545.

Crawford, Y., Zhou, M., Hu, Z., Joly, J., Snedecor, B., Shen, A., & Gao, A. (2013). Fast identification of reliable hosts for targeted cell line development from a limited-genome screening using combined ϕC31 integrase and CRE-Lox technologies. *Biotechnology Progress*, 29, 1307–1315.

Curran, B. P. G., & Bugeja, V. (2015). The biotechnology and biotechnology of yeast. In R. Rapley, & D. Whitehouse (Eds.), *Molecular biology and biotechnology* (6th ed.), (pp. 282–314). Cambridge: Royal Society of Chemistry.

Czajkowsky, D. M., Hu, J., Shao, Z., & Pleass, R. J. (2012). Fc-fusion proteins: new developments and future perspectives. *EMBO Molecular Medicine*, 4(10), 1015–1028.

Daegelen, P., Studier, F. W., Lenski, R. E., Cure, S., & Kim, J. F. (2009). Tracing ancestors and relatives of *Escherichia coli* B, and the derivation of B strains REL606 and BL21(DE3). *Journal of Molecular Biology*, *394*, 634–643.

De Jesus, M., & Wurm, F. M. (2011). Manufacturing recombinant proteins in kg-ton quantities using animal cells in bioreactors. *European Journal of Pharmaceutics and Biopharmaceutics*, *78*, 184–188.

De Marco, V., Stier, G., Blandin, S., & De Marco, A. (2004). The solubility and stability of recombinant proteins are increased by their fusion to NusA. *Biochemical and Biophysical Research Communications*, *322*, 766–771.

Dykes, C. W. (1993). Molecular biology in the pharmaceutical industry. In J. M. Walker, & E. B. Gingold (Eds.), *Molecular biology and biotechnology* (pp. 155–176). Cambridge: Royal Society of Chemistry.

Eiteman, M. A., & Altman, E. (2006). Overcoming acetate in *Escherichia coli* recombinant fermentations. *Trends in Biotechnology*, *24*(11), 530–536.

European Molecular Biology Laboratory. (2016). Protein expression and purification core facility. Available from https://www.embl.de/pepcore/pepcore_services/cloning/choice_vector/index.html. Retrieved April 9, 2016.

Geisse, S. (2009). Reflections on more than 10 years of TGE approaches. *Protein Expression and Purification*, *64*, 99–107.

Girod, P. -A., Nguyen, D. -Q., Calabrese, D., Puttini, S., Grandjean, M., Martinet, D., et al. (2007). Genome-wide prediction of matrix attachment regions that increase gene expression in mammals. *Nature Methods*, *4*, 747–753.

Girod, P. -A., Zhan-Zabal, M., & Mermod, N. (2005). Use of the chicken lysozyme 5′matrix attachment region to generate high producer CHO cell lines. *Biotechnology and Bioengineering*, *91*, 1–11.

Gleeson, M. A., White, C. E., Meininger, D. P., & Komives, E. A. (1998). Generation of protease-deficient strain and their use in heterologous protein expression. *Methods in Molecular Biology*, *103*, 81–94.

Graslund, S., Nordlund, P., Weigelt, J., Hallberg, B. M., Bray, J., Gileadi, O., et al. (2008). Protein production and purification. *Nature Methods*, *5*(2), 135–146.

Graumann, K., & Premstaller, A. (2006). Manufacturing of recombinant therapeutic proteins in microbial systems. *Biotechnology Journal*, *1*, 164–186.

Han, M. J., Kim, J. Y., & Kim, J. A. (2014). Comparison of the large-scale periplasmic proteomes of the *Escherichia coli* K12 and B strains. *Journal of Bioscience and Bioengineering*, *117*(4), 437–442.

Harbron, S. (2015). Protein expression. In R. Rapley, & D. Whitehouse (Eds.), *Molecular biology and biotechnology* (6th ed.), (pp. 76–108). Cambridge: Royal Society of Chemistry.

Holms, W. H. (1997). Metabolic flux analysis. In P. M. Rhodes, & P. F. Stanbury (Eds.), *Applied microbial physiology—A practical approach* (pp. 213–248). Oxford: IRL Press at Oxford University Press.

Idiris, A., Tohda, H., Kumagai, H., & Takegawa, K. (2010). Engineering of protein secretion in yeast: strategies and impact on protein production. *Applied Microbiology and Biotechnology*, *86*, 403–417.

Itakura, K., Hirose, T., Crea, R., Riggs, A. D., Heyneker, H. L., Bolivar, F., & Boyer, H. W. (1977). Expression in *Escherichia coli* of a chemically synthesized gene for the hormone somatostatin. *Science*, *198*(4321), 1056–1063.

Jadhav, V., Hackl, M., Druz, A., Shridhar, S., Chung, C. Y., Heffner, K. M., et al. (2013). CHO microRNA engineering is growing up: recent successes and future challenges. *Biotechnology Advances, 31*, 1501–1513.

Jager, V., Bussow, K., & Schirrmann, T. (2015). Transient recombinant protein expression in mammalian cells. In M. Al-Rubeai (Ed.), *Cell engineering Vol. 9—animal cell culture* (pp. 27–64). Switzerland: Springer International Publishing.

Kaufman, R. J., & Sharp, P. A. (1982). Amplification and expression of sequences cotransfected with a modular dihydrofolate reductase complementary DNA gene. *Journal of Molecular Biology, 159*, 601–621.

Kim, J. Y., Kim, Y. -G., & Lee, G. M. (2012). CHO cells in biotechnology for production of recombinant proteins: current state and further potential. *Applied Microbiology and Biotechnology, 93*, 917–930.

Korotkov, K. V., Sandkvist, M., & Hol, W. G. J. (2012). The type II secretion system: biogenesis, molecular architecture and mechanism. *Nature Reviews Microbiology, 10*, 336–351.

Korpimaki, T., Kurittu, J., & Karp, M. (2003). Surprisingly fast disappearance of beta-lactam selection pressure in cultivation as detected with novel biosensing approaches. *Journal of Microbiological Methods, 53*, 37–42.

Kuczewski, M., Schirmer, E., Lain, B., & Zarbis-Papastoitsis, G. (2011). A single use purification process for the production of a monoclonal antibody produced in a PER.C6 human cell line. *Biotechnology Journal, 6*(1), 56–65.

Kuystermans, D., & Al-Rubeai, M. (2015). Mammalian cell line selection strategies. In M. Al-Rubeai (Ed.), *Animal cell culture* (pp. 327–372). Cham, Switzerland: Springer.

Lai, T., Yang, Y. K., & Ng, S. K. (2013). Advances in mammalian cell line development technologies for recombinant protein production. *Pharmaceuticals, 6*, 579–603.

Lieb, M. (1966). Studies of heat-inducible lambda bacteriophage: 1. Order of genetic sites and properties of mutant prophages. *Journal of Molecular Biology, 16*(1), 149–163.

Liu, M., Feng, X., Ding, Y., Zhao, G., Liu, H., & Xian, M. (2015). Metabolic engineering of *Escherichia coli* to improve recombinant protein production. *Applied Microbiology and Biotechnology, 99*(24), 10367–10377.

Longo, P. A., Kavran, J. M., Kim, M. S., & Leahy, D. J. (2013). Transient mammalian cell transfection with polyethylenimine (PEI). *Methods in Enzymology, 529*, 227–240.

Looser, V., Bruhlmann, B., Bumbak, F., Stebger, C., Costa, M., Camattari, A., Fotiadis, D., & Kovar, K. (2015). Cultivation strategies to enhance productivity of *Pichia pastoris*. *Biotechnology Advances, 33*, 1177–1193.

Majors, B. S., Betenbaug, M. J., Pederson, N. E., & Chiang, G. G. (2008). Enhancement of transient gene expression and culture viability using Chinese hamster ovary cells overexpressing Bcl-xL. *Biotechnology and Bioengineering, 101*(3), 567–578.

Malik, A. (2016). Protein fusion tags for efficient expression and purification of recombinant proteins in the periplasmic space of *E. coli. 3. Biotech, 6*(44), 1–7.

Malik, A., Rudolph, R., & Sohling, B. (2006). A novel fusion protein system for the production of native human pepsinogen in the bacterial periplasm. *Protein Expression and Purification, 47*(2), 662–671.

Malik, A., Jenzsch, M., Lubbert, A., Rudolph, R., & Sohling, B. (2007). Periplasmic production of native proinsulin as a fusion to *E. coli* ecotin. *Proetin Expression and Purification, 55*, 100–111.

Mamat, U., Wilke, K., Bramhill, D., Schromm, A. B., Lindner, B., Kohl, A., et al. (2015). Detoxifying *Escherichia coli* for endotoxin-free production of recombinant proteins. *Microbial Cell Factories, 14*, 57.

Matos, C. F. R. O., Branston, S. D., Albiniak, A., Dhanoya, A., Freedman, R. B., Keshavaraz-Moore, E., & Robinson, C. (2012). High-yield export of a native heterologous protein to the periplasm by the tat translocation pathway in *Escherichia coli*. *Biotechnology and Bioengineering*, *109*(10), 2533–2542.

Mattanovich, D., Branduardi, P., Dato, L., Gasser, B., Sauer, M., & Porro, D. (2012). Recombinant protein production in yeasts. In A. Lorence (Ed.), *Recombinant Gene Expression: Reviews and Protocols, Third Edition, Methods in Molecular Biology* (pp. 329–358). (824). New York: Humana Press.

McDonald, K. A. (2011). Heterologous protein expression. In C. Webb (Ed.), *Comprehensive biotechnology: Vol. 2. Engineering fundamentals of biotechnology* (pp. 441–450). Amsterdam: Elsevier.

Merck Millipore. (2014). UCOE expression technology—Mammalian expression. Available from https://www.merckmillipore.com/GB/en/life-science-research/genomic-analysis/transfection-protein-expression/mammalian-expression/GY2b.qB.ynIAAAFA3sQENF7F,nav

Meredith, T. C., Aggarwal, P., Mamat, U., Lindner, B., & Woodward, R. W. (2006). Redefining the requisite lipopolysaccharide structure in *Escherichia coli*. *ACS Chemical Biology*, *1*, 33–42.

Mieschendahl, M., Petri, T., & Hanggi, U. (1986). A novel prophage independent TRP regulated lambda P(L) expression system. *Bio/Technology*, *4*(9), 802–808.

Nakamura, Y., Gojobori, T., & Ikemura, T. (2000). Codon usage tabulated from international DNA sequence databases: status for the year 2000. *Nucleic Acids Research*, *28*(1), 292.

Natale, P., Bruser, T., & Driessen, A. J. M. (2008). Sec- and Tat-mediated protein secretion across the bacterial cytoplasmic membrane— distinct translocases and mechanisms. *Biochimica et Biophysica Acta*, *1778*, 1735–1756.

Nefarma. (2013). *Biopharmaceuticals and biosimilars*. Available from https://www.nefarma.nl/stream/com-biologicals-and-biosimilars.

Nett, J. H., Stadheim, T. A., Huijuan, Li., Bobrowicz, P., Hamilton, R., Davidson, R. C., et al. (2011). A combinatorial genetic library approach to target heterologous glycosylation enzymes to the endoplasmic reticulum or the Golgi apparatus of *Pichia pastoris*. *Yeast*, *28*, 237–252.

Newstead, S., Kim, J. H., von Heijne, G., Iwata, S., & Drew, D. (2007). High-throughput fluorescent-based optimization of eukaryotic membrane protein overexpression and purification in *Saccharomyces cerevisiae*. *Proceedings of the National Academy of Sciences USA*, *104*, 13936–13941.

Nielsen, J. (2013). Production of biopharmaceutical proteins by yeast: advances through metabolic engineering. *Bioengineering*, *4*(4), 207–211.

O'Callaghan, P. M., & Racher, A. J. (2015). Building a cell culture process with stable foundations: searching for certainty in an uncertain world. In M. Al-Rubeai (Ed.), *Cell engineering Vol. 9—animal cell culture* (pp. 373–406). Switzerland: Springer International Publishing.

Overton, T. W. (2014). Recombinant protein production in bacterial hosts. *Drug Discovery Today*, *19*(5), 590–601.

Parenteau, J., Durand, M., Veronneau, S., Lacombe, A. -A., Morin, G., Guerin, V., et al. (2008). Deletion of many yeast introns reveals a minority of genes that require splicing for function. *Molecular Biology of the Cell*, *19*(5), 1932–1941.

Peebo, K., Vlagepea, K., Nahku, R., Riis, G., Oun, M., Adamberg, K., & Vilu, R. (2014). Coordination of PTA-ACS and TCA cycles strongly reduces overflow metabolism of acetate in *Escherichia coli*. *Applied Microbiology and Biotechnology*, *98*, 5131–5143.

Posfai, G., Plunkett, G., Feher, T., Frisch, D., Keil, G. M., Umenhoffer, K., et al. (2006). Emergent properties of reduced-genome *Escherichia coli*. *Science*, *312*, 1044–1046.

Potvin, G., Ahmad, A., & Zhang, Z. (2012). Bioprocess engineering aspects of heterologous protein production in *Pichia pastoris*: a review. *Biochemical Engineering Journal*, *64*, 91–105.

Prinz, W. A., Aslund, F., Holmgren, A., & Beckwith, J. (1997). The role of thioredoxin and glutaredoxin pathways in reducing protein disulfide bonds in the *Escherichia coli* cytoplasm. *The Journal of Biological Chemistry*, *272*(25), 15661–15667.

Pryor, K. D., & Leiting, B. (1997). High-level expression of soluble protein in *Escherichia coli* using a his$_6$tag and maltose-binding-protein double-affinity fusion system. *Protein Expression and Purification*, *10*(3), 309–319.

Qiao, J., Oumard, A., Wegloehner, W., & Bode, J. (2009). Novel tag-and-exchange (RMCE) strategies generate master cell clones with predictable and stable transgene expression properties. *Journal of Molecular Biology*, *390*(4), 579–594.

Ramon, A., Senorale-Pose, M., & Marin, M. (2014). Inclusion bodies: not that bad. *Frontiers in Microbiology*, *5*, article 56.

Reader, R. A. (2013). FDA biopharmaceutical product approvals and trends in 2012. *BioProcess International*, *11*(3), 18–27.

Reed, G., & Nagodawithana, T. W. (1991). *Yeast technology* (2nd ed.). New York: Van Nostrand Reinhold, pp. 413–437.

Rees, S., Coote, J., Stables, J., Goodson, S., Harris, S., & Lee, M. G. (1996). Bicistronic vector for the creation of stable mammalian cell lines that redisposes all antibiotic resistant cells to express recombinant protein. *Biotechniques*, *20*(1), 102–104106, 108-110.

Rosano, G. L., & Ceccarelli, E. A. (2014). Recombinant protein expression in *Escherichia coli*: advances and challenges. *Frontiers in Microbiology*, *5*, article 172.

Salema, V., & Fernandez, L. A. (2013). High yield purification of nanobodies from the periplasm of *E. coli* as fusions with the maltose binding protein. *Protein Expression and Purification*, *91*(1), 42–48.

Sanchez-Garcia, L., Martin, L., Mangues, R., Ferrer-Miralles, N., Vazquez, E., & Villaverde, A. (2016). Recombinant pharmaceuticals from microbial cells: a 2015 update. *Microbial Cell Factories*, *15*(33), 1–7.

Schwarz, H., Schmittner, M., Duschl, A., & Horejs-Hoeck, J. (2014). Residual endotoxin contaminations in recombinant proteins are sufficient to activate human CD1c$^+$ dendritic cells. *PLOS One*, *9*, e113840.

Stanley, P., Schachter, H., & Taniguchi, N. (2009). N-Glycans. In A. Varki, R. D. Cummings, J. D. Esko et al.,(Eds.), *Essentials of glycobiology* (2nd ed.). New York: Cold Spring Harbor Laboratory Press, chapter 8.

Studier, F. W., & Moffatt, B. A. (1986). Use of bacteriophage T7 RNA polymerase to direct selective high-level expression of cloned genes. *Journal of Molecular Biology*, *189*, 113–130.

Sun, P., Tropea, J. E., & Waugh, D. S. (2010). Enhancing the solubility of recombinant proteins in *Escherichia coli* by using hexa-histidine tagged maltose binding protein as a fusion partner. *Methods in Molecular Biology*, *705*, 259–274.

Tomimoto, K., Fujita, Y., Iwaki, T., Chiba, Y., Jigami, Y., Nakayama, K., Nakajima, Y., & Abe, H. (2013). Protease-deficient *Saccharomyces cerevisiae* strains for the synthesis of human-compatible glycoproteins. *Bioscience, Biotechnology and Biochemistry*, *77*(12), 2461–2466.

Valdez-Cruz, N. A., Caspeta, L., Perez, N. O., Ramirez, O. T., & Trujillo-Roldan, M. A. (2010). Production of recombinant proteins in *E. coli* by the heat inducible expression system based on the phage lamba pL and/or PR promoters. *Microbial Cell Factories*, *9*(18), 1–16.

Valdez-Cruz, N. A., Ramirez, O. T., & Trujillo-Roldan, M. A. (2011). Molecular responses of *Escherichia coli* caused by heat stress and recombinant protein production during temperature induction. *Bioengineered Bugs, 2*(2), 105–110.

Valgepea, K., Adamberg, K., Nahku, R., Lahtvee, P. -J., Arike, L., & Vilu, R. (2010). Systems biology approach reveals that overflow metabolism of acetate in *Escherichia coli* is triggered by carbon catabolite repression of acetyl-CoA synthetase. *BMC Systems Biology, 4*, 166.

Walsh, G. (2010). Biopharmaceutical benchmarks. *Nature Biotechnology, 28*, 917–924.

Walsh, G. (2012). New biopharmaceuticals. *Biopharm International, June 2012*, 1–5.

Walsh, G. (2014a). Biopharmaceutical benchmarks 2014. *Nature Biotechnology, 32*(10), 992–1000.

Walsh, G. (2014b). Protein sources. In G. Walsh (Ed.), *Proteins: biochemistry and biotechnology* (pp. 110–138). Chichester: Wiley.

West, A. G., Gazner, M., & Felsenfeld, G. (2015). Insulators: many functions, many mechanisms. *Genes and Development, 16*, 271–288.

Wulhfard, S., Baldi, L., Hacker, D. L., & Wurm, F. (2010). Valporic acid enhances recombinant mRNA and protein levels in transiently transfected Chinese hamster ovary cells. *Journal of Biotechnology, 148*(2–3), 128–132.

Wurm, F. M. (2004). Production of recombinant therapeutics in cultivated mammalian cells. *Nature Biotechnology, 22*(11), 1393–1398.

Yoon, S. H., Han, M. -J., Jeong, H., Lee, C. H., Xia, X. -X., Lee, D. -H., et al. (2012). Comparative multi-omics systems analysis of *Escherichia coli* strains B and K-12. *Genome Biology, 13*(5), R37.

Yoon, S. H., Jeong, H., Kwon, S. -K., & Kim, J. F. (2009). Genomics, biological features, and biotechnological applications of *Escherichia coli* B12 "Is B for better?!". In S. Y. Lee (Ed.), *Systems biology and biotechnology of Escherichia coli* (pp. 1–17). Heidelberg: Springer.

Yoon, S. H., Kim, S. K., & Kim, J. F. (2010). Secretory production of recombinant proteins in Escherichia coli. *Recent Patents on Biotechnology, 4*, 23–29.

Young, C. L., Britton, Z. T., & Robinson, A. S. (2012). Recombinant protein expression and purification: a comprehensive review of affinity tags and microbial applications. *Biotechnology Journal, 7*, 620–634.

Zhang, N., & An, Z. (2010). Heterologous protein expression in yeasts and filamentous fungi. In R. H. Baltz, A. L. Demain, & J. E. Davies (Eds.), *Manual of industrial microbiology and biotechnology* (3rd ed.). Washington: ASM Press, Chapter 11.

Zhu, J. (2012). Mammalian cell protein expression for biopharmaceutical production. *Biotechnology Advances, 30*(5), 1158–1170.

Index

777